Moshe Shapiro and Paul Brumer

Quantum Control of Molecular Processes

Related Titles

May, V., Kühn, O.

Charge and Energy Transfer Dynamics in Molecular Systems

2011
ISBN: 978-3-527-40732-3

Matta, C. F., Boyd, R. J. (eds.)

The Quantum Theory of Atoms in Molecules
From Solid State to DNA and Drug Design

2007
ISBN: 978-3-527-30748-7

Schleich, W. P., Walther, H. (eds.)

Elements of Quantum Information

2007
ISBN: 978-3-527-40725-5

Wilkening, G., Koenders, L.

Nanoscale Calibration Standards and Methods
Dimensional and Related Measurements in the Micro- and Nanometer Range

2005
ISBN: 978-3-527-40502-2

Rice, S. A., Zhao, M.

Optical Control of Molecular Dynamics

2000
ISBN: 978-0-471-35423-9

Moshe Shapiro and Paul Brumer

Quantum Control of Molecular Processes

Second, Revised and Enlarged Edition

WILEY-VCH

WILEY-VCH Verlag GmbH & Co. KGaA

The Authors

Prof. Moshe Shapiro
Department of Chemistry
University of British Columbia
Vancouver, British Columbia
Canada V6T 1Z1

Prof. Paul Brumer
Department of Chemistry
University of Toronto
Toronto, Ontario
Canada M5S 3H6

Library of Congress Card No.: applied for

British Library Cataloguing-in-Publication Data:
A catalogue record for this book is available from the British Library.

Bibliographic information published by the Deutsche Nationalbibliothek
The Deutsche Nationalbibliothek lists this publication in the Deutsche Nationalbibliografie; detailed bibliographic data are available on the Internet at http://dnb.d-nb.de.

Typesetting le-tex publishing services GmbH, Leipzig
Printing and Binding Fabulous Printers Pte Ltd, Singapore
Cover Design Adam-Design, Weinheim

Printed in Singapore
Printed on acid-free paper

ISBN Print 978-3-527-40904-4

ISBN ePDF 978-3-527-63972-4
ISBN oBook 978-3-527-63970-0
ISBN ePub 978-3-527-63971-7
ISBN Mobi 978-3-527-63973-1

In memory of our parents and
to our wives Rachelle and Abbey, נשי חיל

Contents

Preface to the Second Edition

The growth of the field of the Quantum Control of Molecular Processes, emphasized in the preface of the first edition of this book, continues at an ever-increasing pace. Indeed, the past eight years have seen the explosive growth of both experimental and theoretical studies that are aimed at manipulating atomic and molecular processes at their most fundamental level. This updated book on Quantum Control captures important new directions and challenges in the area.

As in the first edition, our focus is on theories and experiments that utilize and reveal the *underlying physics* that is at the heart of quantum control. For this reason we provide only a minimal discussion of the purely mathematical aspects of control theory (amply covered, for example, in [1]). We also do not focus on modern adaptive feedback studies, which are extensively reviewed elsewhere (as indicated in Chapter 17), but which have yet to reveal conclusive physical insight into control mechanisms. Rather, this book stresses physical ideas that motivate control scenarios, and their resultant implementation.

Amongst other additions in this second edition, we include new approaches to the control of intramolecular dynamics (Chapter 4), a greatly extended treatment of decoherence and its affect on control (Chapter 6), a larger selection of sample coherent control scenarios (in Chapter 7), a detailed analysis of quantum control for collision processes (Chapter 8), a treatment of the very fundamentals of quantum interference in coherent control (Chapter 10), an extended discussion of coherent control via adiabatic passage (Chapter 11), a lengthy treatment of the control of chiral systems (Chapter 14), a Chapter on control with few-cycle "attosecond" pulses (Chapter 16), and an examination of coherent control in the classical limit (Chapter 18). In addition, numerous recent developments are described throughout the book, and several sections have been reorganized. We apologize to those whose work we could not cite; however, keeping this book of reasonable size necessitated (sometimes arbitrary) choices of material.

We express our continuing gratitude to our colleagues, postdocs, and students who have advanced this field of research. In addition, we are grateful for our own longstanding and continuing research collaboration, now lasting forty years, which has proven both fruitful and stimulating. We thank Ms. Susan Arbuckle for undertaking the myriad of annoying, important, and time-consuming tasks that must be completed before a volume can go to press. As always, she did an exceptional job.

We thank Dr. Timur Grinev for outstanding proofreading, and for preparing an extensive table (to be found after Chapter 18) of the symbols used in this book. We thank Dr. Michael Spanner, NRC Ottawa for allowing use of his creative picture of lasers incident on a molecular chain, which graces the cover of this book. Sincere thanks also to NSERC, Canada for ongoing research support.

We have dedicated this book to our wives, who have provided the most essential of ingredients for scientific productivity, a stable and happy home environment.

Finally, we continue to welcome, at our email addresses below, any suggested corrections to this book.

תושלב״ע

Vancouver, March 2011 — *Moshe Shapiro (mshapiro@chem.ubc.ca)*
Toronto, March 2011 — *Paul Brumer (pbrumer@chem.utoronto.ca)*

Preface to the First Edition

Despite its maturity, quantum mechanics remains one of the most intriguing of subjects. Since its emergence over 75 years ago, each generation has discovered, investigated, and utilized different attributes of quantum phenomena. In this book we introduce results from research over the past fifteen years that demonstrates that quantum attributes of light and matter afford the possibility of unprecedented control over the dynamics of atomic and molecular systems. This subject is the result of extensive investigations in chemistry and physics since 1985, and has seen enormous growth and interest over the past years. This growth reflects a confluence of developments – the maturation of quantum mechanics as a tool for chemistry and physics, the development of new laser devices that afford extraordinary facility in manipulating light, and the recognition that coherent laser light can be used to imprint information on atoms and molecules in a manner such that their subsequent dynamics leads towards desirable goals. As such, an appreciation of coherent control requires input from optical physics, physical chemistry, atomic and molecular physics, and quantum mechanics. This book aims to provide this background in a systematic manner, allowing the reader to gain expertise in the area.

We have written this monograph with the mature chemistry or physics graduate student in mind; the development is systematic, starting with the fundamental principles of light–matter interactions and concluding with a wide variety of specific topics. We endeavor to include a sufficient number of steps throughout the book to allow self-study or use in class. To retain the focus on the role of quantum interference in control, we tend to utilize examples from our own research, while including samples from that of others. This focus is partially made possible by the recent appearance of a comprehensive survey of the field by Rice and Zhao [2]. It is our expectation that the two books will complement one another.

This book is organized, after a discussion of light and light–matter interactions in Chapter 1, in order of increasing incident electromagnetic field strength. Chapters 2–8 primarily deal with molecular dynamics and control where the field strengths are such that perturbation theory is applicable. Emphasis is placed on the principle of coherent control, that is, control via quantum interference between simultaneous indistinguishable pathways to the same final state. From the viewpoint of chemistry, the vast majority of control work has thus far been done on photodissociation processes. As a consequence, we provide a thorough introduction to the dynamics of photodissociation in Chapter 2 and discuss its control in Chapters 2, 4 and 6. The extension of quantum control to bimolecular collision processes is provided in Chapter 7 and to the control of chirality (and asymmetric synthesis) in Chapter 8.

Applications of control using moderate fields are discussed in Chapters 9 to 11. These fields allow for new physical phenomena in both bound state and continuum problems, including adiabatic population transfer in both regimes, electromagnetically induced transparency in bound systems, as well as additional unimolecular and bimolecular control scenarios.

Strong fields introduce yet another set of phenomena allowing for the controlled manipulation of matter. Examples of light-induced potentials and the controlled focusing, alignment, and deposition of molecules are discussed in Chapter 12, after the introduction of the quantized electromagnetic field.

All of the quantum control scenarios involve a host of laser and system parameters. To obtain maximal control in any scenario necessitates a means of tuning the system and laser parameters to optimally achieve the desired objective. This topic, optimal control, is introduced and discussed in Chapters 4 and 13. The role of quantum interference effects in optimal control are discussed as well, providing a uniform picture of control via optimal pulse shaping and coherent control.

By definition, quantum control relies upon the unique quantum properties of light and matter, principally the wavelike nature of both. As such, maintenance of the phase information contained in both the matter and light is central to the success of the control scenarios. Chapter 5 deals with "decoherence", that is, the loss of phase information due to the influence of the external environment in reducing the system coherence. Methods of countering decoherence are also discussed.

This book has benefited greatly from the research support that we have received over the past years. First, we acknowledge the ongoing support by the US Office of Naval Research through the research program of Dr. Peter J. Reynolds. We are also grateful to NSERC Canada, Photonics Research Ontario, the Israel Science Foundation and the Minerva Foundation, Germany. Equally importantly, we thank the many students and colleagues who have taken part in the development of coherent control and have contributed so much to the field. We wish the acknowledge Ignacio Franco, Einat Frishman, David Gerbasi, Michal Oren, and Alexander Pegarkov for comments on various parts of the manuscript, Ms. Susan Arbuckle for unstinting assistance with copyediting and indexing, Daniel Gruner for expert assistance on puzzling tex issues and Amnon Shapiro for preparing many of the postscript figure files. On a personal note, P.B. thanks Meir and Malka Cohen-Nehemia for training in the Mitzvah Technique that allowed him to counter the debilitating effects of back pain and body misuse.

None of this work would be possible without our wives, Rachelle and Abbey, who have provided the environment and support so necessary to allow productive science to be done. We are more than grateful to them both, as indicated in the dedication.

Finally, we welcome, at our email addresses below, any suggested corrections or additions to this book.

תושלב"ע

Rehovot, June 2002 *Moshe Shapiro (mshapiro@weizmann.ac.il)*
Toronto, June 2002 *Paul Brumer (pbrumer@chem.utoronto.ca)*

1
Preliminaries of the Interaction of Light with Matter

Influencing the dynamics of atomic and molecular systems, the goal of quantum control, requires exposing the system to an external perturbation to alter its state. The most widely applied approaches use controllable incident external radiation as the source of this perturbation. In this chapter we introduce some of the fundamental concepts of the interaction of radiation with matter. The treatment assumes a classical electromagnetic field interacting with a quantized molecule, and is intended as a summary of the relevant equations used in the early parts of this book. (A summary of the symbols used throughout the book is provided in the Appendix.) A detailed derivation of these equations, starting from fundamentals, is provided in Chapter 9. Further development, in terms of quantized electromagnetic fields, is discussed in Section 10.1.

Consider a material system comprising a set of particles of mass m_i and charge q_i at positions $\boldsymbol{r}_i$ that is interacting with an incident classical electric field $\boldsymbol{E}(\boldsymbol{r}, t)$. Here $\boldsymbol{r}$ denotes the spatial coordinates, and t denotes time. The Hamiltonian H of the system, at time t, is then given by

$$H(t) = H_M + H_{MR}(t) . \tag{1.1}$$

where H_M denotes the Hamiltonian of the material and $H_{MR}(t)$ describes its interaction with an incident external radiation field. The former is of the form

$$H_M = \sum_j \frac{-\hbar^2}{2m_j} \nabla_j^2 + V_C , \tag{1.2}$$

where ∇_j^2 is the Laplacian associated with the coordinates of the jth particle, and V_C is the Coulomb potential between the particles. The $H_{MR}(t)$ term is given below in Eq. (1.5).

Given the Hamiltonian $H(t)$, then, in accord with standard quantum mechanics [3], the dynamical evolution of the wave function $\Psi(t)$ describing the motion of the system in the presence of the field $\boldsymbol{E}(\boldsymbol{r}, t)$ is given by the solution to the time-dependent Schrödinger equation:

$$i\hbar \frac{\partial \Psi(t)}{\partial t} = H(t)\Psi(t) = \left[H_M + H_{MR}(t)\right] \Psi(t) . \tag{1.3}$$

Quantum Control of Molecular Processes, Second Edition. Moshe Shapiro and Paul Brumer

In general, $E(r, t)$ is comprised of modes that are described by plane waves with directions of propagation k, unit vectors $\hat{\epsilon}_k$ describing the polarizations, frequencies ω_k, and complex amplitudes A_k.

Specifically,

$$E(r, t) = i \sum_k k \hat{\epsilon}_k \left\{ A_k \exp\left(-i\omega_k t + i k \cdot r\right) - A_k^* \exp\left(i\omega_k t - i k \cdot r\right) \right\} . \quad (1.4)$$

In the traditional case, where one can neglect the variation of the incident electric field over the size of the material, and assuming that the electric field propagates along the z direction, $H_{MR}(t)$ is given by

$$H_{MR}(t) = -d \cdot E(z, t) , \quad (1.5)$$

where d is the molecular dipole moment,

$$d \equiv \sum_j q_j r_j , \quad (1.6)$$

and where the center of mass of the material is positioned at z.

When considering a coherent *pulse* of light, it is necessary to superimpose a collection of plane waves, as in Eq. (1.4). We adopt the reasonable simplifying assumption that all the modes of the pulse propagate in the same (z) direction and that all the pulse modes have the same polarization direction $\hat{\epsilon}$. We can therefore eliminate the integration over the $\hat{k}$ directions and write Eq. (1.4) (in an infinite volume) as

$$E(z, t) = \hat{\epsilon} \int_0^\infty d\omega \left\{ \epsilon(\omega) \exp\left[i\omega \left(\frac{z}{c} - t \right) \right] + \epsilon^*(\omega) \exp\left[-i\omega \left(\frac{z}{c} - t \right) \right] \right\}$$

$$\equiv E_+(\tau) + E_-(\tau) \equiv E(\tau) = \hat{\epsilon} \int_{-\infty}^{\infty} d\omega \epsilon(\omega) \exp(-i\omega\tau) . \quad (1.7)$$

Here, $\epsilon(\omega) \equiv i k A_k / c$, and τ is the so-called "retarded time",

$$\tau \equiv t - \frac{z}{c} . \quad (1.8)$$

Each mode amplitude in Eq. (1.7) is a complex number,

$$\epsilon(\omega) = |\epsilon(\omega)| \exp[i\phi(\omega)] , \quad (1.9)$$

where $\phi(\omega)$ are frequency-dependent phases. The fact that $E(z, t)$ is real ensures that $\epsilon(-\omega) = \epsilon^*(\omega)$, and hence that

$$\phi(-\omega) = -\phi(\omega) , \quad |\epsilon(-\omega)| = |\epsilon(\omega)| . \quad (1.10)$$

As discussed in later chapters, the phase $\phi(\omega)$ plays a central role in coherent control theory. However, individual phase values depend on an (arbitrary) definition of the origin of time and space. Therefore only the relative phases, which are the only phase factors that can actually be measured, are of any consequence physically.

Equation (1.7) describes a pulse of *coherent* light, where $\mathbf{E}(z, t)$ is represented by an analytic function of time. In cases of partially coherent light, either the phase or amplitude acquires a random component and an analytic expression for $\mathbf{E}(z, t)$ no longer exists. Appropriate descriptions of partially incoherent light interacting with molecules are provided in Section 6.5.

2
Weak-Field Photodissociation

As shown throughout this book, quantum control of molecular dynamics has been applied to a wide variety of processes. Within the framework of chemical applications, control over reactive scattering has dominated. In particular, the two primary chemical processes of interest are photodissociation, in which a molecule is irradiated and dissociates into various products, and bimolecular reactions, in which two molecules collide to produce new products. In this chapter we formulate the quantum theory of photodissociation; that is, the light-induced breaking of a chemical bond. In doing so we provide an introduction to concepts essential for the remainder of this book. The quantum theory of bimolecular collisions is also briefly discussed.

Throughout this chapter we utilize perturbation theory, assuming that the light field is "weak;" "strong" light fields are addressed in Chapter 12. The approach that we advocate applies to both pulsed and continuous wave (CW) excitation sources. This allows us to compare and contrast these photodissociation schemes and to identify aspects of photodissociation that are consistent with, or contrary to, our classical intuition.

A number of issues preliminary to questions of control and process selectivity are also introduced. In particular we ask: What determines the final outcome of a photodissociation process? Although in quantum mechanics the fate of a system can only be known in a probabilistic sense, the linear time dependence of the Schrödinger equation guarantees that the probability of future events is completely determined by the probability of past events. That is, quantum mechanics is a deterministic theory of distributions of various observables. Hence by identifying attributes of the quantum state at earlier times we learn what is required to alter – that is, control – system dynamics in future times.

In addition to the basic concepts of photodissociation, we address a number of more subtle issues, such as the precise definition of the concept of a "lifetime." We show that this attribute is not purely a property of the system. Rather, it is (see Appendix 2.A) intimately related to the way the system was prepared and probed.

Quantum Control of Molecular Processes, Second Edition. Moshe Shapiro and Paul Brumer

2.1 Photoexcitation of a Molecule with a Pulse of Light

Consider a molecule interacting with a pulse of coherent light, where the light is described by a purely classical field of Eq. (1.7) and the molecule is treated quantum mechanically. The dynamics of the radiation-free molecule is fully described by the (discrete or continuous) set of energy eigenvalues and eigenfunctions, denoted respectively as E_n and $|E_n\rangle$, of the material Hamiltonian H_M (Eq. (1.2)),

$$H_M|E_n\rangle = E_n|E_n\rangle \,. \tag{2.1}$$

Here the eigenfunctions are denoted $|E_n\rangle$, with the understanding that the notation will be extended to include additional quantum numbers when energy degeneracies exist.

Given E_n and $|E_n\rangle$, the full time-dependent Schrödinger equation (Eq. (1.3)) can be solved by expanding $|\Psi(t)\rangle$ in terms of $|E_n\rangle$. That is,

$$|\Psi(t)\rangle = \sum_n b_n(t)|E_n\rangle \exp(-iE_nt/\hbar) \;, \tag{2.2}$$

with unknown coefficients $b_n(t)$. To obtain these coefficients we use the orthonormality of the $|E_n\rangle$ basis functions, and substitute Eq. (2.2) into Eq. (1.3), yielding a set of ordinary differential equations for $b_n(t)$:

$$\frac{db_m(t)}{dt} = \frac{1}{i\hbar}\sum_n b_n(t)\exp(i\omega_{m,n}t)\langle E_m|H_{MR}(t)|E_n\rangle \;, \tag{2.3}$$

where the transition frequency, $\omega_{m,n}$, is given by

$$\omega_{m,n} \equiv (E_m - E_n)/\hbar \,. \tag{2.4}$$

Consider first the case where the molecule is initially ($t = -\infty$) in a single state $|\, E_1 \,\rangle$. That is, where

$$b_1(t=-\infty) = 1 \;, \quad \text{and} \quad b_k(t=-\infty) = 0 \quad \text{for} \quad k \neq 1 \;, \tag{2.5}$$

and where the perturbation is weak. The latter condition implies that

$$\int_{-\infty}^{\infty} dt \frac{|\langle E_i|H_{MR}(t)|E_j\rangle \exp(i\omega_{i,j}t)|}{\hbar} \ll 1 \,. \tag{2.6}$$

Under these circumstances we obtain, in first-order perturbation theory, that the expansion coefficients in Eq. (2.2) are given by:

$$\begin{aligned} b_m(t) &= -\frac{d_{m,1}}{i\hbar}\int_{-\infty}^{t} dt' \exp\left[i\omega_{m,1}t'\right]\varepsilon\left(z,t'\right) \\ &= -\frac{d_{m,1}}{i\hbar}\int_{-\infty}^{\infty} d\omega\bar{\epsilon}(\omega)\int_{-\infty}^{t} dt' \exp\left[i(\omega_{m,1}-\omega)t'\right] \;, \end{aligned} \tag{2.7}$$

where

$$d_{m,1} \equiv \langle E_m | \hat{\boldsymbol{\varepsilon}} \cdot \boldsymbol{d} | E_1 \rangle \,, \tag{2.8}$$

with $\hat{\boldsymbol{\varepsilon}} \cdot \boldsymbol{d}$ denoting the projection of the transition dipole operator along the electric field direction. In Eq. (2.7) we have introduced $\varepsilon(z,t) = \int_{-\infty}^{\infty} d\omega \bar{\epsilon}(\omega) \exp(-i\omega t)$ as the length of the $\boldsymbol{E}(z,t) = \varepsilon(z,t)\hat{\boldsymbol{\epsilon}}$ vector with

$$\bar{\epsilon}(\omega) \equiv \epsilon(\omega) \exp(i\omega z/c) = |\epsilon(\omega)| \exp\left[i\left(\phi(\omega) + \omega z/c\right)\right] \,. \tag{2.9}$$

Hence,

$$H_{MR}(t) = -\boldsymbol{E} \cdot \boldsymbol{d} = -\varepsilon(z,t)\hat{\boldsymbol{\varepsilon}} \cdot \boldsymbol{d} \,. \tag{2.10}$$

Equation (2.7) provides the expansion coefficients at any time t. If our interest is in observing or controlling the *final* product states (as it is in photodissociation) then we only require the wave function $\Psi(t)$ as $t \to +\infty$. In this limit we can insert the equality

$$\int_{-\infty}^{\infty} dt' \exp\left[i(\omega_{m,1} - \omega)t'\right] = 2\pi\delta(\omega_{m,1} - \omega) \tag{2.11}$$

into Eq. (2.7) to obtain

$$b_m(+\infty) = \frac{2\pi i}{\hbar}\bar{\epsilon}(\omega_{m,1})d_{m,1} = \frac{2\pi i}{\hbar}|\epsilon(\omega_{m,1})|\, d_{m,1} \exp\left[i\left(\phi(\omega_{m,1}) + \frac{\omega_{m,1}z}{c}\right)\right] \,. \tag{2.12}$$

Equation (2.12) clearly shows that in preparing the state $|E_m\rangle$ the light field has imparted both magnitude and phase to $\Psi(t)$. Similar information is contained in the finite time result as well, but in a somewhat more complex fashion (see Section 2.2).

We note, for use later below, that if $E_m > E_1$ (corresponding to light absorption) then $\omega_{m,1} > 0$ and the phase acquired by b_m from the laser is $\phi(|\omega_{m,1}|)$. Alternatively, when $E_1 > E_m$ (corresponding to stimulated emission) then $\omega_{m,1} < 0$, and by Eq. (1.10) $\phi(\omega_{m,1}) \equiv -\phi(|\omega_{m,1}|)$. Hence we have the rule that light absorption imparts the laser phase evaluated at the frequency of transition to b_m, whereas stimulated emission imparts the negative of the laser phase evaluated at the frequency of emission.

Equation (2.12) also defines the *resonance* (energy conservation) condition: a material energy state $|E_m\rangle$ only absorbs (emits) light, at infinite time, for which $\omega = \omega_{m,1}$ ($\omega = \omega_{1,m}$). Equation (2.11) suggests that it takes an infinite time to establish this resonance condition. However, in the case of pulsed light, no transitions can take place after the pulse is over, no matter how short. Hence the resonance condition must actually be established in the finite time by which the pulse is over. These issues, as well as others related to dynamics during a pulse, are discussed in the next section.

2.2
State Preparation During the Pulse

In order to explore the behavior of the system while the pulse is on we express the integrals over t', in Eq. (2.7), as follows:

$$A(t) \equiv \lim_{T\to\infty} \int_{-T}^{t} dt' \exp\left[i(\omega_{m,1} - \omega)t'\right]$$
$$= \frac{\exp\left[i(\omega_{m,1} - \omega)t\right]}{i\,(\omega_{m,1} - \omega)} - \lim_{T\to\infty} \frac{\exp\left[-i\,(\omega_{m,1} - \omega)\,T\right]}{i\,(\omega_{m,1} - \omega)}\,. \tag{2.13}$$

The *rotating wave approximation* (RWA) chooses desirable ω for absorption and emission by neglecting the highly oscillatory *counter rotating wave* (CRW) term (which has negative and positive ω for absorption and emission, respectively) in favor of the slowly varying *rotating wave* (RW) term (which has positive and negative ω for absorption and emission, respectively).

We can show, using contour integration, that when we insert Eq. (2.13) into Eq. (2.7) and integrate over ω, the contribution from the second term in Eq. (2.13) vanishes. In order to do so, we modify the path along the real axis to include an infinitesimally small semicircle in the upper-half complex ω-plane with $\omega = \omega_{m,1}$ as its center. The contour is then closed by adding a large semicircle in the upper-half plane. The contour, shown in Figure 2.1, thus excludes the $\omega = \omega_{m,1}$ pole. Because of this, and provided we can deform the contour to exclude any existent complex poles of $\bar{\epsilon}(\omega)$ in the upper-half ω plane, the integral over the closed-contour is zero. It remains to be shown that the integral over the large semicircle is also zero. If this is the case, the contribution from the real-line segment – that is, the part appearing in Eq. (2.7) – is also zero.

To show this, write $\omega - \omega_{m,1} = Re^{i\theta}$. Hence, $e^{iT(\omega-\omega_{m,1})} = e^{iRT\cos\theta - RT\sin\theta}$. Since in the upper half of the complex plane, $\theta > 0$, $e^{iT(\omega-\omega_{m,1})}$ vanishes on the large semicircle as $T \to \infty$, a result which holds for all R (since by definition $R > 0$). We can therefore deform the large upper circle portion of the contour

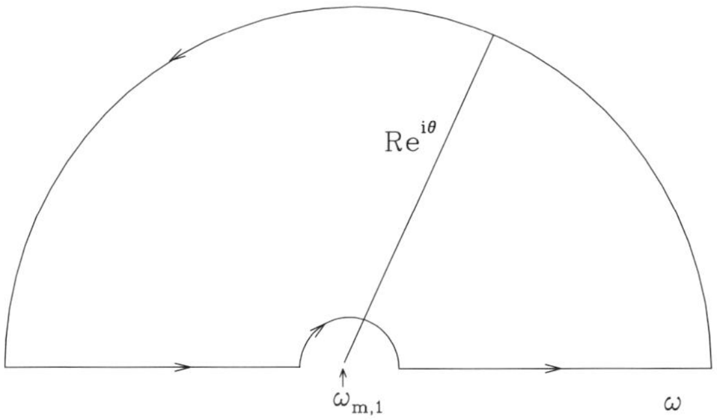

Figure 2.1 Contour to evaluate the ω integral.

to exclude all the poles of $\bar{\epsilon}(\omega)$. Thus, the contribution from the second term in Eq. (2.13), to the integral in Eq. (2.7), is zero, irrespective of the form of $\bar{\epsilon}(\omega)$.

The first term in Eq. (2.13) *does* contribute to $b_m(t)$, since $\exp[i(\omega_{m,1} - \omega)t]$ is nonzero on the large semicircle in the upper plane. Substituting the first term of Eq. (2.13) into Eq. (2.7), we obtain

$$b_m(t) = \frac{i}{\hbar} d_{m,1} \left[\bar{\epsilon}(\omega_{m,1})\, c^+_{m,1}(t) + \bar{\epsilon}(-\omega_{m,1})\, c^-_{m,1}(t) \right] , \tag{2.14}$$

where $c^{\pm}_{m,1}(t)$ are radiative preparation coefficients, defined as

$$c^{\pm}_{m,1}(t) \equiv \frac{1}{\bar{\epsilon}(\pm\omega_{m,1})} \int_0^{\infty} d\omega \bar{\epsilon}(\pm\omega) \frac{\exp[i(\omega_{m,1} \mp \omega)t]}{i(\omega_{m,1} \mp \omega)} . \tag{2.15}$$

The subscript "one", defining the initial state, is suppressed for the remainder of this chapter. The coefficient $c^+_m(t)$ results from the RW term, and $c^-_m(t)$ results from the CRW term. We show below that $c^-_m(t)/c^+_m(t) \to 0$ as $t \to \infty$; that is, only the RW term contributes asymptotically, and that at finite times this ratio is smallest for $\omega \approx \omega_{m,1}$.

To gain insight into the character of the preparation coefficients, we study pulses whose frequency profiles are Gaussian functions:

$$\epsilon(\omega) = \pi^{-\frac{1}{2}} \epsilon_0 \delta_t \exp\left\{-[\delta_t(\omega - \omega_0)]^2\right\} . \tag{2.16}$$

Their time dependence is also described by Gaussian functions:

$$\mathcal{E}(t) = \epsilon_0 \exp\left(-\Gamma^2 t^2\right) \exp(-i\omega_0 t) , \tag{2.17}$$

where $\Gamma \equiv \delta_t/2$. It follows from Eq. (2.15) and Eq. (2.16) that the preparation coefficients for such pulses are given as

$$c^+_m(t) = -\text{sgn}(t) 2\pi \left\{\theta(t) - \frac{1}{2} \exp\left[\beta_+^2\right] W\left[\text{sgn}(t)\beta_+\right]\right\} , \tag{2.18}$$

and

$$c^-_m(t) = \text{sgn}(t)\pi \exp\left[\beta_-^2\right] W\left[\text{sgn}(t)\beta_-\right] , \tag{2.19}$$

where

$$\beta_{\pm} \equiv \delta_t(\omega_{m,1} \pm \omega_0) + it/(2\delta_t) , \tag{2.20}$$

and $W[z]$ is the complex error function

$$W[z] \equiv \exp\left(-z^2\right)\left[1 - \text{erf}(-iz)\right] , \tag{2.21}$$

(see [4, Eqs. (7.1.3) and (7.1.8)]).

Given that $\mathrm{erf}(z) \to 1$ as $z \to \infty$, and that $|\arg z| < \pi/4$, it follows from Eq. (2.18) and Eq. (2.21) that

$$c_m^{\pm}(t) \to 0\,, \quad \text{for} \quad t \ll \frac{-1}{\Gamma}\,,$$

and

$$c_m^{+}(t) \to 2\pi\,, \quad c_m^{-}(t) \to 0\,, \quad \text{for} \quad t \gg \frac{|(\omega_{m,1} - \omega_0)|}{3.7\Gamma^2}\,; \tag{2.22}$$

that is,

$$b_m(t) = (2\pi i/\hbar)\, d_{m,1}\bar{\epsilon}(\omega_{m,1})\,. \tag{2.23}$$

Thus, the CRW coefficients do not contribute after the pulse is over; they are pure transients. Conversely, the RW coefficients $c_m^{+}(t)$ can be nonzero after the pulse is over, with magnitudes that depend on the detuning, as shown below. Equation (2.22) thus gives a criterion, for the Gaussian pulse, as to the time required to establish the resonance condition (Eq. (2.12)). It follows from Eq. (2.22) that the relevant parameter is the pulse duration $1/\Gamma$. This quantity can, in principle, be *shorter* than a single optical cycle, but the resonance condition still holds.

To see the character of the pulse preparation we display the quantity

$$c'_m(t) \equiv c_m^{+}(t) \left| \frac{\epsilon(\omega_{m,1})}{\epsilon(\omega_0)} \right|\,, \tag{2.24}$$

using Eqs. (2.18)–(2.20). The results for $|c'_m(t)|$, $\mathrm{Re}\,c'_m(t)$ and $\mathrm{Im}\,c'_m(t)$ for a Gaussian pulse whose intensity bandwidth $(2(\ln 2)^{1/2}/\delta_t)$ is $120\,\mathrm{cm}^{-1}$, for different detunings $\Delta_0 \equiv \omega_{m,1} - \omega_0$, are presented in Figure 2.2a–c. It is evident from Figure 2.2a that although the amplitude for populating a state with transition frequency $\omega_{m,1}$ at the end of the pulse is proportional to $|\epsilon(\omega_{m,1})|$, the time-dependent path leading to this value varies with $\omega_{m,1}$. For example, for $\omega_{m,1}$ near the line center the $c'_m(t)$ coefficients rise monotonically to their asymptotic values, whereas at off-center energies $(\omega_{m,1} - \omega_0 \neq 0)$ this is not the case. In essence, what happens is that (by Fourier's theorem) the pulse appears to have a much broader frequency profile at short times than it does at long times. This causes all the off-center $c'_m(t)$ coefficients to rise uniformly at short times. As time progresses and as the true nature of the $\epsilon(\omega_{m,1})$ profile becomes apparent, the system responds by depleting the off-center $c'_m(t)$ coefficients until they become proportional to $\epsilon(\omega_{m,1})$. In particular, in the case of extreme detuning where $\epsilon(\omega_{m,1}) \approx 0$, $c'_m(t)$, though these coefficients must vanish at the end of the pulse, they may be nonzero during the pulse. This means that such levels get populated and completely depopulated during the pulse. States of this kind are usually termed "virtual" states, although according to this description they are simply highly detuned (with respect to the pulse center) real states which become transiently populated and depopulated during the pulse.

Understanding how the material phase develops during and after the excitation process will prove to be important for control purposes. In accord with Eq. (2.12),

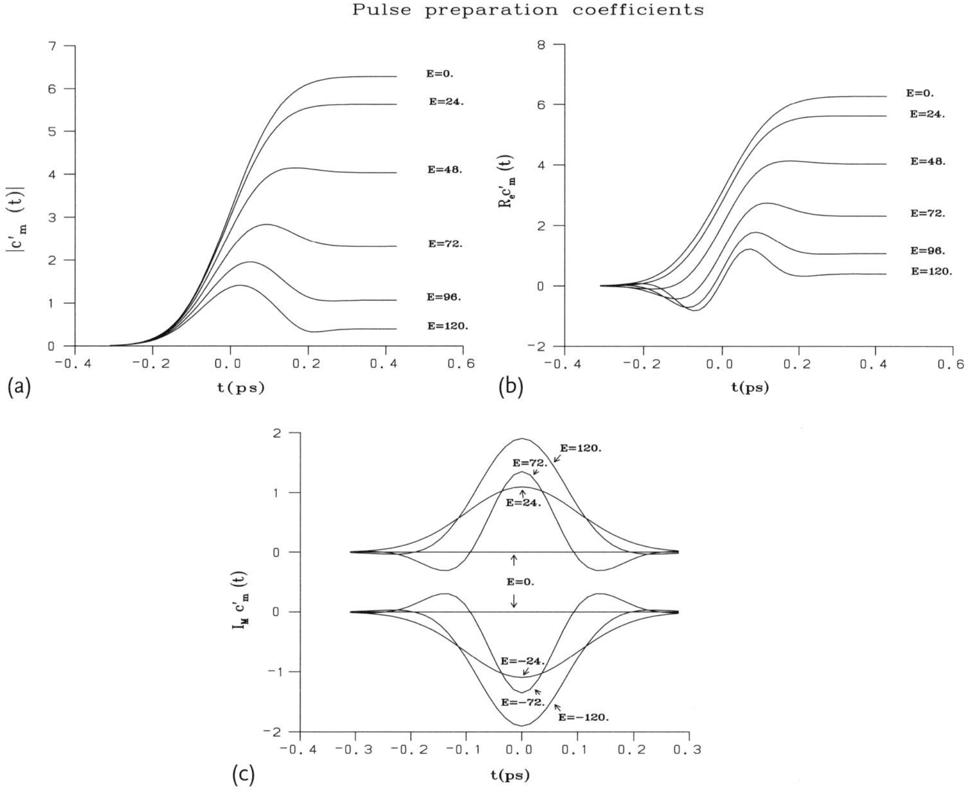

Figure 2.2 Time evolution of the $c'_m(t)$ coefficients at different detunings from the center of the pulse for a Gaussian pulse with FWHM of 120 cm^{-1}. (a) $|c'_m(t)|$, (b) Re$[c'_m(t)]$, (c) Im$[c'_m(t)]$. (E is to be replaced by $\hbar(\omega_{m,1} - \omega_0)$). Note that Re denotes the real part and Im denotes the imaginary part of the argument that follows.

the phase of $b_m(t)$ (see Eq. (2.14)) at the end of the pulse is that of $d_{m,1}\bar{\epsilon}(\omega_{m,1})$, which means (see also Eq. (2.18) and Eq. (2.20)) that $c'_m(t)$ is real at the end of the pulse. As Figure 2.2b shows, $c'_m(t)$ is real at all times for zero detuning ($\omega_{m,1} - \omega_0 = 0$) and complex during the pulse for finite detunings. In fact, the phase of $c'_m(t)$ changes linearly with time at the early stages of the pulse, with a slope given by $\omega_{m,1}$. This time dependence counteracts the natural time evolution of the wave packet of excited states, given by Eq. (2.2). Thus, during the buildup phase of the wave packet, that is, during the early part for which the preparation phase changes linearly, the wave packet hardly moves: It merely grows in size while changing its shape due to the changes in $|c_m(t)|$. As the pulse wanes, the time dependence of the preparation phases becomes less and less pronounced, until they become constant. When this occurs, the $\exp(-iE_nt/\hbar)$ factors in Eq. (2.2) are no longer being offset, and the wave packet of excited states is "released" to move under the full influence of the radiation-free Hamiltonian.

It is also of interest to look at $(d/dt)|c'_m(t)|^2$, which is proportional to the *rate* of populating the mth level. We see from Figure 2.3 that this rate, which depends on the detuning, is far from being constant. This result contradicts Fermi's celebrated "Golden Rule" formula introduced by most textbooks (cf. [5]), according to which $d/dt|c'_m(t)|^2$ *averaged over the pulse modes* is constant. Such averaging is permissible if the action of each pulse mode on the rate is additive. Quite clearly, the effect of the different pulse modes is not additive, since the probability for observing each state involves first calculating $b_m(t)$ – which, according to Eqs. (2.14) and (2.15), is given as an integral over all the pulse modes – and then squaring the result. Contrarily, in the derivation of the "Golden Rule", one first calculates the rate $d/dt|b_m(t)|^2$ for each mode and *then* integrates over the pulse modes. This procedure is permissible only when the $\phi(\omega)$ is in some sense a random variable, which corresponds to a pulse which is "incoherent."

Figures 2.2 and 2.3 only display the RW coefficients. However, it can be shown from Eq. (2.19) that the behavior of CRW coefficients resemble that of the highly detuned RW coefficients, save for the fact that the CRW coefficients rigorously vanish at large times. Thus, the $c_m^-(t)$ coefficients make a noticeable contribution only at short times. This justifies the usual practice of neglecting the CRW terms whenever detuning with respect to some material levels is small.

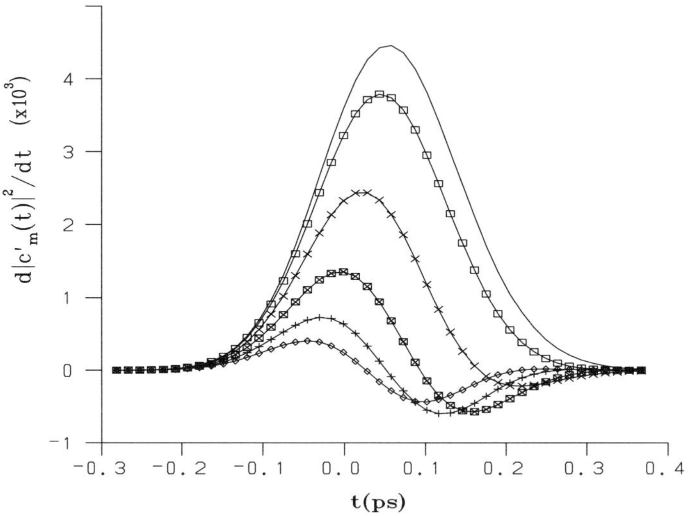

Figure 2.3 Excitation rate $(d|c'_m(t)|^2/dt)$ for a $120\,\mathrm{cm}^{-1}$-wide pulse at different detunings from the pulse center. The line corresponds to $\Delta_0 = 0$, boxes to $\Delta_0 = 24\,\mathrm{cm}^{-1}$, the x's to $\Delta_0 = 48\,\mathrm{cm}^{-1}$, the filled boxes to $\Delta_0 = 72\,\mathrm{cm}^{-1}$, the pluses to $\Delta_0 = 96\,\mathrm{cm}^{-1}$, and the diamonds to $\Delta_0 = 120\,\mathrm{cm}^{-1}$.

The results of Eqs. (2.2), (2.14) and (2.22) are readily summarized as follows: During the pulse the material wave function is

$$|\Psi(t)\rangle = |E_1\rangle \exp(-iE_1t/\hbar) + i/\hbar \sum_m \left[c_m^+(\tau)\bar{\epsilon}(\omega_{m,1}) + c_m^-(\tau)\bar{\epsilon}(-\omega_{m,1})\right] d_{m,1}|E_m\rangle \exp(-iE_mt/\hbar) . \quad (2.25)$$

At the end of the pulse the wave function is given by

$$\left|\Psi\left(t \gg \frac{1}{\Gamma}\right)\right\rangle = |E_1\rangle \exp(-iE_1t/\hbar) + \frac{2\pi i}{\hbar}\sum_m \bar{\epsilon}(\omega_{m,1})\, d_{m,1}|E_m\rangle \exp(-iE_mt/\hbar) . \quad (2.26)$$

In the next section we extend these results to the case of excitation involving a continuous spectrum.

2.3 Photodissociation

2.3.1 General Formalism

Photodissociation results when the energy eigenstates reached by photon absorption are in the continuum. When the spectrum is continuous we have to use the scattering wave functions as the states of matter. These are defined as eigenstates $|E, \boldsymbol{m}\rangle$ of the material Hamiltonian with continuous eigenvalues E, that is,

$$[E - H_M]|E, \boldsymbol{m}\rangle = 0 , \quad (2.27)$$

where $\boldsymbol{m}$ designates a collection of additional quantum numbers that may be necessary to completely specify the state. In particular, if we regard the state $|E, \boldsymbol{m}\rangle$ as representing a collisional or a dissociation process, then $\boldsymbol{m}$ includes the chemical identity as well as all the *internal* (electronic, vibrational, rotational, etc.) quantum numbers of the molecules that participate in the collision, before (or after) the event [6].

The portion of the wave packet excited to a continuous segment of the spectrum is given as

$$|\Psi'(t)\rangle = \frac{2\pi i}{\hbar}\sum_n \int dE\,\bar{\varepsilon}(\omega_{E,1})\,\langle E, \boldsymbol{n}|\hat{\boldsymbol{\varepsilon}}\cdot\boldsymbol{d}|E_1\rangle|E, \boldsymbol{n}\rangle \exp(-iEt/\hbar) . \quad (2.28)$$

Because we now have an integral over E in the expansion, the normalization of the constituent states $|E, \boldsymbol{n}\rangle$ must be different than that of bound states $|E_m\rangle$. It is given as,

$$\langle E', \boldsymbol{m}|E, \boldsymbol{n}\rangle = \delta(E - E')\,\delta_{\boldsymbol{m},\boldsymbol{n}} . \quad (2.29)$$

Likewise, the dimension of the scattering states is different; due to Eq. (2.29), $\psi_{E,\boldsymbol{n}}(R) \equiv \langle R|E, \boldsymbol{n}\rangle$ has dimensions of [length]$^{-1/2}$[energy]$^{-1/2}$.

We wish now to investigate the long-time properties of Eq. (2.28). In order to do so we need to relate the eigenstates of H_M to the eigenstates which describe the freely moving fragments at the end of the process. Take as an example a triatomic molecule ABC, which breaks apart at the end of the process to yield, say, the $A + BC$ channel. (The extension of this treatment to the breakup into different arrangements is discussed at the end of this chapter. Here the term "arrangement" is used to denote the way in which the particles are bound to one another.) Factorizing out the ABC center of mass motion, we partition the remaining part of H_M into three parts

$$H_M = K_{\boldsymbol{R}} + K_{\boldsymbol{r}} + W(\boldsymbol{R}, \boldsymbol{r}) . \tag{2.30}$$

Here $\boldsymbol{R}$ is the radius vector separating A and the BC center of mass, $\boldsymbol{r}$ is the $B{-}C$ separation. $W(\boldsymbol{R}, \boldsymbol{r})$ is the total potential energy of A, B and C, and

$$K_{\boldsymbol{R}} = \frac{-\hbar^2}{2\mu}\nabla^2_{\boldsymbol{R}} , \tag{2.31}$$

$$K_{\boldsymbol{r}} = \frac{-\hbar^2}{2m}\nabla^2_{\boldsymbol{r}} , \tag{2.32}$$

are the kinetic energy operators in $\boldsymbol{R}$ and $\boldsymbol{r}$ in the coordinate representation, with μ and m being the reduced masses,

$$\mu = m_A \frac{(m_B + m_C)}{(m_A + m_B + m_C)} , \quad m = \frac{m_B m_C}{(m_B + m_C)} . \tag{2.33}$$

The asymptotic limit of $W(\boldsymbol{R}, \boldsymbol{r})$ as A departs from BC is denoted as

$$v(r) = \lim_{R\to\infty} W(\boldsymbol{R}, \boldsymbol{r}) . \tag{2.34}$$

It is clear that the $A{-}BC$ *interaction potential*, defined as

$$V(\boldsymbol{R}, \boldsymbol{r}) \equiv W(\boldsymbol{R}, \boldsymbol{r}) - v(r) , \tag{2.35}$$

vanishes as $R \to \infty$:

$$\lim_{R\to\infty} V(\boldsymbol{R}, \boldsymbol{r}) = 0 . \tag{2.36}$$

Defining the BC Hamiltonian as

$$h_{\boldsymbol{r}} \equiv K_{\boldsymbol{r}} + v(r) , \tag{2.37}$$

the triatomic Hamiltonian of Eq. (2.30) can now be broken into three different parts using Eq. (2.35):

$$H_M = K_{\boldsymbol{R}} + h_{\boldsymbol{r}} + V(\boldsymbol{R}, \boldsymbol{r}) . \tag{2.38}$$

We see that it is the interaction potential $V(\boldsymbol{R}, \boldsymbol{r})$ that couples the motion of the A atom to the motion of the BC diatomic. In its absence the two free fragments A and BC described by the free Hamiltonian

$$H_0 \equiv K_R + h_r \tag{2.39}$$

move independently of one another. Because H_0 is a sum of two independent terms, its eigenstates, $|E, \boldsymbol{n}; 0\rangle$, satisfying

$$[E - H_0]|E, \boldsymbol{n}; 0\rangle = 0 \ , \tag{2.40}$$

are given as products

$$|E, \boldsymbol{n}; 0\rangle = |e_n\rangle|E - e_n\rangle \ . \tag{2.41}$$

Here, $|e_n\rangle$, the *internal* states, satisfy the eigenvalue relation,

$$[e_n - h_r]|e_n\rangle = 0 \ , \tag{2.42}$$

with e_n being the internal (electronic, vibrational, rotational) energy of the BC diatomic. $|E - e_n\rangle$, which satisfy the eigenvalue relation,

$$[E - e_n - K_R]|E - e_n\rangle = 0 \ , \tag{2.43}$$

describe the free (translational) motion of A relative to BC.

The $|e_n\rangle$ eigenstates of h_r are often called *channels*. A channel is said to be *open* if $E - e_n > 0$; it is said to be *closed* if $E - e_n < 0$. When a channel is open the solution of Eq. (2.43), written in the coordinate representation,

$$\left[E - e_n + \frac{\hbar^2}{2\mu}\nabla_R^2\right]\langle \boldsymbol{R}|E - e_n\rangle = 0 \ , \tag{2.44}$$

describes a plane wave of kinetic energy $E - e_n$,

$$\langle \boldsymbol{R}|E - e_n\rangle = \left(\frac{\mu k_n}{\hbar^2(2\pi)^3}\right)^{\frac{1}{2}} \exp\left(i\boldsymbol{k}_n \cdot \boldsymbol{R}\right) \ , \tag{2.45}$$

where

$$k_n \equiv |\boldsymbol{k}_n| = \frac{[2\mu\,(E - e_n)]^{\frac{1}{2}}}{\hbar} \tag{2.46}$$

is the wave vector of the free motion of A relative to the BC center of mass. Since the free solutions are continuous they too satisfy the continuous spectrum normalization,

$$\langle E', \boldsymbol{m}; 0|E, \boldsymbol{n}; 0\rangle = \delta\left(E - E'\right)\delta_{\boldsymbol{m},\boldsymbol{n}} \ . \tag{2.47}$$

Note that if a channel is closed, k_n is imaginary (see Eq. (2.46)) and the $\langle \boldsymbol{R}|\boldsymbol{k}_n\rangle$ wave function is proportional to a decaying exponential $\exp(-k_n R)$. Closed channels are therefore not observed in the $R \to \infty$ limit.

The eigenstates $|E, \boldsymbol{n}\rangle$ of the fully interacting Hamiltonian H_M are related to $|E, \boldsymbol{n}; 0\rangle$ in the following way: We first write Eq. (2.27) as an inhomogeneous equation,

$$[E - H_0]|E, \boldsymbol{n}\rangle = V|E, \boldsymbol{n}\rangle , \tag{2.48}$$

and then write the solution $|E, \boldsymbol{n}\rangle$ as a particular solution to Eq. (2.48), $|E, \boldsymbol{n}\rangle = [E - H_0]^{-1} V|E, \boldsymbol{n}\rangle$, plus any solution $|E, \boldsymbol{n}; 0\rangle$ to the homogeneous part of that equation. We have that

$$|E, \boldsymbol{n}\rangle = |E, \boldsymbol{n}; 0\rangle + [E - H_0]^{-1} V|E, \boldsymbol{n}\rangle . \tag{2.49}$$

This is an integral equation. To see this, introduce the spectral resolution as an inverse of an operator,

$$[E - H_0]^{-1} = \int dE' \frac{|E', \boldsymbol{n}; 0\rangle\langle E', \boldsymbol{n}; 0|}{E - E'} , \tag{2.50}$$

and Eq. (2.49) becomes

$$|E, \boldsymbol{n}\rangle = |E, \boldsymbol{n}; 0\rangle + \int dE' \frac{|E', \boldsymbol{n}; 0\rangle\langle E', \boldsymbol{n}; 0|V|E, \boldsymbol{n}\rangle}{E - E'} . \tag{2.51}$$

However, the $\int dE'$ integral is ill-defined in the Riemann sense since the integrand diverges at $E' = E$. Hence, we calculate the integral as a Cauchy integral by considering its value as the limit of a series of well-defined Riemann integrals. That is, we generate a series of well-defined Riemann integrals by adding a small $i\epsilon$ imaginary part to E, thereby avoiding the divergence, calculate the integral and finally letting $\epsilon \to 0$. In what follows, we usually omit writing the $\epsilon \to 0$ step, but it is always implied.

It turns out that the limiting value thus obtained depends on the sign of ϵ. We therefore consider two cases, one in which we add $i\epsilon$ and one in which we subtract $i\epsilon$, with $\epsilon > 0$. The resulting two equations, which replace Eq. (2.49), are

$$|E, \boldsymbol{n}^{\pm}\rangle = |E, \boldsymbol{n}; 0\rangle + \lim_{\epsilon \to 0} [E \pm i\epsilon - H_0]^{-1} V|E, \boldsymbol{n}^{\pm}\rangle . \tag{2.52}$$

Each of the above equations is known as the Lippmann–Schwinger equation. The "+" solutions are called the *outgoing* scattering states, and the "−" solutions are called the *incoming* scattering states. Though each such state is an independent solution of the full Schrödinger equation (Eq. (2.27)), the incoming and outgoing states are not orthogonal to one another, nor do they satisfy the same boundary conditions.

We now use the Lippmann–Schwinger equation to explore the long-time behavior of the wave packet $\Psi'(t)$ that we have created with the laser pulse. We can use either the outgoing or incoming states as the basis set for expanding $\Psi'(t)$. In what follows we shall expose the different boundary conditions and see which type of solution is best suited for which purpose. Substituting Eq. (2.52) in Eq. (2.28), we obtain that

$$|\Psi'(t)\rangle = \frac{2\pi i}{\hbar} \sum_n \int dE \exp(-iEt/\hbar)\, \bar{\varepsilon}(\omega_{E,1}) \langle E, \boldsymbol{n}^{\pm} | \hat{\boldsymbol{\varepsilon}} \cdot \boldsymbol{d} | E_1 \rangle \times \left\{ |E, \boldsymbol{n}; 0\rangle + [E \pm i\epsilon - H_0]^{-1} V | E, \boldsymbol{n}^{\pm}\rangle \right\} . \tag{2.53}$$

Using the spectral resolution of $[E \pm i\epsilon - H_0]^{-1}$ (Eq. (2.50)) we have from Eq. (2.53) that the amplitude for finding a free state $|E', \boldsymbol{m}; 0\rangle$ at time t is given as

$$\langle E', \boldsymbol{m}; 0 | \Psi'(t)\rangle = \frac{2\pi i}{\hbar} \sum_n \int dE \exp(-iEt/\hbar)\, \bar{\varepsilon}(\omega_{E,1}) \langle E, \boldsymbol{n}^{\pm} | \hat{\boldsymbol{\varepsilon}} \cdot \boldsymbol{d} | E_1 \rangle \times \left\{ \langle E' \boldsymbol{m}; 0 | E, \boldsymbol{n}; 0\rangle + \left[E \pm i\epsilon - E'\right]^{-1} \langle E' \boldsymbol{m}; 0 | V | E, \boldsymbol{n}^{\pm}\rangle \right\} . \tag{2.54}$$

Using the normalization of the free states (Eq. (2.47)), we have that

$$\begin{aligned} \langle E', \boldsymbol{m}; 0 | \Psi'(t)\rangle = {} & \frac{2\pi i}{\hbar} \exp(-iE't/\hbar)\, \bar{\varepsilon}(\omega_{E',1}) \langle E', \boldsymbol{m}^{\pm} | \hat{\boldsymbol{\varepsilon}} \cdot \boldsymbol{d} | E_1 \rangle \\ & + \frac{2\pi i}{\hbar} \sum_n \int dE \exp(-iEt/\hbar)\, \bar{\varepsilon}(\omega_{E,1}) \langle E, \boldsymbol{n}^{\pm} | \hat{\boldsymbol{\varepsilon}} \cdot \boldsymbol{d} | E_1 \rangle \\ & \times \left[E \pm i\epsilon - E'\right]^{-1} \langle E' \boldsymbol{m}; 0 | V | E, \boldsymbol{n}^{\pm}\rangle . \end{aligned} \tag{2.55}$$

In the $t \to \infty$ limit the integration over E can be performed analytically by contour integration (see Figure 2.4). To see this we note that in that limit the integrand on a large semicircle in the lower part of the complex-E plane is zero, since, for $E = Re^{i\theta}$, with $\theta < 0$,

$$\begin{aligned} \exp(-iEt/\hbar) &= \exp\left(\frac{-iRe^{i\theta}t}{\hbar}\right) \\ &= \exp\left(\frac{-iR\cos\theta\, t}{\hbar}\right) \exp\left(\frac{R\sin\theta\, t}{\hbar}\right) \xrightarrow{t\to\infty} 0 . \end{aligned} \tag{2.56}$$

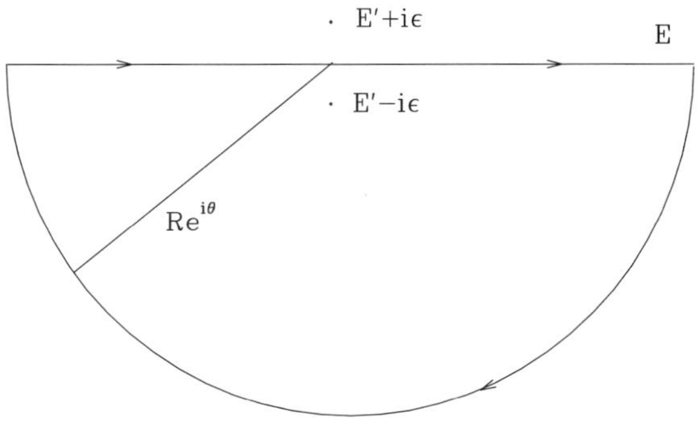

Figure 2.4 Complex energy plane contour integration.

Hence the result of the real E integration remains unchanged by supplementing it with an integration along the above large semicircle in the lower-half of the E-plane. Since in the $-i\epsilon$ case, the integrand has a pole at $E = E' + i\epsilon$ residing outside the closed contour, the whole integral is zero. We obtain that

$$\lim_{t\to\infty} \langle E', \boldsymbol{m}; 0|\Psi'(t)\rangle = \frac{2\pi i}{\hbar}\bar{\varepsilon}(\omega_{E',1}) \exp\left(-iE't/\hbar\right) \langle E', \boldsymbol{m}^-|\hat{\boldsymbol{\varepsilon}}\cdot \boldsymbol{d}|E_1\rangle \,. \tag{2.57}$$

Thus, in terms of the $|E, \boldsymbol{m}^-\rangle$ states, the coefficients of expansion of the excited wave packet directly yield the probability amplitude of observing states $|E, \boldsymbol{m}; 0\rangle$ in the *distant future*.

If instead of the incoming states we use the outgoing states, the closed-contour integration encircles a pole at $E = E' - i\epsilon$, then the integration yields

$$\lim_{t\to\infty} \langle E', \boldsymbol{m}; 0|\Psi'(t)\rangle = \frac{2\pi i}{\hbar} \exp\left(-iE't/\hbar\right) \bar{\varepsilon}\left(\omega_{E',1}\right) \times \sum_n S_{n,m}\left(E'\right) \langle E', \boldsymbol{n}^+|\hat{\boldsymbol{\varepsilon}}\cdot \boldsymbol{d}|E_1\rangle \,, \tag{2.58}$$

where the $S_{n,m}(E')$ matrix,

$$S_{n,m}\left(E'\right) \equiv \delta_{n,m} - 2\pi i\langle E', \boldsymbol{m}; 0|V|E', \boldsymbol{n}^+\rangle \equiv \langle E', \boldsymbol{n}; 0|S|E', \boldsymbol{m}; 0\rangle \,, \tag{2.59}$$

is known as the *S-matrix* or Scattering matrix.

The form of Eq. (2.58) appears more complicated than that of Eq. (2.57) because each $\langle E, \boldsymbol{m}; 0|\Psi'(t)\rangle$ component is made up of contributions from all degenerate $|E, \boldsymbol{n}; 0\rangle$ states. Why use it at all then? The reason is that in ordinary scattering events we use states whose past is well known to us. These are the outgoing states, because when $t \to -\infty$ it is the contour on the semicircle in the upper half of the complex E-plane that vanishes. Thus, supplementing the real-E integration by such a contour keeps the $E = E' - i\epsilon$ pole out of the contour and we obtain that

$$\lim_{t\to-\infty} \langle E', \boldsymbol{m}; 0|\Psi'(t)\rangle = \frac{2\pi i}{\hbar}\bar{\varepsilon}\left(\omega_{E',1}\right) \exp\left(-iE't/\hbar\right) \langle E', \boldsymbol{m}^+|\hat{\boldsymbol{\varepsilon}}\cdot \boldsymbol{d}|E_1\rangle \,. \tag{2.60}$$

That is, the $|E, \boldsymbol{m}^+\rangle$ states expose a simplified structure as $t \to -\infty$. By contrast, the $t \to -\infty$ limit appears more complicated when the wave packet is expanded in the incoming states because now the $E = E' + i\epsilon$ pole is enclosed within the integration contour, and we obtain that

$$\lim_{t\to-\infty} \langle E', \boldsymbol{m}; 0|\Psi'(t)\rangle = \frac{2\pi i}{\hbar} \exp\left(-iE't/\hbar\right) \bar{\varepsilon}\left(\omega_{E',1}\right) \times \sum_n S^-_{n,m}\left(E'\right) \langle E', \boldsymbol{n}^-|\hat{\boldsymbol{\varepsilon}}\cdot \boldsymbol{d}|E_1\rangle \,, \tag{2.61}$$

where the $S^-_{n,m}(E')$ matrix is defined as

$$S^-_{n,m}\left(E'\right) \equiv \delta_{n,m} + 2\pi i\langle E'\boldsymbol{m}; 0|V|E', \boldsymbol{n}^-\rangle \,. \tag{2.62}$$

In the case of optical pulse excitation, we use the incoming solutions $|E, \boldsymbol{n}^-\rangle$ because the origin of the system in the remote past is already known to us and we are interested in the fate of the system in the distant future.

The above results deal with photodissociation. A similar formulation can be applied to an inelastic scattering event, for example, the scattering of $A + BC$ from an initial state $|E, \boldsymbol{m}; 0\rangle$ into a final state $|E', \boldsymbol{n}; 0\rangle$ of $A + BC$. This problem can be phrased as asking for the amplitude of making a transition from free-state $|E, \boldsymbol{m}; 0\rangle$ at $t \to -\infty$ to free-state $|E', \boldsymbol{n}; 0\rangle$ at $t \to \infty$. Since $|E, \boldsymbol{m}^+\rangle$ is known to have evolved from $|E, \boldsymbol{m}; 0\rangle$ and $|E', \boldsymbol{n}^-\rangle$ is known to evolve to state $|E, \boldsymbol{n}; 0\rangle$, the answer is given by the expansion coefficients of an outgoing state in terms of the incoming states,

$$|E, \boldsymbol{m}^+\rangle = \sum_n \int dE' |E', \boldsymbol{n}^-\rangle\langle E', \boldsymbol{n}^- | E, \boldsymbol{m}^+\rangle . \tag{2.63}$$

It is easy to show by the same contour integration as above that

$$\langle E', \boldsymbol{n}^- | E, \boldsymbol{m}^+\rangle = \delta\left(E - E'\right) S_{\boldsymbol{n},\boldsymbol{m}}(E) , \tag{2.64}$$

where $S_{\boldsymbol{n},\boldsymbol{m}}(E)$ is the S-matrix of Eq. (2.59).

The above treatment assumes that the product comprises a single arrangement channel, that is, the formation of $A + BC$ as the final product. The extension of this formalism to multiple product arrangements, for example, where $A + BC$ and $AB + C$ are products of ABC photodissociation, or where $A + BC$ collide to form $A + BC$ and $AB + C$ requires:

1. The addition of a channel label $q = 1, 2, \ldots$ to the descriptor of the state, so that $|E, \boldsymbol{m}, q; 0\rangle$ corresponds to arrangement q; and
2. Rewriting Eqs. (2.30)–(2.52) to partition the Hamiltonian in a fashion appropriate to the final arrangement of interest.

Thus, for example, for the $AB + C$ arrangement, the vector $\boldsymbol{R}$ defines the AB to C distance, $\boldsymbol{r}$ defines the $A - B$ separation, and so on.
For example, and for later use below, the amplitude for reactive scattering from $A + BC$ (labeled q) to $AB + C$ (labeled q') is given by [6, 7] (see Eq. (2.59) and Eq. (2.64))

$$\begin{aligned} S_{\boldsymbol{n},q':\boldsymbol{m},q}\left(E'\right) &\equiv \delta_{\boldsymbol{n},\boldsymbol{m}}\delta_{q,q'} - 2\pi i\langle E', \boldsymbol{m}, q; 0| V_q | E', \boldsymbol{n}, q'^+\rangle \\ &\equiv \langle E', \boldsymbol{n}, q'; 0| S | E', \boldsymbol{m}, q; 0\rangle \end{aligned} \tag{2.65}$$

where V_q is the part of the full potential that goes to zero as the distance between A and BC goes to infinity.

The boundary conditions that the state $|E, \boldsymbol{n}, q^-\rangle$ goes to $|E, \boldsymbol{n}, q; 0\rangle$ in the distant future and that $|E, \boldsymbol{n}, q^+\rangle$ originates from the $|E, \boldsymbol{n}, q; 0\rangle$ that existed in the remote past, is summarized in the expression:

$$\lim_{t \to \mp\infty} e^{-i(E \pm i\epsilon)t/\hbar} |E, \boldsymbol{n}, q^\pm\rangle = e^{-i(E \pm i\epsilon)t/\hbar}\, |E, \boldsymbol{n}, q; 0\rangle \ . \tag{2.66}$$

2.3.2 Electronic States

Thus far, we have suppressed the eigenfunctions associated with the electronic state of the system. We continue to do so below, unless necessary. A complete treatment would consider the total material Hamiltonian H_{MT} as a sum of two terms:

$$H_{MT} = H_M + H_{el} = (K + V_{NN}) + H_{el} \ . \tag{2.67}$$

Here H_M is the nuclear part, written as the sum of a kinetic term (K) and a nuclear–nuclear interaction V_{NN}, and H_{el} is the electronic part. Since H_{el} involves interactions between the electrons and nuclei, H_{MT} is not separable in the nuclear and electronic degrees of freedom. However, adopting the Born–Oppenheimer approximation substantially simplifies matters. In this approximation we:

1. Define the Born–Oppenheimer potential seen by the nuclei in electronic state $|e\rangle$ as $W_e = \langle e|V_{NN} + H_{el}|e\rangle$, where

$$H_{el}|e\rangle = E_e|e\rangle \ . \tag{2.68}$$

2. Assume that

$$K|e\rangle = |e\rangle K \ . \tag{2.69}$$

Thus, as a consequence of Eqs. (2.68) and (2.69), we have that:

$$\langle e|H_{MT}|e'\rangle = \delta_{ee'}\langle e|H_{MT}|e\rangle \ . \tag{2.70}$$

The total eigenfunctions are now of the form $|E_i\rangle|g\rangle$, $|E, \boldsymbol{n}^-\rangle|e\rangle$, and so on, where $|g\rangle$ denotes the ground electronic state. If we consider optical transitions between the electronic states then the matter–radiation Hamiltonian $H_{MR}(t)$ (Eq. (1.5)) should be replaced by

$$H'_{MR}(t) = \sum_{ee'} |e\rangle\langle e|H_{MR}(t)|e'\rangle\langle e'| \ , \tag{2.71}$$

where the sums are over all electronic states (including the ground electronic state) and where $H_{MR}(t)$ is given by Eq. (2.5). As a result, the matrix element $\langle E, \boldsymbol{n}^-|\hat{\boldsymbol{\varepsilon}} \cdot \boldsymbol{d}|E_i\rangle$ would be replaced by $\langle E, \boldsymbol{n}^-|\langle e|\hat{\boldsymbol{\varepsilon}} \cdot \boldsymbol{d}|g\rangle|E_i\rangle \equiv \langle E, \boldsymbol{n}^-|d_{e,g}|E_i\rangle$, where

$$d_{e,g} = \langle e|\hat{\boldsymbol{\varepsilon}} \cdot \boldsymbol{d}|g\rangle \tag{2.72}$$

is an operator in the space of nuclear motion. Similarly, Eq. (2.28) becomes

$$|\Psi'(t)\rangle = \frac{2\pi i}{\hbar} \sum_n \int dE \bar{\epsilon}(\omega_{E,1})|E, \boldsymbol{n}^-\rangle\langle E, \boldsymbol{n}^-|d_{e,g}|E_1\rangle \exp(-iEt/\hbar) \ . \tag{2.73}$$

2.3.3 Energy-Resolved Quantities

We now focus attention on energy-resolved quantities, such as the photodissociation cross section. Measurements of these quantities can be performed using CW excitation sources that excite only a single energy, or by using pulsed sources and extracting energy-resolved information after the pulse is over.

The photodissociation probability $P_n(E|i)$ into the state characterized by $\boldsymbol{n}$ at energy E, is given by the square of $A_{\boldsymbol{n}}(E|i)$, the photodissociation amplitude for observing the free state $\exp(-iEt/\hbar)|E,\boldsymbol{n};0\rangle$ in the long time limit. That is,

$$P_{\boldsymbol{n}}(E|i) = |A_{\boldsymbol{n}}(E|i)|^2 , \tag{2.74}$$

with $A_{\boldsymbol{n}}(E|i)$ defined as

$$A_{\boldsymbol{n}}(E|i) = \lim_{t\to\infty} \exp(iEt/\hbar)\,\langle E,\boldsymbol{n};0|\Psi(t)\rangle . \tag{2.75}$$

Because the bound and continuum wave functions usually belong to different electronic manifolds, they are orthogonal to one another and the only term which contributes to $A_{\boldsymbol{n}}(E|i)$ is derived from Ψ', the excited part of the wave packet. It follows from the boundary conditions on $|E,\boldsymbol{n}^-\rangle$ (Eq. (2.57)), that the $t\to\infty$ limit of Eq. (2.73) can be written as

$$|\Psi'(t\to\infty)\rangle = \frac{2\pi i}{\hbar}\sum_n \int dE\,\bar{\epsilon}(\omega_{E,i})\,|E,\boldsymbol{n};0\rangle\langle E,\boldsymbol{n}^-|d_{e,g}|E_i\rangle \exp(-iEt/\hbar) . \tag{2.76}$$

Hence, using the orthonormality of the $|E,\boldsymbol{n};0\rangle$ functions (Eq. (2.47)), we obtain from Eq. (2.75) that

$$A_{\boldsymbol{n}}(E|i) = \frac{2\pi i}{\hbar}\bar{\epsilon}(\omega_{E,i})\,\langle E,\boldsymbol{n}^-|d_{e,g}|E_i\rangle . \tag{2.77}$$

The photodissociation cross section $\sigma_{\boldsymbol{n}}(E|i)$ is defined as the photon energy absorbed in a transition to a final fragment state, divided by the incident intensity of light per unit energy. The energy absorbed is $\hbar\omega_{E,i}P_{\boldsymbol{n}}(E|i)$. The incident intensity (flux) per unit frequency $I(\omega_{E,i})$ is the radiation energy-density per unit frequency times the velocity of light, $|\epsilon(\omega_{E,i})|^2c$. Hence the incident intensity per unit energy, $I(E)$, is $|\epsilon(\omega_{E,i})|^2c/\hbar$. Thus, from Eqs. (2.75) and (2.74) we have that

$$\sigma_{\boldsymbol{n}}(E|i) = \frac{\hbar\omega_{E,i}P_{\boldsymbol{n}}(E|i)}{I(E)} = \frac{4\pi^2\omega_{E,i}}{c}\left|\langle E,\boldsymbol{n}^-|d_{e,g}|E_i\rangle\right|^2 . \tag{2.78}$$

This formula forms the basis for many of the computations of detailed photodissociation cross sections and of angular distribution of photofragments reported in the literature [8, 9].

Note that by using incoming states, we have shown that the coefficient of the state $|E, \boldsymbol{n}^-\rangle$ at time $t = 0$ in Eq. (2.76) is exactly $A_{\boldsymbol{n}}(E|i)$, the long-time photodissociation amplitude. Thus, we obtain the crucial insight that the probability of obtaining a product in the state $|E, \boldsymbol{n}; 0\rangle$ is given solely by the probability of preparing the state $|E, \boldsymbol{n}^-\rangle$ at the time of preparation.

Note also that the form of the pulse does not appear in the expression (Eq. (2.78)) for the cross section. This is because resolving the energy, embodied in the orthogonality expression (Eq. (2.47)), extracts a single frequency component of $\bar{\epsilon}(\omega)$, whose contribution is cancelled in the division by the incident light intensity. Therefore, as shown in Appendix 2.A, we can use any convenient pulse shape to compute energy-resolved quantities. This is not the case if we want to follow the realtime dynamics of the system, where the pulse shape is intimately linked with the observables. Indeed, this link prevents a pulse-free definition of concepts such as the lifetime of a state. This issue is addressed in Appendix 2.A.

2.A
Appendix: Molecular State Lifetime in Photodissociation

In this Appendix we consider, in greater detail, the time dependence of photodissociation under various excitation conditions. To make a connection with wave packet methodologies we first rewrite the preparation coefficients (Eq. (2.15)) for continuum states as the *finite time* Fourier transform of the pulse,

$$\bar{\epsilon}(\omega_{E,i})c_E^+(t) + \bar{\epsilon}(-\omega_{E,i})c_E^-(t) = \int d\omega\bar{\epsilon}(\omega)\frac{\exp\left[i\left(\omega_{E,i} - \omega\right)t\right]}{i\left(\omega_{E,i} - \omega\right)}$$
$$= \int_{-\infty}^{t} dt'\varepsilon(z, t')\exp\left(i\omega_{E,i}t'\right) . \tag{2.A.1}$$

Using Eq. (2.A.1) in Eq. (2.73) we can write the excited portion of the wave packet, valid for any t, as

$$|\Psi'(t)\rangle = \frac{i}{\hbar}\sum_{n}\int dE\int_{-\infty}^{t} dt'\exp\left[i\left(E - E_i\right)t'/\hbar\right]\varepsilon\left(z, t'\right)|E, \boldsymbol{n}^-\rangle$$
$$\times\langle E, \boldsymbol{n}^-|d_{e,g}|E_i\rangle\exp\left(-iEt/\hbar\right) . \tag{2.A.2}$$

Using the spectral resolution of the evolution operator in the subspace spanned by the continuum wave functions,

$$\exp\left(-iH_c t/\hbar\right) = \sum_{n}\int dE\exp\left(-iEt/\hbar\right)|E, \boldsymbol{n}^-\rangle\langle E, \boldsymbol{n}^-| , \tag{2.A.3}$$

we can rewrite Eq. (2.A.2) as

$$|\Psi'(t)\rangle = \frac{i}{\hbar}\int_{-\infty}^{t} dt'\varepsilon\left(z, t'\right)\exp\left[-iH_c\left(t - t'\right)/\hbar\right]d_{e,g}|E_i(t')\rangle , \tag{2.A.4}$$

where we have defined $|E_i(t')\rangle = \exp(-iE_it'/\hbar)|E_i\rangle$.

One interesting special case is the ultrashort pulse limit, $\varepsilon(t) \approx \epsilon_0\delta(t)$, which is equivalent to choosing a completely white pulse, $\epsilon(\omega) = \epsilon_0/2\pi$. Substituting the ultrashort pulse form in Eq. (2.A.4) (or Eq. (2.28)) we obtain that

$$|\Psi'(t)\rangle = \frac{i\epsilon_0}{\hbar}\exp\left(-iH_ct/\hbar\right)d_{e,g}|E_i\rangle\ . \tag{2.A.5}$$

Equation (2.A.5) has often been used [10, 11] to describe photodissociation in the following way: At $t = 0$ the light pulse creates the wave packet $d_{e,g}|E_i\rangle$ which subsequently evolves under the action of $\exp(-iH_ct/\hbar)$. Strictly speaking, for Eq. (2.A.5) to hold, and hence for this qualitative interpretation to be valid, $\epsilon(\omega_{E,i})$ must vary more slowly with energy than any of the other variables in Eq. (2.28). In particular, it must vary more slowly than the energy (or frequency) dependence of the photodissociation amplitudes $\langle E, \boldsymbol{n}^-|d_{e,g}|E_i\rangle$.

An easy way to decide whether this is the case is to compare the bandwidth of the laser with the frequency width of the absorption spectrum of the molecule, where the latter is determined by $|\langle E, \boldsymbol{n}^-|d_{e,g}|E_i\rangle|^2$. Typically, for the case of direct dissociation, the absorption spectrum extends over a few thousands of cm^{-1}. In order to have a bandwidth broader than this, $\varepsilon(t)$ must, by Eq. (1.7), be as short as ≈ 1–5 fs. Since most pulses used in real photodissociation experiments are much longer, Eq. (2.A.5) is not a valid description of many photodissociation experiments.

One can correct Eq. (2.A.5), by viewing "realtime" dynamics as composed of a superposition of wavelets created by a *sequence* of δ pulses [12, 13]. This follows by writing the pulse $\varepsilon(t)$ in Eq. (2.A.4) as $\varepsilon(t) = \int dt'\varepsilon(t')\delta(t - t')$. Each $\delta(t - t')$ evolves according to Eq. (2.A.5) with a different starting time t'.

In most direct photodissociation cases the laser's bandwidth is much narrower than that of the absorption spectrum, and the limit opposite to Eq. (2.A.5) is realized. Under such circumstances we can approximate, in Eq. (2.28), the narrow range of energies accessed by the laser by a *single* continuum energy level E_0, and write $|\Psi'(t)\rangle$ as

$$|\Psi'(t)\rangle \approx 2\pi i\sum_n |E_0, \boldsymbol{n}^-\rangle\langle E_0, \boldsymbol{n}^-|d_{e,g}|E_i\rangle \left\{\int \frac{dE}{\hbar}\bar{\epsilon}\left(\omega_{E,i}\right)\exp\left(-iEt/\hbar\right)\right\}\ . \tag{2.A.6}$$

We recognize the term in the curly bracket as $\boldsymbol{E}_+(t)$ (Eq. (1.7)). Hence we see that under these circumstances the time dependence of the wave packet created by the laser is the time dependence of the laser pulse. In classical terms we would say that the dissociation is "faster" than the optical excitation process: Every photon absorbed leads to immediate dissociation, so that the only time evolution that can be observed is that of the laser pulse.

To obtain the lifetime of the state we integrate out over the spatial dependence of $\Psi'(\boldsymbol{R}, \boldsymbol{r}, t)$. This is most easily done by considering the *autocorrelation* function, defined as

$$F(t, t_0) \equiv \langle\Psi'(t_0)|\Psi'(t)\rangle\ . \tag{2.A.7}$$

It follows from Eq. (2.28) that $F(t, t_0)$ can be calculated as,

$$F(t, t_0) = \frac{-1}{\hbar^2} \sum_n \int dE\, c_E^+(t) c_E^{+*}(t_0) \left| \epsilon(\omega_{E,i}) \langle E, \boldsymbol{n}^- | d_{e,g} | E_i \rangle \right|^2 \times \exp\left[-iE(t - t_0)/\hbar\right] , \tag{2.A.8}$$

where we have ignored the CRW terms. If both t and t_0 refer to times after the pulse (i.e., $t, t_0 \gg 1/\Gamma$) we obtain, using Eq. (2.28), that

$$F(t, t_0) = -\left(\frac{2\pi}{\hbar}\right)^2 \sum_n \int dE \left| \epsilon(\omega_{E,i}) \langle E, \boldsymbol{n}^- | d_{e,g} | E_i \rangle \right|^2 \times \exp\left[-iE(t - t_0)/\hbar\right] . \tag{2.A.9}$$

We see that after the pulse the autocorrelation function is essentially the Fourier transform of the pulse-modulated absorption spectrum,

$$I_\epsilon(E) \equiv \sum_n \left| \epsilon(\omega_{E,i}) \langle E, \boldsymbol{n}^- | d_{e,g} | E_i \rangle \right|^2 . \tag{2.A.10}$$

Quite clearly the decay of the autocorrelation function, hence the "lifetime of the state", is a function of the pulse profile. There simply is no molecular lifetime measurement which is independent of the measurement apparatus. The only measurement that provides only molecular information is that where the laser width far exceeds the spectral bandwidth (Eq. (2.A.5)), giving the inverse of the spectral bandwidth.

As we shall see below (see Section 3.A), this temporal behavior is separate from the issue of product selectivity (i.e., enhancing one product over another). In fact, in the weak-field regime, it is not possible to alter the yield of the $\boldsymbol{n}$ quantum numbers in a one-photon dissociation process by "shaping" (e.g., shortening) the pulse. In order to do this we need to either interfere two colors, use multiphoton processes, or work with strong pulses. This is discussed in further detail in the chapters that follow.

3
Weak-Field Coherent Control

3.1
Traditional Excitation

Traditional molecular excitation and subsequent system evolution, discussed in Chapter 2, affords little opportunity for us to control the outcome of molecular events. According to perturbation theory, as developed in Section 2.3, the branching ratio (i.e., the relative probability to populate different product channels) at the end of the process depends entirely (see Eq. (2.77) above) on the ratio between squares $|\langle E, \boldsymbol{n}^- | \hat{\boldsymbol{\varepsilon}} \cdot \boldsymbol{d} | E_1 \rangle|^2 / |\langle E, \boldsymbol{m}^- | \hat{\boldsymbol{\varepsilon}} \cdot \boldsymbol{d} | E_1 \rangle|^2$ of the *purely material* transition dipole matrix elements. Thus, the electric field profile does not appear in the branching ratio expression. Since the electric field profile of the pulse is the means by which one could hope to influence and possibly control the outcome of the event, it seems that there is no way that we can change the natural branching ratio [14, 15].

This state of affairs holds true even if we consider pulsed excitation in the strong-field domain, provided that the excitation involves only a single precursor state $|E_1\rangle$. In order to see this [16] we generalize the perturbation theory expressions of Section 2.3.3 to the strong-field domain. We recall that the probability of populating a "free" state $|e_m\rangle|\boldsymbol{k}_m\rangle$ at any given time is given as

$$P_m(E)(t) = |\langle e_m | \langle \boldsymbol{k}_m | \Psi(t) \rangle|^2 . \tag{3.1}$$

Using the expansion of the wave packet

$$|\Psi(t)\rangle = b_1(t) \exp(-i E_1 t/\hbar) |E_1\rangle + \sum_n \int dE\, b_{E,n}(t) \exp(-i E t/\hbar) |E, \boldsymbol{n}^-\rangle ,$$

we have that

$$\langle e_m | \langle \boldsymbol{k}_m | \Psi(t) \rangle = b_1(t) \langle e_m | \langle \boldsymbol{k}_m | E_1 \rangle \exp(-i E_1 t/\hbar) + \sum_n \int dE\, b_{E,n}(t) \langle e_m | \langle \boldsymbol{k}_m | E, \boldsymbol{n}^- \rangle \exp(-i E t/\hbar) , \tag{3.2}$$

since $\langle e_m | E_1 \rangle = 0$, (e.g., the two states belong to different electronic states), it follows from Eq. (3.2) that in the long-time limit,

$$P_m(E) = |\langle e_m | \langle \boldsymbol{k}_m | \Psi(t \to \infty) |^2 = |b_{E,m}(t \to \infty)|^2 . \tag{3.3}$$

Quantum Control of Molecular Processes, Second Edition. Moshe Shapiro and Paul Brumer

It follows from Eq. (2.3) (extended to continuum states) that we can write, in complete generality, that

$$b_{E,n}(t) = \frac{i}{\hbar}\langle E, \boldsymbol{n}^- | \hat{\boldsymbol{\varepsilon}} \cdot \boldsymbol{d} | E_1 \rangle \int_{-\infty}^{t} dt' \varepsilon(t') \exp(i\omega_{E,1} t') b_1(t') \,. \tag{3.4}$$

Hence,

$$\frac{P_n(E)}{P_m(E)} = \left| \frac{b_{E,n}(\infty)}{b_{E,m}(\infty)} \right|^2 = \left| \frac{\langle E, \boldsymbol{n}^- | \hat{\boldsymbol{\varepsilon}} \cdot \boldsymbol{d} | E_1 \rangle}{\langle E, \boldsymbol{m}^- | \hat{\boldsymbol{\varepsilon}} \cdot \boldsymbol{d} | E_1 \rangle} \right|^2 . \tag{3.5}$$

Thus, the branching ratios at a fixed energy E are *independent of the laser power and pulse shape.* This result, which coincides with that of perturbation theory, holds true as long as there is only *one* initial state $|E_1\rangle$ that is excited to the continuum.

This argument motivates the idea that the way to control photodissociation is to use more than one initial state, or in greater generality, to use multiple excitation pathways. In this chapter we demonstrate that such a strategy allows us to actively influence and control which photodissociation product is formed. These ideas, which introduce the notion of "coherent control", will be later shown to hold true for any dynamical process, not just for photodissociation.

3.2 Photodissociation from a Superposition State

We introduce the basic principles of coherent control through a series of examples. In particular, we extend the treatment of Chapter 2 to the photodissociation of a nonstationary *superposition* of bound states, $|\chi(0)\rangle = \sum_{j=1}^{N} a_j |E_j\rangle \exp(-iE_j t/\hbar)$. Numerous experimental techniques can be used to create such a state. Whatever the method of preparation, the amplitude and phase of the coefficients a_j are functions of the experimentally controllable parameters used in creating the superposition.

Repeating the full treatment of weak-field photodissociation given in Chapter 2, but now for an initial superposition state, gives the same result as trivially replacing, in the final result (Eq. (2.28)), the single initial state $|E_1\rangle \exp(-iE_1 t/\hbar)$ by the superposition $\sum_{j=1}^{N} a_j |E_j\rangle \exp(-iE_j t/\hbar)$. Thus the system wave function at time t is given by

$$\begin{aligned} |\Psi(t)\rangle &= \sum_{j=1}^{N} a_j |E_j\rangle \exp\left(-iE_j t/\hbar\right) \\ &+ \frac{i}{\hbar} \sum_{j=1}^{N} a_j \sum_{n} \int dE \left[c_{E,j}^{+}(t)\bar{\epsilon}(\omega_{E,j}) + c_{E,j}^{-}(t)\bar{\epsilon}(-\omega_{E,j}) \right] \\ &\times \langle E, \boldsymbol{n}^- | d_{e,g} | E_j \rangle | E, \boldsymbol{n}^- \rangle \exp\left(-iEt/\hbar\right) \,, \end{aligned} \tag{3.6}$$

where $\omega_{E,j} \equiv (E - E_j)/\hbar$. Here the definition of the c_E preparation coefficients (Eq. (2.15)) has been extended to include many states. That is,

$$c^{\pm}_{E,j}(t) \equiv \frac{1}{\bar{\epsilon}(\pm\omega_{E,j})} \int_0^{\infty} d\omega \bar{\epsilon}(\pm\omega) \frac{\exp\left[i\left(\omega_{E,j} \mp \omega\right)t\right]}{i\left(\omega_{E,j} \mp \omega\right)} . \tag{3.7}$$

As in Eq. (2.22), $c^{+}_{E,j} \to 2\pi$ and $c^{-}_{E,j} \to 0$ for $t \gg 1/\Gamma$, so after the pulse Eq. (3.6) assumes the form

$$\begin{aligned} |\Psi(t)\rangle &= \sum_{j=1}^{N} a_j |E_j\rangle \exp\left(-i E_j t/\hbar\right) \\ &+ \frac{2\pi i}{\hbar} \sum_{j=1}^{N} a_j \sum_{n} \int d E \bar{\epsilon}(\omega_{E,j}) \langle E, \boldsymbol{n}^- | d_{e,g} | E_j \rangle | E, \boldsymbol{n}^- \rangle \\ &\times \exp\left(-i E t/\hbar\right) . \end{aligned} \tag{3.8}$$

The probability $P_{\boldsymbol{n}}(E)$ of being in the final state $|E, \boldsymbol{n}; 0\rangle$ is

$$P_{\boldsymbol{n}}(E) = |A_{\boldsymbol{n}}(E)|^2 , \tag{3.9}$$

where the probability amplitude $A_{\boldsymbol{n}}(E)$ is given by (using Eqs. (2.47) and (2.75))

$$\begin{aligned} A_{\boldsymbol{n}}(E) &= \lim_{t\to\infty} \exp\left(i E t/\hbar\right) \langle E, \boldsymbol{n}; 0 | \Psi(t)\rangle \\ &= \frac{2\pi i}{\hbar} \sum_{j=1}^{N} a_j \bar{\epsilon}(\omega_{E,j}) \langle E, \boldsymbol{n}^- | d_{e,g} | E_j \rangle . \end{aligned} \tag{3.10}$$

Of particular interest is the probability of being in a complete subspace of states, denoted by the label q; that is, all $\boldsymbol{m}$ associated with a fixed q, where $\boldsymbol{n} = (\boldsymbol{m}, q)$. Of greatest concern in chemistry is the case where q labels the chemical identity of the product of a chemical reaction, hence below we often explicitly refer to q in this manner. However, it should be clear that the theory applies to any other choice of q out of the set of $\boldsymbol{n}$ quantum numbers.

The probability of forming product with a particular value of q is given as

$$P_q(E) = \sum_{\boldsymbol{m}} P_{\boldsymbol{m},q}(E) = \sum_{\boldsymbol{m}} |A_{\boldsymbol{m},q}(E)|^2 . \tag{3.11}$$

Inserting $A_{\boldsymbol{m},q}(E)$ from Eq. (3.10) gives

$$P_q(E) = \left(\frac{2\pi}{\hbar}\right)^2 \sum_{i,j=1}^{N} \left[a_i a_j^* \bar{\epsilon}(\omega_{E,i}) \bar{\epsilon}^* \left(\omega_{E,j}\right)\right] d_q(j i) , \tag{3.12}$$

where

$$d_q(j i) = \sum_{\boldsymbol{m}} \langle E_j | d_{g,e} | E, \boldsymbol{m}, q^- \rangle \langle E, \boldsymbol{m}, q^- | d_{e,g} | E_i \rangle , \tag{3.13}$$

and $d_{g,e} = d^*_{e,g}$. The branching ratio between two channels at energy E, $R_{q,q'}(E)$, which we control below is then,

$$R_{q,q'}(E) = \frac{P_q(E)}{P_{q'}(E)} \ . \tag{3.14}$$

Consider then the nature of $P_q(E)$ (Eq. (3.12)). The diagonal terms ($i = j$) give the standard probability, at energy E, of photodissociation out of a bound state $|E_j\rangle$ to produce a product in channel q. The off-diagonal terms ($i \neq j$) correspond to interference terms between these photodissociation routes. These interference terms describe the constructive enhancement, or destructive cancellation, of product formation in subspace q. Equation (3.12) is important *in practice* because the interference terms have coefficients $(a_i a_j^* \bar{\epsilon}(\omega_{E,i})\bar{\epsilon}^*(\omega_{E,j}))$ whose magnitude and sign depend upon *experimentally controllable* parameters. Thus the experimentalist can manipulate laboratory parameters and, in doing so, alter the interference term and hence control the reaction product yield. This control approach can be extended to the domain of moderately strong fields, as shown in Chapter 12, Eq. (12.20).

It is important to note, for experimental implementation, that the interference term contains a z dependence through $\bar{\epsilon}(\omega_{E,i})\bar{\epsilon}^*(\omega_{E,j})$. Thus, it is necessary to design the step which prepares the a_i coefficients in such a way as to cancel this spatial dependence. Failure to do so results in a spatially dependent interference term and hence in the cancellation of control on the average. Several methods in which the spatial dependence are eliminated are described in Sections 3.4.2 and 3.6 below.

Equation (3.12) displays another important feature. That is, the entire control map, that is, $P_q(E)$ or $R_{q,q'}(E)$ as a function of the control parameters, is a function of very few molecular parameters, that is, the $d_q(j\,i)$. As a consequence, the experimentalist need only determine these few parameters in order to produce the entire control map. This statement constitutes the weak-field version of "Adaptive Feedback Control" [17–20] to be discussed in the context of strong-field optimal control in Chapter 5 below. In the general strong-field regime, a numerical nonlinear search procedure must be performed (see Chapter 5), to achieve a desired optimization. However, in the weak-field regime, because of the simple bilinear dependence of each $P_q(E)$ on the $a_j\bar{\epsilon}(\omega_{E,j})$ experimental parameters, we need only carry out $q \times N^2$ measurements to determine all the $d_q(j\,i)$ coefficients. Once these coefficients are known, the bilinear $P_q(E)$ function can be analytically interpolated to give any desired branching ratio between, and including, the extrema of $R_{q,q'}(E)$.

3.2.1
Bichromatic Control

Experimentally attaining control via Eq. (3.12) requires a light source containing N frequencies ω_i, where $i = 1, \ldots, N$. Both pulsed excitation with a source whose frequency width encompasses these frequencies, as well as excitation with N continuous wave (CW) lasers of frequencies $\omega_i = \omega_{E,i}$, ($i = 1, \ldots, N$) are possible approaches, as depicted in Figure 3.1. Here we focus on $N = 2$, that is, the effect of two CW lasers on a system in a superposition of two states (Figure 3.1c), a scenario which we call "bichromatic control."

CW light at frequency ω is described by

$$\boldsymbol{\epsilon}(\omega) = \hat{\boldsymbol{\epsilon}}\left[\epsilon(\omega)\delta(\omega - \omega_i) + \epsilon(-\omega)\delta(\omega + \omega_i)\right] . \tag{3.15}$$

Using Eqs. (1.7), (1.8), (1.9) and (2.9) we obtain

$$\begin{aligned} \boldsymbol{E}(z,t) &= 2\hat{\boldsymbol{\epsilon}}|\epsilon(\omega_i)|\cos\left[\omega_i\tau - \phi\left(\omega_i\right)\right] \\ &= 2\hat{\boldsymbol{\epsilon}}\,\mathrm{Re}\left[\bar{\epsilon}\left(\omega_i\right)\exp\left(-i\omega_i t\right)\right] . \end{aligned} \tag{3.16}$$

Consider then two parallel CW fields of frequencies ω_1 and ω_2 incident on a molecule. The light–molecule interaction potential (Eq. (1.5)) is then

$$H_{MR}(t) = -\sum_{i=1}^{2} 2\boldsymbol{d}\cdot\hat{\boldsymbol{\epsilon}}\,\mathrm{Re}[\bar{\epsilon}(\omega_i)\exp(-i\omega_i t)] . \tag{3.17}$$

Tuning the ω_1 and ω_2 frequencies such that $\omega_2 - \omega_1 = (E_1 - E_2)/\hbar$, we have that $P_q(E)$ of Eq. (3.12), at energy $E = E_1 + \hbar\omega_1 = E_2 + \hbar\omega_2$, only has two contributions, corresponding to the excitations shown in Figure 3.1c. The quantities $P_q(E = E_1 + \hbar\omega_1)$ and $R_{q,q'}(E)$ are therefore given by [21–25]:

$$\begin{aligned} (\hbar/2\pi)^2\, P_q(E = E_1 + \hbar\omega_1) &= |a_1|^2|\bar{\epsilon}(\omega_1)|^2 d_q(11) + |a_2|^2|\bar{\epsilon}(\omega_2)|^2 d_q(22) \\ &\quad + 2\mathrm{Re}\left[a_1 a_2^* \bar{\epsilon}(\omega_1)\bar{\epsilon}^*(\omega_2) d_q(12)\right] , \end{aligned} \tag{3.18}$$

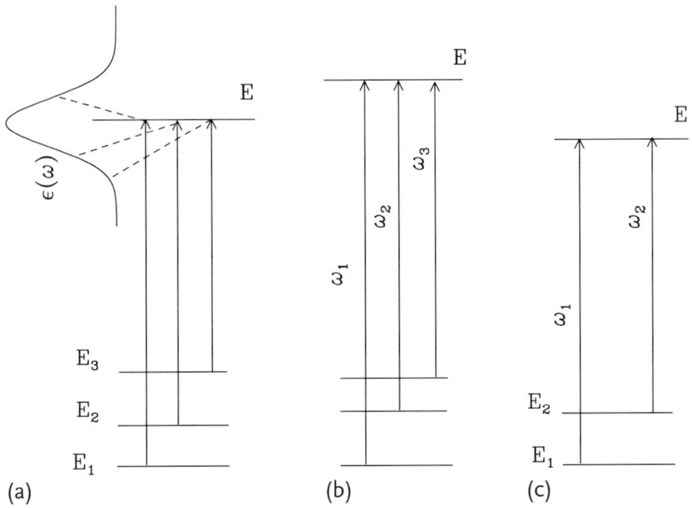

Figure 3.1 Photodissociation of a superposition of N levels using (a) a pulsed light source ($N = 3$ is shown); (b) N CW lasers ($N = 3$ is shown), and (c) $N = 2$ with two CW lasers.

and

$$R_{q,q'}(E = E_1 + \hbar\omega_1)$$
$$= \frac{|d_q(11)| + x^2|d_q(22)| + 2x\cos\left(\theta_1 - \theta_2 + \alpha_q(12)\right)|d_q(12)|}{|d_{q'}(11)| + x^2|d_{q'}(22)| + 2x\cos\left(\theta_1 - \theta_2 + \alpha_{q'}(12)\right)|d_{q'}(12)|}, \tag{3.19}$$

where $\alpha_q(ij)$ and θ_j are defined via

$$d_q(ij) = |d_q(ij)|\exp(i\alpha_q(ij)),$$
$$x = \frac{|\bar{\epsilon}(\omega_2)a_2|}{|\bar{\epsilon}(\omega_1)a_1|},$$
$$\tan\theta_j = \frac{\mathrm{Im}\left[\bar{\epsilon}\left(\omega_j\right)a_j\right]}{\mathrm{Re}\left[\bar{\epsilon}\left(\omega_j\right)a_j\right]}. \tag{3.20}$$

For convenience we now introduce the control variables $\Delta\theta = \theta_1 - \theta_2$ and $s = x^2/[x^2+1]$. The range $0 \le s \le 1$ covers all possible values of relative laser intensities. Varying $\Delta\theta$ or s changes the interference term and thus gives us control over the dissociation probabilities. These changes may be accomplished either by varying the coefficients of the initial superposition state, $\{a_j\}$, or by changing the intensity and relative phases of the dissociation lasers. Note, in particular, that varying $\Delta\theta$ corresponds to just varying a phase. The dependence of the yield on $\Delta\theta$ hence emphasizes the quantum-interference-based nature of the control.

As an example of this approach we consider control over the relative probability of forming $^2P_{3/2}$ vs. $^2P_{1/2}$ atomic iodine, denoted I and I*, in the dissociation of methyl iodide in the regime of 266 nm,

$$\mathrm{CH_3} + \mathrm{I}^*(^2P_{1/2}) \leftarrow \mathrm{CH_3I} \rightarrow \mathrm{CH_3} + \mathrm{I}(^2P_{3/2}) \tag{3.21}$$

This reaction is an example of electronic branching of photodissociation products. The results reported below are for a nonrotating two-dimensional collinear model [26, 27] in which the H_3 center of mass, the C atom, and the I atoms are assumed to lie on a line. Results for a rotating collinear model are discussed in the next section.

Typical results for the control of the I vs. I* channel are shown in Figure 3.2, as contour plots of the yield of CH_3 + I as a function of the control parameters. Two cases are shown: photodissociation out of the two superposition states $|\chi(0)\rangle = a_1|E_1\rangle + a_2|E_2\rangle$ (Figure 3.2a) and $|\chi(0)\rangle = a_1|E_1\rangle + a_3|E_3\rangle$ (Figure 3.2b). Here $|E_1\rangle$ is the ground state and $|E_2\rangle$ and $|E_3\rangle$ correspond to states with one and two quanta of excitation in the C–I bond. The results show a large range of possible control. For example, the yield changes in Figure 3.2b from 30 to 70% as one varies s at small $\theta_1 - \theta_2$. In addition, a comparison of the two figures shows that the topology of the control plot depends strongly on the states which comprise the superposition state.

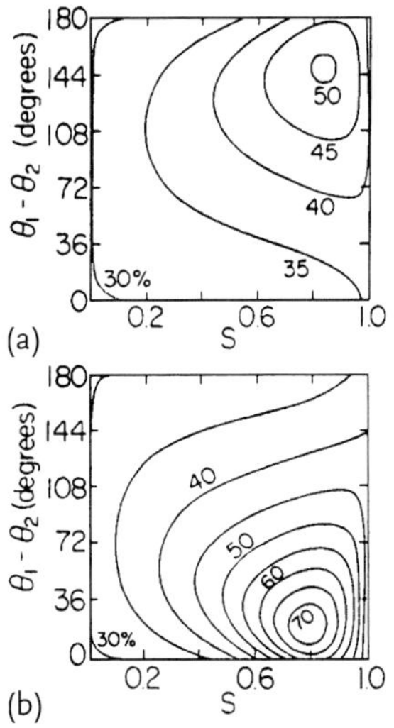

Figure 3.2 Contour plot of the yield of CH_3 + I from the photodissociation of CH_3I from a superposition of states at $\omega_1 = 37\,593.9\,\text{cm}^{-1}$. (a) $|\chi(0)\rangle = a_1|E_1\rangle + a_2|E_2\rangle$, (b) $|\chi(0)\rangle = a_1|E_1\rangle + a_3|E_3\rangle$. Taken from [21, Figure 2]. (Reprinted with permission. Copyright 1986 Elsevier BV.)

3.2.2 Energy Averaging and Satellite Contributions

In general, experiments measure energy-averaged quantities such as

$$P_q = \int dE\, P_q(E)\,,$$
$$R_{q,q'} = \frac{P_q}{P_{q'}}\,, \tag{3.22}$$

since products are not distinguished on the basis of total energy. As such, it is necessary to compute photodissociation to all energies.

For the case considered above, where two states are irradiated with two CW fields of frequencies ω_1 and ω_2, $P_q(E)$ (Eq. (3.12)) is nonzero at three energies: $E = E_1 + \hbar\omega_1 = E_2 + \hbar\omega_2$, $E' = E_1 + \hbar\omega_2$ and $E'' = E_2 + \hbar\omega_1$.

The contribution from the first of these energies $P_q(E = E_1 + \hbar\omega_1)$ is given in Eq. (3.18) and shown on the left hand side of Figure 3.3. The remaining contributions, shown on the right-hand side of Figure 3.3, are

$$P_q(E' = E_1 + \hbar\omega_2) = \left(\frac{2\pi}{\hbar}\right)^2 |a_1\bar{\epsilon}(\omega_2)|^2 d_q(11)\,,$$
$$P_q(E'' = E_2 + \hbar\omega_1) = \left(\frac{2\pi}{\hbar}\right)^2 |a_1\bar{\epsilon}(\omega_1)|^2 d_q(22)\,. \tag{3.23}$$

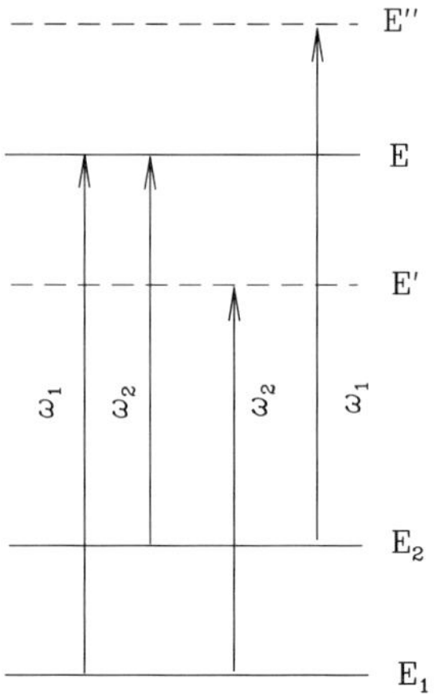

Figure 3.3 Contributions for two levels photodissociated by two frequencies. The interference terms correspond to total energy E. The satellite terms correspond to total energies E' and E''.

Thus, the overall P_q for $N = 2$ is given by

$$P_q = P_q(E = E_1 + \hbar\omega_1) + P_q(E' = E_1 + \hbar\omega_2) + P_q(E'' = E_2 + \hbar\omega_1) \,. \quad (3.24)$$

The latter two terms correspond to traditional photodissociation terms without associated interference contributions and provide *uncontrollable* photodissociation terms which we call "satellites." In this, and all coherent control scenarios discussed below, it is important to attempt to reduce the relative magnitude of the satellite terms in order to increase overall controllability.

We make the general observation that interference between terms of different energies contains oscillatory $\exp[i(E_1 - E_2)t/\hbar]$ terms which average out to zero with time. (This is not to say that the oscillatory interference term can not be put to good use. See, for example, a proposal for generating THz radiation [28] using such oscillatory terms).

Results demonstrating control over electronic branching ratios in the photodissociation of CH_3I, including the satellite terms [29], are shown in Figure 3.4. In this computation, CH_3I was treated as a rotating pseudotriatomic molecule [26, 27]. The bound states $|E_i\rangle$ in this case are described by the vibrational, rotational and magnetic numbers (v, J, M_J). Figure 3.4 shows contour plots of the $CH_3 + I^*$ product in the photodissociation of CH_3I at two different total energies, obtained by using $\omega_1 = 39\,639\,\mathrm{cm}^{-1}$ and $\omega_1 = 42\,367\,\mathrm{cm}^{-1}$, where the initial state is a superposition of $(\mathrm{v}_1, J_1) = (0, 2)$ and $(\mathrm{v}_2, J_2) = (1, 2)$. The M_J magnetic quantum number is averaged over, and all satellite terms are included. The results show that the I^* yield varies over the same large range as that seen in the nonrotating model study. In addition, a comparison of Figure 3.4a with 3.4b, which correspond to results at different excitation frequencies, shows that there is considerable dependence of the control contour topology on total energy.

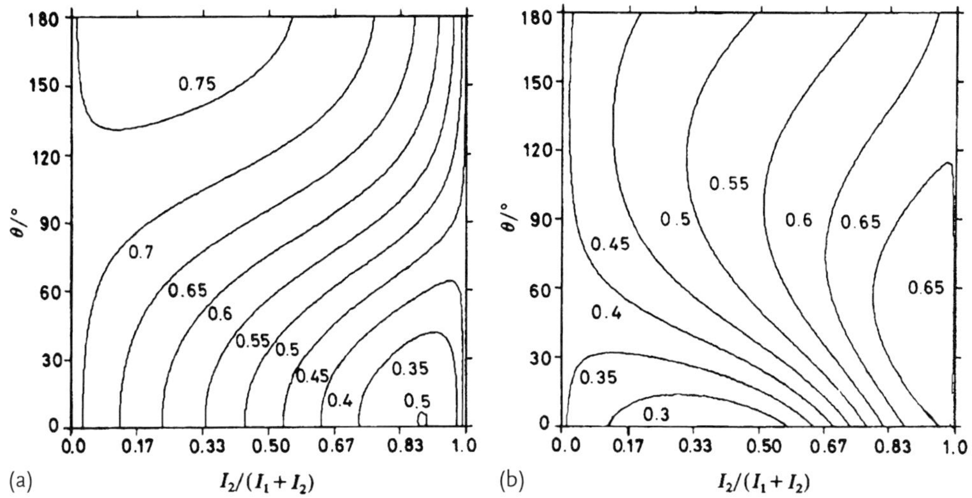

Figure 3.4 Contour plot of the yield of I^* (i.e., fraction of I^* as product) in the photodissociation of CH_3I starting from an M-averaged initial state. (a) $\omega_1 = 39\,639\,\text{cm}^{-1}$, (b) $\omega_1 = 42\,367\,\text{cm}^{-1}$. In both cases $\omega_2 = (E_1 - E_2)/\hbar + \omega_1$. Here, $\theta \equiv \theta_1 - \theta_2$ and $I_2/[I_1 + I_2]$ is equivalent to S in previous figures. Taken from [29, Figure 4]. (Reprinted with permission. Copyright 1986 Royal Society of Chemistry.)

Note that the entire theory would only need to be slightly modified to accommodate control of scattering into different angles, that is, the differential cross section, into channel q. A specific example of this type of control is discussed in Section 3.5, where we apply this approach to manipulating electric currents in semiconductors.

3.3 The Principle of Coherent Control

Control of the type discussed above, in which quantum interference effects are used to constructively or destructively alter product properties, is called coherent control [21]. Photodissociation of a superposition state, the scenario described above, will be seen to be just one particular implementation of a general principle of coherent control; that is, that *coherently driving a state with phase coherence through multiple, coherent, indistinguishable, optical excitation routes to the same final state allows for the possibility of control.* This procedure has a well-known analogy: The interference between paths as a beam of either particles or light passes through a double slit. In this case, interference between two coherent beams leads to spatial patterns of enhanced or reduced probabilities on an observation screen. In the case of coherent control, the overall coherence of a pure state plus a laser source allows for the constructive or destructive manipulation of probabilities in product channels. Active control results because the excitation process explicitly imparts experimentally controllable phase and amplitude information to the molecule.

Several comments are in order:

1. The origin of interference effects in material systems that are subject to external electromagnetic radiation is rooted in fundamental quantum mechanical concepts such as entanglement and decoherence. These issues are fully discussed in Section 10.2, which provides a fully general perspective on the origin of quantum interference in these systems.
2. One caveat to the analogy between the double slit experiment and coherent control is in order. Specifically, in the coherent control case the interference term (see, for example Eq. (3.18)) contains the incident external field. As such, this is a *driven* interference contribution, which has characteristics that are distinct from free evolution in the traditional double slit experiment. The consequences of this distinction for control of systems near the classical limit are the subject of Chapter 18.
3. As mentioned above, photodissociation control can only arise from energetically degenerate states. Another way of seeing this is to note that products of states of different energies E and E' appearing in the square of the wave packet of Eq. (3.8) cannot contribute to any measurement where the total energy is resolved. Such a measurement, which filters out all the wave packet components save those belonging to a given value of E, eliminates all the $E \neq E'$ products. Alternatively, two states of different energies are in principle distinguishable. Hence they can not interfere with one another.
4. Weak-field coherent control (CC) scenarios lead to simple analytic expressions for reaction probabilities in terms of a few molecular parameters and a few control parameters. Hence, the entire dependence of product probabilities on the control parameters can be easily generated experimentally once the molecular terms are determined from fitting the control expression to a small number of experimentally determined yields.
5. Numerous other scenarios can be designed which rely upon the essential coherent control principle. Several are discussed in the following sections.

As discussed in Appendix 3.A, there is an alternative approach to the control of chemical reactions, called "mode-selective chemistry", which does not rely upon quantum interferences. When applied to photodissociation, this approach would entail attempting to excite specific bonds in the molecule (e.g., the A–B bond in the A–B–C molecule) in order to produce a specific product (e.g., the $A + B{-}C$ product in the given example). Mode-selective chemistry, though very useful under favorable circumstances [30–37], is of limited scope, because in most cases the chemical bond we wish to excite is strongly coupled to other bonds (i.e., the "local mode" corresponding to the excitation of one bond is not an eigenstate of the system Hamiltonian). As a consequence, most excitations result in the production of a highly delocalized wave packet that entails the excitation of many bonds. This phenomenon, to be discussed more fully in Appendix 3.A, is called intramolecular vibrational redistribution (IVR).

3.4
Interference between *N*-Photon and *M*-Photon Routes

Another important example of coherent control introduces the possibility of quantum interference arising through competitive optical routes in the excitation of a single bound state to an energy E. Specifically, we consider the photodissociation of a single state via two pathways, an N photon and an M photon dissociation route. As will become evident in our discussion of selection rules (Section 3.4.2.1), the N vs. M scenarios are of two types: N and M of the same parity (e.g., both N and M odd or both even) or of opposite parity (one of (N, M) being odd and the other being even). The latter allows for control over the photodissociation differential cross sections (i.e., scattering into different angles) only, whereas the former allows for control over both the integral and differential cross sections. For simplicity we focus on the two lowest-order cases $(N, M) = (1, 2)$ and $(N, M) = (1, 3)$. We begin by deriving the expressions for one, two and three photon absorption, which will serve as input into these control scenarios.

3.4.1
Multiphoton Absorption

In Chapter 2 we developed the theory of the absorption of a single photon using a first-order perturbation approximation. We now consider processes involving the absorption of many photons. Such multiphoton processes can be treated using high-order perturbation solutions to the time-dependent Schrödinger equation (Eq. (1.3)). The computation becomes somewhat simpler if we work in the interaction representation [38]. Defining

$$V^I(t) = \exp(iH_M t/\hbar)\, H_{MR}(t) \exp(-iH_M t/\hbar) \;,$$
$$|\psi^I(t)\rangle = \exp(iH_M t/\hbar)\, |\Psi(t)\rangle \;, \tag{3.25}$$

we obtain, upon substitution into Eq. (1.3), that

$$i\hbar \frac{\partial |\psi^I(t)\rangle}{\partial t} = V^I(t)|\psi^I(t)\rangle \;. \tag{3.26}$$

The solution to Eq. (3.26), where $|\Psi^I(-\infty)\rangle = |E_i\rangle$ is given by

$$|\psi^I(t)\rangle = |E_i\rangle - \frac{i}{\hbar}\int_{-\infty}^{t} dt_1\, V^I(t_1)|\psi^I(t_1)\rangle \;. \tag{3.27}$$

Iterating this equation gives the series solution

$$|\psi^I(t)\rangle = \left[1 + \sum_{n=1}^{\infty} \mathcal{V}^{(n)}(t)\right] |E_i\rangle \;, \tag{3.28}$$

with

$$\mathcal{V}^{(1)}(t) = -\frac{i}{\hbar}\int_{-\infty}^{t} dt_1\, V^I(t_1)\,,$$

$$\mathcal{V}^{(2)}(t) = \left(-\frac{i}{\hbar}\right)^2 \int_{-\infty}^{t} dt_1\, V^I(t_1) \int_{-\infty}^{t_1} dt_2\, V^I(t_2)\,,$$

$$\vdots$$

$$\mathcal{V}^{(n)}(t) = \left(-\frac{i}{\hbar}\right)^n \int_{-\infty}^{t} dt_1\, V^I(t_1) \int_{-\infty}^{t_1} dt_2\, V^I(t_2) \cdots \int_{-\infty}^{t_{n-1}} dt_n\, V^I(t_n)\,. \qquad (3.29)$$

Using Eq. (3.25), the photodissociation probability amplitude becomes

$$\begin{aligned}\lim_{t\to\infty} \langle E, \boldsymbol{m}, q^- | \Psi(t)\rangle &= \lim_{t\to\infty} \exp(-iEt/\hbar)\, \langle E, \boldsymbol{m}, q^- | \psi^I(t)\rangle \\ &= \lim_{t\to\infty} \exp(-iEt/\hbar)\, \langle E, \boldsymbol{m}, q^- | \sum_{n=1}^{\infty} \mathcal{V}^{(n)}(t) | E_i\rangle\,,\end{aligned} \qquad (3.30)$$

where we have used the fact that $|E, \boldsymbol{m}, q^-\rangle$ are eigenstates of H_M, and that $\langle E_i | E, \boldsymbol{m}, q^-\rangle = 0$. Henceforth we drop the term $\exp(-iEt/\hbar)$ since it provides an overall, and therefore irrelevant, phase factor.

Multiphoton excitation using light pulses that are moderately strong may be treated directly using Eq. (1.5). Here we focus solely on CW excitation with an adiabatically switched interaction potential,

$$H_{MR}(t) = \lim_{\epsilon\to 0} -\boldsymbol{d}\cdot\boldsymbol{E}(z,t)e^{-\epsilon|t|} = \lim_{\epsilon\to 0} -2\hat{\boldsymbol{\varepsilon}}\cdot\boldsymbol{d}\,\mathrm{Re}\left[\bar{\epsilon}(\omega)e^{-i\omega t}e^{-\epsilon|t|}\right]\,. \qquad (3.31)$$

The adiabatic switching is introduced via the slowly varying $e^{-\epsilon|t|}$ term that guarantees that the interaction vanishes as $t \to \pm\infty$. It is the exact time-dependent analog of the procedure used in the derivation of the Lippmann–Schwinger equation (Eq. (2.52)) in the energy domain.

Inserting Eq. (3.31) into Eq. (3.29), and carrying out the integration gives the first-order probability amplitude in the long-time limit:

$$\begin{aligned}&\langle E, \boldsymbol{m}, q^- | \mathcal{V}^{(1)}(t\to\infty) | E_i\rangle \\ &\quad = \frac{i}{\hbar}\langle E, \boldsymbol{m}, q^- | \int_{-\infty}^{\infty} dt_1 \exp(iH_M t_1/\hbar) \\ &\qquad\qquad 2\hat{\boldsymbol{\varepsilon}}\cdot\boldsymbol{d}\,\mathrm{Re}\left[\bar{\epsilon}(\omega)e^{-i\omega t_1}\right]\exp(-iH_M t_1/\hbar)\,|E_i\rangle\,.\end{aligned} \qquad (3.32)$$

Remembering that $|E_i\rangle$ and $|E, \boldsymbol{m}, q^-\rangle$ are eigenstates of H_M, we have that

$$\langle E, \boldsymbol{m}, q^-|\mathcal{V}^{(1)}(t \to \infty)|E_i\rangle$$
$$= \frac{i}{\hbar}\langle E, \boldsymbol{m}, q^-| \int_{-\infty}^{\infty} dt_1 \hat{\boldsymbol{\varepsilon}} \cdot \boldsymbol{d} \left[\bar{\epsilon}(\omega) e^{i(E-E_i-\hbar\omega)t_1/\hbar} + \bar{\epsilon}^*(\omega) e^{i(E-E_i+\hbar\omega)t_1/\hbar}\right] |E_i\rangle \,. \tag{3.33}$$

After integration over t_1 we obtain, using the well-known equality $\int \exp(i\omega t)dt = 2\pi\delta(\omega)$, that

$$\langle E, \boldsymbol{m}, q^-|\mathcal{V}^{(1)}(t \to \infty)|E_i\rangle$$
$$= 2\pi i\langle E, \boldsymbol{m}, q^-|\hat{\boldsymbol{\varepsilon}} \cdot \boldsymbol{d}|E_i\rangle \left[\bar{\epsilon}(\omega)\delta(E - \hbar\omega - E_i) + \bar{\epsilon}^*(\omega)\delta(E + \hbar\omega - E_i)\right] \,. \tag{3.34}$$

When $E > E_i$ only the first term in the square brackets can be nonzero and we obtain that

$$\langle E, \boldsymbol{m}, q^-|\mathcal{V}^{(1)}(t \to \infty)|E_i\rangle = 2\pi i\bar{\epsilon}(\omega)\delta(E - \hbar\omega - E_i)\langle E, \boldsymbol{m}, q^-|\hat{\boldsymbol{\varepsilon}} \cdot \boldsymbol{d}|E_i\rangle \,. \tag{3.35}$$

This term is the probability amplitude for the one-photon absorption. It is the CW analog of the expression obtained in Section 2.3 for a general pulse. The second term in the square brackets of Eq. (3.34) contributes when $E < E_i$, in which case the expression describes the process of photoemission.

In a similar way we can obtain an expression for the matrix element resulting from the second-order perturbation term:

$$\langle E, \boldsymbol{m}, q^-|\mathcal{V}^{(2)}(t \to \infty)|E_i\rangle = \frac{-4}{\hbar^2} \int_{-\infty}^{\infty} dt_1 \langle E, \boldsymbol{m}, q^-|e^{iH_M t_1/\hbar} e^{-\epsilon|t_1|}$$
$$\times \hat{\boldsymbol{\varepsilon}} \cdot \boldsymbol{d}\mathrm{Re}\left[\bar{\epsilon}(\omega)e^{-i\omega t_1}\right] e^{-iH_M t_1/\hbar} \int_{-\infty}^{t_1} dt_2 e^{iH_M t_2/\hbar} \hat{\boldsymbol{\varepsilon}} \cdot \boldsymbol{d}\mathrm{Re}\left[\bar{\epsilon}(\omega)e^{-i\omega t_2}\right]$$
$$\times e^{-\epsilon|t_2|} e^{-iH_M t_2/\hbar}|E_i\rangle \tag{3.36}$$

We can time-integrate the exponential operators as if their arguments were numbers, provided that we maintain the proper ordering of all noncommuting operators. Using the fact that $|E_i\rangle$ is an eigenstate of H_M, we obtain that the integral over t_2 is given as

$$\int_{-\infty}^{t_1} dt_2 e^{iH_M t_2/\hbar} \hat{\boldsymbol{\varepsilon}} \cdot \boldsymbol{d}\mathrm{Re}\left[\bar{\epsilon}(\omega)e^{-i\omega t_2}\right] e^{-\epsilon|t_2|} e^{-iH_M t_2/\hbar}|E_i\rangle$$
$$= \frac{i\hbar}{2}\left\{e^{-i(E_i - i\epsilon - H_M + \hbar\omega)t_1/\hbar}\left[E_i - i\epsilon - H_M + \hbar\omega\right]^{-1}\bar{\epsilon}(\omega)\hat{\boldsymbol{\varepsilon}} \cdot \boldsymbol{d}|E_i\rangle\right.$$
$$\left. + e^{-i(E_i - i\epsilon - H_M - \hbar\omega)t_1/\hbar}\left[E_i - i\epsilon - H_M - \hbar\omega\right]^{-1}\bar{\epsilon}^*(\omega)\hat{\boldsymbol{\varepsilon}} \cdot \boldsymbol{d}|E_i\rangle\right\} \,, \tag{3.37}$$

where the contribution of the lower limit has vanished due to the $e^{-\epsilon|t_2|}$ term.

When the above expression is substituted in Eq. (3.36), and using the fact that $|E, \boldsymbol{m}, q^-\rangle$ is an eigenstate of H_M, we obtain that

$$\begin{aligned}
\langle E, \boldsymbol{m}, q^- | \mathcal{V}^{(2)}(t \to \infty) | E_i \rangle &= \frac{-i}{\hbar} \int_{-\infty}^{\infty} dt_1 \\
&\times \langle E, \boldsymbol{m}, q^- | \left[\bar{\epsilon}^2(\omega) e^{-i(E_i - i\epsilon + 2\hbar\omega - E)t_1/\hbar} + |\bar{\epsilon}(\omega)|^2 e^{-i(E_i - i\epsilon - E)t_1/\hbar} \right] \\
&\hat{\boldsymbol{\varepsilon}} \cdot \boldsymbol{d} [E_i - i\epsilon - H_M + \hbar\omega]^{-1} \hat{\boldsymbol{\varepsilon}} \cdot \boldsymbol{d} | E_i \rangle \\
&+ \langle E, \boldsymbol{m}, q^- | \left[|\bar{\epsilon}(\omega)|^2 e^{-i(E_i - i\epsilon - E)t_1/\hbar} + (\bar{\epsilon}^*(\omega))^2 e^{-i(E_i - i\epsilon - E - 2\hbar\omega)t_1/\hbar} \right] \\
&\hat{\boldsymbol{\varepsilon}} \cdot \boldsymbol{d} [E_i - i\epsilon - H_M - \hbar\omega]^{-1} \hat{\boldsymbol{\varepsilon}} \cdot \boldsymbol{d} | E_i \rangle \,. \qquad (3.38)
\end{aligned}$$

Using the equality $\int_{-\infty}^{\infty} \exp(i\omega t) = 2\pi\delta(\omega)$ again gives:

$$\begin{aligned}
\langle E, \boldsymbol{m}, q^- | \mathcal{V}^{(2)}(t \to \infty) | E_i \rangle &= -2\pi i \\
&\times \Big\{ \langle E, \boldsymbol{m}, q^- | \left[\bar{\epsilon}^2(\omega) \delta(E_i + 2\hbar\omega - E) + |\bar{\epsilon}(\omega)|^2 \delta(E_i - E) \right] \\
&\hat{\boldsymbol{\varepsilon}} \cdot \boldsymbol{d} [E_i - i\epsilon - H_M + \hbar\omega]^{-1} \hat{\boldsymbol{\varepsilon}} \cdot \boldsymbol{d} | E_i \rangle \\
&+ \langle E, \boldsymbol{m}, q^- | \left[|\bar{\epsilon}(\omega)|^2 \delta(E_i - E) + (\bar{\epsilon}^*(\omega))^2 \delta(E_i - 2\hbar\omega - E) \right] \\
&\hat{\boldsymbol{\varepsilon}} \cdot \boldsymbol{d} [E_i - i\epsilon - H_M - \hbar\omega]^{-1} \hat{\boldsymbol{\varepsilon}} \cdot \boldsymbol{d} | E_i \rangle \Big\} \,. \qquad (3.39)
\end{aligned}$$

When $E > E_i$, the only second-order perturbation theory contribution comes from the first term of Eq. (3.39). Thus,

$$\begin{aligned}
\langle E, \boldsymbol{m}, q^- | \mathcal{V}^{(2)}(t \to \infty) | E_i \rangle &= -2\pi i \bar{\epsilon}^2(\omega) \delta(E_i + 2\hbar\omega - E) \\
&\times \langle E, \boldsymbol{m}, q^- | \hat{\boldsymbol{\varepsilon}} \cdot \boldsymbol{d} [E_i - i\epsilon + \hbar\omega - H_M]^{-1} \hat{\boldsymbol{\varepsilon}} \cdot \boldsymbol{d} | E_i \rangle \,. \qquad (3.40)
\end{aligned}$$

The $\delta(E_i + 2\hbar\omega - E)$ term clearly identifies this second-order perturbation theory term as a process in which the molecule undergoes a transition from level E_i to level E by absorbing two photons of frequency ω. (The second and third terms of Eq. (3.39), surviving when $E = E_i$, represent (elastic) light scattering. The fourth term, surviving when $E < E_i$, represents two-photon stimulated emission).

In a similar way we can show that the probability amplitude for three-photon absorption, obtained from the third-order perturbation theory term, is given as

$$\begin{aligned}
&\langle E, \boldsymbol{m}, q^- | \mathcal{V}^{(3)}(t \to \infty) | E_i \rangle \\
&= 2\pi i \bar{\epsilon}^3(\omega) \delta(E_i + 3\hbar\omega - E) \langle E, \boldsymbol{m}, q^- | \\
&\hat{\boldsymbol{\varepsilon}} \cdot \boldsymbol{d} [E_i - i\epsilon + 2\hbar\omega - H_M]^{-1} \hat{\boldsymbol{\varepsilon}} \cdot \boldsymbol{d} [E_i - i\epsilon + \hbar\omega - H_M]^{-1} \hat{\boldsymbol{\varepsilon}} \cdot \boldsymbol{d} | E_i \rangle \,. \qquad (3.41)
\end{aligned}$$

For notational simplicity we have neglected the electronic degrees of freedom in deriving Eq. (3.41). To properly incorporate them we can repeat this derivation with $H_{MR}(t)$ replaced by $H'_{MR}(t)$ (Eq. (2.71)), with H_M replaced by H_{MT} (Eq. (2.67)), and with eigenstates including the electronic component. Adopting the Born–Oppenheimer approximation and its consequences (Eqs. (2.68)–(2.70)), and noting that at energies corresponding to visible and UV (ultraviolet) light the relevant dipole matrix elements within electronic states are small compared to those between different electronic states, gives, instead of Eqs. (3.35), (3.40) and (3.41),

$$\langle E, \boldsymbol{m}, q^-|\mathcal{V}^{(1)}(t \to \infty)|E_i\rangle = 2\pi i\bar{\epsilon}(\omega)\delta(E - \hbar\omega - E_i)\langle E, \boldsymbol{m}, q^-|d_{e,g}|E_i\rangle , \quad (3.42)$$

$$\begin{aligned}&\langle E, \boldsymbol{m}, q^-|\mathcal{V}^{(2)}(t \to \infty)|E_i\rangle \\ &= -2\pi i\bar{\epsilon}^2(\omega)\delta(E - 2\hbar\omega - E_i) \\ &\quad \times \sum_{e'}\langle E, \boldsymbol{m}, q^-|d_{e,e'}[\hbar\omega + E_i - i\epsilon - H_{e'}]^{-1} d_{e',g}|E_i\rangle \\ &\equiv -2\pi i\bar{\epsilon}^2(\omega)\delta(E - 2\hbar\omega - E_i)\langle E, \boldsymbol{m}, q^-|D|E_i\rangle , \end{aligned} \quad (3.43)$$

$$\begin{aligned}&\langle E, \boldsymbol{m}, q^-|\mathcal{V}^{(3)}(t \to \infty)|E_i\rangle = 2\pi i\bar{\epsilon}^3(\omega)\delta(E - 3\hbar\omega - E_i) \\ &\quad \times \sum_{e'e''}\langle E, \boldsymbol{m}, q^-|d_{e,e''}[2\hbar\omega + E_i - i\epsilon - H_{e''}]^{-1} \\ &d_{e'',e'}[\hbar\omega + E_i - i\epsilon - H_{e'}]^{-1} d_{e',g}|E_i\rangle \\ &\quad \equiv 2\pi i\bar{\epsilon}^3(\omega)\delta(E - 3\omega\hbar - E_i)\langle E, \boldsymbol{m}, q^-|T|E_i\rangle . \end{aligned} \quad (3.44)$$

Here the matrix elements $H_{e'} \equiv \langle e'|H_{MT}|e'\rangle$ are operators with respect to nuclear wave functions in the ground and excited electronic states, and $d_{e,g} \equiv \hat{\boldsymbol{\varepsilon}} \cdot \boldsymbol{d}_{e,g}$, and so on. The two- and three-photon absorption operators D and T are defined by the above identities.

3.4.2 One- vs. Three-Photon Interference

Given the above results we can now consider [39] a molecule initially in state $|E_i\rangle$, subjected to two copropagating CW fields of frequencies ω_1 and ω_3, with $\omega_3 = 3\omega_1$. The interaction potential is given by

$$H_{MR}(t) = -2\boldsymbol{d} \cdot \mathrm{Re}\left[\hat{\boldsymbol{\varepsilon}}_3\bar{\epsilon}_3 \exp(-i\omega_3 t) + \hat{\boldsymbol{\varepsilon}}_1\bar{\epsilon}_1 \exp(-i\omega_1 t)\right] , \quad (3.45)$$

where $\bar{\epsilon}_i = \bar{\epsilon}(\omega_i)$.

We assume the following physics:

1. the dipole transitions within electronic states are negligible compared to those between electronic states;
2. the fields are sufficiently weak to allow the use of perturbation theory; and
3. $E_i + 2\hbar\omega_1$ is below the dissociation threshold, with dissociation occurring from the excited electronic state.

In accord with Eq. (3.44) the lowest-order expression for the one-photon or three-photon dissociation amplitude $A_{\boldsymbol{m},q}(E = E_i + \hbar\omega_3)$ is

$$A_{\boldsymbol{m},q}(E = E_i + \hbar\omega_3) = \frac{2\pi i}{\hbar}\left[\delta\left(\omega_3 - \omega_{E,i}\right)\bar{\epsilon}_3\langle E, \boldsymbol{m}, q^-|d_{e,g}|E_i\rangle \right. \\ \left. + \delta\left(3\omega_1 - \omega_{E,i}\right)\bar{\epsilon}_1^3\langle E, \boldsymbol{m}, q^-|T|E_i\rangle\right] , \tag{3.46}$$

where T is the three-photon transition operator, given according to Eq. (3.44) as

$$T = \sum_{e'e''} d_{e,e'}(E_i - H_{e'} + 2\hbar\omega_3)^{-1} d_{e',e''}(E_i - H_{e''} + \hbar\omega_3)^{-1} d_{e'',g} . \tag{3.47}$$

The probability to produce fragments q at a fixed energy E is therefore

$$P_q(E) = \sum_m |A_{\boldsymbol{m},q}\left(E_i + \hbar\omega_3\right)|^2 = P_q^{(1)}(E) + P_q^{(3)}(E) + P_q^{(13)}(E) , \tag{3.48}$$

where the one-photon photodissociation probability is

$$P_q^{(1)}(E) = \left(\frac{2\pi}{\hbar}\right)^2 |\bar{\epsilon}_3|^2 \sum_m |\langle E, \boldsymbol{m}, q^-|d_{e,g}|E_i\rangle|^2 , \tag{3.49}$$

the three-photon photodissociation probability is

$$P_q^{(3)}(E) = \left(\frac{2\pi}{\hbar}\right)^2 |\bar{\epsilon}_1|^6 \sum_m |\langle E, \boldsymbol{m}, q^-|T|E_i\rangle|^2 , \tag{3.50}$$

and the one-photon three-photon interference term is

$$P_q^{(13)}(E) = \left(\frac{2\pi}{\hbar}\right)^2 \left[\bar{\epsilon}_3(\bar{\epsilon}_1^*)^3 \sum_m \langle E_i|T|E, \boldsymbol{m}, q^-\rangle\langle E, \boldsymbol{m}, q^-|d_{e,g}|E_i\rangle + \text{c.c.}\right] . \tag{3.51}$$

As in our discussion of the photodissociation of a superposition state (Section 3.2), we define a "molecular" interference amplitude $|F_q^{(13)}|$ and a "molecular" phase $\alpha_q(13)$ as

$$|F_q(13)| \exp\left[i\alpha_q(13)\right] = \sum_m \langle E_i|T|E, \boldsymbol{m}, q^-\rangle\langle E, \boldsymbol{m}, q^-|d_{e,g}|E_i\rangle . \tag{3.52}$$

Recognizing that $\bar{\epsilon}_i$ is a complex number, $\bar{\epsilon}_i = |\bar{\epsilon}_i|e^{i\phi_i}$ we can write the above interference term as

$$P_q^{(13)}(E) = 2\left(\frac{2\pi}{\hbar}\right)^2 |\bar{\epsilon}_3\bar{\epsilon}_1^3| \cos\left(\phi_3 - 3\phi_1 + \alpha_q(13)\right)|F_q(13)| . \tag{3.53}$$

The branching ratio $R_{qq'}(E)$ for channels q and q', (see Eq. (3.14)) can now be written as

$$R_{qq'}(E) = \frac{F_q(11) + 2x\cos\left[\phi_3 - 3\phi_1 + \alpha_q(13)\right]\epsilon_0^2|F_q(13)| + x^2\epsilon_0^4 F_q(33)}{F_{q'}(11) + 2x\cos\left[\phi_3 - 3\phi_1 + \alpha_{q'}(13)\right]\epsilon_0^2|F_{q'}(13)| + x^2\epsilon_0^4 F_{q'}(33)} , \tag{3.54}$$

where

$$F_q(11) = \left(\frac{\hbar}{\pi|\bar{\epsilon}_3|}\right)^2 P_q^{(1)}(E) ;$$

$$F_q(33) = \left(\frac{\hbar}{\pi|\bar{\epsilon}_1|^3}\right)^2 P_q^{(3)}(E) ; \quad \text{and} \quad x = \frac{|\bar{\epsilon}_1|^3}{\epsilon_0^2|\bar{\epsilon}_3|} , \tag{3.55}$$

where ϵ_0 is defined as a single unit of electric field; x is therefore a dimensionless parameter.

The numerator and denominator of Eq. (3.54) each display the canonical form for coherent control, that is, a form similar to Eq. (3.19) in which there are independent contributions from more than one route, modulated by an interference term. Since the interference term is controllable through variation of the laboratory parameters (x and $\phi_3 - 3\phi_1$), so too is the branching ratio $R_{qq'}(E)$. Thus, the principle upon which this control scenario is based is the same as that in Section 3.2 but the interference is introduced in an entirely different way.

3.4.2.1 One- vs. Three-Photon Interference: Three-Dimensional Formalism

With the qualitative principle of interfering pathways exposed, it remains to demonstrate the quantitative extent to which the one- vs. three-photon scenario alters the yield ratio in a realistic system. To this end we consider the photodissociation of IBr,

$$I + Br \leftarrow IBr \rightarrow I + Br^* \tag{3.56}$$

where Br $=$ Br($^2P_{3/2}$) and Br* $=$ Br($^2P_{1/2}$); the IBr potential curves used in the calculation are shown in Figure 3.5. Details of the computation are discussed below to demonstrate the role of selection rules in the one- vs. three-photon scenario and the extent of achievable control.

Because we want to consider the one- vs. three-photon control of IBr in three dimensions [40], we replace the notation $|E_i\rangle$ for the initial state by $|E_i, J_i, M_i\rangle$, where E_i is (as before) the energy of the state, J_i is its angular momentum, and M_i is the angular momentum projection along the z-axis. Where no confusion might arise, we continue to use $|E_i\rangle$ for simplicity.

The collection of final channel quantum numbers $\boldsymbol{n}$ in the photodissociation amplitude $\langle E, \boldsymbol{n}, q^- | d_{e,g} | E_i, J_i, M_i\rangle$ must now include the scattering angles $\hat{\boldsymbol{k}} = (\phi_k, \theta_k)$. In the case of IBr, the quantum number of primary interest $q = 1, 2$ labels the Br($^2P_{3/2}$) or Br*($^2P_{1/2}$) electronic states. Further, the continuum states $|E, \boldsymbol{n}, q^-\rangle$ are conveniently treated by partial wave expansion in states of total angular momentum J and associated projection M_J along the z-axis.

The required $P^{(1)}$, $P^{(3)}$ and $P^{(13)}$ terms of Eq. (3.48) are conveniently expressed in terms of $d_q(E_j, J_j, M_j; E_i, J_i, M_i; E)$, the angle-averaged products of two amplitudes,

$$d_q(E_j, J_j, M_j; E_i, J_i, M_i; E) = \int d\hat{\boldsymbol{k}} \langle E_j, J_j, M_j | d_{g,e} | E, \hat{\boldsymbol{k}}, q^- \rangle \langle E, \hat{\boldsymbol{k}}, q^- | d_{e,g} | E_i, J_i, M_i \rangle . \tag{3.57}$$

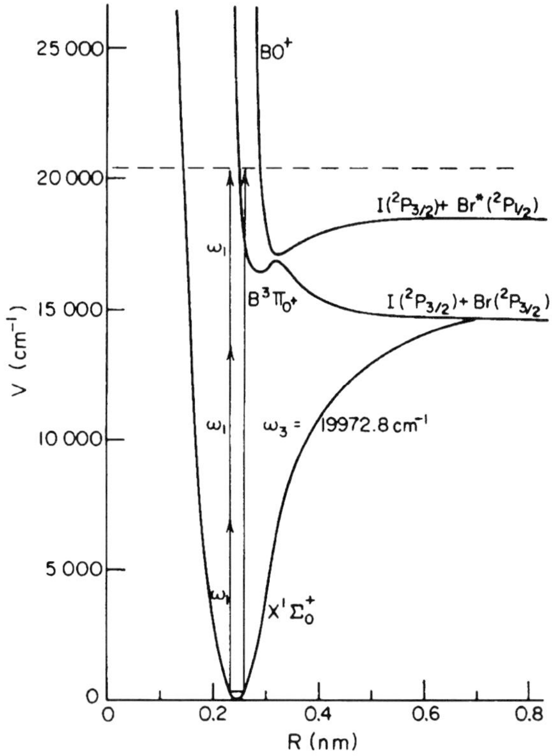

Figure 3.5 IBr potential energy curves relevant to one-photon vs. three-photon dissociation. Taken from [40, Figure 1]. (Reprinted with permission. Copyright 1991 American Institue of Physics.)

Evaluation of this integral [40] shows that $d_q(E_j, J_j, M_j; E_i, J_i, M_i; E)$ is proportional to δ_{M_i,M_j}, and contains nonzero contributions only from terms where $|E_j, J_j, M_j\rangle$ and $|E_i, J_i, M_i\rangle$ are excited to continuum states of the same J.

The probability $P^{(1)}$ (Eq. (3.49)) is given in terms of d_q, by

$$P_q^{(1)}(E, E_i, J_i, M_i) = (2\pi/\hbar)^2 \bar{\epsilon}_3^2 d_q(E_i, J_i, M_i; E_i, J_i, M_i; E) \,. \tag{3.58}$$

Assuming that only two electronic states $|g\rangle$ and $|e\rangle$ are of importance, the terms $P^{(13)}$ and $P^{(3)}$ can also be written in terms of d_q. To do so, we express $\langle E, \boldsymbol{n}, q^-|T|E_i\rangle$ of Eq. (3.52) for two electronic states in terms of $\langle E, \boldsymbol{n}, q^-|\hat{\boldsymbol{\varepsilon}}\cdot \boldsymbol{d}|E_i\rangle$ by inserting appropriate resolutions of the identity

$$\langle E, \boldsymbol{n}, q^-|T|E_i\rangle = \sum_{j,\boldsymbol{n}',q'} \int dE' \frac{\langle E, \boldsymbol{n}, q^-|d_{e,g}|E_j\rangle\langle E_j|d_{g,e}|E', \boldsymbol{n}', q'^-\rangle\langle E', \boldsymbol{n}', q'^-|d_{e,g}|E_i\rangle}{\left(E_j - E_i - 2\hbar\omega_1\right)\left(E' - E_i - \hbar\omega_1\right)} \,. \tag{3.59}$$

Here, as noted above, $|E_i\rangle$ denotes $|E_i, J_i, M_i\rangle$ and the j-summation indicates a sum over all bound states $|E_j\rangle$ of the ground $X^1\Sigma_0^+$ potential surface. Computationally, of all the bound eigenstates of $X^1\Sigma_0^+$, the contribution to $\langle E, \boldsymbol{n}, q^-|T|E_i\rangle$ is dominated by those states with energy E_j which nearly satisfy the two-photon resonance condition $E_j \approx E_i + 2\hbar\omega_1$.

From Eq. (3.59) and the definition of d_q, $|F_q^{(13)}|\exp(i\alpha_q^{(13)})$ in the cross term $P^{(13)}$ (Eq. (3.53)), is given by

$$|F_q^{(13)}|\exp\left(i\alpha_q^{(13)}\right) = \sum_{E_j,J_j,q'} \int dE' \frac{d_q\left(E_j, J_j, M_i; E_i, J_i, M_i; E\right) d^*_{q'}\left(E_j, J_j, M_i; E_i, J_i, M_i; E'\right)}{\left(E_j - E_i - 2\hbar\omega_1\right)\left(E' - E_i - \hbar\omega_1\right)}, \tag{3.60}$$

where the J_j and the E_j summation indicates a summation over all bound eigenstates of the $X^1\Sigma_0^+$ state. Note that the term $d_q(E_j, J_j, M_i; E_i, J_i, M_i; E)$ arose from two dipole matrix elements, one coupling $|E_i, J_i, M_i\rangle$ to the continuum in the one photon route and one coupling $|E_j, J_j, M_j\rangle$ to the continuum as the third step in the three photon route. Thus, in accord with the discussion above, the angular momentum J of the continuum accessed by these two routes must be equal for $F_q^{(13)}$ to be nonzero.

Finally, using Eqs. (3.55), (3.50) and (3.59), the probability of three-photon photodissociation is given by

$$P_q^{(3)}(E, E_i, J_i, M_i) = \sum_{E_j,E_l,J_j,J_l,\bar{q},q'} \int dE' \int d\bar{E} \times \frac{d_q\left(E_l,J_l,M_i; E_j,J_j,M_i; E\right) d_{q'}\left(E_j,J_j,M_i; E_i,J_i,M_i; E'\right) d^*_{\bar{q}}\left(E_l,J_l,M_i; E_i,J_i,M_i; \bar{E}\right)}{\left(E_j - E_i - 2\hbar\omega_1\right)\left(E_l - E_i - 2\hbar\omega_1\right)\left(E' - E_i - \hbar\omega_1\right)\left(\bar{E} - E_i - \hbar\omega_1\right)}. \tag{3.61}$$

Given these results, the branching ratio $(R_{qq'})$ (Eq. (3.54)) can be easily written in terms of d_q using Eqs. (3.55), (3.58), (3.60) and (3.61).

Equations (3.50) and (3.59) are quite complex, and simple qualitative rules for tabulating the terms have been developed [40]. Figures 3.6 and 3.7 provide a means of demonstrating these rules and exposing features of the angular momentum as it affects control.

Consider, as an example, the case of $J_i = M_i = 0$. The angular momentum values associated with the one- and three-photon absorption routes are summarized in Figure 3.6a. In the figure, the solid horizontal line represents angular momentum states of the ground potential surface $(X^1\Sigma_0^+)$ and the dashed horizontal line represents those of the excited potential surfaces ($B^3\Pi_{0+}$ and BO^+). The succession of transitions follows from the usual dipole selection rules for linearly polarized light $(\Delta J = \pm 1, \Delta M = 0)$ for transitions between $\Omega = 0$ states (Ω is the total electronic angular momentum along the molecular axis). Note that there are two J

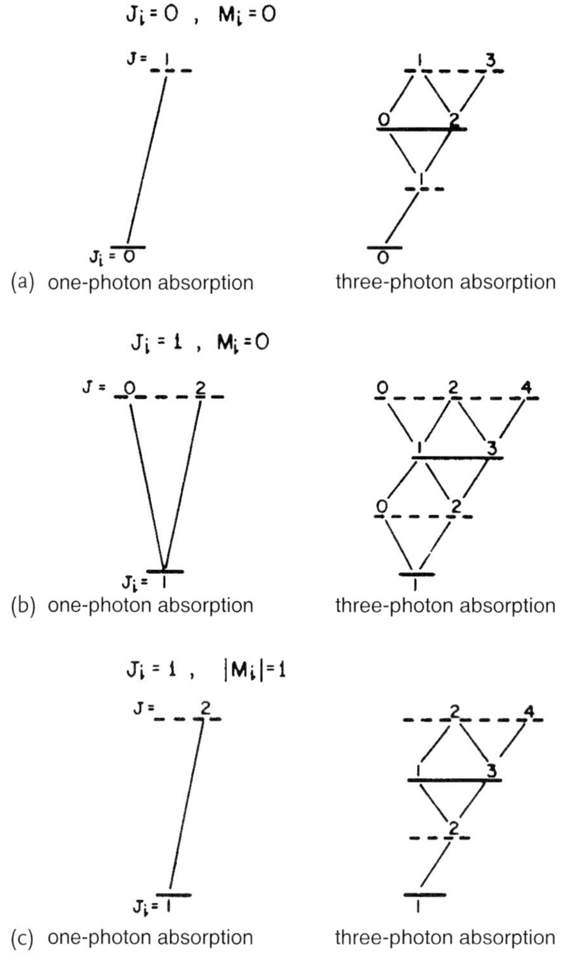

Figure 3.6 Angular momentum levels available for one- and three-photon absorption from (a) $J_i = 0$ and (b) $J_i = 1$, $M_i = 0$ and (c) $J_i = 1$, $|M_i| = 1$. Taken from [40, Figure 2]. (Reprinted with permission. Copyright 1991 American Institute of Physics.)

states which arise from the three photon absorption, that is, $J = 1$ and $J = 3$. Only the $J = 1$ term interferes with the one photon route, which also generates an excited state with $J = 1$. That is, there is no interference term involving the $J = 3$ route, which is therefore an uncontrollable satellite.

To use these diagrams to tabulate the dipole matrix elements which contribute to $P^{(1)}$, $P^{(3)}$, and $P^{(13)}$ consider, for example, contributions to $P^{(3)}$ (Eq. (3.61)) for the case of $J_i = M_i = 0$. The relevant diagram is then that shown on the right-hand side of Figure 3.6a. Detailed consideration of Eq. (3.61) shows that the following diagrammatic method gives the appropriate dipole contributions to $P^{(3)}$. One starts at the bottom of the diagram (here $J = 0$) and proceeds to a state at the top of the

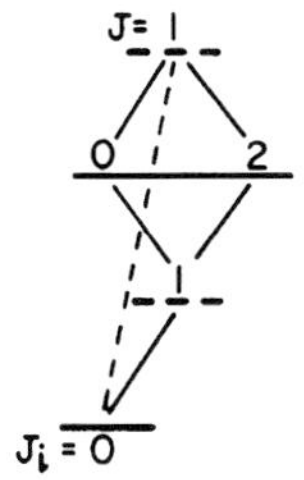

Figure 3.7 Angular momentum levels (for the $J_i = 0$ case) connected by both one- and three-photon routes to the continuum J state which is shared by both types of absorption. Taken from [40, Figure 3]. (Reprinted with permission. Copyright 1991 American Institute of Physics.)

diagram, and then back down to the initial state. Each pair of levels encountered in this route contributes one dipole matrix element to the product in Eq. (3.61). A sample case would be the route $0 \to 1 \to 2 \to 1 \to 0 \to 1 \to 0$ which contributes a term of the form (where the value of the angular momentum is indicated and where the electronic states are distinguished as being unprimed (ground state) or primed (excited state)): $\langle 0|\hat{\boldsymbol{\varepsilon}}\cdot\boldsymbol{d}|1'\rangle\langle 1'|\hat{\boldsymbol{\varepsilon}}\cdot\boldsymbol{d}|2\rangle\langle 2|\hat{\boldsymbol{\varepsilon}}\cdot\boldsymbol{d}|1'\rangle\langle 1'|\hat{\boldsymbol{\varepsilon}}\cdot\boldsymbol{d}|0\rangle\langle 0|\hat{\boldsymbol{\varepsilon}}\cdot\boldsymbol{d}|1'\rangle\langle 1'|\hat{\boldsymbol{\varepsilon}}\cdot\boldsymbol{d}|0\rangle$. Equation (3.61) would include all terms, of which this is one, with products of six-dipole transition matrix elements that can arise from this figure in the manner prescribed. Similarly, contributions to $P^{(1)}$ arise from the only diagram contributing to the left hand side of Figure 3.6a, that is, $|\langle 0|\hat{\boldsymbol{\varepsilon}}\cdot\boldsymbol{d}|1'\rangle|^2$.

Finally, contributions to the interference term $P^{(13)}$ can be obtained from Figure 3.7, which results from superimposing the two angular momentum ladders in Figure 3.6a for those continuum angular momentum states that contribute to the interference term. To obtain the contributions to $P^{(13)}$, one constructs all matrix element products that involve *four*-dipole transitions in making a complete circuit from ground state $J = 0$, through the continuum level, and back down to the ground state.

Computational results were obtained [40] using the one-photon vs. three-photon scenario for the case of IBr photodissociation (Eq. (3.56)). Two different cases were examined, those corresponding to fixed initial M_i values and those corresponding to averaging over a random distribution of M_i, for fixed J_i. Results typical of those obtained are shown in Figures 3.8 and 3.9, where we provide a contour plot of the yield of $Br^*(^2P_{1/2})$ for the case of excitation from $J_i = 1$, $M_i = 0$ and $J_i = 42$ while averaging over M_i, as a function of laser control parameters (relative intensity $s = x^2/(1 + x^2)$ and relative phase).

The range of control in each case is impressive, with essentially no loss of control due to the M_j-averaging. Note also the similarity in behavior of the lower J_i and higher J_i cases, although at a different range of S. A related strong field study of two-photon vs. four-photon control in the photodissociation of Cl_2 has been carried out by Bandrauk and coworkers [41, 42]. Strong-field extensions of the one- vs. three-photon scenario have also been discussed [43, 44].

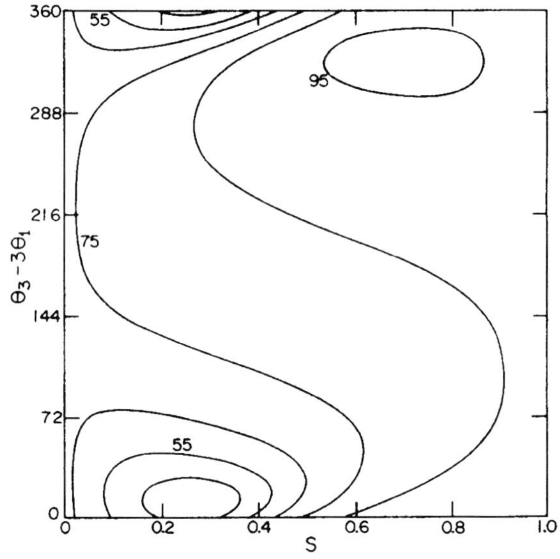

Figure 3.8 Contour plot of the yield of Br^* ($^2P_{1/2}$) (percentage of Br^* as product) in the photodissociation of IBr from an initial bound state in $X\,^1\Sigma_0^+$ with $\nu = 0$, $J_i = 1$, $M_i = 0$. Results arise from simultaneous (ω_1, ω_3) excitation ($\omega_3 = 3\omega_1$) with $\omega_1 = 6657.5\,\text{cm}^{-1}$. The abscissa is labeled by the amplitude parameter $S = x^2/(1+x^2)$, and the ordinate by the relative phase parameter $\theta_3 - 3\theta_1$, equivalent to $\phi_3 - 3\phi_1$ in the text. Taken from [40, Figure 4]. (Reprinted with permission. Copyright 1991 American Institute of Physics.)

The three-photon vs. one-photon scenario has been experimentally realized by Elliott *et al.* in atoms [45, 46], and by Gordon and coworkers [47–51] in a series of experiments on HCl and CO.

A series of experiments on one- vs. three-photon control was carried out by Gordon and coworkers. In the case of HCl, shown in Figure 3.10, the molecule was excited to an intermediate $^3\Sigma^-(0^+)$ vib-rotational resonance, using a combination of three ω_1 ($\lambda_1 = 336$ nm) photons and one ω_3 ($\lambda_3 = 112$ nm) photon. The ω_3 beam was generated from an ω_1 beam by tripling in a Kr gas cell. Ionization of the intermediate state takes place by absorption of one additional ω_1 photon. The relative phase of the light fields was varied by passing the ω_1 and ω_3 beams through a second Ar or H_2 ("tuning") gas cell of variable pressure.

The HCl experiments verified the predictions of coherent control theory concerning the sinusoidal dependence of the ionization rates on the relative phase of the two exciting lasers. The HCl experiment also verified the prediction of the dependence of the strength of the sinusoidal modulation of the ionization current on the relative laser field intensities. Similar demonstrations for ammonia, trimethylamine, triethylamine, cyclooctatetraene, and 1,1-dimethylhydrazine by Bersohn and coworkers [52] have been reported.

Gordon has also demonstrated [49] control of ionization in H_2S in a jet with a large distribution of *j*-states. Although in this case both a dissociation and an ion-

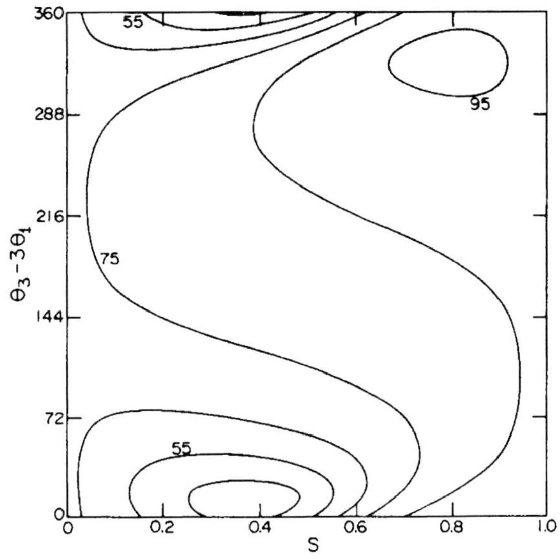

Figure 3.9 As in Figure 3.8, but for v $= 0$, $J_i = 42$, $\omega_1 = 6635.0\,\text{cm}^{-1}$. Results are M-averaged with $\epsilon_0 = 0.125$ a.u. Here $\theta_3 - 3\theta_1$ is equivalent to $\phi_3 - 3\phi_1$ in the text. Taken from [40, Figure 4]. (Reprinted with permission. Copyright 1991 American Institute of Physics.)

ization channel are possible, that is, $H_2S^+ \leftarrow H_2S \rightarrow H + HS$, no discrimination between the possible outcomes of the photoexcitation has been observed: The signals of all final channels oscillate in phase as the relative phase $\phi_3 - 3\phi_1$ is varied.

By contrast, in the $HI^+ \leftarrow HI \rightarrow H + I$ case, control over the production of different channels, specifically the HI^+ vs. the $H + I$ channels, has been observed [50, 51]. The three-photon and one-photon excitation routes possible in this system are shown in Figure 3.11 for both the ionization process and the dissociation process.

The experimental results shown in Figure 3.12 are highly significant as, contrary to the H_2S case, the modulations in the I^+ signal are seen to be *out of phase* with those of the HI^+ signal. Thus, control over different reaction products has been demonstrated. That is, by changing $\phi_3 - 3\phi_1$ (the phase difference between the ω_3 and the ω_1 laser fields) through the change in the pressure of the H_2 gas in the tuning cell, different I^+/HI^+ ratios are attained. Similarly [53] demonstrated one-plus-three-photon control over the photodissociation of $(CH_3)_2S$ to produce measurable product ions CH_3S^+ and $CH_3SCH_2^+$ in two different product channels. Differences in the molecular phases $\alpha_q(13)$ and $\alpha'_q(13)$ allowed control over products of the photodissociation.

The quantitative nature of the observed control depends upon the values of $F_q^{(13)}$ and the "molecular phase", α_q. In particular, the value of $\alpha_q - \alpha_{q'}$ dictates the shift between the peaks in $P_q(E)$ and $P_{q'}(E)$. For example, a molecular case where $\alpha_q - \alpha_{q'} \approx 0$ (e.g., in the $H_2S^+ \leftarrow H_2S \rightarrow H + HS$ case discussed above) shows less discrimination between channels than does a molecular case where $\alpha_q - \alpha_{q'} = \pi$.

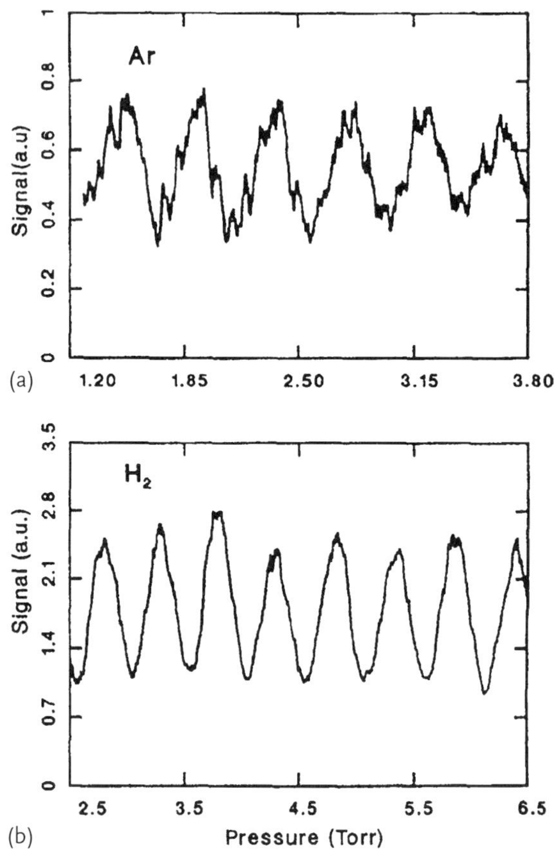

Figure 3.10 Ionization signal for the HCl $R(2)$ transition as a function of pressure in the tuning cell, using either (a) Ar or (b) H_2 to control the relative phases of ω_1 and ω_3. Taken from [47, Figure 2]. (Reprinted with permission. Copyright 1991 American Institute of Physics.)

Hence, the relationship between the nature of the dynamics and the α_q values is of interest. This is a topic studied in detail by Gordon, Seideman and coworkers [54], and is discussed in further detail in Chapter 7.

A crucial aspect of the experiments is that the two copropagating ω_1 and ω_3 beams satisfy the "phase-matching" condition $\boldsymbol{k}_3 = 3\boldsymbol{k}_1$. As a result, Eq. (3.54) no longer depends upon the spatial coordinate z and the interference term is independent of the position in space.

The above results show that it is possible to control the integral cross section into channel q via one-photon vs. three-photon absorption. A similar result is obtained for any N-photon vs. M-photon absorption scenario where N and M are of the same parity. In addition, these scenarios allow for control over differential cross sections as well. To see this, consider rewriting Eqs. (3.49)–(3.52) so that it applies to the probability of observing the product in channel q, but at a fixed scattering angle. Then the sum over the channel indices $\boldsymbol{m}$ no longer includes an integral over scat-

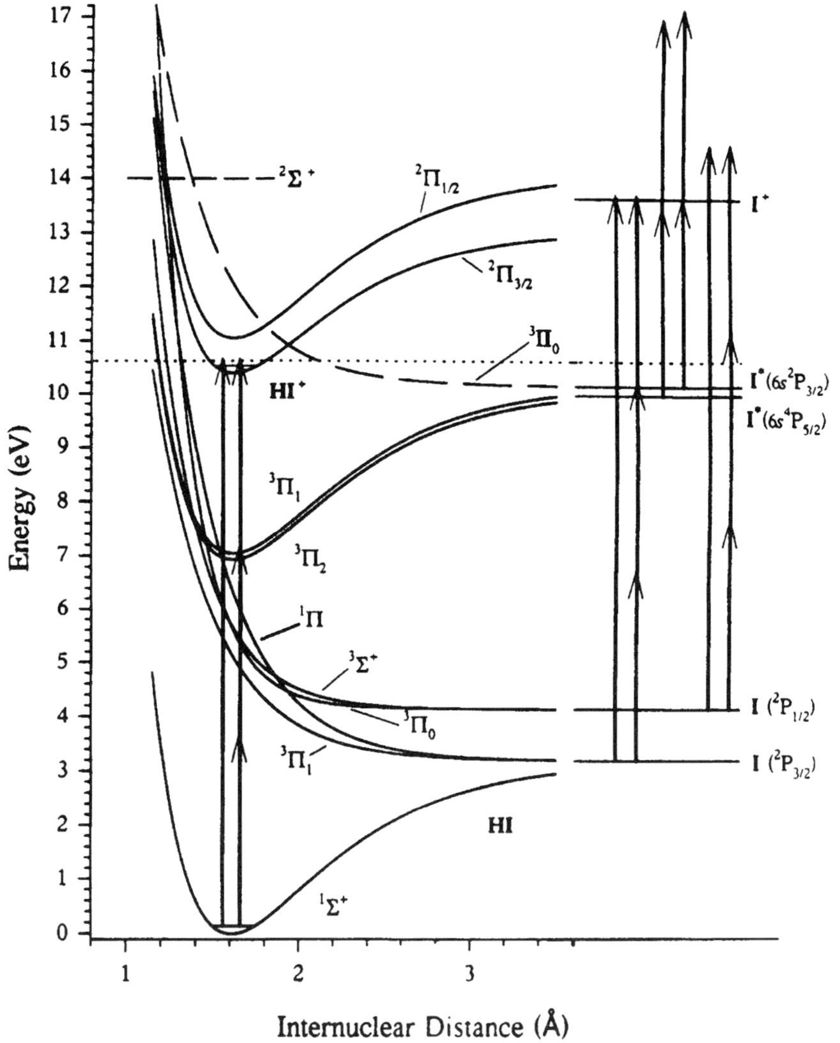

Figure 3.11 Potential energy diagram for HI, with arrows showing the one- and three-photon paths whose interference is used to control the ratio of products formed in the branching reactions $HI \rightarrow HI^+ + e$ and $HI \rightarrow H + I$. Taken from [50, Figure 2]. (Reprinted with permission. Copyright 1995 American Association for the Advancement of Science.)

tering angles. The resultant cross term $P^{(13)}$ is nonzero so that varying properties of the lasers will indeed alter the differential cross section into channel q.

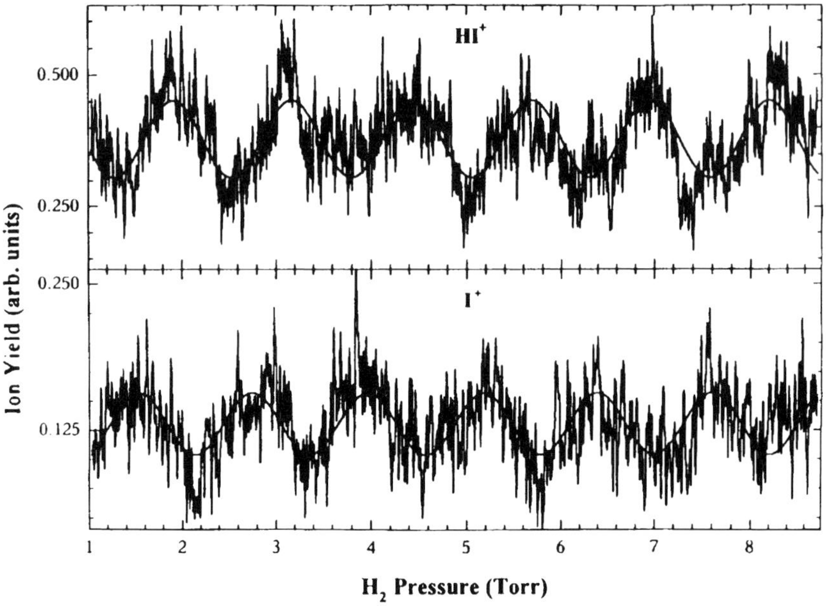

Figure 3.12 Modulation of the HI^+ and I^+ signals as a function of the difference between the one- and three-photon phases (proportional to the H_2 pressure in the cell used to phase shift the beams). Taken from [50, Figure 3]. (Reprinted with permission. Copyright 1995 American Association for the Advancement of Science.)

3.4.3
One- vs. Two-Photon Interference: Symmetry Breaking

Although scenarios for interference between an N-photon route and an M-photon route allow for control over both the differential and integral photodissociation cross sections where (N, M) are of the same parity, this is not the case when N and M are of different parity. In this case only control over the differential cross section is possible. However, the control is such that it leads to the breaking of the usual backward-forward symmetry. This is but one example of the breaking of symmetry afforded via coherent control techniques. A more spectacular example, that of chiral (asymmetric) synthesis, is presented in Chapter 14 below.

In order to understand why control over the total cross section is lost and how the backward-forward symmetry is broken, we analyze in some detail the simplest case in this class, namely the interference between a one-photon and a two-photon absorption process [55]. Consider irradiating a molecule by a field composed of two modes, ω_2 and ω_1, with $\omega_2 = 2\omega_1$, for which the light–matter interaction is

$$H_{MR}(t) = -2\boldsymbol{d} \cdot \mathrm{Re}\left[\hat{\boldsymbol{\varepsilon}}_2\bar{\epsilon}_2 \exp(-i\omega_2 t) + \hat{\boldsymbol{\varepsilon}}_1\bar{\epsilon}_1 \exp(-i\omega_1 t)\right] . \tag{3.62}$$

It follows from Eqs. (3.30) and (3.40) that the amplitude for the combined one-photon, two-photon absorption process is

$$A_{q,\boldsymbol{m}}(E = E_i + \hbar\omega_2) = \frac{2\pi i}{\hbar}\delta(\omega_2 - \omega_{E,i})\left[\bar{\epsilon}_2\langle E, \boldsymbol{m}, q^-|d_{e,g}|E_i\rangle - \bar{\epsilon}_1^2\langle E, \boldsymbol{m}, q^-|D|E_i\rangle\right] , \quad (3.63)$$

where $\langle E_i|D|E, \hat{\boldsymbol{k}}, q^-\rangle$ is the two-photon dissociation amplitude, defined in Eq. (3.43). Here we have implicitly assumed that $E_i + \hbar\omega_1$ is below the threshold for photodissociation.

Suppressing for the moment all channel indices $\boldsymbol{m}$ (which can be readily included), save for the final direction $\hat{\boldsymbol{k}}$, we square the amplitude to obtain $P_q(E, \hat{\boldsymbol{k}})$, the probability of photodissociation into channel q at angles $\hat{\boldsymbol{k}} \equiv (\theta_k, \phi_k)$:

$$P_q(E, \hat{\boldsymbol{k}}) = |A_q(\hat{\boldsymbol{k}}, E_i + \hbar\omega_2)|^2 = P_q^{(1)}(E, \hat{\boldsymbol{k}}) + P_q^{(12)}(E, \hat{\boldsymbol{k}}) + P_q^{(2)}(E, \hat{\boldsymbol{k}}) , \quad (3.64)$$

where

$$P_q^{(1)}(E, \hat{\boldsymbol{k}}) = \left(\frac{2\pi}{\hbar}\right)^2 |\bar{\epsilon}_2|^2 |\langle E, \hat{\boldsymbol{k}}, q^-|d_{e,g}|E_i\rangle|^2 ,$$

$$P_q^{(2)}(E, \hat{\boldsymbol{k}}) = \left(\frac{2\pi}{\hbar}\right)^2 |\bar{\epsilon}_1|^4 |\langle E, \hat{\boldsymbol{k}}, q^-|D|E_i\rangle|^2 ,$$

$$P_q^{(12)}(E, \hat{\boldsymbol{k}}) = -2\left(\frac{2\pi}{\hbar}\right)^2 |\bar{\epsilon}_2\bar{\epsilon}_1^2| \cos\left[\phi_2 - 2\phi_1 + \alpha_q^{(12)}(\hat{\boldsymbol{k}})\right] |F_q^{(12)}(\hat{\boldsymbol{k}})| , \quad (3.65)$$

with the amplitude $|F_q^{(12)}(\hat{\boldsymbol{k}})|$ and phase $\alpha_q^{(12)}(\hat{\boldsymbol{k}})$ defined by

$$|F_q^{(12)}(\hat{\boldsymbol{k}})| \exp(i\alpha_q^{(12)}(\hat{\boldsymbol{k}})) = \langle E_i|D|E, \hat{\boldsymbol{k}}, q^-\rangle\langle E, \hat{\boldsymbol{k}}, q^-|d_{e,g}|E_i\rangle . \quad (3.66)$$

The interference term $P_q^{(12)}(E, \hat{\boldsymbol{k}})$ is generally nonzero, so that control over the differential cross section is possible.

Consider, however, the integral cross section into channel q, that is,

$$P_q(E) = \int d\hat{\boldsymbol{k}}\, P_q(E, \hat{\boldsymbol{k}}) , \quad (3.67)$$

and focus on the contribution from $P_q^{(12)}(E, \hat{\boldsymbol{k}})$. That is, consider

$$P_q^{(12)}(E) = \int d\hat{\boldsymbol{k}} |F_q^{(12)}(\hat{\boldsymbol{k}})| \exp(i\alpha_q^{(12)}(\hat{\boldsymbol{k}}))$$
$$= \int d\hat{\boldsymbol{k}} \langle E_i, J_i, M_i|D|E, \hat{\boldsymbol{k}}, q^-\rangle\langle E, \hat{\boldsymbol{k}}, q^-|d_{e,g}|E_i, J_i, M_i\rangle , \quad (3.68)$$

where we have explicitly inserted the angular momentum characteristics of the initial state. Using the definition of D and inserting unity in terms of the states

$|E_j, J_j, M_j\rangle$ of the intermediate electronic states gives

$$P_q^{(12)}(E) = \sum_{j,e'} \int d\hat{\boldsymbol{k}} \left[\hbar\omega_1 + E_i - E_j\right]^{-1}$$
$$\times \langle E_i, J_i, M_i | d_{g,e'} | E_j, J_j, M_j \rangle \langle E_j, J_j, M_j | d_{e',e} | E, \hat{\boldsymbol{k}}, q^- \rangle$$
$$\times \langle E, \hat{\boldsymbol{k}}, q^- | d_{e,g} | E_i, J_i, M_i \rangle \,. \tag{3.69}$$

For convenience consider the case of diatomic dissociation. Examination of the selection rules shows that when the transition dipole operators $\boldsymbol{d}_{eg}$ and $\boldsymbol{d}_{e'e}$ are parallel to the nuclear axis, the two-photon amplitude is nonzero only if $J_j - J_i = \pm 2, 0$. However, in this case the one photon matrix element $\langle E_i, J_i, M_i | d_{g,e'} | E_j, J_j, M_j \rangle$ is nonzero only if $J_j - J_i = \pm 1$. Since these two conditions are contradictory, $P_q^{12}(E)$ is zero. Hence coherent control over integral cross sections is not possible using the one- vs. two-photon scenario.

This result holds true even when the transition dipole operators are perpendicular to the nuclear axis. Thus, lack of control over the integral cross section in the one- vs two-photon scenario will also occur in polyatomic molecules and for any *N*- vs. *M*-photon process where *N* and *M* are of different parities. The loss of integral control emanates from the fact that the total parity of any molecular wave function is reversed each time a photon is absorbed, since the parity of each photon is negative and the total parity of the photon + molecule system must be conserved. Thus, the parity of a molecular state resulting from a given initial state absorbing an odd number (*N*) of photons is opposite that resulting from the absorption of an even number (*M*) of photons by the same initial state. The integrated interference term which reflects the overlap integral between such states is zero.

However, these features do not prevent control over the differential cross sections for *N* and *M* of different parity, because no integration over angles is required. In fact, because the continuum state $|E, \hat{\boldsymbol{k}}^-\rangle$ accessed via multiphoton pathways of opposite parity has contributions from angular momentum states of opposite parity, the probability of seeing products in a given direction $\hat{\boldsymbol{k}}$ is not the same as the probability of observing products in the opposite direction $-\hat{\boldsymbol{k}}$. That is, the "forward-backward" symmetry has been broken. This is shown below to be but one example of symmetry breaking induced by many coherent control scenarios. One particularly interesting example is control over right- vs. left-handed enantiomers, discussed in detail in Chapter 14.

The experimental implementation of the one-plus-two-photon absorption scenario have taken a variety of forms [56, 57]. Several theoretical papers in addition to [55] have also analyzed this phenomenon [58–60].

For example, Corkum and coworkers [61] have carried out one- vs. two-photon absorption in crafted quantum wells in an experiment depicted schematically in Figure 3.13. As shown in Figure 3.14, by varying $\phi_2 - 2\phi_1$, the relative phase between the second harmonic and twice that of the fundamental frequency (at 10.6 μm), the experimentalists were able to direct the electronic current to move in either the forward or backward direction, or to generate a current that was equally probable in both directions.

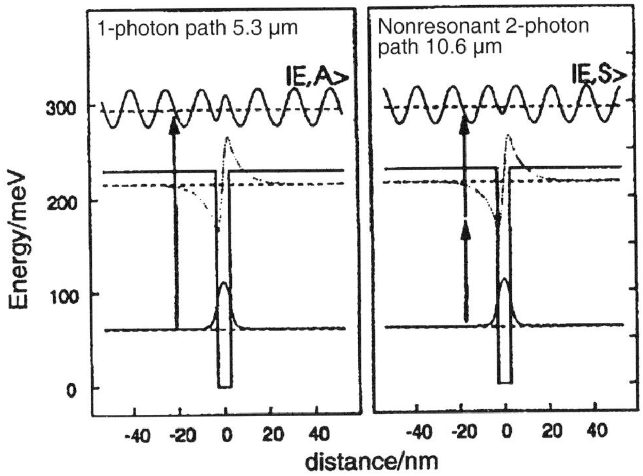

Figure 3.13 Energy band diagram of a 55 Å GaAs/$Ga_{0.74}Al_{0.26}As$ quantum well and wave functions of the states implied in a 5.3 μm single-photon pathway and a 10.6 μm two-photon process. Neither dephasing nor reflections of the electronic waves on the neighbor quantum wells are considered in this simplified figure. Taken from [61, Figure 1]. (Reprinted with permission. Copyright 1995 American Physical Society.)

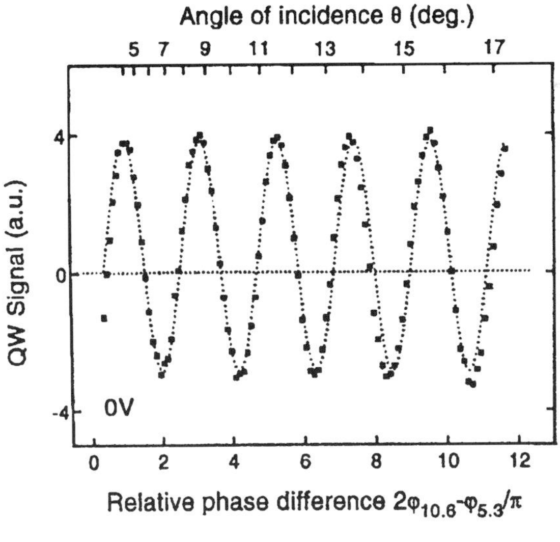

Figure 3.14 An experiment showing the integrated quantum well response versus the relative laser phase. Dashed line: sinusoidal fit. Taken from [61, Figure 4]. (Reprinted with permission. Copyright 1995 American Physical Society.)

Related results were obtained with molecules. For example, following the theoretical predictions of Charron *et al.* [62] on the photodissociation of H_2^+ (displayed in Figure 3.15), Dimauro *et al.* [63] performed an experiment (shown schematically in Figure 3.16) to control product directionality in HD^+ dissociation to $H + D^+$

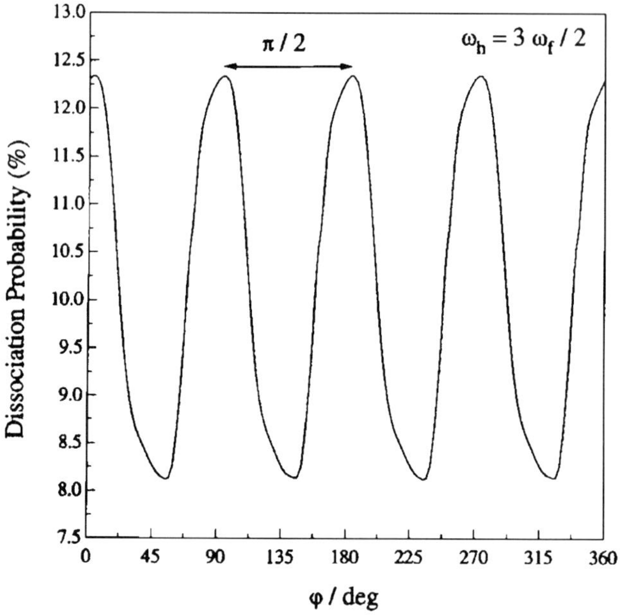

Figure 3.15 The computed H^+ current resulting from the photodissociation of H_2^+ as a function of $\varphi = \phi_2 - 2\phi_1$, the difference between the second harmonic phase and twice the phase of the fundamental photon. Taken from [62, Figure 6]. (Reprinted with permission. Copyright 1995 American Institute of Physics.)

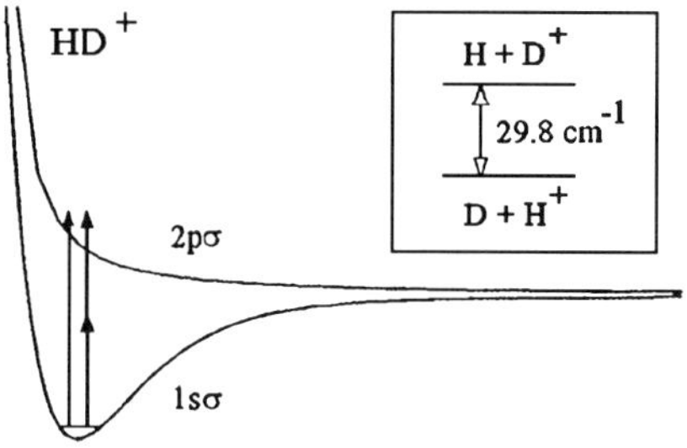

Figure 3.16 Potential curves for the 1sσ and 2pσ states of HD^+. In the homonuclear case (H_2^+), the two states are asymptotically degenerate; the degeneracy is lifted in the heteronuclear case by 29.8 cm^{-1} (inset). Taken from [63, Figure 1]. (Reprinted with permission. Copyright 1995 American Physical Society.)

and $H^+ + D$. Here a combination of a one-photon process, induced by the second harmonic, and a two-photon process, induced by the fundamental frequency, were used to excite the molecule to a repulsive 2pσ state yielding either the $H + D^+$ or the $D + H^+$ products. The results of the experiment are shown in Figure 3.17. We see that the angle at which the ions appear can be varied by changing the $\phi_2 - 2\phi_1$ relative phase.

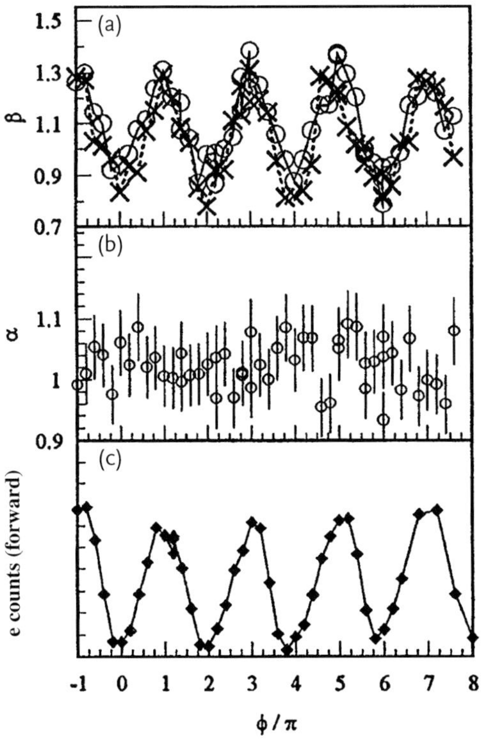

Figure 3.17 The forward/backward yield ratios of protons and deuterons in the dissociation of HD^+ vs. $\phi = \phi_2 - 2\phi_1$, the difference between the phase of the second harmonic photon and twice the phase of the fundamental photon. The ratio of H^+ is $\beta_{H^+} = H_f^+/H_b^+$ (circles), and the ratio of D^+ is $\beta_{D^+} = D_f^+/D_b^+$ (crosses). Taken from [63, Figure 3]. (Reprinted with permission. Copyright 1995 American Physical Society.)

It is interesting to note (see Figure 3.17b) that the ratio between the H^+ and D^+ ions does not vary with the relative phase. This is partly in agreement with the analysis presented above that, within lowest-order perturbation theory, the ratio between integral cross sections of different channels cannot be controlled by an *N*- vs. *M*-photon scenario when *N* and *M* possess different parities. This means that, within the confines of perturbation theory, when we average over all angles we should find a phase-independent H^+/D^+ branching ratio. This argument does not, however, explain why this lack of discrimination should hold for each and every angle: as can be seen from Figure 3.17a, the H^+/D^+ ratio is independent of the dissociation angle. Moreover, an argument based on low-order perturbation theory is not expected to hold in the long wavelength regime where multiple photon transitions are involved, and isotopic discrimination is therefore expected to occur [62].

We conclude that in the short wavelength regime what is being affected in the dissociation of HD^+ is the motion of the (lone) *electron*. The electron is seen to direct itself towards the forward or backward directions in the *laboratory frame* as the $\phi_2 - 2\phi_1$ relative phase is varied. Since the experiment monitored only dissociative events where the electron is still bound to the atomic fragment, the electron simply "rides" on whatever ion happens to be pointing in its preferred laboratory direction. If, while the molecule is rotating and dissociating, the electron finds the D^+ nucleus pointing in its preferred direction, it attaches itself to the deuteron and the neutral D atom will emerge in that direction (with the H^+ ion emerging in the opposite direction). The situation is reversed if the proton happens to be moving in the direction preferred by the electron.

These conclusions, that even if ionization does not occur it is often the electron that is being controlled rather than the nuclei, follow the work of Aubanel and Bandrauk [64], who have shown such electronic control in the photodissociation of Cl_2. The case for electronic control is naturally stronger when the lasers are intense enough to ionize the molecule. In that case the interference between the one-photon and two-photon processes has been shown to affect the ionization yield [44]. Additional theoretical and experimental work on the control of *atomic* phenomena in $\omega + 2\omega$ and $\omega + 3\omega$ scenario has been reviewed in detail by Ehlotzky [65].

The results in this section describe symmetry breaking due to one-versus-two-photon excitation that is on resonance with two degenerate excited states of opposite parity. The extension to off-resonant excitation is discussed in Section 18.1.2.

3.5 Polarization Control of Differential Cross Sections

Rather than attempting coherent control with two different frequencies, it appears that using two different polarizations of the same frequency would be much easier to implement experimentally. It turns out that this scenario is akin to the one-vs. two-photon control in the sense that integral control is not possible. Although differential cross sections can be controlled [66], one can not break the forward-backward symmetry in this case.

In order to see this we consider the photodissociation of a *single* bound state $|E_1\rangle$ by a single CW source with arbitrary polarization vector $\hat{\boldsymbol{\eta}}$,

$$\hat{\boldsymbol{\eta}} = \eta_1 \exp(i\alpha_1)\,\hat{\boldsymbol{\eta}}_1 + \eta_2 \exp(i\alpha_2)\,\hat{\boldsymbol{\eta}}_2\,, \tag{3.70}$$

where $\hat{\boldsymbol{\eta}}_1$ and $\hat{\boldsymbol{\eta}}_2$ are two orthonormal vectors (see Figure 3.18).

We can regard the two components $\hat{\boldsymbol{\eta}}_1$ and $\hat{\boldsymbol{\eta}}_2$ as inducing two independent excitation routes. Choosing $\hat{\boldsymbol{\eta}}_1$ and $\hat{\boldsymbol{\eta}}_2$ parallel and perpendicular to the quantization (z) axis, respectively, the differential cross section is composed of three terms; one corresponds to photodissociation of $|E_1\rangle$ by the $\hat{\boldsymbol{\eta}}_1$ component, one by the $\hat{\boldsymbol{\eta}}_2$ component, and one being the cross term between these two contributions. Excitation by the parallel component allows $\Delta M_J = 0$ transitions, while excitation by the perpendicular $\hat{\boldsymbol{\eta}}_2$ component allows $\Delta M_J = \pm 1$ transitions. The interference term is therefore comprised of a product of two bound-continuum matrix elements, where the two continua differ in M_J by ± 1. If this cross term is nonzero then control over

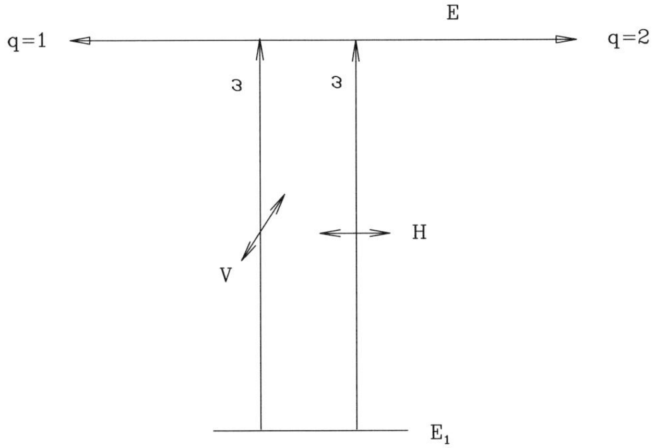

Figure 3.18 Photodissociation by CW source of arbitrary polarization. The two components of the polarization vector are shown.

the differential cross section is possible. However, producing the *integral* cross section necessitates integrating the differential cross section over $\hat{\boldsymbol{k}}$ and, under these circumstances, the cross term vanishes.

Contrary to the one- vs two-photon case, the states comprising the $|E, \hat{\boldsymbol{k}}^-\rangle$ state are of the same parity. Thus, in the differential cross section, the backward-forward symmetry is not broken. Rather, control is manifested in our ability to sharpen or broaden the angular distribution about a given recoil direction.

3.6 Pump-Dump Control: Few Level Excitation

Control of the dynamics via a pump-dump scenario was first introduced by Tannor and Rice [67, 68] with insight afforded by localized wave packets [69], an approach that is associated with many-level excitation, and which is discussed in Section 5.1. Here we consider the few levels case shown qualitatively in Figure 3.19. It can be regarded as a useful extension of the scenario outlined in Section 3.2, in which the initial superposition of bound states is prepared with one laser pulse and subsequently dissociated with another. It is also related to a two- vs. two-photon scenario (Section 7.1) since, as shown in Figure 3.20, pump-dump excitation contains 2 vs. 2 photon pathways to a collection of energies E.

Consider then a system irradiated by two pulses, termed the "pump and dump" pulse. These pulses are assumed to be temporally separated by a time delay Δ_d. The analysis below shows that under these circumstances control over the photodissociation yields is obtained by varying the central frequency of the pump pulse and the time delay between the two pulses.

Consider a molecule, initially ($t = 0$) in an eigenstate $|E_1\rangle$ of the molecular Hamiltonian H_M that is subjected to two transform limited light pulses. The elec-

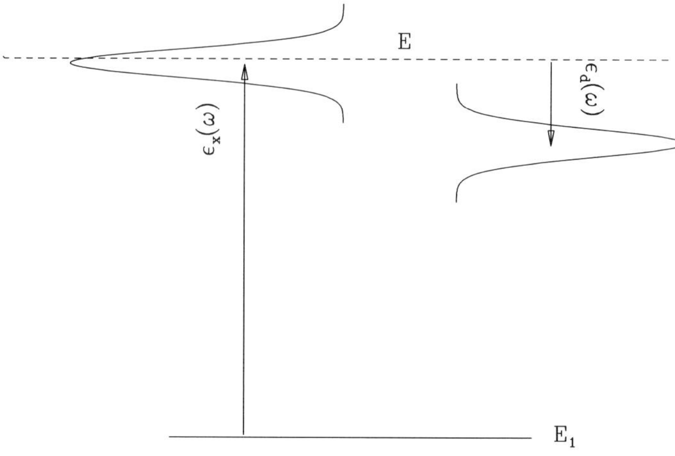

Figure 3.19 Pump-dump control scenario

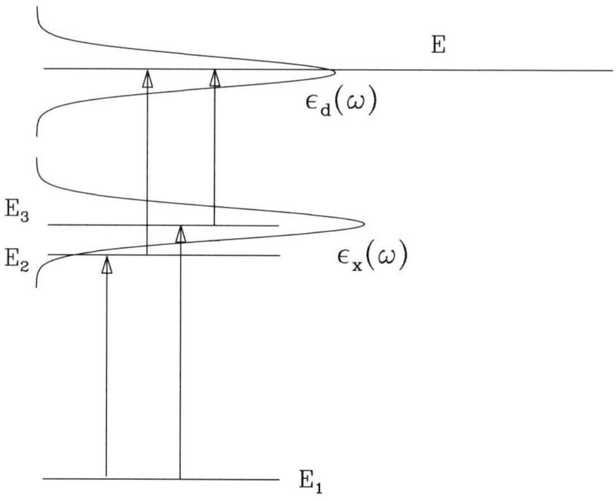

Figure 3.20 Interfering two two-photon pathways to energy E contained in the pump-dump control scheme of Figure 3.19. The ϵ_x, ϵ_d labels indicate whether the excitation or dissociation laser is causing the indicated transition.

tric field consists of two temporally separated pulses $\boldsymbol{E}_x(\tau)$, $\boldsymbol{E}_d(\tau)$ (where τ is the retarded time $(t - z/c)$). For both pulses the electric field is of the form $\boldsymbol{E}(\tau) = 2\hat{\boldsymbol{\varepsilon}}\varepsilon(\tau)\cos(\omega_0\tau)$, which is a parametrization of Eq. (1.7). Here ω_0 is the carrier frequency and $\varepsilon(\tau)$ describes the pulse envelope. Thus, the molecule is subjected to the field,

$$\boldsymbol{E}(\tau) = \boldsymbol{E}_x(\tau) + \boldsymbol{E}_d(\tau) \tag{3.71}$$

For convenience we use Gaussian pulses that peak at $t = t_x$ and t_d, respectively. In particular, the excitation pulse is of the form

$$E_x(\tau) = \frac{1}{2}\hat{\epsilon}_x \epsilon_x \exp[-i(\omega_x \tau + \delta_x)] \exp\left[-\frac{(\tau - t_x)^2}{\tau_x^2}\right] , \tag{3.72}$$

where the Gaussian pulse is spread with width τ_x about time t_x and carries an overall phase δ_x. The associated frequency profile is given by the Fourier transform of Eq. (3.72):

$$\epsilon_x(\omega) = \frac{\sqrt{\pi}}{2}\epsilon_x \tau_x \exp[-i(\omega_x - \omega) t_x] \exp\left[-\frac{\tau_x^2 (\omega_x - \omega)^2}{4}\right] \exp(-i\delta_x) . \tag{3.73}$$

Further, we define $\bar{\epsilon}_x(\omega)$as in Eq. (2.9), with $\phi(\omega) = (\omega - \omega_x)t_x - \delta_x$.

The analogous quantities for the dissociation laser, $E_d(\tau)$, $\epsilon_d(\omega)$ and $\bar{\epsilon}_d(\omega)$ are defined similarly, with the parameters t_d and ω_d replacing t_x and ω_x, and so on The pump pulse $E_x(\tau)$ induces a transition to a linear combination of the eigenstates $|E_i\rangle$ of the excited electronic state. The pump pulse may be chosen to encompass any number of states. Here we choose the pump pulse sufficiently narrow to excite only two of these states, $|E_2\rangle$ and $|E_3\rangle$. The dump pulse $E_d(\tau)$ dissociates the molecule by further exciting it to the continuous part of the spectrum. Both fields are chosen sufficiently weak for perturbation theory to be valid.

Since the two pulses are temporally distinct, it is convenient to deal with their effects consecutively. After the first pulse is over, the superposition state prepared by the $E_x(\tau)$ pulse, whose width is chosen to encompass just the two levels $|E_2\rangle$ and $|E_3\rangle$, is given in first-order perturbation theory as

$$|\phi(t)\rangle = |E_1\rangle e^{-iE_1t/\hbar} + b_2|E_2\rangle e^{-iE_2t/\hbar} + b_3|E_3\rangle e^{-iE_3t/\hbar} , \tag{3.74}$$

where (Eq. (2.23))

$$b_k = \frac{2\pi i}{\hbar}\langle E_k|\hat{\boldsymbol{\varepsilon}}_x \cdot \boldsymbol{d}|E_1\rangle \bar{\epsilon}_x(\omega_{k,1}) , \quad k = 2, 3 , \tag{3.75}$$

with $\omega_{k,1} \equiv (E_k - E_1)/\hbar$.

After a delay time of $\Delta_d \equiv t_d - t_x$ the system is subjected to the $E_d(\tau)$ pulse. It follows from Eq. (3.74) that after this delay time each preparation coefficient has picked up an extra factor of $e^{-iE_k\Delta_d/\hbar}$, $k = 2, 3$. Hence, the phase of b_2 relative to b_3 at that time increases by $[-(E_2 - E_3)\Delta_d/\hbar = \omega_{3,2}\Delta_d]$. Thus, the natural two-state time evolution controls the relative phase of the two terms, replacing the externally controlled relative laser phase of the two-frequency control scenario of Section 3.2.

After the action and subsequent decay of the $E_d(\tau)$ pulse, the system wave function is:

$$|\psi(t)\rangle = |\phi(t)\rangle + \sum_{n,q}\int dE\, b_{E,m,q}(t)|E, \boldsymbol{m}, q^-\rangle e^{-iEt/\hbar} . \tag{3.76}$$

In accord with Eqs. (2.74)–(2.77), the probability of observing the q product at total energy E in the remote future is therefore

$$P_q(E) = \sum_m |b_{E,\boldsymbol{m},q}(t=\infty)|^2$$
$$= \left(\frac{2\pi}{\hbar}\right)^2 \sum_m \left| \sum_{k=2,3} b_k \langle E, \boldsymbol{m}, q^- | d_{e,g} | E_k \rangle \bar{\epsilon}_d(\omega_{EE_k}) \right|^2 , \tag{3.77}$$

where $\omega_{EE_k} = (E - E_k)/\hbar$, b_k is given by Eq. (3.75), and where $\bar{\epsilon}_d(\omega)$ is given via an expression analogous to Eq. (3.73).

Expanding the square and using the Gaussian pulse shape (Eqs. (3.72) and (3.73)) gives,

$$P_q(E) = \left(\frac{2\pi}{\hbar}\right)^2 \left[|b_2|^2 d_q(2,2) \bar{\epsilon}_2^2 + |b_3|^2 d_q(3,3) \bar{\epsilon}_3^2 + 2|b_2 b_3^* \bar{\epsilon}_2 \bar{\epsilon}_3^* d_q(3,2)| \cos\left(\omega_{3,2}\Delta_d + \alpha_q(3,2) + \chi\right)\right] , \tag{3.78}$$

where $\bar{\epsilon}_i = |\bar{\epsilon}_d(\omega_{EE_i})|$, $\omega_{32} = (E_3 - E_2)/\hbar$ and the phases χ, $\alpha_q(3,2)$ are defined via

$$\langle E_1 | d_{e,g} | E_g \rangle \langle E_g | d_{g,e} | E_2 \rangle \equiv |\langle E_1 | d_{e,g} | E_g \rangle \langle E_g | d_{g,e} | E_2 \rangle| e^{i\chi} ,$$
$$d_q(ki) \equiv |d_q(ki)| e^{i\alpha_q(k,i)} = \sum_m \langle E_k | d_{g,e} | E, \boldsymbol{m}, q^- \rangle \langle E, \boldsymbol{m}, q^- | d_{e,g} | E_i \rangle . \tag{3.79}$$

Integrating Eq. (3.78) over E to encompass the full width of the second pulse yields the final expressions for the quantities we wish to control: P_q, the probability of forming channel q, and $R_{q,q'}$, the ratio of product probabilities into q vs. q' (see Eq. (3.22)).

Examination of Eq. (3.78) makes clear that $R_{q,q'}$ can be varied by changing the delay time $\Delta_d = (t_d - t_x)$ or the ratio $x = |b_2/b_3|$; the latter is most conveniently done by detuning the initial excitation pulse. Note that, once again, as in the scenarios above, the z dependence of P_q vanishes due to cancellation between the excitation and dump steps. In addition, the phases δ_x, δ_d do not appear in the final $R_{q,q'}$ expression, so that the relative phases of the two pulses do not affect the result.

This approach was applied to realistic systems such as the control of the Br to Br* branching ratio in the photodissociation of IBr [70], and the control of Li_2 photodissociation [71], discussed later below. To gain insight into the control afforded by this scenario we also applied it to a somewhat artificial model of the photodissociation of a hypothetical collinear DH_2 complex [72]:

$$H + HD \leftarrow DH_2 \rightarrow D + H_2 \tag{3.80}$$

The first pulse is used to excite a pair of states in an electronic state supporting bound states and the second pulse to dissociate the system by de-exciting it back to the ground state, above the dissociation threshold.

The model potentials used in the DH_2 [72] are shown in Figure 3.21 and typical control results are shown in Figure 3.22. Specifically, Figure 3.22 contours of equal DH yield as a function of $E_x - E_{AV}$ and Δ_d. Here $(E_x - E_{AV})$ measures the deviation of the central excitation energy of the pump pulse from the mean energy E_{AV} of the

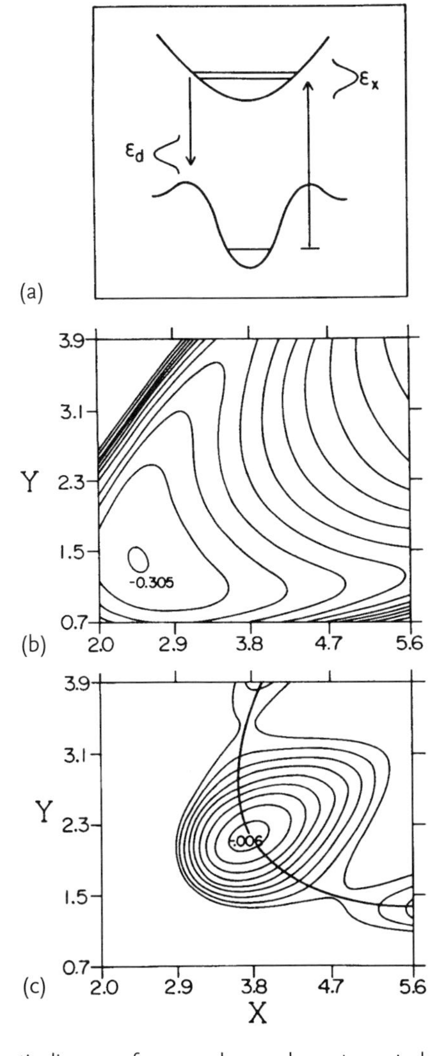

Figure 3.21 (a) Schematic diagram of a pump-dump scheme to control the model DH + H ← DH_2 → D + H_2 branching photodissociation reaction. Here ϵ_x is the excitation pulse and ϵ_d is the dump pulse. (b) Ground potential surface. Contour lines are spaced by 0.02 a.u., increasing outwards from the indicated minimum. (c) Excited potential surface. Contour lines are spaced by 0.0098 a.u., increasing outwards from the indicated minimum. The reaction coordinate S is shown as a thick line that is chosen here as to coincide with the minimum energy path connecting the DH+H and the D+H_2 products. Taken from [72, Figure 1]. (Reprinted with permission. Copyright 1989 American Institute of Physics.)

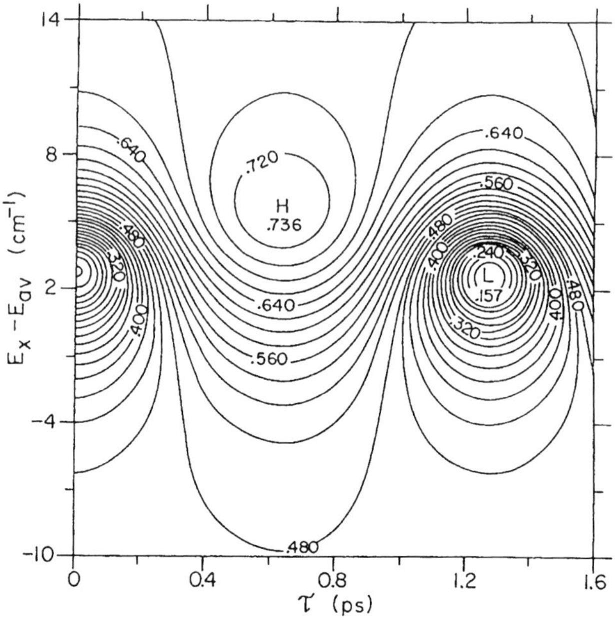

Figure 3.22 Contour plot of the DH yield as a function of the detuning of the exciting pulse $E_x - E_{AV}$, and the delay variable $\tau \equiv \Delta_d$. The actual delay is $(8.44 + 2.11n)$ ps $+ \tau$, where n is an arbitrary positive integer which is chosen high enough to eliminate any overlap between the pulses. Here the initially created superposition state is between levels 56 and 57 ($E_1 = 0.323\,849$ a.u., $E_2 = 0.323\,968$ a.u.) of the excited surface. The letters H and L denote the positions of the absolute maxima and minima, whose magnitudes are explicitly shown. Taken from [25, Figure 6]. (Reprinted with permission. Copyright 2000 Elsevier.)

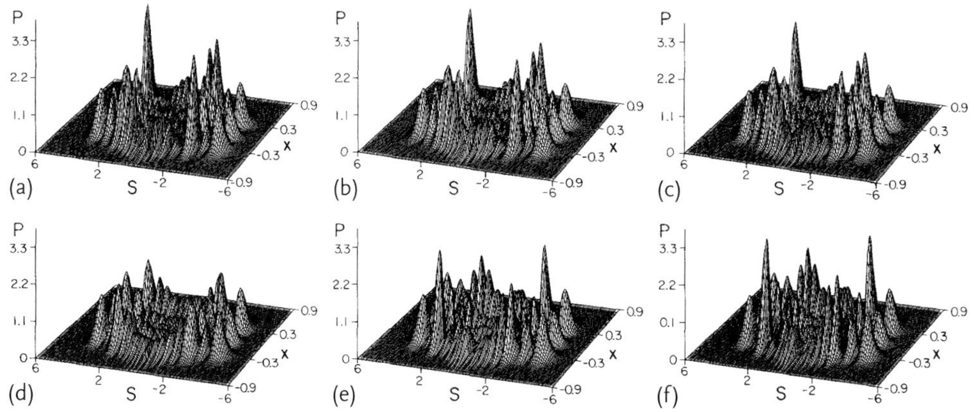

Figure 3.23 Time evolution of the square of the wave function for a superposition state comprised of levels 56 and 57. The probability is shown as a function of S and its orthogonal coordinate x at times (a) 0 ps, (b) 0.0825 ps, (c) 0.165 ps, (d) 0.33 ps, (e) 0.495 ps, (f) 0.66 ps, which correspond to equal fraction of the period $2\pi/\omega_{2,1}$. Taken from [72, Figure 6]. (Reprinted with permission. Copyright 1989 American Institute of Physics.)

pair of bound states which it excites. The DH yield is shown to vary significantly, from 16 to 72%, as the control parameters are varied. This is an extreme range of control, especially if one considers that the product channels only differ by a mass factor.

It is highly instructive to examine the nature of the superposition state prepared in the initial excitation, (Eq. (3.74)) and its time evolution during the delay between pulses. An example is shown in Figure 3.23 where we plot the wave function for the collinear model of model DH_2 photodissociation. Specifically, the axes are the reaction coordinate S and the coordinate x orthogonal to it. The wave function is shown evolving over one half of its total possible period. An examination of Figure 3.23 in conjunction with Figure 3.22 shows that de-exciting this superposition state at the time of Figure 3.23b would yield a substantially different product yield than de-exciting at the time of Figure 3.23e. However, Figure 3.23 shows that there is clearly no particular preference of the wave function for either large positive or large negative S at these particular times, which would be the case if the reaction control were a result of some spatial characteristics of the wave function. Rather, the results make clear that the essential control characteristics of the wave function are encrypted in the quantum amplitude and phase of the created superposition state.

The controlled photodissociation of Li_2 into different atomic states provides another example of pump-dump control [71]. At the energy of interest three product pairs, Li(2s) + Li(2p), Li(2s) + Li(3p), and Li(2s) + Li(3s) may be produced. The pho-

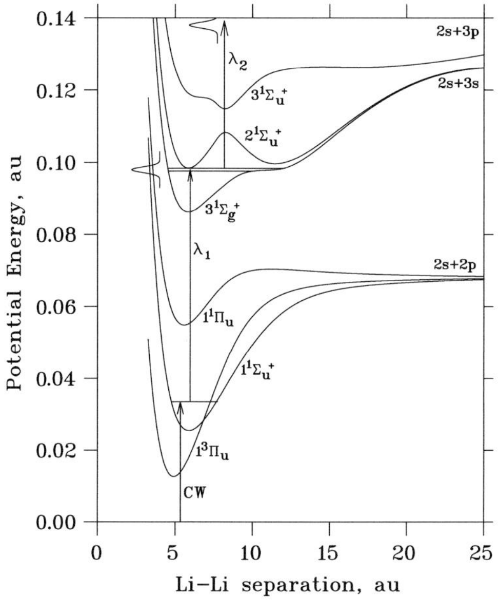

Figure 3.24 Schematic of the pulse sequencing control scheme used in this work. The realistic Li_2 potentials used in this work are shown. From [71, Figure 1]. (Reprinted with permission. Copyright 1998 American Institute of Physics.)

todissociation computations were carried out using realistic potential curves [73] shown in Figure 3.24. A schematic description of a suggested multipulse scheme (a preliminary experiment along these lines has been reported by Leone *et al.* [74, 75]) is also shown in this figure. In this case, a CW laser prepares a single vib-rotational state of the A $(1^1\Sigma_u^+)$ electronic state, chosen as $v = 14$, $J = 22$, which serves as the starting point for the subsequent pump-pump control. The pump pulse "lifts" the nuclear wave function in this electronic state to the $E(3^1\Sigma_g^+)$ state, forming a coherent superposition of ro-vibrational states. The subsequent pulse excites the system above the dissociation energy for the product formation.

Figures 3.25 and 3.26 show the control resulting from the use of a narrowband pump laser to prepare a superposition of the ($v = 14$, $J = 21$) and ($v = 14$, $J = 23$) vib-rotational states in the excited E electronic manifold. The pump pulse center was tuned to the midfrequency between these two states; that is, $\lambda_1 = 805.6$ nm, $\lambda_2 = 1045$ nm, $\Delta_{1\omega} = 36\,\text{cm}^{-1}$ and $\Delta_{2\omega} = 45\,\text{cm}^{-1}$. Here Δ_{iw} is a measure of the frequency width of the ith pulse, that is, $\Delta_{iw} = 4\sqrt{\ln 2}\tau_i$. The yield of Li(2s) + Li(3s) was found to vary from 0.46 to 1.03%, and to be relatively invariant to changes in Δ_d. For this reason its behavior is ignored below.

Figure 3.25 shows the percent yield of forming Li(2s) + Li(2p) (solid curve) and Li(2s) + Li(3p) (dashed curve) as a function of delay time Δ_d. The results clearly show an extensive variation of the relative yield of the different products as a function of τ with the Li(2s) + Li(2p) yield varying from 82.2 to 20.4%, and the Li(2s) + Li(3p) yield varying from 17.2 to 78.7%. This corresponds to the change in the yield ratio $R = P(q = 2\text{p})/P(q = 3\text{p})$ from 4.8 to 0.26. The yield of the Li(2p) product is seen to change more dramatically with Δ_d, a consequence of the fact

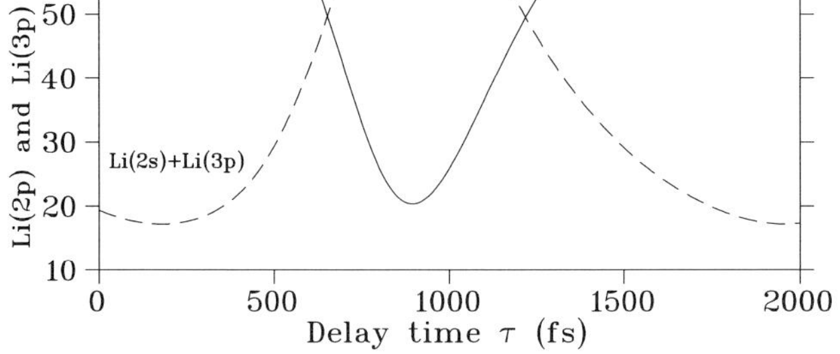

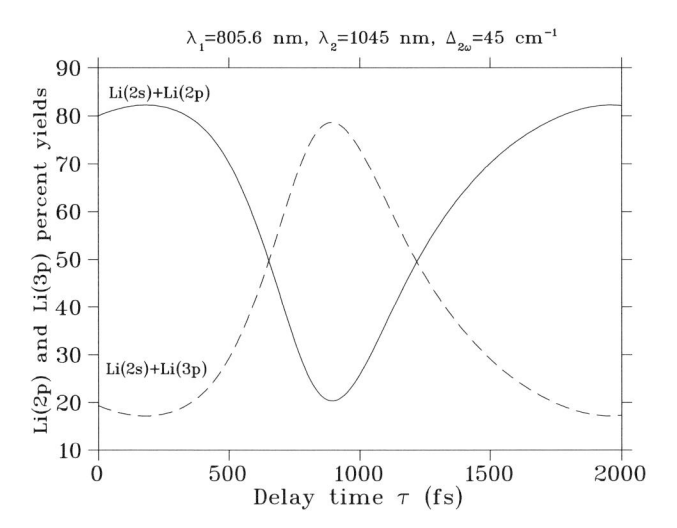

Figure 3.25 (a) Yields of Li(2p) (solid curve) and Li(3p) (dashed curve) as a function of the delay time $\tau \equiv \Delta_d$. Here the initially created superposition state is composed of the ($v = 14, J = 21$) and ($v = 14, J = 23$) states in the E electronic manifold. From [71, Figure 2]. (Reprinted with permission. Copyright 1998 American Institute of Physics.)

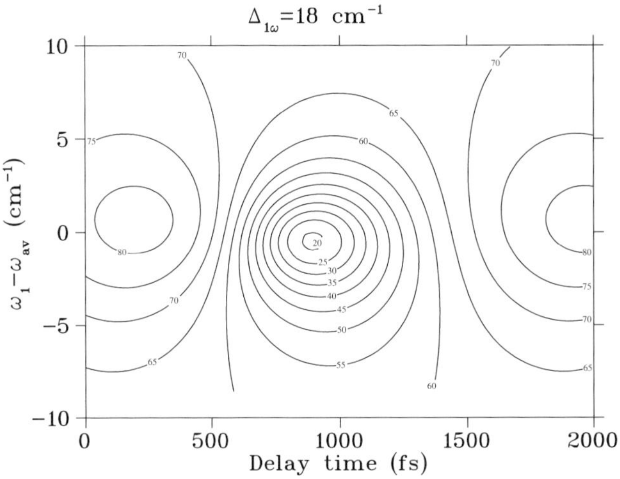

Figure 3.26 Contour plot of the Li(2p) yield as a function of the detuning of the pump pulse $\omega_1 - \omega_{AV}$ and the delay time $\tau \equiv \Delta_d$. The initially created superposition state is between levels v = 14, J = 21 and J = 23. Laser parameters are given in the text. From [71, Figure 4]. (Reprinted with permission. Copyright 1998 American Institute of Physics.)

that the $d_q(12)$ matrix element associated with the Li(2p) product is approximately three times larger than that associated with the Li(3p) product.

In accord with Eq. (3.78), the product probability is seen to be periodic, with an approximate period of $T = 2\pi/\omega_{2,1} = 1773$ fs, which corresponds to the rotational spacing $\omega_{2,1} \equiv (E_{v=14,J=23} - E_{v=14,J=21})/\hbar$ of 18.8 cm^{-1}.

The product ratios can also be controlled by shifting the central frequency of the pump laser, thus altering the b_j coefficients of the superposition state (Eq. (3.74)). Sample results are shown in Figure 3.26 where we display a contour plot of the Li(2s) + Li(2p) product yield as a function of $\tau \equiv \Delta_d$ and of the detuning of ω_1 from $\omega_{AV} = (1/2\hbar)(E_{v=14,J=21} + E_{v=14,J=23} - 2E_1)$, the halfway transition frequency to the two vib-rotational levels. In this case $\Delta_{1\omega} = 18$ cm^{-1} and $\Delta_{2\omega} = 45$ cm^{-1} with the remaining parameters of the second laser chosen as in Figure 3.25. In Figure 3.26, λ_1 is varied from 806.2 to 805 nm, corresponding to the centering of the first pulse on levels $E_{v=14,J=21}$ and $E_{v=14,J=23}$, respectively. The results indicate a large range of control, from 81 to 20% as one varies τ at fixed pulse detuning. Substantial control is also attained by changing ω_1 at fixed τ (e.g., at 890 fs). The results clearly show, however, that the highest level of control is attained when the energy of the first pulse is centered close to E_{AV} ($\lambda_1 = 805.6$ nm). This is because, under these circumstances, the bound-bound dipole matrix elements $\langle E_j|\hat{\boldsymbol{\varepsilon}}\cdot\boldsymbol{d}|E_s\rangle$, $j = 1, 2$ contributing via Eq. (3.75) have very similar values for the (v = 14, J = 21) and (v = 14, J = 23) levels.

Varying other parameters, such as the width of the pulses, also has substantial effect on product control. For example, the effect of exciting more vib-rotational levels in the E electronic state by using a broader pump pulse is shown in Figure 3.27,

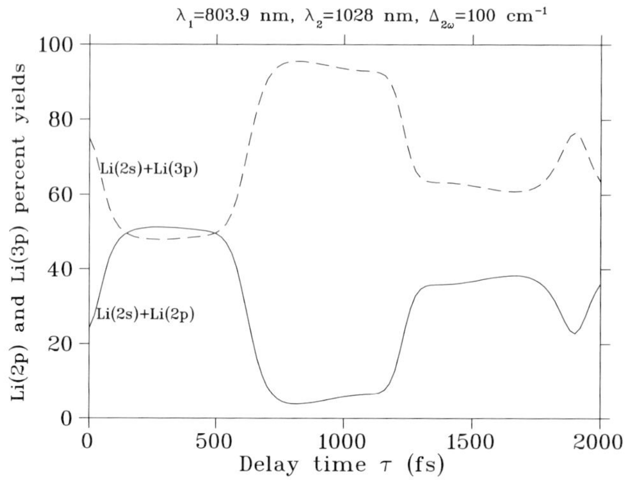

Figure 3.27 Control results with a broader first pulse. The yield ratios of Li(2p) (solid curve) and Li(3p) (dashed curve) as a function of $\tau \equiv \Delta_d$ for the integrated probability $P(q) = \int_0^\infty dE\, P(E, q)$ with wavelengths $\lambda_1 = 803.9$ nm and $\lambda_2 = 1028$ nm and pulse widths $\Delta_{1\omega} = 60\,\text{cm}^{-1}$ and $\Delta_{2\omega} = 100\,\text{cm}^{-1}$. The superposition state consists of v = 14, $J = 21$ and $J = 23$, and v = 15, $J = 21$ and $J = 23$ levels. From [71, Figure 8]. (Reprinted with permission. Copyright 1998 American Institute of Physics.)

where $\Delta_{1\omega} = 60\,\text{cm}^{-1}$ and $\Delta_{2\omega} = 100\,\text{cm}^{-1}$. The superposition state prepared by the first pulse consists of the v = 14, 15 and $J = 21, 23$ levels, where the pulse is centered at $\lambda_1 = 803.88$ nm corresponding to the frequency halfway between the (v = 14, $J = 21$) and the (v = 15, $J = 23$) levels. The resultant behavior of P_q is more complicated than the previous cases since more than two terms are included in Eq. (3.78). Interestingly, we find that overall control is reduced with increasing $\Delta_{1\omega}$. That is, in this case the yield in the Li(2s) + Li(2p) channel only changes from 51.4 to 2.6%, a much smaller control range than in the comparable (narrower pump pulse) case. Adding additional levels to the initial superposition state by pumping with a wider pulse ($\Delta_{1\omega} = 100\,\text{cm}^{-1}$) resulted in slightly more complicated behavior of $P(q)$ as a function of τ, and an even further small reduction in the extent of control.

The pump-dump scheme described above has also been applied to the control of the

$$\text{D} + \text{OH} \leftarrow \text{HOD} \rightarrow \text{H} + \text{OD}$$

dissociation reaction, proceeding via the B^1A' excited state of HOD. In this case both asymptotic channels have identical potential energy surfaces so that control over the relative yield is challenging. To consider the extent of possible control we excite an initial superposition of the (0, 2, 0) and (1, 0, 0) states of ground state HOD ((0, 2, 0) denotes two quanta in the bend mode and (1, 0, 0) denotes one quantum of excitation of the OD stretch). A subsequent pulse dissociates HOD via the B^1A' continuum. A typical result is displayed in Figure 3.28, which shows contours of

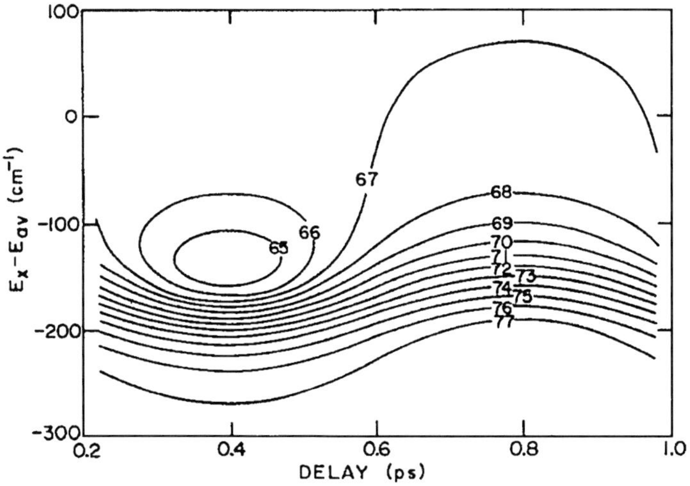

Figure 3.28 Percentage yield of the H + OD channel in the photodissociation of the DOH(0,2,0 + 1,0,0) superposition state. The excitation pulse band width is 50 cm^{-1}, the dissociation pulse bandwidth is 50 cm^{-1}, and the enter frequency is 71 600 cm^{-1}. The ordinate is the detuning of the excitation pulse ω_x from the energy center of the (0,2,0) and (1,0,0) states. Taken from [76, Figure 3]. (Reprinted with permission. Copyright 1993 American Institute of Physics.)

constant percentage yield of H+OD that is, 100P(OD+H)/[P(OD+H)+P(OH+D)] as a function of the time delay Δ_d and of the detuning of the pump laser pulse $E_x - E_{AV}$. Features of this result are of note. First, significant variations of yield ratio accompany changes in $E_x - E_{AV}$. Second, the dependence of the yield ratio on the time delay is weak. The former feature merely reflects a natural preference, on the part of either of the two excited states $|E_2\rangle$, $|E_3\rangle$ to favor production of OD over OH. Changing $E_x - E_{AV}$ changes the relative contribution of each of these two states thereby changing the yield ratio. Thus, changes in yield ratio with changes in $E_x - E_{AV}$ is not due to coherent control. Rather, quantum interference effects are reflected in variations of the yield ratio with changes in Δ_d. The fact that this is weak is indeed a reflection of the similarity of the two product channels.

The approach discussed thus far relies heavily on the interference generated between a very small number of energy levels. An alternative fundamental perspective on control, based on localized wave packets, which are a superposition of many levels, was originally introduced by Tannor and Rice [67, 68] and Tannor, Kosloff and Rice [77]. Their approach, which founded the fundamentals of pump-dump control and of optimal control, is the subject of Chapter 5.

3.A
Appendix: Mode-Selective Chemistry

Coherent control is quite distinct from the mode-selective approach to controlling chemistry, an approach which has been advocated for some time [30, 35–37, 78]. Examining the difference between mode selectivity and coherent control, as done in this section, affords considerable insight into both.

The essence of the mode-selective approach is concisely stated in the Pimentel report of the National Academy of Sciences [79] as "an effort to excite a particular degree of freedom in a molecule so that the molecules react as if this particular degree of freedom was at a very high temperature whereas the rest of the molecular degrees of freedom are cold." For example, in the case of A + BC $\leftarrow$ ABC $\rightarrow$ AB + C one might excite the AB bond in an attempt to enhance production of A + BC. In addition, the Pimentel report identifies that problem which is widely regarded as the major difficulty to overcome in order to achieve control via mode selectivity. That is [79], "apparently the problem is that vibrational redistribution (IVR) takes place within these molecules", presumably preventing maintenance of the selective control based upon the idea of initially localizing energy in some part of the molecule. For this reason, advocates of this approach argued for faster excitation methods. Below we show that rapid laser excitation is unnecessary and that the mode-selective approach affords nothing new in controlling reactions of isolated molecules, assuming one photon processes.

Consider the general case of a pure state wave packet prepared by any of a variety of preparation schemes that are only required to be off at long times. The prepared time-dependent wave packet is

$$|\Psi(t)\rangle = \int dE \sum_{\boldsymbol{n},q} b_{E,\boldsymbol{n},q} |E, \boldsymbol{n}, q^-\rangle e^{-iEt/\hbar} . \tag{3.A.1}$$

Some mode-selective chemistry advocates argued that it is the time dependence of the wave packet, that is, the phase relationships between energy levels, that afford the possibility of effectively controlling reactions. Here we show that in the one precursor state situation this is not the case.

The probability $P_{\boldsymbol{n}',q'}(E')$ of forming the product with energy E', quantum numbers $\boldsymbol{n}'$ and product arrangement q' (Eq. (2.74)) is

$$P_{\boldsymbol{n}',q'}(E') = \left| \lim_{t\to\infty} e^{iE't/\hbar} \langle E', \boldsymbol{n}', q'; 0|\Psi(t)\rangle \right|^2 = \left| b_{E',\boldsymbol{n}',q'} \right|^2 . \tag{3.A.2}$$

Thus, the probability of forming the product state $P_{\boldsymbol{n},q}(E)$ is given by the sum over the squares of the $b_{E,\boldsymbol{n},q}$ coefficients, a quantity which in the one precursor state situation is *totally independent of any coherence established between nondegenerate energy levels in the initially created state.* Indeed as shown below, the product yield is exactly the same as that which would be obtained if the initially created state were a totally time-independent energy-incoherent mixture,

$$\rho(0) = \int dE \sum_{\boldsymbol{n},q} \sum_{\boldsymbol{n}',q'} b_{E,\boldsymbol{n},q} b^*_{E,\boldsymbol{n}',q'} |E, \boldsymbol{n}, q^-\rangle\langle E, \boldsymbol{n}', q'^-| \tag{3.A.3}$$

with the associated probability

$$P_{n,q}(E) = \langle E, \boldsymbol{n}, q; 0|\rho|E, \boldsymbol{n}, q; 0\rangle \; . \tag{3.A.4}$$

This being the case, the coherence amongst energy levels, hence the time dependence of the initial state, has no influence whatsoever on the product yield. This implies, for example, that in the weak-field regime, the product yield obtainable with a subfemtosecond laser pulse with frequency spectrum $I(\omega)$ is exactly equal to the sum of a set of, for example, *microsecond* pulses with an appropriate set of frequencies and intensities which have the same cumulative $I(\omega)$. Shorter pulses alter the product probabilities only to the extent that they excite a larger number of states as the frequency spectrum broadens with diminishing pulse duration. Clearly, an observation of increased yield of a particular product upon use of a shorter pulse is then due solely to the fact that the shorter pulse encountered some states with a preference for a particular product.

One further point is worth noting. The mode-selective approach seeks to excite modes which enhance production of a desired product. The associated assumption is that the correct modes are simple and intuitive, for example, such as exciting local bond modes (e.g., AB in ABC) to selectively dissociate the molecule. However, our discussion (Sections 3.2, 3.4) makes clear that the proper modes to excite to produce product in channel q, for example, are the $|E, \boldsymbol{n}, q^-\rangle$ states, since these states are guaranteed to correlate with the q product. In some instances the appropriate $|E, \boldsymbol{n}, q^-\rangle$ may correspond to bond excitation. This is the case with hydrogen containing molecules such as HOOH [31], C_2HD [36, 78], and HOD [80, 81], where excitation of the OH bond is known to enhance the breaking of the OH in preference to the OD bond and vice versa [34, 81]. However, in general the structure of $|E, \boldsymbol{n}, q^-\rangle$ is considerably different than that of simple bond excitation.

The delta function energy normalization for example, (e.g., Eq. (2.47)) of the scattering states $|E, \boldsymbol{n}, q^-\rangle$ and $|E, \boldsymbol{n}, q; 0\rangle$ necessitates that care be taken in deriving Eq. (3.A.4). The nuances are discussed below. For notational simplicity we drop the $(\boldsymbol{n}, q)$ labels which, as discrete indices, can readily be included.

Consider the stationary density matrix

$$\rho = \int dE |E^-\rangle\langle E^-| g(E) \, , \tag{3.A.5}$$

where $g(E)$ is a square integrable function and $|E^-\rangle$ is a scattering state at energy E. Equation (3.A.5) is the analogue of Eq. (3.A.3) where we suppress the $(\boldsymbol{n}, q)$ variables. In light of the delta function normalization $\langle E^-|E_1^-\rangle = \delta(E - E_1)$, and $\mathrm{Tr}(\rho) = \infty$, the operator ρ is then said to be "non trace-class" [82] and, as a consequence, standard operations such as taking the average values are meaningless.

To bypass these problems we initially introduce broadening in energy, replacing the diagonal density matrix (Eq. (3.A.5)) by the nondiagonal

$$\rho(0) = \lim_{\epsilon \to 0} \int dE_0 \iint dE_1 dE_2 \, f_\epsilon(E_1, E_0) f_\epsilon(E_2, E_0) |E_1^-\rangle\langle E_2^-| g(E_1) g^*(E_2) \, , \tag{3.A.6}$$

where the argument of $\rho(0)$ denotes time zero. Here $f_\epsilon(E_1, E_0)$ is a function with nonzero values within the E_1 range $(E_0 - \epsilon, E_0 + \epsilon)$. We choose $f_\epsilon(x, x')$ such that

$$\lim_{\epsilon \to 0} |f_\epsilon(x, x')|^2 = \delta(x - x') . \tag{3.A.7}$$

In this limit, Eq. (3.A.6) is essentially a stationary mixture of projectors $|E_0^-\rangle\langle E_0^-|$. Note that a small degree of coherence between energy levels has been introduced into Eq. (3.A.6) to account for the continuous character of E.

Consider now $\rho(t \to \infty)$, given $\rho(0)$ (Eq. (3.A.6)). At long times $\rho(t \to \infty)$ is of the same form as Eq. (3.A.6), but with $|E_1^-\rangle\langle E_2^-|$ being $|E_1\rangle\langle E_2|$ states, that is, products of eigenstates of the asymptotic Hamiltonian. Specifically, if $|E_1\rangle\langle E_2|$ denote such asymptotic states then at long times

$$\rho(t) = \lim_{\epsilon \to 0} \int dE_0 \iint dE_1 dE_2 \, f_\epsilon(E_1, E_0) f_\epsilon(E_2, E_0) |E_1\rangle \times \langle E_2| g(E_1) g^*(E_2) e^{-i(E_1 - E_2)t/\hbar} . \tag{3.A.8}$$

Then the probability of observing product states within the range $E - \Delta$ to $E + \Delta$ is

$$\lim_{\epsilon \to 0} \int_{E-\Delta}^{E+\Delta} d\bar{E} \langle \bar{E}|\rho|\bar{E}\rangle = \lim_{\epsilon \to 0} \int_{E-\Delta}^{E+\Delta} d\bar{E} \int dE_0 |f_\epsilon(\bar{E}, E_0)|^2 |g(\bar{E})|^2$$
$$= \int_{E-\Delta}^{E+\Delta} dE_0 |g(E_0)|^2 . \tag{3.A.9}$$

Here, the first equality arises from using the delta function normalization of $|E\rangle$ and the second equality, from Eq. (3.A.7).

Equation (3.A.9) is the relevant result, indicating that the product probability is independent of any coherence established initially between nondegenerate states. It reiterates the discussion above and emphasizes that product control resides in control over the wave functions $|E, \boldsymbol{n}, q^-\rangle$ that exist at fixed energies.

4
Control of Intramolecular Dynamics

The majority of the theory and applications of quantum control discussed in the previous chapters dealt with unimolecular processes whose products lie in the continuum. Such processes have well-defined product states in their long-time limit, with control naturally directed towards manipulating the ratio of populations at asymptotic times.

4.1
Intramolecular Dynamics

By contrast, intramolecular bound-state dynamics, the subject of this chapter, is distinctly different. To see this, consider the dynamics of a bound system described by total Hamiltonian H whose exact energy eigenstates and eigenvalues are denoted $|\gamma\rangle$ and E_γ (note that H includes both nuclear and electronic contributions, as do its eigenvalues and eigenfunctions). The dynamical evolution of the wave function of the bound system satisfies the time-dependent Schrödinger equation. If the system is an initial pure state $|\psi(0)\rangle$, then the system at time t is in the state:

$$|\psi(t)\rangle = \sum_\gamma d_\gamma |\gamma\rangle e^{-iE_\gamma t/\hbar} \tag{4.1}$$

where $d_\gamma = \langle\gamma|\psi(0)\rangle$. As is evident from Eq. (4.1) and unlike continuum systems, the dynamics of bound-state systems need not approach any particular long-time limit. Hence control at a variety of specified times may be of interest, and the system does not naturally display "products" over which control is to be exercised. Rather, intramolecular processes, as described below, are often characterized in terms of the dynamics within zeroth-order pictures that are motivated by chemical or physical considerations, and control is expressed in terms of this zeroth-order perspective. Examples include intramolecular vibrational redistribution amongst molecular bonds, and radiationless processes [83] such as intersystem crossing and internal conversion. In the latter cases, energy flows between electronic surfaces, a result of the coupling between zeroth-order Born–Oppenheimer (BO) electronic surfaces via non-BO terms in the system Hamiltonian. Frequently, radiationless transitions appear decay-like and proceed irreversibly from one electronic surface

Quantum Control of Molecular Processes, Second Edition. Moshe Shapiro and Paul Brumer

to another. However oscillatory dynamics, which manifest as beats, are also well known.

Numerous scenarios to control intramolecular dynamics can be envisioned. In this chapter we focus on one physically realizable approach. That is, control over intramolecular dynamics is shown here to result from the preparation of initial states that evolve preferentially to the desired target state. The significant role of *overlapping resonances* in introducing quantum interference into the control mechanism [84–87], and the role of Hilbert space partitioning in defining the control problem are the subject of the first two subsections below.

4.1.1
Time Evolution and the Zero-Order Basis

Consider a system described by an Hamiltonian H that can be partitioned into the sum of two components H_A and H_B, plus the interaction H_{AB} between them:

$$H = H_A + H_B + H_{AB} \,. \tag{4.2}$$

The eigenstates $|\gamma\rangle$ and eigenvalues E_γ of the full Hamiltonian are defined by:

$$H|\gamma\rangle = E_\gamma|\gamma\rangle \,. \tag{4.3}$$

The "zeroth-order" eigenstates and eigenvalues of the sum of the decoupled Hamiltonians are defined as

$$(H_A + H_B)\,|\kappa\rangle = E^{(\kappa)}|\kappa\rangle \,. \tag{4.4}$$

Interest is in the time evolution of the system between zeroth-order states. Here we consider the system initially prepared in a superposition of zeroth-order states confined to a subspace S:

$$|\Psi(t=0)\rangle = \sum_\kappa c_\kappa|\kappa\rangle \,, \tag{4.5}$$

where $\{c_\kappa\}$ are coefficients set by the system preparation. All sums over $|\kappa\rangle$, here and below, are assumed to be confined to a subspace S. Our goal will be to determine the $\{c_k\}$, and their associated characteristics, that optimize evolution out of the S subspace at a fixed target time.

The time evolution of $|\Psi(0)\rangle$ can be obtained using the expansion of the zeroth-order states $|\kappa\rangle$ in terms of the exact eigenstates $|\gamma\rangle$:

$$|\kappa\rangle = \sum_\gamma |\gamma\rangle\langle\gamma|\kappa\rangle \equiv \sum_\gamma |\gamma\rangle a_{\kappa,\gamma} \tag{4.6}$$

where $a_{\kappa,\gamma} = \langle\gamma|\kappa\rangle$. Inserting Eq. (4.6) into Eq. (4.5) gives, at time t:

$$|\Psi(t)\rangle = \sum_{\kappa,\gamma} c_\kappa a_{\kappa,\gamma} e^{-iE_\gamma t/\hbar}|\gamma\rangle \,. \tag{4.7}$$

The structure of $|\langle\gamma|\kappa\rangle|^2$ as a function of E_γ defines a *resonance shape* that provides insight, in the frequency domain, into the population flow out of the zeroth-order $|\kappa\rangle$ states. Equation (4.7) provides the Fourier transform that converts this frequency dependence into the time evolution of the superposition of $|\kappa\rangle$ states comprising $|\Psi(0)\rangle$.

Given this time evolution, the amplitude for finding the system in a state $|\kappa\rangle$ at time t is

$$\langle\kappa|\Psi(t)\rangle = \sum_{\kappa'} c_{\kappa'} M_{\kappa,\kappa'}(t) , \tag{4.8}$$

where

$$M_{\kappa,\kappa'}(t) \equiv \sum_{\gamma} a^*_{\kappa,\gamma} a_{\kappa',\gamma} e^{-\mathrm{i} E_\gamma t/\hbar} = \langle\kappa| \left(\sum_{\gamma} e^{-\mathrm{i} E_\gamma t/\hbar} |\gamma\rangle\langle\gamma| \right) |\kappa'\rangle \tag{4.9}$$

is the (κ, κ')-element of the overlap matrix $\boldsymbol{M}(t)$ defined by the term in brackets in Eq. (4.9).

From Eq. (4.8), the probability of finding the system in the *collection* of states $|\kappa\rangle$ that are contained in the initial set S at time t is given by

$$P(t) = \sum_{\kappa} |\langle\kappa|\Psi(t)\rangle|^2 = \boldsymbol{c}^\dagger \boldsymbol{M}^\dagger(t)\boldsymbol{M}(t)\boldsymbol{c} = \boldsymbol{c}^\dagger \boldsymbol{K}(t)\boldsymbol{c} . \tag{4.10}$$

That is, $\boldsymbol{c}$ is a κ-dimensional vector whose components are the c_κ coefficients, and $\boldsymbol{K}(t) \equiv \boldsymbol{M}^\dagger(t)\boldsymbol{M}(t)$. Equation (4.10) allows us to address the question of enhancing or restricting the flow of probability out of S by finding the optimal combination of c_κ that achieves the desired goal at a specified time T. In particular, maximizing or minimizing $P(T)$, subject to the constraint $\boldsymbol{c}^+\boldsymbol{c} = 1$, requires solving the simple linear eigenvalue problem: $\boldsymbol{K}(T)\boldsymbol{c} = \lambda(T)\boldsymbol{c}$. Experimentally, the resultant required superposition state can often be prepared using modern laser techniques [88, 89].

4.1.2
Partitioning of the Hilbert Space

As noted above, issues regarding intramolecular dynamics often relate to the flow of energy and population from a *subspace* of the Hilbert space describing the molecule to the remainder of the Hilbert space (for example, the flow of energy and population out of a particular electronic state). This being the case, it proves advantageous to subdivide the Hilbert space into the component Q from within which the dynamics is initiated, to the remainder P. The Feshbach Partitioning Technique [90, 91] offers a useful approach for carrying out this procedure.

Feshbach partitioning of Hilbert Space introduces two projection operators

$$Q \equiv \sum_{\kappa} |\kappa\rangle\langle\kappa| , \quad P \equiv \sum_{\beta} |\beta\rangle\langle\beta| , \tag{4.11}$$

that satisfy the following properties:

$$Q^2 = Q\,, \quad P^2 = P\,,$$
$$[Q, P] = 0\,,$$
$$P + Q = \mathrm{I}\,, \tag{4.12}$$

where I is the identity operator. Hence, $Q + P$ spans the entire Hilbert Space and the subspaces defined by these projection operators are orthogonal.

Using Eqs. (4.11) and (4.12), the eigenstates of the full Hamiltonian can be expanded in its Q and P components as:

$$|\gamma\rangle = \sum_{\kappa} |\kappa\rangle\langle\kappa|\gamma\rangle + \sum_{\beta} |\beta\rangle\langle\beta|\gamma\rangle\,. \tag{4.13}$$

Similarly, the Schrödinger equation can be expressed as

$$[E_\gamma - H][P + Q]|\gamma\rangle = 0\,. \tag{4.14}$$

Multiplying Eq. (4.14) by P and then by Q, and using Eq. (4.12), one obtains the following set of coupled equations:

$$\left[E_\gamma - PHP\right] P|\gamma\rangle = PHQ|\gamma\rangle\,,$$
$$\left[E_\gamma - QHQ\right] Q|\gamma\rangle = QHP|\gamma\rangle\,. \tag{4.15}$$

The states $|\kappa\rangle$ and $|\beta\rangle$ are solutions to the decoupled (homogeneous) equations arising from Eq. (4.15). That is,

$$\left[E_\beta - PHP\right] P|\beta\rangle = 0\,,$$
$$\left[E_\kappa - QHQ\right] Q|\kappa\rangle = 0\,. \tag{4.16}$$

The P, Q partitioning is traditionally used for continuum problems where $E_\gamma = E_\beta$ (see, for example, Section 7.3.1). This is not the case for bound systems so it is possible here to express $P|\gamma\rangle$ in terms of the particular solution of the (inhomogeneous) equation for $P|\gamma\rangle$ (Eq. (4.15)),

$$P|\gamma\rangle = [E_\gamma - PHP]^{-1} PHQ|\gamma\rangle\,. \tag{4.17}$$

Substituting Eq. (4.17) into the equation for $Q|\gamma\rangle$ (Eq. (4.15)) results in

$$[E_\gamma - QHQ]Q|\gamma\rangle = QHP[E_\gamma - PHP]^{-1} PHQ|\gamma\rangle\,. \tag{4.18}$$

By rearranging terms in this equation, one obtains

$$[E_\gamma - \mathcal{H}]Q|\gamma\rangle = 0\,, \tag{4.19}$$

where $\mathcal{H}$ is an effective Hamiltonian, defined as

$$\mathcal{H} = QHQ + QHP[E_\gamma - PHP]^{-1} PHQ\,. \tag{4.20}$$

The term between square brackets can be written as

$$\left[E_\gamma - PHP\right]^{-1} = \sum_\beta \frac{1}{E_\gamma - E_\beta}|\beta\rangle\langle\beta| \tag{4.21}$$

by using the spectral resolution of the inverse operator. The matrix elements of $\mathcal{H}$ are given by

$$\langle\kappa|\mathcal{H}|\kappa'\rangle = E_\kappa\delta_{\kappa,\kappa'} + \Delta_{\kappa,\kappa'}\,, \tag{4.22}$$

where

$$\Delta_{\kappa,\kappa'} = \frac{1}{2\pi}\sum_\beta \frac{\Gamma_{\kappa,\kappa'}}{E_\gamma - E_\beta}\,,$$

$$\Gamma_{\kappa,\kappa'} = 2\pi V(\kappa|\beta)V(\beta|\kappa')\,, \tag{4.23}$$

with $V(\kappa|\beta) = \langle\kappa|QHP|\beta\rangle$ being the coupling term. By solving Eq. (4.19) one obtains the energy eigenvalues, E_γ, and the values for the overlap integrals, $a_{\kappa,\gamma}$.

For a small system the energy eigenvalues and the overlap integrals can also be obtained [92] by directly diagonalizing the full Schrödinger equation (Eq. (4.3)) in the zeroth-order basis. However, partitioning the system into P and Q subspaces has computational advantages for cases where the dimension of the P space is large, since one only needs to diagonalize an effective Hamiltonian $\mathcal{H}$ with dimension given by the Q space [87]. Such larger systems can take advantage of a "QP algorithm" [86] developed especially for treating the dynamics of such systems. This method allows the computation of energy eigenvalues and eigenfunctions. This is particularly useful for wave packet control problems since it allows examination of a wide variety of initial wave packets, something that can not be done with typical time-dependent methods. (See, however, [93] for a time-dependent QP approach.)

4.1.3
Initial State Control and Overlapping Resonances

Consider the dynamical evolution of a state $|\kappa\rangle$ initially in the Q space. This state is nonstationary, is termed a resonance, and has an energy width here given by $|\langle\gamma|\kappa\rangle|^2$ as a function of E_γ. The lifetime of the state, flowing out of the Q space, is qualitatively given by the inverse of the energy width of the resonance. In the case of *overlapping resonances*, the energy width of two resonances, say $|\kappa\rangle$ and $|\kappa'\rangle$, exceed their energy spacing $|E_\kappa - E_{\kappa'}|$. In this case, $|\kappa\rangle$ and $|\kappa'\rangle$, share at least one common $|\gamma\rangle$, for which $\langle\gamma|\kappa\rangle$ and $\langle\gamma|\kappa'\rangle$ are both nonzero. A sample pair of overlapping resonances associated with flow out of the S_2 state of pyrazine is shown in Figure 4.1. Such overlapping resonances were found to be ubiquitous in this system [85, 86], and is most likely the case in many large molecules.

Overlapping resonances explicitly appear in the above formulation through the off-diagonal ($\kappa \neq \kappa'$) term $M_{\kappa,\kappa'}(t)$ (Eq. (4.9)). Specifically, if there are no overlapping resonances then the off-diagonal $M_{\kappa,\kappa'}(t) = 0$.

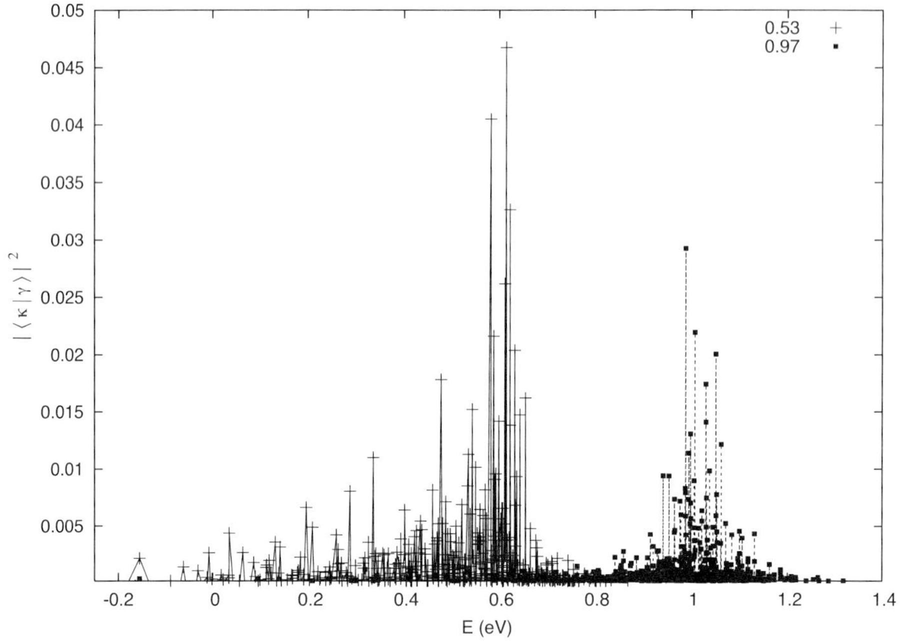

Figure 4.1 Two typical resonance line-shapes in the S_2 state of pyrazine: one centered on a $|\kappa\rangle$ of energy $\approx$ 0.53 eV and the second centered at $\approx$ 0.97 eV. From [86, Figure 3]. (Reprinted with permission. Copyright 2006 American Institute of Physics.)

To further expose the role of overlapping resonances in the time evolution of an initial zeroth-order state, we rewrite Eq. (4.10) as

$$P(t) = \sum_{\kappa} |c_\kappa|^2 g_\kappa + \sum_{\kappa',\kappa,\kappa' \neq \kappa} c^*_{\kappa'} c_\kappa f_{\kappa',\kappa} , \tag{4.24}$$

where

$$g_\kappa = \sum_{\kappa'} |M_{\kappa',\kappa}|^2 , \tag{4.25}$$

and

$$f_{\kappa',\kappa} = \sum_{\kappa''} M^*_{\kappa'',\kappa'} M_{\kappa'',\kappa} . \tag{4.26}$$

Note that, as expressed in Eq. (4.24), the Q space population assumes the generic coherent control form, given as the sum of noninterfering pathways, represented by g_κ, and interfering pathways, represented by $f_{\kappa',\kappa}$. The latter depends upon the off-diagonal $M_{\kappa'',\kappa}$. As noted above, if the zeroth-order states included in $P(t)$ do not display overlapping resonances then $M_{\kappa'',\kappa}$, and hence the interference term $f_{\kappa',\kappa}$, are zero. The resultant $P(t)$ then no longer depends upon the phases of the coefficients c_κ and this type of *phase control* is lost. By contrast, control via the phases of c_κ is expected if overlapping resonances participate in Eq. (4.24) through $M_{\kappa'',\kappa}$.

Note that overlapping resonances also contribute to the g_κ term, as can be seen from their effect on the nature of the decay from the individual $|\kappa\rangle$. These resonances distort the line-shape, and hence the corresponding time dependence.

4.1.3.1 **Internal Conversion in Pyrazine**

As an example of this control methodology, consider the dynamics of pyrazine, a molecule that has long provided a model for studies of internal conversion. In this case, laser excitation from the ground S_0 electronic state to the excited S_2 state (chosen as the Q space) is followed by internal conversion to the S_1 state (chosen as the P space). Of particular interest is the population $P_2(t)$ in S_2 as a function of time. A (realistic) 24 mode model of the S_1 and S_2 Hamiltonian [94] expresses the vibrational motion of pyrazine in terms of a set of oscillators, described by the (dimensionless) normal coordinates, Q_i, coupled by linear and quadratic functions of Q_i. The Hamiltonian is given as

$$\begin{aligned} H = & \sum_i \begin{pmatrix} 1 & 0 \\ 0 & 1 \end{pmatrix} \frac{\omega_i}{2} \left(-\frac{\partial^2}{\partial Q_i^2} + Q_i^2 \right) + \begin{pmatrix} -\Delta & 0 \\ 0 & \Delta \end{pmatrix} \\ & + \sum_{i \in G_1} \begin{pmatrix} a_i & 0 \\ 0 & b_i \end{pmatrix} Q_i + \sum_{i,j \in G_2} \begin{pmatrix} a_{ij} & 0 \\ 0 & b_{ij} \end{pmatrix} Q_i Q_j \\ & + \begin{pmatrix} 0 & \lambda_{10a} \\ \lambda_{10a} & 0 \end{pmatrix} Q_{10a} + \sum_{i,j \in G_4} \begin{pmatrix} 0 & c_{ij} \\ c_{ij} & 0 \end{pmatrix} Q_i Q_j \, . \end{aligned} \tag{4.27}$$

Here ω_i are the normal mode frequencies of the lowest states of S_0, 2Δ is the energy difference between the zero points of the S_2 and S_1 potential energy surfaces, a_i, b_i, a_{ij} and b_{ij} are "intrasurface" coupling constants, and λ_{10a} and c_{ij} are "intersurface" coupling constants. These constants, provided in [94] were determined by fitting the above functional form to the *ab initio* derived points on the potential energy surface. The linear couplings only occur within the A_g symmetry set of modes, denoted as G_1. The quadratic couplings occurs between the set of modes whose products are either of A_g symmetry, denoted as G_2, or of B_{1g} symmetry, denoted as G_4. Parameters are provided in [94].

A sample result of uncontrolled internal conversion dynamics is shown as the middle curve in Figure 4.2. Here the initial state $|\Psi(0)\rangle$ is taken as the ground vibrational state of pyrazine in the S_0 manifold, placed onto the S_2 state. The state (Eq. (4.5)), comprising 176 zeroth-order vibrational states $|\kappa\rangle$ on S_2, evolves from the S_2 to S_1 state on a 20 fs time scale. Also shown, as the curves bracketing this central curve, are the results of optimizing the coefficients c_κ in this superposition of 176 states, here to either maximize or minimize $P_2(t)$ at $T = 20$ fs. The resultant control at 20 fs is almost total, with $P_2(T)$ ranging from essentially zero to ≈ 0.92. Hence, internal conversion can be significantly enhanced, or highly suppressed,

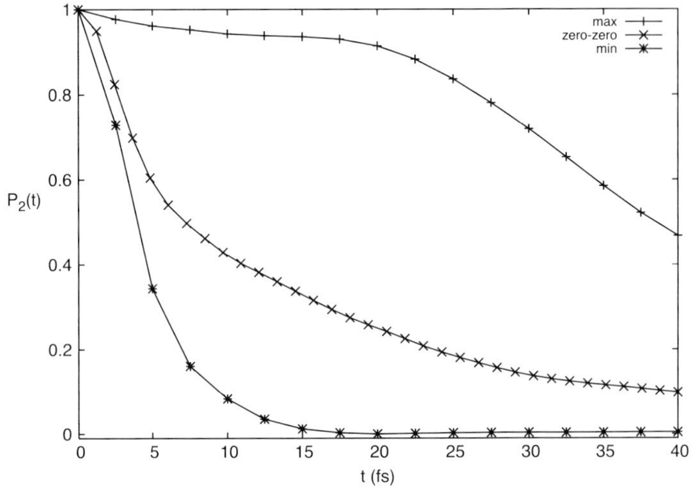

Figure 4.2 Maximal and minimal $P_2(t)$ for a superposition of 176 zeroth-order states, with the optimization time $T = 20$ fs. Also shown, as the middle curve, is the uncontrolled superposition described in the text. From [85, Figure 5]. (Reprinted with permission. Copyright 2006 American Institute of Physics.)

by varying the superposition state coefficients. Results for other superpositions are shown in [85].

Significantly, the results shown here are for a relatively sophisticated model of pyrazine that includes all 24 vibrational modes [94]. By contrast, control results using a 4 mode model of pyrazine [84] show incorrect oscillatory behavior between the Q and P spaces, warning against approximations that simplify the structure of the P space.

The initial superposition state is assumed, in accord with the literature on shaped pulses [95], to be preparable. For example, a desired state can be prepared via shaped laser pulsed excitation of the ground electronic state coupled with a wave function diagnostic based upon time resolved spontaneous emission.

4.1.3.2 **Intramolecular Vibrational Redistribution: OCS**

As a second example of this control methodology, consider the control of the flow of vibrational energy within a single molecule. As an example, consider the flow of energy between bonds in the collinear diatomic OCS [87]. Here, the classical Hamiltonian is given by

$$H = \frac{P_1^2}{2\mu_{13}} + \frac{P_2^2}{2\mu_{23}} - \frac{P_1 P_2}{m_C} + V(R_1, R_2, R_3)\,, \tag{4.28}$$

where

$$\mu_{13} = \frac{m_O m_C}{m_O + m_C}$$
$$\mu_{23} = \frac{m_S m_C}{m_S + m_C}\,, \tag{4.29}$$

are reduced masses; R_1 and R_2 are the CS and CO bond distances ($R_3 = R_1 + R_2$), P_1 and P_2 are the corresponding momenta, and the potential $V(R_1, R_2, R_3)$ is described in detail in [87]. Note that the dynamics of this system is far from simple, insofar as the classical dynamics displays extensive regions of chaotic behavior [87, 96].

To explore control of the intramolecular energy redistribution, we note that the initial OCS wave function can be expanded in the zeroth-order states as

$$|\Psi\rangle = \sum_{m,n} |\eta^m_{\text{CS}}\rangle|\xi^n_{\text{CO}}\rangle d_{mn} , \tag{4.30}$$

where $|\eta^m_{\text{CS}}\rangle$ and $|\xi^n_{\text{CO}}\rangle$ are eigenstates of the uncoupled CS and CO bond Hamiltonians with quantum numbers m and n. Our interest is in the flow, for example, out of the CS bond. In that case the Q subspace is chosen to represent all wave functions containing only excitation in the CS bond, that is, $|\kappa\rangle$ are $|\eta^m_{\text{CS}}\rangle|\xi^0_{\text{CO}}\rangle$ for all m, whereas the P subspace spans the space represented by all other zeroth-order excitations, that is, the $|\beta\rangle$ are $|\eta^m_{\text{CS}}\rangle|\xi^n_{\text{CO}}\rangle$, $n \neq 0$, describing excitation in the CS bond. Initiating excitation within Q and watching the flow into P then corresponds to an experiment wherein excitation flows out of the CS bond. Here $P-Q$ coupling consists of a static term via the potential coupling and a dynamic term proportional to $P_1 P_2$ (Eq. (4.28)).

Consideration of various initial states comprising a superposition of vibrational states showed that control over IVR is best obtained when the vibrational states are at the onset of dissociation. This is the regime in which overlapping resonances predominate and control is found to be most effective. In larger molecules, such overlapping vibrational resonances are expected to be obtained at far lower energies compared to dissociation.

Figure 4.3 shows the time-evolution of the Q space population, $P(t)$, for an initial wave function constructed from the nine zeroth-order Q space states, and optimized for maximal or minimal energy flow at $T = 100$ fs. Results in Figure 4.3a correspond to an initial superposition optimized to minimize the population flow from the Q to the P space, while Figure 4.3b shows results optimized to enhance the flow of population. As is clearly seen, the initial falloff in Figure 4.3a is much slower than that in Figure 4.3b. To quantify this decay, the initial $P(t)$ falloff was fit to an exponentially decreasing function,

$$P(t) = P_\infty + (1 - P_\infty)e^{-t/t_\delta} , \tag{4.31}$$

where t_δ is the decay time, and P_∞ is the average around which $P(t)$ fluctuates for the first 1.0 ps. Note that the t_δ values can only be regarded as approximate since the falloff is, in general, not exponential, and t_δ depends on the time scale over which the exponential is fit. (Here the fit is over 400 fs). In Figure 4.3a, the decay time is $t_\delta \simeq 57.35$ fs, while Figure 4.3b it is $t_\delta \simeq 8.60$ fs, about seven times smaller. Furthermore, we note that in Figure 4.3a, only about 24% of the population has been transferred from Q to P during the first 50 fs, while, in contrast, approximately 82% of the population has being transferred to the P in Figure 4.3b during the same

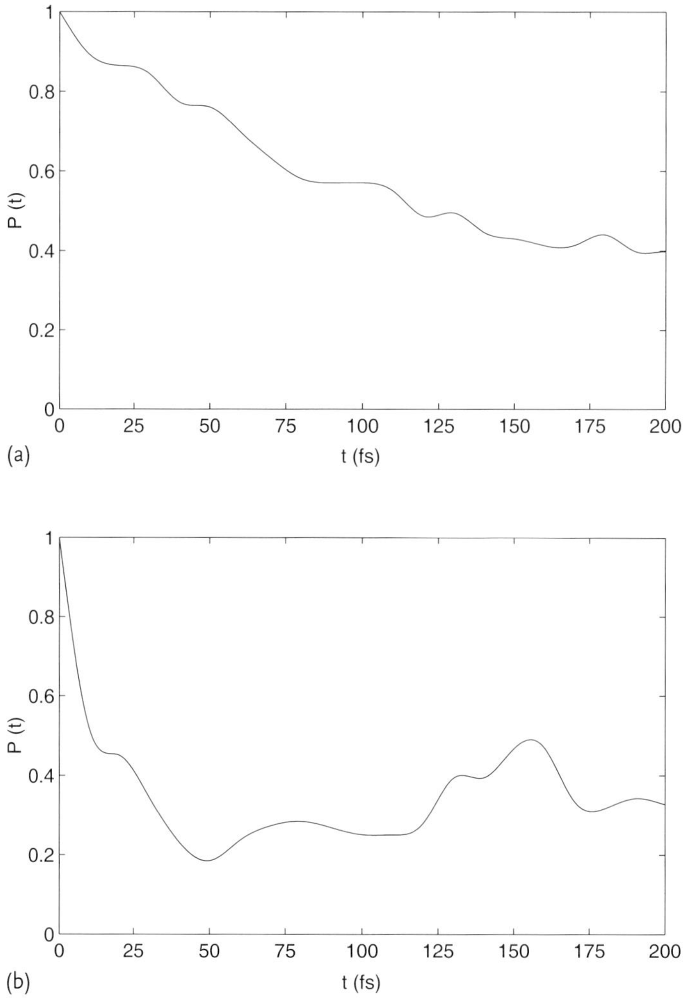

Figure 4.3 IVR control in OCS: (a) IVR suppression, and (b) IVR enhancement. From [87, Figure 3]. (Reprinted with permission. Copyright 2007 American Institute of Physics.)

time. Moreover, the population that asymptotically remains localized along the CS bond is also larger in the case of IVR suppression ($P_\infty \simeq 0.4$) than in that of enhancement ($P_\infty \simeq 0.3$).

A pictorial, and enlightening, view of the results is provided in Figures 4.4 and 4.5, where the wave packets associated with IVR suppression and enhancement are shown. As can be seen in Figure 4.4, for the case of IVR suppression, the wave packet remains highly localized along the R_{CS} mode, with minimum spreading along the R_{CO} mode. In particular, it undergoes a slight oscillation along the

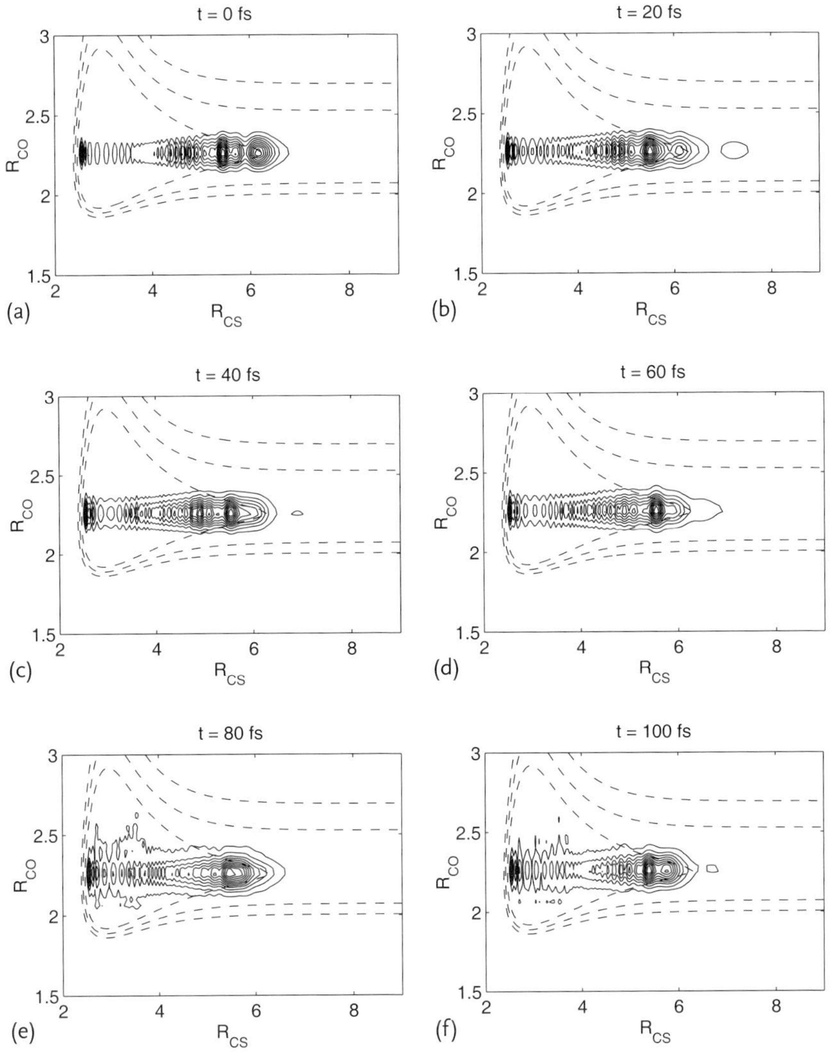

Figure 4.4 Wave packet evolution corresponding to IVR suppression. Dashed lines represent equipotential energy contours, with the innermost corresponding to the wave packet energy, $E_{+} = 0.09\,849$ a.u. From [87, Figure 6]. (Reprinted with permission. Copyright 2007 American Institute of Physics.)

R_{CS} mode, concentrating most of the probability around the region of large R_{CS}, in a clear correspondence to what happens with a classical counterpart. For the case of IVR enhancement, the effect is the opposite. As can be seen in Figure 4.5, the spreading of the wave packet along the R_{CO} mode coordinate is relatively fast.

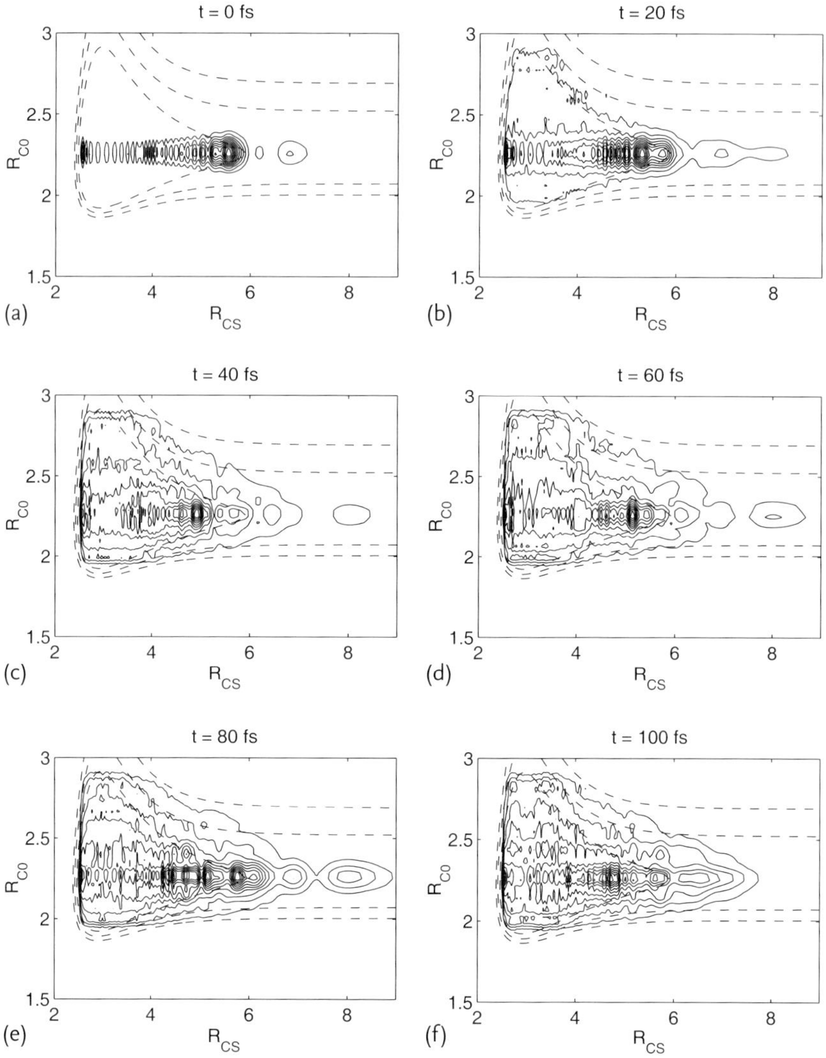

Figure 4.5 Wave packet evolution corresponding to IVR enhancement. Dashed lines represent equipotential energy contours, with the innermost corresponding to the wave packet energy, $E_- = 0.09743$ a.u. From [87, Figure 7]. (Reprinted with permission. Copyright 2007 American Institute of Physics.)

Experimental studies of the control of internal conversion have been carried out on large systems, such as those of interest in light-induced biological processes. Such studies have all utilized adaptive feedback methods that are introduced in Chapter 17.

5
Optimal Control Theory

5.1
Pump-Dump Excitation with Many Levels: the Tannor–Rice Scheme

The approach discussed thus far relies heavily on the interference generated between a small number of energy levels. An alternative perspective on control, based on localized wave packets, was pioneered by Tannor and Rice [67, 68] and Tannor, Kosloff and Rice [77]. As an example of their approach, consider the potential energy profile shown in Figure 5.1, where the product $A + BC$ emerges on the potential surface when leaving through one exit and $AB + C$ emerges from the other. An initial wave packet, localized on the ground electronic state, is laser-excited to the upper electronic state where it evolves. A second laser then causes stimulated emission back to the ground electronic state. Assuming only small spreading of the wave packet, a properly timed second laser pulse induces stimulated emission to either deposit the excited wave function preferentially in the $A+BC$ or $AB+C$ exit channel region, enhancing production of that product. In addition to timing, they introduced the idea of optimizing the shape of the laser pulses using the calculus of variations so as to alter the final state, hence controlling the outcome of the dynamics. This "pump-dump" approach was the underlying basis for developments in optimal control theory, discussed in Section 5.2 below.

Assuming that the states are spatially localized implies that they are comprised of a large number of energy eigenstates. Further, since excitations to produce superpositions of large numbers of states require broadband coherent laser excitation, the required lasers are in the femtosecond domain. Finally, spatial localization of wave packets suggests behavior near the classical limit. In this regime the formalism of Section 3.6 becomes unreasonably complicated, involving interferences between a myriad of energy eigenstates and the direct wave packet propagation approach based on perturbation theory (Section 3.4.1), now proves more useful.

According to this approach, the amplitude $A_{m,q}(E)$ for a pump-dump transition into the final state $|E, \boldsymbol{m}, q; 0\rangle|g\rangle$ at time t, having started in state $|E_i\rangle|g\rangle$, and using second order perturbation theory with respect to the light–matter interaction

Quantum Control of Molecular Processes, Second Edition. Moshe Shapiro and Paul Brumer

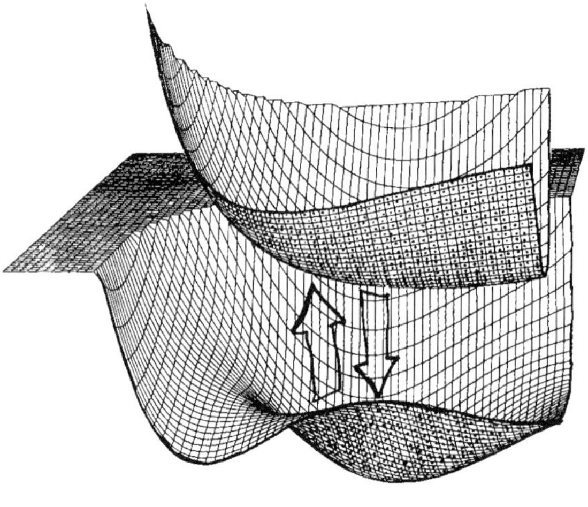

Figure 5.1 A schematic illustration of the original Tannor–Rice scenario for enhancing the yield in a given arrangement channel. The case of enhancing the right channel is shown. From [98, Figure 1]. (Reprinted with permission. Copyright 1989 Elsevier Science.)

(Eq. (3.30) restricted to $n = 2$ at finite time), is:

$$A_{m,q}(E) = \langle E, \boldsymbol{m}, q; 0|\langle g|\Psi(t)\rangle\rangle = \exp(-iEt/\hbar)\,\langle E, \boldsymbol{m}, q^-|\langle g|\psi^I(t)\rangle\rangle$$
$$= \exp(-iEt/\hbar)\,\langle E, \boldsymbol{m}, q^-|\langle g|\mathcal{V}^{(2)}(t)|g\rangle|E_i\rangle\,. \tag{5.1}$$

Here we have explicitly included the electronic state labels ($|g\rangle$ is the ground state) and have chosen the electronic ground state energy as the zero of energy. Note that $\mathcal{V}^{(1)}$ is neglected since it does not contribute significantly to transitions between states of the ground electronic state $|g\rangle$. The $\mathcal{V}^{(2)}$ operator is given by an equation analogous to Eq. (3.29) in which the electronic state terms are explicitly included. That is, H_M is replaced by H_{MT} (Eq. (2.67)) and V^I (Eq. (3.25)) is written in terms of $H'_{MR}(t)$ (Eq. (2.71)). We obtain that

$$A_{m,q}(E) = \left(\frac{-i}{\hbar}\right)^2 \langle E, \boldsymbol{m}, q^-|\langle g|e^{-iEt/\hbar} \int_{-\infty}^{t} dt_1 e^{iH_{MT}t_1/\hbar} H'_{MR}(t_1) e^{-iH_{MT}t_1/\hbar}$$
$$\times \int_{-\infty}^{t_1} dt_2 e^{iH_{MT}t_2/\hbar} H'_{MR}(t_2) e^{-iH_{MT}t_2/\hbar}|g\rangle|E_i\rangle\,. \tag{5.2}$$

Using Eqs. (2.67) and (2.71) allows us to rewrite the pump-dump amplitude as

$$A_{m,q}(E) = \left(\frac{-i}{\hbar}\right)^2 \sum_e \langle E, \boldsymbol{m}, q^-| \int_{-\infty}^{t} dt_1 e^{-iE(t-t_1)/\hbar} \langle g|H_{MR}(t_1)|e\rangle$$
$$\times \int_{-\infty}^{t_1} dt_2 e^{-iH_e(t_1-t_2)/\hbar} \langle e|H_{MR}(t_2)|g e^{-E_i t_2/\hbar}|E_i\rangle\,, \tag{5.3}$$

where $H_e \equiv \langle e|H_{MT}|e\rangle$ is an operator on the nuclear coordinates describing the nuclear motion on electronic state $|e\rangle$. Likewise, $\langle g|H_{MT}|g\rangle$ describes the nuclear motion on $|g\rangle$. Given that $H_{MR}(t) = -\varepsilon(z,t)\hat{\boldsymbol{\varepsilon}}\cdot\boldsymbol{d}$, (Eq. (2.10)), and assuming that the laser is resonant with only one electronic state, we obtain that

$$A_{m,q}(E) = \left(\frac{-i}{\hbar}\right)^2 \langle E, \boldsymbol{m}, q^-| \int_{-\infty}^{t} dt_1 e^{-iE(t-t_1)/\hbar} d_{g,e}\varepsilon(z,t_1) \times \int_{-\infty}^{t_1} dt_2 e^{-iH_e(t_1-t_2)/\hbar} d_{e,g}\varepsilon(z,t_2) e^{-iE_i t_2/\hbar}|E_i\rangle \,. \quad (5.4)$$

Equation (5.4) has the qualitative interpretation that $|E_i\rangle$ evolves on the ground state surface until time $t = t_2$ when it makes a transition to the excited electronic surface. It then propagates for a time t_2 to t_1 at which time it makes a transition back to the ground electronic state. The wave function then evolves on the ground state surface until time t when we measure its overlap with the product state. Since the fields are spread out over time we have to integrate over both t_1 and t_2.

As in Section 3.6, $\varepsilon(z,t)$ is often comprised of two temporally distinct pulses where the timing between the two pulses serves as a control parameter. It is interesting to examine, for example, the limiting case where $\varepsilon(z,t)$ comprises two (temporal) delta function pulses. In this case the pulses have an infinitely wide profile in frequency space and hence encompass a complete set of levels. This is the extreme opposite of the case discussed in Section 3.6 where the laser pulse only encompasses two levels. Here, neglecting the spatial dependence, the field is of the form

$$\varepsilon(z,t) = \mathcal{E}_x \exp(i\omega_x t + i\phi_x)\delta(t-t_x) + \mathcal{E}_d \exp(i\omega_d t + i\phi_d)\delta(t-t_d) \quad (5.5)$$

with $t_d > t_x$. Noting that

$$\int_{-\infty}^{t} dt_1 \int_{-\infty}^{t_1} dt_2\, f(t_1,t_2)\delta(t_1-t_d)\delta(t_2-t_x) = f(t_d,t_x)\,, \quad (5.6)$$

and inserting Eq. (5.5) in Eq. (5.4), we obtain for $-\infty < t_x < t$ and $t_x < t_d < t$ that the dissociation probability is,

$$P_q(E) = \sum_m |A_{m,q}(E)|^2 = (\hbar)^{-4}\mathcal{E}_x^2\mathcal{E}_d^2 \sum_m \left|\langle E, \boldsymbol{m}, q^-|d_{g,e}e^{-iH_e(t_d-t_x)/\hbar}d_{e,g}|E_i\rangle\right|^2 \,. \quad (5.7)$$

Thus, in this case, the ratio of product into various arrangement channels is controlled entirely by the time delay $(t_d - t_x)$ between pulses.

Sample results for pulses of finite duration are shown in Figure 5.2 where the magnitude of the excited state wave function before the second pulse, and of the

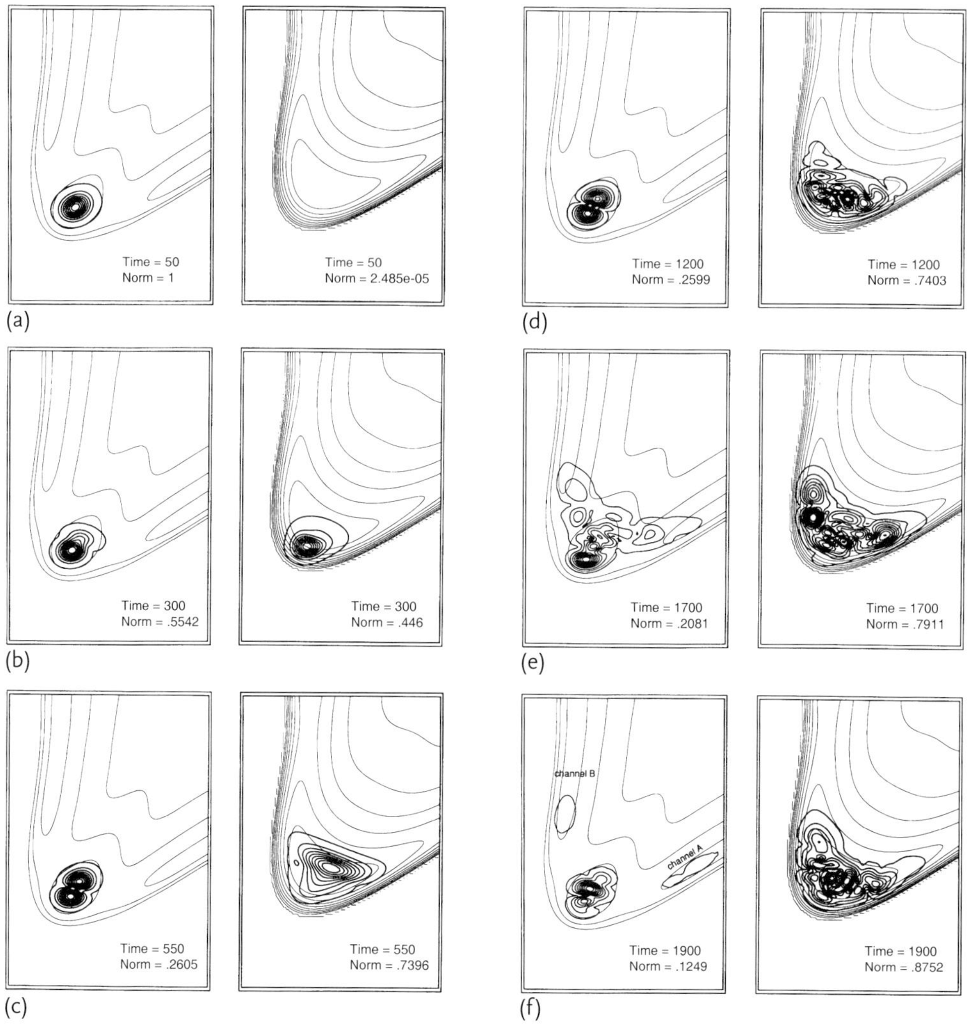

Figure 5.2 Magnitude of the wave function for a model collinear problem. The left panel in each figure shows results on the ground potential surface and the right panel shows results on the excited potential surface. Times shown are at various intervals from $t = 50$ a.u., to $t = 1900$ a.u. Note only a slight increase in probability density in outgoing channel A on the ground potential surface at the last time shown. From [98, Figure 2]. (Reprinted with permission. Copyright 1989 Elsevier Science.)

ground state wave function after the second pulse, are shown. In this case the delay time of 1900 a.u. (1 a.u. of time $= 2.418\,88 \times 10^{-17}$ s) corresponds to the time it would take a compact excited state wave packet to move to the channel A configuration, assuming that the excited state wave packet does not spread appreciably. As shown in Figure 5.2 this in fact is not the case, as the excited state wave packet spreads considerably while losing its original shape. Thus, a sizeable portion of the probability is seen to leak into the "unwanted" channel B.

The failure of the simple model which assumes that the excited state wave packet remains a compact object can be corrected, as discussed below, by optimally shaping the pump and dump pulses. Alternatively, as shown in Figure 5.3, one can simply scan the delay times while leaving the pulses with their original shapes. We see that the timing between the pulses strongly affects the relative yield of each of the two products. In this way it is possible to find the "optimal" delay time to enhance a given channel, a topic to be discussed at length below.

Numerous computational studies of many-level pump-dump scenarios have been considered by Rice, Tannor and Kosloff, and are discussed in detail in [69] and [100]. Further, a number of experiments have been carried out demonstrating control of molecular processes using pulsed light sources. In addition, there are a host of experiments of Zewail *et al.* [101–106] which use a first pulse to initiate a dynamical process, and a second pulse to interrogate the process. By the formalism above, these too are pulsed control experiments. Often, however, they only involve a single product arrangement channel, so that our primary challenge, that is, enhancing one channel over another, is not addressed.

The initial experiment demonstrating control in accordance with the Tannor–Rice scenario is due to Gerber and coworkers [107–109] in which control was demonstrated over the two channel ionization:

$$Na_2^+ + e^- \leftarrow Na_2 \rightarrow Na^+ + Na + e^- \tag{5.8}$$

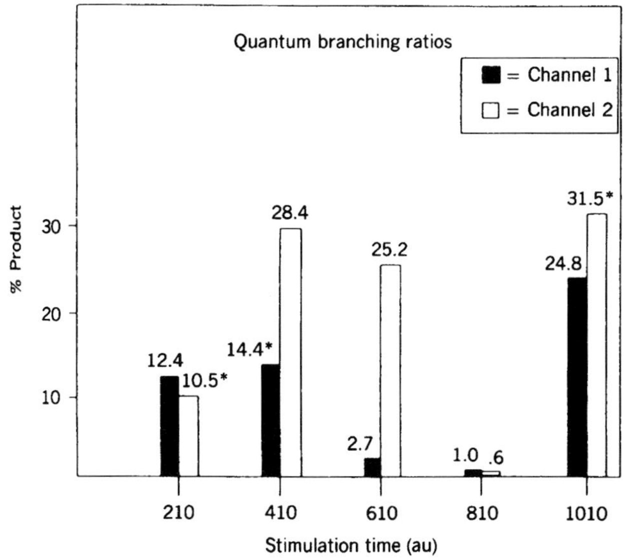

Figure 5.3 Branching ratio between two product channels in a collinear model as a function of the time of the second pulse. Note the strong dependence of the branching ratio on the pulse time. Taken from [99, Figure 5]. (Reprinted with permission. Copyright 1997 John Wiley & Sons, Inc.)

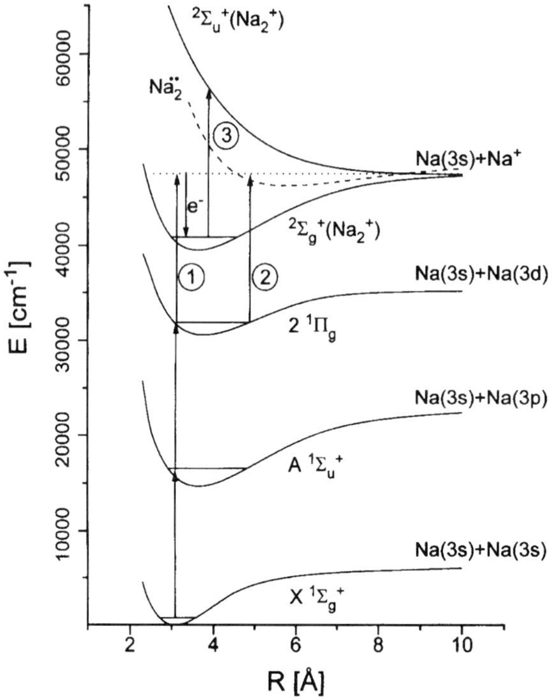

Figure 5.4 Potential energy surfaces and excitation scheme involved in the Tannor–Rice controlled Na_2 ionization. Taken from [109, Figure 3]. (Reprinted with permission. Copyright 1997 John Wiley & Sons, Inc.)

Specifically, as shown in Figure 5.4, Na_2 is pumped from the ground electronic state to the $2^1\Pi_g$ state in a two-photon process by an initial pulse. The wave packet propagates on this potential curve until an additional pulse carries it to the ionized state. The Franck–Condon factors favor production of the Na_2^+ product if excitation

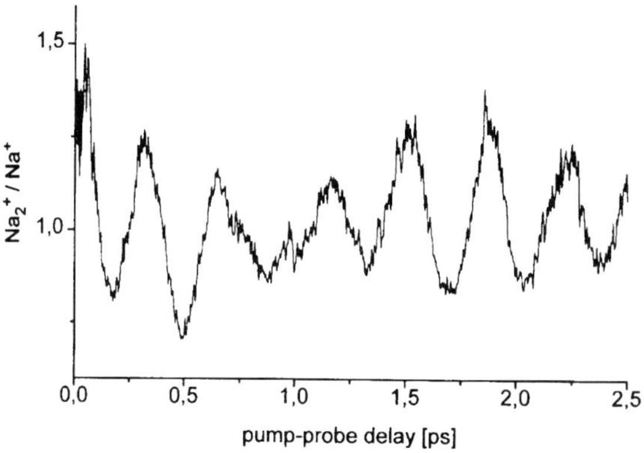

Figure 5.5 Ratio of Na^+ to Na_2^+ product as a function of the time delay between pulses. Taken from [109, Figure 5]. (Reprinted with permission. Copyright 1997 John Wiley & Sons, Inc.)

occurs when the packet is at the inner turning point of the $2^1\Pi_g$ curve, whereas the excited Na_2 decays to the $Na^+ + Na + e^-$ product if excitation occurs at the outer turning point. The experimental results on the ratio of the Na^+ and Na_2^+ signals, as a function of the delay time between the two 80 fs long pulses are shown in Figure 5.5. Clearly, the ratio is a strong function of the delay time.

A number of other early experiments confirming the Tannor–Rice scenario are discussed in detail in the monograph of Rice and Zhao [100].

5.2 Optimal Control Theory

Optimal Control Theory (OCT), was initially developed by Tannor, Rice, Kosloff *et al.* [67–69, 77, 98, 100, 110–116], by Rabitz *et al.* [117–124], by Jakubetz, Manz *et al.* [125–129], by Shapiro, Brumer, *et al.* [130–132], and by Wilson *et al.* [133].

Tannor and Rice's original idea, discussed above, had considerable obvious appeal. However, it soon became apparent that for most realistic surfaces, intuitive ideas do not necessarily tell us which pulses do the best job. In particular, once the wave packet is created, it does not remain localized, but quickly disperses. An example of this can be seen in Figure 5.2 where the wave packet in the excited state is seen to spread, resulting in less than ideal channel A/channel B branching ratios. As a consequence, intuition (which was based on simulations [67, 68, 77]) was abandoned in favor of a more systematic pulse optimization scheme.

In what follows we outline the general principles of OCT. Specific cases are analyzed in Chapter 17.

5.2.1 General Principles of Optimal Control Theory

Optimal Control Theory aims to maximize or minimize certain transition probabilities, called *objectives*, such as the production of a specified wave function Φ at a specified time t_f, given a wave function $\Psi(t_0)$ at time t_0. The general principles of OCT are best understood via a case study [98, 110], illustrated in Figure 5.2, in which the objective is to concentrate the wave function Φ in one of the exit channels of a bifurcating chemical reaction,

$$\text{(channel B)} \qquad A + BC \leftarrow ABC \rightarrow AB + C\,, \qquad \text{(channel A)} \tag{5.9}$$

Mathematically speaking, our objective is to maximize

$$J = \langle \Psi(t_f)|\mathsf{P}|\Psi(t_f)\rangle\,, \tag{5.10}$$

where $\mathsf{P} \equiv |\Phi\rangle\langle\Phi|$ is a projection operator onto the product state of interest.

The maximization of J is subject to a set of physical and practical constraints that make the optimization problem meaningful. For example, we usually wish to optimize J using a laser pulse of a fixed total energy I. This results in the (practical) constraint equation (known as a "penalty") where

$$\int_{t_0}^{t_f} dt|\varepsilon(t)|^2 - I = 0 \,. \tag{5.11}$$

Dynamics is introduced as an additional constraint. In our case this is the requirement that Ψ be a solution of the time-dependent Schrödinger equation,

$$\left(i\hbar\frac{\partial}{\partial t} - H(t)\right)|\Psi\rangle = 0 \,. \tag{5.12}$$

For a problem involving a ground and excited electronic state, it is convenient to rewrite the Schrödinger equation in matrix notation. Specifically [67, 68, 70, 72, 77, 98], $|\Psi\rangle$ is replaced by $\underline{|\Psi\rangle}$, a two-component state-vector

$$\underline{|\Psi\rangle} \equiv \begin{pmatrix} |\psi_e(t)\rangle \\ |\psi_g(t)\rangle \end{pmatrix} , \tag{5.13}$$

where $|\psi_g(t)\rangle$ and $|\psi_e(t)\rangle$ are (time-dependent) wave functions on the ground and excited electronic surfaces, respectively. The Hamiltonian $\underline{\underline{H}}(t)$ is now a 2×2 matrix (for a full justification of this treatment see Eq. 15.36):

$$\underline{\underline{H}}(t) \equiv \begin{pmatrix} H_e & -d_{ge}\varepsilon^*(t) \\ -d_{eg}\varepsilon(t) & H_g \end{pmatrix} , \tag{5.14}$$

where d_{ge}, d_{eg} is the transition dipole moment introduced in Eq. (2.72).

The problem of maximizing the objective J subject to the above constraints can be transformed into an unconstrained problem by using Lagrange multipliers. According to this standard procedure, we multiply Eq. (5.11) by an unknown number λ and Eq. (5.12) by an unknown two-component state-vector

$$\underline{|\chi(t)\rangle} \equiv \begin{pmatrix} |\chi_e(t)\rangle \\ |\chi_g(t)\rangle \end{pmatrix} , \tag{5.15}$$

and add the result to Eq. (5.10). Note that the latter is an extension of the simple Lagrange multiplier procedure to constraints on continuous functions. The resulting unconstrained objective

$$\bar{J} = \underline{\langle\Psi(t_f)|}\cdot \mathsf{P} \cdot \underline{|\Psi(t_f)\rangle} + \lambda\left[\int_{t_0}^{t_f} dt|\varepsilon(t)|^2 - I\right] + i\int_{t_0}^{t_f} dt\left\{\underline{\langle\chi|}\cdot\left[i\hbar\frac{\partial}{\partial t} - \underline{\underline{H}}\right]\cdot\underline{|\Psi\rangle} + \text{c.c.}\right\} , \tag{5.16}$$

is maximized by imposing the $\delta\bar{J} = 0$ (with respect to changes in $\underline{|\Psi\rangle}$) condition. When this is done, it follows from Eq. (5.16) that the Lagrange multipliers $\underline{|\chi\rangle}$

satisfy [98]

$$i\hbar \frac{\partial |\underline{\chi}\rangle}{\partial t} = \underline{\underline{\mathrm{H}}} \cdot |\underline{\chi}\rangle \,, \tag{5.17}$$

with the boundary condition

$$\underline{|\chi(t_f)\rangle} = \mathsf{P} \cdot \underline{|\Psi(t_f)\rangle} \,. \tag{5.18}$$

The optimum field, defined via the requirement $\delta \bar{J} = 0$ is then related to the above solutions and to λ as

$$\varepsilon(t) = O(t)/\lambda \,, \tag{5.19}$$

where

$$O(t) = i \left[\langle \chi_e | d_{eg} | \psi_g \rangle - \langle \psi_e | d_{ge} | \chi_g \rangle \right] \,. \tag{5.20}$$

Once $O(t)$ is known, λ can be obtained by substituting Eq. (5.19) in Eq. (5.11)) to obtain

$$\lambda = \pm \left(\frac{1}{I} \int_{t_0}^{t_f} dt |O(t)|^2 \right)^{1/2} \,. \tag{5.21}$$

Likewise, $\varepsilon(t)$ can be written explicitly, using Eq. (5.19) and Eq. (5.21) as

$$\varepsilon(t) = O(t) \left(\frac{1}{I} \int_{t_0}^{t_f} dt |O(t)|^2 \right)^{-1/2} \,. \tag{5.22}$$

The above set of equations can be solved by an iterative procedure which starts by guessing some $\varepsilon(t)$ function. One then determines $\underline{\underline{\mathrm{H}}}(t)$ using Eq. (5.14), and $\underline{|\Psi(t)\rangle}$ by propagating Eq. (5.13), using Eq. (5.12), from t_0 to t_f. The final value of $\underline{\chi}$, which by Eq. (5.18), is just $\mathsf{P} \cdot \underline{|\Psi(t_f)\rangle}$, is used to obtain $\underline{\chi}(t)$ from Eq. (5.17) for all $t < t_f$. An improved guess for $\varepsilon(t)$ is obtained from Eq. (5.20) and Eq. (5.22), and the whole procedure is repeated until convergence.

An example of the success of this procedure is given in Figure 5.6 where the probability density for the model studied in Figure 5.2 is shown for a field optimized to deposit the system in channel A. We see that at the time shown (1950 a.u. (= 47.2 fs)) essentially all the probability density in the ground state (whose absolute norm is still quite small at 0.0046) is deposited in channel A, while none exists in channel B. This result should be contrasted with Figure 5.2 where no pulse shaping was carried out and a nonoptimal delay between the pump and dump pulse was applied.

This procedure was applied to a number of cases by Kosloff *et al.* [98] who studied a model for the theoretical

$$\mathrm{H} + \mathrm{HD} \leftarrow \mathrm{HHD} \rightarrow \mathrm{H}_2 + \mathrm{D}$$

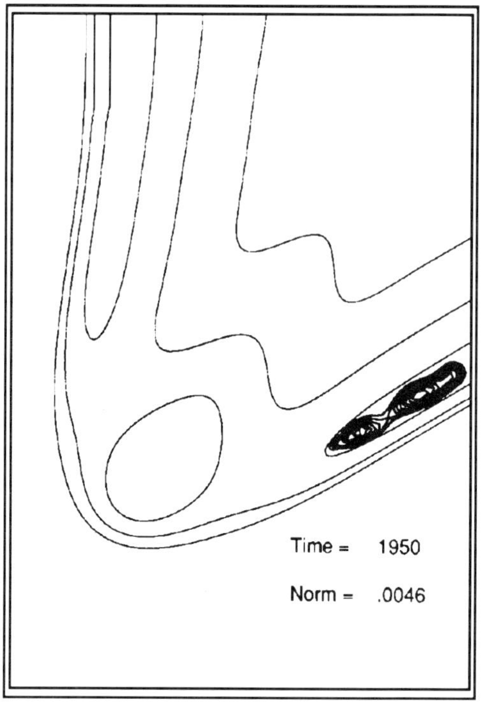

Figure 5.6 Magnitude of the wave function for the same model collinear problem as in Figure 5.2, but for a field optimized to deposit the excited state wave packet in channel A. From [98, Figure 3]. (Reprinted with permission. Copyright 1989 Elsevier Science.)

dissociation reaction, Amstrup *et al.* [111] who studied the control of the HgAr and I_2 photodissociation, Hartke *et al.* [127] who simulated the dynamics of control of the Br^*/Br branching ratio in the photodissociation of Br_2, and Jin *et al.* [114] who studied the control of population inversion between two displaced harmonic oscillators. In all cases it was possible to show a theoretically appreciable increase in the yield of the desired product by optimizing the pulse shape. Although the increase in yield relative to the zero order guess is dramatic, the final selectivity is often similar to that obtained in the linear regime by the coherent control procedure. This has been shown explicitly in the HHD case for the Rice–Tannor potential [72].

Both coherent control and OCT attempt to extremize the same transition probabilities. At first, the two approaches appear to be different: coherent control is based on using multiple interfering pathways, and relies heavily on the (linear) superposition principle of quantum mechanics, whereas OCT is a (nonlinear) theory in which one attempts to perform a completely general optimization of some external inputs in order to reach the desired objectives. However, in similar problems the control fields calculated by the two methods are often quite similar. This can be traced back to the finding, discussed in Section 3.3 below, that the presence of interfering pathways is essential to all forms of quantum control, whether included explicitly in the algorithm (as in coherent control) or not (as in OCT). In the absence of interference the degree of control attainable is greatly reduced.

General global optimization is a difficult procedure [98], one that involves searching in a complex function space (see however [114] for a description of an improved search routine using conjugate gradient methods). The search is difficult because each step necessitates the solution of the time-dependent Schrödinger equation, which although done very efficiently [134–138], is still time consuming. An additional complication arises because of uncertainties in the parameters of the system to be optimized, the most obvious such parameter being the Hamiltonian itself (interaction potentials are only rarely known to sufficient accuracy). By comparison, if one can remain in the perturbative-regime, then coherent control theory teaches that the experimentalist does not require a theoretical evaluation of the molecular matrix elements appearing in the control equations (e.g. Eq. (3.19)). Rather, one can evaluate these parameters, which are coefficients of a quadratic expression, by making a preliminary scan of the external laser parameters and fitting the result to the quadratic form.

There are now many direct experimental confirmations [18–20, 95, 139–144], to be discussed in Section 17.3, of the effect of the pulse shaping. In addition, the role of the phase between two pulses, predicted in both OCT [98, 110] and in coherent control studies [70, 72, 145] has been confirmed experimentally by Fleming *et al.* [146, 147], Girard *et al.* [148–151], Kinrot *et al.* [152], Kauffmann *et al.* [153] and Ohmori *et al.* [154–157] in the so-called "wave packet interferometry" experiments [158]. For example, Fleming [146, 147] and Kauffmann [153] describe experiments where the fluorescence from I_2 in the B-state is influenced by constructive or destructive interference between two wave packets induced by two phase related excitation pulses. This study relates to a large volume of work on wave packet interferometry in atoms [159–163], as well as to various femtochemistry experiments, where similar effects were seen in absorption [101–106].

Within OCT the very existence of a unique solution is not guaranteed. This point has been investigated in the context of square integrable functions by Peirce and Rabitz [118] and for unbounded systems [112, 113] where the existence of an optimal solution under certain conditions was established. Yao *et al.* [123] have shown that nonlinearities may give rise to multiple solutions, each producing exactly the same physical effect on the molecule. This result can be viewed from two alternate perspectives. From a positive perspective, this implies that the experimenter can obtain the optimal result using equipment that may be readily available in his laboratory. From a negative perspective, this result makes it very difficult to *understand* the physics behind optimization since many different electric fields give essentially the same final result.

Theoretical strategies for designing optimal pulses that are least sensitive to errors in the molecular Hamiltonians have also been proposed [122]. Basically one confines the search to pulses that prohibit the wave packet from making excursions into regions of uncertainties [122]. Such pulses may not always yield the best results as far as *objectives*. In fact, studies [98, 110] show great sensitivity of the outcome to even slight changes in pulse shapes. A "phase-space" representation in frequency and time of optimal pulses shows surprisingly small difference between pulses designed to select the AA + B channel vs. the AB + A channel. This means that one has to tailor these pulses very carefully in order to achieve the desired product, a task now achievable due to the progress made in pulse shaping techniques [164–171]. This problem can be made easier by smoothing the pulses. It is often [111] possible to do almost as well as in the full optimization by restricting the last stages of the iteration to a search in a space spanned by two Gaussian pulses.

These problems can be solved experimentally in the context of Optimal Control by using experimental data directly, as in the Automated Feedback Control approach to OC [17], to be discussed in detail in Section 17.3. Essentially, in this approach one obtains the optimal laser pulses required to achieve a given task by linking a computer to a pulse shaping device [164, 166]. By letting the computer vary the laser pulse parameters and performing successive comparisons of the measured outcomes with the preset target, one can get closer and closer (often using "Genetic" search algorithms [172, 173]) to the objective of interest. The strength (and weakness) of this approach is that in adopting it one treats the system as a "black box", avoiding altogether the theoretical treatment of the dynamics.

It is worth mentioning that a different approach to the control problem in finite N-state space was introduced by Harel and Akulin [174]. Rather than search for the optimal $\varepsilon(t)$ field yielding the desired time evolution operator $U(t)$ associated with an optimized $N \times N$ Hamiltonian matrix $\underline{\underline{H}}(t) = \underline{\underline{H}}_0 - \underline{\underline{d}}\varepsilon(t)$, where $\underline{\underline{d}}$ is the $N \times N$ transition dipole matrix, these authors have shown that it is possible to control the system using just two *fixed-amplitude* instantaneous pulses $\mathcal{E}_A\delta(t - t_n)$ and $\mathcal{E}_B\delta(t - t_{n'})$ applied alternately at N^2 optimally chosen time points t_n and $t_{n'}$. The end result is the representation of the evolution operator at time t as

$$U(t) = e^{-iBt_{N2}} e^{-iAt_{N2-1}} \cdots e^{-iBt_2} e^{-iAt_1} , \tag{5.23}$$

where $A \equiv \underline{\underline{H}}_0 - \underline{\underline{d}}\mathcal{E}_A$ and $B \equiv \underline{\underline{H}}_0 - \underline{\underline{d}}\mathcal{E}_B$. The evolution operator is controlled by the choice of the t_n time points, constrained such that $t = \sum_{n=1}^{N^2} t_n$.

6
Decoherence and Its Effects on Control

Thus far we have dealt with the idealized case of isolated molecules which are neither subject to external collisions nor display spontaneous emission. Further, we have assumed that the molecule is initially in a pure state (i.e., described by a wave function) and that the externally imposed electric field is coherent, that is, that the field is described by a well-defined function of time (e.g., Eq. (1.7)). Under these circumstances the molecule is in a pure state before and after laser excitation, and remains so throughout its evolution. However, if the molecule is initially in a mixed state (e.g., due to prior collisional relaxation), or if the incident radiation field is not fully coherent (e.g., due to random fluctuations of the laser phase or of the laser amplitude), or if collisions cause the loss of quantum phase after excitation, then phase information is degraded, interference phenomena are muted and laser control is jeopardized. Issues related to control in such "open systems" (as distinct from isolated, or closed, systems) are major challenges, motivating this detailed discussion of decoherence and control.

6.1
Decoherence

Loss of quantum information (either of the phase or of the amplitude of a state) due to the interaction of a system with its environment is termed decoherence. Examples include the obvious case where a system is actually embedded in an external environment, for example a molecule in solution, or more subtle cases, for example where the system is chosen as the center of mass of an object and the environment is the 10^{23} variables associated with the motion of the atoms that comprise the object.

The current view is that certain forms of decoherence can cause the loss of quantum interference in just such a way that the system then obeys classical mechanics [175, 176]. This view does not obviate the possibility that classical mechanics is, in fact, the limit of quantum mechanics when $\hbar \to 0$, (i.e., when the system action becomes very large) [177]. Rather, it proposes an alternate route to classical mechanics for systems in interaction with their environment. Clearly, decoherence effects that change the dynamics from quantum to classical mechanics destroy

Quantum Control of Molecular Processes, Second Edition. Moshe Shapiro and Paul Brumer

quantum phases and hence alter coherent control. Aspects of the issue of control in the classical limit are treated in Chapter 18.

In this chapter we focus on how decoherence effects can alter control (rarely positively, often negatively), as well as on approaches to reestablish coherent control.

Consider then a total Hamiltonian H_{tot} that we subdivide into a system s interacting with an environment. This subdivision is physically motivated: the system s is *defined* as that component in which we have an interest, and the environment is the remainder. As will be evident from the decoherence discussion below, any size environment can cause loss of phase information. The total Hamiltonian H_{tot} is then of the form

$$H_{tot} = H_s + H_e + H_{int} , \tag{6.1}$$

where H_s is the system Hamiltonian, H_e is the Hamiltonian of the environment, and H_{int} is the interaction between them.

We intend to focus solely on the properties of the system, that is, the quantities of interest. To do so it is convenient to invoke the density matrix formulation of quantum mechanics [178]. That is, the system + environment is described by a density operator $\rho_{tot}(t)$ whose dynamics is given by the quantum Liouville-von Neumann equation:

$$i\hbar \frac{d\rho_{tot}(t)}{dt} = \left[H_{tot}, \rho_{tot}(t)\right] . \tag{6.2}$$

Equation (6.2) reduces to the Schrödinger equation for the case of a pure state, that is, where $\rho_{tot}(t) = |\psi(t)\rangle\langle\psi(t)|$. To obtain system properties from known values of $\rho_{tot}(t)$, say the average value $\overline{A}$ of the operator $A(t)$, requires evaluating $\overline{A(t)} = \text{Tr}[\rho_{tot}(t)A(t)]$. Operators that pertain solely to the system, denoted $A_s(t)$, are obtained by averaging over (i.e., ignoring) environmental variables, that is, $A_s(t) = \text{Tr}_e[\rho_{tot}(t)A(t)]$. In particular, the density matrix $\rho_s(t)$ pertaining solely to the system dynamics is given by $\rho_s(t) = \text{Tr}_e[\rho_{tot}(t)]$.

A typical measure of system decoherence [179, 180] for a system starting out as a pure state is obtained as follows. Consider the ideal case of an initial normalized separable state $\rho_{tot}(0) = \rho_s(0)\rho_e(0)$. Here ρ_s is associated with the system and ρ_e with the environment. Initially, $\text{Tr}[\rho_s(0)] = 1$ and the purity of the state, $S(0) \equiv \text{Tr}[\rho_s^2(0)] = 1$. As time evolves the interaction term H_{int} causes the system and environment to become *entangled*, that is, $\rho_{tot}(t)$ can no longer be written as a separable product of system and bath density matrices. In this case, focusing solely on $\rho_s(t)$ means a loss of quantum information associated with the system–bath entanglement. This is reflected in the value of numerous measures, including the purity, $\text{Tr}[\rho_s^2(t)]$, which is now less than unity. Hence, decoherence and entanglement are intimately related to one another: tracing over the environment of an entangled system–environment state is manifested as system decoherence.

Written in terms of any arbitrary system basis $\{|i\rangle\}$ the system density matrix and purity are then given by:

$$\begin{aligned} S(t) = \mathrm{Tr}\left(\rho_s^2(t)\right) &= \sum_{i,j} \langle i|\rho_s(t)|j\rangle\langle j|\rho_s(t)|i\rangle \\ &= \sum_i |\langle i|\rho_s(t)|i\rangle|^2 + \sum_{i\neq j} \langle i|\rho_s(t)|j\rangle\langle j|\rho_s(t)|i\rangle \,, \end{aligned} \tag{6.3}$$

As seen below, typical environmentally induced decoherence results in the disappearance of the off-diagonal elements $\langle j|\rho_s(t)|i\rangle$ at long times. The specific basis $|i\rangle$ in which this "bath-induced diagonalization" occurs depends upon the nature of the system–bath coupling [181]. Subsequent relaxation leads to a limiting result:

$$\mathrm{Tr}\left(\rho_s^2(t \to \infty)\right) = \sum_i |\langle i|\rho_s(t \to \infty)|i\rangle|^2 \,. \tag{6.4}$$

It is useful to note the extreme limiting case where, in addition to loss of off-diagonal matrix elements, relaxation has led to a uniform distribution of the final system density matrix populations, that is, $|\langle j|\rho_s(t \to \infty)|j\rangle|^2 = 1/N^2$, where N is the total number of system states participating in ρ_s. In this case, the statistical limit of the purity is given by the sum over N such terms, that is,

$$\mathrm{Tr}\left(\rho_s^2\right) = 1/N \,. \tag{6.5}$$

Two points are worth emphasizing:

1. The discussion above makes clear that decoherence arises from loss of information in averaging over the environment subsequent to *unitary* evolution of the system + environment. An alternate process that causes loss of off-diagonal elements in the system density matrix is *dephasing* (also called "fake decoherence" by the decoherence community [176, 182]). This behavior is associated with a density matrix $\rho_s(t)$ that arises from some type of averaging process, e.g., the evolution of a *mixture* of system states (i.e., where $\mathrm{Tr}(\rho_s^2(0)) < 1$). In this case, loss of off-diagonal contributions to $\rho_s(t)$ can arise from the different contributions to $\rho_s(t)$ within the mixture of states. Both decoherence and dephasing are detrimental to control, due to loss of required coherence of the system.
   Decoherence, as defined above, is then a quantum mechanical effect arising within unitary dynamics. It often has no classical analog. By contrast, dephasing can have an analog in the classical evolution of an ensemble; for example, the loss of synchronization amongst a collection of vibrating oscillators that start with different initial conditions.
2. Although the purity, as a trace, is basis set independent, the loss of coherence viewed as the decay of the off-diagonal elements of the system density matrix is a property that is basis set dependent. In particular, different types of couplings between the system and environment will result in the decay of coherence in different bases [176, 181]. For example, [183] describes model cases for

the decay of decoherence in mechanisms of electron transfer for interactions that cause localization vs. those that cause energy eigenstate dephasing. (Note, however, that the authors use the term "coherence" here to denote the presence of tunneling contributions, and not to the role of off diagonal elements of $\rho_s(t)$.)

The most well-known system–bath example is the density matrix of a system interacting with a large thermal environment, for example, via system–bath collisions, viewed in the basis of the system energy eigenstates. The density matrix ρ_s^T of a system that has reached thermal equilibrium at temperature T is $\rho_s^T = A\exp(-H_s/(k_B T))$. Here A is a normalization factor and k_B is Boltzmann's constant. In the energy representation. $\rho_{i,j}^T = \langle E_i|\rho_s^T|E_j\rangle = A\exp(-E_i/(k_B T))\delta_{i,j}$, where $|E_i\rangle \equiv |\phi_s^{(i)}\rangle$ are eigenstates of H_s of energy $E_i \equiv E_s^{(i)}$. That is, the system shows population in each of the energy levels, but no off-diagonal terms. Nonzero off-diagonal terms would represent quantum coherences between the energy eigenstates of the Hamiltonian. To see this, contrast the equilibrium result with that of the density operator associated with a pure state

$$|\psi\rangle = \sum_k a_k|E_k\rangle\,, \tag{6.6}$$

where the density operator is given by

$$\rho_s = |\psi\rangle\langle\psi| = \sum_{k,m} a_k a_m^*|E_k\rangle\langle E_m|\,. \tag{6.7}$$

In this case,

$$\langle E_k|\rho_s|E_m\rangle = \rho_{k,m} = a_k a_m^*\,, \tag{6.8}$$

where here and below $\rho_{m,k}$ denotes the m, k element of the density matrix representing ρ_s.

Thus, the fact that there are well-defined phase relationships between the eigenstates of the Hamiltonian, contained in the wave function, is manifested in the nonzero off-diagonal elements in the energy representation. The absence of off-diagonal matrix elements for the thermally equilibrated case makes clear that collisions have destroyed matter coherence manifested as quantum correlations between energy eigenstates.

6.1.1 Master Equations

Since it is the dynamics of the system that is of interest, it would be convenient to pre-average over the environment variables and obtain an equation of motion for $\rho_s(t)$, the system component of the density matrix. Formal work of this kind [182, 184] yields the so-called generalized master Equation. Deriving the generalized master Equation, and extracting the various approximations utilized, goes well astray of the central focus of this book. For this reason we just sketch some models below and direct the reader to suitable reviews [182, 184–186] that provide an appropriate overview.

For example, we consider below one class of such approximate master equations. Specifically, if the correlation time of the environment is much shorter than the typical time scale for the variation of the system, then the generalized master equation is of the form

$$\frac{\partial \rho_s(t)}{\partial t} = -i\hbar^{-1}\left[H_s, \rho_s(t)\right] + F(\rho_s(t)) \,, \tag{6.9}$$

where $F(\rho_s)$ is a functional of ρ_s. Its functional form depends upon the nature of the environment and on the coupling H_{int} and the resultant equations are generally quite complex. As such, a variety of approximations to the generalized master equations have yielded equations that are used to model the effect of the environment on the system dynamics. Of note is the general Lindblad form [185, 187–189]

$$F(\rho_s(t)) = \sum_j \left[2A_j \rho_s A_j^\dagger - \left\{A_j^\dagger A_j, \rho_s\right\}\right] \,, \tag{6.10}$$

where $\{A, B\} = AB + BA$ is the anticommutator, and the operators, A_j, chosen to represent the physics, are called Lindblad operators. Mathematical advantages of this Lindblad form are discussed in [185], a key advantage being that the Lindblad form assures "positivity", that is, that $\langle\psi|\rho_s(t)|\psi\rangle \geq 0$ for any system state $|\psi\rangle$.

The relation of Eq. (6.10) to the system–environment Hamiltonian is complicated [185]. Given a system–environment Hamiltonian H_{int} of the general form $H_{int} = \sum_\alpha S_\alpha B_\alpha$, where S_α operates on the system and B_α operates on the bath, the most general master equation that ensures positivity has been shown [190] to be of the form of Eq. (6.9) with

$$F\left[\rho_s(t)\right] = \frac{1}{2}\sum_{\alpha,\beta} \gamma_{\alpha,\beta}\left(\left[S_\alpha, \rho_s S_\beta^\dagger\right] + \left[S_\alpha \rho_s, S_\beta^\dagger\right]\right) \,. \tag{6.11}$$

The form in Eq. (6.10) then results by diagonalizing the matrix with elements $\gamma_{\alpha\beta}$. The resultant Lindblad operators A_j are linear combinations of the original S_α operators. Note that the collection of S_α operators is essentially arbitrary [185] as long as it forms a complete set of system operators with $\mathrm{Tr}_s(S_\alpha^\dagger S_\beta) = \delta_{\alpha,\beta}$. An example of this form is discussed later below.

An alternative, exact, approach to propagating $\rho_s(t)$, as yet to be fully utilized, is based upon a linear mapping approach [191, 192]. The structure is as follows. Consider a system that is M-dimensional. Under the assumption of complete positivity [188], the reduced density matrix evolves according to the canonical Kraus representation, which is general for the case where the overall density matrix is initially a product of the density matrices for the system and the environment [193]. This canonical form is given by

$$\rho_s(t) = \sum_{\alpha=1}^{d} \boldsymbol{E}^\alpha(t)\rho_s(0)\boldsymbol{E}^{\alpha\dagger}(t) \,. \tag{6.12}$$

Here the sum is over the d "Kraus operators" [188] $\boldsymbol{E}^{\alpha}$ where $\sum_{\alpha=1}^{d} \boldsymbol{E}^{\alpha\dagger}\boldsymbol{E}^{\alpha} \leq I$. Specifically, $\sum_{\alpha=1}^{d} \boldsymbol{E}^{\alpha\dagger}\boldsymbol{E}^{\alpha} = I$ for trace-preserving quantum operations and $\sum_{\alpha=1}^{d} \boldsymbol{E}^{\alpha\dagger}\boldsymbol{E} < I$ for nontrace-preserving quantum operations such as quantum measurements. Note, significantly, that the operator-sum representation for open systems, Eq. (6.12), has at most M^2 Kraus operators, that is, $2 < d \leq M^2$ and that the operators are fixed by the given system, bath and any incident external fields. Hence, in this representation, the effect of the environment is embedded in the d Kraus operators. Once they are known, Eq. (6.12) provides $\rho_s(t)$ at any time, without the need for propagating the dynamics. A method for obtaining the Kraus operators, given the Hamiltonian H_{tot} for the system + environment has been proposed [194].

Note that Eq. (6.12) is general insofar as it applies to isolated system dynamics as well. In this case Eq. (6.12) reduces to a single term, that is,

$$\rho_s(t) = \boldsymbol{E}^{\alpha}\rho_s(0)\boldsymbol{E}^{\alpha\dagger} = U(t)\rho_s(0)U^{\dagger}(t) , \tag{6.13}$$

where $U(t)$ is the propagator associated with the H_s Hamiltonian [for example, $U(t) = \exp(-iH_s/\hbar)$ in the case where H_s is time independent]. One control related application of the Kraus operator formalism is provided in Section 6.3.

6.2
Sample Computational Results on Decoherence

6.2.1
Electronic Decoherence

As a relevant example of decoherence, consider the case of electron dynamics in an isolated molecule, where decoherence arises in focusing on the electrons as the system and in ignoring the environment to which the electrons are coupled, for example, the vibrations.

Consider, for example, excitation from an initial state comprising a product of the ground electronic state $|g\rangle$ and ground vibrational state $|v_g\rangle$ to an excited electronic state $|e\rangle$. If the pulse is sufficiently fast, then in accord with Eq. (2.A.5), the prepared state at time zero is of the form (where A is a normalization factor)

$$|\psi(0)\rangle = A\left(|g\rangle|v_g\rangle + c|e\rangle|v_g\rangle\right) = A\left[|g\rangle + c|e\rangle\right]|v_g\rangle , \tag{6.14}$$

where $c = (i\epsilon_0/\hbar)d_{e,g}$. Note that initially the wave function is factorizeable into electronic and vibrational components and hence (by definition) the state in Eq. (6.14) is unentangled. However, since the vibrational state $|v_g\rangle$ is not an eigenstate of the nuclear Hamiltonian associated with $|e\rangle$ it will evolve to $|v_g(t)\rangle \neq |v_g\rangle$. The associated density matrix of the system is then

$$|\psi(t)\rangle\langle\psi(t)| = A^2\Big(|g\rangle|v_g\rangle\langle g|\langle v_g| + |c|^2|e\rangle|v_g(t)\rangle\langle e|\langle v_g(t)| \\ + c^*|g\rangle|v_g\rangle\langle e|\langle v_g(t)|e^{i\delta t/\hbar} + c|e\rangle|v_g(t)\rangle\langle g|\langle v_g|e^{-i\delta t/\hbar}\Big) , \tag{6.15}$$

where δ is the difference in energy between the minima in the nuclear potential on the ground and excited electronic states.

Since our interest is in the dynamics of the electrons, ignoring the environment means tracing over the vibrational states. Assuming the Condon approximation, the resultant system density matrix is

$$\rho_s(t) = A^2 \Big[|g\rangle\langle g| + |c|^2|e\rangle\langle e| + c|e\rangle\langle g|\langle v_g|v_g(t)\rangle e^{-i\delta t} + c^*|g\rangle\langle e|\langle v_g(t)|v_g\rangle e^{i\delta t} \Big] . \tag{6.16}$$

The electronic coherence, manifested in the off-diagonal matrix elements $|g\rangle\langle e|$ and $|e\rangle\langle g|$ is seen to be modulated by the overlap $\langle v_g|v_g(t)\rangle$. As the vibrational wave packet $|v_g(t)\rangle$ evolves on the upper electronic surface the overlap $\langle v_g|v_g(t)\rangle$ will decrease as $|v_g\rangle$ and $|v_g(t)\rangle$ become different. The result is "electronic decoherence", that is, the reduction of the off-diagonal contributions $|g\rangle\langle e|$ and $|e\rangle\langle g|$ [195, 196].

Numerous studies, for example, [195, 197–199], suggest that this electronic decoherence occurs in $\sim$ 10 fs for typical large molecules. An example is shown in Figure 6.1 where electronic decoherence is computed [197] for polyacetylene chains of varying length N. Decoherence times are seen to reduce from $\approx$ 100 to $\approx$ 7 fs as N is increased from 4 to 100. This is due to both the vibrational overlap effect as well as the vibrationally induced transitions of populations between electronic states.

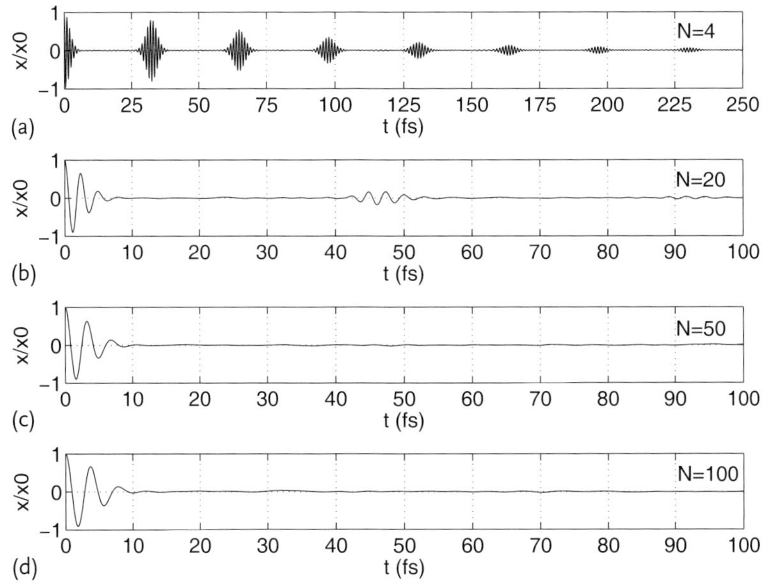

Figure 6.1 Time dependence of the electronic polarization $\langle \mu(t)\rangle = \langle \psi(t)|\hat{\varepsilon}\cdot d|\psi(t)\rangle$ for neutral flexible PA chains with N sites. The initial state is a superposition, with equal coefficients, between the ground and first excited electronic state. The nuclei are taken to be initially in the ground state configuration. From [197, Figure 10]. (Reprinted with permission. Copyright 2008 American Institute of Physics.)

Hence, rapid electronic decoherence for reasonably large molecules is expected. When this is not the case, as in some light-harvesting systems [200, 201] it is regarded as unexpected and intriguing behavior. Significantly, in the polyacetylene case [197, 198, 202] the speed and extent of electronic decoherence was sufficient to destroy control of electrical currents by one- vs. two-photon control, that is, the scenario described in Section 3.4.3.

6.2.2
Vibrational Decoherence in Condensed Phases

Consider now coherent control as it would apply in condensed phases, for example, in a liquid. Here, molecules of some species B in solution would be subjected to laser irradiation. Since we are ultimately interested in the fate of the B molecules, B is the system and the remaining molecules in the solution, as well as the laser, are the environment. Decoherence effects can then arise from the collisions of the solvent with the molecule of interest, or from incoherence properties of the laser that cause some loss of quantum phase information. Both of these effects are discussed below, along with proposed methods to reduce the effects of collisional or laser induced decoherence effects.

As an example consider the pump-dump scenario described in Section 3.6. In this case, control relies upon the maintenance of temporal coherence in the intermediate state created in the first pump step. If, for example, interaction with an environment causes decoherence on a very short time scale then each of the two excited states $|E_2\rangle$ and $|E_3\rangle$ are independently excited and do not interfere. The resultant probability of finding product in channel q at energy E [that is, Eq. (3.78)] would be given by

$$P_q(E) = \left(\frac{2\pi}{\hbar}\right)^2 \left[|b_2|^2 d_q(2,2)\bar{\epsilon}_2^2 + |b_3|^2 d_q(3,3)\bar{\epsilon}_3^2\right] ; \tag{6.17}$$

that is, control is lost.

Clearly, this is an extreme case, where decoherence in the energy eigenbasis occurs almost instantaneously. Experimental studies of the preparation of coherent superpositions of states in solution show that phase coherences of molecules exist in solution for time scales greater than 100 fs [203, 204]. Similar time scales for vibrational decoherence are shown in the computational example, below.

As a model consider the canonical example of a diatomic vibrator immersed in a bath comprising a collection of harmonic oscillators, the latter being characteristic of a general integrable bath. Specifically, results are available [205] for the case of a Morse oscillator coupled linearly to $(f-1)$ harmonic oscillators so that the Hamiltonian is given by:

$$H = \frac{p^2}{2\mu} + D\left[1 - e^{-a(q-q_e)}\right]^2 + \sum_{j=1}^{f-1}\left[\frac{P_j^2}{2} + \frac{\omega_j^2}{2}\left(Q_j + \frac{c_j}{\omega_j^2}(q-q_e)\right)^2\right] . \tag{6.18}$$

Here $\{p, q\}$ are the diatom system phase space variables and $\{P_j, Q_j\}$ are the mass-scaled momenta and coordinates of the bath. The Morse potential parameters are those of molecular iodine and the remainder of the system–bath parameters, including the system–bath coupling constant η that quantifies the spectral content of the bath, are discussed and given in [205], as is the converged semiclassical procedure used to carry out the dynamics.

Several of the cases examined were: (a) a coherent initial iodine state $|\Phi_0\rangle$ given by

$$\langle q|\Phi_0\rangle = \left(\frac{\gamma}{\pi}\right)^{\frac{1}{4}} \exp\left[-\frac{\gamma}{2}(q-q_i)^2 + \frac{i}{\hbar}p_i(q-q_i)\right], \tag{6.19}$$

where (q_i, p_i) are the average position and momentum of an initial wave packet of width γ; (b) a superposition of two different coherent states (Schrödinger "cat"-states) given by

$$\langle q|\Phi_0\rangle = \sum_{j=a,b} \left(\frac{\gamma}{4\pi}\right)^{\frac{1}{4}} \exp\left[-\frac{\gamma}{2}\left(q-q_j\right)^2 + \frac{i}{\hbar}p_j\left(q-q_j\right)\right]; \tag{6.20}$$

and (c) superposition states of the form

$$\langle q|\Phi_0\rangle = \sum_n c_n \langle q|E_n\rangle, \tag{6.21}$$

where $|E_n\rangle$ is the nth bound eigenstate of the I_2 Morse oscillator.

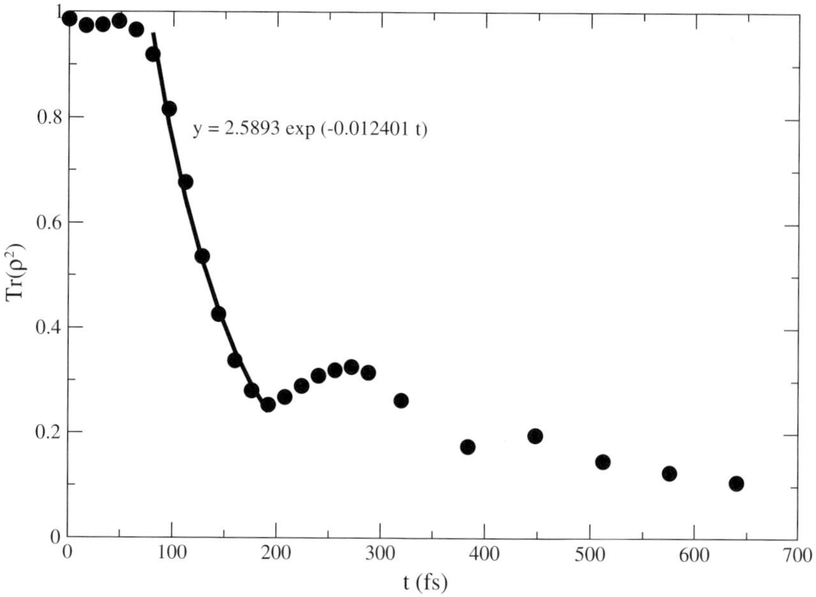

Figure 6.2 Purity as a function of time for a multilevel coherent state. $T = 300$ K, $\eta = 0.25$. The group of points corresponding to the first abrupt decay of this function have been fitted to an exponential curve described by the given equation. The ordinate here and in the related figures below is for the system ρ, that is, ρ_s. From [205, Figure 1]. (Reprinted with permission. Copyright 2004 American Institute of Physics.)

Figure 6.2 shows the purity $\mathrm{Tr}(\rho_s^2)$ as a function of time for an initial multilevel coherent state placed at $q_i = 2.4$, $p_i = 0.0$. This coherent state populates the first 61 vibrational levels of the Morse oscillator, peaking around state 24. Figure 6.2 shows that the purity of the system falls off very slowly until $\approx$ 100 fs, when it abruptly decays, showing short recurrences at $\approx$ 250 and $\approx$ 450 fs. The decay between $t \approx$ 100–250 fs, can be fitted well to an exponential. The recurrence frequency is slightly larger than the harmonic vibrational period of the Morse oscillator $\sim$ 156 fs, suggesting a correlation between the vibrational motion and the quantum coherences. The value of $\mathrm{Tr}(\rho_s^2)$ at the largest time shown is $\approx$ 0.1. If, at this time, ρ were a statistically mixed state with equal probabilities, it would imply that the total number of states in the system were $N = 10$. Clearly this is inconsistent with the actual number of states that are in the wave packet and with Figure 6.3, discussed below, which shows the involvement of far more states. Nonetheless, this "effective number of levels" provides an idea as to how far the system is from complete relaxation. The time scale of these coherences is shown in greater detail in Figure 6.4, which plots the real and imaginary parts of a sample matrix element near the diagonal: $\rho_{23,27}$ and $\rho_{27,23}$. Here the overall decay of this matrix element is evident, but on a far longer time scale than projected by the purity measure.

A more general view of the time evolution of the off-diagonal density matrix elements (in the energy representation) is given in Figure 6.3, where the elements

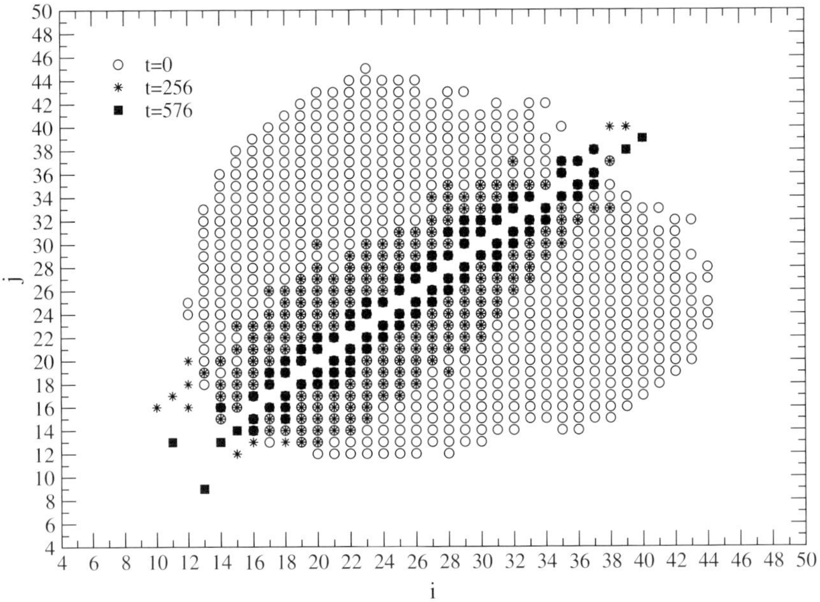

Figure 6.3 Map of the off-diagonal density matrix elements that obey $|\rho_{ij}| > 0.02$ at three different times. For clarity note that those elements $|\rho_j| > 0.02$ at time t also satisfy this condition at earlier times. Note the collapse of the density to the diagonal with increasing time. From [205, Figure 3]. (Reprinted with permission. Copyright 2004 American Institute of Physics.)

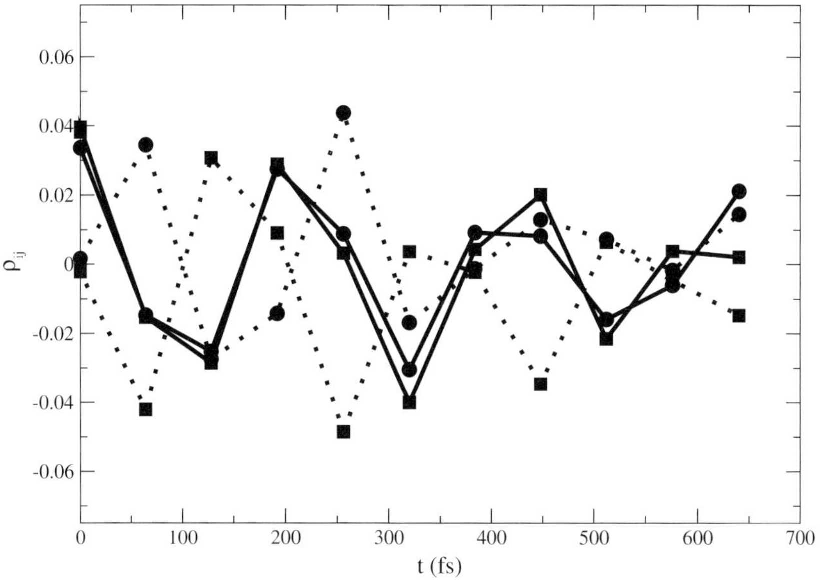

Figure 6.4 Sample $\rho_{i,j}$ as a function of time. Solid lines: $\mathrm{Re}(\rho_{i,j})$; Dotted lines: $\mathrm{Im}(\rho_{i,j})$. Circles: $i = 27$, $j = 23$; Squares: $i = 23$, $j = 27$. From [205, Figure 2] (Reprinted with permission. Copyright 2004 American Institute of Physics.)

$\rho_{ij} = \langle E_i|\rho_s(t)|E_j\rangle$ of the density matrix are shown in the basis of the system energy eigenstates $|E_i\rangle$. Here, each point depicts a density matrix element whose absolute value is larger than the cutoff of 0.02. Results are shown at three different times. The collapse of these matrix elements towards the diagonal is visibly impressive. Generally speaking, the higher frequency coherences arising from levels that are widely separated from one another in energy decay much faster than levels that are close to one another.

The combination of Figures 6.2, 6.3 and 6.4 plus auxiliary studies of the matrix elements make it clear that the primary decoherence process seen in Figure 6.2 is the disappearance of the high frequency off-diagonal components of the density matrix contributing to the $i \neq j$ in Eq. (6.3).

Despite the fact that exponential decay is observed over a small time region, the total falloff in Figure 6.2 is distinctly nonexponential. Nevertheless one can extract some sort of dephasing time by noting that $\mathrm{Tr}(\rho_s^2)$ has fallen below $1/e$ by $t \approx 200$ fs. By contrast, the off-diagonal matrix element in Figure 6.4 appears to be falling off, with oscillations, over a much longer time scale. Specifically, we would very crudely estimate that the peaks of the oscillations would reach $1/e$ of their initial value by ≈ 700 fs.

The overall decoherence rate depends upon various conditions, including the initial state, bath temperature and system–bath coupling, as discussed and demonstrated in [205].

6.2.3 Decoherence: Towards the Classical Limit

It is widely argued that certain types of decoherence lead to the disappearance of quantum effects, and to the emergence of classical mechanics. We consider here an approximate model, due to Caldeira–Leggett and to Zurek [206–209], that is widely used as a paradigm for such behavior, leading to the loss of off-diagonal elements of the density matrix in the coordinate representation. Here, as above, the system interacts with a bath comprising harmonic oscillators, but this model is approximate and assumes weak system–bath coupling and a high temperature limit. As a concrete example, we consider the vibrational motion of a model molecule with two degrees of freedom coupled to an harmonic bath.

It proves convenient to carry out the computations in the Wigner representation, where the physics of the classical limit is more evident. The Wigner representation ρ^W of ρ is defined, for an N degree of freedom system, by:

$$\rho^W \equiv \rho^W(\boldsymbol{q}, \boldsymbol{p}) = (\pi\hbar)^N \int d\boldsymbol{v}\, e^{\frac{-2i\boldsymbol{p}\cdot\boldsymbol{v}}{\hbar}} \langle \boldsymbol{q} - \boldsymbol{v}|\rho|\boldsymbol{q} + \boldsymbol{v}\rangle \,, \tag{6.22}$$

and the Wigner representation of any operator A, denoted by the superscript W, is given by

$$A^W = 2^N \int d\boldsymbol{v}\, e^{\frac{-2i\boldsymbol{p}\cdot\boldsymbol{v}}{\hbar}} \langle \boldsymbol{q} - \boldsymbol{v}|A|\boldsymbol{q} + \boldsymbol{v}\rangle \,. \tag{6.23}$$

This Wigner representation of the density $\rho^W(\boldsymbol{q}, \boldsymbol{p})$ proves particularly useful since it satisfies a number of properties that are similar to the classical phase space distribution $\rho_{cl}(\boldsymbol{q}, \boldsymbol{p})$. For example, if $\rho = |\psi\rangle\langle\psi|$, that is if ρ represents a pure state, then $\int d\boldsymbol{p}\rho^W = |\psi(\boldsymbol{q})|^2$; that is, the integral over $\boldsymbol{p}$ gives the probability density in coordinate space. Similarly, integrating ρ^W over $\boldsymbol{q}$ gives the probability density in momentum space. These features are shared by the classical density $\rho_{cl}(\boldsymbol{p}, \boldsymbol{q})$ in phase space. Note, however, that ρ^W is not a probability density, as evidenced by the fact that it can be negative, a reflection of the quantum features of the dynamics [210].

Transforming Eq. (6.9) in accord with Eqs. (6.22) and (6.23) gives the Wigner representation of the master equation:

$$\frac{\partial \rho_s^W(t)}{\partial t} = -i\hbar^{-1}\left[H_s, \rho_s(t)\right]^W + F_{CL}^W(\rho_s) \,. \tag{6.24}$$

The Caldeira–Leggett and Zurek model for an oscillator system with two degrees of freedom and with Hamiltonian H_s in contact with a bath is a particular example of Eq. (6.24) given by the master equation [211]:

$$\frac{\partial \rho_s^W(t)}{\partial t} = \left[\left\{H_s, \rho_s^W\right\} + \sum_{(l_1+l_2)odd} \frac{(\hbar/2i)^{(l_1+l_2-1)}}{l_1!l_2!} \frac{\partial^{(l_1+l_2)} V(x,y)}{\partial x^{l_1}\partial y^{l_2}} \frac{\partial^{(l_1+l_2)}\rho_s^W}{\partial p_x^{l_1}\partial p_y^{l_2}}\right] + F_{CL}^W(\rho_s) \,. \tag{6.25}$$

Here

$$F_{CL}^{W}(\rho_s) = D\left(\frac{\partial^2\rho_s^W}{\partial p_x^2} + \frac{\partial^2\rho_s^W}{\partial p_y^2}\right)$$

denotes the Caldeira–Leggett form of $F(\rho_s)$ in the Wigner representation. The term in square brackets in Eq. (6.25) is the Wigner representation $[H_s, \rho_s(t)]^W$ of $[H_s, \rho_s(t)]$. Here (p_x, p_y, x, y) are the system momenta and coordinates, $V(x, y)$ is the system potential in the Hamiltonian H, and $\rho_s^W = \rho_s^W(p_x, p_y, x, y; t)$. The first term on the right hand side of Eq. (6.25),

$$\{H_s, \rho_s^W\} = \frac{\partial H_s}{\partial x}\frac{\partial \rho_s^W}{\partial p_x} - \frac{\partial H_s}{\partial p_x}\frac{\partial \rho_s^W}{\partial x} + \frac{\partial H_s}{\partial y}\frac{\partial \rho_s^W}{\partial p_y} - \frac{\partial H_s}{\partial p_y}\frac{\partial \rho_s^W}{\partial y},$$

is the classical Poisson bracket that generates classical dynamics, the second term is responsible for the difference between quantum and classical mechanics, and the third term induces decoherence. Note that if the $F_{CL}^{W}(\rho_s)$ term cancels the effect of the second term in brackets in Eq. (6.25), then $\rho_s^W(t)$ evolves classically. Evidently, this limit, advocated by the decoherence community as the origin of observed classical behavior, depends on the magnitude of D, often obtained by a fit to experimental data.

Numerical calculations on Eq. (6.25) can be compared to classical mechanics by computing the classical phase space density $\rho_{cl}(x, y, p_x, p_y)$, which is obtained as the solution to the Fokker–Planck equation:

$$\begin{aligned}\frac{\partial}{\partial t}\rho_{cl}(x, y, p_x, p_y) &= \{H_s, \rho_{cl}(x, y, p_x, p_y)\} \\ &+ D\left(\frac{\partial^2}{\partial p_x^2}\rho_{cl}(x, y, p_x, p_y) + \frac{\partial^2}{\partial p_y^2}\rho_{cl}(x, y, p_x, p_y)\right).\end{aligned} \tag{6.26}$$

For example, consider the specific case of the nonlinear oscillator Hamiltonian [212]

$$H_s = \frac{1}{2}\left(p_x^2 + p_y^2 + \alpha x^2 y^2\right) + \frac{\beta}{4}(x^4 + y^4), \tag{6.27}$$

with parameters that can be related to typical molecules: $\beta = 0.01$, $\alpha = 1.0$. The extent to which quantum effects are diminished in the presence of decoherence is demonstrated in the figures that follow. In particular, Figure 6.5 shows the classical and quantum expectation values in the absence of decoherence (i.e., $D = 0$ in Eqs. (6.25) and (6.26)) for four moments associated with y. All figures show qualitatively similar behavior; that is, after an initial period of agreement, the quantum results continue to oscillate while the classical results show smooth relaxation [213]. Note in particular that the quantum results do not always simply oscillate about the classical ones (e.g., see results for $\langle y^2\rangle$), and that the quantum fluctuations about the mean are substantial (e.g., 30% in the case of average energy in the y-oscillator $\langle E_y \rangle$).

Results for the same moments, after introducing decoherence, are shown in Figure 6.6. A comparison of Figures 6.5 and 6.6 shows substantially improved correspondence between classical and quantum results upon introducing decoherence. Remarkably, this is true even for $\langle y^2 \rangle$, where the long term quantum average in the closed system deviated significantly from the long term classical average. Qualitatively similar results have been obtained for reactive scattering [214].

These computational results demonstrate the way in which decoherence tends to eliminate quantum effects in the system. As a consequence, quantum control processes must be effectively shielded from detrimental types of decoherence effects if quantum control is to survive. Control that persists in the classical limit is discussed in Chapter 18.

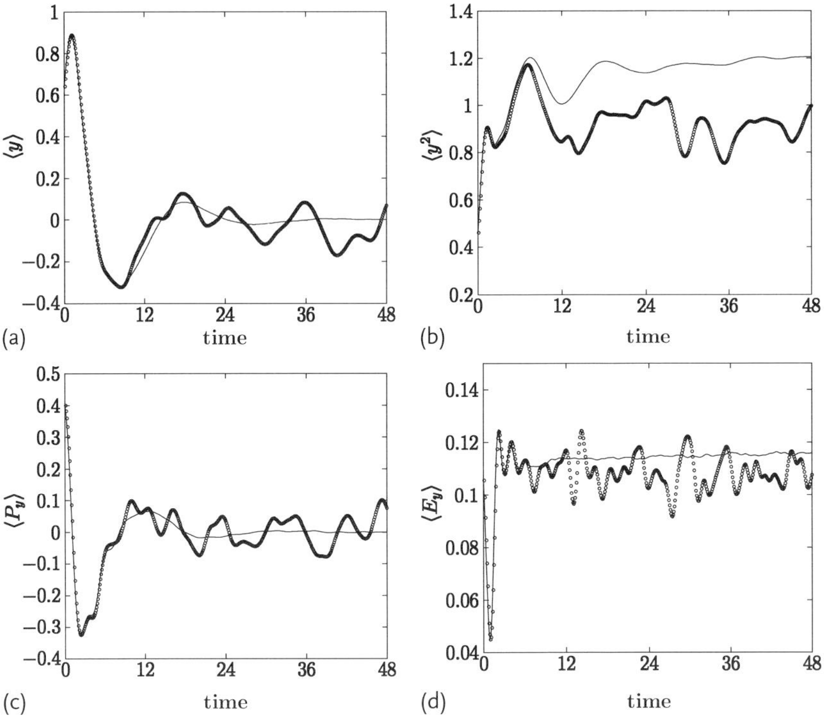

Figure 6.5 Time dependence of four statistical moments ($\langle y \rangle$, $\langle y^2 \rangle$, $\langle P_y \rangle$, and $\langle E_y \rangle$), for the system in the absence of decoherence. Dark dots denote quantum results, thin solid lines are classical results. From [211, Figure 1] (Reprinted with permission. Copyright 1999 American Physical Society.)

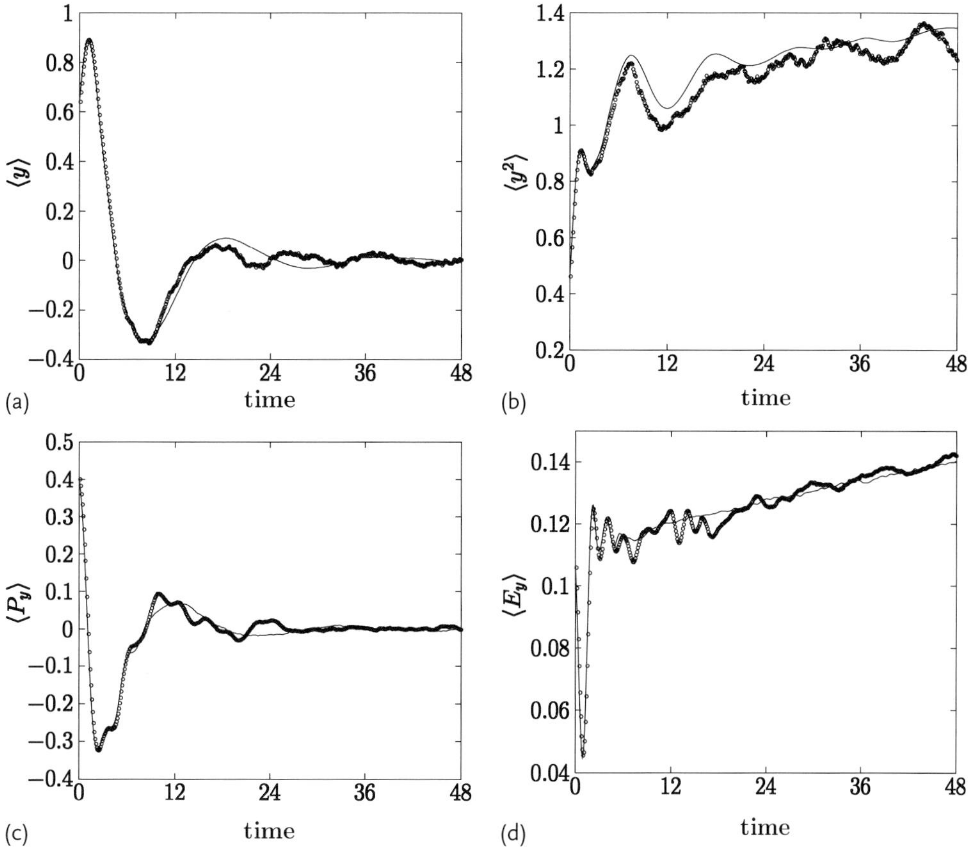

Figure 6.6 As in the previous figure, but in the presence of decoherence. From [211, Figure 2]. (Reprinted with permission. Copyright 1999 American Physical Society.)

6.3 Environmental Effects on Control: Some Theorems

6.3.1 Environment Can Limit Control

Few quantitative results are available on the effect of the environment on molecular control scenarios. A particularly clear case is discussed in this section. Here we attempt to control the extent to which the population is transferred between the two subspaces of the Hilbert space of the system by preparing a superposition of states in the subspace in which the dynamics are initiated [215]. This is shown to be far more difficult in an open system than in an isolated system [216].

Consider a system whose Hamiltonian (H_s) eigenstates are partitioned into bases for three subspaces: H_0, H_1, H_2, of dimensionality M_0, M_1 and M_2, respectively.

The basis vectors $|i, n_i\rangle$ span the subspace H_i, where $i = 0, 1, 2$ labels the subspace and $n_i = 1, \dots, M_i$ labels the states in H_i. The total dimension of the system is then $M = M_0 + M_1 + M_2$. We assume that the system initially resides in M_0, and that the dynamical evolution of the system will cause the population to flow into the other two subspaces. The specific control question under consideration is: Under what conditions can one prevent dynamics from flowing into the subspace H_2 by preparing a initial superposition state in H_0, say by laser excitation, that goes solely into H_1? Doing so would invoke interference contributions that destructively interfere to cancel the evolution into H_2. The general discussion below, applicable to open systems, is a generalization of an earlier closed system result [216].

We assume that the initial density matrix is a product of the density matrix for the system and for the environment, and denote elements of the system density matrix $\langle i, n_i|\rho_s|j, j, n_j\rangle$ by $\rho^s_{(i,n_i),(j,n_j)}$. Under generic conditions [193] the resultant system density matrix ρ_s evolves in accord with the Kraus representation (Eq. (6.12)). In order to prevent dynamics from going into H_2, we require

$$\rho^s_{(2,k_2)(2,k_2)} = 0 \tag{6.28}$$

for all states $|2, k_2\rangle$, $k_2 = 1, \dots, M_2$.

The Kraus representation implies that matrix elements of a propagated initial pure system state $|\psi_0\rangle\langle\psi_0|$ in M_0, becomes

$$\rho^s_{(2,k_2)(j,n_j)}(t) = \sum_{\alpha=1}^{d} \langle 2, k_2|\, \boldsymbol{E}^\alpha(t)\, |\psi_0\rangle \langle\psi_0|\, \boldsymbol{E}^{\alpha\dagger}(t) \left|j, n_j\right\rangle . \tag{6.29}$$

Given that the diagonal element $\rho_{(2,k_2)(2,k_2)}(t)$ is $\sum_{\alpha=1}^{d} |\langle 2, k_2|\boldsymbol{E}^\alpha(t)|\psi_0\rangle|^2$, satisfying $\rho_{(2,k_2)(2,k_2)}(t) = 0$, means requiring

$$\langle 2, k_2|\, \boldsymbol{E}^\alpha(t)\, |\psi_0\rangle = 0\,, \quad \alpha = 1, \dots, d\,. \tag{6.30}$$

This being the case, demanding zero population in H_2 also implies that all elements $\rho_{(2,k_2)(j,n_j)}(t)$, which include coherences within the H_2 Hilbert subspace, are also zero.

Equation (6.30) has a nontrivial solution if $M_0 > d M_2$, and control via initial state preparation is achievable no matter what the form or dynamics of the d-dimensional Kraus operators. Further, this equation provides the initial state that allows for the desired control. However, this control condition is, as expected, far more stringent in the open system than in the closed system case, where $d = 1$. Note, however, that the bath effects are still limited insofar as the minimum of d in an open system can be[1] as small as two, and the maximum of d is M^2, which can be far smaller than the dimensionality of the bath. Further work is necessary, however, to establish conditions for the value of d for a given system + bath.

If $M_0 \leq d M_2$, control via initial state preparation is also far more difficult in the open system than in the (already difficult) closed system case [216]. Define a

1) Alicki, R. has shown that for the case of pure dephasing, d can be as small as two (private communication, 2010).

$dM_2 \times M_0$ dimensional matrix W with matrix elements $W_{(2k,\alpha),(0n)} = \langle 2, k|\boldsymbol{E}^{\alpha}|0, n\rangle$. In the case where $M_0 \le dM_2$, the rank of W is equal to M_0 unless all $M_0 \times M_0$ dimensional submatrices of W are singular. Hence, nontrivial solutions to Eq. (6.30) exist if $\det(W^{(k)}_{M_0,M_0}) = 0$, where (k) numbers all of the submatrices. This condition also implies that a set of columns of W are linearly dependent. Hence control is possible for a *specific* class of W, expected to be difficult to obtain physically.

Note that the definition of the subspaces H_i and associated dimensions M_i can, in some instances, also be manipulated. For example, in bound-state systems (such as that in the example below) these subspaces can be defined by accessing only specific system eigenstates using a pulsed laser field. This facility might prove additionally useful in attempting to satisfy the $M_0 > dM_2$ bound of this theorem.

These results allow numerous applications, as yet to be carried out. Consider, for example, electronic energy transfer from donor to acceptor molecules, where both may be part of one larger molecule. These systems are ubiquitous, ranging from relatively small systems [217] to large structures such as carotenoid-to-bacteriochlorophyll energy transfer in photosynthesis [201]. The most interesting cases take place in condensed matter environments, that is, open systems. Studies of the dynamics and spectroscopy of such systems can be carried out by laser excitation of the system from a lower electronic manifold (here H_0) to the donor (here H_1). Electronic energy transfer from the donor (H_1) to the acceptor (H_2) is then measured. Here, the subspaces H_i, and associated dimensions M_i, are determined by a combination of the molecular state densities and the width of the laser pulses that prepare a preliminary superposition of states on H_0 and that subsequently excite the system into H_1 and H_2.

In some systems of interest, excitation from H_0 to the donor is contaminated by partial excitation of the acceptor as well, with a concomitant reduction in the quality of the data on the subsequent electronic energy transfer dynamics. A considerable improvement would result from being able to excite H_1 with reduced population transfer to H_2 from H_0. Results of the open system theorem indicate that this is (a) difficult to achieve physically if $M_0 \le dM_2$, and (b) attainable if $M_0 > dM_2$. In the latter case, one could ensure significantly reduced, and ideally zero, acceptor population at the target time. The extent to which this is achievable is dependent upon the particular system.

The quantum conditions obtained above are completely general, applying to both systems that are controlled, as well as to uncontrolled system evolution. The result establishes an inequality between the dimension of the subspace M_2 which we do not desire to populate, the initial subspace M_0 from which dynamics evolves, and the dimensionality of the Kraus representation. Indeed, it is the involvement of the dimension $d > 1$ of the Kraus representation that implies that this type of control is more difficult in open systems than in closed systems.

6.3.2 Environment Can Enhance Control

6.3.2.1 Environmentally Assisted Transport

The above scenario assumes a circumstance where several subspaces are interconnected dynamically, and the goal is to suppress population flow to a specific subspace by utilizing interfering pathways. An opposite case, where interfering pathways exist in the uncontrolled dynamics so as to prevent the flow of population to a subspace, have been the subject of considerable attention in the framework of the transport of electronic energy in light-harvesting complexes [218–220]. Models for these processes assume the form of excitation on a collection of sites that are interconnected by coupling terms. Specifically, the Hamiltonian for a collection of N sites is of the form

$$H = \sum_{j=1}^{N} \hbar\omega_j \sigma_j^+ \sigma_j^- + \sum_{j \neq k} \hbar v_{j,k} \left(\sigma_j^- \sigma_k^+ + \sigma_j^+ \sigma_k^- \right) , \tag{6.31}$$

with $\sigma_j^+ = |j\rangle\langle 0|$ and $\sigma_j^- = |0\rangle\langle j|$ being transfer operators and where $|j\rangle$ indicates that site j is singly excited. Here $|0\rangle$ denotes a state where no site is excited. The $\hbar\omega_j$ are the site energies and $v_{j,k}$ are the so-called coherent tunneling amplitudes that modulate energy flow between sites j and k. Note that the Hamiltonian assumes a maximum of single excitations on each site.

This collection of N sites is assumed to interact with a complex environment that results in dissipative processes (i.e., loss of energy) and dephasing processes, and is coupled to a "sink" that is numbered site $N+1$, where population is lost. The master equation is then of the Lindblad form:

$$\frac{\partial \rho_s(t)}{\partial t} = -i\hbar^{-1}[H, \rho_s(t)] + L_{deph} + L_{diss} + L_{sink} , \tag{6.32}$$

where each of the L_i are of the Lindblad type (see Eq. (6.10))

$$L_{diss} = \sum_{j=1}^{N} \Gamma_j \left[-\left\{ \sigma_j^+ \sigma_j^-, \rho_s \right\} + 2\sigma_j^- \rho_s \sigma_j^+ \right] , \tag{6.33}$$

$$L_{deph} = \sum_{j=1}^{N} \gamma_j \left[-\left\{ \sigma_j^+ \sigma_j^-, \rho_s \right\} + 2\sigma_j^+ \sigma_j^- \rho_s \sigma_j^+ \sigma_j^- \right] , \tag{6.34}$$

$$L_{sink} = \Gamma_{N+1} \left[-\left\{ \sigma_k^+ \sigma_{N+1}^- \sigma_{N+1}^+ \sigma_k^-, \rho_s \right\} + 2\sigma_{N+1}^+ \sigma_k^- \rho_s \sigma_k^+ \sigma_{N+1}^- \right] , \tag{6.35}$$

where $\{A, B\}$ is the anticommutator $AB + BA$. Qualitatively, the selected structure of the Lindblad operators on the right hand side of the equation can be understood by examining their effect on ρ_s. For example, the term $\sigma_j^- \rho_s \sigma_j^+$ projects the j component of ρ_s onto the state $|0\rangle\langle 0|$, that is, a dissipative effect on ρ_s. Similarly, the term $\sigma_j^+ \sigma_j^- \rho \sigma_j^+ \sigma_j^-$ in L_{deph} does not change the population of state j, but its effect on $\partial\rho_s/\partial t$ can be found to be a loss of phase. Finally, L_{sink} contains terms like $2\sigma^{+}_{N+1}\sigma^{-}_{k}\rho\sigma^{+}_{k}\sigma^{-}_{N+1}$, which eliminates population in site k and moves it to the sink at site $N+1$.

Computations on this system show that dynamics are enhanced by interaction with the environment. This is readily seen by examining some simple cases, where the goal is to drive the population to the sink. Consider the case where $\omega_j = \gamma_j = \Gamma_j = 0$ for all $j = 1, \ldots, N$. Then the only relaxation process in the system is loss of the population to the sink. Results for the population $p_{sink}(t)$ for this case, with $N = 10$ and excitation initially on site 1, is shown as the lowest curve in Figure 6.7. The value of $p_{sink}(t)$ at long times is seen to be ≈ 0.1, far smaller than a classical random walk model where $p_{sink}(t \to \infty) = 1$. The restricted dynamical behavior is attributed, in an exactly soluble model [218], to destructively interfering quantum mechanical pathways that suppress transfer to the $N+1$ site/sink.

Consider then the effect of including dephasing at the site of the initial excitation, that is, $\gamma_1 \neq 0, \gamma_i = 0$ for $i \neq 1$. The results are shown as the upper curve in Figure 6.7, where rapid population transfer to the sink, with an asymptotic value of $p_{sink}(t \to \infty) = 1$ is clearly seen. That is, the introduction of dephasing has destroyed the interference between pathways that suppressed dynamics to the sink, allowing the sink to be accessed. Indeed, the resultant dynamics to the sink occurs far faster than would be the case if energy differences between sites was introduced, as shown in the middle curve in Figure 6.7. Note, however, that increasing the decoherence further tends to then reduce transport, so that there is an optimal degree of decoherence that aids transport.

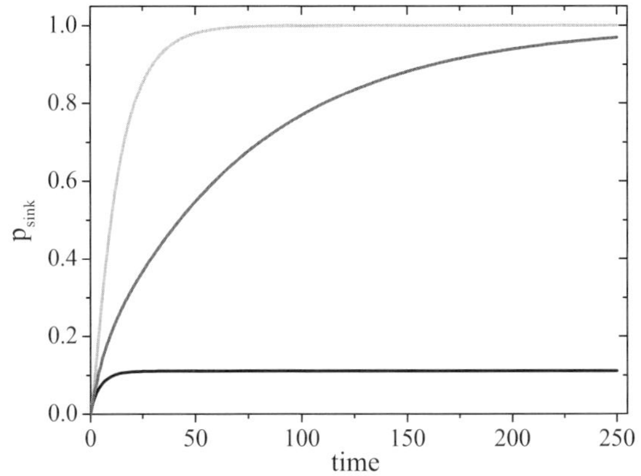

Figure 6.7 p_{sink} vs. time is shown for a fully connected graph of $N = 10$ nodes with $\Gamma_i = 0$ for $i = 1, \ldots, N$, $J = 1$, and $\Gamma_{N+1} = 1$. At $t = 0$ one excitation is in the site 1 and the energy transfer evolves. The cases of an energy mismatch (middle line), that is, $\omega_1 = 1, \omega_i = 0$ for $i \neq 1$, a dephasing mismatch (top line), that is, $\gamma_1 = 1, \gamma_i = 0$ for $i \neq 1$, and the basic case (lowest line) $\gamma_i = \omega_i = 0$ are shown. For the latter, the population in the sink asymptotically reaches $p_{sink} = 1/9$. Both of the upper curves reach unit transfer asymptotically, but at considerably different rates. From [218, Figure 1]. (Reprinted with permission. Copyright 2009 American Institute of Physics.)

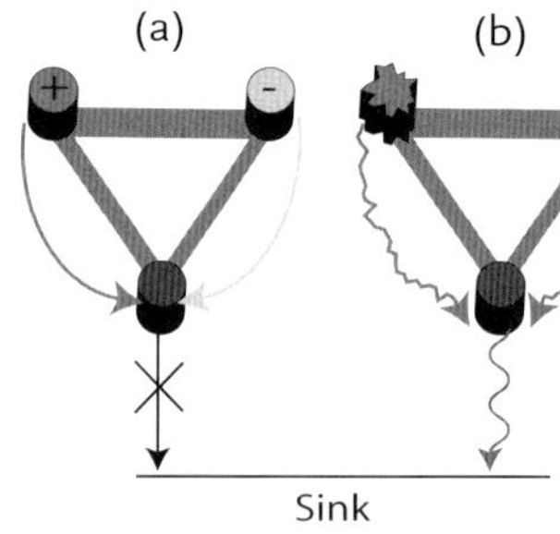

Figure 6.8 A fully connected three-site network. In (a) the excitation wave function is a delocalized superposition over the two top sites with equal probability of being found at either site. However, as the wave function is antisymmetric with respect to the interchange on these two sites, this state has no overlap with the dissipative site at the bottom due to the destructive interference of the tunneling amplitudes from each site in the superposition. The network can therefore store an excitation in this state indefinitely. In (b) pure dephasing causes the loss of this phase coherence and the two amplitudes no longer cancel, leading to total transfer of the excitation to the sink. From [218, Figure 2]. (Reprinted with permission. Copyright 2009 American Institute of Physics.)

This mechanism, in which decoherence aids transport, has been proposed as being relevant to the observation of long coherence times for electronic energy transfer in light harvesting systems [218–220]. A cartoon depiction of this environmentally assisted transport, where the suppression of transport is due to symmetry effects, is shown in Figure 6.8. An analogue of this in a Mach–Zender interferometer is described in [218]. Similar interference based "trapping" behavior is known to occur in optical [221] and condensed matter systems [222, 223].

The introduction of random behavior through coupling to the environment can also result from the use of partially coherent light, as discussed in Section 6.5. In that section, partial coherence is shown to result in loss of control. However, as in the case above, in some specific instances the introduction of partially coherent light can improve control. Such is the case, for example, in multiladder systems [224–226]. Another example of environmentally assisted control is discussed in the next section.

6.3.3 Environmentally Assisted One-Photon Phase Control

The discussion in Section 3.1 included the result that control over product ratios via weak-field excitation from a single initial state, in isolated molecules, is not possible. Here we reformulate this result for finite time or asymptotic time dynamics in both open and isolated systems. For an isolated system at finite time, the time dependence of the preparation of the state, discussed in Section 2.2, would lead us to expect that the state at finite time would depend upon the laser phase. This result is formalized below. In addition, the extension to open systems shows that coupling to an environment, in this case, might well aid the possibility of control via the laser phase.

For simplicity we focus on one-photon induced transitions between two electronic states, and consider measurements of the excited state dynamics. Consider a quantum system with Hamiltonian H_0, eigenenergies $\{E_n\}$ and eigenstates $\{|E_n\rangle\}$, where the labels $\boldsymbol{n}$ denote all quantum numbers needed to specify the state. The system is irradiated with a weak laser pulse with a frequency spectrum given by Eq. (1.9):

$$\epsilon(\omega) = |\epsilon(\omega)|e^{i\phi(\omega)}\,, \tag{6.36}$$

where $\phi(\omega)$ is the spectral phase, a quantity central to the present study. With the system initially in a pure state $|E_i\rangle$ of the ground electronic state, the excited wave function following one-photon absorption is given in first-order perturbation theory by

$$|\Psi\rangle = \sum_n d_{ni}\epsilon(\omega_{ni})|E_n\rangle\,, \tag{6.37}$$

where $d_{nm} = \langle E_n|\boldsymbol{d}\cdot\hat{\varepsilon}|E_m\rangle$ are the dipole transition matrix elements, and $\omega_{nm} = (E_n - E_m)\hbar$. The expectation value $\langle O(t)\rangle$, corresponding to an arbitrary measurement of a system property O at some later time t on the excited electronic state, is given by:

$$\begin{aligned}\langle O(t)\rangle &= \sum_n O_{nn}\,|d_{ni}|^2\,|\epsilon(\omega_{ni})|^2 \\ &\quad + \sum_{n\neq m} O_{nm}d_{mi}d^*_{ni}\epsilon\,(\omega_{mi})\,\epsilon^*\,(\omega_{ni})\,e^{i\omega_{nm}t}\,,\end{aligned} \tag{6.38}$$

where $O_{nm} = \langle E_n|O|E_m\rangle$. The dependence of $\langle O(t)\rangle$ on the laser phase $\phi(\omega)$ can be made more explicit by writing Eq. (6.38) as

$$\begin{aligned}\langle O(t)\rangle &= \sum_n O_{nn}\,|d_{ni}|^2\,|\epsilon(\omega_{ni})|^2 \\ &\quad + \sum_{n\neq m} |O_{nm}d_{mi}d_{ni}\epsilon\,(\omega_{mi})\,\epsilon\,(\omega_{ni})| \\ &\quad \times \cos\left[\omega_{nm}t + \theta_{nm} + \phi\,(\omega_{mi}) - \phi\,(\omega_{ni})\right]\,,\end{aligned} \tag{6.39}$$

where the hermiticity of O was used, θ_{nm} is the phase of the matrix element $(O_{nm} = |O_{nm}|e^{i\theta_{nm}})$, and d_{nm} is assumed real. In the case where the initial state is described by a density matrix $\rho(0) = \sum_i w_i|E_i\rangle\langle E_i|$, Eq. (6.39) is replaced by a w_i-weighted sum over the right hand side of the equation. In the case of a continuous spectrum, all sums become integrals.

Equation (6.39) displays the effect of the laser phase $\phi(\omega)$ on observables after one-photon induced state preparation. The first sum in Eq. (6.39) represents the incoherent contribution to the observable arising from level populations, and is both time-independent and $\phi(\omega)$-independent. By contrast, the second sum, which is the coherent interference term, is both time-dependent and $\phi(\omega)$-dependent. Interest is in the control achievable through changes in laser phase since varying in laser intensity, which affects the $|\epsilon(\omega)|$, constitutes passive control associated with changing the laser power spectrum. By contrast, the $\phi(\omega)$ terms explicitly alter the interference contribution.

Given Eq. (6.39), two distinct classes of measurements can be identified:

1. *Class A* – Measurements of properties O where $[H_0, O] = 0$, that is, operators O that are constants of the motion under laser-free system evolution. In this case $O_{nm} = 0$ so that $\langle O(t) \rangle$ then only involves the first sum in Eq. (6.39) and does not depend on $\phi(\omega)$.
2. *Class B* – Measurements whose corresponding operators O do not commute with H_0: $\langle O(t) \rangle$ then involves the second sum in Eq. (6.39) and depends on $\phi(\omega)$.

Hence, Class A observables can not be coherently phase controlled via a one-photon transition, while one-photon coherent phase control of Class B observables is possible. Since Class A measurements rigorously lead to time-independent results, any observables exhibiting time dependence must fall into Class B and can therefore be sensitive to $\phi(\omega)$. Below, we examine several examples in each class.

6.3.4
Isolated Molecules

Consider first the case of isolated molecules. A relevant example is that treated in Section 3.1. There, weak-field photofragmentation of an isolated molecule was considered, and the observable was the ratio of the photodissociation products. Such product states are eigenstates of H_s [227]. Hence, the operator

$$P_n(E) = \sum_{n \in \alpha} |E, \boldsymbol{n}^-\rangle\langle E, \boldsymbol{n}^-| \,, \tag{6.40}$$

corresponding to measurements of product populations in a particular channel α at fixed energy E, is diagonal in the energy basis. The ratio of populations of two different product channels $P_{\boldsymbol{n}}(E)/P_{\boldsymbol{m}}(E)$ examined in Eq. (3.5) is therefore of Class A, and is independent of $\phi(\omega)$ and laser intensity in the one-photon limit. One-photon control is therefore not possible.

Many additional Class A measurements clearly exist. For example, they also include all time-averaged quantities $\overline{O}$, with

$$\overline{O} = \lim_{T\to\infty} \left(\frac{1}{T}\right) \int_0^T dt \langle \hat{O}(t) \rangle = \sum_n O_{nn} \left|d_{nn_0}\right|^2 \left|\epsilon(\omega_{nn_0})\right|^2 \,. \tag{6.41}$$

Consider, by contrast, Class B cases for isolated molecules, that is, all measurements whose corresponding operators are time dependent under laser-free evolution. A particular example amongst the large class of such measurements is that of spatially distinct measurements, such as that of cis vs. trans molecular isomers. Since the associated measurement operator does not commute with H_s, it constitutes a Class B measurement, and coherent control is indeed possible in this case.

6.3.5 Nonisolated Systems

As a more general type of Class B measurements, consider the case where a system, as part of the system + environment, is measured. The full Hamiltonian is given by

$$H = H_{tot} + H_{MR} , \tag{6.42}$$

where H_{tot} is given in Eq. (6.1),

$$H_{tot} = H_s + H_e + H_{se} , \tag{6.43}$$

and H_{MR} is the laser interaction (Eq. (2.10)) driving the one-photon transition. The system space is defined to be that on which the measurement operators O act. For O to be a Class B measurement requires that $[O, H_{tot}] \neq 0$. This condition can be achieved in two ways: either $[O, H_s] \neq 0$, or if $[O, H_s] = 0$, then $[O, H_{se}] \neq 0$ and the coupling H_{se} is dissipative (i.e., the subsystem can exchange energy with the environment, so that $[H_{se}, H_s] \neq 0$). Note that virtually any O that commutes with H_s will not commute with H_{se}.

The generality of this statement implies that in an open system, virtually every subsystem property can, in principle, be coherently phase controlled via a one-photon transition. Note, in particular that in the case where $[O, H_s] = 0$, such control would be *environmentally assisted*. Whether such control is significant in practice depends upon the nature of the $O_{n,m}$ matrix elements in Eq. (6.38), where $|\boldsymbol{n}\rangle$ and $|\boldsymbol{m}\rangle$ are eigenstates of the full (subsystem + environment) Hamiltonian.

Demonstrating one-photon phase control appears to require a careful treatment of both the laser preparation and the dynamics of the total system. For example, the efforts of Spanner to computationally demonstrate one-photon phase control on a model open system using both secular and nonsecular Redfield theory [228], as well as a detailed study of model bacteriorhodopsin isomerization using a TDSCF approach [229], failed to show one-photon phase control. By contrast, experimental results observing one-photon phase control in dyes and in retinal isomerization have been reported [230, 231], as have environmental effects on control in solvents [232]. Katz *et al.* [233] have successfully shown such control in a computation that models one photon absorption modeling full subsystem–environment dynamics. Specifically, in the latter case they showed that the phase of a weak-field chirped pulse can significantly control the branching ratio in a dynamical process provided the environment can stabilize the outcome on a short time scale.

The model problem examined in [233] consists of a diatomic with three electronic states, the ground state (labeled by subscript g), the first excited state accessible by electromagnetic radiation from the ground state, (a "bright state", labeled with subscript b), and a state that is inaccessible by excitation with light from the ground state (a "dark state", labeled with subscript d). The system is immersed in a bath comprising a collection of two-level systems, and coupled by a system–bath interaction that induces vibrational relaxation of the diatomic. The particular focus is on control of the relative population N_d/N_b of the dark vs. bright states of the diatomic by varying the phase characteristics of weak incident radiation that excites the diatomic from state g to state b. The arrangement of potentials are shown in Figure 6.9, along with an inset showing that the computation is being carried out in the linear regime.

Here, the frequency dependence of the incident electromagnetic radiation was taken to be

$$\epsilon(\omega) = A \exp\left[\frac{-(\omega-\omega_0)^2}{2\Gamma^2} + \frac{i\chi(\omega-\omega_0)^2}{2}\right], \tag{6.44}$$

where A is the peak amplitude, ω_0 is the pulse frequency center, Γ is the frequency bandwidth, and χ is the so-called chirp. Note that the field intensity $|\epsilon(\omega)|^2$ is independent of χ, so that by varying χ one changes phase characteristics of the pulse without changing the intensity profile of the pulse.

Results of the studies for various values of the strength λ of the system–bath coupling are shown in Figure 6.10. Clearly, control with varying chirp χ is evident, displaying, as shown in the inset figure, a nonmonotonic dependence of control on λ. The mechanism of control is as follows [233]: the population transfer between the bright and dark states is stabilized by interaction with the environment to diminish the flow of population between the bright and dark states. The transferred wave

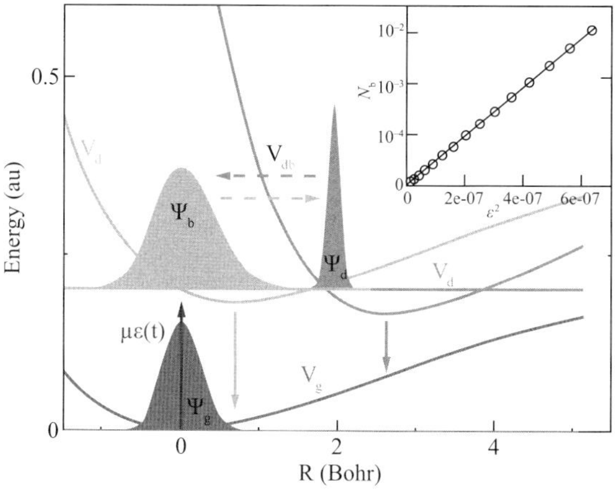

Figure 6.9 The general scheme: the ground state $V_g(r)$, the bright excited state $V_b(r)$ and the dark excited state $V_d(r)$. Superimposed are snapshots of three wave functions: the initial ground state density, the bright state after the pulse is over, and the dark state density after some dynamics. The ratio between long-time emission of the bright and dark states is an experimental indication of the population ratio between these states. From [233, Figure 1].

function is reflected from the outer turning point of the dark state and collides again with the crossing region. During this period dissipation takes place. As a result the system energy approaching the crossing point is reduced. For conditions where the initial energy is slightly below the crossing point, the nonadiabatic crossing is in the tunneling regime and a decrease in energy due to interaction with the environment will exponentially reduce the back transfer to the bright state. Further relaxation will stabilize the product. The chirp control mechanism is understood to behave as a timing control. The negative chirp enhances the population transfer in the first nonadiabatic transfer event, increasing the probability of producing the stabilized dark state. An observed dependence of the control on the strength of the system–bath coupling is also evident. Too strong a coupling will quench the system in a relaxed state, independent of the phases of the incident laser, whereas too weak a coupling will not quench the system sufficiently. The result is a nonmonotonic dependence of the phase control on the coupling strength λ.

6.4 Decoherence and Control

Below we consider a number of approaches to combating decoherence in solutions and other media where collisions are present. Other alternative approaches, which we do not address, include, for example, designing "decoherence-free subspaces" [188, 235]. In these approaches one deals with the explicit *design* of the system + environment, where a particular subspace is free from decoherence effects. These approaches are, for example, of particular interest to the development of subspaces in which to carry out quantum computation, an approach in which the

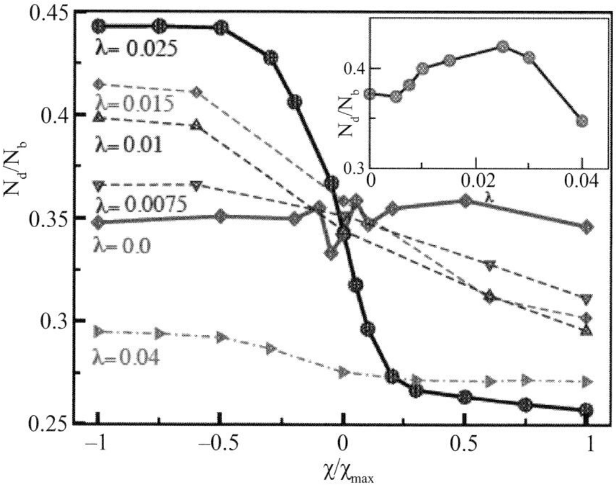

Figure 6.10 Branching ratio at 7 ps between the bright and dark states as a function of the chirp rate (where $\chi_{max} = 0.0184\ \text{fs}^{-2}$). The insert shows results for $\chi/\chi_{max} = -1.0$ for several values of the system–bath coupling parameter λ. From [234, Figure 3].

computational machinery follows the laws of quantum mechanics [188]. Similarly, we do not address efforts to utilize external fields to modify the environment or the system–environment coupling (see, for example, [236, 237]). By contrast, we deal below with the need to curb decoherence effects in natural preexistent systems. This area of research is in its infancy.

6.4.1 The Optical Bloch Equation

Given the significance of collisional effects in solution, we consider the simplest of master equations for relaxation and concomitant decoherence. That is, the equation of motion of $\rho_s(t)$ in the energy representation is given the form:

$$\frac{\partial \rho_{i,j}(t)}{\partial t} = -i\hbar^{-1}\left[H_s, \rho_s(t)\right]_{i,j} - \frac{1}{T_{i,j}}\rho_{i,j}(t) , \tag{6.45}$$

with $T_{i,i} = T_1$ and $T_{i,j} = T_2$ for $i \neq j$. Here T_1 and T_2 are phenomenological relaxation times in the energy representation. In this model, where (in Eq. (6.9)) $F(\rho)_{i,j} = -\rho_{i,j}(t)/T_{i,j}$, the coherent terms $\rho_{i,j}(t), i \neq j$ decay with a rate $1/T_2$ and populations $\rho_{i,i}(t)$ decay with rate $1/T_1$. If the system is in the presence of a radiation field, then H_s in Eq. (6.45) is augmented by the dipole–electric field interaction H_{MR} (Eq. 2.10). The result is the so-called optical Bloch equations.

The simplest optical Bloch equations result from a system comprised of two eigenstates $|E_1\rangle$, $|E_2\rangle$ of the molecule Hamiltonian H_M that experience the electric field–dipole interaction

$$H_{MR} = -\mathcal{E}\hat{\boldsymbol{\epsilon}} \cdot \boldsymbol{d}\cos(\omega t + \phi) . \tag{6.46}$$

The Hamiltonian H_s in Eq. (6.45) is $H_s = H_M + H_{MR}$. Equation (6.45) then becomes (where we suppress the t dependence of $\rho(t)$)

$$\frac{\partial \rho_{i,j}}{\partial t} = -i\hbar^{-1}\sum_k \left[H_{i,k}\rho_{k,j} - \rho_{i,k}H_{k,j}\right] - \frac{1}{T_{i,j}}\rho_{i,j} , \tag{6.47}$$

where

$$H_{i,k} = E_i\delta_{i,k} - \mathcal{E}\cos(\omega t + \phi)d_{i,k}(1 - \delta_{i,k}) , \tag{6.48}$$

and $d_{i,k} = \langle E_i|\hat{\boldsymbol{\epsilon}} \cdot \boldsymbol{d}|E_k\rangle$. Noting that $\rho_{i,j} = \rho^*_{j,i}$, we define

$$\begin{aligned} R_1 &= 2\mathrm{Im}(\rho_{1,2}) = \mathrm{Im}(\rho_{1,2} - \rho_{2,1}) , \\ R_2 &= 2\mathrm{Re}(\rho_{1,2}) = \mathrm{Re}(\rho_{1,2} + \rho_{2,1}) , \\ R_3 &= \rho_{11} - \rho_{22} . \end{aligned} \tag{6.49}$$

Then, with $H_{1,2} = H_{2,1}$ for bound states subjected to Eq. (6.46), Eq. (6.47) becomes

$$\frac{dR_1}{dt} = \Delta R_2 - \frac{1}{T_2} R_1 + 2H_{1,2} R_3/\hbar \ ,$$
$$\frac{dR_2}{dt} = -\Delta R_1 - \frac{1}{T_2} R_2 \ ,$$
$$\frac{dR_3}{dt} = -\frac{1}{T_1} R_3 - 2H_{1,2} R_1/\hbar \ . \tag{6.50}$$

Here $\Delta \equiv (E_2 - E_1)/\hbar - \omega \equiv \omega_{2,1} - \omega$, that is, the detuning of ω from the $|E_1\rangle$ to $|E_2\rangle$ transition. Equation (6.50) constitutes the standard form of the two-level optical Bloch equation.

It is worth emphasizing that the Bloch equation structure identifies two different timescales characterized by T_1 and T_2. Time evolution of the (system + environment) is manifested in both timescales. However, within the T_2 time, dynamics is still describable as being due to system coherence. After that time, as the system relaxes over the T_1 time, the system no longer evolves coherently. Rather, one sees time evolution of the *populations* of the system states, which is a projection of the overall coherent dynamics of the total (system + environment).

6.4.1.1 Decoherence Effects in One-Photon vs. Three-Photon Absorption

As an example of the effect of decoherence on coherent control, consider the case of one-photon plus three-photon control introduced in Section 3.4.2 . There, consideration was given to using two electric fields of frequency ω_1 and $\omega_3 = 3\omega_1$ and amplitude ϵ_1 and ϵ_3 to photodissociate a molecule. Interference effects between the two pathways to the photodissociation continua resulted in coherent control by varying the amplitude and relative phase of the two fields.

Here we consider [238] the simpler problem of controlling the population of level $|2\rangle$ in a two-level system by one- vs. three-photon excitation. Of particular interest is the effect of decoherence on population control. Of note is that related experiments on one versus three photon control in condensed phases have been carried out[52] on large molecules.

Consider a two-level bound system interacting with a CW electromagnetic field and assume that the energy levels undergo random Stark shifts without a change of state during collisions with an external bath, for example, elastic collisions between atoms in a gas. The CW field is treated classically, and the ground and the excited energy eigenstates states, of energy E_1 and E_2 are denoted $|1\rangle$ and $|2\rangle$, respectively.

In the one vs. three control scenario, the interaction between the system and electric field is given by Eq. (3.45):

$$H_{MR}(t) = -2\mathbf{d} \cdot \mathrm{Re}\left[\hat{\varepsilon}_3 \bar{\epsilon}_3 \exp^{-i\omega_3 t} + \hat{\varepsilon}_1 \bar{\epsilon}_1 \exp^{-i\omega_1 t}\right]$$
$$\equiv -2\hat{\varepsilon} \cdot \mathbf{d}\mathrm{Re}\left[\epsilon_f e^{-i\omega_f t} e^{i\phi_f} + \epsilon_h e^{i\omega_h t} e^{-i\phi_h}\right] \ . \tag{6.51}$$

Here $\epsilon_f = \epsilon_f e^{i\phi_f}, \epsilon_h = \epsilon_h e^{i\phi_h}$, and we have assumed that $\hat{\varepsilon}_f = \hat{\varepsilon}_h = \hat{\varepsilon}$. To avoid confusion with subscripts associated with the numerical labeling of density matrix elements, here (unlike Section 3.4.2) we use subscripts f, h to denote the fundamental and its third harmonic. The field frequencies ω_f and $\omega_h = 3\omega_f$ are chosen so that the third harmonic and the three fundamental photons are on resonance with the transition from state $|1\rangle$ to state $|2\rangle$ and the z dependence of the fields is neglected. In the standard scenario (Section 3.4.2) control is obtained by changing the relative phase and amplitudes of two fields, which results in the alteration of the degree of interference between the two pathways to the excited state.

To model the decoherence effects, we utilize the Bloch equations (Eq. (6.45)) and set $1/T_1 = 0$. Within the rotating wave approximation that is, neglecting highly oscillatory contributions – (see Eq. (2.13)), the slowly varying density matrix elements of states $|1\rangle$ and $|2\rangle$, $\sigma_{ii} = \rho_{ii}$, $(i = 1, 2)$, and $\sigma_{21} = \rho_{21} e^{3i(\omega_f t + \phi_f)}$ obey the following set of equations:

$$\frac{\partial \sigma_{11}}{\partial t} = -\text{Im}\left[\left(\frac{\mu_{1,2}^{(3)}\epsilon_f^3}{\hbar} + \frac{d_{1,2}\epsilon_h e^{i\phi}}{\hbar}\right)\sigma_{21}\right] ,$$

$$\frac{\partial \sigma_{22}}{\partial t} = \text{Im}\left[\left(\frac{\mu_{1,2}^{(3)}\epsilon_f^3}{\hbar} + \frac{d_{1,2}\epsilon_h e^{i\phi}}{\hbar}\right)\sigma_{21}\right] ,$$

$$\frac{\partial \sigma_{21}}{\partial t} = -\gamma_p \sigma_{21} + \frac{i}{2}\left(\frac{\mu_{1,2}^{(3)}\epsilon_f^3}{\hbar} + \frac{d_{1,2}\epsilon_h e^{-i\phi}}{\hbar}\right)(\sigma_{11} - \sigma_{22}) , \tag{6.52}$$

with

$$\mu_{12}^{(3)} \equiv \frac{1}{(2\hbar)^2} \sum_{n,m} \frac{d_{1n} d_{nm} d_{m2}}{(\omega_{n1} - \omega_f)(\omega_f - \omega_{2m})} . \tag{6.53}$$

Here γ_p is the dephasing rate, ω_{nm} is the frequency difference between levels $|n\rangle$ and $|m\rangle$ and $d_{nm} \equiv \langle n|\hat{\varepsilon} \cdot \boldsymbol{d}|m\rangle$. The $\mu_{1,2}^{(3)}$ denotes the effective three-photon matrix element for the fundamental field for the $|1\rangle \rightarrow |2\rangle$ transition. The controllable relative phase is $\phi = \phi_h - 3\phi_f$.

We define the one- and three-photon Rabi frequencies by

$$\Omega_h = |\Omega_h| e^{i\theta_h} = \frac{d_{1,2}\epsilon_h}{\hbar}$$

and

$$\Omega_f = |\Omega_f| e^{i\theta_f} = \frac{\mu_{1,2}^{(3)}}{\hbar}\epsilon_f^3 .$$

Note that, although $d_{1,2}$ and $\mu_{1,2}^{(3)}$ are real for a bound system, all equations below allow complex matrix elements for the case of transitions to the continuum. Since ϵ_h and ϵ_f are real and positive, θ_h and θ_f are determined by $d_{1,2}$ and $\mu_{1,2}^{(3)}$:

$$e^{i\theta_h} = \frac{d_{1,2}}{|d_{1,2}|} , \tag{6.54}$$

$$e^{i\theta_f} = \frac{\mu_{1,2}^{(3)}}{|\mu_{1,2}^{(3)}|} . \tag{6.55}$$

To amalgamate these Rabi frequencies and the relative laser phase of ϕ, define the effective Rabi frequency Ω_{eff}:

$$\Omega_{eff} e^{i\theta} \equiv \Omega_h e^{i\phi} + \Omega_f = |\Omega_h e^{i\phi} + \Omega_f| e^{i\theta} , \tag{6.56}$$

where Ω_{eff} is real and positive. Here Ω_{eff} and θ are related to Ω_h and Ω_f as

$$\Omega_{eff} = \sqrt{|\Omega_h|^2 + |\Omega_f|^2 + 2|\Omega_h \Omega_f| \cos \Phi} , \tag{6.57}$$

$$\tan\theta = \frac{\sin(\phi + \theta_h) + \frac{|\Omega_f|}{|\Omega_h|} \sin\theta_f}{\cos(\phi + \theta_h) + \frac{|\Omega_f|}{|\Omega_h|} \cos\theta_f} , \tag{6.58}$$

where $\Phi = \phi + \theta_h - \theta_f$.

By rewriting Eq. (6.52) in terms of Ω_{eff} and θ and introducing $u = 2\mathrm{Re}(\sigma_{12} e^{-i\theta})$, $v = 2\mathrm{Im}(\sigma_{12} e^{-i\theta})$, $w = \sigma_{22} - \sigma_{11}$, the equations assume the form

$$\frac{du}{dt} = -\gamma_p u , \tag{6.59}$$

$$\frac{dv}{dt} = -\gamma_p v + \Omega_{eff} w , \tag{6.60}$$

$$\frac{dw}{dt} = -\Omega_{eff} v . \tag{6.61}$$

These are now of standard optical Bloch form (Eq. (6.50)) but with Ω_{eff} as the Rabi frequency, and here with the detuning $\Delta = 0$. The fact that one can rewrite these equations with a single such frequency supports a perspective that regards one vs. three control as the effect of varying the properties of a *single* incident field on the molecule, with amplitude and phase control caused by controlling the properties of that field.

Substituting Eq. (6.61) into Eq. (6.60) gives a simple equation for w:

$$\frac{d^2 w}{dt^2} + \gamma_p \frac{dw}{dt} + \Omega_{eff}^2 w = 0 . \tag{6.62}$$

In the case where initially the ground state is populated and the coherence is zero (i.e., $w(0) = -1$, $u(0) = v(0) = 0$), the resultant excited state population $\rho_{22} = \sigma_{22}$ is given by

$$\rho_{22}(t) = -\frac{e^{-\frac{\gamma_p t}{2}}}{2} \left[\cos(st) + \frac{\gamma_p}{2s} \sin(st)\right] + \frac{1}{2} \quad \text{for} \quad \gamma_p < 2\Omega_{eff} , \tag{6.63}$$

$$\rho_{22}(t) = \frac{\left[-\lambda_2 e^{\lambda_1 t} + \lambda_1 e^{\lambda_2 t}\right]}{2(\lambda_2 - \lambda_1)} + \frac{1}{2} \quad \text{for} \quad \gamma_p > 2\Omega_{eff}, \tag{6.64}$$

$$\rho_{22}(t) = -\frac{e^{-\frac{\gamma_p t}{2}}}{2}\left(1 + \frac{\gamma_p t}{2}\right) + \frac{1}{2} \quad \text{for} \quad \gamma_p = 2\Omega_{eff}, \tag{6.65}$$

where $s = 1/2\sqrt{4\Omega_{eff}^2 - \gamma_p^2}$, and $\lambda_{1,2} = 1/2[-\gamma_p \pm \sqrt{\gamma_p^2 - 4\Omega_{eff}^2}]$. The general behavior of the solution is seen to be determined by the relative size of the dephasing time and the Rabi oscillation period. Analogous analytic results can be obtained for σ_{12} which decays with rate γ_p. If the external field is intense enough so that $\gamma_p < 2\Omega_{eff}$, then ρ_{22} shows oscillations that are exponentially damped with time. On the other hand, if dephasing dominates over the Rabi oscillation, so that $\gamma_p \geq 2\Omega_{eff}$ or ρ_{22} increases monotonically.

The behavior of the excited state population for several parameter values is sketched in Figure 6.11. For $\gamma_p < 2\Omega_{eff}$, the introduction of dephasing increases the period of the oscillation and causes the amplitudes to decay as $e^{-\gamma_p t/2}$. Although this is a CW laser field case, we can extract the result for the field being switched off at a specific time, that is, a square pulsed laser which is on from $t = 0$ to $t = t_f$, by examining the population at time t_f.

Typical behavior of ρ_{22} and of the one- vs. three-photon phase control profile (i.e., ρ_{22} as a function of generic phase control variable Φ) for several values of Rabi frequencies and γ_p are shown in Figure 6.12. Here we assume that the fields are abruptly turned off at the times indicated in the figure captions to produce a square pulse and the intensities are chosen so that $|\Omega_h| = |\Omega_f|$, to enhance the interference effects. The effective Rabi frequency is then $\Omega_{eff} = |\Omega_h|\sqrt{2(1 + \cos\Phi)}$. While $\Phi = 0$ leads to a complete constructive interference of the two transition amplitudes, $\Phi = \pi$ leads to a complete destructive interference, that is, no excitation from the ground to the excited state.

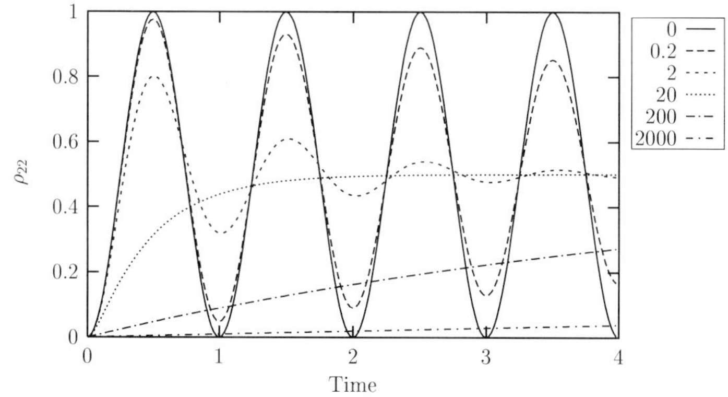

Figure 6.11 Excited state population as a function of time for the various dephasing rates γ_p shown inside the box, with $\Omega_{eff} = 2\pi$. The variables are in dimensionless units. Taken from [238, Figure 1]. (Reprinted with permission. Copyright 2005 Elsevier BV.)

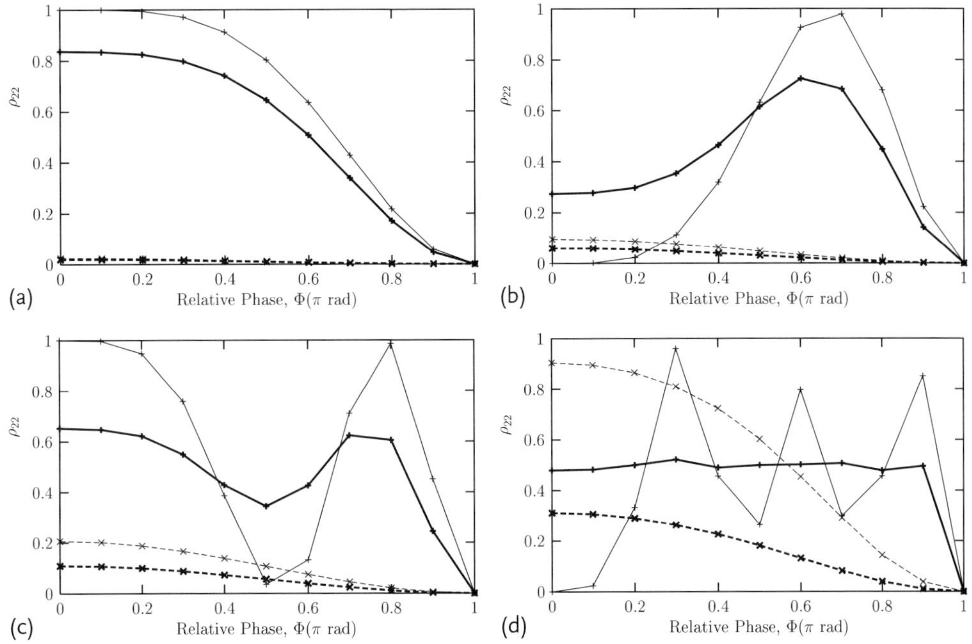

Figure 6.12 Excited state population versus relative phase for two different $|\Omega_h|$: Solid lines and dashed lines denote the case at $|\Omega_h| = 2\pi$ and $|\Omega_h| = \pi/5$, respectively. Thin lines and thick lines denote the case at $\gamma_p = 0$ and $\gamma_p = \pi$, respectively. Fields are turned off at (a) $t = 0.25$, (b) $t = 0.25 \times 2$, (c) $t = 0.25 \times 3$, and (d) $t = 0.25 \times 8$. All the variables are in dimensionless units. Data points are connected by straight lines as a guide. From [238, Figure 2]. (Reprinted with permission. Copyright 2005 Elsevier BV.)

The control profiles for small Ω_h and $\gamma_p = 0$ (thin dashed lines in Figure 6.12) are seen to be monotonically decreasing from the maximum excitation at $\Phi = 0$, to zero excitation at $\Phi = \pi$; that is, they follow a "cos Φ rule". By contrast, for strong intensity (thin solid lines in Figure 6.12) in which there are many Rabi cycles during the pulse, the control curve is not necessarily monotonic since the final excited populations are determined by the time at which the fields are turned off. Introducing dephasing is seen to lead to a decreased range of control whose magnitude depends on the relative strength of the dephasing and on the effective Rabi frequencies, according to Eqs. (6.63)–(6.65).

Figure 6.12 demonstrates that phase control profiles are strongly dependent on the pulse duration. For weak intensities, as the pulse duration increases, the degree of control improves and the control curve continues to approximately follow a cos Φ law. This behavior is seen both in the absence and in the presence of dephasing, although dephasing reduces the yield for a given pulse duration. In the strong-field case the control profile varies strongly with pulse duration. In particular, with $\gamma_p = 0$, if the pulse duration is smaller than the oscillation period ($= 1/2$) of the $\Omega_{eff}(\Phi = 0)$ case, then ρ_{22} decreases with increasing Φ, as shown in Figure 6.12a. For the pulse duration greater than that period, the control profile no longer follows cos Φ and the maximal yields start to appear at $\Phi \neq 0$, as shown in Figure 6.12b–d. In all strong intensity cases, the addition of dephasing results in a decay of ρ_{22} with a rate of $e^{-\gamma_p t/2}$ for a given Ω_{eff}. Thus the degree of the control worsens in the presence of dephasing as the pulse duration increases. Note that the introduction of dephasing leads to a degree of control C that converges to 0.5, where C is defined as the difference between the maximum and minimum excited state populations.

One feature that these results make clear is that stronger fields tend to counter decoherence effects, at least at shorter times. This competition between excitation and decoherence is used below to counter decoherence effects in a model scenario.

6.4.2
Countering Collisional Effects

Consider now a scenario that is capable of maintaining coherent control in the presence of collisions for systems that are described by the optical Bloch equations. In particular, we reconsider the bichromatic control scenario discussed in Section 3.2.1, however, for the case that the molecules are in solution at temperature T. Further, we irradiate the system so as to saturate the $|E_1\rangle$ to $|E_2\rangle$ transition and simultaneously photodissociate the system.

The initial state, prior to dissociation, is a mixed state described in the energy representation by a 2×2 density matrix with elements $\rho_{i,j}$, $(i, j = 1, 2)$. Photodissociation of this mixed state can be written as a generalization [239] of Eq. (3.12). In Eq. (3.12) we assumed an initial state of the form of Eq. (6.6) (with $k = 1, 2$), so that the corresponding density matrix would be Eq. (6.8). That is, Eq. (3.12) could be rewritten as

$$P_q(E) = \left(\frac{2\pi}{\hbar}\right)^2 \sum_{i,j=1}^{N} \left[\rho_{i,j}\bar{\epsilon}(\omega_{E,i})\bar{\epsilon}^*(\omega_{E,j})\right] d_q(j,i) \; . \qquad (6.66)$$

Equation (6.66) is, in fact, the correct generalization to the case where the initial state is mixed and is represented by $\rho_{i,j}$. Below we neglect the z dependence in $\bar{\epsilon}$, replacing it by ϵ (see Eq. (2.9)).

To utilize Eq. (6.66) we determine $\rho_{i,j}$ for the case where two levels $|E_1\rangle$ and $|E_2\rangle$ are continuously subjected to radiation and to collisions, using the optical Bloch approach. We note that if, as in the bichromatic control cases discussed in Chapter 3, $|E_1\rangle$ and $|E_2\rangle$ have the same parity, a one-photon absorption cannot couple these states. We must therefore consider saturating this transition using two-photon absorption through an off-resonant intermediate bound state $|E_0\rangle$ with dipole matrix elements $d_{0j} = \langle E_0|\boldsymbol{d} \cdot \hat{\boldsymbol{\epsilon}}|E_j\rangle$, and with $E_2 > E_0 > E_1$. For simplicity we assume that $2\omega = (E_2 - E_1)/\hbar$, so that the transition is two-photon resonant, and we sketch the extension [240] of Eq. (6.50) to two-photon absorption. To carry out this extension, we write the Bloch equations (Eq. (6.47)) for three levels $|E_0\rangle$, $|E_1\rangle$ and $|E_2\rangle$, and adopt the adiabatic approximation for off-resonant transitions to level

$|E_0\rangle$. This approximation is equivalent [240] to setting $d\rho_{2,0}/dt = d\rho_{10}/dt = 0$. Substituting the result into the remaining equations gives the set of equations for $d\rho_{i,j}/dt$, with $i, j = 1, 2$. This set can be rewritten, with the help of a modified version of Eq. (6.49), where $\rho_{1,2}$ is replaced by $\rho_{1,2}\exp(-2i\phi)$, as

$$\begin{aligned}\frac{dR_1}{dt} &= -\frac{D_{2,1}R_3}{2} - \frac{R_1}{T_2}\,,\\ \frac{dR_2}{dt} &= -\frac{R_2}{T_2}\,,\\ \frac{dR_3}{dt} &= \frac{D_{2,1}R_1}{2} - \frac{(R_3 - R_3^e)}{T_1}\,,\end{aligned} \tag{6.67}$$

where

$$D_{2,1} = \frac{\mathcal{E}^2 d_{2,0} d_{0,1}}{2\hbar^2(\omega - \omega_{01})}\,. \tag{6.68}$$

Here we have recognized that the quantity R_3 relaxes to the thermodynamic population difference R_3^e at temperature T, with:

$$R_3^e = \frac{1 - \exp\left(\frac{-\hbar\omega_{2,1}}{k_B T}\right)}{1 + \exp\left(\frac{-\hbar\omega_{2,1}}{k_B T}\right)}\,. \tag{6.69}$$

At long times ($t \gg T_2, T_1$) the system saturates, that is, $dR_3/dt = 0$ and Eq. (6.67) implies that $dR_1/dt = 0$. The equations for R_i can then be readily solved and, in conjunction with Eq. (6.49), gives $\rho_{1,2}$ at saturation:

$$\begin{aligned}\rho_{1,2} &= R_3^e T_2 D_{2,1} \frac{\exp\left[i\left(2\phi - \frac{\pi}{2}\right)\right]}{4 + D_{2,1}^2 T_1 T_2}\,,\\ \rho_{1,1} &= 0.5\left(1 + \frac{R_3^e}{1 + D_{2,1}^2 T_1 T_2/4}\right),\\ \rho_{2,2} &= 0.5\left(1 - \frac{R_3^e}{1 + D_{2,1}^2 T_1 T_2/4}\right).\end{aligned} \tag{6.70}$$

This ρ_{ij} then serves as the state to be pumped to dissociation. Consider excitation of this mixed state with a Gaussian pulse, within the rotating wave approximation. The pulse is of the form

$$\varepsilon(t) = \mathcal{E} e^{-i(\omega_L t + \delta)} e^{-\frac{(t-t_0)^2}{\tau^2}}\,, \tag{6.71}$$

with Fourier transform

$$\begin{aligned}\epsilon(\omega) &= \frac{\mathcal{E}\tau}{\sqrt{\pi}} e^{-\frac{\tau^2(\omega_L - \omega)^2}{4}} e^{-i(\omega_L - \omega)t_0} e^{-i\delta}\\ &\equiv \epsilon_\omega e^{-i(\omega_L - \omega)t_0} e^{-i\delta}\,.\end{aligned} \tag{6.72}$$

Note that this control arrangement differs from that in Section 3.2.1 insofar as the two frequencies $\omega_1 = (E - E_1)/\hbar$ and $\omega_2 = (E - E_2)/\hbar$ that dissociate the system are components of a single pulse. The pulse time is chosen such that $t_0 \gg T_1, T_2$.

Inserting the long-time $\rho_{i,j}$ into Eq. (6.66) gives (denoting $\epsilon(\omega_{\epsilon,i})$ by ϵ_i) the probability $P_q(E)$ of forming the product in channel q at energy E as

$$P_q(E) = \frac{4\pi^2}{\hbar^2}\left[P_{1,1}(E,q) + P_{2,2}(E,q) + P_{1,2}(E,q)\right], \tag{6.73}$$

where

$$P_{1,1}(E,q) = \rho_{1,1}\epsilon_1^2 d_q(1,1) = 0.5\left(1 + \frac{R_3^e}{1 + D_{2,1}^2 T_1 T_2/4}\right)\epsilon_1^2 d_q(1,1),$$

$$P_{2,2}(E,q) = \rho_{2,2}\epsilon_2^2 d_q(2,2) = 0.5\left(1 - \frac{R_3^e}{1 + D_{2,1}^2 T_1 T_2/4}\right)\epsilon_2^2 d_q(2,2)$$

and

$$\begin{aligned} P_{1,2}(E,q) &= 2|\rho_{1,2}||d_q(1,2)|\epsilon_1\epsilon_2 \cos\left(\alpha_q(1,2) + \omega_{2,1}t_0 + 2\phi - \frac{\pi}{2}\right) \\ &= \frac{0.5\, T_2 D_{2,1} R_3^e}{1 + D_{2,1}^2 T_1 T_2/4}\epsilon_1\epsilon_2|d_q(1,2)|\cos\left(\alpha_q(1,2) + \omega_{2,1}t_0 + 2\phi - \frac{\pi}{2}\right), \end{aligned} \tag{6.74}$$

where $\alpha_q(1,2)$ is the phase of $d_q(1,2)$ (see Eq. (3.20)). From these equations it is evident that coherent control can be achieved in solution by, for example, varying τ to alter the quantum interference term.

A number of simple qualitative observations are evident. First, $P_q(E)$ depends upon the parameters associated with the saturation through the combinations $D_{2,1}T_1$ and $D_{2,1}T_2$ (or their ratio and product T_1/T_2 and $D_{2,1}^2 T_1 T_2$ used below). Control vanishes if $P_{1,2}(E,q) = 0$, which occurs if either the temperature $T \to \infty$ (i.e., $R_3^e \to 0$) or $D_{2,1}T_2 \to 0$. Both these limits correspond to complete loss of coherence. Examination of Eq. (6.74) shows that this is not the case, however, for $D_{2,1}T_1 \to 0$, consistent with the fact that T_1 relates to population, rather than phase, relaxation. Physically [241] however, in collisional environment, $T_1 > T_2$ so that the limit $T_1 \to 0$ also implies loss of control. Note also that control vanishes under extremely large pumping rates, $D_{2,1} \to \infty$ for which $\rho_{1,1} = \rho_{2,2}$ and $\rho_{1,2} \to 0$.

Sample computational results are shown in Figures 6.13–6.16 for the case of the photodissociation of CH_3I into $CH_3 + I$ vs. $CH_3 + I^*$. In particular, we show control over the ratio $I^*/(I + I^*)$ for the collision-free case in Figure 6.13, and at three successively higher temperatures in the subsequent three figures. The abscissa is $S = \epsilon_1^2/(\epsilon_1^2 + \epsilon_2^2)$ and the ordinate is the angle $\chi_{1,2} = \omega_{2,1}t_0 + 2\phi - \pi/2$. The results clearly show control for the two lower temperatures, including $T = \hbar\omega_{2,1}/k$, and the total loss of control at the highest temperature, as manifested in the lack of dependence of the yield ratio on the phase angle $\chi_{1,2}$.

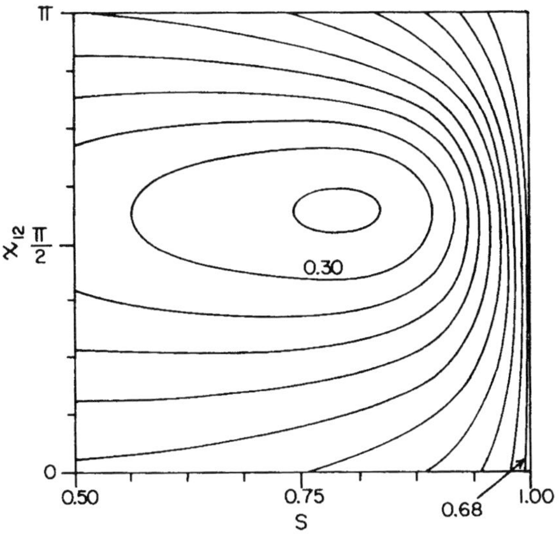

Figure 6.13 Contour plot of the I* yield [I*/(I + I*)] for two color photodissociation of a pure CH_3I superposition state composed of bound states with vibrational and rotational quantum numbers $(\nu, J) = (0, 2)$ and $(1, 2)$ excited with frequencies $\omega_1 = 41\,579\,\text{cm}^{-1}$ and $\omega_2 = 41\,163\,\text{cm}^{-1}$. Contours increase in increments of 0.04 from the "center well". Taken from [239, Figure 1]. (Reprinted with permission. Copyright 1989 American Institute of Physics.)

Thus we see that, although collisional effects do reduce the degree of control relative to the collision-free case, saturation pumping of superposition in the bichromatic control scenario can be used to overcome collisional effects up to some reasonable temperature.

6.4.3
Additional Control Studies

A number of additional theoretical studies, several using optimal or pulse control (see Section 5.1), have been carried out to study control in the presence of solvent and decoherence effects. These include work in the Wilson group on a two-level oscillator model coupled to a background bath [242] and on electronic population transfer in a molecule in solution [243]. This, and other related work [244–247], has generally concluded that some degree of control is indeed possible in solution, depending upon the extent of the coupling between system and solvent, and the degree to which one can manipulate the incident pulses. Some specific theoretical approaches have been proposed for condensed phase control, including the use of STIRAP based methods [244, 245, 248]. However, no quantitative rules have yet emerged on the extent to which control is possible.

In addition to model systems, there exists [249] one fully converged computation on control of intramolecular dynamics in the presence of decoherence. This semiclassical [250, 251] computation deals with controlled proton transfer between

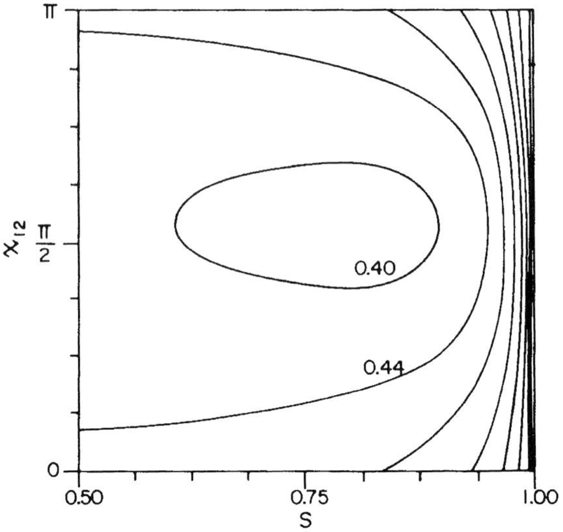

Figure 6.14 As in Figure 6.13, but at $k_B T = 0.2\hbar\omega_{2,1}$. Taken from [239, Figure 2]. (Reprinted with permission. Copyright 1989 American Institute of Physics.)

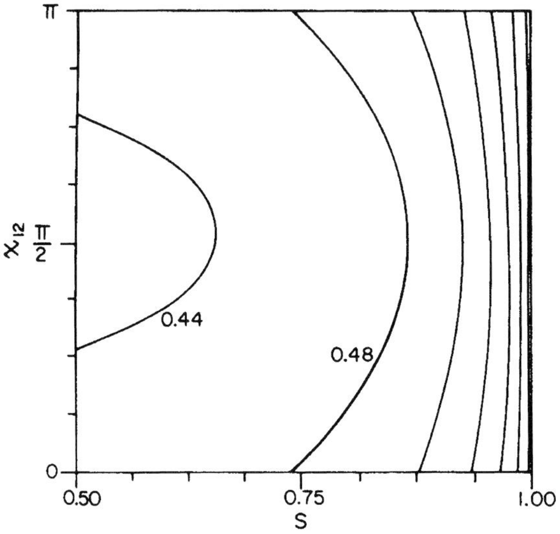

Figure 6.15 As in Figure 6.13, but at $k_B T = \hbar\omega_{2,1}$. Taken from [239, Figure 3]. (Reprinted with permission. Copyright 1989 American Institute of Physics.)

the keto and enol forms of 2-(2′-hydroxyphenyl)-oxazole (see Figure 6.17). Here the proton is "the system" and the remainder of the molecule, comprising 35 coupled degrees of freedom plus 16 out-of-plane vibrational modes, serves as "the environment". The results show that despite extensive dephasing, the proton transfer dynamics is easily controlled using the bichromatic control scenario. For exam-

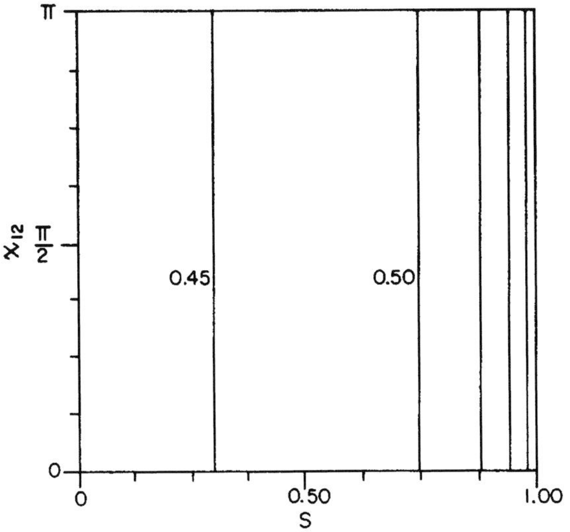

Figure 6.16 As in Figure 6.13, but at $k_B T = 1000\hbar\omega_{2,1}$. Contours are spaced by 0.05. Taken from [239, Figure 4]. (Reprinted with permission. Copyright 1989 American Institute of Physics.)

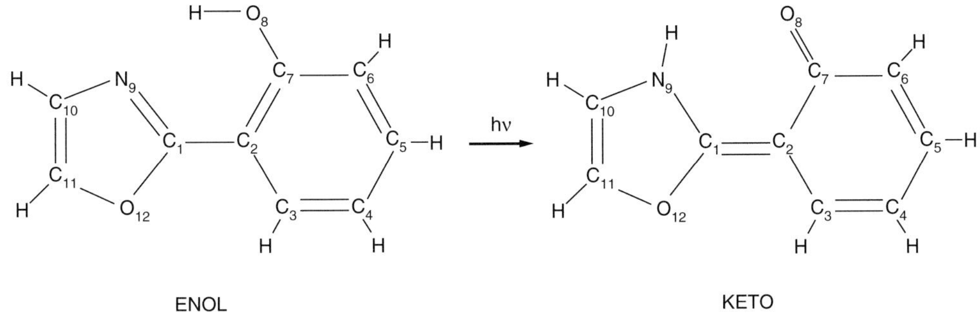

Figure 6.17 Keto and enol forms of 2-(2′-hydroxyphenyl)-oxazole. From [249, Figure 1]. (Reprinted with permission. Copyright 2002 American Physical Society.)

ple, consider the case where the initial superposition state involves the ground vibrational state of the oxazole-hydroxyphenyl in-the-plane bending mode, that is, bending motion of the $C_1C_2C_7$ angle, and the first excited state associated with a vibrational mode. Figure 6.18 shows a contour plot of the percentage yield of the reactant at 200 fs after excitation of the system. Here, the degree of yield control is maximum in the $0.2 < S < 0.8$ range, where the amount of the reactant can be reduced from more than 80% to less than 40% by changing the relative phase of the two incident lasers from the range of 120–180 to 0°.

The extent to which these results are significant is associated with the advent of intrinsic decoherence experienced by the proton during the course of the dynamics. That is, if there is little decoherence then the system would effectively be

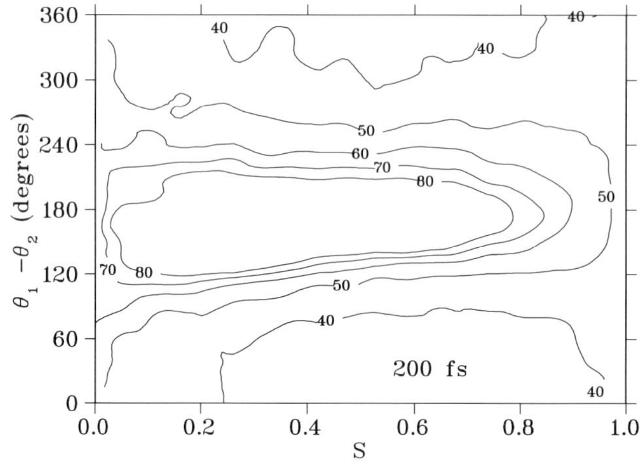

Figure 6.18 Contour plot of the reactant for bichromatic coherent control at 200 fs after photoexcitation of 2-(2'-hydroxyphenyl)-oxazole. Here $\theta_i \equiv \phi(\omega_i)$, where $\phi(\omega_i)$ is the phase of the electric field, of frequency ω_i, incident on the system (see Eq. (3.16)). From [249, Figure 3]. (Reprinted with permission. Copyright 2002 American Physical Society.)

a small molecule. To this end it is necessary to ascertain the extent of decoherence via, for example, the purity $\text{Tr}[\rho_s^2(t)]$, where $\rho_s(t)$ is the density matrix of the proton. A computation of the purity in this case [249] shows that as decoherence sets in, $\text{Tr}[\rho_s^2(0)]$ decays, reaching $\text{Tr}[\rho_s^2(t)] = 0.38$ by 200 fs. Hence decoherence is rapid and effective during the time scale of the control, shown above. Nonetheless, bichromatic control is possible. A similar result has been reported for control of cis-trans isomerization in model retinal [229].

In addition to the various control results reported in this chapter, we note some additional experimental studies of control of dynamics in solution [95, 252–254] indicate that control in the presence of collisional effects is indeed possible. For example, coherent control of the dynamics of I_3^- in ethanol and acetonitrile has been demonstrated. Specifically, I_3^- was excited with 30 fs UV laser pulse to the first excited state. The resultant wave function was comprised of a localized wave function on the ground electronic state and a corresponding depletion of wave function density, that is, a "hole", on the ground electronic state. In this instance the target of the control was the nature of the spectrum associated with the coherences associated with the symmetric stretch. By manipulating various attributes of the exciting pulse (intensity, frequency, and chirp of the excitation pulse) aspects of the spectrum were controlled, despite the decoherence associated with collision effects.

However, the vast majority of experiments done on control in condensed phases are done on femtosecond time scales in an effort to "beat out" decoherence effects [95]. Typical experiments are now done using the automatic feedback control approach introduced in Section 5.2 and described in further detail in Section 17.3. The result of such a procedure is a pulse shape that enhances the probability of the selected target. The success of many of these experiments has encouraged *conjec-*

tures on the nature of the coherent control in such systems. What is required, however, is a detailed analysis of the mechanism that underlies the laser pulse resulting from feedback control. Examples from such analyses are discussed in Section 17.4, where, often, the results show that computations are vital to the successful understanding of the control scenario.

6.4.4
State Stability against Decoherence

Several control scenarios, described in Chapter 3, rely upon the initial creation of system superposition states. Section 6.2.2 above clearly demonstrated that such superpositions are sensitive to decoherence induced by interaction with the environment. Clearly, general principles characterizing the rate of decoherence are important. For example, in the Caldeira–Leggett model (Eq. (6.25)), states with greater oscillatory structure have been shown to decohere faster [180]. This result, however, is specific to the Caldeira–Leggett model. One emerging approach, where the stability of states against decoherence is linked to overlapping resonances and to interference effects, is discussed below.

This approach is a major extension of the Q, P partitioning of Hilbert space described in Section 4.1.2. There, population flowed between the Q and P spaces (e.g., between the S_2 and S_1 electronic states in pyrazine) and the Hilbert space was partitioned into a *sum* of the Q and P spaces. By contrast, in the case of decoherence population does not flow between the system (to be chosen as the Q space) and the environment (to be chosen as the P space). Hence partitioning, in this case, is into a *product* of Hilbert spaces.

The interrelationship between overlapping resonances and the ability to design states that are resistant to decoherence, shown below, is general. It proves useful, however, to introduce the approach via the specific example of simple spin–boson problems. Recall that decoherence, as discussed in Section 6.1, is the loss of quantum information due to the neglect of the entanglement of a system with its environment. As such, decoherence can occur effectively even with a small bath, as in the case in the example below.

Consider then the control of decoherence in a spin one-half system interacting with a single-mode thermal oscillator environment. The most general system–environment Hamiltonian for this case is given by

$$\frac{H}{\hbar} = \omega a^\dagger a + \frac{\omega_0}{2} S_z + g_r(S_+ a + S_- a^\dagger) + g_{nr}(S_+ a^\dagger + S_- a) + g_{ph} S_z(a + a^\dagger) , \tag{6.75}$$

where a and $a^\dagger$ are the environment annihilation and creation operators, ω is the frequency of the bath mode, $\hbar\omega_0$ is the energy difference between the two spin-half states $|+\rangle$ and $|-\rangle$, and the spin operators are defined as $S_\pm = |\pm\rangle\langle\mp|$, $S_z = |+\rangle\langle+|-|-\rangle\langle-|$. Note the three system–environment coupling terms with coupling parameters g_r, g_{nr}, and g_{ph}. The g_{ph} term corresponds to decoherence of the spin due to the operator S_z, which induces a relative phase between the spin states $|+\rangle$

and $|-\rangle$, whereas g_r and g_{nr} mediate energy exchange between the spin and the bath. We focus below on the region of strong coupling, where g_{nr} and g_r are both greater than ω_0.

The class of Hamiltonians (Eq. (6.75)) possesses a plane defined by $g_r = g_{nr}$ in the three-dimensional g_{nr}, g_r, g_{ph} parameter space upon which the system displays a decoherence free subspace (DFS). A DFS is a collection of system states that do not interact with the environment. That is, in these special cases the interaction term in the total Hamiltonian factorizes into a product of spin $\mathcal{E}_I$ and bath operators B_I. Here $B_I = (a + a^\dagger)$ and the spin operator $\mathcal{E}_I$ can be written in the $(|+\rangle, |-\rangle)$ spin basis as:

$$\mathcal{E}_I = \begin{pmatrix} g_{ph} & g \\ g & -g_{ph} \end{pmatrix}, g = g_r = g_{nr} . \tag{6.76}$$

The DFS corresponds to the eigenstates of the $\mathcal{E}_I$ matrix. For other values of $g_{nr} \neq g_r$ the system does not have a DFS. Of interest is the effect of decoherence on an initial spin superposition state $|\psi\rangle = c_+|+\rangle + c_-|-\rangle$, where $c_\pm$ are complex amplitudes and the bath is in thermal equilibrium at temperature T. The bath is described in terms of the oscillator eigenstates $|n\rangle$ by the bath density matrix $\rho_B = \sum_n p_n|n\rangle\langle n|$, $p_n = e^{-E_n/k_BT}/(\sum_n e^{-E_n/k_BT})$ and $E_n = (n + 1/2)\hbar\omega$.

The temporal evolution of the system can be described by expanding the total (system + bath) evolution operator U in terms of $|\gamma\rangle$, the eigenstates of the full Hamiltonian (i.e., $(E_\gamma - H)|\gamma\rangle = 0$) with

$$U = \sum_\gamma e^{\frac{-iE_\gamma t}{\hbar}} |\gamma\rangle\langle\gamma| . \tag{6.77}$$

The eigenstates $|\gamma\rangle$ can be expanded in the zeroth-order basis of the "bare" spin–bath states $\{|\pm, n\rangle \equiv |\pm\rangle|n\rangle\}$, (i.e., the eigenstates of H without the spin–bath interaction) and gives, for the time-evolved density matrix of the total spin–bath system:

$$\begin{aligned} \rho(t) &= \sum_n p_n U [|\psi\rangle\langle\psi| \otimes |n\rangle\langle n|] U^\dagger \\ &= \sum_{\gamma,\gamma',n} p_n e^{-i(E_\gamma - E_{\gamma'})t} |\gamma\rangle\langle\gamma'|\langle\gamma|\psi, n\rangle\langle\psi, n|\gamma'\rangle . \end{aligned} \tag{6.78}$$

Further, tracing over the bath states gives the following reduced system density matrix elements $(\rho_s)_{kl}$ in the basis of spin states $|k\rangle$ and $|l\rangle$ (where k, l are $+, -$):

$$\begin{aligned} (\rho_s)_{kl} &= \langle k|\rho_s|l\rangle \\ &= \sum_{m,n,\gamma,\gamma'} p_n e^{-i(E_\gamma - E_{\gamma'})t} \langle k, m|\gamma\rangle\langle\gamma|\psi, n\rangle\langle\psi, n|\gamma'\rangle\langle\gamma'|l, m\rangle . \end{aligned} \tag{6.79}$$

Expanding $|\psi\rangle$ in terms of $c_\pm$, we can rewrite the above expression as

$$(\rho_s)_{kl} = |c_+|^2 Q_{kl}(t) + |c_-|^2 R_{kl}(t) + c_+c_-^* P_{kl}(t) + c_-c_+^* T_{kl}(t) , \tag{6.80}$$

where P, Q, R, T are complex matrices with elements

$$P_{kl}(t) = \sum_{m,n,\gamma,\gamma'} p_n e^{-i(E_\gamma - E_{\gamma'})t} \langle k, m|\gamma\rangle\langle\gamma'|l, m\rangle\langle\gamma|+, n\rangle\langle -, n|\gamma'\rangle\ ,$$

$$Q_{kl}(t) = \sum_{m,n,\gamma,\gamma'} p_n e^{-i(E_\gamma - E_{\gamma'})t} \langle k, m|\gamma\rangle\langle\gamma'|l, m\rangle\langle\gamma|+, n\rangle\langle +, n|\gamma'\rangle\ ,$$

$$R_{kl}(t) = \sum_{m,n,\gamma,\gamma'} p_n e^{-i(E_\gamma - E_{\gamma'})t} \langle k, m|\gamma\rangle\langle\gamma'|l, m\rangle\langle\gamma|-, n\rangle\langle -, n|\gamma'\rangle\ ,$$

$$T_{kl}(t) = \sum_{m,n,\gamma,\gamma'} p_n e^{-i(E_\gamma - E_{\gamma'})t} \langle k, m|\gamma\rangle\langle\gamma'|l, m\rangle\langle\gamma|-, n\rangle\langle +, n|\gamma'\rangle\ . \qquad (6.81)$$

Consider the decoherence of the spin state, quantified in terms of the purity $S = \mathrm{Tr}(\rho_s^2)$. As noted in Section 6.1, $S = 1$ denotes a pure state whereas $S = 1/2$ here describes a completely mixed, or decohered, state. Given the above results, the purity $S(t)$ can be written as

$$S(t) = \sum_{k,l\in+,-} \langle k|\rho_s|l\rangle\langle l|\rho_s|k\rangle = \sum_{k,l\in+,-} |(\rho_s)_{kl}|^2 \qquad (6.82)$$

so that, using $P_{kl}^* = T_{lk}$:

$$\begin{aligned} S(t) = {} & |c_+|^4 \sum_{k,l\in+,-} |Q_{kl}|^2 + |c_-|^4 \sum_{k,l\in+,-} |R_{kl}|^2 \\ & + |c_+|^2|c_-|^2 \sum_{k,l\in+,-} \left[|P_{kl}|^2 + |T_{kl}|^2 + \left(Q_{kl}R_{kl}^* + Q_{kl}^* R_{kl}\right)\right] \\ & + 2\mathrm{Re}\left\{ \sum_{k,l\in+,-} \left[c_+^2 c_-^{*2} P_{kl}P_{lk} + |c_+|^2 c_+^* c_- \left(Q_{kl}P_{kl}^* + P_{lk}^* Q_{kl}^*\right)\right.\right. \\ & \left.\left. + |c_-|^2 c_+^* c_- \left(R_{kl}P_{kl}^* + P_{lk}^* R_{kl}^*\right)\right]\right\} . \end{aligned} \qquad (6.83)$$

The first three of these terms depend solely on the magnitude of the $c_\pm$ coefficients, whereas the remainder of the terms depend upon the phases of the coefficients as well. The latter are clearly contributions to the purity that rely upon quantum interference between the spin states. With the P, Q, R, T matrices known, one can vary $S(t)$ with respect to c_+ and c_- to determine the state with maximum purity $S(t)$ at a fixed time τ. In particular, the relative phase of c_+ and c_- is important if either Q_{kl} or R_{kl} is nonzero, and P_{kl} is nonzero. Note that Eq. (6.83) of the typical coherent control form, with direct terms dependent on the magnitude $|c_i|$ of the coefficients and an interference term dependent on the c_i. The latter depend on the overlapping resonances contributions.

6.4.5
Overlapping Resonances and Decoherence Control: Qualitative Motivation

Since the bare states are not eigenstates of H, they will evolve in the presence of the system–bath coupling, manifested as a broadening of the bare states into resonances. As in Section 4.1.3, two such resonances are said to overlap when their widths are larger than the level spacing between them, that is, when the two bare states have nonzero overlap with the same eigenstate $|\gamma\rangle$ of the total Hamiltonian. Here the bare states, in this system–bath problem, are the unperturbed spin–boson states $|\pm, n\rangle$ for all n and overlapping resonances occur when two states $|k, m\rangle$ and $|l, n\rangle$ with $k \neq l$ and/or $m \neq n$ $(k, l \in +, -)$ have nonzero overlap with the same eigenstate $|\gamma\rangle$. In the case of the Hamiltonians given by Eq. (6.75), overlapping resonances were found in, but not limited to, the strong coupling regime characterized by g_r and g_{nr} greater than ω_0. All of the matrices P, Q, R, T depend on such overlapping resonance terms. Their dependencies differ, however, when $k \neq l$. In that case overlapping resonances contribute to T_{kl} (and P_{kl}), but not to Q_{kl} and R_{kl}.

Note that the number and character of the overlapping resonances are functions of the parameters of the total Hamiltonian, for example, here g_r, g_{nr}, g_{ph}, ω, and ω_0. For example, in the extreme case where there is no system–bath coupling (i.e., $g_r = g_{nr} = g_{ph} = 0$), the total Hamiltonian is diagonal in the $|\pm, n\rangle$ bare state basis. The overlap between the spin–boson bare states then vanishes, and in turn, so does the overlap between the states $|\pm\rangle$. That is, as expected, in absence the system of coupling to the bath, the spin states are not broadened and thus do not overlap.

The rather complex contribution of the overlapping resonances to the system purity is discussed below. However, qualitative motivation for the role of overlapping resonances in controlling decoherence is readily provided by considering the survival probability of an initial separable (spin + bath) state. Consider the dynamics of a wave function which, at time $t = 0$ is given by

$$|\Psi_0\rangle = |\psi\rangle|n\rangle \; . \tag{6.84}$$

Evolution of the state yields $|\Psi(t)\rangle$.

Preventing decoherence, from this viewpoint, means preserving the $|\psi\rangle$ component of $|\Psi_0\rangle$ during the course of the dynamics. Consider then the "survival probability" $I(t) = |\langle\Psi_0|\Psi(t)\rangle|^2$ of finding the state $|\Psi_0\rangle$ at time t. Expanding $|\Psi(t)\rangle$ in terms of $|\gamma\rangle$ gives:

$$|\Psi(t)\rangle = \sum_\gamma |\gamma\rangle\langle\gamma|\Psi_0\rangle e^{-iE_\gamma t/\hbar} \; , \tag{6.85}$$

so that

$$I(t) = \sum_{\gamma,\gamma'} |\langle\Psi_0|\gamma\rangle|^2 |\langle\Psi_0|\gamma'\rangle|^2 e^{i(E_\gamma - E'_\gamma)t/\hbar} \; . \tag{6.86}$$

To gain insight, we consider the simplified case where both t and the density of the energy spectrum $\{E_\gamma\}$ are large. In that case only the $\gamma = \gamma'$ term contributes to Eq. (6.86), so that $I(t)$ becomes $I(t) = \sum_\gamma |\langle\gamma|\Psi_0\rangle|^4$.

In the decoherence case, for an initial product state $|\Psi_0\rangle = [c_+|+\rangle + c_-|-\rangle]|n\rangle$, the survival probability is then:

$$I(t) = \sum_{\gamma} |\langle\gamma|[c_+|+\rangle + c_-|-\rangle]|n\rangle|^4 = \sum_{\gamma} |c_+\langle\gamma|+, n\rangle + c_-\langle\gamma|-, n\rangle|^4 \, . \tag{6.87}$$

Clearly, both terms inside the absolute value sign only contribute if both $\langle\gamma|+, n\rangle$ and $\langle\gamma|-, n\rangle]$ are nonzero, that is, if the resonances $|+, n\rangle$ and $|-, n\rangle$ overlap. In the event that they do not overlap, control over the survival probability is limited to two parameters, $|c_+|$ and $|c_-|$. In the presence of overlapping resonances, however, a third parameter, the relative phase of c_+ and c_- is important, guaranteeing improved control over the survival probability. Specifically, increased $I(t)$ would imply less decoherence, and vice versa. Finally, note further that the structure of Eq. (6.87) is clearly that of interfering quantum contributions to the amplitude associated with the γ contribution to $I(t)$.

Given this qualitative motivation, exact contributions to the standard purity measure of decoherence are discussed below in terms of their relation to overlapping resonances. From Eq. (6.83), it is apparent that $S(t)$ is strongly dependent on overlapping resonances through Q_{kl}, R_{kl}, P_{kl}, and T_{kl}. To quantify this dependence, we identify overlapping resonances contributions associated with the plus and minus spin states. That is, first define

$$A_{+-} = \sum_{m \neq n} A_+^{m,n} A_-^{m,n} \, , \quad A_\pm^{m,n} = \left[\sum_{\gamma} |\langle\pm, m|\gamma\rangle\langle\gamma|\pm, n\rangle| \right] . \tag{6.88}$$

For a pair of bath states $|m\rangle$ and $|n\rangle$, $A_+^{m,n}$ determines the overlap (summed over all eigenstates $|\gamma\rangle$) between the bare states $|+, m\rangle$ and $|+, n\rangle$, when the spin state is $|+\rangle$. If, for the same pair of $|m\rangle$ and $|n\rangle$, there exists nonzero overlap $A_-^{m,n}$ between the bare state $|-, m\rangle$ and $|-, n\rangle$, then A_{+-} provides a measure of the overlap between the states $|+\rangle$ and $|-\rangle$, that arises through the overlap between the states $|\pm, m\rangle$ and $|\pm, n\rangle$. Note that Eq. (6.88) is somewhat similar to the off-diagonal T_{+-} and P_{+-} terms at $t = 0$.

Consider then the relationship between overlapping resonances as reflected in A_{+-} and decoherence as embodied in the system purity S. As an example, consider the dynamics where the bath is at a temperature of 25 mK (corresponding to $\approx$ 20 occupied bath states), and choose to vary g_r, keeping the other Hamiltonian parameters fixed. To assess this decoherence one can find, for each value of g_r, the maximum possible purity $S_{\max}$ in the system at a fixed time $t = t_0$ by optimizing the complex coefficients $c_\pm$. Results are shown in Figure 6.19 where both $S_{\max}$ and the overlapping resonance measure A_{+-} are plotted vs. g_r at $t = 0.1$ ns. Larger A_{+-} is seen to correlate well with larger $S_{\max}$; that is, the larger the overlapping resonances contribution, the more resistant is the optimal superposition to decoherence. Note, significantly, that the optimally resistant state need not be, and is often not, an eigenstate of the spin Hamiltonian, but is rather superposition of eigenstates.

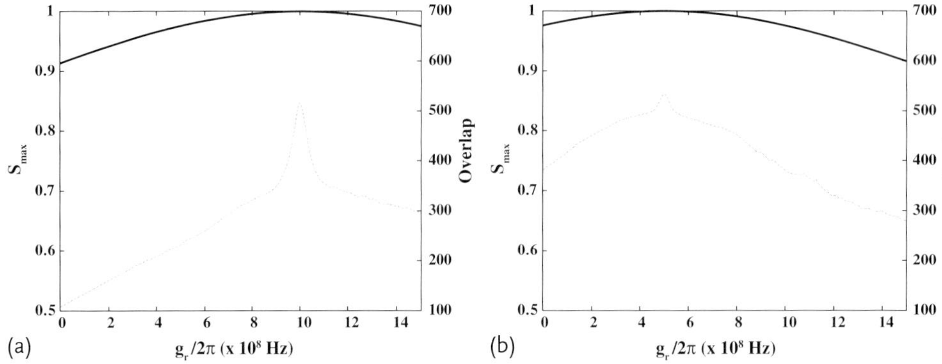

Figure 6.19 Dependence of the maximum achievable purity $S_{\max}$ of the system (solid line) and the overlapping resonance A_{+-} (dashed line; right y-axis) on the coupling strength $g_r/(2\pi)$. Here (a) $g_{ph} = 2\pi \times 500$ MHz, $g_{nr} = 2\pi$ GHz and (b) $g_{ph} = 2\pi$ GHz, $g_{nr} = 2\pi \times 500$ MHz. The other parameters are $\omega = 2\pi$ GHz, $\omega_0 = 2\pi \times 100$ MHz, and the temperature $T = 25$ mK. The purity is calculated at $t = 0.1$ ns. From [194, Figure 2]. (Reprinted with permission. Copyright 2010 American Institute of Physics.)

Prominent in the graph (and another presented below) is the DFS point $g_r = g_{nr}$, where the A_{+-} curve displays a sharp peak, commensurate with $S_{\max}$ reaching its maximum value of unity. That is, maximal purity correlates with maximum A_{+-}. The deviation of $S_{\max}$ from unity as one moves away from this point is significant, particularly in scenarios such as quantum computation, where extremely high degrees of coherence are necessary.

The results clearly show that states take advantage of overlapping resonances (and hence of quantum interference) to increase the state purity. This is the case even if the system–bath coupling is very large, as is clear from Figure 6.19b, where, for example, $g_r \sim 2\pi \times 1.5$ GHz ($g_r \gg \omega_0$), but where the purity remains as large as 0.92 for $g_{ph} = 2\pi$ GHz.

Additional results showing how overlapping resonances between spin states is manifested in decoherence is shown in Figure 6.20. Here, for a given value of g_r, the $|c_+|$ and $|c_-|$ that maximize S are obtained and then the relative phase θ between c_+ and c_- is varied to obtain S_{diff}, defined as the difference between the maximum and minimum purity so attained. This difference is plotted in Figure 6.20 as a function of g_r, along with the corresponding value of A_{+-}. Results show that as A_{+-} increases, so does the extent to which the purity varies with θ. Hence, in the presence of large overlapping resonance, one can *actively* control the purity of the system by taking advantage of the phase dependent quantum interference contribution to S. Indeed, in accord with Eq. (6.83), an observed variation of the decoherence with the phase of a superposition state provides evidence of the presence of overlapping resonances.

In summary, the results show clearly that overlapping resonances provide one unifying approach to considering the resistance to decoherence of system superpo-

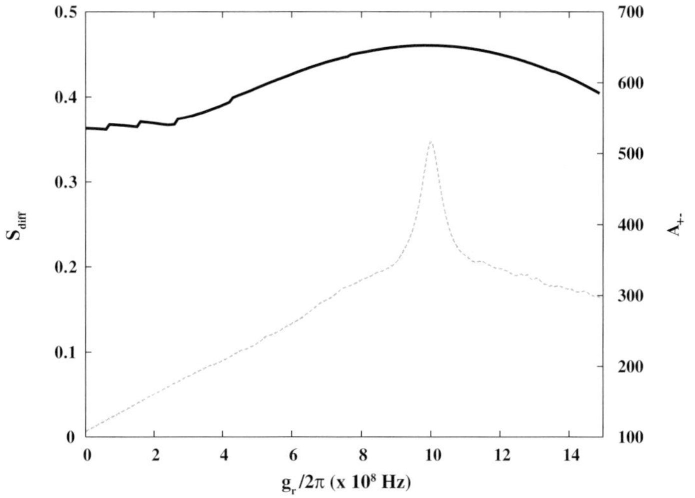

Figure 6.20 Variation of S_{diff} (solid line) and the overlap A_{+-} (gray dashed line; right y-axis) with g_r for $g_{nr} = 2\pi$ GHz and $g_{ph} = 2\pi \times 500$ MHz. The remaining parameters are as in Figure 6.19. From [194, Figure 4]. (Reprinted with permission. Copyright 2010 American Institute of Physics.)

sition states. The larger the overlapping resonance contribution to the Hamiltonian, the larger the opportunity for larger resistance to decoherence.

6.4.6 Control of Dephasing

Experimental results on the control of decoherence and dephasing in molecular processes are limited. However, the control of dephasing in coupled vibrational-rotational dynamics in a collection of diatomic molecules has been demonstrated. In particular, Branderhorst *et al.* [255] have considered the control of the dephasing of the vibrational degree of freedom of a diatomic molecule, regarded as the system, coupled to the rotational degrees of freedom of the diatomic, regarded as the bath.

To appreciate that this is control of dephasing, as distinct from decoherence, recall that, as a central force problem, the wave function for the internal degrees of freedom of a diatomic molecule with rotational quantum number j can be written as a separable product of a j-dependent vibrational wave function and a spherical harmonic $Y_{j,m}$ of degree j [256]. The vibrational wave functions are j-dependent since they arise from vibrational motion in an effective potential that includes the centrifugal contribution. That is, if the internuclear potential is $V(R)$, then the vibrational motion occurs in the effective potential $V_{eff}(R)$:

$$V_{eff}(R) = V(R) + \hbar^2 \frac{j(j+1)}{(2mR^2)}, \tag{6.89}$$

where m is the reduced mass of the diatomic. Since the rotational and vibrational degrees of freedom remain separable throughout the dynamics there is no vibrational decoherence due to vibration-rotation coupling, in accord with our discussion above (Section 6.1). Rather, the vibrational motion of different diatomics in the overall system differ since they experience different effective potentials associated with different values of j. As the diatomics vibrate, post laser excitation, the vibrations of the diatomics go out of phase, resulting in dephasing.

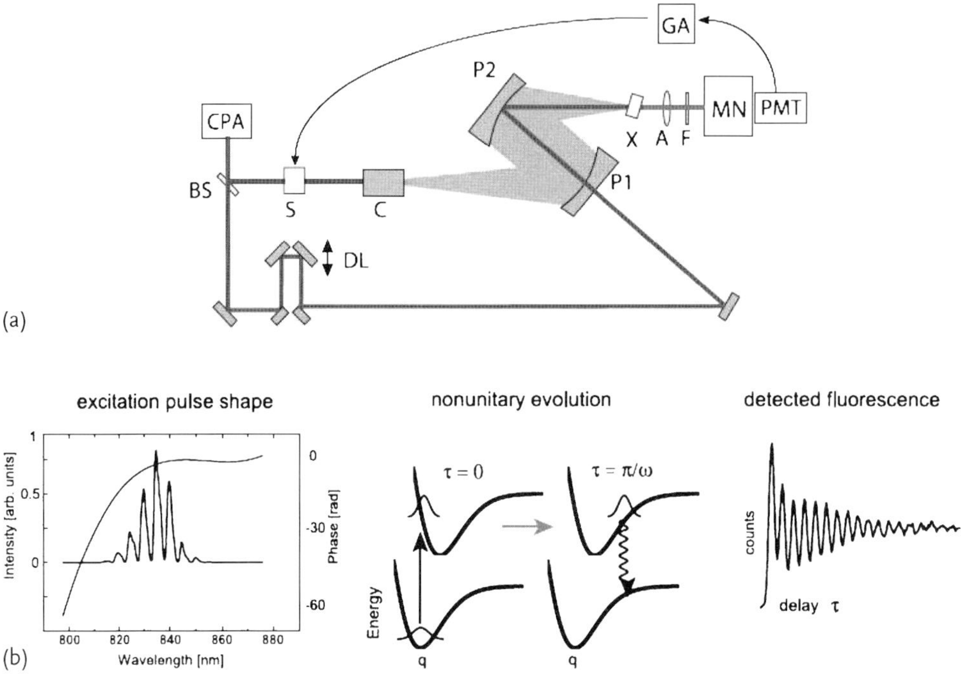

Figure 6.21 (a) Apparatus for control and measurement of the vibrational wave packet states of potassium dimers and (b) excitation of a shaped vibrational wave packet and its detection at the outer turning point. A pulse from a chirped-pulse-amplified mode-locked Ti:sapphire laser system (CPA) is shaped via a Fourier plane pulse shaper (S) and incident on a cell (C) containing a gas of K2 molecules. Absorption generates a wave packet in the lowest excited electronic state. Fluorescence from this state is collected on a nonlinear crystal (X) where it is mixed with a portion of the unshaped laser pulse that has been sent through an adjustable delay line (DL). The resulting frequency-upconverted radiation is imaged (depicted as A) through a filter (F) onto a monochromator (MN) and detected by a photomultiplier tube (PMT). Quantum beats are observed in the time- and frequency-resolved fluorescence as the vibrational wave packet oscillates in the excited electronic state potential, as illustrated in the middle panel of (b). The resulting detected signal is shown in the right panel of (b). The time- and frequency-resolved signal shows quantum beats as a function of the time delay τ between the gate pulse and excitation pulse. The optimal pulse profile that minimized the dephasing is shown in (b) on the left. From [255, Figure 1]. (Reprinted with permission. Copyright 2008 American Association for the Advancement of Science.)

The experiment [255] was carried out by laser excitation to the $A^1\Sigma_u^+$ excited electronic state, of a collection of K_2 molecules at 400° C, and fluorescence from the outer vibrational turning point of the dynamics was detected (see Figure 6.21). Initially, the laser excited molecules start off near the inner turning point of the vibrational motion. As vibrational dynamics on different *j*-dependent potentials occurs, fluorescence from the outer turning point reflects the dephasing of the collection of molecules. Estimated time scales of dephasing of vibrations due to coupling to vibrations is 3 ps, shorter than the spreading of the anharmonic vibrational wave packet ($\approx$ 12 ps) and far shorter than collision induced dephasing at the low pressures considered experimentally. Hence, the primary observed effect, with variations in laser characteristics, is due to the vibrational dephasing described above. Sample control results, showing successful control with changes in the laser chirp are shown in Figure 6.22. Hence, the experimental results show that changing the phase of the incident excitation pulse suffices to control the ensemble dephasing. An effort to analyze these results in detail [257], however, led to the conclusion that either phase control of dephasing was only possible at intermediate times, or that subtle *j*-dependent features of the excitation (i.e., rotational distortion of the vibrational wave function during the pump pulse) are necessary to account for the experimental results.

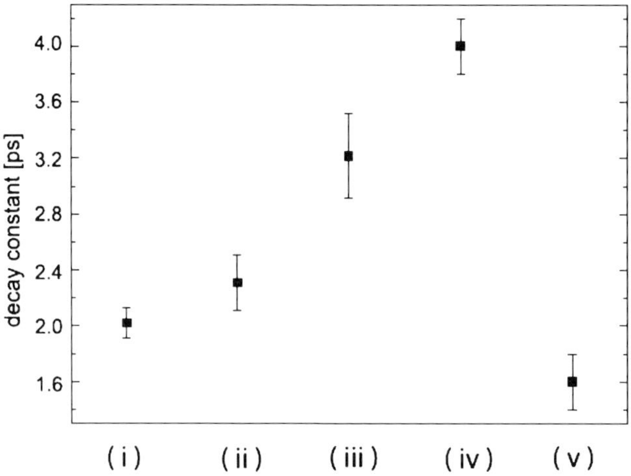

Figure 6.22 The decay time of the quantum beats for several pulse shapes: (i) transform-limited excitation pulses, (ii) optimized pulses constrained to have a flat phase, (iii) optimized pulses with no constraints, (iv) pulses with -3.1×10^3 fs^2 negative chirp, and (v) pulses with 3.1×10^3 fs^2 positive chirp. The error bars indicate $\pm$ standard deviation of the fit. From [255, Figure 3]. (Reprinted with permission. Copyright 2008 American Association for the Advancement of Science.)

6.5 Countering Partially Coherent Laser Effects in Pump-Dump Control

An alternate source of decoherence is in the nature of the laser used to irradiate the system. Specifically, if the laser has random components then, in general, it inputs a degree of randomness into the system, reducing the phase information content and hence decohering the system.

For example, consider Eq. (3.19) corresponding to control of photodissociation of an initial two-level superposition state that is excited to the continuum. If there are sufficiently strong external fluctuations, or excessive jitter in the laser phase, then this results in a complete average over $\phi_1 - \phi_2$ and Eq. (3.19) becomes

$$R_{q,q'}(E) \equiv \frac{P_q(E)}{P_{q'}(E)} = \frac{|d_q(1,1)| + x^2|d_q(2,2)|)}{|d_{q'}(1,1)| + x^2|d_{q'}(2,2)|} . \tag{6.90}$$

Thus, there is no longer any dependence on any phase characteristics and all phase control would be lost.

Equations (3.19) and (6.90) represent two extremes, where the result is either fully coherent or fully incoherent. It is enlightening to consider [258] the effect of partially coherent laser sources through the example of the pump-dump scenario of Section 3.6 in the case where the laser is not fully coherent.

To characterize partially coherent pulses [259, 260] consider a Gaussian laser pulse with time profile $\varepsilon_x(t)$ and phase $\delta_x(t)$, that is,

$$\varepsilon_x(t) = \frac{\mathcal{E}_{x0}}{2} e^{\frac{-(t-t_x)^2}{\tau_x^2}} e^{-i[\omega_x t + \delta_x(t)]} . \tag{6.91}$$

We adopt a modified phase diffusion model for the partially coherent laser source in which the phase δ_x is allowed to be time dependent and random. Thus, the molecule–laser interaction is modeled by the interaction with an ensemble of lasers, each of different phase. This ensemble is described by a Gaussian correlation for the stochastic phases with a decorrelation time scale τ_{xc}:

$$\langle e^{i\delta_x(t_2)} e^{-i\delta_x(t_1)} \rangle = e^{\frac{-(t_2-t_1)^2}{2\tau_{xc}^2}} . \tag{6.92}$$

Here the angle brackets denote an average over the ensemble of laser phases.

To examine photodissociation given this field requires, as shown below, the frequency-frequency correlation function $\langle \epsilon_x(\omega_2)\epsilon_x^*(\omega_1)\rangle$ where $\epsilon_x(\omega)$ is the Fourier transform of $\varepsilon(t)$ for $\delta_x(t)$ equal to a constant δ_x. Given Gaussian pulses (Eqs. (6.91)-(6.92)) we have [259]

$$\langle \epsilon_x(\omega_2)\epsilon_x^*(\omega_1)\rangle = \frac{\mathcal{E}_{x0}^2 \tau_x T_x}{8} e^{i(\omega_2-\omega_1)t_x} e^{-\left[\tau_x^2 \frac{(\omega_2-\omega_1)^2}{8}\right]} e^{-\left[T_x^2 \frac{(\omega_2+\omega_1-2\omega_x)^2}{8}\right]} , \tag{6.93}$$

where

$$T_x = \frac{\tau_x \tau_{xc}}{(\tau_x^2 + \tau_{xc}^2)^{\frac{1}{2}}} . \tag{6.94}$$

Since the pump-dump control scenario involves two lasers we adopt a similar description for the dump pulse $\varepsilon_d(t)$, with appropriate change in parameter labels, that is,

$$\varepsilon_d(t) = \frac{\mathcal{E}_{d0}}{2} e^{\frac{-(t-t_d)^2}{\tau_d^2}} e^{-i(\omega_d t + \delta_d(t))} , \tag{6.95}$$

$$\langle e^{i\delta_d(t_2)} e^{-i\delta_d(t_1)} \rangle = e^{\frac{-(t_2-t_1)^2}{2\tau_{dc}^2}} , \tag{6.96}$$

$$\langle \epsilon_d(\omega_2)\epsilon_d^*(\omega_1)\rangle = \frac{\mathcal{E}_{d0}^2 \tau_d T_d}{8} e^{i(\omega_2-\omega_1)t_d} e^{-\left[\tau_d^2 \frac{(\omega_2-\omega_1)^2}{8}\right]} e^{-\left[T_d^2 \frac{(\omega_2+\omega_1-2\omega_d)^2}{8}\right]} , \tag{6.97}$$

where $T_d = \tau_d \tau_{dc}/(\tau_d^2 + \tau_{dc}^2)^{1/2}$. Here τ_{xc} and τ_{dc} define the degree of field coherence, the two limiting cases being a coherent source ($\tau_{ic} \to \infty, i = x, d$) and fully incoherent source ($\tau_{ic} \to 0, i = x, d$).

The frequency spectrum of the pulse provides a primary means of characterizing the laser–molecule interaction and is given, for the excitation pulse, by

$$I_x(\omega) = \frac{T_x}{\sqrt{2\pi}} e^{-T_x^2 \frac{(\omega-\omega_x)^2}{2}} , \tag{6.98}$$

whose full width at half maximum (FWHM) is $2\sqrt{2\ln 2}/T_x$. For a fully coherent pulse $T_x = \tau_x$ giving a FWHM of the intensity spectrum equal to $\Delta_x/\sqrt{2}$, where $\Delta_x \equiv 4\sqrt{\ln 2}/\tau_x$ is the FWHM of the frequency profile of the pulse.

Similar definitions apply to characterize the dump pulse. Further, for consistency we define the partial coherence parameter $\Delta_{xc} = 4\sqrt{\ln 2}/\tau_{xc}$ and similar parameters (Δ_d, Δ_{dc}) for the dump pulse. Note that fitting Eq. (6.98) to the measured pulse frequency spectrum provides a means of obtaining T_x and ω_x from experimental data [259].

Consider now the pump-dump scenario where, for generality, we assume a pump pulse whose spectral width is sufficiently large to encompass a large number of levels. As in Section 3.6, a molecule with Hamiltonian H_M is subjected to two temporally separated pulses $\varepsilon(t) = \varepsilon_x(t) + \varepsilon_d(t)$. The probability of forming channel q at energy E is now given by the extension of Eq. (3.77) to many level excitations by $\varepsilon_x(t)$, that is,

$$P_q(E) = \frac{(2\pi)^2}{\hbar^2} \sum_m |\sum_j b_j \langle E, \boldsymbol{m}, q^- | \boldsymbol{d} \cdot \hat{\boldsymbol{\epsilon}} | E_j \rangle \epsilon_d(\omega_{E,E_j})|^2 , \tag{6.99}$$

where $\omega_{E,E_j} = (E - E_j)/\hbar$ and b_j is given by (Eq. (3.75)):

$$b_j = \frac{2\pi i}{\hbar} \langle E_j | \hat{\boldsymbol{\varepsilon}} \cdot \boldsymbol{d} | E_1 \rangle \epsilon_x(\omega_{j,1}) \equiv c_j \epsilon_x(\omega_{j,1}) . \tag{6.100}$$

Equation (6.100) also defines c_j as the field independent component of b_j.

Expanding the square allows us to write the probability in a canonical form:

$$P_q(E) = \frac{(2\pi)^2}{\hbar^2} \sum_{i,j} b_i b_j^* \epsilon_d(\omega_{E,E_i}) \epsilon_d^*(\omega_{E,E_j}) d_q(i,j)$$
$$\equiv \frac{(2\pi)^2}{\hbar^2} \sum_{i,j} c_i c_j^* \epsilon_x(\omega_{ig}) \epsilon_x^*(\omega_{jg}) \epsilon_d(\omega_{E,E_i}) \epsilon_d^*(\omega_{E,E_j}) d_q(i,j) \,, \quad (6.101)$$

where $d_q(i,j)$ is defined by Eq. (3.79). To incorporate effects due to a partially coherent source we independently average Eq. (6.101) over the ensemble of phases of the excitation and dump pulses, giving

$$P_q(E) = \frac{2\pi}{\hbar^2} \sum_{i,j} c_i c_j^* \langle \epsilon_x(\omega_{i,g}) \epsilon_x^*(\omega_{j,g}) \rangle \langle \epsilon_d(\omega_{E,E_i}) \epsilon_d^*(\omega_{E,E_j}) \rangle d_q(i,j) \,. \quad (6.102)$$

Note that despite the excitation of multiple levels, the only correlation function required is between ϵ at two frequencies. Assuming the phase diffusion model described above, then the frequency-frequency correlation functions is given by Eq. (6.93). The probability $P(q)$ of forming product in channel q is then obtained by integrating over the pulse width:

$$P(q) = \int dE\, P_q(E) = \frac{2\pi}{\hbar^2} \sum_{i,j} c_i c_j^* \langle \epsilon_x(\omega_{i,g}) \epsilon_x^*(\omega_{j,g}) \rangle$$
$$\int dE \langle \epsilon_d(\omega_{E,E_i}) \epsilon_d^*(\omega_{E,E_j}) \rangle d_q(i,j) \,. \quad (6.103)$$

It is often the case that the Franck–Condon factors contained in $d_q(i,j)$ vary sufficiently slowly over the range of E encompassed by the dump pulse to be regarded as constant [261]. (This assumption, called the "slowly-varying continuum approximation", is discussed in detail in Chapter 12). Under these circumstances we can use the following generalized Parseval's equality to show that $P(q)$ is independent of the coherence properties of the dump pulse. Specifically we have

$$\int d\omega \langle \epsilon^*(\omega - \omega_1) \epsilon(\omega - \omega_2) \rangle = \frac{1}{2\pi} \int d\omega$$
$$\iint dt_1 dt_2 \langle \varepsilon^*(t_1) \varepsilon(t_2) \rangle e^{i\omega(t_2 - t_1)} e^{i(\omega_1 t_1 - \omega_2 t_2)}$$
$$= \iint dt_1 dt_2 \langle \varepsilon^*(t_1) \varepsilon(t_2) \rangle \delta(t_2 - t_1) e^{i(\omega_1 t_1 - \omega_2 t_2)}$$
$$= \int dt \langle \varepsilon^*(t) \varepsilon(t) \rangle e^{i(\omega_1 - \omega_2)t} \,. \quad (6.104)$$

(The conventional Parseval equality $\int d\omega \langle |\epsilon(\omega)|^2 \rangle = \int dt \langle \varepsilon^*(t) \varepsilon(t) \rangle$ is the special case of $\omega_1 = \omega_2$.) Since the right-hand side is independent of the phase of $\varepsilon(t)$ then the frequency integrated correlation function is independent of the degree of coherence of the dump pulse.

Assuming that $d_q(i,j)$ is independent of E over the pulse width allows us to write Eq. (6.103) as

$$P(q) = \frac{(2\pi)^2}{\hbar^2}\sum_{i,j} c_i c_j^* \langle \epsilon_x(\omega_{ig})\epsilon_x^*(\omega_{jg})\rangle F(\omega_{j,i}) d_q(i,j)\,, \tag{6.105}$$

with $F(\omega_{j,i}) = \int dt \langle |\varepsilon_d(t)|^2\rangle e^{i(E_j - E_i)t/\hbar}$. For the Gaussian dump pulse,

$$F(\omega) = \left(\frac{\epsilon_{d0}}{2}\right)^2 \sqrt{\frac{\pi}{2}} \exp\left[-\tau_d^2\omega^2/8 + i\omega t_d\right]\,. \tag{6.106}$$

Given Eqs. (6.93), (6.106), and (3.13), Eq. (6.105) assumes the form

$$P(q) = \frac{(2\pi)^2}{\hbar^2}\sum_{i,j} c_i c_j^* |\langle \epsilon_x(\omega_{ig})\epsilon_x^*(\omega_{jg})\rangle F(\omega_{j,i}) d_q(i,j)| \exp\left[i\omega_{i,j}\Delta_t + i\alpha_q(i,j)\right]\,, \tag{6.107}$$

where $\Delta_t = (t_d - t_x)$. For the simplest case, excitation which encompasses only two levels, $P(q)$ is given by

$$\begin{aligned} P(q) = \frac{(2\pi)^2}{\hbar^2} \Big[& |c_1|^2 \langle|\epsilon_x(\omega_{1g})|^2\rangle d_q(1,1) F(\omega_{11}) + |c_2|^2 \langle|\epsilon_x(\omega_{2g})|^2\rangle d_q(2,2) F(\omega_{22}) \\ & + 2|c_1 c_2^* d_q(1,2) \langle\epsilon_x(\omega_{1g})\epsilon_x^*(\omega_{2g})\rangle F(\omega_{2,1})| \cos(\omega_{2,1}\Delta_t + \alpha_q(1,2) + \beta)\Big]\,, \end{aligned} \tag{6.108}$$

with $\langle E_1|\boldsymbol{d}\cdot\hat{\boldsymbol{\epsilon}}|E_g\rangle\langle E_g|\boldsymbol{d}\cdot\hat{\boldsymbol{\epsilon}}|E_2\rangle \equiv |\langle E_1|\boldsymbol{d}\cdot\hat{\boldsymbol{\epsilon}}|E_g\rangle\langle E_g|\boldsymbol{d}\cdot\hat{\boldsymbol{\epsilon}}|E_2\rangle| \exp(i\beta)$.

Since partial laser phase coherence affects both the direct terms as well as the cross terms, the extent of control is dependent on the laser properties through the relative magnitudes of $|\langle\epsilon_x(\omega_{1g})\epsilon_x^*(\omega_{2g})\rangle|$ and $\langle|\epsilon_x(\omega_{ig})|^2\rangle$, $i = 1, 2$. To expose the dependence on the coherence of the pump field denote the terms $|c_k c_j^* F(\omega_{jk}) d_q(k,j)(E)|$ by $a_{k,j}^{(q)}$ and consider the ratio of the $k \neq j$ term in Eq. (6.107) to the associated diagonal terms. That is, consider the contrast ratio:

$$C_{k,j}^{(q)} = \frac{a_{k,j}^{(q)} \langle\epsilon_x(\omega_{kg})\epsilon_x^*(\omega_{jg})\rangle}{a_{k,k}^{(q)}\langle|\epsilon_x(\omega_{kg})|^2\rangle + a_{j,j}^{(q)}\langle|\epsilon_x(\omega_{jg})|^2\rangle}\,. \tag{6.109}$$

For the Gaussian model of a partially coherent source Eq. (6.109) assumes the form:

$$C_{k,j}^{(q)} = \exp\left[-\omega_{k,j}^2(\tau_x^2 - T_x^2)/8\right]\left[\frac{a_{k,j}^{(q)}\exp\left(-\omega_{jk}T_x^2\delta_{E_{jk}}/2\hbar\right)}{a_{k,k}^{(q)} + a_{j,j}^{(q)}\exp\left(-\omega_{jk}T_x^2\delta_{E_{jk}}/\hbar\right)}\right]\,. \tag{6.110}$$

Here $\delta_{E_{jk}} \equiv E_x - E_{av} \equiv \hbar\omega_x - (E_k + E_j)/2$. The second term in Eq. (6.110), in brackets, is a function of $T_x^2\delta_{E_{jk}}$. If two bound-state levels dominate the pump excitation then this term contributes a scaling characteristic to control plots. That is, if we plot contours of constant dissociation probability as a function of Δ_t and $\delta_{E_{jk}}$ then, barring the first term, plots with different T_x will appear similar, with a new range scaled by $\delta_{E'} = (T_x/T'_x)^2 \delta_{E_{jk}}$.

The first term in Eq. (6.110) can be rewritten as

$$A_{k,j} = \exp\left[\frac{-\omega_{k,j}^2(\tau_x^2 - T_x^2)}{8}\right] = \exp\left[\frac{-\omega_{k,j}^2 \tau_x^2}{8}\left(1 - \frac{1}{(\tau_x/\tau_{xc})^2 + 1}\right)\right], \tag{6.111}$$

which affords additional insight into the dependence of control, achieved by varying Δ_t, on coherence characteristics of the pump laser. Note first that Eq. (6.111) implies that for fixed τ_x control comes predominantly from nearby molecular states, that is, those with small $\omega_{k,j}$. Second, for fixed pulse duration τ_x, control is expected to decrease with decreasing τ_{xc}, that is, with decreasing pulse coherence. This is reasonable since decreasing τ_x leads to the preparation of mixed molecular states with increasing degrees of state impurity and hence loss of phase information. Somewhat unexpected, however, is the prediction of improved control with decreasing pulse duration τ_x at fixed (τ_x/τ_{xc}), embodied in the $\exp(-\omega_{k,j}^2 \tau_x^2)$ term.

As an example, consider the dependence of the photodissociation the model collinear DH_2, discussed in Section 3.6, as a function of Δ_t and of the detuning $\delta_E \equiv E_x - E_{av} \equiv \hbar\omega_x - (E_1 + E_2)/2$, where E_1 and E_2 are two selected neighboring energy levels for variable Δ_{xc}. For small Δ_{xc} only two levels are excited but as Δ_{xc} increases the excitation encompasses a larger number of levels. Each panel of Figure 6.23 shows a contour plot of the fractional yield of DH (i.e., $P(DH)/(P(DH) + P(H_2))$) as a function of Δ_t and δ_E, for different values of Δ_{xc}. Figure 6.23a, for comparison, shows the case of a coherent pulse ($\Delta_{xc} = 0$); the range of control is large with the yield ratio varying from 0.19 to 0.92. Figure 6.23b,c shows control in the same system but with differing amounts of pump laser incoherence. Increasing incoherence is clearly accompanied by a reduction in the range of control, with considerable loss of control by $\Delta_{xc} = 80\,\text{cm}^{-1}$ control (Figure 6.23c). Note that regions where the yield depends on δ_E but not on Δ_t does not constitute interference based control. Rather this dependence is a consequence of the (generally uninteresting) predisposition of various bound levels of the excited electronic state to preferentially dissociate to particular products. The pattern shown in Figure 6.23, where the control plot is dominated by a single well-defined peak and valley, is characteristic of the excitation of essentially two levels, that is, two molecular levels under the laser excitation envelope have significant Franck–Condon factors. More complicated behavior is seen upon excitation of additional levels. Figure 6.24 shows results typical of those obtained with the excitation of a large number of levels. The resultant dependence of the yield ratio on δ_E and Δ_t is far more complex than the behavior seen in previous figures. However, the reduction of control with increasing Δ_{xc} is clearly evident; control is lost by $\Delta_{xc} = 1000\,\text{cm}^{-1}$ (not shown).

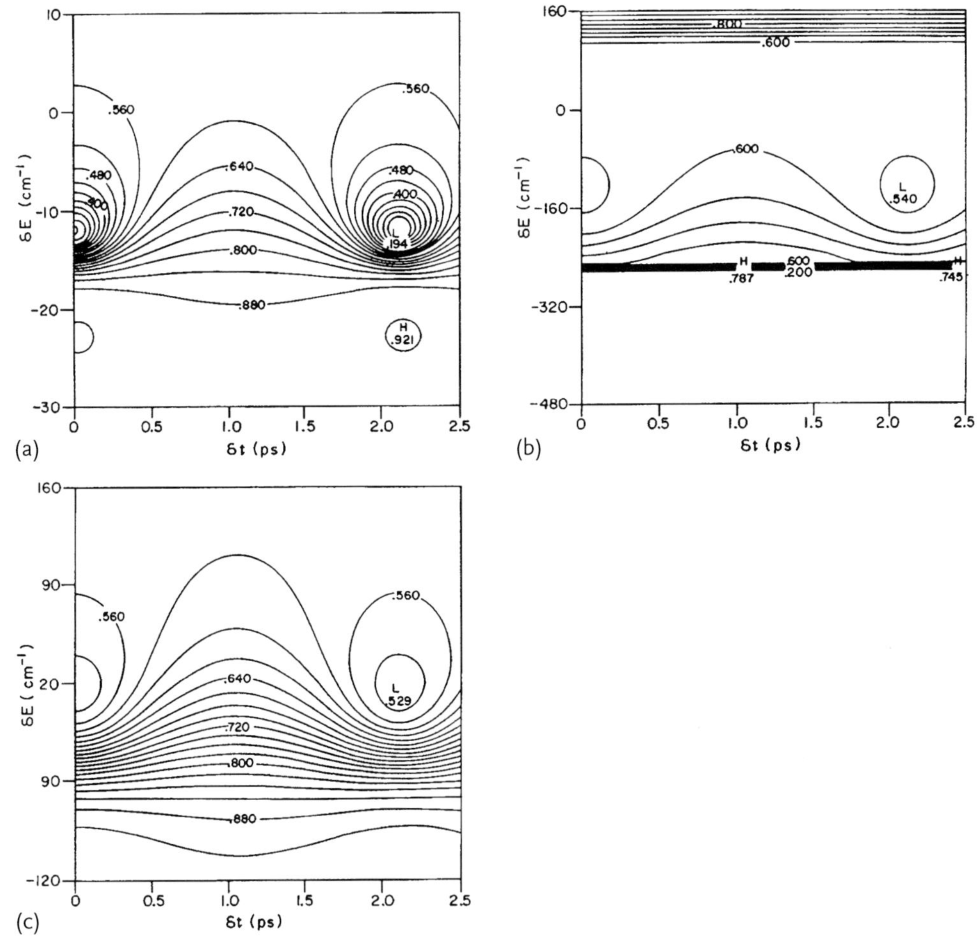

Figure 6.23 Contour plots of the fraction of DH yield as a function of the detuning, δE and the time delay Δ_t for partially coherent pulsed excitation. Here $\Delta_d = 80\,\text{cm}^{-1}$, $\Delta_x = 20\,\text{cm}^{-1}$, (a) $\Delta_{xc} = 0\,\text{cm}^{-1}$, (b) $\Delta_{xc} = 40\,\text{cm}^{-1}$ and (c) $\Delta_{xc} = 80\,\text{cm}^{-1}$. Taken from [258, Figure 1]. (Reprinted with permission. Copyright 1996 American Institute of Physics.)

From these and related studies we find that control in the model DH_2 system is lost when $(\Delta_{xc}/\Delta_x) = (\tau_x/\tau_{xc}) \approx 4$–$5$. This precludes the use of typical nanosecond lasers, where $(\Delta_{xc}/\Delta_x) > 10^2$, for pump-dump control.

A more complex analysis of the effect of laser phase diffusion has been applied to the case of one-photon vs. three-photon absorption (i.e., simultaneous absorption of $3\omega_1$ and ω_3 with $3\omega_1 = \omega_3$) (Section 3.4.2) by Lambropoulos and coworkers [262]. They assumed that the ω_3 photon was made by third harmonic generation from the ω_1 laser. As discussed in Section 3.4.2, current experiments vary the relative phase of two laser beams by passing ω_3 and ω_1 through a gas. If the laser frequency is somewhat unstable then the relative phase of the two beams will ac-

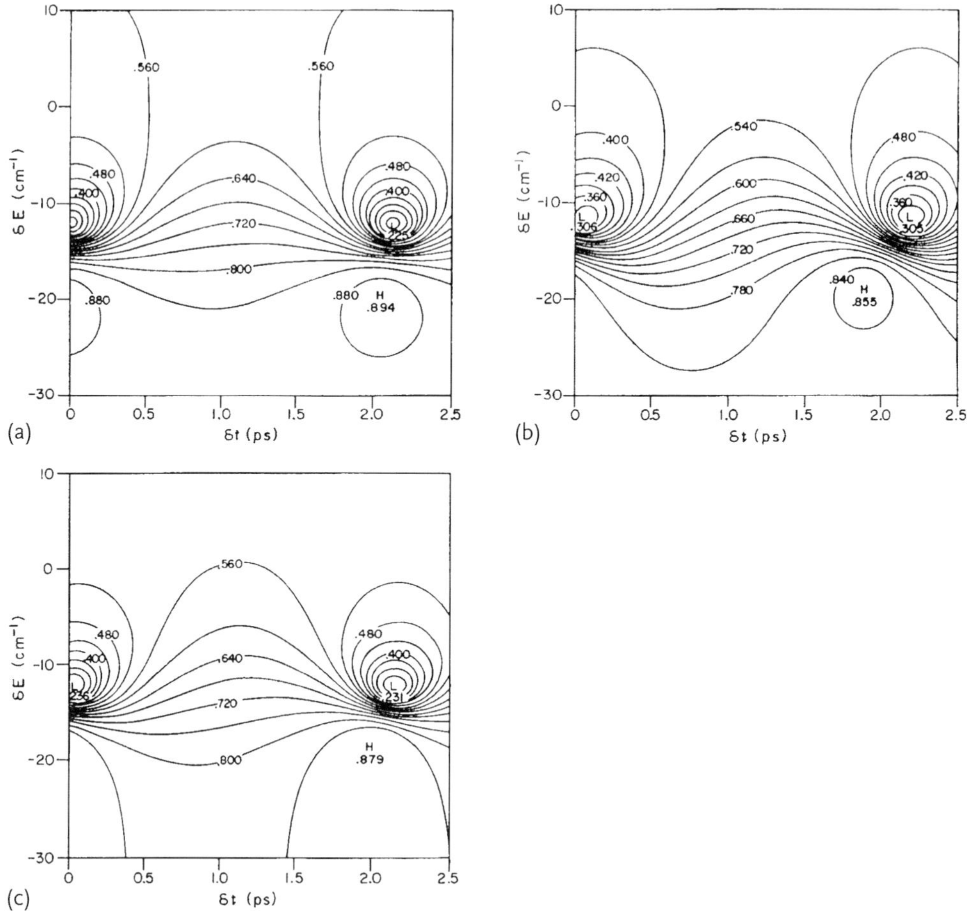

Figure 6.24 Same as Figure 6.23, but for excitation of denser set of bound states [258]. Here $\Delta_x = 200\,\text{cm}^{-1}$ and (a) $\Delta_{xc} = 0\,\text{cm}^{-1}$, (b) $\Delta_{xc} = 200\,\text{cm}^{-1}$ and (c) $\Delta_{xc} = 800\,\text{cm}^{-1}$. Taken from [258, Figure 2]. (Reprinted with permission. Copyright 1996 American Institute of Physics.)

quire a fluctuating phase which is a source of phase loss in the system. The phase fluctuations of the ω_1, denoted $\delta\phi$ are assumed proportional to the fluctuations of ω_1, that is, $\delta\phi = \alpha\delta\omega_1$. For typical experimental circumstances it is reasonable to model the $\delta\omega_1$ distribution by the Gaussian probability distribution

$$P(\delta\omega_1) = (2\pi\gamma\beta)^{-1/2} \exp\left[-(\delta\omega_1)^2/(2\gamma\beta)\right] , \tag{6.112}$$

where γ and β parametrize the laser frequency fluctuations:

$$\langle\delta\omega_1\delta\omega_1(t-\tau)\rangle = \gamma\beta\exp^{-\beta|t|} . \tag{6.113}$$

If the one- vs. three-photon scenario is applied to the simple case of He photoionization (i.e., only a single product channel is considered) then in accord with

Eq. (3.54) the ionization probability would be given by:

$$P_q(E) = F_q(11) - 2x \cos\left[\phi_3 - 3\phi_1 + \delta_q(13)\right] \epsilon_0^2 |F_q(13)| + x^2 \epsilon_0^4 F_q(33) . \tag{6.114}$$

Assuming that the direct one-photon and direct three-photon parts are unity and the phase assumes a time-dependent fluctuation then the essential part of the Eq. (6.114) is of the form:

$$P_q(\phi_0, t) = 1 + \cos(\phi_0 + \delta\phi(t)) = 1 \pm \cos(\alpha\delta\omega_1(t)) , \tag{6.115}$$

where $\phi_3 - 3\phi_1 + \delta_q(13) \equiv \phi_0 + \delta\phi(t)$ and where the plus and minus refer to the cases where $\phi_0 = 0$ and π, respectively. The average probability is then obtained as

$$\begin{aligned}\langle P_q(E)\rangle &= 1 \pm \int_{-\infty}^{\infty} P(\delta\omega_1) \cos(\alpha\delta\omega_1) d[\delta\omega_1] \\ &= 1 \pm \exp\left(-\alpha^2\gamma\beta/2\right) .\end{aligned} \tag{6.116}$$

A useful criteria for the effect of the partially coherent laser on the control is given by the "contrast ratio", that is, the ratio of the photoionization signal at its maximum and minimum,

$$\zeta_0 = \log_{10}\left[1 + \exp\left(-\alpha^2\gamma\beta/2\right)\right] - \log_{10}\left[1 - \exp\left(-\alpha^2\gamma\beta/2\right)\right] . \tag{6.117}$$

Numerical studies [262] show that Eq. (6.117) provides a zeroth-order approximation to the results of a full computation, which underestimates the degree of possible control in a realistic system. In addition, these results show that even for phase diffusion fields which have widths on the order of wavenumbers, control is still extensive (e.g., $\zeta_0 \approx 5$). Examination of the experimental results on one-photon vs. three-photon control show, however, contrast ratios on the order of 30% [52, 263]. That is, the main experimental limitation, thus far, is due to experimental issues other than the partial laser coherence.

Finally, we note for completeness, that there are some circumstances where weak laser amplitude noise can, in conjunction with a deterministic field, enhance control to a specified target [224–226].

6.6 Countering CW Laser Jitter

For CW lasers, laser decoherence appears via the jitter and drift of the laser phase ϕ in the field $\boldsymbol{E}(z, t)$ for example, Eq. (3.16) with a concomitant reduction in control (see Section 6.5). However, suitable design of the control scenario can result in a method which is immune to the effects of laser jitter. In particular, to do so we rely upon the way in which the laser phase enters into control scenarios.

6.6.1
Laser Phase Additivity

The role of the laser phase in controlling molecular dynamics was clear in the examples shown in Chapter 3. For example, in the one- vs. three-photon scenario the relative laser phase $(\phi_3 - 3\phi_1)$ enters directly into the interference term (see, for example, Eq. (3.53)), as does the relative phase $(\phi_1 - \phi_2)$ in the bichromatic control scenario (Eq. (3.19)). These results embody two useful general rules about the contribution of the laser phase to coherent control scenarios. The first is that the interference term contains the *difference* between the laser phase imparted to the molecule by one route, and that imparted to the molecule by an alternate route. Second, the phase imparted to the state $|E_m\rangle$ by a light field of the form

$$\hat{\epsilon} \int_{-\infty}^{\infty} d\omega |\epsilon(\omega)| \exp[i\phi(\omega)] \exp(-i\omega\tau)$$

in an excitation from level $|E_1\rangle$ to $|E_m\rangle$ is $\pm\phi(|\omega_{m,1}|)$. The plus sign applies to light absorption $(E_m > E_1)$ and the minus sign to stimulated emission $(E_1 > E_m)$. This observation proves very useful in designing schemes which are insensitive to laser phase.

One example is the two-photon plus two-photon scheme discussed in Section 7.1. An alternative, which we sketch here and discuss in further detail in relation to strong-field scenarios (Section 13.2) is called incoherent interference control.

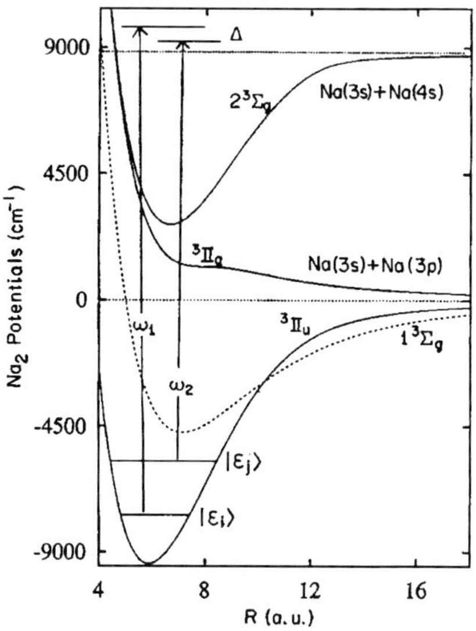

Figure 6.25 Sample scenario for the incoherent interference control of the photodissociation of Na_2. Taken from [269, Figure 1]. (Reprinted with permission. Copyright 1995 American Institute of Physics.)

6.6.2 Incoherent Interference Control

Figure 6.25 shows a level scheme where a CW field with frequency ω_1 excites the level $|E_i\rangle$ to the photodissociative continuum. Simultaneously, a stronger CW laser field of frequency ω_2 couples the continuum to the initially empty state $|E_j\rangle$. The phases associated with these two fields are ϕ_1 and ϕ_2, respectively. The effect of the strong field is to cause Rabi cycling of the population between $|E_j\rangle$ and the continuum. Thus, in this arrangement, the population can be transferred from $|E_i\rangle$ to the continuum by a variety of routes, as shown in Figure 6.26. The method is the multichannel generalization of "laser induced continuum structure" (LICS) [264–268].

The first panel of Figure 6.26 shows the bichromatic control scenario. The second panel in Figure 6.26 shows the simplest path to the continuum, consisting of one-photon absorption of ω_1. The subsequent panels show the three-photon process to the continuum (absorption of ω_1 followed by stimulated emission and reabsorption of ω_2, and so on), and a five-photon process (absorption of ω_1 followed by stimulated emission and reabsorption of ω_2, twice). This series goes on *ad-infinitum*, resulting in an infinite number of interfering pathways.

In accord with Section 6.6.1 the phase imparted to the continuum state by the first route in Figure 6.26 is ϕ_1, and by the second is $\phi_1 - \phi_2 + \phi_2$. The $-\phi_2$ contribution to the latter phase is due to the stimulated emission step, and the following $+\phi_2$ is due to the absorption. Hence both routes impart the overall phase ϕ_1 to the continuum state. It is clear that this is also the case for all additional

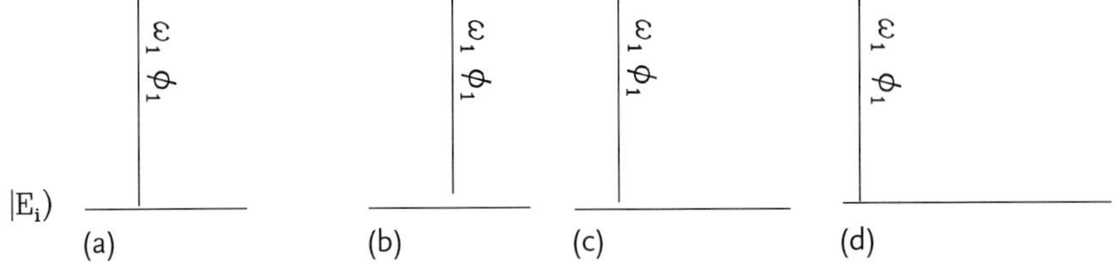

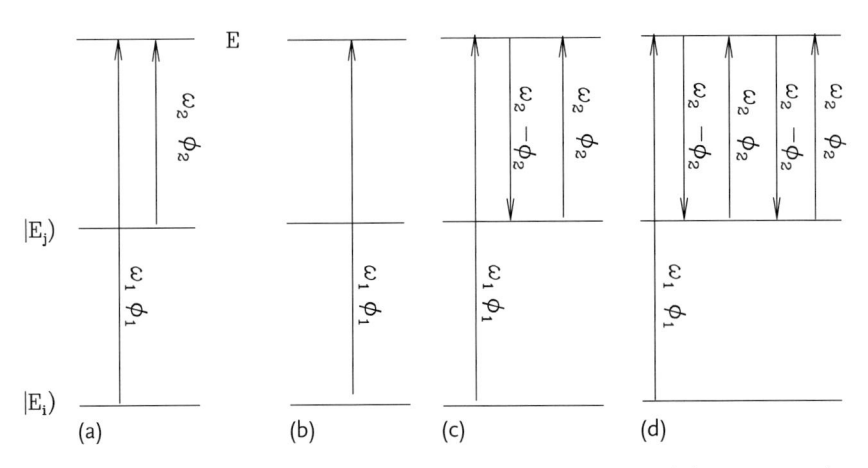

Figure 6.26 Interfering pathways from $|E_i\rangle$ to the continuum associated with the scenario in the previous figure. The frequency and phase of the lasers are ω_i and ϕ_i. (a) Bichromatic control. (b) One-photon absorption. (c) A three-photon process in which the initially unpopulated state $|E_j\rangle$ is coupled to the continuum at energy E and interferes with the one-photon absorption from state $|E_i\rangle$. (d) The same as in (c) for a five-photon process. Notice that in the processes depicted in (c) and (d) the phase ϕ_2 gets cancelled at the completion of each stimulated-emission-followed-by-absorption cycle.

routes to the continuum, since they must contain an equal number of stimulated emission and absorption steps.

Examination of all previous described scenarios makes clear, however, that it is the *relative* phase imparted to the routes which affects control. In the case described here, the relative phase of the routes is $\phi_1 - \phi_1 = 0$, so that control is independent of the laser phase. As a consequence, even lasers with extreme laser jitter and drift can be used in this scenario. Note, however, the additional consequence that in this scenario, control is achieved by varying the frequencies ω_1 and ω_2.

An experimental realization with the pulsed laser version of this approach [270] is discussed in detail in Section 13.2 where control over both relative populations into different channels, as well as over the photodissociation yield into both channels is computed and demonstrated.

7
Case Studies in Coherent Control

Chapter 3 provided an introduction to the principles of coherent control and included a number of examples of scenarios that embody that principle. In this chapter we consider several other scenarios that shed further light on these principles and suggest a number of useful experimental scenarios.

7.1
Two-Photon vs. Two-Photon Control

The *M*- vs. *N*-photon scenarios, where both routes are nonresonant, was discussed in Section 3.4. Here we consider resonantly enhanced routes and show, in particular that a resonantly enhanced two-photon vs. two-photon excitation (see Figure 7.1) provides a means of maintaining control in a molecular system in thermal equilibrium. The resonant character of the excitations ensure that only a particular initial state, out of the thermal distribution of molecular levels, participates in the photodissociation. Hence coherence is established by the excitation, and maintained throughout the process. The design of the proposed control scenario also relies upon the phase additivity arguments in Section 6.6.1 to provide a method for overcoming the reduction in the interference term due to phase jitter in the laser source. In addition, we show that this approach allows one to reduce the contributions from uncontrolled satellite routes. Thus, this scenario is capable of overcoming three of the major decoherence mechanisms discussed in Chapter 6, thereby enabling the execution of coherent control in “natural” thermal environments.

The two-photon control scenario that we describe is completely general, but we focus here on the photodissociation of diatomic molecules. First consider photodissociation along *one* of the paths shown in Figure 7.1a; that is, two-photon dissociation of a molecule where the first laser is resonant with an intermediate bound level. The molecule, initially in a state $|E_i, J_i, M_i\rangle$, is subjected to two CW fields. Once again J_i, M_i denote the angular momentum and its projection along the z-axis. The matter–radiation interaction term is of the form

$$H_{MR}(t) = -2\boldsymbol{d} \cdot \mathrm{Re}\left[\hat{\boldsymbol{\varepsilon}}_2 \bar{\epsilon}_2 \exp(-i\omega_2 t) + \hat{\boldsymbol{\varepsilon}}_1 \bar{\epsilon}_1 \exp(-i\omega_1 t)\right] . \tag{7.1}$$

Quantum Control of Molecular Processes, Second Edition. Moshe Shapiro and Paul Brumer

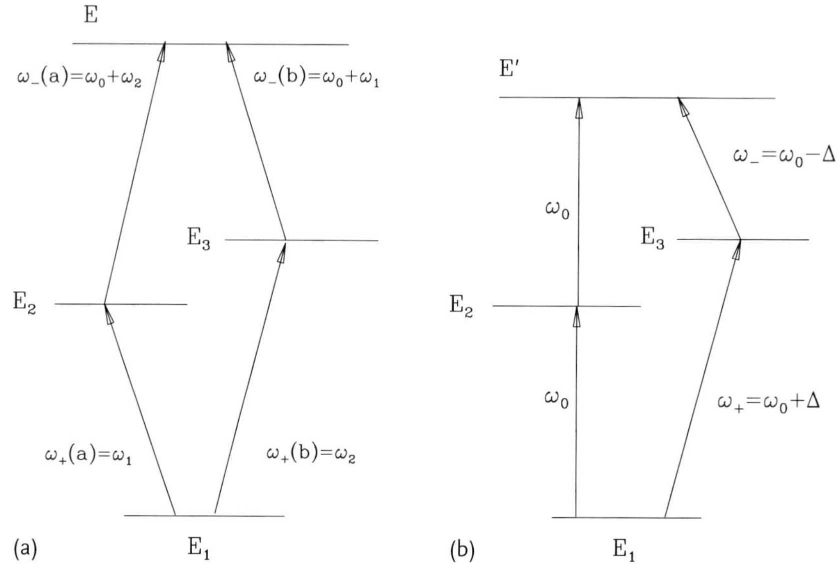

Figure 7.1 Resonantly enhanced two-photon vs. two-photon control scenarios: (a) using four frequencies, (b) using three frequencies.

As a result of irradiation by this field the molecule photodissociates, yielding a number of different product channels labeled by q. Absorption of the first photon of frequency ω_1 lifts the system to an energy close to the energy E_m of an intermediate bound state $|E_m, J_m, M_m\rangle$. A second photon of frequency ω_2 carries the system to the dissociating states $|E, \hat{\boldsymbol{k}}, q^-\rangle$.

Extending Eq. (3.35) to the case of two different frequencies, ω_1 and ω_2, it is possible to show [271] that the probability amplitude $D_{\hat{k},q,i}(E, E_i J_i M_i, \omega_2, \omega_1)$ for resonantly enhanced two-photon dissociation is

$$D_{\hat{k},q,i}(E, E_i J_i M_i, \omega_2, \omega_1)$$
$$= \sum_{E_m, J_m, e'} \frac{\langle E, \hat{\boldsymbol{k}}, q^- | d_{e,e'} \bar{\epsilon}_2 | E_m, J_m, M_i\rangle \langle E_m, J_m, M_i | d_{e',g} \bar{\epsilon}_1 | E_i J_i M_i\rangle}{\hbar\omega_1 - (E_m + \Delta_m - E_i) + i\Gamma_m/2} . \quad (7.2)$$

Here $E = E_i + (\omega_1 + \omega_2)\hbar$, Δ_m and Γ_m are respectively the radiative shift and the full width at half maximum (FWHM) of the intermediate state (the derivation of these quantities is discussed in detail in Section 7.3.1 below). We assume that the lasers are linearly polarized and that their electric field vectors are parallel to one another.

The term $(\hbar\omega_1 - (E_m + \Delta_m - E_i) + i\Gamma_m/2)$ in the denominator of Eq. (7.2) allows us to tune ω_1 so that only a select few of the thermally populated levels $|E_i, J_i, M_i\rangle$ are excited. That is, making this denominator small; that is, achieving this resonance condition establishes coherence, despite the thermal environment.

The probability of producing the fragments in channel q is obtained by integrating the square of Eq. (7.2) over the scattering angles $\hat{\boldsymbol{k}}$:

$$P_q(E, E_i J_i M_i, \omega_2, \omega_1) = \int d\hat{\boldsymbol{k}} \left| D_{\hat{\boldsymbol{k}},q,i}(E, E_i J_i M_i, \omega_2, \omega_1) \right|^2 . \tag{7.3}$$

We consider now the scenarios, shown in Figure 7.1, of simultaneously exciting a molecule by two resonantly enhanced two-photon routes. For example (Figure 7.1b), a molecule is irradiated with three interrelated frequencies, $\omega_0, \omega_+, \omega_-$ with $\omega_\pm = \omega_0 \pm \Delta$. The interference occurs between the two two-photon dissociation routes leading to identical final energies $E = E_i + 2\hbar\omega_0 = E_i + \hbar(\omega_+ + \omega_-)$, where ω_0 and ω_+ are chosen to be resonant with intermediate bound-state levels. The associated field amplitudes are $\bar{\epsilon}_0, \bar{\epsilon}_+$ and $\bar{\epsilon}_-$, whose phases are denoted ϕ_0, ϕ_+, ϕ_-. These fields generate two independent routes to the continuum at energy E. Thus, the probability of photodissociation at energy E into arrangement channel q is given by the square of the sum of the D matrix elements from pathway "a" (absorption of $\omega_0 + \omega_0$) and pathway "b" (absorption of $\omega_+ + \omega_-$). That is, the probability into channel q is

$$\begin{aligned} P_q(E, E_i J_i M_i; \omega_0, \omega_+, \omega_-) &= \int d\hat{\boldsymbol{k}} \Big| D_{\hat{\boldsymbol{k}},q,i}(E, E_i J_i M_i, \omega_0, \omega_0) \\ &\quad + D_{\hat{\boldsymbol{k}},q,i}(E, E_i J_i M_i, \omega_+, \omega_-) \Big|^2 \\ &\equiv P_q(a) + P_q(b) + P_q(ab) . \end{aligned} \tag{7.4}$$

Here $P_q(a)$ and $P_q(b)$ are the independent photodissociation probabilities associated with routes a and b, respectively, and $P_q(ab)$ is the interference term between them. This is discussed below.

The interference term $P_q(ab)$ can be written as

$$P_q(ab) = 2|F_q(ab)| \cos[\alpha_q(ab) + \delta\phi] , \tag{7.5}$$

where

$$\delta\phi = 2\phi_0 - \phi_+ - \phi_- , \tag{7.6}$$

and the amplitude $|F_q(ab)|$ and the "molecular phase" $\alpha_q(ab)$ are defined via the cross term in Eq. (7.4). (See also Section 3.2, Eq. (3.20).)

Consider now the quantity of interest, the channel branching ratio $R_{qq'}$. Noting that in the weak-field case $P_q(a)$ is proportional to $|\bar{\epsilon}_0|^4$, $P_q(b)$ to $|\bar{\epsilon}_+\bar{\epsilon}_-|^2$, and $P_q(ab)$ to $|\bar{\epsilon}_0^2\bar{\epsilon}_+\bar{\epsilon}_-|$, we can write

$$R_{qq'} = \frac{d_q(a) + x^2 d_q(b) + 2x|d_q(ab)| \cos[\alpha_q(ab) + \delta\phi] + \left(B_q/|\bar{\epsilon}_0|^4\right)}{d_{q'}(a) + x^2 d_{q'}(b) + 2x|d_{q'}(ab)| \cos[\alpha_{q'}(ab) + \delta\phi] + \left(B_{q'}/|\bar{\epsilon}_0|^4\right)} , \tag{7.7}$$

where

$$d_q(a) = \frac{P_q(a)}{|\bar{\epsilon}_0|^4} , \quad d_q(b) = \frac{P_q(b)}{|\bar{\epsilon}_+\bar{\epsilon}_-|^2} ,$$

and

$$|d_q(ab)| = \frac{|F_q(ab)|}{|\bar{\epsilon}_0^2 \bar{\epsilon}_+ \bar{\epsilon}_-|}$$

and

$$x = \left| \frac{\bar{\epsilon}_+ \bar{\epsilon}_-}{\bar{\epsilon}_0^2} \right| .$$

The terms with B_q, $B_{q'}$, described below, correspond to resonantly enhanced photodissociation routes to energies other than $E = E_i + 2\hbar\omega_0$. That is, they are satellite terms that do not coherently interfere with the a and b pathways. These uncontrollable terms should be minimized, and we discuss how this can be done below. Here we just note that the product ratio in Eq. (7.7) depends upon both the laser intensities and relative laser phase, which are therefore control parameters in this scenario.

This scenario, embodied in Eq. (7.7), also provides a means by which control can be improved by eliminating effects due to laser jitter. Specifically, the $\delta\phi$ term of Eq. (7.6) can be subject to the phase fluctuations arising from laser instabilities, substantially reducing control. One can, however, design an experimental implementation of the two-photon plus two-photon scenario which readily compensates for this problem. For example, consider generating the two frequencies $\omega_\pm = \omega_0 \pm \Delta$ as the "signal" and the "idler" beams in a nonlinear down-conversion process that occurs when a beam of frequency $2\omega_0$ passes through an optical parametric oscillator (OPO) [272]. The $2\omega_0$ beam is assumed generated by second harmonic generation from the laser ω_0 with the phase ϕ_0. Because ω_+ and ω_- are generated in this way the phase difference $\delta\phi$ between the $(\omega_0 + \omega_0)$ and $(\omega_+ + \omega_-)$ routes is a constant [273]. That is, fluctuations in ϕ_0 cancel and have no effect on $\delta\phi$ of Eq. (7.6), nor on the interference term $P_q(ab)$ of Eq. (7.5).

The above control scheme also allows for the systematic reduction of the satellite contributions B_q and $B_{q'}$ (Eq. (7.7)). These terms, explicitly given in [271], include contributions from the resonantly enhanced photodissociation processes due to absorption of $(\omega_0 + \omega_-)$, $(\omega_0 + \omega_+)$, $(\omega_+ + \omega_0)$ or $(\omega_+ + \omega_+)$ that lead to photodissociation at energies $E_- = E_i + \hbar(\omega_0 + \omega_-)$, $E_+ = E_i + \hbar(\omega_0 + \omega_+)$, $E_{++} = E_i + \hbar(\omega_+ + \omega_+)$, respectively . Other nonresonant pathways are possible but are negligible by comparison. Controllable reduction of this background term is indeed possible because B_q and $B_{q'}$ are functions of $|\bar{\epsilon}_+|$ or $|\bar{\epsilon}_-|$, while the photodissociation products resulting from paths a and b depend on the product $|\bar{\epsilon}_+\bar{\epsilon}_-|$. Thus, changing $|\bar{\epsilon}_+|$ (or $|\bar{\epsilon}_-|$) while keeping $|\bar{\epsilon}_+\bar{\epsilon}_-|$ fixed will not affect the yield from the controllable paths a and b, but will affect B_q. To this end we introduce the parameters $\bar{\epsilon}_b^2 = \bar{\epsilon}_+\bar{\epsilon}_-$ with $|\bar{\epsilon}_-|^2 = \eta|\bar{\epsilon}_b|^2$, $|\bar{\epsilon}_+|^2 = |\bar{\epsilon}_b^2|/\eta$. The terms B_q and $B_{q'}$ are the only terms dependent upon η and can be reduced by appropriate choice of this parameter. Numerical examples are provided below.

To examine the range of control afforded by this scheme consider the photodissociation of Na_2 (Figure 7.2). For simplicity we focus on the regime below

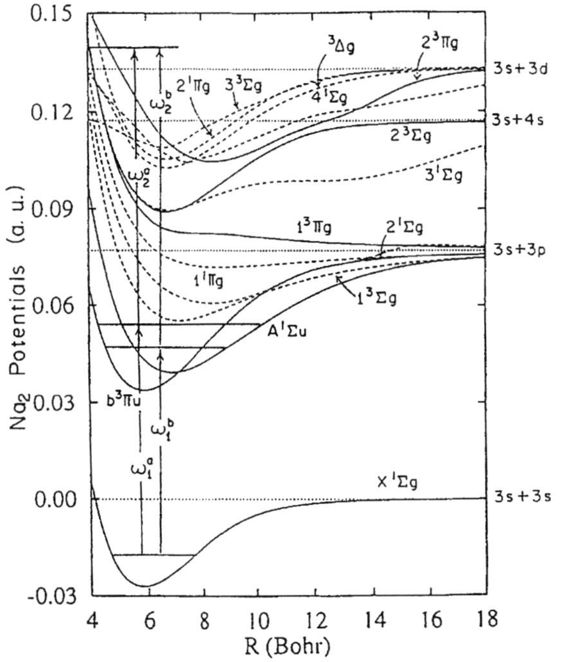

Figure 7.2 Na_2 potential energy surfaces included in the two-photon vs. two-photon control scenario. The arrows indicate the resonantly enhanced two-photon vs. two-photon pathways included in the computation discussed below. From [275, Figure 2]. (Reprinted with permission. Copyright 1994 Wiley-VCH Verlag GmbH+Co. KGaA.)

the Na(3d) + Na(3s) threshold, where dissociation is to the two product channels Na(3s) + Na(3p) and Na(3s) + Na(4s). Results above this energy, where the channel Na(3s) + Na(3d) is open, are described in the literature [269, 274].

Resonantly enhanced two-photon dissociation of Na_2 from a bound state of the ground electronic state occurs [271] by initial excitation to an excited intermediate bound state $|E_m, J_m, M_m\rangle$. The latter is a superposition of states of the $A^1\Sigma_u^+$ and $b^3\Pi_u$ electronic curves, a consequence of spin–orbit coupling. The continuum states reached in the two-photon excitation can have either a singlet or a triplet character, but, despite the multitude of electronic states involved in the computation, the predominant contributions to the products Na(3s) + Na(3p) and Na(3s) + Na(4s) are found to come from the $^3\Pi_g$ and $^3\Sigma_g^+$ electronic states, respectively. The resonant character of the two-photon excitation allows the selection of a single initial state from a thermal ensemble; here results for $v_i = J_i = 0$, where v_i, J_i denote the vibrational and rotational quantum numbers of the initial state, are discussed.

The ratio $R_{qq'}$ depends on a number of laboratory control parameters including the relative laser intensities x, relative laser phase, and the ratio of $|\bar{\epsilon}_+|$ and $|\bar{\epsilon}_-|$ via η. In addition, the relative cross sections can be altered by modifying the detuning. Typical control results are shown in Figures 7.3 and 7.4, which provide contour plots of the Na(3s) + Na(3p) yield as a function of the ratio of the laser am-

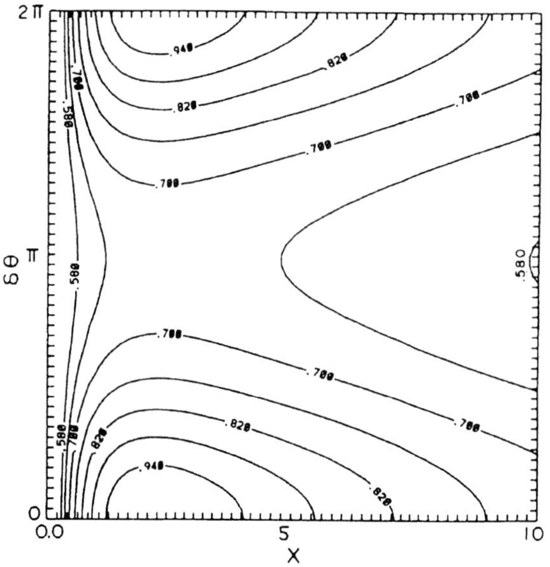

Figure 7.3 Contours of equal Na(3p) yield. Ordinate is the relative laser phase and the abscissa is the field intensity ratio *x*. Here for $\lambda_0 = 623.367$ nm, $\lambda_+ = 603.491$ nm, $\lambda_- = 644.596$ nm and $\eta = 1$. From [276, Figure 1]. (Reprinted with permission. Copyright 1992 Elsevier BV.)

plitudes x, and of the relative laser phase $\delta\phi$ of Eq. (7.6). Consider first Figure 7.3 resulting from excitation with $\lambda_0 = 2\pi c/\omega_0 = 623.367$ nm and $\lambda_+ = 2\pi c/\omega_+ = 603.491$ nm, which are in close resonance with the intermediate states $v = 13$ and 18, $J_m = 1$ of ${}^1\Sigma_u^+$, respectively. The corresponding $\lambda_- = 2\pi c/\omega_-$ is 644.596 nm. The yield of Na(3p) is seen to vary from 58 to 99%, with the Na(3p) atom predominant in the products. Although this range is large, variation of the η parameter should allow for improved control by minimizing the background contributions. This improvement is, in fact, not significant in this case since reducing η decreases B_q from the $(\omega_0 + \omega_-)$ route but increases the contribution from the $(\omega_0 + \omega_+)$ route. By contrast, the background can be effectively reduced for the frequencies shown in Figure 7.4. Here $\lambda_0 = 631.899$ nm, $\lambda_+ = 562.833$ nm and $\lambda_- = 720.284$ nm. In this example, with $\eta = 5$, a larger range of control is achieved, from 30% Na(3p) to 90% as $\delta\phi$ and x are varied.

This control scenario is not limited to the specific frequency scheme discussed above. Essentially all that is required is that two or more resonantly enhanced photodissociation routes lead to interference and that the cumulative laser phases of the two routes be independent of laser jitter. As one sample extension, consider the case (Figure 7.1a) where paths a and b are composed of totally different photons, $\omega_+(a)$ and $\omega_-(a)$, and $\omega_+(b)$ and $\omega_-(b)$, with $\omega_+(a)+\omega_-(a) = \omega_+(b)+\omega_-(b)$. Both these sets of frequencies can be generated, for example, by passing $2\omega_0$ light through two nonlinear crystals, hence yielding two pathways whose relative phase is independent of laser jitter in the initial $2\omega_0$ source. A sample control re-

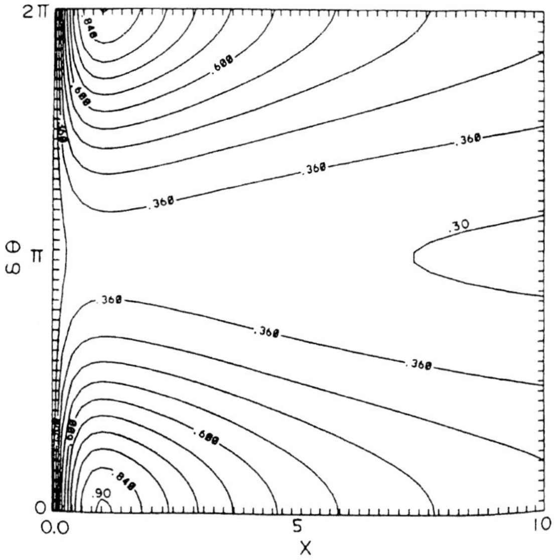

Figure 7.4 As in Figure 7.3 but for $\lambda_0 = 631.899$ nm, $\lambda_+ = 562.833$ nm, $\lambda_- = 720.284$ nm, and $\eta = 5$. From [276, Figure 2]. (Reprinted with permission. Copyright 1992 Elsevier BV.)

sult is shown in Figure 7.5 where $\lambda_+(a) = 599.728$ nm, $\lambda_-(a) = 652.956$ nm, $\lambda_+(b) = 562.833$ nm and $\lambda_-(b) = 703.140$ nm, and $\eta_a = 1/2$, $\eta_b = 10$, where η_a and η_b are the analogue of η for the a and b paths, respectively. Here the range of the control is from 14 to 95%, which is a substantial fraction of total product control and an improvement over the three-frequency approach.

An alternative method of generating the $\omega_+(a), \omega_-(a), \omega_+(b)$ and $\omega_-(b)$ frequencies necessary for the two-photon vs. two-photon scenario is shown in Figure 7.1a. This method has been implemented experimentally [277] in an attempt to control the branching $Na_2 \rightarrow Na + Na(3p)$, $Na + Na(3d)$ two-photon dissociation reaction. One starts with two dye lasers of frequencies $\omega_+(a) = \omega_1$ and $\omega_+(b) = \omega_2$ with two, totally uncorrelated phases ϕ_1 and ϕ_2. By mixing these two beams (using two nonlinear crystals) with a third frequency ω_0 (in the actual experiment this was the fundamental frequency of a Nd:YAG laser operating at $\lambda_0 = 1.06$ μm), one generates two additional frequencies $\omega_-(b) = \omega_1 + \omega_0$ and $\omega_-(a) = \omega_2 + \omega_0$. As depicted in Figure 7.1a, one can interfere a two-photon route composed of absorption of ω_1 and $\omega_2 + \omega_0$ with the ω_2 and $\omega_1 + \omega_0$ two-photon route. Using the phase additivity, which is realized quite well in the nonlinear mixing process, the phase of the first route is $\phi_1 + (\phi_2 + \phi_0)$ and the phase of the second route is $\phi_2 + (\phi_1 + \phi_0)$. That is, the overall phase associated with each route is the same. This is the case, even though the four phases that make up the two two-photon routes are totally uncorrelated. By adding a delay line to either of the two-photon pathways one can introduce in a controlled way any desired *relative phase* $\delta\phi$ between the two routes, which then serves as a laboratory knob in the control experiment.

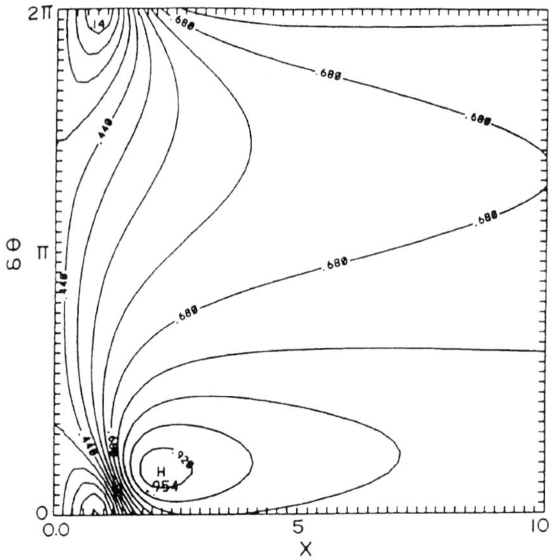

Figure 7.5 As in Figure 7.3 but for the Na(4s) product in the four-field case with $\lambda_+(a) = 599.728$ nm, $\lambda_-(a) = 652.956$ nm, $\lambda_+(b) = 562.833$ nm, $\lambda_-(b) = 703.140$ nm, $\eta_a = 1/2$ and $\eta_b = 10$. From [276, Figure 3]. (Reprinted with permission. Copyright 1992 Elsevier BV.)

A simpler, though much more limited implementation of two-photon vs. two-photon control, entails the use of just two frequencies, ω_1 and ω_2. It is a special case of the scenario using four frequencies, depicted in Figure 7.1a, where $\omega_0 = 0$ (i.e., no mixing with ω_0 is performed). Because any phase change in either ω_1 or ω_2 will automatically affect the two routes, it is not possible in this two-frequency scenario to introduce an externally controlled relative phase change between the two routes. It is, however, possible to alter the $\alpha_q(ab)$ "molecular phase" by detuning either ω_1 or ω_2 off their respective resonances. Due to the presence of the $i\Gamma_m/2$ term in the denominator of Eq. (7.2), the detuning of just one frequency from resonance results in a phase change (in addition to the much more noticeable amplitude change) of one two-photon matrix element relative to another.

7.1.1 Experimental Implementation

This type of two-photon vs. two-photon phase control has been implemented experimentally in molecules by Pratt [278], who studied the photoionization of the $A^2\Sigma^+, v' = 1$ state of NO, and in atoms by Elliott *et al.* [279], who demonstrated control over the branching ratio for photoionization of Ba into the $6s_{1/2}$, $5d_{3/2}$ and $5d_{5/2}$ states of Ba^+. Quite unexpectedly, the interference between the two-photon routes turned out to be destructive when both lasers were resonant with their respective transitions. A theoretical treatment of this effect was provided by Luc–Koenig *et al.* [280].

An experimental implementation of a three-color two-photon vs. two-photon control scheme of Figure 7.1b was performed by Georgiades *et al.* [281]. Their experimental setup is shown in Figure 7.6 where the ω_0 beam is generated by a Ti:sapphire laser and the $\omega_\pm$ beams are generated from an optical parametric oscillator (OPO) which is pumped by a frequency doubled Ti:sapphire laser beam at $2\omega_0$. Thus, the conditions set out in Figure 7.1b, that $\omega_+ + \omega_- = 2\omega_0$, are satisfied. Using a piezoelectric transducer it is possible to change the optical path of the ω_0 beam relative to the $\omega_\pm$ beams. The three beams serve to excite the $6D_{5/2}, F'' = 6$ state of Cs whose population is measured by monitoring the $6D_{5/2}, F'' = 6 \rightarrow 6P_{3/2}, F' = 5$ fluorescence. As shown in Figure 7.7, modulation of the fluorescence signal as a function of the relative phase between the $\omega_+ + \omega_-$ and the $2\omega_0$ routes is obtained, indicating a modulation in the population of the $6D_{5/2}, F'' = 6$ state.

An experimental application of the two-photon vs. two-photon control scenario was presented by Meshulach and Silberberg [140]. These authors considered the off-resonance two-photon excitation between two levels using a broadband pulse. Following Eq. (7.2) of Section 7.1 we write the two-photon transition amplitude between two bound states of energies E_g and E_f subject to two CW sources of frequencies ω_1 and ω_2 as

$$b^{(2)}_{E_f}(\omega_2, \omega_1) = \bar{\epsilon}(\omega_2)\bar{\epsilon}(\omega_1) \sum_i \frac{\langle E_f | d | E_i \rangle \langle E_i | d | E_g \rangle}{\hbar\omega_1 - \left(E_i + \Delta_i - i\Gamma_i/2 - E_g\right)} . \tag{7.8}$$

Consider the case in which ω_1 and ω_2 are both derived from the same pulse. We know from the resonance condition (Eq. (2.12)) that after the pulse is over only those frequencies whose sum is in exact resonance with the transition of interest,

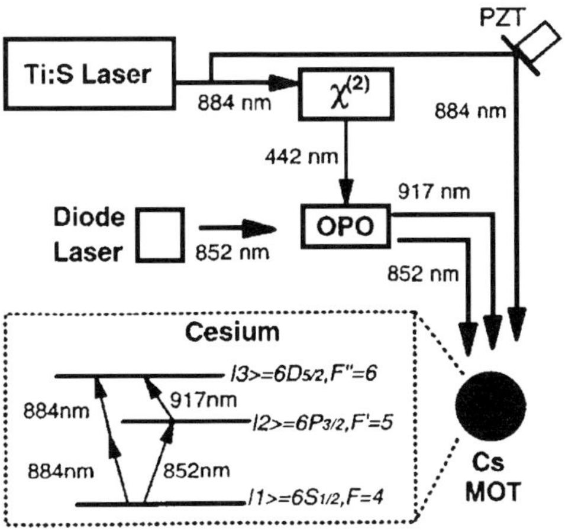

Figure 7.6 Experimental arrangement for the $(\omega_+ + \omega_-)$ vs. $2\omega_0$, two-photon vs. two-photon control of the population of Cs in the $6D_{5/2}, F'' = 6$ state. PZT is the piezoelectric transducer, OPO is optical parametric oscillator, and MOT is the magneto-optical trap of Cs atoms. The inset shows the Cs levels of interest. From [281, Figure 1]. (Reprinted with permission. Copyright 1996 Optical Society of America.)

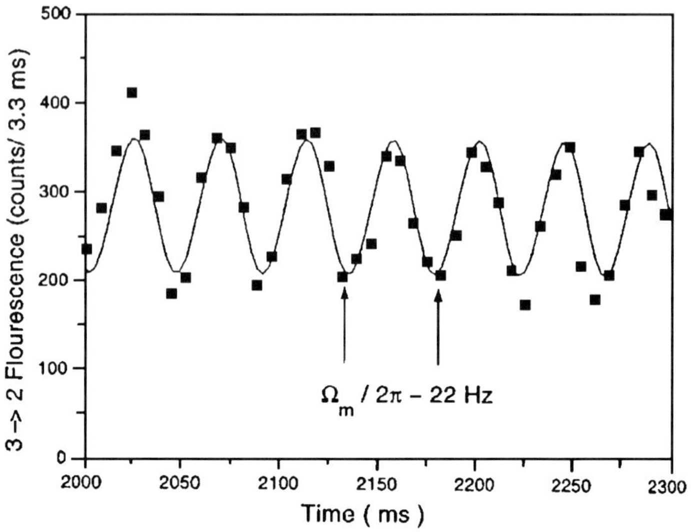

Figure 7.7 Fluorescence due to the $|3\rangle \rightarrow |2\rangle$ transition, where $|3\rangle = |6D_{5/2}, F'' = 6\rangle$ and $|2\rangle = |6P_{3/2}, F' = 5\rangle$ as the PZT is scanned in time, thereby inducing a proportional phase change. The solid curve is a fit of a constant plus a sinusoidal function. From [281, Figure 2]. (Reprinted with permission. Copyright 1996 Optical Society of America.)

$\omega_{f,g} \equiv \omega_1 + \omega_2 = (E_f - E_g)/\hbar$, are effective. Because of the presence of many modes in the pulse, there are multiple paths of different ω_1 all having the same $\omega_1 + \omega_2$ $(= \omega_{f,g})$, that contribute to the two-photon transition amplitude. Integrating over all these paths, we have (denoting ω_1 as ω) that,

$$b^{(2)}_{E_f} = \int_0^\infty d\omega \bar{\epsilon}(\omega_{f,g} - \omega)\bar{\epsilon}(\omega) \sum_i \frac{\langle E_f | d | E_i \rangle \langle E_i | d | E_g \rangle}{\hbar\omega - \left(E_i + \Delta_i - i\Gamma_i/2 - E_g\right)} . \tag{7.9}$$

Assuming that there is no intermediate resonance within the pulse bandwidth, that is, that $E_i - E_g \gg \hbar\omega$, the denominator can be approximated as $\hbar\omega - E_i - \Delta_i + E_g + i\Gamma_i/2 \approx \hbar\omega_{f,g}/2 - (E_i + \Delta_i - E_g) + i\Gamma_i/2$ and taken out of the integration to obtain that

$$\begin{aligned} b^{(2)}_{E_f} &= \langle E_f | M^{(2)} | E_g \rangle \int_0^\infty d\omega \bar{\epsilon}(\omega)\bar{\epsilon}(\omega_{f,g} - \omega) \\ &= \langle E_f | M^{(2)} | E_g \rangle \int_{-\omega_{f,g}/2}^{\infty} d\Omega\, \bar{\epsilon}\left(\omega_{f,g}/2 + \Omega\right) \bar{\epsilon}\left(\omega_{f,g}/2 - \Omega\right) , \end{aligned} \tag{7.10}$$

where $\Omega \equiv \omega - \omega_{f,g}/2$, and

$$\langle E_f | M^{(2)} | E_g \rangle \equiv \sum_i \frac{\langle E_f | d | E_i \rangle \langle E_i | d | E_g \rangle}{\hbar\omega_{f,g}/2 - (E_i + \Delta_i - E_g) + i\Gamma_i/2} .$$

An alternative way of seeing this result is to use the fact that the Fourier transform of a product of functions is the *convolution* integral,

$$\int dt\epsilon'(t)\epsilon''(t)\exp(-i\omega t) = 2\pi \int d\omega'\epsilon'(\omega')\epsilon''(\omega - \omega') . \tag{7.11}$$

Hence,

$$b^{(2)}_{E_f} = \frac{\langle E_f|M^{(2)}|E_g\rangle}{2\pi} \int dt\varepsilon^2(z,t)\exp(-i\omega_{f,g}t) . \tag{7.12}$$

We see that in the absence of an intermediate level, the two-photon transition amplitude is proportional to the Fourier transform of the square of the field at the transition frequency. Thus in complete analogy with the one-photon case (Eq. (2.12)), in the absence of an intermediate level, a single ($\omega = \omega_{f,g}$) frequency component of the Fourier transform of the field-squared determines the two-photon transition amplitude.

Denoting the phase of $\bar{\epsilon}(\omega)$ as $\phi(\omega)$, we can write $\bar{\epsilon}(\omega) = |\epsilon(\omega)|\exp[i\phi(\omega)]$, hence

$$\begin{aligned} b^{(2)}_{E_f} = \langle E_f|M^{(2)}|E_g\rangle \int d\Omega \left|\epsilon\left(\omega_{f,g}/2 + \Omega\right)\epsilon\left(\omega_{f,g}/2 - \Omega\right)\right| \\ \cdot \exp\left[i\phi\left(\omega_{f,g}/2 - \Omega\right) + \phi\left(\omega_{f,g}/2 + \Omega\right)\right] . \end{aligned} \tag{7.13}$$

Quite clearly, the integral in Eq. (7.13) is maximized when the integrand is positive over the entire integration range. This is the case for a *transform limited* pulse for which, in the dipole approximation, ϕ can be taken as a constant independent of ω. Hence, in the *absence* of an intermediate resonance, for a fixed frequency bandwidth, the transform limited pulse yields the highest two-photon absorption probabilities.

Equation (7.13) is especially interesting because it shows that any *antisymmetric* phase modulation about the $\omega_{f,g}/2$ point,

$$\phi\left(\omega_{f,g}/2 - \Omega\right) = -\phi\left(\omega_{f,g}/2 + \Omega\right) ,$$

that is, a modulation for which the phase sum about the midpoint is zero, should have the same two-photon absorption probability as a transform-limited pulse with the same power spectrum! Thus, the modulated pulse can have a temporal shape that is drastically different than the transform-limited shape, yet it will be just as effective in bringing about a two-photon absorption process.

Silberberg *et al.* [140] demonstrated this effect by introducing an extra *phase modulation* of the type

$$\phi(\Omega) = \alpha\cos(\beta\Omega + \gamma) . \tag{7.14}$$

When $\gamma = \pi/2$, ϕ is an antisymmetric sine function and $b^{(2)}_{E_f}$ should remain the same as the transform-limited case. When $\gamma = 0$, ϕ is a symmetric cosine function and $b^{(2)}_{E_f}$ should go down. There may in fact be particular values of α and β that give rise to a complete destructive interference between all the paths that contribute to the integral.

An experimental demonstration [140] of the above destructive interference is given in Figure 7.8. We see that when $\gamma = n\pi, n = 0, 1, \ldots$ and the modulation period is 220 fs, the two-photon amplitude goes to zero for pulses whose modulation depths are given by $\alpha \approx 1.2, 2.9, 4.4, \ldots$ Such pulses may therefore be termed "dark" pulses. In contrast, when an antisymmetric modulation ($\gamma = \pi/2$) is applied, there is indeed little loss in the two-photon absorption probability relative to the transform limited case. (The small degree of loss observed with increasing modulation depth is attributed [140] to imperfections in the pulse shaper.)

The situation changes significantly when the two-photon process is resonantly enhanced. For a *single* intermediate resonance, it follows from Eq. (7.9) that

$$b_{E_f}^{(2)} = \int_0^\infty d\omega \bar{\epsilon}(\omega_{f,g} - \omega)\bar{\epsilon}(\omega) \frac{\langle E_f|d|E_i\rangle\langle E_i|d|E_g\rangle}{\hbar(\omega - \omega_{i,g}) + i\Gamma_i/2}, \tag{7.15}$$

where $\omega_{i,g} = (E_i + \Delta_i - E_g)/\hbar$. Writing

$$\frac{1}{\hbar(\omega - \omega_{i,g}) + i\Gamma_i/2} = \frac{\hbar(\omega - \omega_{i,g})}{\hbar^2(\omega - \omega_{i,g})^2 + \Gamma_i^2/4} - i\frac{\frac{\Gamma_i}{2}}{\hbar^2(\omega - \omega_{i,g})^2 + \Gamma_i^2/4},$$

we see that whereas the imaginary part is *symmetric* about the $\omega = \omega_{i,g}$ point, the real part is *antisymmetric* about that point. The analysis becomes even simpler

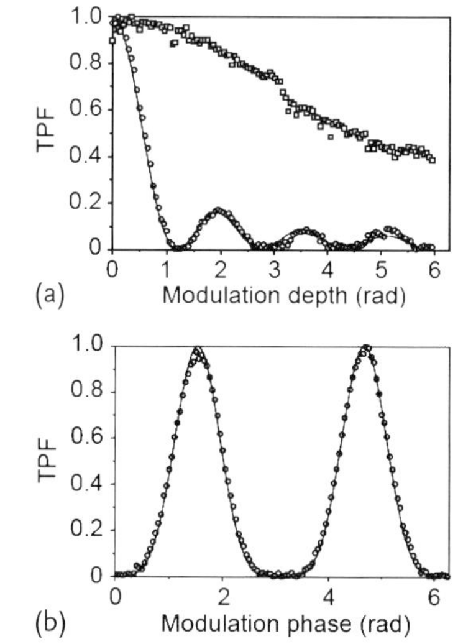

Figure 7.8 Experimental (○, or, □) and calculated (—) results of two-photon excitation probabilities in Cs with a pulse whose phase is modulated according to Eq. (7.14). (a) As a function of the modulation depth α. Lower trace – $\gamma = 0$; upper trace – $\gamma = \pi/2$. (b) As a function of the modulation constant γ for the $\alpha = 1.2$ modulation depth. Taken from [140, Figure 3]. (Reprinted with permission. Copyright 1998 Nature Publishing Group.)

when the resonance is narrow, because in that case

$$\lim_{\Gamma_i \to 0} \mathrm{Im}\left[\frac{1}{\omega - \omega_{i,g} + i\frac{\Gamma_i}{2\hbar}}\right] = -\pi\delta(\omega - \omega_{i,g}) ,$$

and

$$\left(\mathrm{Im} b^{(2)}_{E_f}\right)^2 = \left|\frac{\pi}{\hbar}\epsilon(\omega_{f,i})\epsilon(\omega_{i,g})\langle E_f|d|E_i\rangle\langle E_i|d|E_g\rangle\right|^2 . \tag{7.16}$$

We see that the phase of the pulse does not affect the square of the imaginary part. In contrast, the square of the real part does not reach its maximal value for pulses, such as a transform limited pulse for which ϕ is constant, whose phase function is symmetric about the $\omega_{i,g}$ point, because in that case the contribution of the $\omega > \omega_{i,g}$ frequencies tends to cancel the contribution of the $\omega < \omega_{i,g}$ frequencies. Hence it is possible, as demonstrated experimentally by Silberberg *et al.* [282] in the two-photon excitation of Rb, illustrated in Figure 7.9, to do *better* than a transform limited pulse by either eliminating a portion of the frequency profile of the pulse (as shown in Figure 7.10), or by introducing some asymmetry about the $\omega = \omega_{i,g}$ point into the phase function $\phi(\omega)$ (e.g., via a $\pi/2$ phase jump window – as shown in Figure 7.11).

The same principles outlined above can be used to *discriminate* between various accessible final states using a pulse whose bandwidth by far exceeds the spacing between such states [283]. In absorption, this follows immediately from Eq. (7.13), where we see that an antisymmetric phase modulation about $\omega_{f,g}$ that leaves $\phi(\omega_{f,g}/2 - \Omega) + \phi(\omega_{f,g}/2 + \Omega) = 0$ will no longer be antisymmetric with respect to another final state whose transition frequency is $\omega_{f',g} \neq \omega_{f,g}$.

Similar ideas apply to stimulated Raman transitions, where the second (virtual) step constitutes the stimulated emission (rather than the absorption) of a photon. The step being that of emission means that $E_g + \hbar\omega > E_f$, or that $\omega_{f,g} - \omega < 0$. Using the reality condition of the pulse (Eq. (1.10)), according to which each elec-

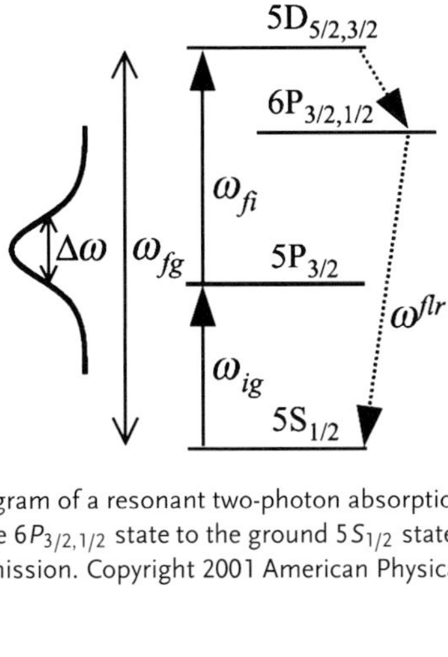

Figure 7.9 Energy level diagram of a resonant two-photon absorption in Rb. Monitored is the fluorescence signal from the $6P_{3/2,1/2}$ state to the ground $5S_{1/2}$ state. Taken from [282, Figure 1]. (Reprinted with permission. Copyright 2001 American Physical Society.)

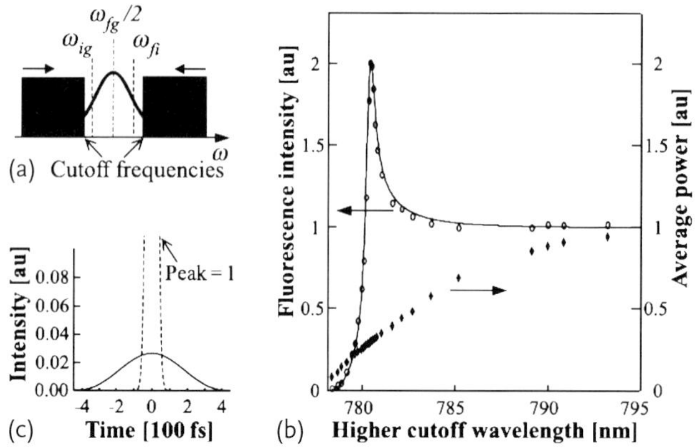

Figure 7.10 (a) The spectral cut-off region (shaded in black). (b) Calculated (—) and measured (○) (monitored by the fluorescence signal from the $6P_{3/2,1/2}$ state) two-photon absorption probability, and transmitted power (◇), as a function of the position of the higher spectral cut-off point. (c) The temporal intensity of the optimal pulse (—) relative to that of the transform limited pulse. (— — —) The peak intensity of the optimally truncated pulse is lower by a factor of 38 relative to the transform limited pulse, yet at that point the two-photon absorption probability is double that of the transform limited pulse. Taken from [282, Figure 2]. (Reprinted with permission. Copyright 2001 American Physical Society.)

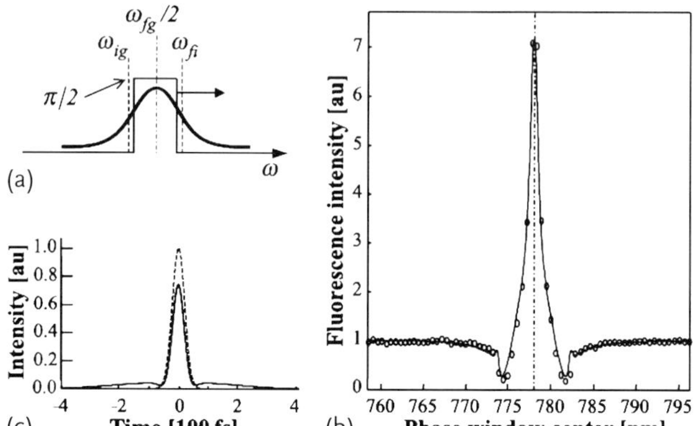

Figure 7.11 (a) The $\pi/2$ phase jump window. (b) Calculated (—) and experimental (○) (monitored by the fluorescence signal from the $6P_{3/2,1/2}$ state) two-photon absorption probability as a function of the center of the $\pi/2$ phase jump window. (c) The temporal intensity of the optimal pulse (—) and the transform limited pulse (— — —). Taken from [282, Figure 3]. (Reprinted with permission. Copyright 2001 American Physical Society.)

tric field component of negative frequency is the complex conjugate of its positive frequency counterpart, we can write that $\bar{\epsilon}(\omega_{f,g} - \omega) = \bar{\epsilon}^*(\omega - \omega_{f,g}) = |\epsilon(\omega - \omega_{f,g})| \exp[-i\phi(\omega - \omega_{f,g})]$. Hence, in the absence of an intermediate resonance, the stimulated Raman amplitude is given as

$$b_{E_f}^{(2)} = \int_0^\infty d\omega |\epsilon(\omega - \omega_{f,g})\epsilon(\omega)| \exp\left[i\left(\phi(\omega) - \phi(\omega - \omega_{f,g})\right)\right] \langle E_f | M^{(2)} | E_g \rangle \; . \tag{7.17}$$

It is clear that any phase modulation that will keep

$$\phi(\omega) - \phi(\omega - \omega_{f,g}) = 0 \tag{7.18}$$

will cause the modulated pulse to be as efficient as a transform limited pulse in bringing about a stimulated Raman transition. This means that we need to make sure that for every frequency ω there is a lower frequency component at $\omega - \omega_{f,g}$ with an identical phase in order to maximize the *constructive* interference. As pointed out above, the condition of Eq. (7.18) does not necessarily hold for other levels for which $\omega_{f',g} \neq \omega_{f,g}$. The various ω components of the integrand of Eq. (7.17) are therefore expected to interfere in a random-like manner with each other, thereby reducing the stimulated Raman transition probabilities to those levels. One can therefore *tune in* any Raman transition of interest by merely modulating the laser pulse in accordance with Eq. (7.18). In this way it is possible to achieve an effective spectral resolution that is by far finer than the pulse bandwidth, with the tuning of the desired transition performed most conveniently via a computer that controls the voltages applied to the pixels of a pulse shaper. (See Chapter 5 and [164, 166–171].)

It is of interest to examine the extent to which these results hold in the general N-photon case. Following the discussion of N-order perturbation theory in Section 3.4.1, we can write the N-photon transition amplitude between states $|E_g\rangle$ and $|E_f\rangle$ under the action of a broadband pulse of light as

$$b_{E_f}^{(N)} = \int \left(\prod_{j=1}^{N-1} d\omega_j \right) \bar{\epsilon}(\omega_N) \sum_{i_1,\ldots,i_{N-1}} \langle E_f | d | E_{i_{N-1}} \rangle$$
$$\prod_{k=1}^{N-1} \bar{\epsilon}(\omega_k) \cdot \frac{\langle E_{i_k} | d | E_{i_{k-1}} \rangle}{\sum_{j=1}^{k} \hbar\omega_j - \left(E_{i_k} + \Delta_{i_k} - i\Gamma_{i_k}/2 - E_g\right)} \; , \tag{7.19}$$

where $i_1, \ldots, i_{N-1}$ go over all the intermediate states with $i_0 = g$ and $\omega_N = \omega_{f,g} - \sum_{j=1}^{N-1} \omega_j$. In the absence of intermediate resonances, for example, when $|\omega_{f,g}| \ll |E_i - E_g|/\hbar$ for all $i \neq f$ or g, we can replace the

$$\sum_{j=1}^{k} \hbar\omega_j - \left(E_{i_k} + \Delta_{i_k} - \frac{i\Gamma_{i_k}}{2} - E_g \right)$$

term in the denominator with some constant term $\hbar\overline{\omega} - \left(E_k + \Delta_k - i\Gamma_k/2 - E_g\right)$, and write that

$$b_{E_f}^{(N)} = \langle E_f | M^{(N)} | E_g \rangle \int \left(\prod_{j=1}^{N-1} d\omega_j \right) \prod_{k=1}^{N} \bar{\epsilon}(\omega_k) \; , \tag{7.20}$$

where

$$\langle E_f | M^{(N)} | E_g \rangle \equiv \sum_{i_1,\ldots,i_{N-1}} \langle E_f | d | E_{i_{N-1}} \rangle \prod_{k=1}^{N-1} \frac{\langle E_{i_k} | d | E_{i_{k-1}} \rangle}{\hbar\overline{\omega} - \left(E_{i_k} + \Delta_{i_k} - i\Gamma_{i_k}/2 - E_g\right)} \, .$$

By a repeated application of Eq. (7.11) it follows that $b_{E_f}^{(N)}$ of Eq. (7.20) can be written as,

$$b_{E_f}^{(N)} = \frac{\langle E_f | M^{(N)} | E_g \rangle}{(2\pi)^{N-1}} \int dt \varepsilon^N(t) \exp(-i\omega_{f,g} t) \, . \tag{7.21}$$

Equation (7.21) is very useful if we want to control the strength of transitions between *nondegenerate* states, essentially using variants of the pulse modulations of the $N = 2$ case above [284]. In the highly nonlinear category, the $N = 3$ case, arising from the application of three external fields (which can all be derived from a *single* broadband pulse), leading to a variety of the so-called "four-wave mixing" scenarios [285], has been most widely applied [286–290]. Of particular interest is the CARS (coherent anti-Stokes Raman spectroscopy) setup, which is a variant of the Stimulated Raman process in which the extra field is introduced to improve the detection, using the directionality of the emitted fourth beam [285].

The reason highly nonlinear processes such as CARS are so useful lies in their N-order power law dependence which makes it possible by focusing the pulse on a given plane to sharply discriminate against objects that are not in the focusing plane. In this way it is possible to perform nonlinear *microscopy* which for the above reasons obviates the need to dye the sample. The introduction of coherent control techniques to nonlinear microscopies allows one to focus on different layers of a (still living!) cell, while at the same time changing the (frequencies of the) pulse modulation [284, 286, 291] so as to tune in a few well chosen transitions and identify the chemical makeup of the probed region [284, 286, 291, 292].

Besides providing insight for developing strategies for altering the magnitude of N-photon transitions between *nondegenerate* states, Eq. (7.21) shows that, in the absence of an intermediate resonance, it is not possible to optically control the populations of *degenerate* states, using N-photon processes from a *single* precursor state. This is because in the absence of intermediate resonances, the transition amplitude $b_{E_f}^{(N)}$ factorizes into a product of the material part $\langle E_f | M^{(N)} | E_g \rangle$ and a radiative part $\int dt \epsilon^N(t) \exp(-i\omega_{f,g} t)$. Since the only information regarding the final state appearing in the radiative part is its energy (contained in the $\omega_{f,g}$ factor), there is no way that the shape of the pulse can affect the populations of states with the *same* energy. In particular, if we consider control in a multichannel continuum, then the amplitude at time t for observing channel $\boldsymbol{n}$ in the far future is given in this case as,

$$b_{E,\boldsymbol{n}}^{(N)}(t) = \frac{\langle E, \boldsymbol{n}^- | M^{(N)} | E_g \rangle}{(2\pi)^{N-1}} \int_0^t dt' \varepsilon^N(t') \exp(-i\omega_{E,g} t') b_{E_g}(t') \, , \tag{7.22}$$

where $b_{E_g}(t)$ is the amplitude of the initial state, and

$$\langle E, \boldsymbol{n}^- | M^{(N)} | E_g \rangle = \sum_{i_1,\ldots,i_{N-1}} \langle E, \boldsymbol{n}^- | d | E_{i_{N-1}} \rangle \prod_{k=1}^{N-1} \frac{\langle E_{i_k} | d | E_{i_{k-1}} \rangle}{\hbar\varpi - \left(E_{i_k} + \Delta_{i_k} - i\Gamma_{ik}/2 - E_g\right)} , \quad (7.23)$$

where $|E_{i_0}\rangle \equiv |E_g\rangle$. We see that the branching ratio between different final channels is given as

$$R_{\boldsymbol{n},\boldsymbol{m}}(t) \equiv \left| \frac{b^{(N)}_{E,\boldsymbol{n}}(t)}{b^{(N)}_{E,\boldsymbol{m}}(t)} \right|^2 = \left| \frac{\langle E, \boldsymbol{n}^- | M^{(N)} | E_g \rangle}{\langle E, \boldsymbol{m}^- | M^{(N)} | E_g \rangle} \right|^2 , \quad (7.24)$$

and the pulse attributes, as well as the detailed evolution history of the initial state coefficient $b_{E_g}(t')$, have completely disappeared. Thus it is not possible by manipulating the laser field(s) to control the branching ratio into different scattering channels in a fixed-N multiphoton process which starts with only one precursor state. As in our general discussion, the remedy in this case is to interfere M-photon processes with N-photon processes leading to the same final state $|E, \boldsymbol{n}^-\rangle$, with $M \neq N$, or to start from several precursor states.

7.2 Control over the Refractive Index

Photodissociation is but one of many processes which are amenable to control. A host of other processes that have been studied are discussed later in this book, such as asymmetric synthesis, control of bimolecular reactions, strong-field effects, and so on Also of interest is control of nonlinear optical properties of materials [272], particularly for device applications. In this section we describe an application of the bichromatic control scenario discussed in Section 3.2.1 pertaining to the control of refractive indices.

The real and imaginary parts of the refractive index n quantify the scattering and absorption (or amplification) properties of a material. The refractive index is best derived from the susceptibility tensor $\underline{\underline{\chi}}$ of the material, defined below, which describes the response of a macroscopic system to incident radiation [293]. Specifically, an incident electric field $\boldsymbol{E}(\boldsymbol{r}, t)$, where $\boldsymbol{r}$ denotes the location in the medium, tends to displace charges, thereby polarizing the medium. The change in $\boldsymbol{d}^{ind}(\boldsymbol{r}, t)$, the induced dipole moment, from point $\boldsymbol{r}$ to point $\boldsymbol{r} + d\boldsymbol{r}$ is given in terms of the polarization vector $\boldsymbol{P}(\boldsymbol{r}, t)$, defined as

$$d\boldsymbol{d}^{ind}(\boldsymbol{r}, t) = \boldsymbol{P}(\boldsymbol{r}, t) d\boldsymbol{r} . \quad (7.25)$$

It is customary to relate the polarization to the external field by defining a susceptibility $\underline{\underline{\chi}}(t)$ tensor via the relation

$$P(r, t) = \epsilon_0 \int_0^\infty d\tau \underline{\underline{\chi}}(r, \tau) \cdot E(r, t - \tau) . \tag{7.26}$$

$\underline{\underline{\chi}}$ is a complex tensor whose real part relates to light scattering and whose imaginary part describes absorption or amplification of light. In the weak-field (linear) domain it is independent of the field $E(r, t)$. As the field gets stronger $\underline{\underline{\chi}}$ may become dependent on the field, in which case we say that $\underline{\underline{\chi}}$ has nonlinear contributions. Below, for convenience, we suppress the spatial dependence of $\underline{\underline{\chi}}$.

For CW light, $E(z, t) = (\mathcal{E}/2)\hat{\epsilon} \exp(-i\omega t + ik \cdot r) + \text{c.c.}$, where c.c. denotes the complex conjugate of the preceding term, and Eq. (7.26) becomes

$$P(r, t) = \frac{\epsilon_0 \mathcal{E}}{2} \left[\underline{\underline{\chi}}(\omega) \exp(-i\omega t + ik \cdot r) + \underline{\underline{\chi}}(-\omega) \exp(i\omega t - ik \cdot r) \right] \cdot \hat{\epsilon} , \tag{7.27}$$

where $\underline{\underline{\chi}}(\omega)$ is the Fourier transform of $\underline{\underline{\chi}}(t)$.

To see the relationship to the refractive index we focus on the case where the tensor $\underline{\underline{\chi}}$ reduces to a scalar. This is the case, for example, if the electronic response of the medium is isotropic, in which case $\underline{\underline{\chi}}(\omega)$ reduces to a scalar $\chi(\omega)$, or where the field is, for example, along the laboratory z-axis and only the single $\chi_{zz}(\omega)$ component is of interest. In the former case, the complex refractive index $n(\omega)$ is given by

$$n^2(\omega) = 1 + \chi(\omega) . \tag{7.28}$$

Analogous expressions hold for radiation incident on *individual* molecules, but here the polarizability $\underline{\underline{\alpha}}(t)$ replaces the susceptibility $\underline{\underline{\chi}}(t)$. That is, the induced molecular dipole $d^{ind}(t)$ is given as

$$d^{ind}(t) = \int_0^\infty d\tau \underline{\underline{\alpha}}(\tau) \cdot E(r, t - \tau) . \tag{7.29}$$

Therefore, given the system and field $E(t)$, to obtain either the susceptibility or the polarizability requires that we compute the induced polarization, or induced dipole, respectively.

The susceptibility and polarizability differ in that the latter deals with single molecules, and the former with an entire medium. Further, the polarizability can be defined for the system in a particular quantum state, whereas the susceptibility generally refers to the bulk system, often in thermodynamic equilibrium. Typically, then [293] the susceptibility includes the number of particles per unit volume ρ

and an average over populated system energy levels. Below we compute the polarizability and susceptibility for neither of these cases. Rather, we extend the standard definition of the polarizability to include a molecule initially in a superposition state. In this case the polarizability and susceptibility are virtually identical, and are related by

$$\epsilon_0 \underline{\underline{\chi}}(\omega) = \rho \underline{\underline{\alpha}}(\omega) , \tag{7.30}$$

where $\underline{\underline{\alpha}}(\omega)$ is the Fourier transform of $\underline{\underline{\alpha}}(t)$.

7.2.1 Bichromatic Control

Here we show that an application of bichromatic control (Section 3.2.1) allows us to control both the real and imaginary parts of the refractive index. In doing so we consider isolated molecules [294, 295], or molecules in a very dilute gas, where collisional effects can be ignored and time scales over which radiative decay occurs can be ignored.

Consider then the case of bichromatic control where a system prepared in a superposition of bound states $|E_m\rangle$,

$$|\Phi(t)\rangle = c_1 |E_1\rangle \exp(-iE_1 t/\hbar) + c_2 |E_2\rangle \exp(-iE_2 t/\hbar) , \tag{7.31}$$

is subjected to two CW fields,

$$\boldsymbol{E}(t) = \sum_{i=1}^{2} 2\hat{\boldsymbol{\epsilon}} \operatorname{Re}\left[\bar{\epsilon}(\omega_i) \exp(-i\omega_i t)\right] . \tag{7.32}$$

In accord with Section 3.2.1, $\omega_{2,1} \equiv (\omega_2 - \omega_1) = (E_1 - E_2)/\hbar$, so that excitation of $|E_1\rangle$ by ω_1 and of $|E_2\rangle$ by ω_2 lead to the same energy $E = E_1 + \hbar\omega_1 = E_2 + \hbar\omega_2$. Here, however, there is no dissociation at E. We focus attention on obtaining $\underline{\underline{\chi}}(\omega_1)$ and $\underline{\underline{\chi}}(\omega_2)$.

Given the interaction Hamiltonian of Eq. (3.17),

$$H_{MR}(t) = -\sum_{i=1}^{2} 2\boldsymbol{d} \cdot \hat{\boldsymbol{\epsilon}} \operatorname{Re}\left[\bar{\epsilon}(\omega_i) \exp(-i\omega_i t)\right] , \tag{7.33}$$

we have, according to perturbation theory, that the wave function $|\psi(t)\rangle$ resulting from the interaction of the superposition state of Eq. (7.31) with the field is

$$\begin{aligned} |\psi(t)\rangle &= c_1 \exp(-iE_1 t/\hbar) |E_1\rangle + c_2 \exp(-iE_2 t/\hbar)\, E_2\rangle \\ &\quad + \sum_{m>2} c_m^{(1)}(t) \exp(-iE_m t/\hbar) |E_m\rangle . \end{aligned} \tag{7.34}$$

with expansion coefficients that are given as

$$c_m^{(1)}(t) = \frac{1}{\hbar} c_1 \sum_{k=1}^{2} d_{m,1}^{\varepsilon_k} \left(\bar{\epsilon}(\omega_k) \frac{e^{i(\omega_{m,1}-\omega_k)t}}{\omega_{m,1} - \omega_k - i\gamma} + \bar{\epsilon}(\omega_k)^* \frac{e^{i(\omega_{m,1}+\omega_k)t}}{\omega_{m,1} + \omega_k - i\gamma} \right)$$
$$+ \frac{1}{\hbar} c_2 \sum_{k=1}^{2} d_{m,2}^{\varepsilon_k} \left(\bar{\epsilon}(\omega_k) \frac{e^{i(\omega_{m,2}-\omega_k)t}}{\omega_{m,2} - \omega_k - i\gamma} + \bar{\epsilon}(\omega_k)^* \frac{e^{i(\omega_{m,2}+\omega_k)t}}{\omega_{m,2} + \omega_k - i\gamma} \right) . \tag{7.35}$$

Here γ is the average radiative line half width at half maximum (HWHM) of bound levels (which has been introduced phenomenologically), $\omega_{m,n} = (E_m - E_n)/\hbar$, and $d_{j,m}^{\varepsilon_k} = \langle E_j | \boldsymbol{d} \cdot \hat{\varepsilon}_k | E_m \rangle$. In obtaining this result we have assumed that: (i) the CW fields are turned on at $t \to -\infty$, at which time the system is in its initial superposition state (Eq.(7.31)), (ii) the medium has no permanent dipole moment, and (iii) $\langle E_1 | \boldsymbol{d} | E_2 \rangle = 0$ due to the fact that $|E_1\rangle$ and $|E_2\rangle$ are assumed to have the same parity.

The expectation value of the induced dipole is given according to perturbation theory by

$$\langle \boldsymbol{d}^{ind}(t) \rangle = \sum_m c_m^{(1)}(t) [c_1^* \boldsymbol{d}_{1,m} \exp(-i\omega_{m,1}t) + c_2^* \boldsymbol{d}_{2,m} \exp(-i\omega_{m,2}t)] + \text{c.c.} \,, \tag{7.36}$$

where $\boldsymbol{d}_{j,m} = \langle E_j | \boldsymbol{d} | E_m \rangle$. The sum above is over all $|E_m\rangle$ states, including $m = 1, 2$. Inserting Eq. (7.35) in Eq. (7.36) gives 32 terms contributing to $\langle \boldsymbol{d}^{ind}(t) \rangle$. Half of these terms are proportional to $|c_1|^2$ or $|c_2|^2$ and hence correspond to the independent effects of ω_1 and ω_2. The other half are proportional to $c_i c_j^* (i \neq j)$ and are interference terms resulting from the irradiation of the initial coherent superposition of two $|E_i\rangle$ states. Of these sixteen interference terms, eight do not oscillate with frequency ω_1 or ω_2 and hence do not contribute to the susceptibility at these frequencies. Identifying the terms that contribute at ω_1 and ω_2 gives the susceptibilities, for ω_i, $i = 1, 2$, as

$$\underline{\underline{\chi}}(\omega_i) = \underline{\underline{\chi}}^n(\omega_i) + \underline{\underline{\chi}}^{in}(\omega_i) \tag{7.37}$$

with

$$\frac{\epsilon_0 \underline{\underline{\chi}}^n(\omega_i)}{\rho} \equiv \underline{\underline{\alpha}}^n(\omega_i) = \frac{|c_1|^2}{\hbar} \sum_m \boldsymbol{d}_{1,m} \otimes \boldsymbol{d}_{m,1} \left(\frac{1}{\omega_{m,1} - \omega_i - i\gamma} + \frac{1}{\omega_{m,1} + \omega_i - i\gamma} \right)$$
$$+ \frac{|c_2|^2}{\hbar} \sum_m \boldsymbol{d}_{2,m} \otimes \boldsymbol{d}_{m,2} \left(\frac{1}{\omega_{m,2} - \omega_i - i\gamma} + \frac{1}{\omega_{m,2} + \omega_i - i\gamma} \right) ,$$
$$\frac{\epsilon_0 \underline{\underline{\chi}}^{in}(\omega_1)}{\rho} \equiv \underline{\underline{\alpha}}^{in}(\omega_1) = \frac{c_1^* c_2 \bar{\epsilon}(\omega_2)}{\hbar \bar{\epsilon}(\omega_1)} \sum_m \left(\frac{\boldsymbol{d}_{1,m} \otimes \boldsymbol{d}_{m,2}}{\omega_{m,1} - \omega_1 - i\gamma} + \frac{\boldsymbol{d}_{m,2} \otimes \boldsymbol{d}_{1,m}}{\omega_{m,2} + \omega_1 - i\gamma} \right) ,$$
$$\frac{\epsilon_0 \underline{\underline{\chi}}^{in}(\omega_2)}{\rho} \equiv \underline{\underline{\alpha}}^{in}(\omega_2) = \frac{c_1 c_2^* \bar{\epsilon}(\omega_1)}{\hbar \bar{\epsilon}(\omega_2)} \sum_m \left(\frac{\boldsymbol{d}_{2,m} \otimes \boldsymbol{d}_{m,1}}{\omega_{m,2} - \omega_2 - i\gamma} + \frac{\boldsymbol{d}_{m,1} \otimes \boldsymbol{d}_{2,m}}{\omega_{m,1} + \omega_2 - i\gamma} \right) . \tag{7.38}$$

$\underline{\underline{\chi}}^n$ is the noninterfering component of $\underline{\underline{\chi}}$ and $\underline{\underline{\chi}}^{in}$ is the interfering component of $\underline{\underline{\chi}}$. Here the $\otimes$ symbol denotes the *outer product* of two vectors. For example, $\boldsymbol{a} \otimes \boldsymbol{b}$, where $\boldsymbol{a}$ and $\boldsymbol{b}$ each have x, y, and z components, is a 3×3 matrix whose elements are $a_x b_x$, $a_x b_y$, and so on.

Below we assume that all of the incident light is linearly polarized along the z-axis and denote the laboratory zz component of $\underline{\underline{\chi}}$ as χ_{zz}. Hence, as above, the desired index of refraction n is obtained from the susceptibility as $n(\omega_i) = \sqrt{1 + \chi_{zz}(\omega_i)}$.

Examination of Eq. (7.38) shows that $\underline{\underline{\chi}}(\omega)$ is comprised of two terms that are proportional to $|c_i|^2$ and that are associated with the traditional contribution to the susceptibility from state $|E_1\rangle$ and $|E_2\rangle$ independently, plus two field-dependent terms, proportional to $a_{i,j} = c_i^* c_j \bar{\epsilon}(\omega_j)/\bar{\epsilon}(\omega_i)$, which results from the coherent excitation of both $|E_1\rangle$ and $|E_2\rangle$ to the same total energy $E = E_1 + \hbar\omega_1 = E_2 + \hbar\omega_2$. As a consequence, changing $a_{i,j}$ alters the interference between excitation routes and allows for coherent control over the susceptibility. As in all bichromatic control scenarios, this control is achieved by altering the parameters in the state preparation in order to affect c_1, c_2 and/or by varying the relative intensities of the two laser fields. Note that control over $\underline{\underline{\chi}}(\omega_i)$ is expected to be substantial if $\bar{\epsilon}(\omega_j)/\bar{\epsilon}(\omega_i)$ is large. However, under these circumstances control over $\underline{\underline{\chi}}(\omega_j)$ is minimal since the corresponding interference term is proportional to $\bar{\epsilon}(\omega_i)/\bar{\epsilon}(\omega_j)$. Hence, effective control over the refractive index is possible only at one of the ω_1 or ω_2.

Sample control results for $n(\omega)$, both off-resonance and near-resonance, for gaseous N_2 are shown below. Control is shown as a function of the relative laser phase $\delta\phi = \theta_{1,2} + \phi_2 - \phi_1$ where $\theta_{1,2}$ is the initial phase of $c_1^* c_2$ and $\phi_i (i = 1, 2)$ is the phase of $\bar{\epsilon}(\omega_i)$.

Figures 7.12 and 7.13 show the dependence of the real and imaginary parts of $n(\omega_1) = n'(\omega_1) + in''(\omega_1)$ on $|F_2/F_1| \equiv |\bar{\epsilon}(\omega_2)/\bar{\epsilon}(\omega_1)|$ for various different values of $\delta\phi$. Results are shown for the N_2 molecule in the $v_1 = 0$ ground vibrational level, and a superposition of rotational states, $|\Phi\rangle = \sqrt{0.8}|J_1 = 0, M_1 = 0\rangle + \sqrt{0.2}|J_2 = 2, M_2 = 0\rangle$, using $\omega_1 = 3 \times 10^{15}$ Hz, and $\omega_2 = 2.997\,75 \times 10^{15}$ Hz. The quantities (v_i, J_i, M_i) denote quantum numbers for vibration, rotation, and for the projection of the angular momentum along the z-axis.

Consider first the case of $\delta\phi = -\pi/2$. Here $n''(\omega_1) = 0$ (corresponding to no absorption of the field) and $n'(\omega_1)$ is seen to grow linearly on the log–log plot for $|F_2/F_1| > 10$, that is, once the interference term in Eq. (7.38) dominates. Extensive control over n' is evident; for example, n' has changed by well over 10% by $|F_2/F_1| \sim 10^4$. This is in sharp contrast with the tiny refractive index changes associated, for example, with the optical Kerr effect or self focusing [285] (which change the index of refraction of N_2 by as little as 10^{-6} and which require laser intensities of $> 10^{12}$ W/cm^2).

Figures 7.12 and 7.13 display a broad range of behavior of n' and n''. For example, for the case of $\delta\phi = 0$ and π the n' increases for $|F_2/F_1| > 1100$. For $\delta\phi = 0$ this increase is accompanied by positive n'', and hence by the absorption of the

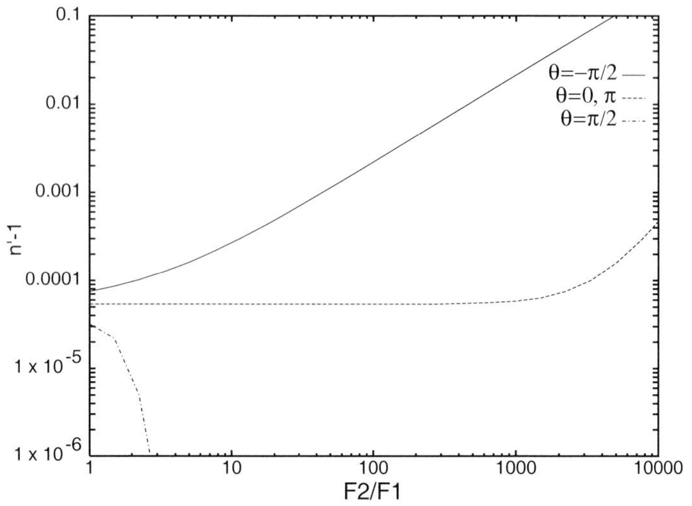

Figure 7.12 Dependence of the real part of $n(\omega)$ on F_2/F_1 in N_2 (in the superposition state described in the text) for different values of the relative laser phase $\delta\phi$. Here $\delta\phi = -\pi/2$ (solid), $\delta\phi = 0$ and π (dashed), and $\delta\phi = \pi/2$ (dot-dash). From [294, Figure 1], where $\delta\phi$ was denoted θ. (Reprinted with permission. Copyright 2000 American Physical Society.)

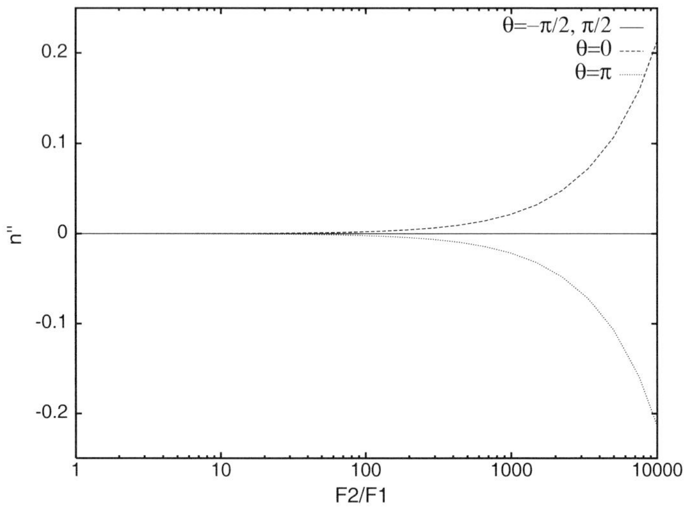

Figure 7.13 Dependence of imaginary part of $n(\omega)$ on F_2/F_1 in N_2 (in the superposition state described in the text) for different values of the relative laser phase $\delta\phi$. Here $\delta\phi = -\pi/2$ and $\pi/2$ (solid), $\delta\phi = 0$ (dashed) and $\delta\phi = \pi$ (dotted). From [294, Figure 1] where $\delta\phi$ was denoted θ. (Reprinted with permission. Copyright 2000 American Physical Society.)

field by the molecules. By contrast, the case of $\delta\phi = \pi$ shows negative n''; that is, the field is amplified. Also of interest is the case of $\delta\phi = \pi/2$ which shows rapidly decreasing n' with increasing $|F_2/F_1|$, accompanied by zero n''. Qualitatively similar results are attained for thermally distributed initial populations.

Consider now near-resonant excitation of the superposition state used in Figure 7.12. Excitation with $\omega_1 = 1.900\,884 \times 10^{16}$ Hz, and $\omega_2 = 1.900\,659 \times 10^{16}$ Hz excites the system, on resonance, to the $|v = 0, J = 1, M = 0\rangle$ bound state, of energy E_b, of the $b^1\Pi_u$ electronic state of N_2. Figures 7.14 and 7.15 show n' and n'' as a function of the detuning Δ ($E = E_1 + \hbar\omega_1 = E_2 + \hbar\omega_2 = E_b - \hbar\Delta$) for $\delta\phi = -\pi/2$, $F_2/F_1 = 1000$. Here large values of the index of refraction are seen to be associated with negligible absorption at $\Delta > 20$ GHz. Further, this significant change in the index of refraction leads to a substantial change in the speed with which light travels through the medium. Specifically, we can calculate the group velocity of light [296] as $v_g = c/[n' + \omega_1 dn'/d\omega_1]$ which, in this regime, can be estimated to be 150 m/s. More dramatic examples of slow light have been demonstrated by Lau, Harris and coworkers [296] using the electromagnetically induced transparency (EIT) effect (discussed in Section 11.1), in a Bose–Einstein condensate where they initially obtained a group velocity of 17 m/s.

The sensitivity to the control parameters is evident by changing $\delta\phi$ to $\pi/2 + 10^{-6}$, shown in Figure 7.15. Here the resultant n'' is negative, corresponding to amplification of the beam. One should note then that a rich range of behavior is possible in near-resonance cases as the control variables $a_{i,j}$ are altered. Numerous examples are provided in [295].

An interference-based scheme of increasing the refractive index characterized by low absorption near resonances, was originally proposed by Scully [297]. The

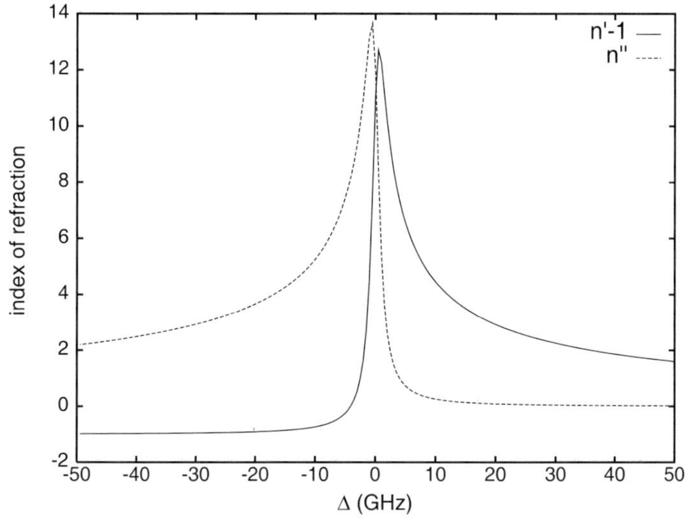

Figure 7.14 The real and imaginary parts of the index of refraction of N_2 as a function of the detuning Δ for the initial superposition state described in the text with $F_2/F_1 = 1000$, for $\delta\phi = -\pi/2$. From [294, Figure 3]. (Reprinted with permission. Copyright 2000 American Physical Society.)

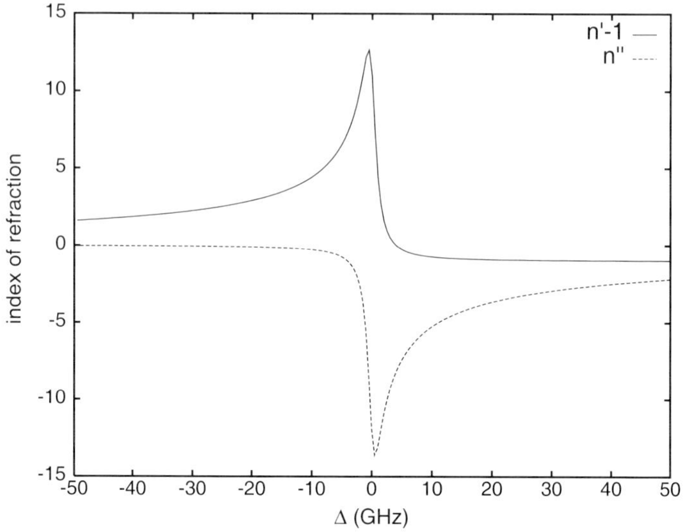

Figure 7.15 The real and imaginary parts of the index of refraction of N_2 as a function of the detuning Δ for the initial superposition state described in the text with $F_2/F_1 = 1000$, for $\delta\phi = \pi/2 + 10^{-6}$. From [294, Figure 3]. (Reprinted with permission. Copyright 2000 American Physical Society.)

scheme is similar to the one above, but relies upon excitation of two very closely spaced levels $|E_1\rangle$ and $|E_2\rangle$ using a pulsed laser. Once again, control is extensive.

7.3 The Molecular Phase in the Presence of Resonances

Spectroscopy has long had, as its central goal, the use of light to extract molecular information, such as potential surfaces, system properties, and so on Gordon and Seideman [298–300] have noted that because coherent control depends upon interference effects it introduces a new spectroscopic tool to extract previously unattainable information regarding the molecular continuum. In particular, they theoretically analyzed the physical origin of, and information contained in, the molecular phase term $\alpha_q(13)$ (Eq. (3.52)). Recall that $\alpha_q(13)$ is defined as the phase of cumulative molecular matrix elements and appears prominently in the one-photon vs. three-photon interference term (Eq. (3.53)). Further, Gordon and Seideman have focused experimentally and theoretically upon the information contained in the "phase lag" $\delta(q, q') = \alpha_q(13) - \alpha_{q'}(13)$ between two product channels q and q'. This would correspond, for example, to the phase difference between the two curves shown in Figure 7.16.

A formal analysis of the physical origins of the molecular phase and phase lag in the one- vs. three-photon scenario is provided below. Appreciating these results requires that we recall the idea of a resonance [227]; that is, a state that results from

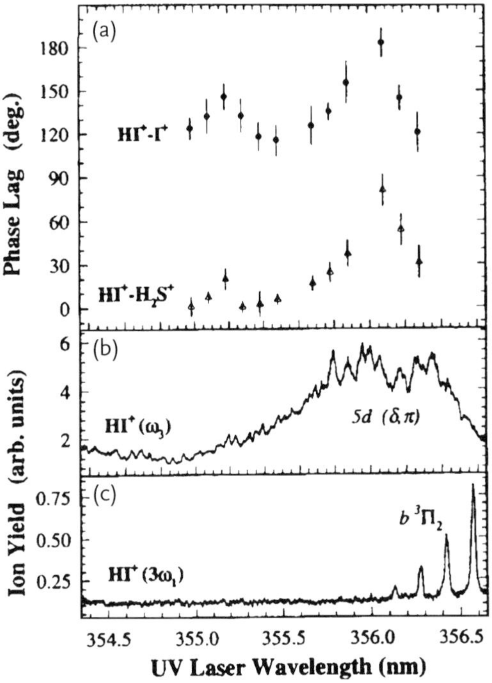

Figure 7.16 The phase lag spectrum (a) for the photodissociation and photoionization of HI (circles) and for the photoionization of a mixture of HI and H_2S (triangles). The bottom two panels (b,c) are the one- and three-photon ionization spectra of HI. From [300, Figure 7]. (Reprinted with permission. Copyright 2001 American Chemical Society.)

a bound state $|\phi_s\rangle$ coupled to a continuum. From the time-dependent viewpoint, such states decay into the continuum over some lifetime. The complementary picture in energy space (depicted in Figure 7.17) is that the resonance is characterized by a rapid change, for example, a peak, in the cross section for the process as a function of energy. The energy range over which this change occurs defines the width of the resonance. If the resonances are sufficiently far apart in energy from one another so that their energy regions of influence do not overlap, they are called "isolated" resonances. Otherwise, a set of closely spaced resonances that affect the same energy region are called "overlapping" resonances.

The results of the analysis of Seideman and Gordon [300] show that a nonzero molecular phase $\alpha_q(13)$ can arise from a number of circumstances, discussed below. These include (i) a multichannel scattering problem that displays coupling between the continua associated with different product channels, (ii) the presence of a resonance connecting to a product channel, or (iii) the presence of a resonance at an energy lower than the continuum that contributes to the phase of one of the interfering pathways. Thus, experimental evidence for a nonzero molecular phase provides information on the nature of the continuum. Further, the functional form

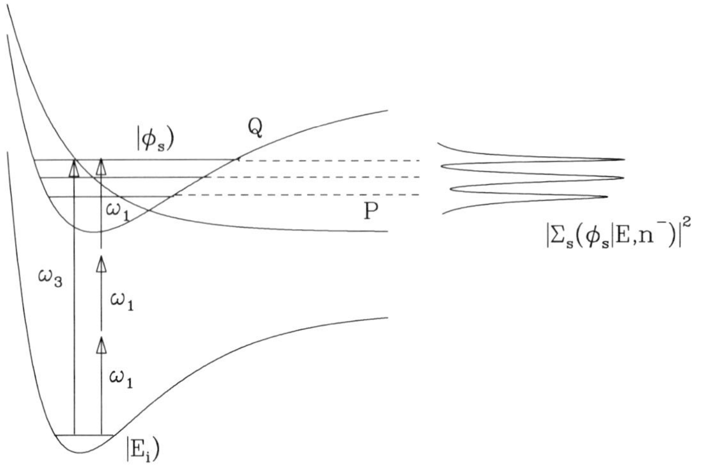

Figure 7.17 A schematic illustration of the formation of a resonance from a bound state $|\phi_s\rangle$ and the way it is probed by a competing three-photon vs. one-photon transition from an initial state $|E_i\rangle$.

of the energy dependence of the molecular phase provides further useful information on the character of the dynamics.

In several other simpler cases, discussed below, the molecular phase vanishes. We note in passing that, in accord with Eq. (3.53), the vanishing of the molecular phase does not imply that control is lost. However, a significant phase lag, from the viewpoint of control, is advantageous.

To see the origin of the molecular phase lag in the one- vs. three-photon control scenario we reconsider the formalism discussed in Section 3.4.2. However, for notational simplicity we denote the set of scattering eigenstates of the full Hamiltonian at energy E and fragment quantum numbers $\boldsymbol{n}$ in channel q as $|E, \boldsymbol{n}^-\rangle$; that is, we subsume the q within the labels $\boldsymbol{n}$.

7.3.1
Theory of Scattering Resonances

In order to understand how resonances affects the molecular phase we briefly outline the basics of the theory of scattering resonances. We consider bound states $|\phi_s\rangle$ interacting with a set of continuum states denoted $|E, \boldsymbol{n}^-; 1\rangle$, where, as for the full scattering states $|E, \boldsymbol{n}^-\rangle$ (see Eq. (2.66)), the states $|E, \boldsymbol{n}^-; 1\rangle$ approach the free asymptotic solutions at infinite time:

$$\lim_{t\to\infty} e^{-i(E-i\epsilon)t/\hbar}|E, \boldsymbol{n}^-; 1\rangle = e^{-i(E-i\epsilon)t/\hbar}|E, \boldsymbol{n}^-; 0\rangle\ . \qquad (7.39)$$

The interaction between the bound and continuum parts is depicted schematically in Figure 7.17. The emergence of resonances is derived most naturally via the use of "partitioning" technique [301], which focuses attention on either the bound or continuum subspace. This method was introduced in Section 4.1.2. Here we formulate this partitioning for continuum problems.

As in Section 4.1.2 one defines two projection operators Q and P that satisfy Eq. (4.12):

$$QQ = Q\,, \quad PP = P\,, \quad PQ = QP = 0\,, \quad P + Q = \mathrm{I}\,, \tag{7.40}$$

where I is the identity operator. Here the Q and P operators are chosen to project out the subspaces spanned by the bound states and the continuum states, respectively. Further, as Eq. (7.40) indicates, they are orthogonal, for example, they may project onto two different electronic states, or any two spaces previously known to be orthogonal.

The full scattering incoming states $|E, \boldsymbol{n}^-\rangle$, introduced in Chapter 2, are eigenstates of the Schrödinger equation $[E - i\epsilon - H]|E, \boldsymbol{n}^-\rangle = 0$, where the $-i\epsilon$ serves to remind us of the incoming boundary conditions (Eq. (2.66)). This equation can be rewritten as

$$[E - i\epsilon - H][P + Q]|E, \boldsymbol{n}^-\rangle = 0\,. \tag{7.41}$$

Multiplying this equation once by P and once by Q, we obtain, using Eq. (7.40), two coupled equations:

$$[E - i\epsilon - PHP]P|E, \boldsymbol{n}^-\rangle = PHQ|E, \boldsymbol{n}^-\rangle\,. \tag{7.42}$$

$$[E - i\epsilon - QHQ]Q|E, \boldsymbol{n}^-\rangle = QHP|E, \boldsymbol{n}^-\rangle\,. \tag{7.43}$$

We define two basis sets, $|E, \boldsymbol{n}^-; 1\rangle$ and $|\phi_s\rangle$, which are the solutions of the *homogeneous* (decoupled) parts of Eqs. (7.42) and (7.43). That is,

$$[E - i\epsilon - PHP]|E, \boldsymbol{n}^-; 1\rangle = 0\,, \tag{7.44}$$

and

$$[E_s - QHQ]|\phi_s\rangle = 0\,. \tag{7.45}$$

Implicit in Eqs. (7.44) and (7.45) is that $|E, \boldsymbol{n}^-; 1\rangle \in P$ and $|\phi_s\rangle \in Q$ and as such they are orthogonal to one another. We, in fact, assume that each basis set spans the entire subspace to which it belongs, hence we can write an explicit representation of Q and P as

$$Q = \sum_s |\phi_s\rangle\langle\phi_s|\,, \tag{7.46}$$

$$P = \sum_n \int dE|E, \boldsymbol{n}^-; 1\rangle\langle E, \boldsymbol{n}^-; 1|\,. \tag{7.47}$$

Using Eqs. (7.46) and (7.47) we can therefore write $|E, \boldsymbol{n}^-\rangle = [P + Q]|E, \boldsymbol{n}^-\rangle$ in terms of Q and P as

$$|E, \boldsymbol{n}^-\rangle = \sum_s |\phi_s\rangle\langle\phi_s|E, \boldsymbol{n}^-\rangle + \sum_{\boldsymbol{n}'} \int dE'|E', \boldsymbol{n}'^-; 1\rangle\langle E', \boldsymbol{n}'^-; 1|E, \boldsymbol{n}^-\rangle\,. \tag{7.48}$$

In order for this expansion to be useful requires that we have explicit expressions for $\langle \phi_s | E, \boldsymbol{n}^- \rangle$ and for $\langle E', \boldsymbol{n}'^-; 1 | E, \boldsymbol{n}^- \rangle$. These are obtained below.

We first solve for $P|E, \boldsymbol{n}^-\rangle$ by writing it as a sum of the homogeneous solution of Eq. (7.44) and a particular solution of Eq. (7.42), obtained by inverting $[E - i\epsilon - PHP]$:

$$P|E, \boldsymbol{n}^-\rangle = P|E, \boldsymbol{n}^-; 1\rangle + [E - i\epsilon - PHP]^{-1} PHQ|E, \boldsymbol{n}^-\rangle \,. \tag{7.49}$$

Substituting this solution into Eq. (7.43) we obtain that

$$\begin{aligned}[E - i\epsilon - QHQ]Q|E, \boldsymbol{n}^-\rangle &= QHP|E, \boldsymbol{n}^-; 1\rangle \\ &\quad + QHP[E - i\epsilon - PHP]^{-1} PHQ|E, \boldsymbol{n}^-\rangle \,. \end{aligned} \tag{7.50}$$

Reordering terms in this equation gives

$$[E - i\epsilon - Q\mathcal{H}Q]Q|E, \boldsymbol{n}^-\rangle = QHP|E, \boldsymbol{n}^-; 1\rangle \,, \tag{7.51}$$

where

$$Q\mathcal{H}Q \equiv QHQ + QHP[E - i\epsilon - PHP]^{-1} PHQ \,. \tag{7.52}$$

Equation (7.51) can be solved to yield

$$Q|E, \boldsymbol{n}^-\rangle = [E - i\epsilon - Q\mathcal{H}Q]^{-1} QHP|E, \boldsymbol{n}^-; 1\rangle \,. \tag{7.53}$$

An explicit representation of Eq. (7.53) is obtained by using the well-known identity (obtained by a similar contour integration to that depicted in Figure 2.1),

$$[E - i\epsilon - PHP]^{-1} = \mathsf{P}_v[E - PHP]^{-1} + i\pi\delta(E - PHP) \,, \tag{7.54}$$

with P_v denoting a Cauchy principal value integral

$$\mathsf{P}_v \int_a^b dE' \frac{f(E')}{E - E'} \equiv \lim_{\epsilon \to 0} \int_a^{E-\epsilon} dE' \frac{f(E')}{E - E'} + \int_{E+\epsilon}^b dE' \frac{f(E')}{E - E'} \,. \tag{7.55}$$

The $\mathsf{P}_v[E - PHP]^{-1}$ operator above is given, using the spectral resolution of an operator, as

$$\mathsf{P}_v[E - PHP]^{-1} \equiv \sum_n \mathsf{P}_v \int \frac{dE'}{E - E'} |E', \boldsymbol{n}^-; 1\rangle\langle E', \boldsymbol{n}^-; 1| \,. \tag{7.56}$$

Using Eqs. (7.54) and (7.56) we can write $Q\mathcal{H}Q$ of Eq. (7.52) as

$$Q\mathcal{H}Q = QHQ + QHP\mathsf{P}_v[E - PHP]^{-1} PHQ + i\pi QHP\delta(E - PHP)PHQ \,. \tag{7.57}$$

Assuming for simplicity the case of "noninteracting" overlapping resonances in which (by definition) $Q\mathcal{H}Q$ is diagonal (the case of overlapping and interacting resonances is dealt with in Section 11.1), we can use Eqs. (7.56) and (7.57) to write the representation of $[E - i\epsilon - Q\mathcal{H}Q]$ in the $\{|\phi_s\rangle\}$ basis as

$$\langle\phi_s|[E - i\epsilon - Q\mathcal{H}Q]|\phi_s\rangle = \left[E - E_s - \Delta_s(E) - \frac{i\Gamma_s(E)}{2}\right], \tag{7.58}$$

where

$$V(s|E, \boldsymbol{n}) \equiv \langle\phi_s|QHP|E, \boldsymbol{n}^-; 1\rangle \equiv \langle\phi_s|H|E, \boldsymbol{n}^-; 1\rangle, \tag{7.59}$$

$$\Gamma_s(E) \equiv \sum_n 2\pi|V(s|E, \boldsymbol{n})|^2, \tag{7.60}$$

$$\Delta_s(E) \equiv \mathsf{P}_v \sum_n \int \frac{dE'}{E - E'}|V(s|E', \boldsymbol{n})|^2. \tag{7.61}$$

It follows from Eqs. (7.58), (7.60), and (7.61) that the $\langle\phi_s|E, \boldsymbol{n}^-\rangle$ overlap integrals are given as,

$$\langle\phi_s|E, \boldsymbol{n}^-\rangle = \frac{V(s|E, \boldsymbol{n})}{E - E_s - \Delta_s(E) - i\Gamma_s(E)/2}. \tag{7.62}$$

Using Eqs. (7.49) and (7.54) we obtain the expression

$$\begin{aligned}\langle E', \boldsymbol{m}^-; 1|E, \boldsymbol{n}^-\rangle &= \delta(E - E')\delta_{n,m} \\ &+ \sum_s V(E', \boldsymbol{m}|s)\left[\mathsf{P}_v \frac{1}{E - E'} + i\pi\delta(E - E')\right]\langle\phi_s|E, \boldsymbol{n}^-\rangle.\end{aligned} \tag{7.63}$$

Given Eqs. (7.62) and (7.63) we can express, via Eq. (7.48), the full scattering wave function $|E, \boldsymbol{n}^-\rangle$ in terms of $|\phi_s\rangle$ and $|E, \boldsymbol{n}^-; 1\rangle$.

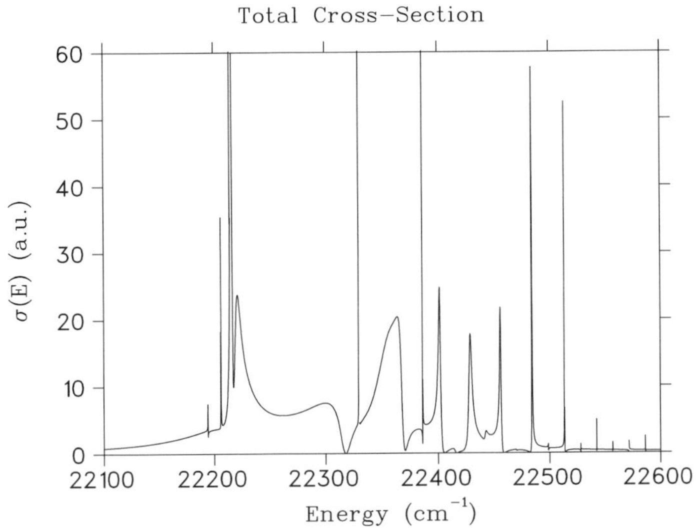

Figure 7.18 Absorption Spectrum resulting from a sum of overlapping resonances.

The form given in Eq. (7.62) gives rise to a Lorentzian shape that is depicted schematically in Figure 7.17. A spectrum resulting from the excitation to a collection of overlapping resonances is shown in Figure 7.18.

7.3.2
Three-Photon vs. One-Photon Coherent Control in the Presence of Resonances

We now apply the above partitioning method to the three-photon vs. one-photon coherent control treated in detail in Section 3.4.2. Consider a molecule initially in a bound energy eigenstate $|E_i\rangle$ subjected to two co-propagating pulses,

$$\mathbf{E}(t) = 2\int d\omega_3 \mathrm{Re}\left[\hat{\boldsymbol{\varepsilon}}_3\bar{\epsilon}_3(\omega_3)\exp(-i\omega_3 t) + \hat{\boldsymbol{\varepsilon}}_1\bar{\epsilon}_1(\omega_1)\exp(-i\omega_1 t)\right] , \quad (7.64)$$

where $3\omega_1 = \omega_3$. After the pulse is over some of the molecules have absorbed either a single photon or three photons from the field and the excited wave packet is given by

$$|\Psi(t)\rangle = \sum_n \int dE|E, \boldsymbol{n}^-\rangle A_{\boldsymbol{n}}(E)\exp(-iEt/\hbar) , \quad (7.65)$$

where $E = E_i + \hbar\omega_3 = E_i + 3\hbar\omega_1$. Assuming electric field–dipole interaction, $H_{MR}(t) = -\boldsymbol{d}\cdot\mathbf{E}(t)$, and that $\langle E_i|E, \boldsymbol{n}^-\rangle = 0$, $A_{\boldsymbol{n}}(E)$, the continuum preparation coefficients, are given by

$$A_{\boldsymbol{n}}(E) = \frac{2\pi i}{\hbar}\langle E, \boldsymbol{n}^-|\left[\bar{\epsilon}_3(\omega_{E,i})d_{e,g} + \bar{\epsilon}_1^3(\omega_{E,i}/3)\,T_{e,g}\right]|E_i\rangle . \quad (7.66)$$

Here, $T_{e,g}$ is the three-photon transition operator, given in Eq. (3.44) as

$$T_{e,g} = \sum_{e'e''} d_{e,e'}(E_i - H_{e'} + 2\hbar\omega_1 - i\gamma_2)^{-1}d_{e',e''}(E_i - H_{e''} + \hbar\omega_1 - i\gamma_1)^{-1}d_{e'',g} . \quad (7.67)$$

It follows from Eq. (7.66) and Eq. (7.48) that $A_{\boldsymbol{n}}(E)$ can be written as,

$$A_{\boldsymbol{n}}(E) = \frac{2\pi i}{\hbar}\left\{\sum_s \langle E, \boldsymbol{n}^-|\phi_s\rangle \mathcal{T}_{s,i}(\bar{\epsilon}_3, \bar{\epsilon}_1) + \sum_m \int dE'\langle E, \boldsymbol{n}^-|E', \boldsymbol{m}^-; 1\rangle \mathcal{T}_{m,i}(E', \bar{\epsilon}_3, \bar{\epsilon}_1)\right\} , \quad (7.68)$$

where

$$\mathcal{T}_{s,i}(\bar{\epsilon}_3, \bar{\epsilon}_1) \equiv \langle\phi_s|\left[\bar{\epsilon}_3(\omega_{E,i})d_{e,g} + \bar{\epsilon}_1^3(\omega_{E,i}/3)\,T_{e,g}\right]|E_i\rangle ,$$
$$\mathcal{T}_{m,i}(E', \bar{\epsilon}_3, \bar{\epsilon}_1) \equiv \langle E', \boldsymbol{m}^-; 1|\left[\bar{\epsilon}_3(\omega_{E,i})d_{e,g} + \bar{\epsilon}_1^3(\omega_{E,i}/3)\,T_{e,g}\right]|E_i\rangle .$$

Substituting the complex conjugate of Eq. (7.63) into Eq. (7.68) gives the amplitude as

$$A_{\boldsymbol{n}}(E) = (2\pi i/\hbar)\mathcal{T}_{n,i}(E,\bar{\epsilon}_3,\bar{\epsilon}_1) + (2\pi i/\hbar)\sum_s \langle E,\boldsymbol{n}^-|\phi_s\rangle$$
$$\times\left[\mathcal{T}_{s,i}(\bar{\epsilon}_3,\bar{\epsilon}_1) + \sum_m \int \frac{dE'}{E+i\epsilon-E'} V^*(s|E',\boldsymbol{m})\mathcal{T}_{m,i}(E',\bar{\epsilon}_3,\bar{\epsilon}_1)\right]. \tag{7.69}$$

Inserting Eqs. (7.62) and (7.63) into Eq. (7.69) we obtain that

$$A_{\boldsymbol{n}}(E) = \frac{2\pi i}{\hbar}\langle E,\boldsymbol{n}^-;1|\left[\bar{\epsilon}_3(\omega_{E,i})d_{e,g} + \bar{\epsilon}_1^3(\omega_{E,i}/3)\,T_{e,g}\right]|E_i\rangle$$
$$+\frac{2\pi i}{\hbar}\sum_s \frac{V^*(E,\boldsymbol{n}|s)}{E-E_s-\Delta_s(E)+i\Gamma_s(E)/2}\Big\{\langle\phi_s|\left[\bar{\epsilon}_3(\omega_{E,i})d_{e,g}+\bar{\epsilon}_1^3(\omega_{E,i}/3)\,T_{e,g}\right]|E_i\rangle$$
$$+\sum_m -i\pi V^*(s|E,\boldsymbol{m})\langle E,\boldsymbol{m}^-;1|\left[\bar{\epsilon}_3(\omega_{E,i})d_{e,g} + \bar{\epsilon}_1^3(\omega_{E,i}/3)T_{e,g}\right]|E_i\rangle$$
$$+\mathrm{P}_v\int\frac{dE'}{E-E'}V^*(s|E',\boldsymbol{m})\langle E',\boldsymbol{m}^-;1|\left[\bar{\epsilon}_3(\omega_{E,i})d_{e,g}+\bar{\epsilon}_1^3(\omega_{E,i}/3)\,T_{e,g}\right]|E_i\rangle\Big\}. \tag{7.70}$$

The probability $P_{\boldsymbol{n}}(E)$ of observing the final state $\boldsymbol{n}$ at energy E, is given as $P_{\boldsymbol{n}}(E) = |A_{\boldsymbol{n}}(E)|^2$. In accord with Eq. (3.48) we identify the components associated with one-photon excitation $P_{\boldsymbol{n}}^{(1)}(E)$, three-photon excitation $P_{\boldsymbol{n}}^{(3)}(E)$, and one-photon/three-photon interference $P_{\boldsymbol{n}}^{(13)}(E)$ so that:

$$P_{\boldsymbol{n}}(E) = P_{\boldsymbol{n}}^{(1)}(E) + P_{\boldsymbol{n}}^{(3)}(E) + P_{\boldsymbol{n}}^{(13)}(E). \tag{7.71}$$

From Eq. (7.70) we have that the interference term is

$$P_{\boldsymbol{n}}^{(13)}(E) = 2\bar{\epsilon}_3(\omega_{E,i})\bar{\epsilon}_1^3(\omega_{E,i}/3)\left(\frac{2\pi}{\hbar}\right)^2$$
$$\times\,\mathrm{Re}\left\{\mathrm{T}(i|E,\boldsymbol{n}) + \sum_{s'}\frac{\mathrm{V}(s'|E,\boldsymbol{n})\left[\mathrm{T}(i|s') + i\Gamma_{s'}^{\mathrm{T}}(i|E)/2 + \Delta_{s'}^{\mathrm{T}}(i|E)\right]}{E-E_{s'}-\Delta_{s'}(E)-i\Gamma_{s'}(E)/2}\right\}$$
$$\times\left\{d(E,\boldsymbol{n}|i) + \sum_s\frac{V^*(E,\boldsymbol{n}|s)\left[d(s|i) - i\Gamma_s^d(E|i)/2 + \Delta_s^d(E|i)\right]}{E-E_s-\Delta_s(E)+i\Gamma_s(E)/2}\right\}, \tag{7.72}$$

where

$$d(E,\boldsymbol{n}|i) \equiv \langle E,\boldsymbol{n}^-;1|d_{e,g}|E_i\rangle\,,\quad d(s|i)\equiv\langle\phi_s|d_{e,g}|E_i\rangle\,,$$
$$\mathrm{T}(i|E,\boldsymbol{n}) \equiv \langle E_i|T_{e,g}^*|E,\boldsymbol{n}^-;1\rangle\,,\quad \mathrm{T}(i|s')\equiv\langle E_i|T_{e,g}^*|\phi_{s'}\rangle\,,$$

$$\Delta_s^d(E|i) \equiv \sum_m \mathrm{P}_v \int \frac{dE'}{E-E'} V^*(s|E', \boldsymbol{m}) d(E', \boldsymbol{m}|i) ,$$
$$\Gamma_s^d(E|i) \equiv 2\pi \sum_m V^*(s|E, \boldsymbol{m}) d(E, \boldsymbol{m}|i) ,$$
$$\Delta_s^{\mathrm{T}}(i|E) \equiv \sum_m \mathrm{P}_v \int \frac{dE'}{E-E'} \mathrm{T}(i|E', \boldsymbol{m}) V(E', \boldsymbol{m}|s) ,$$
$$\Gamma_s^{\mathrm{T}}(i|E) \equiv 2\pi \sum_m \mathrm{T}(i|E, \boldsymbol{m}) V(E, \boldsymbol{m}|s) . \tag{7.73}$$

These equations are completely general. They allow us to look at a variety of limiting cases, discussed below.

7.3.2.1 Case (a): an Indirect Transition to an Isolated Resonance

In this case the photodissociation occurs by excitation to a single resonance, followed by a transition from the resonance to the continuum. In that case the sum over s reduces to a single term and the direct optical transitions to the continuum are suppressed. That is,

$$d(E, \boldsymbol{n}|i) = \mathrm{T}(i|E, \boldsymbol{n}) = 0 , \tag{7.74}$$

and therefore

$$\Gamma_s^d(E|i) = \Delta_s^d(E|i) = \Delta_s^{\mathrm{T}}(i|E) = \Gamma_s^{\mathrm{T}}(i|E) = 0 . \tag{7.75}$$

Equation (7.72) therefore becomes

$$P_{\boldsymbol{n}}^{(13)}(E) = 2\bar{\epsilon}_3(\omega_{E,i})\bar{\epsilon}_1^3(\omega_{E,i}/3)\left(\frac{2\pi}{\hbar}\right)^2 \frac{|V(s|E, \boldsymbol{n})|^2 \mathrm{Re}\left[\mathrm{T}(i|s) d(s|i)\right]}{(E - E_s - \Delta_s(E))^2 + \Gamma_s^2(E)/4} . \tag{7.76}$$

$P_{\boldsymbol{n}}^{(13)}(E)$ is seen to be real, since both $|\phi_s\rangle$ and $|E_i\rangle$ are bound and hence can be chosen real. Hence the molecular phase, that is, the phase of this term, is zero.

7.3.2.2 Case (b): a Purely Direct Transition to the Continuum

In this case the resonances are not optically coupled to the initial state $|E_i\rangle$, that is, $d(s|i) \equiv \langle \phi_s | d_{e,g} | E_i \rangle = \mathrm{T}(i|s') \equiv \langle E_i | T_{e,g}^* | \phi_{s'} \rangle = 0$, and only direct transitions to the continuum survive. We obtain that

$$P_{\boldsymbol{n}}^{(13)}(E) = 2\bar{\epsilon}_3(\omega_{E,i})\bar{\epsilon}_1^3(\omega_{E,i}/3)\left(\frac{2\pi}{\hbar}\right)^2 \mathrm{Re}\left[\mathrm{T}(i|E, \boldsymbol{n}) d(E, \boldsymbol{n}|i)\right] . \tag{7.77}$$

In this case there are two different possibilities. If the physics is such that $|E, \boldsymbol{n}^-; 1\rangle$ is a solution of a single channel problem then its coordinate space representation can always be written [227] as

$$\langle r|E, \boldsymbol{n}^-; 1\rangle = e^{i\delta} \langle r|E, \boldsymbol{n}^-; 1\rangle_R , \tag{7.78}$$

where $\langle r|E, \boldsymbol{n}^-; 1\rangle_R$ is a real function and δ is a phase that is independent of r. In this case the phase of the $\mathrm{T}(i|E, \boldsymbol{n})$ exactly cancels the phase of $d(E, \boldsymbol{n}|i)$ and the phase of the $\mathrm{T}(i|E, \boldsymbol{n})d(E, \boldsymbol{n}|i)$ products vanishes. As a result, the molecular phase is zero. If, on the other hand, the scattering is multichannel, then $|E, \boldsymbol{n}^-; 1\rangle$ is a solution of a multichannel problem, the factorization in Eq. (7.78) no longer holds, and the molecular phase is nonzero and is a function of $\boldsymbol{n}$. In this case phase control is possible.

7.3.2.3 Case (c): an Indirect Transition to a Set of Overlapping Resonances

Here the dynamics occurs by excitation to a set of overlapping resonances, with subsequent decay into the continuum. In this case there is a sum over the resonances in Eq. (7.72), but there is no direct transition to the continuum. That is, Eq. (7.74) still holds and Eq. (7.72) becomes

$$P_{\boldsymbol{n}}^{(13)}(E) = 2\bar{\epsilon}_3(\omega_{E,i})\bar{\epsilon}_1^3(\omega_{E,i}/3)\,(2\pi/\hbar)^2\,\mathrm{Re}\left[\mathcal{D}(E,\boldsymbol{n})\mathcal{T}^*(E,\boldsymbol{n})\right]\,, \tag{7.79}$$

where

$$\mathcal{D}(E,\boldsymbol{n}) \equiv \sum_s \frac{V(s|E,\boldsymbol{n})d(s|i)}{E - E_s - \Delta_s(E) - i\Gamma_s(E)/2}\,,$$
$$\mathcal{T}^*(E,\boldsymbol{n}) \equiv \sum_s \frac{V^*(E,\boldsymbol{n}|s)\mathrm{T}(i|s)}{E - E_s - \Delta_s(E) + i\Gamma_s(E)/2}\,. \tag{7.80}$$

Here no factorization of the $V(s|E, \boldsymbol{n})$ terms out of the sum is possible, the molecular phase $\alpha_{\boldsymbol{n}}(E)$ is now a function of $\boldsymbol{n}$ and phase control is possible. Note that the energy dependence of Eq. (7.79) is distinctly different than that in Eq. (7.77) providing insight into the nature of the continuum.

7.3.2.4 Case (d): a Sum of Direct and Indirect Transition to an Isolated Resonance

In this case the product is reached either via a single resonance or directly via the continuum. Hence the sum over s in Eq. (7.72) reduces to a single term, but the direct optical transitions to the continuum are *not* suppressed. This gives the case of a Fano type interference [302] to an isolated resonance, where the two pathways to the continuum interfere with one another. In this case Eq. (7.72) becomes

$$\begin{aligned}
P_{\boldsymbol{n}}^{(13)}(E) &= 2\bar{\epsilon}_3(\omega_{E,i})\bar{\epsilon}_1^3(\omega_{E,i}/3)\,(2\pi/\hbar)^2 \\
&\times \mathrm{Re}\left\{ d(E,\boldsymbol{n}|i)\mathrm{T}(i|E,\boldsymbol{n}) + \frac{d(E,\boldsymbol{n}|i)V(s|E,\boldsymbol{n})\left[\mathrm{T}(i,s) + i\Gamma_s^{\mathrm{T}}(i|E)/2 + \Delta_s^{\mathrm{T}}(i|E)\right]}{E - E_s - \Delta_s(E) - i\Gamma_s(E)/2}\right. \\
&+ \frac{\mathrm{T}(i|E,\boldsymbol{n})V^*(E,\boldsymbol{n}|s)\left[d(s|i) - i\Gamma_s^d(E|i)/2 + \Delta_s^d(E|i)\right]}{E - E_s - \Delta_s(E) + i\Gamma_s(E)/2} \\
&+ \left.\frac{|V(s|E,\boldsymbol{n})|^2\left[d(s|i) - i\Gamma_s^d(E|i)/2 + \Delta_s^d(E|i)\right]\left[\mathrm{T}(i,s) + i\Gamma_s^{\mathrm{T}}(i|E)/2 + \Delta_s^{\mathrm{T}}(i|E)\right]}{\left[E - E_s - \Delta_s(E)\right]^2 + \Gamma_s^2(E)/4}\right\}.
\end{aligned} \tag{7.81}$$

Once again, the molecular phase $\alpha_n(E)$ is nonzero and is a function of $\boldsymbol{n}$. Hence phase control is possible.

Thus, we see, in accord with extensive work by Gordon and Seideman, that the presence of a nonzero molecular phase provides insight into features of the continuum [300]. Further, the detailed nature of the energy dependence of the molecular phase assists in distinguishing between the various cases discussed above.

As an example consider, once again, the one- vs. three-photon excitation of HI which undergoes two competitive processes:

$$\begin{aligned} &\mathrm{HI} \to \mathrm{HI}^+ + e^- \\ &\mathrm{HI} \to \mathrm{H} + \mathrm{I} \end{aligned} \tag{7.82}$$

As noted in Section 3.4.2.1, Gordon and coworkers have measured (see Figure 3.12) the modulation of the HI^+ and I^+ (from the H + I channel) as the relative laser phase of ω_1 and ω_3 is varied. Experiments of this kind provide the phase lag $\delta(q, q')$ plotted as a function of energy in Figure 7.16. This figure shows the phase lag data for the above two channels in the region of the overlapping $5d\pi$ and $5d\delta$ Rydberg resonances of HI at 356 nm. Also shown is the phase lag for the mixture of HI^+ and H_2S^+, where the phase lag of the latter species is known to vanish. Both phase lag curves shown show a maximum at 356.1 nm, in the region of the resonance, as well as another maximum at 355.2 nm. The peaks are atop an almost zero background, indicating that in this region the continua are elastic. Thus, this figure demonstrates both the resonance and nonresonant contributions to the phase lag.

7.4 Control of Chaotic Dynamics

Studies in classical nonlinear mechanics over the past few decades have shown that systems can be categorized as lying between two limits: that of integrable dynamics and that of chaotic dynamics [96, 303–305]. In the integrable case the dynamics of a system of N degrees of freedom possesses N conserved integrals of motion and is stable with respect to small external perturbations. In the chaotic case the system dynamics usually possesses only symmetry based integrals of motion, such as the total energy and angular momentum, and the dynamics is extremely sensitive to initial conditions and external perturbations. Even in the absence of external perturbations a classical chaotic system "loses memory" of the initial state exponentially fast. This categorization extends to quantum mechanics in the sense that a system is said to be quantum mechanically chaotic if its classical counterpart is classically chaotic. Numerous computational studies [306] have shown that quantum systems do display characteristics of classical chaos if they are sufficiently close to the classical limit, a manifestation of the correspondence principle [177]. It is expected that the vast majority of realistic systems are sufficiently complex so as to display some degree of classically chaotic behavior.

Considering the sensitivity of classical chaotic systems to external perturbations, and the ubiquitous nature of chaotic dynamics in larger systems, it is important to establish that quantum mechanics allows for control in chaotic systems as well.

One simple molecular system that displays quantum chaos is the rotational excitation of a diatomic molecule using pulsed microwave radiation [307]. Under the conditions adopted below this system is a molecular analogue of the "delta-kicked rotor", that is, a rotor that is periodically kicked by a delta function potential, which is a paradigm for chaotic dynamics [308, 309]. The observed energy absorption of such systems is called "quantum chaotic diffusion".

If the orientation of a diatomic molecule is described by two angles θ and ϕ [310], then the corresponding Hamiltonian is

$$H = \frac{\hat{J}^2}{2I} + \boldsymbol{d} \cdot \boldsymbol{E}_0 \cos\theta \sum_n \Delta\left(t/T - n\right) , \tag{7.83}$$

where $\hat{J}$ is the angular momentum operator in three dimensions:

$$\hat{J}^2 = -\hbar^2 \left[\frac{1}{\sin\theta} \frac{\partial}{\partial\theta} \left(\sin\theta \frac{\partial}{\partial\theta} \right) + \frac{1}{\sin^2\theta} \frac{\partial^2}{\partial\phi^2} \right] . \tag{7.84}$$

Here $\boldsymbol{d}$ is the molecular electric dipole moment, $\boldsymbol{E}_0$ is the amplitude of the driving field whose polarization direction defines the z direction, I is the moment of inertia of the molecule about an axis perpendicular to the symmetry axis, and $\Delta(t/T - n)$ is the pulse shape function of the form

$$\Delta\left(t/T - n\right) = 1 + 2 \sum_{m=1}^{m=7} \cos\left[2m\pi \left(t/T - n - \frac{1}{2} \right) \right] . \tag{7.85}$$

Eigenstates of the Hamiltonian H are $|n_J, m_J\rangle$, where n_J is the angular momentum quantum number with projection m_J along the z-axis.

As shown by Fishman [307], the kicked CsI molecule is a particularly appropriate candidate for this study since it has a large dipole moment, increasing the molecule–field coupling strength, and the rotation-vibration coupling is small at low excitation energies so that one may consider solely rotational excitation. We consider then the dynamics of CsI in the indicated pulsed field, in a parameter range known to display classical chaos [311].

To demonstrate control of chaotic dynamics we assume that an initial superposition state of the form

$$|\psi(0)\rangle = \cos\alpha |j_1, 0\rangle + \sin\alpha \exp(-i\beta) |j_2, 0\rangle \tag{7.86}$$

has been previously prepared. This system is now subjected to pulsed microwave irradiation, and the rotational energy absorption is measured. In particular, we define the dimensionless rotational energy $\tilde{E} \equiv \sum_j P_j j(j+1)\tau^2/2$, $\tau = \hbar T/I$, where P_j is the occupation probability of the $|j, 0\rangle$ state, as a measure of the absorbed energy.

To anticipate the result of pulsed excitation of a superposition state, note from Eqs. (7.83) and (7.85) that the Hamiltonian is strictly periodic in time. We denote the time evolution operator associated with one period T as $\hat{F}$. Although it is not possible to give an explicit form of $\hat{F}$ in the kicked molecule case, the existence of this formal solution yields a stroboscopic description of the dynamics,

$$|\psi(nT)\rangle = \hat{F}|\psi((n-1)T)\rangle = \hat{F}^n|\psi(0)\rangle \,, \qquad (7.87)$$

where n is an integer.

The operator $\hat{F}$ can be formally diagonalized by a unitary transformation $\underline{\underline{U}}$ so that,

$$\langle j_a, 0|\hat{F}|j_b, 0\rangle = \sum_{j_c} \exp(-i\phi_{j_c}) U^*_{j_c, j_a} U_{j_c, j_b} \,, \qquad (7.88)$$

where $U_{j_c,j_a} \equiv \langle j_c, 0|\hat{U}|j_a, 0\rangle$ ($j_a = 0, 1, 2, \ldots$) is the eigenvector with eigenphase ϕ_{j_c}. Moreover, since the basis states $|j, 0\rangle$ are time-reversal invariant, one can prove that the matrix elements U_{j_c,j_b} can be chosen as real numbers [312]; that is,

$$U^*_{j_c,j_a} = U_{j_c,j_a} \,, \quad j_a, j_c = 0, 1, 2, \ldots \qquad (7.89)$$

Further, evaluating $\tilde{E}$ at $t = NT$ with Eqs. (7.86), (7.88) and (7.89) gives

$$\begin{aligned} \frac{2\tilde{E}}{\tau^2} &= \langle\psi(0)|\hat{F}^{-N}\frac{\hat{J}^2}{\hbar^2}\hat{F}^N|\psi(0)\rangle \\ &= \cos^2\alpha \sum_{j j_a j_b} j(j+1) U_{j_a j_1} U_{j_b j} U_{j_a j} U_{j_b j_1} e^{iN(\phi_{j_a} - \phi_{j_b})} \\ &+ \sin^2\alpha \sum_{j j_a j_b} j(j+1) U_{j_a j_2} U_{j_b j} U_{j_a j} U_{j_b j_2} e^{iN(\phi_{j_a} - \phi_{j_b})} \\ &+ \frac{1}{2}\sin(2\alpha)\left(e^{-i\beta}\sum_{j j_a j_b} j(j+1) U_{j_a j_1} U_{j_b j_2} U_{j_a j} U_{j_b j} e^{iN(\phi_{j_a} - \phi_{j_b})} + \text{c.c.}\right). \end{aligned} \qquad (7.90)$$

Evidently, the first two terms are incoherent since they do not depend on the value of the phase β in Eq. (7.86). They represent quantum dynamics associated with each of the states $|j_1, 0\rangle$ and $|j_2, 0\rangle$ independently. The last two terms represent interference effects due to initial-state coherence between $|j_1, 0\rangle$ and $|j_2, 0\rangle$. Hence, the absorption of rotational energy in this system, that is, quantum chaotic diffusion, can be controlled by manipulating the quantum phase β in the initial state, which corresponds to manipulating the interference term in Eq. (7.90).

In Figure 7.19 we present a representative example of phase control in this system. In the chosen parameter region the underlying classical dynamics of rotational excitation is strongly chaotic[311] and the excitation is far off-resonance, with many levels excited. We choose $j_1 = 1$ and $j_2 = 2$ to create the initial superposition state $(|1, 0\rangle \pm |2, 0\rangle)/\sqrt{2}$; that is, $\alpha = \pi/4$ and $\beta = 0, \pi$ in Eq. (7.86). (Such states can be prepared experimentally by, for example, STIRAP, a technique discussed in detail in Section 11.1). The results, shown in Figure 7.19, display striking

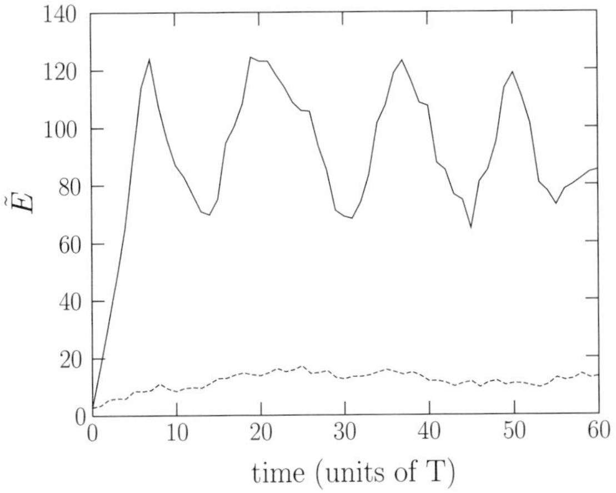

Figure 7.19 The dimensionless rotational energy of the kicked diatomic molecule $\tilde{E} = \sum_j P_j j(j+1)\tau^2/2$ versus time (in units of T). Solid line and dashed lines are for the initial states $(|1,0\rangle+|2,0\rangle)/2^{1/2}$ and $(|1,0\rangle-|2,0\rangle)/2^{1/2}$, respectively, for $\tau = 1.2$, $k = 4.8$. From [311, Figure 2]. (Reprinted with permission. Copyright 2001 American Institute of Physics.)

phase control. That is, $(|1,0\rangle - |2,0\rangle)/\sqrt{2}$ shows almost no energy absorption at all, whereas the $(|1,0\rangle + |2,0\rangle)/\sqrt{2}$ case shows extraordinarily fast energy absorption [311] before it essentially stops at $t \approx 10T$. Note (i) that this huge difference is achieved solely by changing the initial relative phase between the two participating states $|1,0\rangle$ and $|2,0\rangle$ in the initial superposition state, and (ii) that by contrast, each of $|1,0\rangle$ or $|2,0\rangle$ individually would give very similar diffusion behavior lying between the solid and dashed lines in Figure 7.19. This shows that the two par-

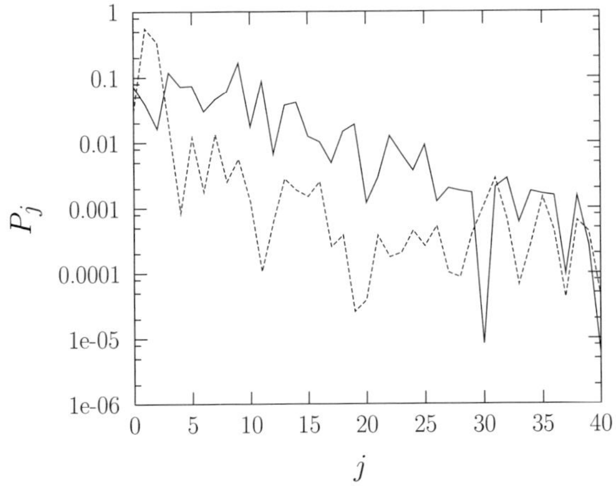

Figure 7.20 Occupation probability P_j versus the rotational quantum number j at $t = 60\,T$. Solid line and dashed lines are for the initial states $(|1,0\rangle + |2,0\rangle)/2^{1/2}$ and $(|1,0\rangle - |2,0\rangle)/2^{1/2}$, respectively, for $\tau = 1.2$, $k = 4.8$. From [311, Figure 3]. (Reprinted with permission. Copyright 2001 American Institute of Physics.)

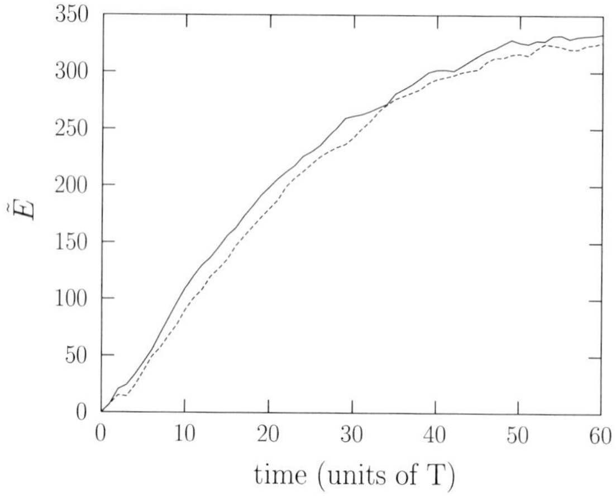

Figure 7.21 As in 7.19 except that τ is 50 times smaller: $\tau = 0.024$, $k = 240$. From [311, Figure 5]. (Reprinted with permission. Copyright 2001 American Institute of Physics.)

ticipating states $|1,0\rangle$ and $|2,0\rangle$ can either constructively or destructively interfere with one another, even though the underlying classical dynamics is strongly chaotic. A detail of the respective wave functions at $t = 60\ T$ is shown in Figure 7.20 in terms of the occupation probability P_j versus j. One sees vividly that changing β from 0 to π alters the occupation probability of many states by almost an order of magnitude.

The quantum dynamics of the kicked molecule depends on two parameters, $\tau \equiv \hbar T/I$ and $k \equiv \boldsymbol{d} \cdot \boldsymbol{E}_0/\hbar$. However, as shown elsewhere [311] the classical dynamics depends solely in the product $k\tau$. Thus, by decreasing the magnitude of $\tau \equiv \hbar T/I$ while keeping $k\tau$ fixed, we can approach the classical limit while keeping the underlying classical dynamics unaffected. This is a useful tool to show that the demonstrated phase control is indeed quantal in nature. Specifically we show, in Figure 7.21, the CsI quantum dynamics after reducing the effective Planck constant τ by 50 times, while keeping $k\tau$ constant. Here, with $\tau = 0.024$ and $k = 240$, the energy diffusion only shows slight dependence on β. That is, the phase control disappears, clearly demonstrating the quantum nature of the control.

8
Coherent Control of Bimolecular Processes

The results described above deal with control of unimolecular processes, that is, processes that begin with a single molecule that subsequently undergoes excitation and dynamics. However, the vast majority of chemical reactions occur via bimolecular processes, for example,

$$A + B \to C + D \tag{8.1}$$

where A, B, C, D are, in general, molecules of mass M_A, M_B, M_C and M_D. Here C and D can be identical to A and B (nonreactive scattering) or different from A and B (reactive scattering). We label $A+B$ as arrangement q and $C+D$ as arrangement q'. Below we describe coherent control of bimolecular collisions, demonstrating that coherent control is possible in bimolecular scattering [313–320]. Developments are discussed below both for control within the framework of scattering at fixed total energy (Section 8.1) and for time-dependent scattering (Section 8.2). The latter will be seen to shed considerable light on conditions for control in the former case. In both cases, focus is on control via quantum interference.

8.1
Fixed Energy Scattering: Entangled Initial States

In accord with Eq. (2.65), the amplitude for scattering between the asymptotic states $|E, q, \boldsymbol{m}; 0\rangle$ of $A + B$ (labeled q) and $|E, q', \boldsymbol{n}; 0\rangle$ of $C + D$ (labeled q') is given by the matrix element element $\langle E, q', \boldsymbol{n}; 0|S|E, q, \boldsymbol{m}; 0\rangle$. The probability of making this transition is therefore

$$P_E(\boldsymbol{n}, q'; \boldsymbol{m}, q) = \left|\langle E, q', \boldsymbol{n}; 0|S|E, q, \boldsymbol{m}; 0\rangle\right|^2 . \tag{8.2}$$

Alternatively, using Eq. (2.65) gives the probability in terms of the potential. Specifically, we define the cross section $\sigma_E(\boldsymbol{n}, q'; \boldsymbol{m}, q)$ for forming $|E, q', \boldsymbol{n}; 0\rangle$, having initiated the scattering in $|E, q, \boldsymbol{m}; 0\rangle$, as:

$$\sigma_E(\boldsymbol{n}, q'; \boldsymbol{m}, q) = \left|\langle E, q', \boldsymbol{n}^-|V_q|E, q, \boldsymbol{m}; 0\rangle\right|^2 . \tag{8.3}$$

Quantum Control of Molecular Processes, Second Edition. Moshe Shapiro and Paul Brumer

Here $|E, q', \boldsymbol{n}^-\rangle$ denotes the incoming scattering solutions associated with product in state $|E, q', \boldsymbol{n}; 0\rangle$ and V_q is the component of the total potential that vanishes as the A to B distance becomes arbitrarily large. The cross section for scattering into arrangement q', independent of the product internal state $\boldsymbol{n}$, is then

$$\sigma_E(q'; \boldsymbol{m}, q) = \sum_n \left|\langle E, q', \boldsymbol{n}^- | V_q | E, q, \boldsymbol{m}; 0\rangle\right|^2 . \tag{8.4}$$

Assorted other cross sections may be defined, depending upon which of the elements of $\boldsymbol{n}$ are summed over. For example, by not including the scattering angles θ, ϕ in the sum we obtain $\sigma_E(q', \theta, \phi; \boldsymbol{m}, q)$, corresponding to scattering into the q' product channel and into scattering angles (θ, ϕ). Similarly, $\sigma_E(q', \theta; \boldsymbol{m}, q)$ is the traditional differential cross section $\sigma_E(q', \theta; \boldsymbol{m}, q) = \int_0^{2\pi} d\phi\, \sigma_E(q', \theta, \phi; \boldsymbol{m}, q)$ into angle θ.

Note that Eqs. (8.2)–(8.4) describe motion in the center of mass coordinate system, that is, they arise in scattering theory after separating out the motion of the center of mass of $A-B$, a feature discussed in greater detail in Section 8.1.1.

Control of bimolecular collisions at fixed energy is achieved by constructing an initial state $|E, q, \{a_{\boldsymbol{m}}\}\rangle$ composed of a superposition of N energetically degenerate asymptotic states $|E, q, \boldsymbol{m}; 0\rangle$:

$$|E, q, \{a_{\boldsymbol{m}}\}\rangle = \sum_m a_{\boldsymbol{m}} |E, q, \boldsymbol{m}; 0\rangle . \tag{8.5}$$

The cross section associated with using Eq. (8.5) as the initial state, obtained by replacing $|E, q, \boldsymbol{m}; 0\rangle$ by Eq. (8.5) in Eq. (8.3), is

$$\begin{aligned}
\sigma_E(\boldsymbol{n}, q'; \{a_{\boldsymbol{m}}\}, q) &= \left|\langle E, q', \boldsymbol{n}^- | V_q \sum_m a_{\boldsymbol{m}} |E, q, \boldsymbol{m}; 0\rangle\right|^2 \\
&= \sum_m |a_{\boldsymbol{m}}|^2 \left|\langle E, q', \boldsymbol{n}^- | V_q | E, q, \boldsymbol{m}; 0\rangle\right|^2 \\
&\quad + \sum_{m'} \sum_{m \neq m'} a_{\boldsymbol{m}} a^*_{\boldsymbol{m}'} \langle E, q, \boldsymbol{m}'; 0 | V_q | E, q', \boldsymbol{n}^-\rangle \langle E, q', \boldsymbol{n}^- | V_q | E, q, \boldsymbol{m}; 0\rangle \\
&\equiv \sum_m |a_{\boldsymbol{m}}|^2 \sigma(\boldsymbol{n}, q'; \boldsymbol{m}, q) + \sum_{m'} \sum_{m \neq m'} a_{\boldsymbol{m}} a^*_{\boldsymbol{m}'} \sigma(\boldsymbol{n}, q'; \boldsymbol{m}', \boldsymbol{m}, q) ,
\end{aligned} \tag{8.6}$$

where $\sigma(\boldsymbol{n}, q'; \boldsymbol{m}', \boldsymbol{m}, q)$ is defined via Eq. (8.6). The total cross section into arrangement q' is then given by

$$\sigma_E(q'; \{a_{\boldsymbol{m}}\}, q) = \sum_n \sigma_E(\boldsymbol{n}, q'; \{a_{\boldsymbol{m}}\}, q) . \tag{8.7}$$

Note that Eq. (8.6), and hence Eq. (8.7), are of a standard coherent control form, that is, direct contributions from each individual member of the superposition, proportional to $|a_{\boldsymbol{m}}|^2$, plus interference terms that are proportional to $a_{\boldsymbol{m}} a^*_{\boldsymbol{m}'}$. It is clear that if we control the a_m, through assorted preparation methods, then we can control the interference term, and hence the scattering cross section.

8.1.1 Issues in the Preparation of the Scattering Superposition

To describe how the required superposition state (Eq. (8.5)) can be constructed in the laboratory requires some introductory remarks. Note first that Eqs. (8.2)–(8.7) and the $|E, q, \boldsymbol{m}; 0\rangle$ states are understood to be in the center of mass coordinate system and describe the relative translational motion as well as the internal state of A and B. In typical $A-B$ scattering, separating out the center of mass motion comes about in a straightforward way. That is, let $\boldsymbol{r}_A$ and $\boldsymbol{r}_B$ denote the laboratory position of A and B and $\hbar\boldsymbol{k}^A$, $\hbar\boldsymbol{k}^B$ denote their laboratory momenta. The relative momentum $\boldsymbol{k}$, relative coordinate $\boldsymbol{r}$, center of mass momentum $\boldsymbol{K}$ and position $\boldsymbol{R}_{\rm cm}$ are defined as

$$K = \boldsymbol{k}^A + \boldsymbol{k}^B \, ; \quad \boldsymbol{R}_{\rm cm} = \frac{(m_A \boldsymbol{r}_A + m_B \boldsymbol{r}_B)}{(m_A + m_B)} \, ;$$
$$\boldsymbol{k} = \frac{(m_B \boldsymbol{k}^A - m_A \boldsymbol{k}^B)}{(m_A + m_B)} \, ; \quad \boldsymbol{r} = \boldsymbol{r}_A - \boldsymbol{r}_B \, . \tag{8.8}$$

In the traditional case where A and B are initially in internal states $|\phi_A(i)\rangle$ and $|\phi_B(j)\rangle$, of energies $e_A(i)$ and $e_B(j)$, and the initial A and B translational motion are described by plane waves of momenta $\boldsymbol{k}_i^A$ and $\boldsymbol{k}_j^B$ then the incident wave function $|\psi_{in}\rangle$ is the product

$$\begin{aligned} |\psi_{in}\rangle &= |\phi_A(i)\rangle|\phi_B(j)\rangle \exp(i\boldsymbol{k}_i^A \cdot \boldsymbol{r}_A) \exp(i\boldsymbol{k}_j^B \cdot \boldsymbol{r}_B) \, , \\ &= |\phi_A(i)\rangle|\phi_B(j)\rangle \exp(i\boldsymbol{k} \cdot \boldsymbol{r}) \exp(i\boldsymbol{K} \cdot \boldsymbol{R}_{\rm cm}) \, . \end{aligned} \tag{8.9}$$

The second equality follows from Eqs. (8.8). Since the interaction potential V_q between A and B depends only upon the relative coordinates of $A-B$, the center of mass momentum is conserved in the collision, allowing us to separate out the center of mass motion and to describe the dynamics in the center of mass coordinate system, that is, in terms of $|\phi_A(i)\rangle|\phi_B(j)\rangle \exp(i\boldsymbol{k} \cdot \boldsymbol{r})$. This state is, in fact, $\langle \boldsymbol{r}|E, q, \boldsymbol{m}; 0\rangle$, where the relative motion is in the coordinate representation.

Note that the scattering in Eq. (8.9) occurs at fixed value of the center of mass momentum $\boldsymbol{K}$. Scattering may also occur from a state comprised of different $\boldsymbol{K}$ values. For example, the incident wave function may be of the form

$$|\psi_{in}\rangle = \sum_{lm} d_{l\boldsymbol{m}} |E, q, \boldsymbol{m}; 0\rangle |\boldsymbol{K}_l\rangle \, , \quad (\boldsymbol{K}_{l'} \neq \boldsymbol{K}_l) \, . \tag{8.10}$$

Since the center of mass momentum is conserved and can be measured, components of the wave function with different values of $|\boldsymbol{K}_l\rangle$ contribute independently to the reaction cross section and do not interfere with one another. That is, the cross section for scattering into $|E, q', \boldsymbol{n}; 0\rangle$ in this case is given by

$$\sigma_E(\boldsymbol{n}, q'; \{d_{l\boldsymbol{m}}\}, q) = \sum_l \left| \langle E, q', n^- | V_q \sum_m d_{l\boldsymbol{m}} | E, q, \boldsymbol{m}; 0\rangle \right|^2 . \tag{8.11}$$

Consider now preparation of the generalized superposition states (Eq. (8.5)) where for simplicity we limit consideration to a superposition of two states. To do so we examine the scattering of A and B, each previously prepared in the laboratory in a superposition state. The wave functions of A and B in the laboratory frame, ψ_A and ψ_B, are of the general form:

$$|\psi_A\rangle = a_1|\phi_A(1)\rangle \exp(i\boldsymbol{k}_1^A \cdot \boldsymbol{r}_A) + a_2|\phi_A(2)\rangle \exp(i\boldsymbol{k}_2^A \cdot \boldsymbol{r}_A) \tag{8.12}$$

$$|\psi_B\rangle = b_1|\phi_B(1)\rangle \exp(i\boldsymbol{k}_1^B \cdot \boldsymbol{r}_B) + b_2|\phi_B(2)\rangle \exp(i\boldsymbol{k}_2^B \cdot \boldsymbol{r}_B) . \tag{8.13}$$

The incident wave function is then the product

$$\begin{aligned} |\psi_{in}\rangle = |\psi_A\rangle|\psi_B\rangle &= \left[a_1|\phi_A(1)\rangle \exp(i\boldsymbol{k}_1^A \cdot \boldsymbol{r}_A) + a_2|\phi_A(2)\rangle \exp(i\boldsymbol{k}_2^A \cdot \boldsymbol{r}_A)\right] \\ &\quad \times \left[b_1|\phi_B(1)\rangle \exp(i\boldsymbol{k}_1^B \cdot \boldsymbol{r}_B) + b_2|\phi_B(2)\rangle \exp(i\boldsymbol{k}_2^B \cdot \boldsymbol{r}_B)\right] \\ &= \sum_{i,j=1}^{2} A_{ij} \exp(i\boldsymbol{k}_{ij} \cdot \boldsymbol{r}) \exp(i\boldsymbol{K}_{ij} \cdot \boldsymbol{R}_{\text{cm}}) , \end{aligned} \tag{8.14}$$

where $A_{ij} = a_i b_j |\phi_A(i)\rangle|\phi_B(j)\rangle$, $\boldsymbol{k}_{ij} = (M_B \boldsymbol{k}_i^A - M_A \boldsymbol{k}_j^B)/(M_A + M_B)$, and $\boldsymbol{K}_{ij} = \boldsymbol{k}_i^A + \boldsymbol{k}_j^B$.

As constructed, Eq. (8.14) is composed of four independent noninterfering incident states since each has a different center of mass wave vector $\boldsymbol{K}_{ij}$. However, we can set conditions so that interference, and hence control, is allowed. That is, we can require the equality of the center of mass motion of two components, plus energy degeneracy:

$$\begin{aligned} \boldsymbol{K}_{12} &= \boldsymbol{K}_{21} , \\ \frac{\hbar^2 k_{12}^2}{2\mu} + e_A(1) + e_B(2) &= \frac{\hbar^2 k_{21}^2}{2\mu} + e_A(2) + e_B(1) , \end{aligned} \tag{8.15}$$

with $\mu = M_A M_B/(M_A + M_B)$. Equation (8.14) then becomes

$$\begin{aligned} \psi_{in} &= \left[A_{12} \exp(i\boldsymbol{k}_{12} \cdot \boldsymbol{r}) + A_{21} \exp(i\boldsymbol{k}_{21} \cdot \boldsymbol{r})\right] \exp(i\boldsymbol{K}_{12} \cdot \boldsymbol{R}_{\text{cm}}) \\ &\quad + A_{11} \exp(i\boldsymbol{k}_{11} \cdot \boldsymbol{r}) \exp(i\boldsymbol{K}_{11} \cdot \boldsymbol{R}_{\text{cm}}) + A_{22} \exp(i\boldsymbol{k}_{22} \cdot \boldsymbol{r}) \exp(i\boldsymbol{K}_{22} \cdot \boldsymbol{R}_{\text{cm}}) , \end{aligned} \tag{8.16}$$

where the term in the first bracket, due to Eq. (8.15), is a linear superposition of two *degenerate* states. We therefore expect that the scattering cross section will be composed of noninterfering contributions from three components with differing

K_{ij}, but where the first term allows for control via the interference between the A_{12} and A_{21} terms. The two remaining terms, proportional to A_{11} and A_{22}, are uncontrolled satellite contributions.

For example, if we design the experiment so that $\boldsymbol{k}_1^A = -\boldsymbol{k}_2^B$ and $\boldsymbol{k}_2^A = -\boldsymbol{k}_1^B$ then $\boldsymbol{K}_{12} = \boldsymbol{K}_{21} = 0$, and $\boldsymbol{k}_{12} = \boldsymbol{k}_1^A, \boldsymbol{k}_{21} = -\boldsymbol{k}_1^B$, so that the degeneracy requirement (Eq. (8.15)) becomes

$$\frac{\hbar^2 (k_1^A)^2}{2\mu} + e_A(1) + e_B(2) = \frac{\hbar^2 (k_1^B)^2}{2\mu} + e_A(2) + e_B(1) \,. \tag{8.17}$$

Note also that we can implement Eq. (8.17) for the case of atom + diatomic molecule scattering by setting $|\phi_A(1)\rangle = |\phi_A(2)\rangle = |\phi_A(g)\rangle$, where $|\phi_A(g)\rangle$ is the, for example, ground electronic state of atom A. In this case the degeneracy condition (Eq. (8.17)) is

$$\frac{\hbar^2}{2\mu}\left[(k_1^A)^2 - (k_2^A)^2\right] = \left[e_B(1) - e_B(2)\right] \,. \tag{8.18}$$

In general, these conditions demand a method of preparing $|\psi_A\rangle$ and $|\psi_B\rangle$ that correlate the internal states $|\phi_A(i)\rangle$ and $|\phi_B(i)\rangle$ with their associated momenta $\boldsymbol{k}_i^A, \boldsymbol{k}_i^B$ so as to obtain Eq. (8.17). Since the overall phase of the wave function is irrelevant to the state of the system, the dynamics is not sensitive to the overall phase of $|\psi_A\rangle|\psi_B\rangle$. However, the phases of the interference term must be well defined, or the control will average to zero.

8.1.2
Identical Particle Collisions

The situation simplifies considerably for the case of identical particle collisions, that is, when $B = A$. Specifically, consider

$$A + A' \rightarrow C + D \tag{8.19}$$

Here we have used A' to denote the molecule A, but in a superposition state that is not necessarily the same as A. If we prepare each of the two initial A and A' superposition states from the same molecular bound states, for example, $|\phi_A(1)\rangle = |\phi_{A'}(1)\rangle$ and $|\phi_A(2)\rangle = |\phi_{A'}(2)\rangle$ then the requirement for conservation of energy in the center of mass (Eq. (8.15)) becomes

$$k_{12}^2 = k_{21}^2 \,. \tag{8.20}$$

For the case of $A + A'$ collisions, this condition is always satisfied.

This scenario opens up a wide range of possible experimental studies of control in bimolecular collisions. Specifically, we need only prepare A and A' in a controlled superposition of two states (e.g., by resonant laser excitation of $|\phi_A(1)\rangle$) to produce a superposition with $|\phi_A(2)\rangle$, direct them antiparallel in the laboratory and vary the coefficients in the superposition to affect the reaction probabilities. Control originates in quantum interference between two degenerate states associated with the contributions of $|\phi_A(1)\rangle|\phi_{A'}(2)\rangle$ and $|\phi_A(2)\rangle|\phi_{A'}(1)\rangle$. This is accompanied by two uncontrolled scattering contributions corresponding to the contributions of $|\phi_A(1)\rangle|\phi_{A'}(1)\rangle$ and $|\phi_A(2)\rangle|\phi_{A'}(2)\rangle$. Control is achieved by varying the four coefficients $a_i, b_i, i = 1, 2$. Stimulated rapid adiabatic passage (STIRAP) [321, 322] to be discussed in detail in Section 11.1, provides one choice for such state preparation.

The control approach described above can be generalized to a superposition of N levels in each of the two A and A' reactants. Specifically, choosing all $\boldsymbol{k}_i^A = \boldsymbol{k}^A$ and with $\boldsymbol{k}_i^{A'} = -\boldsymbol{k}^A$ we have

$$|\psi_A\rangle = \exp(i\boldsymbol{k}^A \cdot \boldsymbol{r}_A)\left[\sum_{i=1}^{N} a_i|\phi_A(i)\rangle\right] ,$$

$$|\psi_{A'}\rangle = \exp(-i\boldsymbol{k}^A \cdot \boldsymbol{r}_{A'})\left[\sum_{j=1}^{N} b_j|\phi_{A'}(j)\rangle\right] . \tag{8.21}$$

The scattering wave function is then

$$|\psi_{in}\rangle = |\psi_A\rangle|\psi_{A'}\rangle = \exp(i\boldsymbol{k}\cdot\boldsymbol{r})\left[\sum_{i=1}^{N} a_i|\phi_A(i)\rangle\right]\left[\sum_{j=1}^{N} b_j|\phi_{A'}(j)\rangle\right] . \tag{8.22}$$

Since $M_A = M_{A'}$, $\boldsymbol{k} = (\boldsymbol{k}^A - \boldsymbol{k}^{A'})/2 = \boldsymbol{k}^A$. The kinetic energy $k^2/2\mu$ is the same for each term in Eq. (8.22) so that degenerate states in the center of mass frame correspond to states $|\phi_A(i)\rangle|\phi_{A'}(j)\rangle$ in Eq. (8.22) which are of equal internal energy $e_A(i) + e_{A'}(j)$. Expanding the product in Eq. (8.22) gives N^2 terms, N terms of which are of differing energy $2e_A(i), i = 1, \ldots, N$ and $(N^2 - N)$ states of energy $e_A(i) + e_{A'}(j), i \neq j$. Of the latter terms, each is accompanied by another term of equal energy (i.e., $e_A(i) + e_{A'}(j) = e_A(j) + e_{A'}(i)$). Hence the N^2 terms are comprised of N direct terms plus $(N^2 - N)/2$ degenerate pairs which are a source of interference, and hence control. Here control is achieved by altering the $2N$ coefficients a_i, b_i in the initially prepared state (Eq. (8.22)), for example, by shaped pulsed laser excitation of A and A'.

Computational examples of this approach have been restricted to control over rotational excitation in $H_2 + H_2$, a consequence of limitations on the ability to perform quantum computations on AB + AB scattering. A careful analysis of the scattering [323, 324] requires consideration both of the interference effects as well as the nature of the identical particle scattering. Typical results are shown in Figure 8.1 for various low energy scattering cases. Specifically, we show the differential cross section into scattering angles θ and final states $j_1' = j_2' = 2$, arising from scattering of para H_2 + para H_2, where each H_2 is in an initial superposition, with either a plus or minus sign, of $j_1 = 4$ and $j_2 = 0$. The cross term contributing to the scattering in these cases is

$$|\psi_{j_1 j_2}^{\pm}\rangle = \frac{1}{\sqrt{2}}[|j_1\rangle|j_2\rangle \pm |j_2\rangle||j_1\rangle]|m_1 = 0, \mathrm{v}_1 = 0\rangle|m_2 = 0, \mathrm{v}_2 = 0\rangle , \tag{8.23}$$

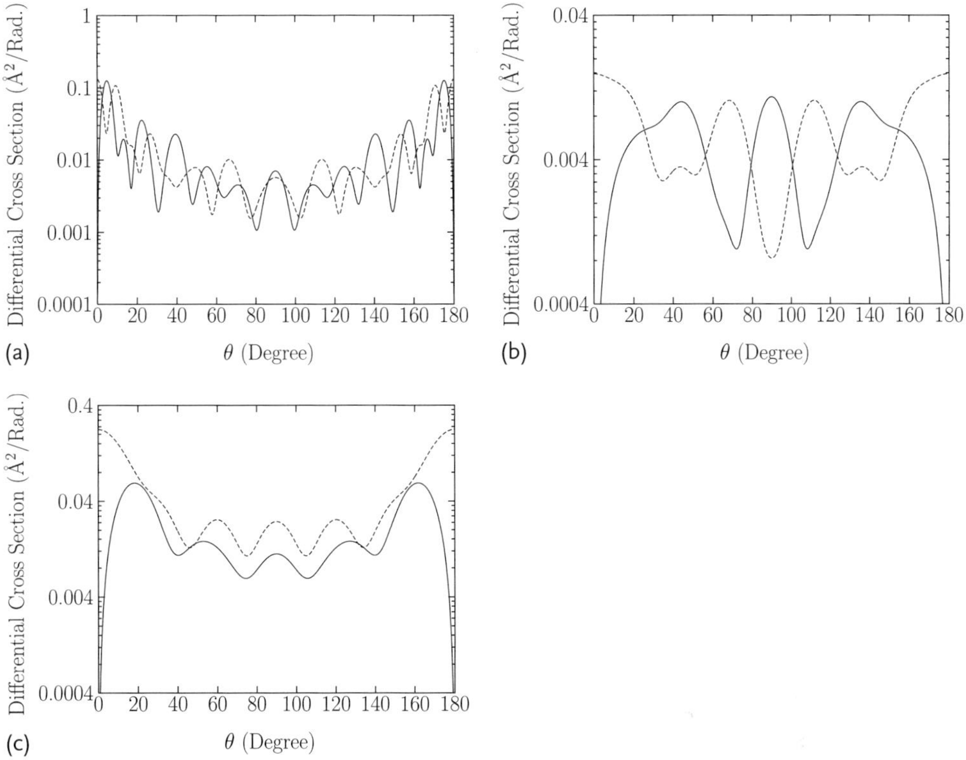

Figure 8.1 Inelastic differential cross section for para H_2 + para H_2, where the collision energy is (a) 400 cm^{-1}, (b) 40 cm^{-1}, and (c) 4 cm^{-1}. Dashed and solid lines are for the incoming free entangled states $|\psi^{+}_{j_1 j_2}\rangle$ and $|\psi^{-}_{j_1 j_2}\rangle$. Here $j_1 = 4$, $j_2 = 0$, $j_1' = j_2' = 2$. From [323, Figure 2]. (Reprinted with permission. Copyright 2003 American Institute of Physics.)

where v_i, m_i denote the vibrational state and angular momentum projection along the z-axis. Note that results for $m_i = m_2$ are essentially independent of the value of m_i.

The results of this computation (Figure 8.1 – note the logarithmic ordinate scale) clearly show that the phase of the j_1, j_2 superposition has a significant effect on the differential cross section. This translates into considerable control over the total inelastic cross section, that is, the integral of the differential cross section over θ. For example, in Figure 8.1c, the total inelastic cross section is 0.057 for $|\psi^{+}_{j_1 j_2}\rangle$ and 0.032 for $|\psi^{-}_{j_1 j_2}\rangle$.

Results on more complex scattering, for example, reactive scattering, await further computational (or experimental) developments.

8.1.3
Sample Control Results

8.1.3.1 *m* Superpositions

For the case of nonidentical particle scattering, insight into control can be gained by starting with a superposition of degenerate states of the fragments. In atom +diatomic molecule scattering, one possibility is to use the $(2j+1)$ diatomic molecule rotational states $|j, m\rangle$, where m is the projection of diatomic molecule angular momentum j along a spacially fixed axis. In this case we show below that control over the differential cross section is possible but control over the total cross section is not. The argument is related to that given in our discussion of polarization control of photodissociation processes (Section 3.5).

Consider first superimposing two m states of a diatomic molecule in atom +diatomic molecule scattering. For this case the initial state (Eq. (8.5)) assumes the form:

$$|E, q, \{a_{\mathbf{m}}\}\rangle = \sum_{i=1,2} a_i |q\mathrm{v}\, j\, m_i\rangle |E_q^{kin}\rangle\,, \tag{8.24}$$

where $|q\mathrm{v}\, j\, m_i\rangle$ is an eigenstate of the diatomic molecule internal Hamiltonian in the q channel, of energy $e_q(\mathrm{v}\, j)$, with v denoting the diatomic molecule vibrational quantum number. $|E_q^{kin}\rangle$ is a plane wave of energy $E_q^{kin} = E - e_q(\mathrm{v}\, j)$, describing the free motion of the atom relative to the diatomic molecule in the q arrangement.

Consider scattering into a final state $\langle E, q', \mathbf{n}; 0|$ with $\mathbf{n}$ defined by v', j', λ' and scattering angles θ, φ from the initial state $|E, q, \{a_{\mathbf{m}}\}\rangle$ in Eq. (8.24). Here λ' is the helicity; that is, the projection of the product of the diatomic molecule angular momentum onto the final relative translational velocity vector. Then the resultant differential cross section can be written as [325]

$$\sigma(q'\mathrm{v}'\, j'\lambda' \leftarrow q, \mathrm{v}, j, m_1, m_2|\theta, \varphi) = \left| \sum_{i=1,2} a_i f_{q'\mathrm{v}'\, j'\lambda' \leftarrow q\mathrm{v}\, j\, m_i}(\theta, \varphi) \right|^2\,, \tag{8.25}$$

where f, the so-called scattering amplitude, is given by

$$f_{q'\mathrm{v}'\, j'\lambda' \leftarrow q\mathrm{v}\, j\, m_i}(\theta, \varphi) = \frac{i^{j-j'+1} e^{i m_i \varphi}}{2k_q(\nu\, j)} \sum_J (2J+1) d^J_{\lambda' m_i}(\pi - \theta) \times \left[S^J_{q'\mathrm{v}'\, j'\lambda', q\mathrm{v}\, j\, m_i} - \delta_{q'q}\delta_{\mathrm{v}'\mathrm{v}}\delta_{j'j}\delta_{\lambda' m_i} \right], \tag{8.26}$$

with $d^J_{\lambda', m_i}(\theta)$ being the reduced Wigner rotation matrices [326]. $S^J_{q'\mathrm{v}'\, j'\lambda', q\mathrm{v}\, j\, m_i}$ of the above are the elements of the scattering S-matrix in the so-called helicity representation [327, 328], that is, where helicity is one of the final quantum numbers. The quantity $k_q(\mathrm{v}\, j) = \sqrt{2\mu_q[E - e_q(\mathrm{v}\, j)]}/\hbar$, with μ_q being the atom + diatomic molecule reduced mass in the q channel.

Expanding the square in Eq. (8.25) gives the reactive differential scattering cross section as (where we drop the initial state labels for convenience)

$$\sigma^R(\mathrm{v}' j'\lambda'|\theta,\varphi) = |a_1|^2\sigma_{11}^R(\mathrm{v}' j'\lambda'|\theta) + |a_2|^2\sigma_{22}^R(\mathrm{v}' j'\lambda'|\theta) + 2\mathrm{Re}\left\{a_1 a_2^* \sigma_{12}^R(\mathrm{v}' j'\lambda'|\theta,\varphi)\right\} , \tag{8.27}$$

where

$$\begin{aligned}\sigma_{ii}^R(\mathrm{v}' j'\lambda'|\theta) &= \left|f_{q'\mathrm{v}' j'\lambda' \leftarrow q\mathrm{v} j m_i}(\theta,\varphi)\right|^2 \\ &= [2k_q(\mathrm{v} j)]^{-2}\sum_{J,J'}(2J+1)(2J'+1)d^J_{\lambda' m_i}(\pi-\theta)d^{J'}_{\lambda' m_i}(\pi-\theta) \\ &\quad \times S^J_{q'\mathrm{v}' j'\lambda',q\mathrm{v} j m_i}\left[S^{J'}_{q'\mathrm{v}' j'\lambda',q\mathrm{v} j m_i}\right]^* , \quad q \neq q' , \quad i = 1,2 ,\end{aligned} \tag{8.28}$$

and

$$\begin{aligned}\sigma_{12}^R(\mathrm{v}' j'\lambda'|\theta,\varphi) &= f_{q'\mathrm{v}' j'\lambda' \leftarrow q\mathrm{v} j m_1}(\theta,\varphi) f^*_{q'\mathrm{v}' j'\lambda' \leftarrow q\mathrm{v} j m_2}(\theta,\varphi) \\ &= \frac{e^{i(m_1-m_2)\varphi}}{4k_q^2(\mathrm{v} j)}\sum_{J,J'}(2J+1)(2J'+1) \\ &\quad \times d^J_{\lambda' m_1}(\pi-\theta)d^{J'}_{\lambda' m_2}(\pi-\theta)S^J_{q'\mathrm{v}' j'\lambda',q\mathrm{v} j m_1}\left[S^{J'}_{q'\mathrm{v}' j'\lambda',q\mathrm{v} j m_2}\right]^* , \quad q \neq q' .\end{aligned} \tag{8.29}$$

Here, the superscript R denotes reactive scattering into a specific final arrangement channel $q' \neq q$. The total differential cross section, $\sigma^R(\theta,\varphi)$, for reaction out of a state in Eq. (8.24) is given by the sum over final states at energy E as

$$\sigma^R(\theta,\varphi) = \sum_{\mathrm{v}',j',\lambda'} \sigma^R(\mathrm{v}' j'\lambda'|\theta,\varphi) . \tag{8.30}$$

Note that the φ-dependence of the measurable cross sections is due solely to the interference term. Thus, traditional (uncontrolled) scattering is φ-independent. Integration of Eq. (8.27) or (8.30) over angles $\theta \in [0,\pi]$ and $\varphi \in [0,2\pi]$ gives the state-resolved integral reactive cross section $\sigma^R(\mathrm{v}' j'\lambda')$. However, the integral of $\sigma_{12}^R(\mathrm{v}' j'\lambda'|\theta,\varphi)$ over φ is zero so that the interference term, and hence control over the integral cross section, disappears. Indeed, the integral over $0 < \varphi < \pi$ exactly cancels the integral over $\pi < \varphi < 2\pi$. For this reason we consider control over scattering into the hemisphere $0 < \varphi < \pi$, giving the state resolved integral cross section which we denote as $\sigma^R(\mathrm{v}' j'\lambda';\varphi \leq \pi)$. This can also be written as three terms, as in Eq. (8.27), but with the σ_{ij}^R replaced by $\sigma_{ij}^R(\mathrm{v}' j'\lambda';\varphi \leq \pi)$, $(i, j = 1,2)$, where

$$\sigma_{ik}^R(\mathrm{v}' j'\lambda';\varphi \leq \pi) = \int_0^\pi \sin\theta\, d\theta \int_0^\pi d\varphi\, \sigma_{ik}^R(\mathrm{v}' j'\lambda'|\theta,\varphi) . \tag{8.31}$$

Summing over the final v′, j', λ' at energy E, gives the total integral cross section for total scattering into the hemisphere as

$$\sigma^R(\varphi \le \pi) = \sum_{v',j',\lambda';\varphi\le\pi} \sigma^R(v', j', \lambda'; \varphi \le \pi) . \quad (8.32)$$

It is important to stress that state-to-state cross sections $\sigma^R_{ii}(v' j'\lambda'|\theta)$ and $\sigma^R_{ii}(v' j'\lambda'; \varphi \le \pi)$ in Eqs. (8.28) and (8.31), as well as the corresponding total cross sections $\sigma^R_{ii}(\theta)$ and σ^R_{ii}, appear in standard scattering theory (see, for example, [329]), whereas $\sigma^R_{12}(v' j'\lambda'|\theta, \varphi)$ and $\sigma^R_{12}(v' j'\lambda'; \varphi \le \pi)$ are new types of interference terms that allow for control, through the a_i, over atom + diatomic molecule collision process. As in the case of photodissociation, significant control requires substantial σ^R_{12}, which follows from the Schwartz inequality $[|\sigma^R_{12}| \le \sqrt{\sigma^R_{11}\sigma^R_{22}}]$. That is, large σ^R_{12} requires also large σ^R_{11} and σ^R_{22}.

To examine the extent of control over the reaction we rewrite the reactive cross section in the form (where we refer to scattering into a hemisphere, but drop the notation "$\varphi \le \pi$" for convenience),

$$\sigma^R = \frac{\sigma^R_{11} + x^2\sigma^R_{22} + 2x|\sigma^R_{12}|\cos(\delta^R_{12} + \phi_{12})}{1 + x^2} , \quad (8.33)$$

where $x = |a_2/a_1|$, $\phi_{12} = \arg(a_2/a_1)$, and $\delta^R_{12} = \arg(\sigma^R_{12})$ with the branching ratio between the reactive and nonreactive total cross sections given by

$$\frac{\sigma^R}{\sigma^{NR}} = \frac{\sigma^R_{11} + x^2\sigma^R_{22} + 2x|\sigma^R_{12}|\cos(\delta^R_{12} + \phi_{12})}{\sigma^{NR}_{11} + x^2\sigma^{NR}_{22} + 2x|\sigma^{NR}_{12}|\cos(\delta^{NR}_{12} + \phi_{12})} . \quad (8.34)$$

Here NR refers to nonreactive scattering; definitions of the nonreactive cross sections are analogous to their reactive counterparts.

It follows from Eqs. (8.33) and (8.34) that by varying the a_i coefficients (i.e., varying either the relative magnitude, x or the relative phase, ϕ_{12}) in Eq. (8.24) through the initial preparation step, we can directly alter the interference term σ^R_{12} (and/or σ^{NR}_{12}) and hence control the scattering cross sections. Such a preparation might be carried out, for example, by a suitably devised molecular beam experiment where the diatomic molecule is excited with elliptically polarized light to a collection of well-defined m states.

As an example of this approach consider control over the reaction $D + H_2$ (for other isotopes see [318].) Typical results corresponding to $D + H_2 \to H + HD$ at $E = 0.93$ eV, with scattering from an initial superposition of ($\nu = 0, j = 2, m_1 = 1, m_2 = 0$) are shown in Figure 8.2.

Results are reported as contour plots of the cross section ratios vs. the phase ϕ_{12} and the parameter $s = x^2/(1 + x^2)$. The value $s = 0$ corresponds to scattering from the state with $m_1 = 1$, and $s = 1$ corresponds to scattering from the state with $m_2 = 0$. The ratio of cross sections is seen to vary from 0.05 to 0.079, showing a maximum and minimum that are well outside (up to factors of 1.22 and 1.26 respectively) of the range of results for scattering from a single m state. Thus, the

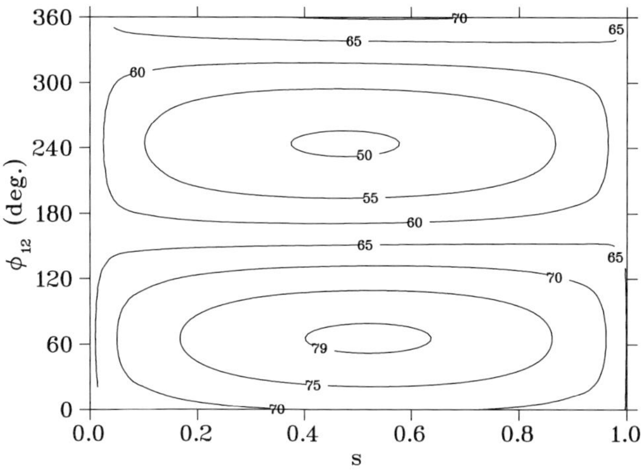

Figure 8.2 Contour plot of the ratio $\sigma^R(\varphi \leq \pi)/\sigma^{NR}(\varphi \leq \pi)$ ($\times 10^3$) for $D + H_2$ at $E = 0.93$ eV as a function of ϕ_{12} and s. The initial state is a superposition of $\nu = 0, j = 2, m_1 = 1, m_2 = 0$. From [318, Figure 2]. (Reprinted with permission. Copyright 1999 Royal Society of Chemistry.)

ratio of cross sections can be increased or decreased by approximately 20% through coherent control effects.

Thus, superposing two m levels provides some degree of control over the differential cross sections. Nonetheless, control is far from extensive. The origin of this behavior is evident from an examination of the σ_{12} compared to σ_{11} and σ_{22}. In particular, the cross term σ_{12} is found to be 4 to 10 times smaller than the σ_{ii} for the reactive case and far smaller for the nonreactive case. As a consequence, the extent of control is rather limited. Similar control, in this case over the total cross section, was found for superpositions of rovibrational states of the diatomic [320].

Two alternatives for improved control suggest themselves. The first is to seek alternate linear superpositions, or possibly different chemical reactions, with larger σ_{12}. The second is to examine the extent of control resulting from the inclusion of more than two degenerate reactant states in the initial superposition, as discussed in Section 8.1.5.

8.1.3.2 **Control in Cold Atoms: Penning vs. Dissociative Ionization**

Recent developments in cold and ultracold chemistry [330, 331] make this an interesting laboratory environment in which to explore and utilize quantum effects (e.g., [332–334]). For example, limited experimental results have been reported [335, 336] on the effects of quantum statistics on ultracold chemical reaction dynamics between Rb_2 and K_2. Below we focus on the role of interference phenomena on collisional control in the cold and ultracold regimes.

Consider then the collision of metastable atoms A^* and ground state target atoms B under such conditions. Two product channels are possible: (i) the ionization of the target atom and the de-excitation of the metastable species, that is, Penning Ionization (PI) [337], or (ii) Associative Ionization (AI), wherein the colliding partners form an ionic dimer while emitting an energetic electron. That is, the two processes are

$$\text{(AI)}\quad AB^+ + e^- \leftarrow A^* + B \rightarrow A + B^+ + e^- \quad \text{(PI)} \tag{8.35}$$

As an example of this process consider the coherent control of the cross section for PI and for AI resulting from collisions between $\text{Ne}^*(3s, {}^3P_2)$ and $\text{Ar}({}^1S_0)$ in the cold and ultracold regimes.

The Initial Superposition State: In this case, coherent control can be achieved by preparing the colliding pair in an initial superposition of internal states, such as,

$$|\psi\rangle = e^{iK\cdot R_{cm}+ik\cdot r}|\phi_{Ar}\rangle \sum_M a_M |\phi^M_{Ne^*}\rangle\,, \tag{8.36}$$

where $|\psi_{Ar}\rangle$ is the initial state of the Ar atom and $|\phi^M_{Ne^*}\rangle$, are Ne* Zeeman sublevels, with $M = \{-2, -1, 0, 1, 2\}$ being the projection of the Ne* electronic angular momentum on the fixed-space quantization axis. The a_M are preparation coefficients, to be optimized to yield a desired objective. As above, $\boldsymbol{R}_{cm}$ is the center of mass coordinate, $\boldsymbol{R}_{cm} \equiv (m_{Ne}\boldsymbol{r}^{Ne} + m_{Ar}\boldsymbol{r}^{Ar})/(m_{Ne} + m_{Ar})$, and $\boldsymbol{r}$ is the internuclear separation vector, $\boldsymbol{r} \equiv \boldsymbol{r}^{Ne} - \boldsymbol{r}^{Ar}$. The (body-fixed) momenta are given as, $\boldsymbol{K} \equiv \boldsymbol{k}^{Ne} + \boldsymbol{k}^{Ar}$, $\boldsymbol{k} \equiv (m_{Ar}\boldsymbol{k}^{Ne} - m_{Ne}\boldsymbol{k}^{Ar})/(m_{Ne}+m_{Ar})$. Here $\boldsymbol{r}^{Ar}$ and $\boldsymbol{k}^{Ar}$ ($\boldsymbol{r}^{Ne}$ and $\boldsymbol{k}^{Ne}$) denote the position and momentum of the Ar (Ne*) atom in the laboratory frame. Note that the fact that the initial superposition state is comprised of degenerate M states, and that the collision partners are atoms, ensures that the conditions for coherent control described above (Section 8.1.1) are satisfied.

The details of the computational method used to obtain the scattering results are described in [338, 339]. Here we focus on the physics of the process and on the character of the control results.

The rates of the PI and AI processes mainly depend on λ, the body-fixed (BF) projection of the electronic angular momentum on $\boldsymbol{r}$, the interatomic axis. If one adopts the "Rotating Atom Approximation" [340], then the axis of quantization of the electrons faithfully follows the internuclear separation vector, establishing a $1:1$ correspondence between the M values and the λ values as the atoms approach one another. Hence, the (even parity) linear combination in the BF frame is written as,

$$|\psi\rangle = |\phi_{Ar}\rangle e^{iK\cdot R_{cm}+ik\cdot r} \sum_{\Omega=0}^{2} |\phi^{\Omega}_{Ne^*}\rangle a_\Omega\,, \tag{8.37}$$

where $\Omega \equiv |\lambda|$, and (due to an assumed even parity) $a_\Omega \equiv (a_M + a_{-M})$.

At the low temperatures considered, (cold collisions at a temperature of 1 mK and ultracold collisions at 1 μK) the PI or AI cross sections are very large since the two atoms are in the vicinity of one another for an extended period of time. Further, in the ultracold case only the s partial wave contributes; three angular momentum states contributing in the cold case. For these energies the relative velocities between the collisional pair are ≈ 1 and ≈ 0.006 m/s respectively, and are experimentally attainable using laser cooling and manipulation techniques. The internal state superposition can be prepared [341] after cooling while the atoms are trapped in the lattice using, for example, stimulated Raman adiabatic passage (STIRAP) or coherent population trapping (CPT) [342].

First, consider control results across a broad spectrum of energies up to temperatures of 1 K. Figure 8.3 shows the cross sections for $\Omega = 0$ and $\Omega = 1$ scattering as a function of energy. Also shown are the maximum and minimum controlled cross sections (optimized over the a_i coefficients) at each energy for the $\Omega = 0$ plus $\Omega = 1$ linear combination. Several resonances [343, 344] are evident, since the collision energy is very close to the dissociation threshold for the Ne*–Ar

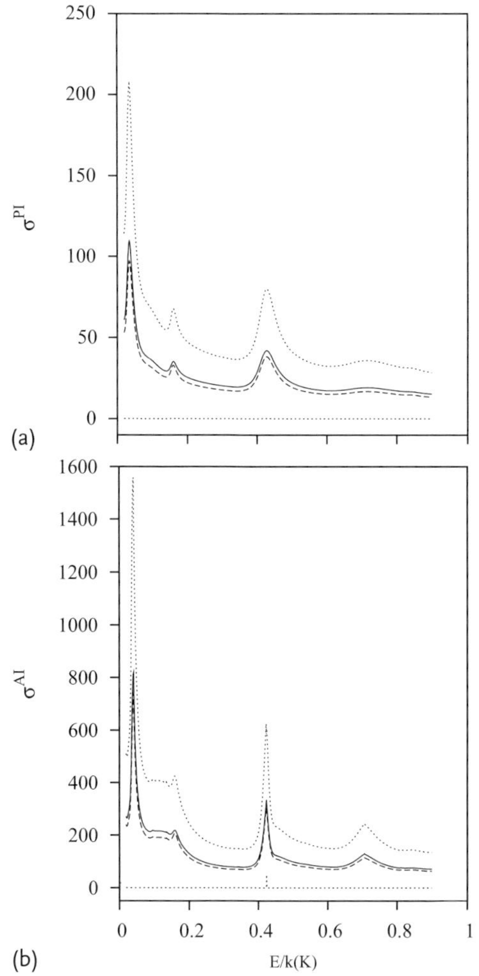

Figure 8.3 PI and AI cross sections in Å^2 for $\Omega = 0, 1$ linear combination. $\Omega = 0$ (solid); $\Omega = 1$ (dashed); maximum and minimum for the linear combination of $\Omega = 0, 1$ (dotted). From [338, Figure 1]. (Reprinted with permission. Copyright 2006 American Physical Society.)

quasimolecule. Noteworthy is that control is extensive, with enhancement and suppression of both cross sections being possible at both resonant and nonresonant energies.

Table 8.1 presents numerical results for the cold collision (1 mK) case. We see that it is possible to actively change the AI and PI cross sections by as much as four orders of magnitude for the $\Omega = 0,1$ linear combination and three orders of magnitude for the $\Omega = 0,2$ linear combination. For both linear combinations, the position of the minima and maxima for σ^{PI} and σ^{AI} occur at close points in the parameter space (not shown here). A similar observation has been noted at higher temperatures (above 1 K), indicating that both PI and AI cross sections can be controlled simultaneously [338, 339]. Sample results for the control of σ^{PI} as a function of the a_i coefficients for the cold collision case are shown in Figure 8.4a.

Ultracold collisions, where only *s* waves contribute to the process, show even more dramatic behavior. As seen in Table 8.2, active changes of up to four orders of magnitude, using the $\Omega = 0, 1$ superposition states, and up to three orders of magnitude, using the $\Omega = 0, 2$ superposition states, are possible. The AI process can also be almost as well controlled. The resulting σ^{AI} cross sections are shown in Figure 8.4b for ultracold collisions as function of the a_i. Note that in all cases, the maxima and minima in the control plots (Figures 8.3 and 8.4) are well separated, making an experiment less sensitive to the control parameters.

Table 8.1 Cross section for cold collision at T = 1 mK. Rows labeled "$\Omega = 0,1$" and "$\Omega = 0,2$" show the minimum and maximum of the σ^{PI} and σ^{AI}, obtained by varying the a_i for the indicated superposition. From [338].

Ω	σ^{PI} (Å^2)	σ^{AI} (Å^2)
0	74.68	346.91
1	64.90	306.25
2	13.75	87.01
0,1	1.27×10^{-2} – 139.57	3×10^{-2} – 653.13
0,2	0.63 – 87.80	0.60 – 433.32

Table 8.2 As in Table 8.1, but for ultra cold collision at T = 1 μK. From [338].

Ω	σ^{PI} (Å^2)	σ^{AI} (Å^2)
0	1357.32	7056.92
1	1174.19	6199.40
2	244.27	1728.65
0,1	0.128 – 2531.38	3.03 – 13 256.30
0,2	9.88 – 1591.71	2.28 – 8783.30

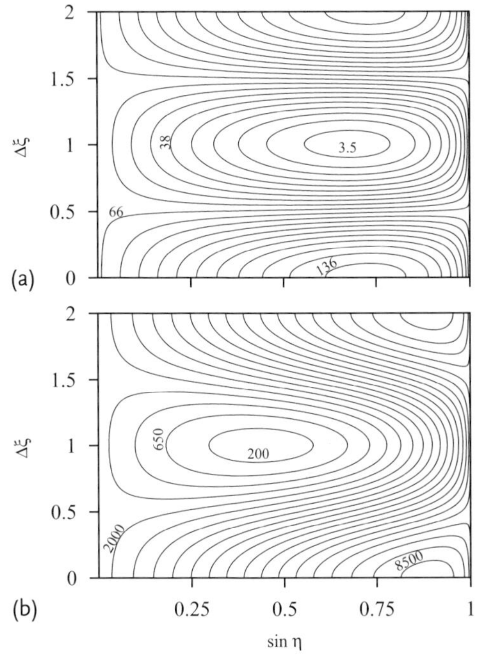

Figure 8.4 Coherent control contours for (a) PI for $\Omega = 0, 1$ in cold collisions at $T = 1$ mK; (b) AI for $\Omega = 0, 2$ in ultra cold collisions at $T = 1\,\mu$K. The parameters η and $\Delta\xi$ are defined via: $a_\Omega = \sin\eta e^{i\xi_\Omega}$ and $a_{\Omega'} = \cos\eta e^{i\xi_{\Omega'}}$, with $\Omega, \Omega' = 0, 1, 2$. From [338, Figure 2]. (Reprinted with permission. Copyright 2006 American Physical Society.)

Proposals to perform the cold collision experiments of this type have been advanced by Bergmann *et al.* [341].

Further, we note an alternative cold atom proposal [345] for creating superposition of scattering states that satisfy the conditions in Eq. (8.15). Here the suggestion is to prepare translational states of the colliding pair with momenta

$$\boldsymbol{k}_1^A = \boldsymbol{k}_2^A \,; \quad \boldsymbol{k}_1^B = \boldsymbol{k}_2^B \,, \tag{8.38}$$

which is possible in cold environments due to the narrow Boltzmann velocity distribution. This ensures equality of the center of mass momenta, and reduces the second requirement in Eq. (8.15) to

$$e_A(1) + e_B(2) = e_A(2) + e_B(1) \,. \tag{8.39}$$

Herrera [345] proposes using static magnetic fields to shift Zeeman levels of alkali metal atoms to achieve this equality. Using, as an example, collisions of Zeeman levels $^7\mathrm{Li}(\mathrm{M}_F = 1)$ and $^{133}\mathrm{Cs}(\mathrm{M}_F = -3)$ degeneracy is achieved at specific magnetic field values, as shown in Figure 8.5.

As an example, consider states $|\phi_A(1)\rangle$ and $|\phi_A(2)\rangle$ (in Eq. (8.14)) to be the Li hyperfine states $|I + 1/2, M_F\rangle$ and $|I - 1/2, M_F\rangle$, denoted $|\phi(1), M_F\rangle$. Here $\phi(1)$

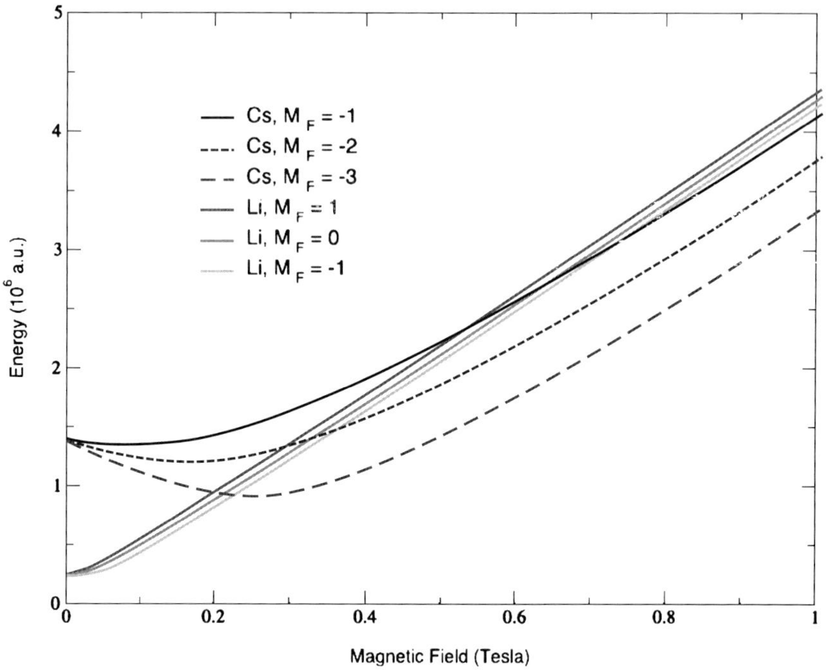

Figure 8.5 Energy difference between atomic Zeeman states as a function of the magnetic field. The curves correspond to different values of M_F for Li and Cs states. At the intersection points, the magnetic field induces degeneracy between the incoming channels $|\phi(1), M_F\rangle_{\text{Li}}|\phi(2), M'_F\rangle_{\text{Cs}}$ and $|\phi(2), M_F\rangle_{\text{Li}}|\phi(1), M'_F\rangle_{\text{Cs}}$, where the index $\phi(i)$ denotes the nuclear spin quantum number. From [345, Figure 1]. (Reprinted with permission. Copyright 2008 American Physical Society.)

denotes the total angular momentum quantum number, and I is the nuclear spin quantum number. Similar notation is used to define the states $|\phi_B(1)\rangle$ and $|\phi_B(2)\rangle$ of Cs. Control is considered as a function of a "relative phase" Φ which consists of the combination of coefficients a_i, b_j in the incident superposition (Eq. (8.14)):

$$\tan\Phi = \frac{\text{Im}(a_2 b_1)/\text{Re}(a_2 b_1) - \text{Im}(a_1 b_2)/\text{Re}(a_1 b_2)}{1 + \text{Im}(a_2 b_1)/\text{Re}(a_2 b_1)\text{Im}(a_1 b_2)/\text{Re}(a_1 b_2)} . \tag{8.40}$$

Results of a computation showing significant control over the cross section of the scattering to specific Zeeman states of the product atoms are shown in Figure 8.6, where a magnetic field has been used to shift the Zeeman states so that they are energetically degenerate. The resultant phase control is considerable, with the ratio of cross sections varying by a factor of six as a function of the coefficients in the incident superposition state.

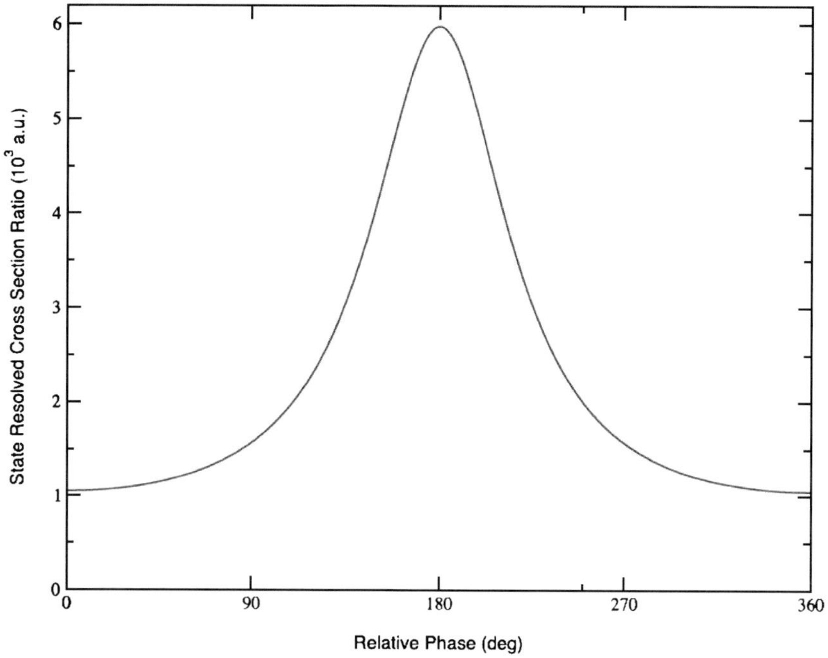

Figure 8.6 Ratio of the state resolved cross sections $\sigma_{i\leftarrow n}/\sigma_{j\leftarrow n}$. The final state labeled i corresponds to the outgoing channel with the projections $M_F = 0$ for Li and $M_F = 2$ for Cs. The final state labeled j corresponds to the outgoing channel with the projections $M_F = 1$ for Li and $M_F = 3$ for Cs. The curve is computed with the magnetic field that induces interference between the incoming channels $|\phi(1), M_F\rangle_{\text{Li}}|\phi(2), M_F'\rangle_{\text{Cs}}$ and $|\phi(2), M_F\rangle_{\text{Li}}|\phi(1), M_F'\rangle_{\text{Cs}}$. The incoming asymptotic states have projections $M_F = 1$ for Li states and $M_F = 3$ for Cs states. The collision energy is 107 cm^{-1}. From [345, Figure 3]. (Reprinted with permission. Copyright 2008 American Physical Society.)

8.1.3.3 Control in Electron Impact Dissociation

A considerably different example [346] of computed control results is shown in Figure 8.7 for electron impact dissociation of H_2^+, that is,

$$e + H_2^+ \rightarrow e + H + H^+ \tag{8.41}$$

Here interest is in the control of the total dissociation cross section and of the product electron energy distribution, using a superposition of vibrational eigenstates of H_2^+. We denote, for Eqs. (8.12) and (8.13), the electron by A and H_2^+ by B, and choose $a_2 = 0$, $|b_1| = |b_2|$, $b_1/b_2 = \exp(i\phi)$. The dissociation cross section $\sigma^{(D)}$ is shown in Figure 8.7a as a function of the relative phase ϕ between the two H_2^+ states $|\phi_B(1)\rangle$ and $|\phi_B(2)\rangle$, here chosen as the ground and first excited vibrational states with rotational angular momentum $J = 0$. The solid curve shows that the cross section ranges from 0.44 to 0.64π a.u. as the relative phase ϕ is varied. By contrast (dotted lines) $\sigma^{(D)}$ for the ground and first excited vibrational state are, in the absence of a superposition of states, 0.5 and 0.58 π. The remaining two panels

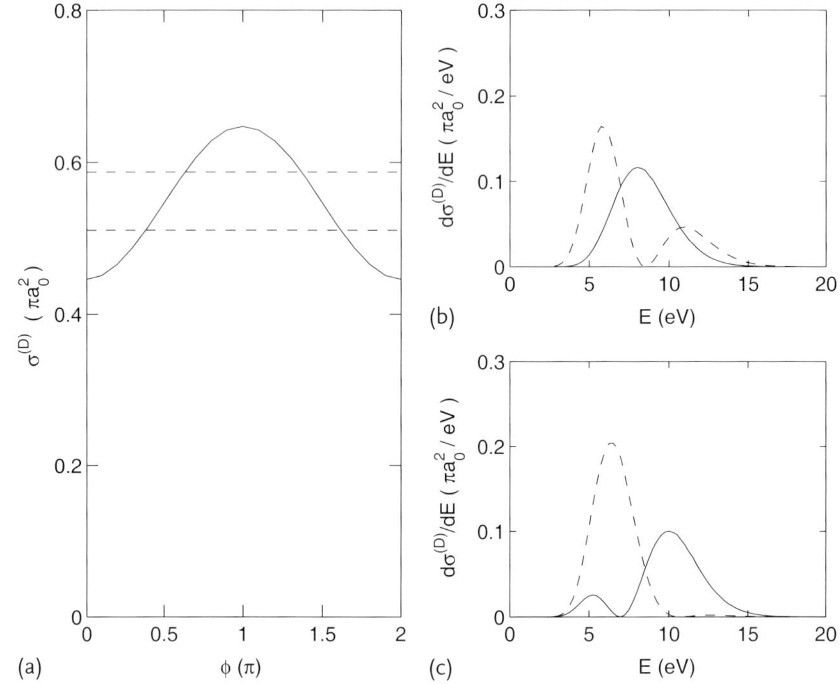

Figure 8.7 Control of dissociative scattering using the entangled initial state described in the text. (a) Dependence of the dissociation cross section $\sigma^{(D)}$ on ϕ. The dashed lines denote $\sigma^{(D)}$ for a single component wave function using ground (v $=$ 0, lower line) and excited (v $=$ 1, upper line) vibrational wave functions. (b) Energy spectrum of the proton fragments for v $=$ 0 (solid) and v $=$ 1 (dashed). (c) Analogous spectra using the two state superposition with $\phi =$ 0 (solid) and $\phi = \pi$ (dashed). From [346, Figure 1]. (Reprinted with permission. Copyright 2007 American Physical Society.)

show the energy resolved cross sections, $d\sigma^{(D)}/dE$, for the states v $=$ 0 and v $=$ 1 individually (Figure 8.7a), and for the superposition states corresponding to the extrema of $\sigma^{(D)}(\phi)$, namely $\phi = 0$ and π (Figure 8.7b). A large degree of control over the proton energy spectrum is evident.

Scattering at fixed total energy is clearly time-independent [227]. As such, we can plot the wave function to ascertain qualitative insight into features that underlie control. In doing so we note that the behavior of the reactive cross section is particularly dependent on the configuration when the reactants are close to one another [347–349]. Hence, insight into control can be obtained by focusing on the character of the wave function at short range. Below we show the incident states in the absence of interparticle interactions and do not include the scattered component; addition of the latter will not change the argument qualitatively.

Consider then the internal states and incident momenta used to generate Figure 8.7. Figure 8.8a,b shows the probability $P(R, x) = |\langle R, x|\Psi\rangle|^2$, of finding the system at nuclear bond length R and electron–ion separation x; Figure 8.8a shows

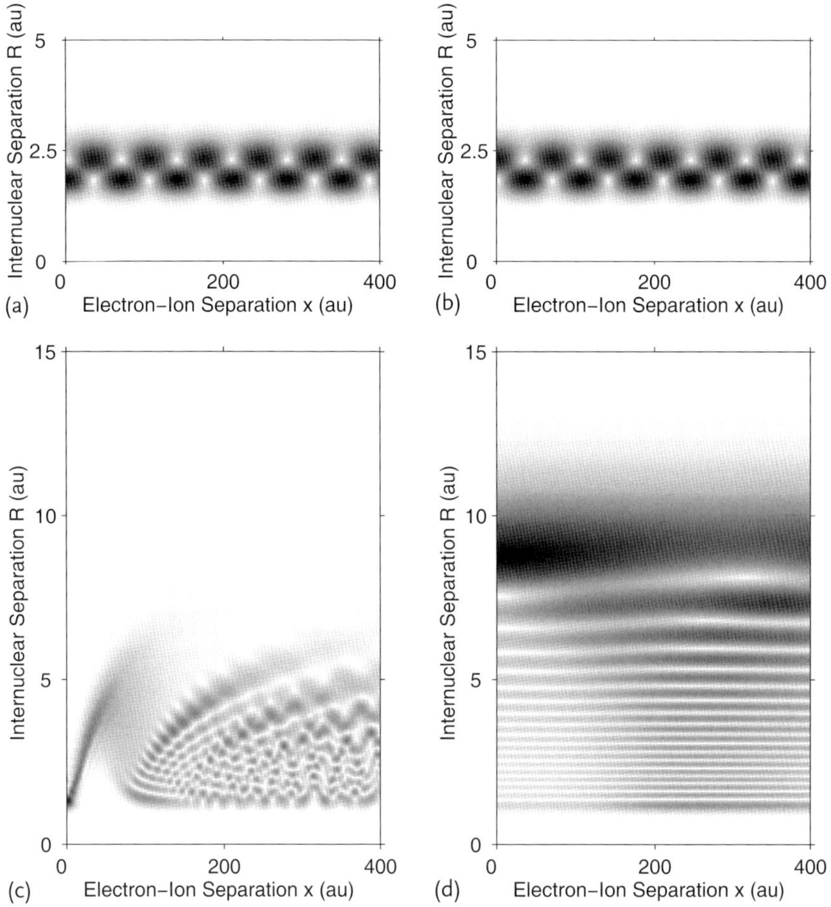

Figure 8.8 Probability density of relative and internal motion $P(R, x) = |\langle R, x|\Psi\rangle|^2$. Panels (a) and (b) correspond to the case shown in Figure 8.7 for $\phi = 0$ and π respectively. Panels (c) and (d) show incident states constructed by hand to minimize and maximize the nuclear bond length at the moment of collision, and hence to (qualitatively) minimize and maximize $\sigma^{(D)}$. From [346, Figure 2]. (Reprinted with permission. Copyright 2007 American Physical Society.)

results for $\phi = 0$ and Figure 8.8b for $\phi = \pi$. As noted above, of particular interest is the character of the internal state near the collision region, $x \approx 0$. Figure 8.8a,b shows that the internal vibrational wave function in this region is controlled by ϕ. Hence, by preparing an initial superposition of vibrational states and by varying ϕ, the wave function is shaped into structures that are not accessible by using individual incident vibrational eigenstates, thus allowing control over $\sigma^{(D)}$ and $d\sigma^{(D)}/dE$. This is then the underlying qualitative mechanism of coherently controlled reactive scattering using the entangled superposition state [Eq. (8.16)].

In standard stationary state scattering theory one usually considers scattering of incident time-independent eigenstates [227], and hence this description may seem obvious. This is not the case. Had a nonentangled wave function of the form

$$
\begin{aligned}
|\Psi_{os}^{(ne)}\rangle &= \left[C_1|\phi_B(1)\rangle + C_2 e^{i\phi_a}|\phi_B(2)\rangle\right] \\
&\times \left[C_3|\boldsymbol{k}_1^A\rangle + C_4 e^{i\phi_b}|\boldsymbol{k}_2^A\rangle\right] \\
&\times \left[C_5|\boldsymbol{k}_1^B\rangle + C_6 e^{i\phi_c}|\boldsymbol{k}_2^B\rangle\right]
\end{aligned} \tag{8.42}
$$

been used, the resultant incident state would contain 8 terms, only two of which are at the same total energy and center of mass momentum, namely the components satisfying Eq. (8.15). From an energy domain perspective this means that only two of the total eight terms exhibit on-shell coherence, and hence only two of the eight terms can interfere. The remaining six terms contribute incoherently to the cross sections. From a time domain perspective, only the superposition of the two on-shell components with the same energy lead to a time-independent wave function. All other components, being at different energies, accumulate a time-dependent phase relative to the on-shell superposition. With respect to the internal configuration near the collision region, this means that the off-shell components will alternate between constructive and destructive interferences at different times, resulting in a time average over all possible phases relative to the on-shell component, and hence an average over internal configurations associated with the internal states at the collision point. These time varying interferences then turn the off-shell contribution into an effective incoherent contribution. Hence, entanglement in the few-state superposition Eq. (8.16) plays a crucial role in limiting the particular internal state wave functions that participate in the scattering event.

The configuration-based qualitative mechanism described above is consistent with alternate information on the $e^-H_2^+$ cross section. In particular, electron impact dissociation cross sections of H_2^+ are monotonic function of bond length R; as $\langle R\rangle$ increases, so does $\sigma^{(D)}$ [350]. Comparing the $x = 0$ structure in Figures 8.8a and 8.8b shows that $\sigma^{(D)}$ control follows the expected $\langle R\rangle$-dependence; the $\phi = \pi$ case has both the larger $\langle R\rangle$ and larger $\sigma^{(D)}$. One can, therefore imagine building a *multistate* entangled coherent superposition in order to achieve better enhancement/suppression than that of Figure 8.7. For example, two superpositions that approximately achieve these goals are

$$
|\Psi_{\max}\rangle = N_{\max}\left[\sum_{\mathrm{v}}(-1)^{\mathrm{v}} e^{-\left(\frac{\mathrm{v}-18}{1.8}\right)^2}|\mathrm{v}\rangle|k_{\mathrm{v}}\rangle\right]|K\rangle \tag{8.43}
$$

and

$$
|\Psi_{\min}\rangle = N_{\min}\left[\sum_{\mathrm{v}} e^{-\left(\frac{\mathrm{v}-7}{6}\right)^2}|\mathrm{v}\rangle|k_{\mathrm{v}}\rangle\right]|K\rangle\ , \tag{8.44}
$$

where $N_{\max}$ and $N_{\min}$ are normalization constants, $k_0 = 4$ au, and the remaining k_{v} are set to ensure that all components satisfy the on-shell requirement, that is, have the same total energy and center of mass momentum. The sum runs over all the vibrational states $|\mathrm{v}\rangle$. The corresponding $P(R, x)$ are shown in Figure 8.8c and d. For these states the values of $\langle R\rangle$ in the collision region are $\langle\Psi_{\max}|R|\Psi_{\max}\rangle_{x=0} =$ 8.75 au and $\langle\Psi_{\min}|R|\Psi_{\min}\rangle_{x=0} = 1.32$ a.u.. This contrasts with the largest and smallest values possible using an incoherent initial state, which would correspond to those of the highest and lowest vibrational states of H_2^+, that is, $\langle \mathrm{v} = 18|R|\mathrm{v} = 18\rangle_{x=0} = 8.12$ a.u. and $\langle \mathrm{v} = 0|R|\mathrm{v} = 0\rangle_{x=0} = 2.05$ a.u.. Hence the coherent superpositions $|\Psi_{\max}\rangle$ and $|\Psi_{\min}\rangle$ access values of $\langle R\rangle$ at $x = 0$ beyond those accessible to any incoherent mixture of states, and hence provide more control over $\sigma^{(D)}$ than any incoherent scenario.

Although in the case of $e^-H_2^+$ scattering there is a clear classical interpretation of the characteristics of the configuration that enhance control, this is need not be the case in general. Rather, whether the optimal configuration is easily understood classically depends upon the system under consideration.

8.1.4
Experimental Implementation: Fixed Total Energy

The development in Section 8.1.1 shows that coherent control of bimolecular processes at fixed total energy requires the production of states (Eq. (8.13)) where the translational and internal states are entangled, that is, composed of components in which two or more translational and internal states are correlated, and the wave function is not separable into a product of internal times translational states. Although such states do result from photodissociation processes (see, for example, Eq. (2.73) which is a sum over $|E, \boldsymbol{n}^-\rangle$ states, each going over in the long-time limit to a translational, internal product state $|E, \boldsymbol{n}; 0\rangle$, hence their sum goes over to a state in which the translational and the internal motions are entangled), they are not necessarily suitable for our purposes [351].

The experiment described above is limited to the attosecond domain. Below we introduce three tentative suggestions for control of molecular collisions. Their realization would require an extension of current laboratory techniques.

For example, entangled states which might be useful for bimolecular control have been prepared in atoms in a relatively straightforward way [352]. Consider, for example, a system with two levels $|E_1\rangle, |E_2\rangle$, initially in the lower state $|E_1\rangle$ and moving with kinetic energy $E_t(1)$. Passing the system through a spatially dependent field with off-resonant frequency $\omega = (E_2 - E_1)/\hbar - \delta$ results in excitation to the state $|E_2\rangle$ with kinetic energy $E_t(2)$. Conservation of energy requires, however, that

$$E_1 + E_t(1) + \hbar\omega = E_2 + E_t(2) , \tag{8.45}$$

or $E_t(2) = E_t(1) - \hbar\delta$. That is, the created superposition state has two internal states correlated with two different translational energy states, precisely as required for bimolecular coherent control. Tuning δ above or below the resonance results in an increase, or decrease, of kinetic energy upon excitation. The extension of this technique to most cases of interest to us will, however, be difficult since the $\hbar\delta$ required is far larger than that in the atomic case.

Similarly, the momentum transfer associated with a collision of photons with atoms is used regularly to cool atoms [353, 354], that is, to alter the translational energy of an atom. Indeed, the momentum of large numbers of photons (over 140 photon momenta) have been successfully transferred coherently to atoms [355]. This suggests the possibility of preparing an initial superposition of internal states of a molecule, followed by the state-specific absorption of photon momenta of one of the internal states in order to form the required entangled superposition of the translational and internal states.

In addition, a number of experiments have shown that it is possible to accelerate or decelerate molecules using time varying electric fields [356]. In this case the molecule is passed through an array of synchronously pulsed electric field stages that interact with the molecular dipole. Since the dipole is a function of the state of the system it may be possible to prepare a superposition of internal states and then selectively accelerate one of the two internal states to produce the desired superposition.

Finally, in the domain of atomic physics, pulsed electromagnetic coils (an atomic coilgun) have been used [357] to stop beams of atoms, and may provide a useful source of translationally cold atoms for subsequent collision studies.

8.1.5
Optimal Control of Bimolecular Scattering

We can readily extend bimolecular control to superpositions composed of more than two states. Indeed, we can introduce a straightforward method to *optimize* the reactive cross section as a function of a_m for any number of states [358]. Doing so is an example of optimal control theory, a general approach to altering control parameters to optimize the probability of achieving a certain goal, introduced in Chapter 5.

Consider scattering from incident state $|E, q, i; 0\rangle$ to final state $|E, q', f; 0\rangle$. The label f includes the angles θ, ϕ into which the products scatter. Hence summations over f imply integrations over these angles. In accord with Eq. (8.2), the probability $P(f, q'; i, q)$ of producing product in final state $|E, q', f; 0\rangle$ having started in the initial state $|E, q, i; 0\rangle$ is

$$P(f, q'; i, q) = |S_{fi}|^2 , \tag{8.46}$$

where $S_{fi} = \langle E, q', f; 0|S|E, q, i; 0\rangle$ and where S is the scattering matrix for the process. The total probability $P(q'; i, q)$ of scattering into arrangement channel q', assuming n open product states, is then given by

$$P(q'; i, q) = \sum_{f=1}^{n} |S_{fi}|^2 . \tag{8.47}$$

To simplify the notation we have not carried an E label in the probabilities: fixed energy E is understood. (Note that Eqs. (8.46) and (8.47) are quite general and can be applied to a host of processes, other than just scattering, since one may, in general, write the probability of transitions between states in terms of a generalized S matrix).

If we now consider scattering from an initial state $|E, q, \{a_i\}; 0\rangle$ comprised of a linear superposition of N states, then the probability of forming $|E, q', f; 0\rangle$ from this initial state is

$$P(f, q'; \boldsymbol{a}, q) = \left| \sum_{i=1}^{N} a_i S_{fi} \right|^2 , \tag{8.48}$$

where $\boldsymbol{a} \equiv \{a_i\}$. The total reactive scattering probability into channel q', $P(q'; \boldsymbol{a}, q)$, is therefore,

$$P(q'; \boldsymbol{a}, q) = \sum_{f=1}^{n} \left| \sum_{i=1}^{N} a_i S_{fi} \right|^2 . \tag{8.49}$$

To simplify the notation we introduce the matrix $\boldsymbol{\sigma} = \boldsymbol{S}_{q'}^{\dagger} \boldsymbol{S}_{q'}$ with elements $\boldsymbol{\sigma}_{ij} = \sum_{f=1}^{n} S_{fj}^{*} S_{fi}$, which allows us to rewrite Eq. (8.49) as

$$P(q'; \boldsymbol{a}, q) = \boldsymbol{a}^{\dagger} \boldsymbol{\sigma} \boldsymbol{a} . \tag{8.50}$$

Here $\dagger$ denotes the Hermitian conjugate and the q' subscript on the $\boldsymbol{S}$ indicates that we are dealing with the submatrix of the S matrix associated with scattering into the product manifold defined by the q' quantum number.

One can optimize scattering into arrangement channel q', with the normalization constraint $\sum_{i=1}^{N} |a_i|^2 = 1$, by requiring

$$\frac{\partial}{\partial a_k^*} \left[P(q'; \boldsymbol{a}, q) - \lambda \sum_{i=1}^{N} |a_i|^2 \right] = \frac{\partial}{\partial a_k^*} [\boldsymbol{a}^{\dagger} \boldsymbol{\sigma} \boldsymbol{a} - \lambda \boldsymbol{a}^{\dagger} \boldsymbol{a}] = 0 , \quad k = 1, \ldots, N , \tag{8.51}$$

where λ is a Lagrange multiplier. Explicitly taking the derivative gives the result that the vector of optimal coefficients $\boldsymbol{a}_{\lambda}$ satisfies the eigenvalue equation

$$\boldsymbol{\sigma} \boldsymbol{a}_{\lambda} = \lambda \boldsymbol{a}_{\lambda} . \tag{8.52}$$

Additional labels may be necessary to account for degeneracies of the eigenvectors $\boldsymbol{a}_{\lambda}$. Optimization is now equivalent to solving Eq. (8.52).

This approach has been used to obtain optimal $\boldsymbol{a}_{\lambda}$ for various isotopic variants of the $H + H_2$ reaction [318]. In general, the range of control was found to increase with j, that is, with the increasing number of available m states. Thus, for example, for scattering into $H' + HD$, where H and H′ are assumed distinguishable, σ^R/σ^{NR} could be varied between 1.15×10^{-2} and 2.62×10^{-2} for initial $j = 2$, and between 1.85×10^{-5} if initial $j = 2$ and 6.29×10^{-3} for initial $j = 10$.

The optimization procedure yields a set of coefficients a_i. Of considerable interest is the question of whether these coefficients merely define a new vector that is just a ket vector in a rotated coordinate system. If so, this would indicate that the optimum solution corresponds to a simple classical reorientation of the diatomic molecule angular momentum vector. Examination of the optimal results [318] indicate that this is not the case. That is, control is the result of quantum interference effects.

8.1.5.1 Optimized Bimolecular Scattering: the Total Suppression of a Reactive Event

In Section 8.1.5 we considered optimizing reactive scattering by varying the coefficients a_i of a superposition of states. In this section we show that when the number of initial open states in the reactant space exceeds the number of open states in the product space it is possible to find a particular set of a_i coefficients such that one can totally *suppress* reactive scattering. This result is proven below and applied to display the total suppression of tunneling [358].

To see this result consider Eq. (8.52). Note that if $\lambda = 0$ is an eigenvalue of this equation with eigenfunctions $\boldsymbol{a}_0$ then by inserting Eq. (8.52) into Eq. (8.50) we have that $P(q'; \boldsymbol{a}_0, q) = 0$. That is, if $\lambda = 0$ is a solution to Eq. (8.52) then the coefficients $\boldsymbol{a}_0$ completely suppress reaction into arrangement channel q'.

Clearly, $\lambda = 0$ is a solution if

$$\det(\boldsymbol{\sigma}) = \det(\boldsymbol{S}_q'^{\dagger} \boldsymbol{S}_q') = 0 , \tag{8.53}$$

which is the case if the number of initial states N participating in the initial superposition is greater than the number M of open product states. To see this result note that, under these circumstances, $\boldsymbol{\sigma}$ is a matrix of order $N \times N$ and $\boldsymbol{S}_{q'}$ is of order $M \times N$. If $N > M$ we can construct an $N \times N$ matrix $\boldsymbol{A}_q'$ by adding a submatrix of $(N - M)$ rows of zeroes to the lower part of $\boldsymbol{S}_{q'}$. This procedure does not change the $(\boldsymbol{S}_{q'}^{\dagger} \boldsymbol{S}_{q'})$ product, hence,

$$\det(\boldsymbol{\sigma}) = \det(\boldsymbol{S}_{q'}^{\dagger} \boldsymbol{S}_{q'}) = \det(\boldsymbol{A}_{q'}^{\dagger} \boldsymbol{A}_{q'}) = \det(\boldsymbol{A}_{q'}^{\dagger})\det(\boldsymbol{A}_{q'}) = 0 . \tag{8.54}$$

The last equality holds because the determinants of $\boldsymbol{A}_{q'}$ and $\boldsymbol{A}_{q'}^{\dagger}$ are zero.

As an example of the kind of results that are possible, consider the optimization of a barrier penetration problem modeled by a set of multichannel Schrödinger equations of the type:

$$\boldsymbol{\Psi}''(r) = -\frac{2\mu}{\hbar^2} \left(\underline{\underline{\hat{\mathrm{E}}}} - \underline{\underline{\mathrm{V}}} \right) \boldsymbol{\Psi}(r) \tag{8.55}$$

where μ is the relevant mass, $\underline{\underline{\mathrm{V}}}$ is a potential matrix and $\underline{\underline{\hat{\mathrm{E}}}}$ is a diagonal matrix with elements $E - e_i$. Sample scattering results for four open channels (namely, channels satisfying the $E - e_i > 0$ condition), were obtained for a model for which $\underline{\underline{\mathrm{V}}}$ is a matrix of Eckart potentials given as

$$V_{ij}(r) = -\frac{a_{ij}\xi}{1 - \xi} - \frac{b_{ij}\xi}{(1 - \xi)^2} + c_{ij} , \quad (i, j = 1, \ldots, 4) , \tag{8.56}$$

where $\xi = -\exp(2\pi r/d)$, with d a distance potential parameter. These potentials are shown in Figure 8.9. Scattering results, obtained from numerically integrating Eq. (8.55) are shown in Figure 8.10. Here reactivity is shown as a function of energy for the case where the number of populated initial states N_{pop} is less than N: here $N_{pop} = 3$. The curves labeled P_i correspond to the standard $P(q'; i, q)$, that is, total reaction probability from each of the individual initial states. The quantities P_1 and P_3, which are open asymptotically at all energies, show a gradual rise with increasing energy, whereas P_2, which is closed on the product side until $E_{th}(3) = 0.008$ a.u., stays rather small until $E = E_{th}(3)$, where it displays a very rapid rise to near unity. Total reaction probability reaches unity above $E_{th}(4) = 0.010$ a.u., the threshold for the opening of the fourth channel.

Of particular relevance here are the solid curves in Figure 8.10 which show the reactivity maximum and minimum obtained from the optimal solutions to Eq. (8.52). The reactivity maximum is seen to be substantially larger than any of the individual P_i and to reach unity at significantly lower energies than any of these solutions. Of greater significance is that the reactivity minimum is, as predicted by the argument presented above, exactly zero for $E < E_{th}(3)$ since the total number of states ($N_{pop} = 3$) in the superposition exceeds the number of open product states ($M = 2$). At $E > E_{th}(3)$ a third product channel opens so that $N_{pop} = M$ and the minimal solution is no longer zero.

Note also that the minimum reactivity curve in Figure 8.10 reflects a variety of different interesting behaviors, depending on the particular energy. Specifically, below the maximum of V_{11} at 0.005 a.u. the zero minimum corresponds to suppression of tunneling through that barrier. Above 0.005 a.u. the zero minimum corresponds to suppression of the reactive scattering that occurs *above* the barrier.

Thus it is clear that the ability to superimpose degenerate scattering states affords great potential to control scattering processes. Note also that, as an obvious

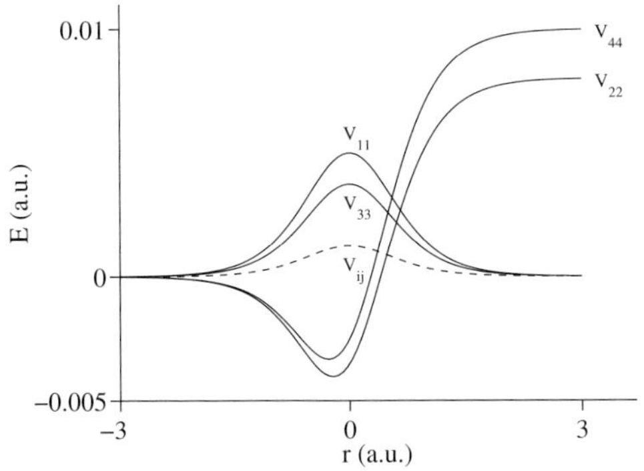

Figure 8.9 Elements of the potential matrix for the model scattering problem. V_{ij} denotes the three off diagonal matrix elements $i \neq j$, which are all chosen to be equal. The nonzero a_{ij}, b_{ij}, c_{ij} elements (Eq. (8.56)) are given (in a.u.) by $b_{11} = 0.02$, $a_{22} = 0.008$, $b_{22} = -0.03$, $b_{33} = 0.015$, $a_{44} = 0.01$, $b_{44} = -0.03$, $b_{ij} = 0.005$, $i \neq j$ with $d = 2.5$ and $\mu = 0.6666$ a.m.u. From [358, Figure 1]. (Reprinted with permission. Copyright 1999 American Institute of Physics.)

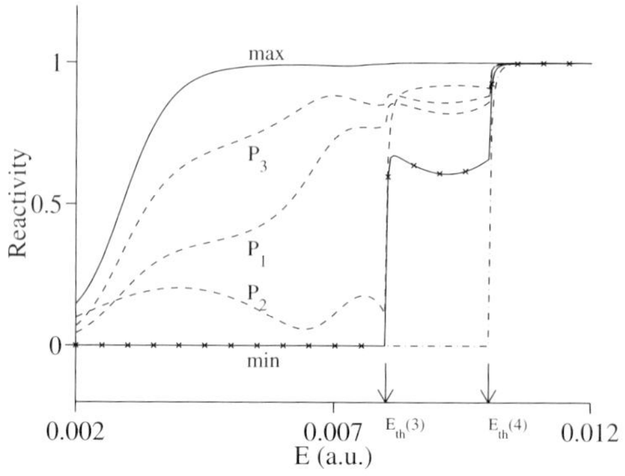

Figure 8.10 Reactivity shown as a function of energy in a model system. Dashed curves labeled P_i correspond to the total reactivity from each of the three individual initial states in the prepared superposition. Solid curves with crosses denote the reactivity obtained by solving Eq. (8.52) for the optimal solutions. The two arrows indicate the threshold energies for opening of the third and fourth channels. The dot-dash curve shows the minimum reactivity resulting from a separate computation which includes four states in the initial superposition. From [358, Figure 2]. (Reprinted with permission. Copyright 1999 American Institute of Physics.)

extension, similar results hold for tunneling in bound systems if the total number of initial degenerate states at the energy of interest exceeds the number of accessible final states at that energy. This approach has also been extended [359] to the design of decoherence free subspaces, that is, subspaces, as mentioned in Chapter 6, within which the dynamics is free of decoherence effects. In brief, one takes a subspace α that undergoes transitions, to a subspace β due to coupling to the environment. In accord with the discussion in this section, these transitions can be suppressed if β is sufficiently larger than α. If this is not the case then α can be augmented with states from another subspace γ to aid in suppressing transitions to β from α.

8.1.6 Sculpted Imploding Waves

Since, in principle, control can be achieved by superposing any degenerate set of eigenstates, rather than using a superposition of internal diatomic molecule states, as above, it is possible to affect bimolecular control by using a superposition of equal energy translational wave functions [358]. To see how this is done note that the relative motion of two particles (say in a molecular beam) is generally well described by a plane wave $\exp(i\boldsymbol{k}\cdot\boldsymbol{r})$. The character of the plane wave can be exposed by a partial wave decomposition. That is, assuming the wave is directed along the z-axis we can write

$$\exp(i\boldsymbol{k}\cdot\boldsymbol{r}) = \exp(ikz) = \exp(ikr\cos\theta) = \sum_l a_l j_l(kr) P_l(\cos\theta) , \qquad (8.57)$$

where $a_l = i^l(2l+1)$ and $j_l(kr)$, $P_l(\cos\theta)$ are the spherical Bessel function and Legendre polynomials, respectively. We see that each incoming plane wave is, in fact, a superposition of energetically degenerate states with fixed coefficients a_l. This suggests the possibility of altering the a_l to produce modified states, that is, a sculpted incoming wave packet, $\langle \boldsymbol{r}|\boldsymbol{k}_{mod}\rangle$ which will display different quantum interferences, hence altering the product cross sections.

This approach is discussed in detail in [360]. As an example of the control afforded by this scenario, consider rotational excitation in a model of the $Ar + H_2(j, m) \rightarrow Ar + H_2(j', m')$ collision. Optimizing the phases χ_l of $a_l = |a_l|\exp(i\chi_l)$ allows a direct study of the effect of varying the interferences between partial wave components on the outgoing flux into any selected product state. Typically, altering χ_l allowed for considerable control. For example, with $j = 2$, $m = 0$, the outgoing flux into $j' = 0$ could be changed by two orders of magnitude, from 5.1×10^{-4} to 3.8×10^{-2}, just by varying the χ_l. These values are to be compared to a flux of 1×10^{-2} associated with scattering from an incident plane wave. Real and imaginary parts of the incident wave functions leading to these maximum and minimum values of the outgoing flux are shown in Figure 8.11. They are distinctly different from one another, and from a plane wave $\exp(i\boldsymbol{k}\cdot\boldsymbol{r})$.

8.2 Time Domain: Fast Timed Collisions

The sections above dealt with stationary state collisional dynamics, that is, scattering at fixed total energy. There control relied upon interference effects between pathways established by constructing entangled initial states. Here we consider an alternate control procedure [346] that takes advantage of interference effects associated with temporal correlations that arise within time-dependent scattering.

8.2.1 Nonentangled Wave Packet Superpositions: Time-Dependent Scattering

Section 8.1.3.3 discussed the qualitative mechanism that underlies successful control when using entangled initial superposition states. This was shown to be the creation of scattering wave functions which, at small interparticle distances, assume spatial configurations that enhance the target cross section. Given this insight, alternative approaches to coherent control of reactive scattering that do not rely on initial state entanglement can be identified. For example, there is an alternative route to ensuring that collisions between two particles occur at the particular phase of the internal state motion that enhances the desired product formation: design *time-dependent* states of translational motion plus superpositions of internal states that localize the colliding partners in advantageous spatial configurations, but restrict the duration of the collision between the wave packet to less than the internal state motion (see Figure 8.12a). Quantum interference would then manifest in this case due to the numerous energetically degenerate sets of states that are

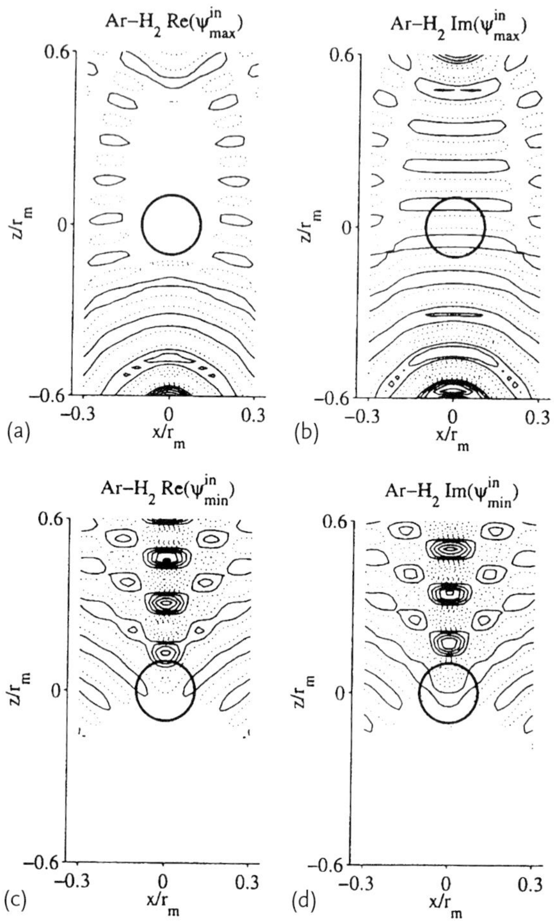

Figure 8.11 Real and imaginary parts of the incident wave function leading to maximum and minimum outgoing flux for Ar + $H_2(\nu = 0, j = 2, m_j = 0)$ Taken from [25, Figure 15]. (Reprinted with permission. Copyright 2000 Elsevier.)

associated with the energy widths of the two incident wave packets associated with the collision partners.

The electron impact dissociation of H_2^+ (see Eq. (8.41)), was used [346] to illustrate this scenario starting with the initial state wave packet:

$$|\Psi_W(t=0)\rangle = |\psi_v\rangle|\psi_B\rangle|\psi_A\rangle \,, \tag{8.58}$$

where

$$|\psi_v\rangle = \left[|v=0\rangle + e^{i\phi}|v=1\rangle\right]/\sqrt{2} \,, \tag{8.59}$$

$$|\psi_B\rangle = \int dp \left(\Delta_p\sqrt{\pi}\right)^{-\frac{1}{2}} e^{-\frac{1}{2}\left(\frac{p-p_0}{\Delta_p}\right)^2} |p\rangle \,, \tag{8.60}$$

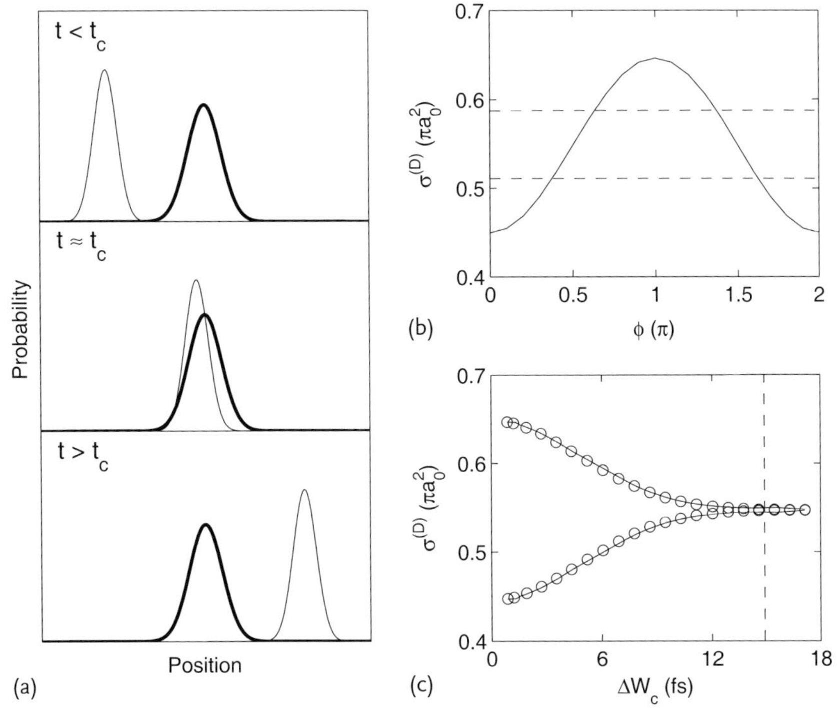

Figure 8.12 Control of reactive scattering via translational wave packets. Panel (a): Schematic of the time evolving wave packets. Altering the internal state of the molecular ion during the finite duration of the collision allows for control. t_c denotes the time of the collision. Panel (b): degree of control using the initial Eq. (8.58), with parameters indicated in the text. Panel (c): loss of control as the duration of the collision between the two wave packet approaches the timescale of the internal state motion. The dashed line denotes the vibrational period of H_2^+, $\tau_{vib} = 14.9$ fs. Taken from [345, Figure 3]. (Reprinted with permission. Copyright 2007 American Physical Society.)

and

$$|\psi_A\rangle = \int dP\,(\Delta_P\sqrt{\pi})^{-\frac{1}{2}}\, e^{-\frac{1}{2}\left(\frac{P-P_0}{\Delta_P}\right)^2} |P\rangle\,, \tag{8.61}$$

where $p = |\boldsymbol{k}^B|$, $P = |\boldsymbol{k}^A|$ and $|\mathrm{v}\rangle$ denotes a vibrational state of the H_2^+. (Recall from Section 8.1.3.3 that here the electron is particle A and the H_2^+ is particle B). The incident wave packet as depicted schematically in Figure 8.12a. Note that this is not simply wave packet scattering off of a single vibrational state. Rather, scattering is off an internal superposition of two vibrational states, $\mathrm{v} = 0$ and $\mathrm{v} = 1$, necessary to incorporate interference and achieve control.

The following measure, the time-dependent scattering probability $W_c(t)$, is introduced to quantify the duration of the collision of the scattering partners:

$$W_c(t) = N\,|\langle \Psi(t)\,|\delta(\boldsymbol{x}-\boldsymbol{y})|\,\Psi(t)\rangle|^2 \tag{8.62}$$

where x and y are the electron and ion positions. Here $|\Psi(t)\rangle$ is the incident wave function and N is a normalization constant such that $\int W_c(t)dt = 1$. This quantity gives the probability of finding the electron and the ion at the same position in space, and hence reflects the probability of a collision occurring at time t. For wave packet collisions, $W_c(t)$ is a localized function that is nonzero during a select time window wherein the two colliding wave packets overlap in space. The duration of the wave packet collision is then defined as

$$\Delta W_c = 2\sqrt{\langle t^2\rangle_{W_c} - \langle t\rangle^2_{W_c}}\,, \tag{8.63}$$

where the angular brackets indicate an average value with $W_c(t)$ used as the distribution function.

The results of the wave packet scenario can be compared to that of the entangled state scenario described in Section 8.1.3 by setting the mean incident momenta of the scattering state to $P_0 = 0$ and $p_0 = 4$ au, the same as used in Section 8.1.3. Using momentum widths of $\Delta_P = 1$ a.u. and $\Delta_p = 0.01$ a.u. then gives a scattering scenario with $\Delta W_c = 0.87$ fs. Since ΔW_c in this case is much smaller than the vibrational period of H_2^+ ($\tau_{vib} = 14.9$ fs), we expect that control is possible using these parameters. Figure 8.12b plots the corresponding reactive cross section as ϕ is varied. Control is indeed present. Further, upon comparing Figures 8.12b and 8.7a, one sees that using the nonentangled state in Eq. (8.58) gives the same degree of control as the entangled state, confirming that the wave packet scenario offers an alternate and equivalent (in terms of the controllability of the total cross section) means of coherent control of reactive scattering, at least for reactive scattering absent of any resonances related to the relative momentum of the incoming scattering partners.

The remaining panel of Figure 8.12 shows the minimum and maximum values of the cross section, which correspond to $\phi = 0$ and π, as the duration of the wave packet collision ΔW_c is increased. The control is seen to go to zero as ΔW_c approaches the H_2^+ vibrational period τ_{vib}. These calculations most clearly demonstrate the configuration-based mechanism underlying the control.

8.2.2
Entangled or Wave Packets?

Both the stationary state and nonstationary state control schemes discussed in this chapter present technological challenges. The wave packet case requires collisions over very short times, and therefore must be run in either a pulsed mode, or with very tightly focused molecular beams so that the molecules move through the collision region in a time smaller than the timescale characteristic of the dynamics of the internal superposition. The stationary state case, on the other hand, requires initial state entanglement, but can be used in a continuous beam regime and/or with arbitrarily large spatial region of overlap of the two beams. Hence, the entangled version allows arbitrarily large spacetime scattering volumes whereas spatial restrictions exist for the wave packet version. This implies that the stationary state approach, reliant on entanglement, will always permit larger total yields since it can be used in conjunction with arbitrarily large volumes and incident fluxes.

Reactive scattering processes may also involve narrow resonances [361, 362]. In this case the entangled scenario can be advantageous since the pulsed wave packet scenario requires a broad superposition of incident momenta that may well wash out narrow resonances. By contrast, the entangled stationary state approach can use as few as two incident momenta, and hence can take advantage of the resonance features. The latter was indeed the case in a computational study of reactive scattering in F + HD [361].

Note that both of these aspects of initial state entanglement, the possibility of (i) efficiently exploiting narrow resonances and of (ii) accessing arbitrarily large scattering volumes, represent nonclassical aspects of controlled reactive scattering. By contrast, control via wave packet scattering has something of a classical analog.

9
The Interaction of Light with Matter: a Closer Look

Chapter 1 provided a capsule introduction to the theory of the interaction of light with matter. In this chapter we expand upon that treatment, deriving, for completeness, the relevant equations from essentially first principles. As in Chapter 1 we consider a system of classical charged particles that interact with a pulse of electromagnetic radiation. We then quantize the particle variables, and develop the semiclassical theory of light interacting with quantized particles. In later chapters (Chapter 10, Chapter 15) we extend this treatment to include the quantization of the electromagnetic field.

9.1
Classical Electrodynamics of a Pulse of Light

9.1.1
The Classical Hamiltonian

Consider a system of charged particles interacting with a pulse of light. The dynamics of the particles, electric field $\boldsymbol{E}(\boldsymbol{r}, t)$, and magnetic field $\boldsymbol{B}(\boldsymbol{r}, t)$ are determined by combining Maxwell's equations for the fields [363]

$$\begin{aligned}
\nabla \cdot \boldsymbol{E}(\boldsymbol{r}, t) &= \frac{1}{\epsilon_0}\rho(\boldsymbol{r}, t) , \\
\nabla \cdot \boldsymbol{B}(\boldsymbol{r}, t) &= 0 , \\
\nabla \times \boldsymbol{E}(\boldsymbol{r}, t) &= -\frac{1}{c}\frac{\partial}{\partial t}\boldsymbol{B}(\boldsymbol{r}, t) , \\
\nabla \times \boldsymbol{B}(\boldsymbol{r}, t) &= \frac{1}{c}\frac{\partial}{\partial t}\boldsymbol{E}(\boldsymbol{r}, t) + \frac{1}{\epsilon_0 c}\boldsymbol{j}(\boldsymbol{r}, t) ,
\end{aligned} \tag{9.1}$$

with Lorentz's equation for a charged particle moving in an electric and magnetic field:

$$m_i \frac{d\boldsymbol{v}_i}{dt} = \boldsymbol{F}(\boldsymbol{r}_i, t) = q_i \left(\boldsymbol{E}(\boldsymbol{r}_i, t) + \frac{\boldsymbol{v}_i}{c} \times \boldsymbol{B}(\boldsymbol{r}_i, t) \right) . \tag{9.2}$$

Quantum Control of Molecular Processes, Second Edition. Moshe Shapiro and Paul Brumer

Here m_i, q_i, $\boldsymbol{r}_i$, and $\boldsymbol{v}_i$ are, respectively, the mass, the charge, the position and the velocity of the ith particle. The quantities ρ and $\boldsymbol{j}$ are the charged particle density and the current of the charged particles, defined by

$$\rho(\boldsymbol{r}, t) = \sum_i q_i \delta[\boldsymbol{r} - \boldsymbol{r}_i(t)] ,$$
$$\boldsymbol{j}(\boldsymbol{r}, t) = \sum_i q_i \boldsymbol{v}_i \delta[\boldsymbol{r} - \boldsymbol{r}_i(t)] . \tag{9.3}$$

The symbol ϵ_0 denotes "the permittivity of free space" and is a constant, as is c, the speed of light. In atomic units (a.u.) $\hbar = 1$, $m_e = 1$, $q_e = -1$, where m_e is the mass, q_e is the charge of the electron, and $\epsilon_0 = 1/(4\pi)$.

It is advantageous to re-express the electrodynamics in terms of the vector and scalar potentials $\boldsymbol{A}(\boldsymbol{r}, t)$ and $\Phi(\boldsymbol{r}, t)$, which are related to the electric and magnetic fields by the following relations,

$$\boldsymbol{E} = -\frac{1}{c}\frac{\partial \boldsymbol{A}}{\partial t} - \nabla \Phi ,$$
$$\boldsymbol{B} = \nabla \times \boldsymbol{A} . \tag{9.4}$$

These equations do not completely specify the vector and scalar potentials. Rather, it is possible to define different forms of the vector and scalar potentials, the so-called *gauges*, that give the same electric and magnetic fields. Specifically, it follows from Eq. (9.4) that given $\boldsymbol{A}$ and Φ we can construct other potentials $\boldsymbol{A}'$ and Φ' as

$$\boldsymbol{A}' = \boldsymbol{A} + \nabla\chi ,$$
$$\Phi' = \Phi - \frac{1}{c}\frac{\partial \chi}{\partial t} , \tag{9.5}$$

where χ is a scalar field, which result in $\boldsymbol{E}$ and $\boldsymbol{B}$ fields that satisfy Maxwell's equations.

The choice that is often made, called the *Coulomb gauge*, is defined by choosing χ such that $\nabla^2\chi = -\nabla \cdot \boldsymbol{A}$, which means, given Eq. (9.5) that,

$$\nabla \cdot \boldsymbol{A}' = 0 . \tag{9.6}$$

Requiring $\nabla^2\chi = -\nabla \cdot \boldsymbol{A}$, still does not completely determine $\boldsymbol{A}$ and Φ because additional gauge transformations, defined by different choices of the scalar fields χ' that satisfy $\nabla^2\chi' = 0$, also satisfy Eq. (9.6). We shall make use of this flexibility later below.

The total energy of a particle plus field system (i.e., the Hamiltonian H) is the sum of the kinetic energy of the particles and the energy of the field. That is

$$H = \sum_i \frac{1}{2} m_i \left(\frac{d\boldsymbol{r}_i}{dt}\right)^2 + \frac{\epsilon_0}{2}\int d^3 r \left[\boldsymbol{E}^2(\boldsymbol{r}, t) + \boldsymbol{B}^2(\boldsymbol{r}, t)\right] . \tag{9.7}$$

Notice that, remarkably, this form does not contain an explicit contribution from the potential energy of the particles. Rather, the potential energy will arise

(Eq. (9.36) below) naturally by accounting for the way in which the particle density and the particle current affect the electric and magnetic field via Maxwell equations, and the way in which the electric and magnetic fields modify the particle position and velocity through the Lorentz equation. Hence, in this sense, the kinetic energy is a more fundamental quantity than is the potential energy of interaction.

To obtain this potential energy contribution, we start with the fact that the electric field can be written as a sum of longitudinal (L) and transverse (R) components

$$\boldsymbol{E} = \boldsymbol{E}_L + \boldsymbol{E}_R \,, \tag{9.8}$$

defined by the relations,

$$\nabla \times \boldsymbol{E}_L = 0 \,, \quad \nabla \cdot \boldsymbol{E}_R = 0 \,. \tag{9.9}$$

It is clear from Eq. (9.9) that the first of Maxwell's equations pertains only to $\boldsymbol{E}_L$. It also follows from the second of Maxwell's equations that $\boldsymbol{B}$ is a purely transverse vector field.

Substituting Eq. (9.8) into Eq. (9.7) gives

$$H = \sum_i \frac{1}{2} m_i \left(\frac{d\boldsymbol{r}_i}{dt}\right)^2 + H_L + H_R \,, \tag{9.10}$$

where we define

$$H_L \equiv \frac{\epsilon_0}{2} \int d^3 r \boldsymbol{E}_L^2(\boldsymbol{r}, t) \,, \tag{9.11}$$

and

$$H_R \equiv \frac{\epsilon_0}{2} \int d^3 r [\boldsymbol{E}_R^2(\boldsymbol{r}, t) + \boldsymbol{B}^2(\boldsymbol{r}, t)] \,. \tag{9.12}$$

Using Eq. (9.4) and Eq. (9.6), according to which $\boldsymbol{E}_L = -\nabla \Phi$, we can write H_L, with the aid of the first of Maxwell equations (Eq. (9.1)), as

$$\begin{aligned} H_L &= \frac{\epsilon_0}{2} \int d^3 r (\nabla \Phi)^2 = -\frac{\epsilon_0}{2} \int d^3 r \Phi \nabla^2 \Phi \\ &= \frac{\epsilon_0}{2} \int d^3 r \Phi \nabla \cdot \boldsymbol{E}_L = \frac{1}{2} \int d^3 r \Phi \rho \,. \end{aligned} \tag{9.13}$$

Using the form of the charge density ρ (Eq. (9.3)), we obtain that

$$H_L = \frac{1}{2} \sum_i q_i \Phi(\boldsymbol{r}_i) \,, \tag{9.14}$$

with $\Phi(\boldsymbol{r})$ being the potential induced by all the charged particles,

$$\Phi(\boldsymbol{r}) = \frac{1}{4\pi\epsilon_0} \sum_j \frac{q_j}{|\boldsymbol{r} - \boldsymbol{r}_j|} \,. \tag{9.15}$$

We see that the H_L gives rise, and is identical to, the electrostatic potential energy of the particles.

Equations (9.14) and (9.15) contain divergent terms which are independent of the particles' positions. These terms occur whenever $\boldsymbol{r} = \boldsymbol{r}_i$ and represent the electrostatic interaction of each particle with itself. However, because these terms are independent of the particles' positions, subtracting them from the electrostatic energy is equivalent to a simple redefinition of the zero-point energy of the particles. Since all forces in nature derive from *changes* in energy, such a redefinition of the zero-point of energy is of no dynamical consequence.

In subtracting the divergent terms, we find that the H_L contribution to H is replaced by the "Coulomb potential" V_C, defined as

$$V_C = \sum_i q_i \Phi_i(\boldsymbol{r}_i) \,, \tag{9.16}$$

where

$$\Phi_i(\boldsymbol{r}) = \frac{1}{4\pi\epsilon_0} \sum_{j<i} \frac{q_j}{|\boldsymbol{r} - \boldsymbol{r}_j|} \,. \tag{9.17}$$

In contrast to $\Phi(\boldsymbol{r})$ (Eq. (9.15)), each $\Phi_i(\boldsymbol{r})$, $i = 1, \ldots, N$ term is the potential due to particles whose indices j are less than that of the ith particle.

The Hamiltonian of the particle + radiation system now assumes the form,

$$H = \sum_i \frac{1}{2} m_i \left(\frac{d\boldsymbol{r}_i}{dt}\right)^2 + V_C + H_R \,. \tag{9.18}$$

9.1.2 The Free Light Field

In the absence of particles, $\boldsymbol{j}(r, t) = 0$, and the fourth Maxwell equation (Eq. (9.1)) takes on the form

$$-\nabla^2 \boldsymbol{A} + \frac{1}{c^2}\frac{\partial^2 \boldsymbol{A}}{\partial t^2} = 0 \,. \tag{9.19}$$

Here we have made use of Eq. (9.4) and the identity $\nabla \times \nabla \times \boldsymbol{A} = \nabla(\nabla \cdot A) - \nabla^2 \boldsymbol{A}$, which simplifies in the Coulomb gauge to, $\nabla \times \nabla \times \boldsymbol{A} = -\nabla^2 \boldsymbol{A}$.

As particular solutions of Eq. (9.19) we can choose plane waves (also called "field-modes")

$$\boldsymbol{A}_k(\boldsymbol{r}, t) = \boldsymbol{A}_k(t) \exp(i\boldsymbol{k} \cdot \boldsymbol{r}) \,, \tag{9.20}$$

where $\boldsymbol{k}$ is an arbitrary vector (the "wave vector") that determines the direction of propagation of the plane wave. Upon substitution of Eq. (9.20) in Eq. (9.19) each expansion coefficient $\boldsymbol{A}_k(t)$ is seen to be a solution of the differential equation for a harmonic oscillator

$$k^2 \boldsymbol{A}_k(t) + \frac{1}{c^2}\frac{d^2 \boldsymbol{A}_k(t)}{dt^2} = 0 \,, \tag{9.21}$$

with solution

$$A_k(t) = A_k \exp(\mp i\omega_k t) . \tag{9.22}$$

That is, each field mode is represented by a classical harmonic oscillator of mode-frequency $\omega_k \equiv ck$.

Each field-mode must also satisfy the Coulomb gauge condition, $\nabla \cdot \boldsymbol{A}_k(\boldsymbol{r}, t) = 0$, which, when substituted into Eq. (9.20), implies that

$$\boldsymbol{k} \cdot \boldsymbol{A}_k = 0 . \tag{9.23}$$

We can ensure the validity of Eq. (9.23) by writing $\boldsymbol{A}_k$ as a product of a unit polarization vector $\hat{\boldsymbol{\epsilon}}_k$, and a complex scalar amplitude A_k

$$\boldsymbol{A}_k \equiv \hat{\boldsymbol{\epsilon}}_k A_k , \tag{9.24}$$

and require that

$$\boldsymbol{k} \cdot \hat{\boldsymbol{\epsilon}}_k = 0 . \tag{9.25}$$

Since, according to Eq. (9.25), the polarization vector is perpendicular to the direction of propagation, each field mode can have only two independent polarization directions.

Because Eq. (9.19) is linear, its general solution can be expressed as a sum over all the field modes and their complex conjugates when the radiation field is in a cavity of a finite volume. That is, since $\boldsymbol{A}$ must be real it satisfies

$$\boldsymbol{A}(\boldsymbol{r}, t) = \sum_k \hat{\boldsymbol{\epsilon}}_k \{A_k \exp(-i\omega_k t + i\boldsymbol{k} \cdot \boldsymbol{r}) + A_k^* \exp(i\omega_k t - i\boldsymbol{k} \cdot \boldsymbol{r})\} . \tag{9.26}$$

If the field is in an infinite volume, then the sum in Eq. (9.26) is replaced by an integral. It follows from Eq. (9.4) that the electric field (which in the absence of particles has only the transverse component) and the magnetic field are given as

$$\boldsymbol{E}_R(\boldsymbol{r}, t) = i \sum_k k\hat{\boldsymbol{\epsilon}}_k \{A_k \exp(-i\omega_k t + i\boldsymbol{k} \cdot \boldsymbol{r}) - A_k^* \exp(i\omega_k t - i\boldsymbol{k} \cdot \boldsymbol{r})\} , \tag{9.27}$$

$$\boldsymbol{B}(\boldsymbol{r}, t) = i \sum_k \boldsymbol{k} \times \hat{\boldsymbol{\epsilon}}_k \{A_k \exp(-i\omega_k t + i\boldsymbol{k} \cdot \boldsymbol{r}) - A_k^* \exp(i\omega_k t - i\boldsymbol{k} \cdot \boldsymbol{r})\} . \tag{9.28}$$

Equations (9.27) and (9.28) represent general time dependent pulses of light.

We see (with the definition of the magnetic field adopted in Eq. (9.1)) that the electric and magnetic fields are two mutually perpendicular vector fields with the same amplitude. Hence, the contribution of each field to the radiation energy is, according to Eq. (9.12), the same. Using this fact and Eq. (9.27) we can write Eq. (9.12) for an infinite cavity as

$$\begin{aligned} H_R = -\epsilon_0 \int d^3r d^3k d^3k' \hat{\boldsymbol{\epsilon}}_k \cdot \hat{\boldsymbol{\epsilon}}_{k'} k k' \\ \times \left\{A_k \exp(-i\omega_k t + i\boldsymbol{k} \cdot \boldsymbol{r}) - A_k^* \exp(i\omega_k t - i\boldsymbol{k} \cdot \boldsymbol{r})\right\} \\ \left\{A_{k'} \exp(-i\omega_{k'} t + i\boldsymbol{k}' \cdot \boldsymbol{r}) - A_{k'}^* \exp(i\omega_{k'} t - i\boldsymbol{k}' \cdot \boldsymbol{r})\right\} . \end{aligned} \tag{9.29}$$

Using the expression

$$\int d^3r \exp[i(\boldsymbol{k}-\boldsymbol{k}')\cdot\boldsymbol{r}] = (2\pi)^3\delta(\boldsymbol{k}-\boldsymbol{k}') , \tag{9.30}$$

we obtain

$$H_R = (2\pi)^3\epsilon_0 \int d^3k\, k^2 \{A_{\boldsymbol{k}}A_{-\boldsymbol{k}}\exp(-2i\omega_k t) + A^*_{\boldsymbol{k}}A^*_{-\boldsymbol{k}}\exp(2i\omega_k t) + 2|A_{\boldsymbol{k}}|^2\} . \tag{9.31}$$

Consider now the integral of H_R over time. Using the time analog of Eq. (9.30), we see that the first two terms in the curly bracket average to zero when we integrate Eq. (9.31) over a cycle of time. The cycle-averaged radiation energy is therefore given by

$$\bar{H}_R = 2(2\pi)^3\epsilon_0 \int d^3k\, k^2|A_{\boldsymbol{k}}|^2 . \tag{9.32}$$

For the finite case of a cavity, of volume V, an analogous derivation, coupled with

$$\int_V d^3r \exp[i(\boldsymbol{k}-\boldsymbol{k}')\cdot\boldsymbol{r}] = V\delta_{\boldsymbol{k},\boldsymbol{k}'} , \tag{9.33}$$

for a finite cavity of volume V, gives

$$\bar{H}_R = 2\epsilon_0 V \sum_{\boldsymbol{k}} k^2|A_{\boldsymbol{k}}|^2 . \tag{9.34}$$

Coherent pulses of light, obtained by superposing a collection of plane waves, were discussed in Chapter 1.

9.2
The Dynamics of Quantized Particles and Classical Light Fields

Consider now the transition from classical mechanics to the quantum mechanics of the particles in the presence of a classical field. (The case of quantized particles in the presence of a quantized field is discussed in Chapter 15).

To quantize the dynamics of the particles first requires that we express the velocities of the particles in terms of canonical momenta. In the presence of electromagnetic fields, the canonical momenta are not merely $m_i d\boldsymbol{r}_i/dt$. Rather, in order to incorporate Lorentz's velocity-dependent forces into Hamilton's formulation of classical mechanics, the canonical momenta are given by [303]

$$\boldsymbol{p}_i = m_i\frac{d\boldsymbol{r}_i}{dt} + \frac{q_i}{c}\boldsymbol{A}(\boldsymbol{r}_i, t) . \tag{9.35}$$

It follows from Eqs. (9.18) and (9.35) that

$$H = \sum_i \frac{1}{2m_i}\left(\boldsymbol{p}_i - \frac{q_i}{c}\boldsymbol{A}(\boldsymbol{r}_i, t)\right)^2 + V_C + H_R . \tag{9.36}$$

Having expressed the Hamiltonian in terms of the canonical momenta, we can readily quantize the particles' dynamics. To do so we replace each particle's canonical momentum by the momentum operator in the coordinate representation,

$$\boldsymbol{p}_j \rightarrow -i\hbar\nabla_j \,. \tag{9.37}$$

The quantized Hamiltonian then assumes the form,

$$H = \sum_j \frac{1}{2\,m_j}\left(-i\hbar\nabla_j - \frac{q_j}{c}\boldsymbol{A}(\boldsymbol{r}_j, t)\right)^2 + V_C + H_R = H_M + H'(t) + H_R \,, \tag{9.38}$$

with H_M, the "material Hamiltonian", given by

$$H_M = \sum_j \frac{-\hbar^2}{2m_j}\nabla_j^2 + V_C \,, \tag{9.39}$$

and where $H'(t)$, the "interaction Hamiltonian", is

$$\begin{aligned} H'(t) &= \sum_j \frac{iq_j\hbar}{m_j c}\nabla_j \cdot \boldsymbol{A}(\boldsymbol{r}_j, t) + \frac{q_j^2}{2m_j c^2}A^2(\boldsymbol{r}_j, t) \\ &= \sum_j \frac{iq_j\hbar}{m_j c}\boldsymbol{A}(\boldsymbol{r}_j, t)\cdot\nabla_j + \frac{q_j^2}{2m_j c^2}A^2(\boldsymbol{r}_j, t) \,. \end{aligned} \tag{9.40}$$

Here we have used the fact that, in the Coulomb gauge,

$$\nabla_j \cdot \boldsymbol{A}(\boldsymbol{r}_j, t)\langle \boldsymbol{R}|\psi\rangle = \boldsymbol{A}(\boldsymbol{r}_j, t)\cdot\nabla_j\langle \boldsymbol{R}|\psi\rangle \,,$$

where $\boldsymbol{R} \equiv \boldsymbol{r}_1, \ldots, \boldsymbol{r}_N$, with N being the total number of particles. Equation (9.40) is often referred to as being in the "velocity gauge".

Given the Hamiltonian of Eq. (9.38), the dynamics of the particles the presence of the field are obtained by solving for the wave function $\Psi(\boldsymbol{R}, t)$ via the time-dependent Schrödinger equation

$$i\hbar\frac{\partial\Psi(\boldsymbol{R}, t)}{\partial t} = \left[\sum_j \frac{1}{2\,m_j}\left(-i\hbar\nabla_j - \frac{q_j}{c}\boldsymbol{A}(\boldsymbol{r}_j, t)\right)^2 + V_C\right]\Psi(\boldsymbol{R}, t)\,. \tag{9.41}$$

Here H_R does not contribute since it is a function of the field variables only.

Equation (9.41) may be further simplified by noting that the variation of $\boldsymbol{A}$ over a typical displacement $\boldsymbol{r}_j$ of a particle is small. For example, for visible light, a typical wavelength of the field is 5000 Å, whereas the particle displacements within a molecule vary over 1–10 Å. It is therefore reasonable to replace all of the $\boldsymbol{r}_j$ displacements in $\boldsymbol{A}$ by the position of the center of mass of the molecule. For a plane wave, only the z projection of the center of mass position is relevant, and we can approximate $\boldsymbol{A}$ as

$$A(\boldsymbol{r}_j, t) \approx A(z, t) . \tag{9.42}$$

For reasons that will become evident below, this is called the "dipole approximation".

Given this approximation, we can transform the Hamiltonian of Eq. (9.40) from the velocity gauge to the so-called "length gauge", in which the matter–radiation interaction term contains only the dot product of the dipole moment and the electric field. In order to do so we choose χ (Eq. (9.5)) as

$$\chi = -\sum_i \boldsymbol{r}_i \cdot A(z, t) . \tag{9.43}$$

Clearly, due to the neglect of the $\boldsymbol{r}_i$ dependence in $\boldsymbol{A}$ in Eq. (9.42), this gauge transformation leaves $A(z, t)$ within the Coulomb gauge (Eq. (9.6)), since $\nabla^2 \chi = 0$.

Using the definition V_C (Eq. (9.16)) and χ (Eq. (9.43)) in the Schrödinger equation (Eq. (9.41)), and noting that $\nabla \cdot \chi = -\boldsymbol{A}$ we obtain that

$$i\hbar \frac{\partial \Psi(\boldsymbol{R}, t)}{\partial t} = \sum_j \left[\frac{-\hbar^2}{2\, m_j} \nabla_j^2 + q_j\, \Phi_j(\boldsymbol{r}_j) + \frac{q_j}{c} \boldsymbol{r}_j \cdot \frac{\partial A(z, t)}{\partial t} \right] \Psi(\boldsymbol{R}, t) . \tag{9.44}$$

Using Eq. (9.4) we can write the last term in the square brackets as $-q_j \boldsymbol{r}_j \cdot \boldsymbol{E}(z, t)$, where we have used the fact that gauge transformations do not change the electric field, which was calculated from the untransformed vector potential.

We obtain that,

$$i\hbar \frac{\partial \Psi(\boldsymbol{R}, t)}{\partial t} = \sum_j \left[\frac{-\hbar^2}{2\, m_j} \nabla_j^2 + q_j\, \Phi_j(\boldsymbol{r}_j) - q_j \boldsymbol{r}_j \cdot \boldsymbol{E}(z, t) \right] \Psi(\boldsymbol{R}, t) . \tag{9.45}$$

In this form, both the vector potential and the gradient operator no longer appear. Instead, a scalar potential, proportional to the scalar product of the transverse field and the displacement of each particle from the origin, has been added to the Coulomb potential.

Equation (9.45) can be written in a more concise form as

$$i\hbar \frac{\partial \Psi(t)}{\partial t} = H(t)\Psi(t) = [H_M + H_{MR}(t)]\Psi(t) , \tag{9.46}$$

where $H(t) = H_M + H_{MR}$ is the total Hamiltonian, H_M is the material Hamiltonian, given in Eq. (9.39) and H_{MR} is the matter–radiation interaction in the dipole approximation, given by Eqs. (1.5) and (1.6).

It is possible to go beyond the dipole approximation in the length gauge and treat the interactions between higher multipoles with the field derivatives, which is relevant when the variation of the field with $\boldsymbol{r}_j$ cannot be neglected [364]. However, we do not pursue these extensions here because, in all the applications discussed in this book, the dipole approximation suffices. Equations (9.46), (1.5) and (1.6) are the central expressions used to describe molecule–light interactions within the classical field picture. Extensions of this approach to include quantized electromagnetic fields are described in Chapter 10 and Chapter 15.

10
Coherent Control with Quantum Light

Prior chapters have dealt with incident light described classically. In this chapter we discuss the implications of the quantization of light on coherent control. To this end we introduce the basics of the quantization of light; a fuller treatment can be found in, for example, [364]. We then utilize (Section 10.2) the quantum light formulation to expose the underlying origins of light-induced, interference-based control. An example of how one can use quantum light to create and control entanglement between light and matter is discussed in Section 10.3.

10.1
The Quantization of the Electromagnetic Field

In complete analogy with the procedure for quantizing the coordinates and momenta of particles (Chapter 9) we first express H_R, the classical Hamiltonian of the electromagnetic field, in terms of the canonical coordinates, Q_k, and momenta P_k. To do so we introduce the relevant canonical variables Q_k, P_k that are related to the field mode amplitudes, A_k and A_k^* (Eq. (9.24)) by the definitions:

$$Q_k \equiv \left(\frac{\epsilon_0 V}{c^2}\right)^{\frac{1}{2}} \left(A_k + A_k^*\right) , \quad P_k \equiv i\omega_k \left(\frac{\epsilon_0 V}{c^2}\right)^{\frac{1}{2}} \left(A_k - A_k^*\right) , \tag{10.1}$$

where the cavity volume V is assumed, in this chapter, to be finite. As a consequence, the number of field modes is discrete. It can be easily verified that both Q_k and P_k satisfy Eq. (9.21), that is, the dynamical equations for a set of harmonic oscillators.

Using Eq. (10.1) to express A_k and A_k^* in terms of Q_k and P_k, gives

$$A_k = c(4\epsilon_0 V\omega_k^2)^{-\frac{1}{2}} (\omega_k Q_k - iP_k) , \quad A_k^* = c(4\epsilon_0 V\omega_k^2)^{-\frac{1}{2}} (\omega_k Q_k + iP_k) , \tag{10.2}$$

and the cycle-averaged energy (Eq. (9.34)) can therefore be written as

$$\bar{H}_R = \frac{1}{2}\sum_k (\omega_k^2 Q_k^2 + P_k^2) . \tag{10.3}$$

Since the energy of the electromagnetic field fluctuates during one cycle, it is the cycle-averaged energy $\bar{H}_R$ that represents the energy of the field after many cycles. Hence, this is the quantity that we now proceed to quantize. We do so, in the coordinate representation, by replacing the classical momenta with the operators

$$P_k \to -i\hbar \frac{\partial}{\partial Q_k} . \tag{10.4}$$

As a result, the radiative Hamiltonian $\bar{H}_R$ (Eq. (10.3)) assumes the quantized form:

$$H_R = \frac{1}{2} \sum_k \left(\omega_k^2 Q_k^2 - \hbar^2 \frac{\partial^2}{\partial^2 Q_k} \right) . \tag{10.5}$$

Introducing the annihilation and creation operators, $\hat{a}_k$ and $\hat{a}_k^\dagger$, defined as

$$\hat{a}_k = (2\hbar\omega_k)^{-\frac{1}{2}} (\omega_k Q_k + i P_k) = (2\hbar\omega_k)^{-\frac{1}{2}} \left(\omega_k Q_k + \hbar \frac{\partial}{\partial Q_k} \right) ,$$
$$\hat{a}_k^\dagger = (2\hbar\omega_k)^{-\frac{1}{2}} (\omega_k Q_k - i P_k) = (2\hbar\omega_k)^{-\frac{1}{2}} \left(\omega_k Q_k - \hbar \frac{\partial}{\partial Q_k} \right) , \tag{10.6}$$

allows us to rewrite the radiative Hamiltonian as

$$H_R = \sum_k \hbar\omega_k \left(\hat{a}_k^\dagger \hat{a}_k + \frac{1}{2} \right) , \tag{10.7}$$

where we have used the commutation relation,

$$\left[\hat{a}_k, \hat{a}_k^\dagger \right] \equiv \hat{a}_k \hat{a}_k^\dagger - \hat{a}_k^\dagger \hat{a}_k = 1 . \tag{10.8}$$

Similarly, comparing Eq. (10.2) and Eq. (10.6) we obtain the operators $\hat{A}_k$ and $\hat{A}_k^\dagger$ as the quantum analogues of A_k and A_k^*:

$$\hat{A}_k = c \left(\frac{\hbar}{2\epsilon_0 V \omega_k} \right)^{\frac{1}{2}} \hat{a}_k , \quad \hat{A}_k^\dagger = c \left(\frac{\hbar}{2\epsilon_0 V \omega_k} \right)^{\frac{1}{2}} \hat{a}_k^\dagger . \tag{10.9}$$

Because Eq. (10.7) is separable, the eigenstates of H_R are products of the eigenstates $|N_k\rangle$ of the different harmonic mode oscillators. That is,

$$H_R |\mathbf{N}\rangle = \mathbf{N} |\mathbf{N}\rangle , \tag{10.10}$$

$$|\mathbf{N}\rangle \equiv \prod_k |N_k\rangle , \tag{10.11}$$

where $|N_k\rangle$ are eigenstates of the operator $\hat{N}_k \equiv \hat{a}_k^\dagger \hat{a}_k$,

$$\hat{N}_k |N_k\rangle = \hat{a}_k^\dagger \hat{a}_k |N_k\rangle = N_k |N_k\rangle . \tag{10.12}$$

It follows from the properties of harmonic oscillators that the eigenvalues N_k are nonnegative integers ($N_k = 0, 1, 2, \ldots$). Hence, the $|N_k\rangle$ states are called *number states* and the $\hat{N}_k$ operator is called the *number operator*. We say that the N_k^{th} level of a given field mode, labeled by $\boldsymbol{k}$, has N_k *photons* in that mode. Note also that we can define $\widehat{e^{i\theta_k}}$, the exponential phase operator of the radiation field, through the relationship $\hat{a}_k = (\hat{N}_k + 1)^{1/2}\widehat{e^{i\theta_k}}$. It is clear from the commutation relations for $\hat{a}_k$ and $\hat{a}_k^{\dagger}$ that $\hat{N}_k$ and $\widehat{e^{i\theta_k}}$ do not commute. As a result, a state with a well-defined number of photons has an ill defined radiative phase.

A second class of radiation states of general interest are the coherent states [365, 366], defined as the eigenstates of the annihilation operator,

$$\hat{a}|\alpha\rangle = \alpha|\alpha\rangle\,, \tag{10.13}$$

with α being any complex number. The coherent states are related to the number states as [364]

$$|\alpha\rangle = \exp(-|\alpha|^2/2)\sum_N \alpha^N/(N!)^{1/2}|N\rangle\,. \tag{10.14}$$

It follows from the normalization of the number states that although the coherent states are normalized, they are not orthogonal to one another. Specifically,

$$|\langle\alpha|\beta\rangle|^2 = \exp(-|\alpha-\beta|^2)\,. \tag{10.15}$$

The free field Schrödinger equation assumes a simple form in the number representation. Using Eq. (10.7), the time-independent Schrödinger equation for the free radiation field, we have:

$$H_R|\boldsymbol{N}\rangle = \left[\sum_k \hbar\omega_k\left(N_k + \frac{1}{2}\right)\right]|\boldsymbol{N}\rangle\,. \tag{10.16}$$

Notice that we have adopted the Schrödinger representation here insofar as all operators, such as $\hat{a}_k$ and $\hat{a}_k^{\dagger}$, have no time dependence; rather, the time dependence is contained in the wave functions. It is, of course, possible to define the Heisenberg representation wherein the operators do vary with time and where the wave functions are time-independent, or a mixed interaction representation where the operators assume the time dependence of the free field. It is in the latter representation that the quantized electric field most resembles the classical form.

10.1.1 Light–Matter Interactions

To obtain the Schrödinger equation for the interaction of a molecule with the quantized radiation field, that is, the Schrödinger equation for the (matter + radiation) system, we need the quantum analogue of H_{MR}, the matter–radiation interaction. In the dipole approximation H_{MR} depends, according to Eq. (1.5), on the transverse electric field. The required quantized electric field is obtained by substituting Eq. (10.9) into Eq. (9.27), to give

$$\hat{\mathsf{E}}(r) = i\sum_k\left(\frac{\hbar\omega_k}{2\epsilon_0 V}\right)^{\frac{1}{2}}\hat{\varepsilon}_k\left[\hat{a}_k e^{ik\cdot r+i\phi_k} - \hat{a}_k^{\dagger}e^{-ik\cdot r-i\phi_k}\right]. \tag{10.17}$$

Here, in accord with Eq. (2.9), we have added an extra phase, ϕ_k, to each plane wave field mode in Eq. (10.17), representing the phase shifts accumulated by the light in the $\boldsymbol{k}$ mode as it travels from the source to the sample.

By substituting Eq. (10.17) into Eq. (1.5) we obtain the quantum analogue of the radiation–matter interaction as

$$H_{MR} = -i \sum_k \left(\frac{\hbar \omega_k}{2\epsilon_0 V} \right)^{\frac{1}{2}} \hat{\varepsilon}_k \cdot \boldsymbol{d} \left[\hat{a}_k e^{i\boldsymbol{k}\cdot\boldsymbol{r}+i\phi_k} - \hat{a}_k^\dagger e^{-i\boldsymbol{k}\cdot\boldsymbol{r}-i\phi_k} \right] , \tag{10.18}$$

where $\boldsymbol{d}$, as before, is the dipole operator.

The quantum analogue of the total (matter + radiation) Hamiltonian now assumes the form

$$H = H_M + H_R + H_{MR} \equiv H_f + H_{MR} , \tag{10.19}$$

where

$$H_f = H_M + H_R \tag{10.20}$$

is called the "radiatively decoupled" Hamiltonian, it being the sum of the independent material and radiative parts. By contrast, H is the total Hamiltonian; its eigenstates are called "fully interacting states".

10.2
Quantum Light and Quantum Interference

The myriad of possible coherent control scenarios and light sources calls for a general perspective on conditions under which interference-based control can be achieved. The key to such a formulation, as shown in this section, is the recognition that the material system of interest is part of an open system comprising matter + radiation. As such, as discussed in Chapter 6 dynamics of the matter subsystem is obtained by tracing over the states of the radiation field. This formulation is carried out below to achieve an all-encompassing perspective on conditions under which interference-based control can be achieved.

We now examine a material system with Hamiltonian H_M interacting via H_{MR} with a quantized radiation field with Hamiltonian H_R. For simplicity we consider a case where there are two light-induced pathways from an initial state $|\Psi^i\rangle$ to the final state $|\Psi^f\rangle$; the generalization to multiple pathways is straightforward. The field, at both times, is not active so each of these states can be written as the product of a matter state $|M_i\rangle$ times a state $|R_i\rangle$ of the radiation field. In particular, given two pathways to the same final state, the overall final wave function $|\Psi^f\rangle$ is of the form

$$|\Psi^f\rangle = |M_1\rangle|R_1\rangle + |M_2\rangle|R_2\rangle . \tag{10.21}$$

The notation is chosen generally, to imply that neither $|M_i\rangle$ nor $|R_j\rangle$ need be eigenstates of H_M and H_R, but rather are general states of the material and radiative systems. Note that in writing Eq. (10.21) we are implicitly stating that the experiment under consideration is such that we could distinguish, should we choose to do so, between $|M_1\rangle$ and $|M_2\rangle$, as well as between $|R_1\rangle$ and $|R_2\rangle$.

Consider then the average value in the final state of any system property A_s, in which case

$$\langle\Psi^f|A_s|\Psi^f\rangle = \langle M_1|A_s|M_1\rangle + \langle M_2|A_s|M_2\rangle + \langle M_1|A_s|M_2\rangle\langle R_1|R_2\rangle + \langle M_2|A_s|M_1\rangle\langle R_2|R_1\rangle \,. \tag{10.22}$$

Note that this applies to the measurement of any property of the material system, including the probability of finding the system $|E, \boldsymbol{m}; 0\rangle$, wherein $A_s = |E, \boldsymbol{m}; 0\rangle\langle E, \boldsymbol{m}; 0|$. Interference contributions, required for coherent control, arise from the last two terms in Eq. (10.22). Significantly, the key to the possibility of control is seen to lie in the magnitude of the *radiation overlap matrix element* $\langle R_1|R_2\rangle$.

Examining the limiting cases associated with Eq. (10.22) is enlightening. If the states of the fields $|R_1\rangle$ and $|R_2\rangle$ are orthogonal, then the interference terms disappear, and coherent control is not possible. At the other extreme, if $|R_1\rangle$ and $|R_2\rangle$ differ only by a constant c, then the interference terms are maximal. The intermediate cases follow, depending on the degree of overlap of the field states. The argument holds for any kind of fields, from quantum states that are highly numbered to classical-like coherent states.

The interpretation of these results merges three qualitative concepts and their role in control: decoherence, entanglement, and "welcher-weg", as summarized below.

When $|R_2\rangle = \mathrm{c}|R_1\rangle$, that is, when interference (and hence the possibility of interference-based control) is maximal, then Eq. (10.21) is of the form

$$|\Psi^f\rangle = |M_1\rangle|R_1\rangle + |M_2\rangle|R_2\rangle = [|M_1\rangle + c|M_2\rangle]|R_1\rangle \,. \tag{10.23}$$

Thus, the wave function $|\Psi^f\rangle$ is a completely unentangled state, being the product of a matter state and the field state. From the decoherence perspective, the material system density matrix (see Chapter 6)

$$\rho_s = \mathrm{Tr}_B\left[|\Psi^f\rangle\langle\Psi^f|\right] = [|M_1\rangle + c|M_2\rangle][\langle M_1| + c\langle M_2|]$$

has robust off-diagonal elements $|M_i\rangle\langle M_j|$ in a matter states, and hence robust matter coherence. Note, significantly, that this is the case independent of the particular form of the radiation field $|R_1\rangle$.

By contrast, when $|R_1\rangle$ and $|R_2\rangle$ are orthogonal, Eq. (10.21) remains in the fully system–field entangled form:

$$|\Psi^f\rangle = |M_1\rangle|R_1\rangle + |M_2\rangle|R_2\rangle \,. \tag{10.24}$$

In this case, the associated system density matrix $\rho_s = \mathrm{Tr}_B[|\Psi^f\rangle\langle\Psi^f|]$ has no off-diagonal elements in $|M_i\rangle\langle M_j|$ and matter coherence is lost, as is control. Other cases of $|R_1\rangle$ and $|R_2\rangle$ fall between these limits.

Finally, we note one alternative perspective – that of welcher-weg (which-way) information [351, 367]. For convenience we focus on the specific case where we are interested in the probability of finding the final state $|E, \boldsymbol{m}; 0\rangle$, that is, where $A_s = |E, \boldsymbol{m}; 0\rangle\langle E, \boldsymbol{m}; 0|$. Consider first Eq. (10.23). This case is consistent with robust matter coherence contributions and we can not know whether the final material state $|E, \boldsymbol{m}; 0\rangle$ was accessed via the $|M_1\rangle|R_1\rangle$ or $|M_2\rangle|R_2\rangle$ route by measuring the state of the radiation field. By contrast, for the case where control is lost (Eq. (10.24)), if we were to measure the state of the radiation field we would be able to tell whether the $|M_1\rangle|R_1\rangle$ or the $|M_2\rangle|R_2\rangle$ route was taken, since observing $|R_i\rangle$ would collapse the amplitude for observing the material state to $|M_i\rangle$. Hence, the superposition of material states would not survive knowledge of the radiation field, and the possibility for control is lost. Intermediate degrees of entanglement between $|M_1\rangle|R_1\rangle$ and $|M_2\rangle|R_2\rangle$ yield intermediate amounts of which-way information, and hence intermediate possibilities for control. This result constitutes the foundation for the assertion of Section 3.4.3 that quantum interference exists only for processes in which we do not possess information as to which path was taken by the dynamics to the final state.

An important observation is that the above holds true even when no measurement of the radiation state is ever performed. As shown by Mandel [368], in a given setup it is enough to have the *potential* of extracting the "which-way information" for the interference not to exist [369].

Sample applications of this general approach follow below.

10.2.1 One-Photon vs. Two-Photon Quantum Field Control

10.2.1.1 Use of Number States

We first discuss the one-photon vs. two-photon symmetry breaking control scenario introduced previously (Section 3.4.3) and examine the outcome of exciting a material system in bound state $|E_i\rangle$ to a final dissociative state $|E, \boldsymbol{m}^-\rangle$ using *number* states. We choose the field to be composed of two modes, $|N_\omega, N_{\omega/2}\rangle$, where $|N_\omega, N_{\omega/2}\rangle$ is a short hand notation for $|0, \ldots, 0, N_\omega, 0, \ldots, 0, N_{\omega/2}, 0, \ldots\rangle$, a number state containing N_ω photons in the ω mode, $N_{\omega/2}$ photons in the $\omega/2$ mode, and zero photons in all the other modes. The matter–radiation state is therefore initially

$$|\Psi^i\rangle = |N_\omega, N_{\omega/2}\rangle|E_i\rangle\ . \tag{10.25}$$

The final matter–radiation state is given (in second-order perturbation theory) as,

$$\begin{aligned}|\Psi^f\rangle = (2\pi i)^{\frac{1}{2}}|E, \boldsymbol{m}^-\rangle \{\bar{\epsilon}(\omega)\langle E, \boldsymbol{m}^-|\mathrm{d}|E_i\rangle|N_\omega - 1, N_{\omega/2}\rangle - \bar{\epsilon}^2(\omega/2)\\ \times\ \langle E, \boldsymbol{m}^-|D|E_i\rangle|N_\omega, N_{\omega/2} - 2\rangle\}\ ,\end{aligned} \tag{10.26}$$

where $\langle E, \boldsymbol{m}^{-}|\mathrm{d}|E_i\rangle$ is the one-photon transition dipole matrix element, $\langle E, \boldsymbol{m}^{-}|D|E_i\rangle$ is the two-photon matrix element, given by Eq. (3.43) as

$$\langle E, \boldsymbol{m}^{-}|D|E_i\rangle = \sum_j \langle E, \boldsymbol{m}^{-}|\mathrm{d}|E_j\rangle\langle E_j|\mathrm{d}|E_i\rangle \left[\hbar\omega/2 + E_i - i\epsilon - E_j\right]^{-1} , \tag{10.27}$$

and the field amplitude $\bar{\epsilon}(\omega)$ is related to N_ω by:

$$\bar{\epsilon}(\omega) = -\left(\frac{\hbar\omega N_\omega}{\epsilon_0 V}\right)^{\frac{1}{2}} \exp\left(-i\omega z/c - i\phi_\omega\right) , \tag{10.28}$$

with z denoting the axis of propagation of the light beams, ϵ_0 being the permittivity of the vacuum, V the cavity volume, and ϕ_ω is an extra phase which we add (e.g., by modifying the optical path transversed by the field) to the inherent phase associated with the time dependence of the quantum state of the radiation. As discussed in the classical field control Section 3.4.3 the $\phi_\omega - 2\phi_{\omega/2}$ phase difference is introduced in order to control the process.

Comparing of Eq. (10.26) with Eq. (10.21) shows that $|R_1\rangle \propto |N_\omega - 1, N_{\omega/2}\rangle$ and $|R_2\rangle \propto |N_\omega, N_{\omega/2} - 2\rangle$. Hence, the radiative overlap matrix element relevant to control $\langle R_1|R_2\rangle = \langle N_\omega - 1, N_{\omega/2}|N_\omega, N_{\omega/2} - 2\rangle = 0$. Interference contributions, and hence the opportunity for control, are therefore lost. That is, due to the orthogonality of the number states the probability to observe the $|E, \boldsymbol{m}^{-}\rangle$ material state is given in this case as

$$\langle\Psi^f|E, \boldsymbol{m}^{-}\rangle\langle E, \boldsymbol{m}^{-}|\Psi^f\rangle = |\bar{\epsilon}(\omega)\langle E, \boldsymbol{m}^{-}|\mathrm{d}|E_i\rangle|^2 + \left|\bar{\epsilon}^2(\omega/2)\langle E, \boldsymbol{m}^{-}|D|E_i\rangle\right|^2 . \tag{10.29}$$

Hence, coherent control cannot take place when the initial radiative states are number states.

10.2.1.2 Use of Coherent States

The situation is dramatically different if the radiation field is comprised of a product of coherent states, $|\alpha_\omega\rangle$, defined as,

$$\hat{a}_\omega|\alpha_\omega\rangle = \alpha_\omega|\alpha_\omega\rangle . \tag{10.30}$$

In this case, the initial matter + radiation state is

$$|\Psi^i\rangle = |\alpha_\omega, \alpha_{\omega/2}\rangle|E_i\rangle . \tag{10.31}$$

Because $|\alpha_\omega\rangle$ is an eigenstate of the annihilation operator, which is the only part of the quantum field operator that contributes in the rotating wave approximation, the absorption of one photon from an $|\alpha_\omega\rangle$ state results in the production of $\bar{\epsilon}(\omega)|\alpha_\omega\rangle$ the initial coherent state multiplied by the field amplitude,

$$\bar{\epsilon}(\omega) = -\left(\frac{\hbar\omega}{\varepsilon_0 V}\right)^{\frac{1}{2}} \alpha_\omega \exp\left(-i\omega z/c - i\phi_\omega\right) . \tag{10.32}$$

Hence, the final matter + radiation state can be written as,

$$|\Psi^f\rangle = (2\pi i)^{\frac{1}{2}} |E, \boldsymbol{m}^-\rangle \left\{\bar{\epsilon}(\omega)\langle E, \boldsymbol{m}^-|\mathrm{d}|E_g\rangle - \bar{\epsilon}^2\omega/2\langle E, \boldsymbol{m}^-|D|E_g\rangle\right\} |\alpha_\omega, \alpha_{\omega/2}\rangle\,. \tag{10.33}$$

The wave function $|\Psi^f\rangle$ is seen to be of the completely unentangled form (as in Eq. (10.23)) and, as discussed above, allows for robust interference and the possibility of control. That is, a measurement of the final radiation states (or lack thereof) would collapse this final state to the material superposition in Eq (10.33), allowing for control. Thus achieving coherent control with coherent states is possible and is essentially identical to the classical field scenario. Indeed, specific coherent states are known to be the quantum description of a classical electromagnetic field [364], as discussed in the case of the pump-dump scenario, below.

10.2.2 Pump-Dump Coherent Control

To demonstrate the origin of interference effects in the classical field scenarios discussed in earlier chapters of this book, we consider as a specific example the pump-dump scenario with few levels, discussed in Section 3.6. There we showed that by using an initial (pump) pulse to excite an initial state $|E_i\rangle$ to two molecular levels $|E_2\rangle$ and $|E_3\rangle$, followed after a delay time $\Delta_d \equiv t_d - t_x$ by a second (dump) pulse to excite to the final continuum state $|E, \boldsymbol{m}, q^-\rangle$, that the probability of observing the qth channel at energy E is given by (Eq. (3.77)):

$$P_q(E) = \sum_{\mathrm{m}} |b_{E,\boldsymbol{m},q}(t=\infty)|^2 = \frac{(2\pi)^2}{\hbar^2} \sum_{\mathrm{m}} \left| \sum_{k=2,3} b_k \langle E, \boldsymbol{m}, q^- |\hat{\varepsilon}_d \cdot \boldsymbol{d}| E_k\rangle \bar{\epsilon}_d(\omega_{E E_k}) \right|^2, \tag{10.34}$$

with

$$b_k = \frac{2\pi i}{\hbar} \langle E_k|\hat{\varepsilon}_x \cdot \boldsymbol{d}|E_1\rangle \bar{\epsilon}_x(\omega_{k,1})\,, \tag{10.35}$$

where the notation and parameters, here and below, are defined in Section 3.6. Here we recast [351] this scenario in terms of general quantized radiation fields, allowing us to expose the origin of interference control in the specific case of classical electromagnetic fields. For notational simplicity we suppress, below, the q index in the states $|E, \boldsymbol{m}, q^-\rangle$.

10.2.2.1 Results with Quantized Light

Consider a molecule subjected to a quantized electromagnetic field. The total Hamiltonian is that given in Eq. (10.19) with the molecule + field coupling given by Eq. (10.18) with, in this case, $\phi_k = 0$. First-order perturbation theory gives [69]

$$e^{-iHt/\hbar} = e^{-iH_f t/\hbar} \left\{ 1 - \sum_k \left(\frac{-\omega_k}{2\hbar\epsilon_0 V} \right)^{\frac{1}{2}} \sum_{m,n} |E_m\rangle\langle E_m|\hat{\varepsilon}\cdot \boldsymbol{d}|E_n\rangle \right.$$
$$\left. \times \langle E_n| \left[\frac{e^{i(\omega_{m,n}-\omega_k)t}\hat{a}_k}{\omega_{m,n}-\omega_k - i\epsilon} - \frac{e^{i(\omega_{m,n}+\omega_k)t}\hat{a}_k^\dagger}{\omega_{m,n}+\omega_k - i\epsilon} \right] \right\} , \tag{10.36}$$

where $t = 0$ marks the end of the first pulse. As usual, the positive constant ϵ is equated to 0 at the end of the computation.

Denoting the initial radiative state of the first pulse as $|R_x\rangle$, the combined matter + radiation wave function after the first pulse can be written in the rotating wave approximation as,

$$|\Psi(0)\rangle = |R_x\rangle|E_1\rangle + d_{2,1}|R_{x,2}\rangle|E_2\rangle + d_{3,1}|R_{x,3}\rangle|E_3\rangle , \tag{10.37}$$

where the $|R_{x,j}\rangle$ states are

$$|R_{x,j}\rangle \equiv \sum_k \left(-\frac{\omega_k}{2\hbar\epsilon_0 V} \right)^{\frac{1}{2}} \frac{\hat{a}_k|R_x\rangle}{\omega_{j1} - \omega_k - i\epsilon} , \quad j = 2, 3 . \tag{10.38}$$

For the second laser pulse, the quantum state of light is assumed to be $|R_d\rangle$ at $t = 0$. Applying first-order perturbation theory a second time gives rise to the wave function $|\Psi(t)\rangle$ for the entire system at any time $t \geq 0$. For the sake of comparison with the classical treatment, we assume that only the $|E_2\rangle$ and $|E_3\rangle$ molecular levels contribute to the photodissociation probabilities. One then finds

$$|\Psi^f\rangle = \lim_{t\to+\infty} |\Psi(t)\rangle = e^{-iH_f t/\hbar}|\Psi(0)\rangle$$
$$= -\sum_n \int dE e^{-iEt/\hbar}|E, \boldsymbol{n}^-\rangle \left[d^{\boldsymbol{n}}_{E,2} d_{2,1}|R_{E,2}\rangle|R_{x,2}\rangle + d^{\boldsymbol{n}}_{E,3} d_{3,1}|R_{E,3}\rangle|R_{x,3}\rangle \right] , \tag{10.39}$$

where $d^{\boldsymbol{n}}_{E,j} \equiv \langle E, \boldsymbol{n}^-|\hat{\varepsilon}\cdot\boldsymbol{d}|E_j\rangle$, and $|R_{E,j}\rangle$ is defined as

$$|R_{E,j}\rangle \equiv \sum_k \left(-\frac{\omega_k}{2\hbar\epsilon_0 V} \right)^{\frac{1}{2}} \frac{\hat{a}_k|R_d\rangle}{\omega_{E,j} - \omega_k + i\epsilon} , \quad j = 2, 3 . \tag{10.40}$$

The continuum part of Eq. (10.39) is thus comprised of two pathways:

$$|\Psi^f\rangle = |\psi_1\rangle + |\psi_2\rangle \tag{10.41}$$

where

$$|\psi_1\rangle \equiv \int dE \exp(-iEt/\hbar) \sum_n |E, \boldsymbol{n}^-\rangle d^{\boldsymbol{n}}_{E,2} d_{2,1}|R_{E,2}\rangle|R_{x,2}\rangle ,$$
$$|\psi_2\rangle \equiv \int dE \exp(-iEt/\hbar) \sum_n |E, \boldsymbol{n}^-\rangle d^{\boldsymbol{n}}_{E,3} d_{3,1}|R_{E,3}\rangle|R_{x,3}\rangle . \tag{10.42}$$

A comparison of Eq. (10.41) with Eq. (10.21) shows that the relevant radiation overlap matrix element in this case is given by:

$$\langle R_1 | R_2 \rangle = \langle R_{x,3} | R_{x,2} \rangle \langle R_{E,3} | R_{E,2} \rangle \,, \tag{10.43}$$

valid for all types of radiation fields. The interference contribution to the probability of producing the product state $|E, \boldsymbol{n}^-\rangle$ is then of the form

$$\begin{aligned} &\sum_{\mathrm{n}} \langle \psi_1 | E, \boldsymbol{n}^- \rangle \langle E, \boldsymbol{n}^- | \psi_2 \rangle + \langle \psi_2 | E, \boldsymbol{n}^- \rangle \langle E, \boldsymbol{n}^- | \psi_1 \rangle \\ &\quad = \sum_{\mathrm{n}} d^{*n}_{E,3} d^{n}_{E,2} d^{*}_{3,1} d_{2,1} \langle R_{x,3} | R_{x,2} \rangle \langle R_{E,3} | R_{E,2} \rangle + \text{c.c.} \end{aligned} \tag{10.44}$$

10.2.2.2 Results with Classical Fields

Significantly, an exact correspondence between classical and quantum treatment of the light fields can be made under certain conditions. Suppose both $|R_x\rangle$ and $|R_d\rangle$ are products of coherent states of light for different frequencies, that is,

$$\begin{aligned} |R_{x,j}\rangle &= \sum_k \left(-\frac{\omega_k}{2\hbar\epsilon_0 V} \right)^{\frac{1}{2}} \frac{\alpha^x_k}{\omega_{j,1} - \omega_k - i\epsilon} |R_x\rangle = (2\pi)^{\frac{1}{2}} E_x(\omega_{j,1}) |R_x\rangle \,, \\ |R_{E,j}\rangle &= \sum_k \left(-\frac{\omega_k}{2\hbar\epsilon_0 V} \right)^{\frac{1}{2}} \frac{\alpha^d_k}{\omega_{E,j} - \omega_k + i\epsilon} |R_d\rangle \equiv (2\pi)^{\frac{1}{2}} E_d(\omega_{E,j}) |R_d\rangle \,, \end{aligned} \tag{10.45}$$

where α^x_k and α^d_k are the eigenvalues of $\hat{a}_k$ for the first and second light pulses, characterizing the coherent states of light. One can then establish the equivalence between Eq. (10.44) and the interference term in Eq. (3.78) by requiring $E_x(\omega_{j,1})$ and $E_d(\omega_{E,j})$ defined in Eq. (10.45) to be the same as the Fourier components $\overline{\epsilon}_x(\omega_{j,1})$ and $\overline{\epsilon}_d(\omega_{E,j})$ of the classical light fields in Eq. (3.71). In this case, Eq. (10.43) becomes:

$$\langle R_1 | R_2 \rangle = (2\pi)^2 \overline{\epsilon}^*_x(\omega_{3,1}) \overline{\epsilon}_x(\omega_{2,1}) \overline{\epsilon}^*_d(\omega_{E,3}) \overline{\epsilon}_d(\omega_{E,2}) \,. \tag{10.46}$$

This result establishes the origin of control with classical light in the nonzero overlap of the radiative components of the associated quantum coherent states of the light fields.

Consistent with this result is the indistinguishability of pathways associated with the classical light fields. Specifically, substituting Eq. (10.45) into Eq. (10.42), shows $|\psi_1\rangle$ to be absolutely indistinguishable from $|\psi_2\rangle$ except for a c-number phase factor. That is, the final wave function $|\Psi^f\rangle$ is of the unentangled form in Eq. (10.23).

By contrast, some quantum states of light, where the radiation overlap matrix elements are much smaller, offer less useful methods to carry out control in this scenario.

These results are also relevant to quantum computation. Since quantum computation relies on coherently controlled evolution of atomic and molecular systems, the analysis here suggests that nonclassical light fields may affect the reliability of a quantum computer by leaking out some which-way information. This is consistent with a recent study suggesting that the quantum nature of light may have important implications for the limits of quantum computation [188].

10.2.3
Phase-Independent Control

The physical picture established in the previous subsection applies also to other weak field coherent control scenarios (e.g., "1 photon + 3 photon" control, "1 photon + 2 photon" control, "$\omega_1 + \omega_2$" vs. "$\omega_3 + \omega_4$" control ($\omega_{1(2)} \neq \omega_{3(4)}$)) with minor changes. However, it remains to examine the situation associated with several control schemes where interference between multiple pathways was found to be insensitive to laser phases [271, 279], but allows for control through laser detunings. These include the scenarios discussed in Section 6.6.2 and the specific two-photon vs. two-photon control (Section 7.1) discussed below.

First consider two-photon vs. two-photon control as in Figure 7.1, but where [279] $\omega_0 = 0$, the so-called "$\omega_1 + \omega_2$" vs. "$\omega_2 + \omega_1$" control. Rigorously describing such laser–molecule interaction requires a general resonant two-photon photodissociation theory [271], by which both level shifts and level widths can be explicitly taken into account. Nevertheless, for the purpose here it suffices to apply second-order perturbation theory with the fully quantized Hamiltonian. Substituting Eq. (10.19) into the following perturbation series:

$$e^{-iHt/\hbar} = e^{-iH_f t/\hbar}\left[1 - \frac{i}{\hbar}\int_0^t dt_1 e^{iH_f t_1/\hbar} H_{MR} e^{-\epsilon t_1} e^{-iH_f t_1/\hbar}\right.$$
$$\left. - \frac{1}{\hbar^2}\int_0^t dt_1 \int_0^{t_1} dt_2 e^{iH_f t_1/\hbar} H_{MR} e^{-\epsilon t_1} e^{-iH_f t_1/\hbar} e^{iH_f t_2/\hbar} H_{MR} e^{-\epsilon t_2} e^{-iH_f t_2/\hbar}\right]. \tag{10.47}$$

If none of the laser frequencies is in resonance with any of the transition frequencies involved, we can neglect the first-order term, leaving only the second-order term. If the initial state $|\Psi^i\rangle$ is a direct product state of the matter wave function $|E_1\rangle$ and the light field wave function $|R^l\rangle$, we obtain, using the rotating wave approximation, that

$$\lim_{t\to+\infty} |\Psi(t)\rangle = e^{-iH_f t/\hbar}\left[|E_1\rangle|R^l\rangle - \frac{1}{2\epsilon_0 V\hbar}\sum_{k'k,j,n}\int dE|E,n^-\rangle\right.$$
$$\left. \times \frac{(\omega_k\omega_{k'})^{1/2} d^n_{E,j} d_{j,1} \hat{a}_{k'} \hat{a}_k}{(\omega_{E,1} - \omega_k - \omega_{k'} + i\epsilon)(\omega_{E,j} - \omega_{k'} + i\epsilon)}|R^l\rangle\right], \tag{10.48}$$

where the intermediate states are assumed to be $|E_j\rangle$, $j = 2, 3$, with $d^n_{E,j}$ and $d_{j,1}$ being the associated transition dipole moments. As in the two pulse control case, Eq. (10.48) indicates that in general the final state is an entangled state between the molecule and the light fields. However, and interestingly, this is not the case in the special "$\omega_1+\omega_2$" vs. "$\omega_2+\omega_1$" scheme. Here there are only two near-resonant and dominant intermediate bound states $|E_2\rangle$ and $|E_3\rangle$ of energy E_2 and E_3, satisfying $\omega_{E,2} = \omega_{3,1}$. Hence,

$$\omega_{E,2} - \omega_{k'} = \omega_{3,1} - \omega_{k'} = \omega_k - \omega_{E,3} \,, \tag{10.49}$$

where in obtaining the second equality we used the two-photon resonance condition, $\omega_{E,1} - \omega_k - \omega_{k'} = 0$. Using Eq. (10.49) and manipulating the order of the sum in Eq. (10.48), the $E, \boldsymbol{m}$-continuum component of the wave function is found to be

$$\lim_{t\to\infty} \langle E, \boldsymbol{m}^- | \Psi(t)\rangle = \frac{e^{-iEt/\hbar}}{2\epsilon_0 \hbar V} \sum_{kk'} \left[\frac{(\omega_k \omega_{k'})^{1/2} \hat{a}_{k'} \hat{a}_k | R^l\rangle}{(\omega_{E,1} - \omega_k - \omega_{k'} + i\epsilon)(\omega_{E,2} - \omega_{k'} + i\epsilon)} \right] \times \left[d^m_{E,3} d_{3,1} - d^m_{E,2} d_{2,1} \right] . \tag{10.50}$$

Here, the first term on the right-hand side of Eq. (10.50) represents the contribution from the first path through the intermediate state $|E_3\rangle$, and the second term represents the contribution from the second path associated with $|E_2\rangle$. Hence, without any restriction on the form of $|R^l\rangle$, the two terms in Eq. (10.50) arising from two excitation pathways are seen to be identical except the c-number coefficients, that is, the final molecular state is of the unentangled form of Eq. (10.23), so that this version of incoherent interference control holds for any type of (classical or quantized) radiation field.

A related argument holds for the incoherent interference control scenario discussed in Section 6.6.2. Here the first pathway to the final state is direct excitation from an initial state to a target state, and the second pathway begins with an excitation from the initial state to the same target state, followed by back and forth transitions (induced by a strong field) between the target state and a third intermediate state (see Figure 6.26). Classifying independent pathways in this way is just a convenient zero-order picture for understanding the associated quantum effects. That is, these "back and forth" transitions are simply fictitious excitation pathways in a perturbation theory interpretation of the excitation from the initial state to a dressed target state. Given that it is impossible to distinguish between fictitious pathways, even in principle, the maximal degree of pathway indistinguishability is automatically guaranteed, as is the associated quantum interference.

Hence, incoherent interference control is markedly different from traditional coherent control, in that the former utilizes a specific kind of quantum interference that results from the absolute indistinguishability of multiple excitation pathways to the same target state. It can therefore be anticipated that incoherent interference control schemes will be applicable in cases involving highly quantum states of light.

10.3
Quantum Field Control of Entanglement

Entanglement is a central attribute of quantum mechanics and widely regarded as a fundamental resource for quantum computing, quantum dense coding and quantum teleportation [188, 370–372]. Three main types of entanglement have been implemented in the laboratory: entanglement between photon states [373], entanglement between matter states [374], and entanglement between atoms and single photon states [375, 376].

In this section we show how to use nonclassical light [272, 377] to control the makeup of entangled states [378–382]. We first examine the creation and control of matter–radiation entangled states and then extend the treatment to the creation of a chain of entangled molecules.

10.3.1
Light–Matter Entanglement

As a platform for examining the effects of nonclassical light [55, 383, 384] we again use the one vs. two coherent control scenario. This time the radiation state chosen is the so-called radiative "cat-state" [385, 386] defined as

$$|C\rangle \equiv \frac{1}{\sqrt{2}}(|\alpha\rangle + |-\alpha\rangle)\,. \tag{10.51}$$

It is called a "cat-state" following Schrödinger's famous "linear combination of live-cat and dead-cat" gedanken experiment [387] meant to highlight the difficulties associated with the notion of a quantum superposition of two radically different states. In a similar fashion the cat-state of Eq. (10.51) is a linear combination of two radically different $|\pm\alpha\rangle$ coherent states.

With this choice of the radiation state, the matter + radiation initial state to be used in the one vs. two coherent control scenario is given as

$$|\Psi^i\rangle = \frac{1}{\sqrt{2}}(|\alpha_\omega\rangle + |-\alpha_\omega\rangle)\,|\alpha_{\omega/2}\rangle|E_i\rangle\,. \tag{10.52}$$

Specializing our objective to controlling the direction of motion of photofragments, that is, identifying $\boldsymbol{m} = \hat{\boldsymbol{k}}$, the final state $|\Psi^f\rangle$ is (using the rotating wave approximation)

$$\begin{aligned}|\Psi^f\rangle = (i\pi)^{\frac{1}{2}} \sum_{\hat{k}} |E,\hat{\boldsymbol{k}}^-\rangle \Big\{ \Big[\bar{\epsilon}(\omega)\langle E,\hat{\boldsymbol{k}}^-|\mathrm{d}|E_i\rangle - \bar{\epsilon}^2(\omega/2)\,\langle E,\hat{\boldsymbol{k}}^-|D|E_i\rangle\Big]\,|\alpha_\omega\rangle \\ + \Big[-\bar{\epsilon}(\omega)\langle E,\hat{\boldsymbol{k}}^-|\mathrm{d}|E_i\rangle - \bar{\epsilon}^2(\omega/2)\,\langle E,\hat{\boldsymbol{k}}^-|D|E_i\rangle\Big]\,|-\alpha_\omega\rangle\Big\}\,|\alpha_{\omega/2}\rangle\,.\end{aligned} \tag{10.53}$$

Thus our choice of an initial cat-state has produced a matter + radiation entangled final state. To see this more clearly we rewrite Eq. (10.53) as

$$|\Psi^f\rangle = (i\pi)^{\frac{1}{2}}|\alpha_{\omega/2}\rangle \sum_{\hat{k}} |E, \hat{\boldsymbol{k}}^-\rangle \Big\{\bar{\epsilon}(\omega)\langle E, \hat{\boldsymbol{k}}^-|\mathrm{d}|E_i\rangle \left[|\alpha_\omega\rangle - |-\alpha_\omega\rangle\right]$$
$$-\bar{\epsilon}^2(\omega/2)\langle E, \hat{\boldsymbol{k}}^-|D|E_i\rangle \left[|\alpha_\omega\rangle + |-\alpha_\omega\rangle\right]\Big\} . \qquad (10.54)$$

We now note that the two-photon amplitude $\langle E, \hat{\boldsymbol{k}}^-|D|E_i\rangle$ is symmetric about a coordinate inversion (i.e., $\hat{\boldsymbol{k}}_f \leftrightarrow \hat{\boldsymbol{k}}_b$) where f denotes the forward direction and b denotes the backward direction, and the one-photon amplitude $\langle E, \hat{\boldsymbol{k}}^-|\mathrm{d}|E_i\rangle$ is antisymmetric with respect to inversion. Writing the product of the one-photon and two-photon amplitude in terms of its absolute value and phase,

$$\langle E, \hat{\boldsymbol{k}}^-|D|E_i\rangle\langle E, \hat{\boldsymbol{k}}^-|\mathrm{d}|E_i\rangle = |B(E, \hat{\boldsymbol{k}}|E_i)|\exp(i\beta_{\hat{k}}) .$$

It follows from the above that

$$|B(E, \hat{\boldsymbol{k}}_f|E_i)| = |B(E, \hat{\boldsymbol{k}}_b|E_i)| ,$$
$$\beta_{\hat{k}} \equiv \beta_{\hat{k}_f} = \pi - \beta_{\hat{k}_b} . \qquad (10.55)$$

We now assume that only the forward, "f", or backward, "b", directions are allowed. This can be achieved, for example, in an ensemble of (ultra) cold molecules [388] (say Ca_2^+) trapped in an optical lattice along the z direction. A pulse polarized in the x direction, composed of an ω cat-state and an $\omega/2$ coherent state, is allowed to propagate along the z direction and to dissociate the molecules it encounters as it moves from a z_i site to the next (at z_{i+1}). Each dissociation event yields "forward" moving fragments or "backward" moving fragments. For a Ca_2^+ molecule located at z_i, the forward direction is defined as the Ca atom moving to a neighboring site at $[z_i, \delta x]$, and the Ca^+ ion moving to the opposite $[z_i, -\delta x]$ site. The "backward" direction is obtained by interchanging the Ca and the Ca^+ positions. Quite clearly one simply controls the direction of motion of the electron as the Ca_2^+ molecule falls apart.

In this "pseudo" one-dimensional setup we can simplify the expressions by tuning the field intensities such that

$$\left|\frac{\epsilon(\omega)}{\epsilon^2(\omega/2)}\right| = \frac{|\langle E^-|D|E_i\rangle|}{|\langle E^-|\mathrm{d}|E_i\rangle|} . \qquad (10.56)$$

With this choice of the intensities we have that

$$|\Psi^f\rangle = (i\pi)^{\frac{1}{2}}|\alpha_{\omega/2}\rangle\,|\epsilon(\omega)\langle E^-|\mathrm{d}|E_i\rangle|$$
$$\times\Big\{|E, f^-\rangle\Big[(|\alpha_\omega\rangle - |-\alpha_\omega\rangle) - e^{i(\phi_\omega - 2\phi_{\omega/2} + \beta)}(|\alpha_\omega\rangle + |-\alpha_\omega\rangle)\Big]$$
$$+ |E, b^-\rangle\Big[(|\alpha_\omega\rangle - |-\alpha_\omega\rangle) + e^{i(\phi_\omega - 2\phi_{\omega/2} + \beta)}(|\alpha_\omega\rangle + |-\alpha_\omega\rangle)\Big]\Big\} .$$
$$(10.57)$$

If we now choose the field's phase difference such that

$$\phi_{\omega} - 2\phi_{\omega/2} = -\beta \, , \tag{10.58}$$

we have from Eq. (10.57) that

$$|\Psi^f\rangle = -2(i\pi)^{\frac{1}{2}} |\epsilon(\omega)\langle E^-|\mathrm{d}|E_i\rangle| \, |\alpha_{\omega/2}\rangle \{|E, f^-\rangle|-\alpha_{\omega}\rangle - |E, b^-\rangle|\alpha_{\omega}\rangle\} \, . \tag{10.59}$$

Equation (10.59) describes a matter + radiation entangled cat-state, where the state is a superposition of a material state of photofragment directed in the *forward* direction being multiplied by the $|-\alpha_{\omega}\rangle$ radiative coherent state, and a material state of photofragments going in the *backward* direction being accompanied by the $|\alpha_{\omega}\rangle$ state.

The makeup of the matter + radiation entangled cat-state produced in this scenario is completely controllable. For example, if we choose the field's phase difference such that

$$\phi_{\omega} - 2\phi_{\omega/2} = \pi - \beta \, , \tag{10.60}$$

then the resulting state will be

$$|\Psi^f\rangle = 2(i\pi)^{\frac{1}{2}} |\epsilon(\omega)\langle E^-|\mathrm{d}|E_i\rangle| \, \alpha_{\omega/2}\rangle \{|E, f^-\rangle|\alpha_{\omega}\rangle - |E, b^-\rangle|-\alpha_{\omega}\rangle\} \, . \tag{10.61}$$

Our ability to control the entangled state comes from the one-photon vs. two-photon interference term which is not zero because, as Eqs. (10.59) and (10.61) show, a measurement of the radiative state ($|\alpha_{\omega}\rangle$ or $|-\alpha_{\omega}\rangle$) does not tell us whether the one-photon or the two-photon pathway was followed. However, as in all two-component entangled states, a measurement of one component (the radiative one) collapses the other (material) component. Thus, if we tune the radiative phase difference as in Eq. (10.58) and we measure that the radiation state is say, $|\alpha_{\omega}\rangle$, then the photofragments must have recoiled in the backward direction. The important aspect here is that we can control the entanglement at will. For example, by tuning the radiative phase difference as in Eq. (10.60) a measurement of an $|\alpha_{\omega}\rangle$ radiation state implies that the photofragments must have recoiled in the forward direction.

10.3.2
Creating Entanglement between a Chain of Molecules and a Radiation Field

The treatment given above can be extended to create a *sequence* of entangled atom-ion pairs. Choosing the phases according to Eq. (10.58) we first dissociate $Ca_2{}^+$ molecule number 1, thereby creating an entangled $|\Psi_1^f\rangle$ state of Eq. (10.59) with respect to the first pair of atoms,

$$\begin{aligned} |\Psi_1^f\rangle = &-2(i\pi)^{\frac{1}{2}} e^{-i\omega z_1/c} |\epsilon(\omega)\langle E^-|\mathrm{d}|E_{i_1}\rangle| \\ &\{|E, f_1^-\rangle|-\alpha_{\omega}\rangle - |E, b_1^-\rangle|\alpha_{\omega}\rangle\} \, |\alpha_{\omega/2}\rangle \, . \end{aligned} \tag{10.62}$$

We now apply this $|\Psi_1^f\rangle$ state to a second Ca_2^+ diatomic in the array for which our initial state is therefore,

$$|\Psi_2^i\rangle = 2(i\pi)^{\frac{1}{2}} e^{-i\omega z_1/c} |\epsilon(\omega)\langle E^-|d|E_{i_1}\rangle| \\ \{-|E, f_1^-\rangle|-\alpha_\omega\rangle + |E, b_1^-\rangle|\alpha_\omega\rangle\} |\alpha_{\omega/2}\rangle|E_{i_2}\rangle \,. \tag{10.63}$$

The final state for molecule 1 and molecule 2 thus becomes

$$|\Psi_{1,2}^f\rangle = 4i\pi e^{-i\omega(z_1+z_2)/c} |\epsilon^2(\omega)\langle E^-|d|E_{i_2}\rangle\langle E^-|d|E_{i_1}\rangle| \\ \{|E, f_2^-\rangle|E, f_1^-\rangle|-\alpha_\omega\rangle + |E, b_2^-\rangle|E, b_1^-\rangle|\alpha_\omega\rangle\} |\alpha_{\omega/2}\rangle \,. \tag{10.64}$$

We have thus entangled *two* atom-ion pairs and the radiation field. Explicitly, if we detect the radiation field to be in the $|-\alpha_\omega\rangle$ state, then both molecules 1 and 2 must have been dissociated in the forward direction, whereas if we detect the radiation field to be in the $|\alpha_\omega\rangle$ state then molecules 1 and 2 must have been dissociated

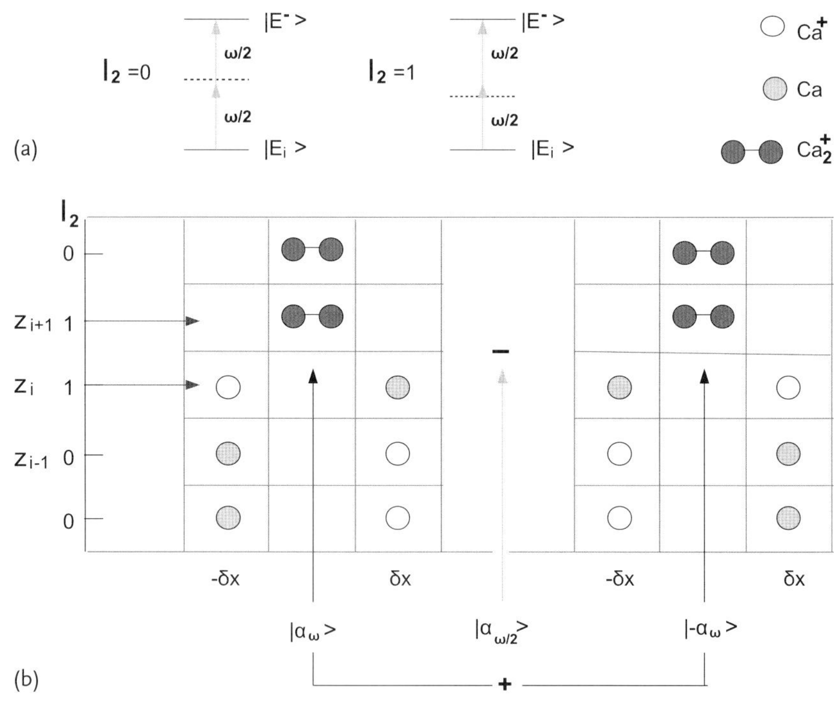

Figure 10.1 The proposed optical lattice setup: The lower panel shows the method of writing an arbitrary word composed of a sequence of entangled qbits using a set of Ca_2^+ molecules placed along a line in an optical lattice, irradiated by a combination of a $|\alpha_\omega\rangle + |-\alpha_\omega\rangle$ and $|\alpha_{\omega/2}\rangle$ field states. Upper panel: The modulation of the intensity I_2 of an auxiliary laser induces an AC Stark shift of the intermediate resonance of the two photon dissociation process. By setting the auxiliary laser to the off ($I_2 = 0$) or on ($I_2 = 1$) states one is able to shift the phase of the two-photon matrix element by π, thereby changing the direction of dissociation.

in the backward direction. Alternatively, if we detect the first Ca atom to be in site $[z_i, \delta x]$, then the second Ca atom must also be in site $[z_{i+1}, \delta x]$ and the radiation field must be in the $|-\alpha_\omega\rangle$ state. Moreover, we can, if we wish, interchange between the above $|\alpha_\omega\rangle$ and $|-\alpha_\omega\rangle$ radiation states (or interchange $f \leftrightarrow b$) by choosing the field phases according to Eq. (10.60).

In a similar manner one can continue to act on more and more molecules in the sequence, creating states in which n atom-ion pairs plus one radiation field are entangled together as

$$|\Psi^f_{1,\dots,n}\rangle = (4i\pi)^{\frac{n}{2}} |\alpha_{\omega/2}\rangle e^{-i\omega(z_1+z_2+\cdots+z_n)/c} \left|\epsilon^n(\omega) \prod_{m=1}^{n} \langle E^-|\mathrm{d}|E_{i_m}\rangle\right| \left\{\left[(-1)^n \prod_{m=1}^{n} |E, f_m^-\rangle\right] |-\alpha_\omega\rangle + \left[\prod_{m=1}^{n} |E, b_m^-\rangle\right] |\alpha_\omega\rangle\right\} . \tag{10.65}$$

Alternatively, as shown in Figure 10.1 the phase β can be varied from site to site using an auxiliary classical field which induces an AC Stark shift in an intermediate Ca_2^+ level. If we assume for simplicity that the two-photon amplitude $\langle E, \hat{\boldsymbol{k}}^- |D| E_i\rangle$ is dominated by a single intermediate resonance,

$$\langle E, \hat{\boldsymbol{k}}^- |D| E_i\rangle \approx \frac{\langle E, \hat{\boldsymbol{k}}^- |\mathrm{d}| E_n\rangle \langle E_n |\mathrm{d}| E_i\rangle}{E_i + \hbar\omega/2 - E_n} , \tag{10.66}$$

then by shifting the intermediate level position E_n from being above $E_i + \hbar\omega/2$ to being below that value, we effectively add a π phase shift to the two-photon process. Thus, the final state to be written, such as,

$$|\Psi^f_{1,\dots,n}\rangle = (4i\pi)^{\frac{n}{2}} |\alpha_{\omega/2}\rangle e^{-i\omega(z_1+z_2+\cdots+z_n)/c} \left|\epsilon^n(\omega) \prod_{m=1}^{n} \langle E^-|\mathrm{d}|E_{i_m}\rangle\right| \{(-1)^n|-\alpha_\omega\rangle|E, f_1^-\rangle|E, f_2^-\rangle|E, b_3^-\rangle \dots + |\alpha_\omega\rangle|E, b_1^-\rangle|E, b_2^-\rangle|E, f_3^-\rangle \dots\} \tag{10.67}$$

is controlled by the auxiliary field (e.g., via an intensity modulating pulse shaper).

The result of this scenario is a sequence of atom–ion pairs entangled with one another as well as with the light field. This constitutes an entirely new entangled species, and a possible resource in quantum information. Explicit implementations of quantum computation with this scenario are being developed. We note that the scenario is expected to be relatively immune to decoherence, since it relies merely on the positional identity of an atom vs. an ion (as distinct from coherence). We note also that symmetry breaking in the one- vs. two-photon scenarios in optical lattices has been shown to be highly robust to decoherence [389].

10.4 Control of Entanglement in Quantum Field Chiral Separation

We now discuss another scenario [382] in which we use quantum fields to control entanglement, this time the process is associated with the enantio-conversion process to be discussed in Section 14.5 in which we devise an adiabatic "enantio-converter" where we transfer populations from an initial left-handed state, say $|1_L\rangle$, to a right-handed state, say $|3_D\rangle$. The process uses delocalized states, denoted as $|2_S\rangle$ and $|2_A\rangle$, as intermediates. The entanglement in the converter can be realized by replacing one of the four light fields by the nonclassical $|\mathcal{C}\rangle$ cat-state. Since the two coherent $|\pm\alpha\rangle$ components of opposite quantum phases drive the system along the two mirror-imaged paths (see Figure 10.2), the quantum field enantio-converter populates the entangled radiation–matter state

$$|\psi\rangle = \frac{|3_L\rangle|\alpha\rangle + |3_D\rangle|-\alpha\rangle}{\sqrt{2}} \,. \tag{10.68}$$

In order to understand better what determines the superposition state we expand the system wave function in all six states,

$$|\Psi(t)\rangle = \sum_{k=1}^{6}\sum_{n=0}^{\infty} c_{k,n}(t) e^{-i(\omega_k + n\omega_0)t}|k, n\rangle \,, \tag{10.69}$$

where $k = \{1_L, 1_D, 2_S, 2_A, 3_L, 3_D\}$ are the molecular states and n denotes the number of photons in the nonclassical dump field of frequency ω_0. The expansion coefficients are obtained by solving the Schrödinger equation,

$$\begin{aligned}
i\dot{c}_{1L,n} &= \left(\Omega_S^p/2\right) c_{2S,n} + \left(\Omega_A^p/2\right) c_{2A,n} \,, \\
i\dot{c}_{1D,n} &= \left(\Omega_S^p/2\right) c_{2S,n} - \left(\Omega_A^p/2\right) c_{2A,n} \,, \\
i\dot{c}_{2S,n} &= \left(\Omega_S^{p*}/2\right) (c_{1L,n} + c_{1D,n}) + \left(\Omega_S^{d*}/2\right) (c_{3L,n} + c_{3D,n}) \,, \\
i\dot{c}_{2A,n} &= \left(\Omega_A^{p*}/2\right) (c_{1L,n} - c_{1D,n}) \\
&\quad + (g_Q/2)\sqrt{n+1}(c_{3L,n+1} - c_{3D,n+1}) \,, \\
i\dot{c}_{3L,n} &= \left(\Omega_S^d/2\right) c_{2S,n} + (g_Q/2)\sqrt{n} c_{2A,n-1} \,, \\
i\dot{c}_{3D,n} &= \left(\Omega_S^d/2\right) c_{2S,n} - (g_Q/2)\sqrt{n} c_{2A,n-1} \,.
\end{aligned} \tag{10.70}$$

The numerical solution for $|\Psi(t)\rangle$ enables us to obtain the density operator $\rho(t) = |\Psi(t)\rangle\langle\Psi(t)|$. The system entanglement is related to the diagonal matrix elements $\rho_{k,k}^{\beta,\beta}(t) \equiv \langle k, \beta|\rho(t)|k, \beta\rangle$ ($k = 1$–6) in the (overcomplete) coherent state $|\beta\rangle$ basis [377], evaluated in terms of the Fock state matrix elements as,

$$\rho_{k,k}^{\beta,\beta}(t) = \sum_{n=0}^{\infty}\sum_{m=0}^{\infty} \langle\beta|n\rangle c_k^n(t)(c_k^m(t))^*\langle m|\beta\rangle \,. \tag{10.71}$$

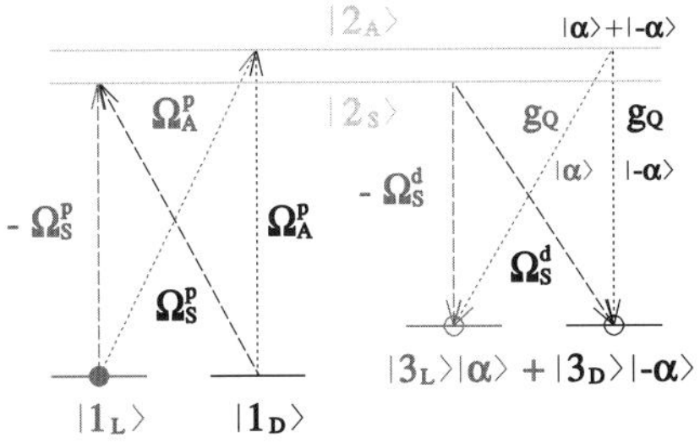

Figure 10.2 The nonclassical enantio-converter. In the classical field converter, an initial (left-handed) $|1_L\rangle$ state is transformed into the $|3_D\rangle$ state. The nonclassical converter uses the $|\alpha\rangle + |-\alpha\rangle$ cat-state to produce the $|3_L\rangle|\alpha\rangle+|3_D\rangle|-\alpha\rangle$ entangled state. Taken from [382]. (Reprinted with permission. Copyright 2007 American Physical Society.)

We can plot the elements as a function of the real and imaginary parts of β, that is, $\beta_x = \text{Re}\beta$ and $\beta_y = \text{Im}\beta$. In Figure 10.3, we display the final state matrix elements $\rho_{L,L}^{\beta,\beta}$, $\rho_{D,D}^{\beta,\beta}$ obtained by solving Eq. (10.70) for the Rabi frequencies, $\Omega_S^p(t) = \Omega_A^p(t) = \Omega_0 f(t-2\tau)$, $\Omega_S^d(t) = \Omega_0 f(t)$ and $g_Q(t) = g_0 f(t)$, with $f(t) = \exp[-t^2/\tau^2]$, $\tau = 30$, $\Omega_0 = 2$ and $g_0 = 0.2$. Figure 10.3a,c shows the results of driving the system with the $|C\rangle$ cat-state, with $|\alpha|^2 = 7$. Starting from the $|1_L\rangle$ state, we prepare the entangled $|\Psi\rangle \approx |3_L\rangle|\alpha\rangle + |3_D\rangle|-\alpha\rangle$ state.

In Figure 10.3 (right column), we show the results of driving the system using the $|n=7\rangle$ Fock state, where we obtain the final entangled state,

$$
\begin{aligned}
|\Psi\rangle &\approx |3_L\rangle\left(|n+1\rangle + |n\rangle\right) + |3_D\rangle\left(|n+1\rangle - |n\rangle\right) \\
&\approx |3_S\rangle|n+1\rangle + |3_A\rangle|n\rangle \,.
\end{aligned}
\tag{10.72}
$$

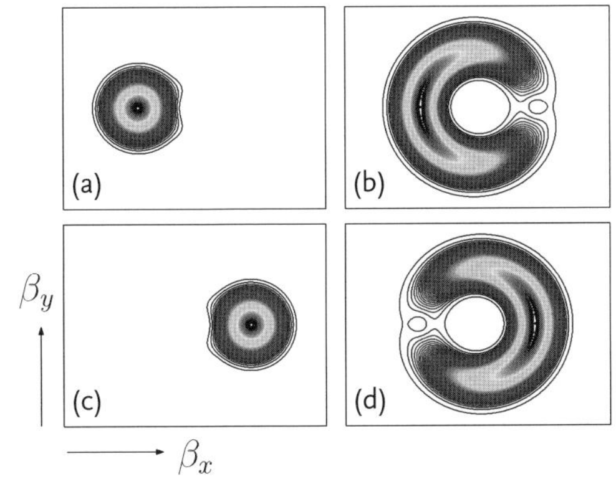

Figure 10.3 Photon distributions $\rho_{L,L}^{\beta,\beta}$ (a,b) and $\rho_{D,D}^{\beta,\beta}$ (c,d) in the quantum field enantio-converter for entangled chiral states. The distributions are presented in the $\Delta\beta_x = \pm 6.4$, $\Delta\beta_y = \pm 5$ range. (a,c) The driving field is given by the cat-state $|C\rangle$ with $|\alpha|^2 = 7$. (b,d) The driving field is given by the $|n = 7\rangle$ Fock state. Taken from [448, Figure 16]. (Reprinted with permission. Copyright 2007 American Physical Society.)

We can see the formation of the $|n = 8\rangle \pm |n = 7\rangle$ states. These states with "banana-like" distributions are related to the multiphase $\sum_n |e^{i2\pi/n}\alpha\rangle$ coherent states generated by the nonlinear Kerr effect [390].

11
Coherent Control beyond the Weak-Field Regime: Bound States and Resonances

In Chapters 2 and 3 we treated n-photon molecular dissociation and coherent control in the weak-field regime and showed that control arises through quantum interference effects. The moderately strong-field regime treated in this chapter is far more complicated because the higher laser powers imply that nth order perturbation theory is no longer valid. The "moderately strong" field regime discussed here, in contradistinction to the "strong" field regime discussed in Chapter 15, is taken here to mean that, although nth order perturbation theory is inappropriate, the (radiation plus matter) wave functions can still be expressed in terms of a moderate number of radiation-free states. In the moderately strong field regime the use of the "field-dressed" basis of states, discussed in Chapter 15, though an option, is not an absolute necessity.

The formalism is best divided into a discussion of control involving bound states, treated in this section, and control involving the continuum, discussed in Chapter 12 and Section 13.1 of this chapter.

11.1
Adiabatic Population Transfer

In this section we show how strong fields can be used to adiabatically control population transfer between bound states and, especially, how to achieve *complete* population transfer between such states. In doing so we describe realistic methods for control, introduce a number of useful methods in strong-field control, and pave the way for a discussion of adiabatic population transfer in problems involving the continuum.

The ability to induce complete population transfer between states is intimately linked to the concept of a "trapped" state; that is, a state that remains invariant under the action of incident CW radiation. These states, which only change when the field changes, often enable one to guide a quantum system from one state to another, a phenomenon known as adiabatic passage (AP), first introduced in the context of magnetic resonance [391].

Adiabatic passage was most commonly exploited in two and three level systems [321, 392–410]. In particular, the three-level "Λ system" (see Figure 11.1), in

Quantum Control of Molecular Processes, Second Edition. Moshe Shapiro and Paul Brumer

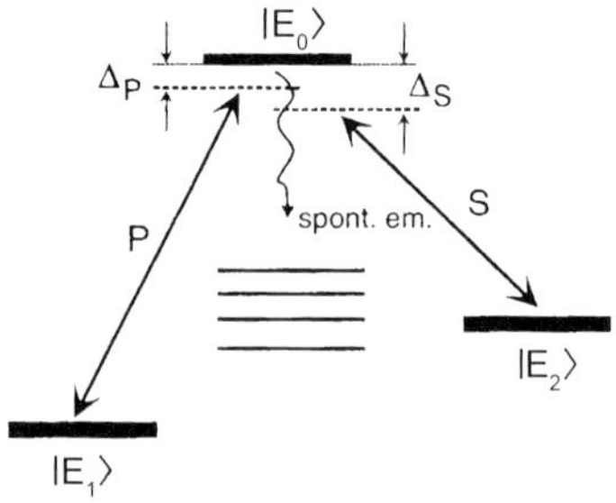

Figure 11.1 The Λ configuration associated with adiabatic passage. Taken from [321, Figure 1]. (Reprinted with permission. Copyright 1998 American Physical Society.)

which one (initially unpopulated) level is higher in energy than the two other levels, was extensively investigated theoretically [321, 322, 401, 402, 408, 411] and experimentally [407–409]. Bergmann *et al.* [407] were the first to experimentally demonstrate, in a process called "stimulated Raman adiabatic passage" (STIRAP), that adiabatic passage in a Λ system enables the *complete* transfer of population from one level to another under certain conditions. A related phenomenon, called "electromagnetically induced transparency" (EIT), in which a medium is made transparent at a certain transition frequency, was investigated by Harris *et al.* [412–415] and others [416, 417]. Moreover, it is possible to use trapped states to cause "lasing without inversion" (LWI), as was shown by Harris [418], Scully [419], Kocharovskaya [420], and others [421, 422]. These phenomena, as seen below, share a common basis insofar as they rely upon interference effects associated with the preparation and evolution of specific superposition states.

11.1.1
Adiabatic States, Trapping, and Adiabatic Following

In order to understand these phenomena we first consider a three-level Λ system, composed of a lowest energy state $|E_1\rangle$, coupled radiatively to an intermediate state $|E_0\rangle$, which in turn is coupled radiatively to a third state $|E_2\rangle$ where $E_0 > E_2 > E_1$. The coupling is due to the combined action of two laser pulses of central frequencies ω_1 and ω_2. We assume (see Figure 11.1) that ω_1, "the pump pulse", is in near-resonance with a transition from $|E_1\rangle$ to the bound state $|E_0\rangle$ and that ω_2, "the dump pulse" is in near-resonance with the transition from $|E_0\rangle$ to $|E_2\rangle$.

In most applications one chooses the two frequencies to fulfill the *two-photon* resonance condition,

$$\omega_1 - \omega_2 = (E_2 - E_1)/\hbar \, . \tag{11.1}$$

Writing the total Hamiltonian in the dipole approximation as

$$H = H_M - 2\boldsymbol{d}_1 \cdot \hat{\boldsymbol{\epsilon}}_1 \mathcal{E}_1(t) \cos(\omega_1 t) - 2\boldsymbol{d}_2 \cdot \hat{\boldsymbol{\epsilon}}_2 \mathcal{E}_2(t) \cos(\omega_2 t) \, , \tag{11.2}$$

where $\hat{\epsilon}_1$ and $\hat{\epsilon}_2$ are the polarization directions, $\mathcal{E}_1(t)$ and $\mathcal{E}_2(t)$ are "slowly varying" electric field amplitudes and $\boldsymbol{d}_1$ and $\boldsymbol{d}_2$ are the transition dipoles, we can solve the time-dependent Schrödinger equation, $i\hbar\partial|\Psi(t)\rangle/\partial t = H|\Psi(t)\rangle$, by expanding the total wave function as

$$|\Psi(t)\rangle = b_1(t)|E_1\rangle e^{-iE_1t/\hbar} + b_0(t)|E_0\rangle e^{-iE_0t/\hbar} + b_2(t)|E_2\rangle e^{-iE_2t/\hbar} , \quad (11.3)$$

where

$$[E_i - H_M]|E_i\rangle = 0 , \quad (i = 0, 1, 2) . \quad (11.4)$$

Doing so gives the three-state version of Eq. (2.3):

$$\frac{db_1}{dt} = i\Omega_1(t)e^{-i\Delta_1 t}b_0(t) , \quad (11.5a)$$

$$\frac{db_0}{dt} = i\Omega_1^*(t)e^{i\Delta_1 t}b_1(t) + i\Omega_2^*(t)e^{i\Delta_2 t}b_2(t) , \quad (11.5b)$$

$$\frac{db_2}{dt} = i\Omega_2(t)e^{-i\Delta_2 t}b_0(t) , \quad (11.5c)$$

where the Rabi frequencies Ω_i and detunings Δ_i are

$$\Omega_i(t) \equiv \langle E_i|d_i \cdot \hat{\epsilon}_i|E_0\rangle\mathcal{E}_i(t)/\hbar , \quad \Delta_i \equiv (E_0 - E_i)/\hbar - \omega_i . \quad (11.6)$$

The two-photon resonance condition (Eq. (11.1)) implies that

$$\Delta_2 = \Delta_1 . \quad (11.7)$$

Defining a vector of coefficients $\underline{b} \equiv (b_1, b_0, b_2)^T$ we write Eq. (11.5) in matrix notation as

$$\frac{d}{dt}\underline{b}(t) = i\underline{\underline{H}} \cdot \underline{b}(t) , \quad (11.8)$$

where

$$\underline{\underline{H}} = \begin{pmatrix} 0 & \Omega_1(t)e^{-i\Delta_1 t} & 0 \\ \Omega_1^*(t)e^{i\Delta_1 t} & 0 & \Omega_2^*(t)e^{i\Delta_2 t} \\ 0 & \Omega_2(t)e^{-i\Delta_2 t} & 0 \end{pmatrix} . \quad (11.9)$$

We first derive the adiabatic approximation to Eq. (11.8). The derivation begins by diagonalizing the $\underline{\underline{H}}$ matrix,

$$\underline{\underline{H}} \cdot \underline{\underline{U}} = \underline{\underline{U}} \cdot \underline{\underline{\lambda}} , \quad (11.10)$$

where $\underline{\underline{\lambda}}$ is a diagonal eigenvalue matrix. We then transform the Schrödinger equation to the adiabatic representation by operating on Eq. (11.8) with $\underline{\underline{U}}^\dagger(t)$, the Hermitian adjoint of $\underline{\underline{U}}$. Defining

$$\underline{a}(t) = \underline{\underline{U}}^\dagger(t) \cdot \underline{b}(t) , \quad (11.11)$$

and using the unitarity property of $\underline{\underline{\mathrm{U}}}$:

$$\underline{\underline{\mathrm{U}}} \cdot \underline{\underline{\mathrm{U}}}^{\dagger} = \underline{\underline{\mathrm{U}}}^{\dagger} \cdot \underline{\underline{\mathrm{U}}} = \underline{\underline{\mathrm{I}}} \,, \tag{11.12}$$

we have that

$$\frac{d}{dt}\underline{\mathrm{a}} = \left\{ i\underline{\underline{\lambda}}(t) + \underline{\underline{\mathrm{A}}} \right\} \cdot \underline{\mathrm{a}} \,, \tag{11.13}$$

where

$$\underline{\underline{\mathrm{A}}} \equiv \frac{d\underline{\underline{\mathrm{U}}}^{\dagger}(t)}{dt} \cdot \underline{\underline{\mathrm{U}}} \,, \tag{11.14}$$

is the "nonadiabatic coupling matrix".

In the adiabatic approximation one neglects $\underline{\underline{\mathrm{A}}}$. This can be done whenever the rate of change of $\underline{\underline{\mathrm{U}}}$ with time is slow, or more specifically, whenever

$$\left(\underline{\underline{\mathrm{A}}}\right)_{i,j} \ll \left|\lambda_i - \lambda_j\right| \,. \tag{11.15}$$

A discussion of the range of validity of the adiabatic approximation, and methods of improving upon it, is given in Section 13.1.

When the adiabatic approximation is adopted, Eq. (11.13) becomes

$$\frac{d}{dt}\underline{\mathrm{a}} = i\underline{\underline{\lambda}}(t) \cdot \underline{\mathrm{a}}(t) \,, \tag{11.16}$$

yielding the adiabatic solutions

$$\underline{\mathrm{a}}(t) = \exp\left\{ i \int_0^t \underline{\underline{\lambda}}(t')dt' \right\} \cdot \underline{\mathrm{a}}(0) \,. \tag{11.17}$$

Given $\underline{\mathrm{a}}(t)$ and $\underline{\underline{\mathrm{U}}}(t)$ one can generate the desired $\underline{\mathrm{b}}(t)$ via Eq. (11.11). For the case in Eq. (11.9) it is easy to show that $\underline{\underline{\lambda}}$ comprises the following eigenvalues:

$$\begin{aligned} \lambda_1 &= 0 \,, \\ \lambda_{2,3}(t) &= \pm\left[|\Omega_1(t)|^2 + |\Omega_2(t)|^2\right]^{\frac{1}{2}} \,, \end{aligned} \tag{11.18}$$

where the plus sign applies to λ_2 and the minus sign applies to λ_3. Because of their time dependence, these adiabatic eigenvalues are also called "quasi-energies".

We now seek the adiabatic states, denoted $|\lambda_i(t)\rangle$, which are the $\underline{\mathrm{b}}(t)$ vectors obtained from the $\underline{\mathrm{a}}(t)$ eigenvectors using Eq. (11.11), that is, $\underline{\mathrm{b}}(t) = \underline{\underline{\mathrm{U}}}(t) \cdot \underline{\mathrm{a}}(t)$. Solving explicitly for $\underline{\mathrm{U}}^{(1)}(t)$, the eigenvector corresponding to $\lambda_1 = 0$, that is

$$\underline{\underline{\mathrm{H}}} \cdot \underline{\mathrm{U}}^{(1)} = 0 \,, \tag{11.19}$$

together with Eq. (11.9), yields

$$\underline{\mathrm{U}}^{(1)} = \frac{1}{[|\Omega_1|^2 + |\Omega_2|^2]^{\frac{1}{2}}} \begin{pmatrix} \Omega_2^* \\ 0 \\ -\Omega_1 e^{-i(\Delta_2-\Delta_1)t} \end{pmatrix} = \begin{pmatrix} \cos\theta(t) \\ 0 \\ -e^{i\chi(t)}\sin\theta(t) \end{pmatrix} . \tag{11.20}$$

Equation (11.20) involves both the "mixing" angle $\theta(t)$, where

$$\theta(t) = \arctan\left(\left|\frac{\Omega_1(t)}{\Omega_2(t)}\right|\right) , \tag{11.21}$$

and the "azimuthal angle" $\chi(t)$, where

$$\chi(t) \equiv (\Delta_1 - \Delta_2)t - \phi_2(t) + \phi_1(t) = \phi_1(t) - \phi_2(t) , \tag{11.22}$$

and where $\phi_i(t)$ are the phases of $\Omega_i(t)$:

$$\Omega_i(t) \equiv |\Omega_i(t)|e^{i\phi_i(t)} , \quad i = 1, 2 . \tag{11.23}$$

The last equality in Eq. (11.22) is a result of the two-photon resonance condition (Eq. (11.7)).

Thus, the adiabatic state associated with the $\lambda_1 = 0$ eigenvector is

$$|\lambda_1(t)\rangle = \cos\theta(t)e^{-iE_1t/\hbar}|E_1\rangle - e^{i\chi(t)}\sin\theta(t)e^{-iE_2t/\hbar}|E_2\rangle . \tag{11.24}$$

It follows from Eqs. (11.6) and (11.20) that $\underline{\mathrm{U}}^{(1)}(t)$ does not change with time for CW fields. It evolves in time, in accord with Eq. (11.20), only when the (pulse envelopes) $\mathcal{E}_1(t)$ and $\mathcal{E}_2(t)$ are themselves time-dependent. The adiabatic eigenvectors with zero eigenvalues are called "trapped" (or "null") states, since, for CW fields, the population stays trapped in the two states $|E_1\rangle$ and $|E_2\rangle$. It is possible to show that the "trapping" is a consequence of quantum interference established between the two routes to level $|E_0\rangle$. As mentioned above, if the pulse envelopes *do* vary in time, the trapped states will also vary in time, following the time evolution in Eq. (11.24). This phenomenon is called "adiabatic following".

We now focus attention on the dynamics of $|\lambda_1(t)\rangle$. It follows from Eqs. (11.20), (11.21), and (11.24) that if the pulse sequence is arranged in the so-called "counter-intuitive" order [321], that is, where the $\mathcal{E}_2(t)$ dump-pulse is applied *before* the $\mathcal{E}_1(t)$ pump-pulse, then the trapped state $\lambda_1(t)$ will smoothly pass from state $|E_1\rangle$ at $t = 0$ to state $|E_2\rangle$ at $t = \infty$. This phenomenon, known as "adiabatic passage" occurs because $\theta(t = 0) = 0$, which means that (see Eq. (11.24)) $|\lambda_1(t = 0)\rangle = |E_1\rangle$, (corresponding to $\underline{\mathrm{U}}^{(1)}(t = 0) = (1\ 0\ 0)^T$), whereas $\theta(t \to \infty) = \pi/2$ and $|\lambda_1(t \to \infty)\rangle = e^{-i\chi(t)}e^{-iE_2t/\hbar}|E_2\rangle$. Thus, the trapped state starts as $|E_1\rangle$ and goes over to the $|E_2\rangle$ state, never populating the intermediate state $|E_0\rangle$. In this way full population transfer from $|E_1\rangle$ to $|E_2\rangle$ results. Further, there are no losses due to spontaneous emission since $|E_0\rangle$ is never populated. (Spontaneous emission from $|E_2\rangle$ to $|E_1\rangle$ is forbidden).

As shown in Figure 11.2, the fact that $|E_0\rangle$ does not contribute to the trapped state enables the description of the adiabatic passage process by the nutation of a "Bloch vector" [423] whose direction is given by an azimuthal angle $\chi(t)$ and a polar angle $2\theta(t)$. During the adiabatic passage process the Bloch vector nutates [321] from the $|b_1|^2 - |b_2|^2 = 1$ up-vertical position to the $|b_1|^2 - |b_2|^2 = -1$ down-vertical position while undergoing a precession at a frequency $\omega_{2,1}$ about the z-axis. Figure 11.3 shows computed results for the time evolution of the Ω associated with the dump

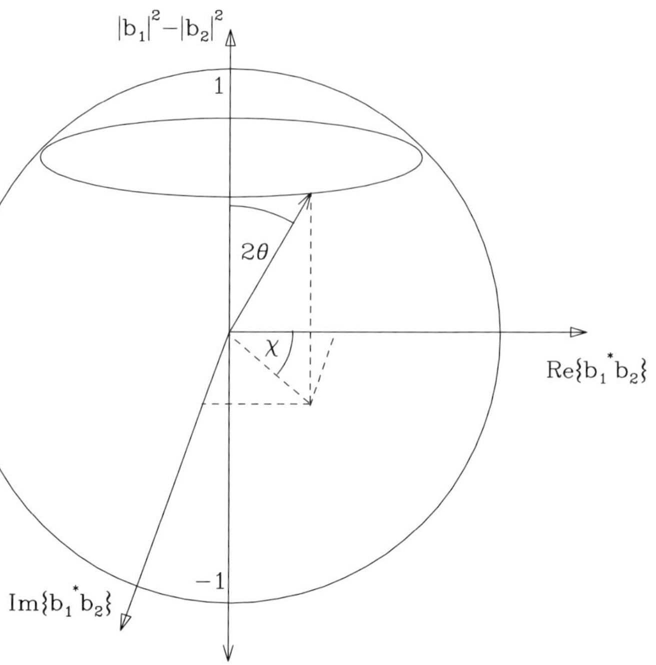

Figure 11.2 The "Bloch vector" of the time evolution of the $b_1(t)$ and $b_2(t)$ complex coefficients of the $b_1(t)|E_1\rangle + b_2(t)|E_2\rangle$ superposition state. The two complex coefficients are constrained by normalization and therefore can be represented as a point on a three-dimensional sphere, shown here as a function of twice the mixing angle θ and the azimuthal angle χ of Eq. (11.22). In this space the z-axis corresponds to the $|b_1|^2 - |b_2|^2$ population difference, the x-axis to $\mathrm{Re}\{b_1^* b_2\}$, and the y-axis to $\mathrm{Im}\{b_1^* b_2\}$. In the case of two-photon resonance ($\Delta_1 - \Delta_2 = 0$) and when the phases of the two lasers fields are the same ($\phi_2(t) = \phi_1(t)$), it follows from Eq. (11.22) that the azimuthal angle is constant and the Bloch vector executes a pure nutational motion from the north pole of the sphere to the south pole during a complete adiabatic population transfer.

and pump pulse as well as the mixing angle θ, the dressed state eigenvalues and the population of $|E_1\rangle$ and $|E_2\rangle$. The smooth and complete transfer of population between the initial and final state is evident. The dependence of the transfer efficiency, gleaned from experimental studies on Ne, is shown as a function of the pulse ordering in Figure 11.4. Specifically, the transfer of population between the ${}^3P_0$ and ${}^2P_2$ states of Ne is seen to be small when the dump and the pump pulses do not overlap, and to maximize at 100% population transfer when the dump laser of frequency ω_2 slightly precedes the pump laser of frequency ω_1.

The two remaining eigenvectors $\underline{U}^{(2)}$, $\underline{U}^{(3)}$ are obtained by solving the second and third eigenvalue equations:

$$\underline{\underline{H}} \cdot \left(\underline{U}^{(2)}, \underline{U}^{(3)}\right) = \left(\lambda_2 \underline{U}^{(2)}, \lambda_3 \underline{U}^{(3)}\right) = \left(\lambda_2 \underline{U}^{(2)}, -\lambda_2 \underline{U}^{(3)}\right) . \tag{11.25}$$

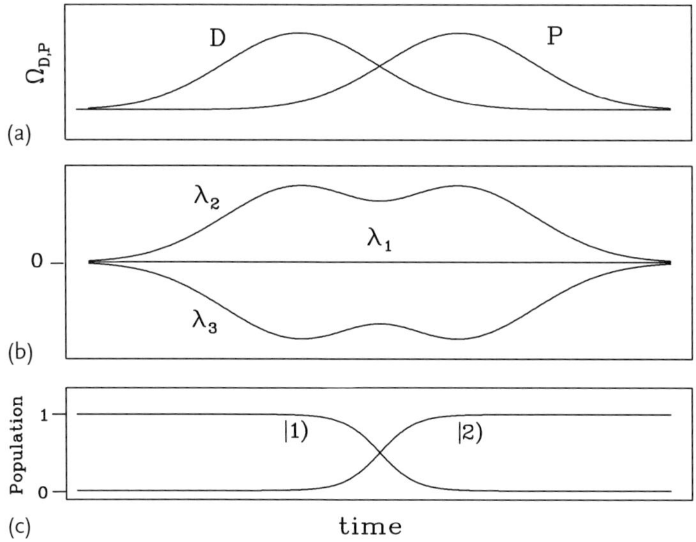

Figure 11.3 The computed time evolution of the Rabi frequencies, adiabatic eigenvalues and population of states $|1\rangle$ and $|2\rangle$. Taken from [447, Figure 3]. (Reprinted with permission. Copyright 2009 American Physical Society.)

Using Eq. (11.9) the solutions are

$$\underline{U}^{(2,3)} = \frac{1}{\sqrt{2}\lambda_2}\begin{pmatrix} \Omega_1^* e^{-i\Delta_1 t} \\ \pm\lambda_2 \\ \Omega_2 e^{-i\Delta_2 t} \end{pmatrix} = \frac{1}{\sqrt{2}}\begin{pmatrix} e^{-i\phi_1(t)-i\Delta_1 t}\sin\theta \\ \pm 1 \\ e^{i\phi_2(t)-i\Delta_2 t}\cos\theta \end{pmatrix}, \tag{11.26}$$

where the plus sign applies to $\underline{U}^{(2)}$ and the minus sign to $\underline{U}^{(3)}$. The associated adiabatic eigenvectors are

$$\begin{aligned} |\lambda_2(t)\rangle &= \frac{1}{\sqrt{2}} e^{\int_0^t i\lambda_2(t')dt'} \Big[e^{-i\phi_1(t)-i\Delta_1 t}\sin\theta\, e^{-iE_1 t/\hbar}|E_1\rangle + |E_0\rangle \\ &\qquad + e^{i\phi_2(t)-i\Delta_2 t}\cos\theta\, e^{-iE_2 t/\hbar}|E_2\rangle \Big] \\ |\lambda_3(t)\rangle &= \frac{1}{\sqrt{2}} e^{-\int_0^t i\lambda_2(t')dt'} \Big[e^{-i\phi_1(t)-i\Delta_1 t}\sin\theta\, e^{-iE_1 t/\hbar}|E_1\rangle - |E_0\rangle \\ &\qquad + e^{i\phi_2(t)-i\Delta_2 t}\cos\theta\, e^{-iE_2 t/\hbar}|E_2\rangle \Big] \end{aligned} \tag{11.27}$$

Assuming that $\phi_1(t) = \phi_2(t) = 0$, we have that in the "counterintuitive" pulse ordering

$$\underline{U}^{(2,3)}(t=0) = \frac{1}{\sqrt{2}}\begin{pmatrix} 0 \\ \pm 1 \\ 1 \end{pmatrix}, \quad \underline{U}^{(2,3)}(t\to\infty) = \frac{1}{\sqrt{2}}\begin{pmatrix} e^{-i\Delta_1 t} \\ \pm 1 \\ 0 \end{pmatrix}. \tag{11.28}$$

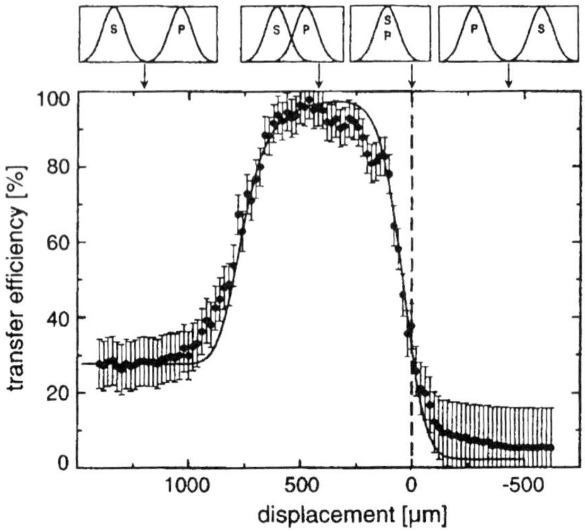

Figure 11.4 Experimental results on the population transfer between two states of Ne as a function of laser pulse ordering. The ordering of the two laser pulses is shown at the top and is controlled by the displacement between the two CW laser beams transversing the path of the Ne beam, thereby introducing an effective time delay between the pulses. *S* of the figure denotes the "Stokes" pulse (the "dump" pulse of the text), and *P* of the figure denotes the "pump" pulse of the text. From [321, Figure 9]. (Reprinted with permission. Copyright 1998 American Physical Society.)

Since at $t = 0$ the first element of $\underline{U}^{(2)}$ and of $\underline{U}^{(3)}$ is zero, then the two "non-trapped" adiabatic states are orthogonal to the initial state $|E_1\rangle$ at the beginning of the process. Hence, the only adiabatic state populated initially is the trapped state $|\lambda_1(t)\rangle$. If there is no coupling between the three adiabatic states, the system will continue to evolve as the trapped state $|\lambda_1(t)\rangle$, executing an adiabatic passage to the $|E_2\rangle$ state as $t \to \infty$. Thus we can achieve the control objective of complete population transfer between two bound states.

A variety of STIRAP pulse configurations and generalizations of STIRAP have been studied. For example, and of interest later below, is when one or more of the levels is either part of the continuum [405, 424, 425] or a resonant state [426–428] (i.e., a bound state coupled to the continuum).

11.1.2
The Multistate Extension of STIRAP

Following the success of three-state STIRAP experiments discussed above, the theory has also been greatly advanced [402, 429–431]. Among the properties examined were the sensitivity of the population transfer to the delay between the dump and the pump pulse [321] and the effect of single-photon [432] and two-photon detuning [431]. Going beyond the RWA [433, 434], parasitic effects due to losses from intermediate state(s) [435–439], and the existence of many intermediate states [408, 440] have also been studied.

Generalizations of the three-state STIRAP to multistate chains [441–450]; to adiabatic momentum transfer [442, 444, 451, 452]; to branched-chain excitation [411, 427, 453, 454]; and to population transfer via a continuum of intermediate states [406, 454–459], which is directly related to laser induced continuum structure (LICS) [264, 266, 460, 461], whose molecular version is discussed in Section 13.2 have also been made.

Numerous additional techniques related to STIRAP have been developed. Examples are the hyper-Raman STIRAP (STIHRAP) [433, 434, 462], Stark-chirped rapid adiabatic passage (SCRAP) [463, 464], adiabatic passage by light-induced potentials (APLIP) [465–468] and photoassociative STIRAP, as a source for cold molecules [469–472]. Various experimental implementations of these ideas, for example, to the formation of dark states in the photoassociation of an atomic BEC to form a high density ultracold molecular gas, have also been reported [388, 473, 474].

11.2 An Analytic Solution of the Nondegenerate Quantum Control Problem

In the context of bound-state dynamics the "nondegenerate quantum control" (NQC) problem is the achievement of complete population transfer between two arbitrary wave packets composed of many nondegenerate energy eigenstates. We emphasize the nondegenerate nature of the constituent states because in this case frequency discrimination allows for an easy and simple addressing of each individual energy eigenstate. As discussed in Chapter 3, such simple means of addressing individual states is not possible if degeneracies exist. We therefore postpone the discussion of quantum control between superpositions of degenerate energy eigenstates in the nonperturbative regime to Section 11.3.

Nondegenerate quantum control is of great interest in many atomic and molecular systems (see for example, [475–477]). There is a "controllability" existence theorem [100, 478], which guarantees that a solution to the NQC problem exists under certain conditions. This theorem provides however no clue as to the nature of the external fields capable of bringing about the desired outcome. Although solutions of the NQC problem for some particular weak-field CC cases [25], the three-state STIRAP [321, 407, 411, 431, 479, 480] and its generalization to a few more final states [408, 427, 453, 481, 482], have been known for quite some time, an explicit construction of the quantum control fields for the general nondegenerate case has only been presented in 2002 [448, 483]. Though this solution is by no means unique, its existence is enough to establish that population transfer between two arbitrary wave packets made up of bound nondegenerate energy eigenstates is controllable by construction. As discussed below, the simplicity of this analytic solution is a key factor in the success of its laboratory implementations that eventually followed.

We proceed now with a detailed description of the solution of the NQC problem. Our task is to execute a complete population transfer between two arbitrary wave packets composed of nondegenerate "bare" energy eigenstates. We do so by going through an intermediate $|0\rangle$ state. The whole process is thus written as

$$|\Psi^i(t)\rangle = \sum_{k=1}^{n} c_k^i e^{-i\omega_k t}|k\rangle \longrightarrow |0\rangle e^{-i\omega_0 t} \longrightarrow |\Psi^f(t)\rangle = \sum_{l=n+1}^{n+m} c_l^f e^{-i\omega_l t}|l\rangle \,. \tag{11.29}$$

Figure 11.5 illustrates the fields used, consisting of two time-delayed pulses composed of discrete frequencies chosen to be in resonance with $\omega_{0,k}$ the transitions from every initial (final) energy eigenstate $|k\rangle$ ($|l\rangle$) to the $|0\rangle$ intermediate state. The electric field of these pulses is given explicitly as

$$\boldsymbol{E}(t) = \mathrm{Re} \sum_{j=1}^{n+m} \hat{\epsilon}_j \mathcal{E}_j(t) e^{-i\omega_{0,j} t} \,. \tag{11.30}$$

In the above, $\mathcal{E}_j(t)$ are the (slowly varying) envelopes of the electromagnetic field modes, $\omega_{0,j}$, each polarized along the $\hat{\epsilon}_j$ direction. Because of the assumed complete resonance, we can equate the field mode frequencies with the transition frequencies $\omega_{0,j} = \omega_0 - \omega_j$.

The first n components of $\boldsymbol{E}(t)$ describe the pump pulse, and the last m components the dump (Stokes) pulse. In keeping with the counterintuitive pulse ordering, the dump pulse is made to precede (while partially overlapping in time) the pump pulse. The time dependence of the *common* dump (pump) pulse envelope is controlled by the function, $0 < f_\nu(t) < 1$, ($\nu = S, P$), according to, $\mathcal{E}_l(t) = f_S(t)\eta_l$ for $l = n+1, \ldots, n+m$ and $\mathcal{E}_k(t) = f_P(t)\eta_k$ for $k = 1, \ldots, n$, with $\eta_j = \max\{\mathcal{E}_j(t)\}(j = 1, \ldots, n+m)$.

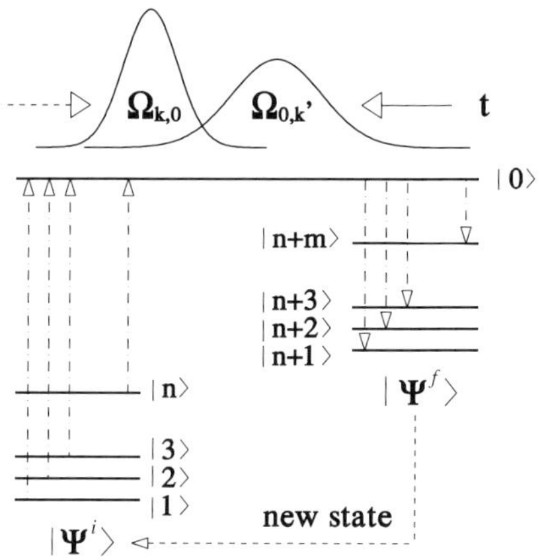

Figure 11.5 The NQC scheme. Population in an initial wave packet $|\Psi^i\rangle$, is transferred via a single state, $|0\rangle$, to the final state $|\Psi^f\rangle$. Taken [483, Figure 1]. (Reprinted with permission. Copyright 2002 American Physical Society.)

The time-dependent radiation–matter Hamiltonian assumes the form

$$H(t) = H_M - \frac{\hbar}{2} \sum_{j=1}^{n+m} \left[\Omega_{0,j}(t) e^{-i\omega_{0,j}t} |0\rangle\langle j| + \text{h.c.} \right] , \tag{11.31}$$

where h.c. denotes the "Hermitian conjugate", and

$$\Omega_{0,j}(t) \equiv d_{0,j} \mathcal{E}_j(t)/\hbar = \frac{d_{0,j} \eta_j f_\nu(t)}{\hbar} \equiv \mathcal{O}_{0,j} f_\nu(t) , \quad \nu = S, P \tag{11.32}$$

are a set of Rabi frequencies, with $d_{0,j}$ being the electric dipole matrix elements between the $|0\rangle$ and $|j\rangle$ states in the polarization $\hat{\epsilon}_j$.

We denote by $c(t) \equiv (c_0, c_1, \ldots, c_{n+m})^{\mathsf{T}}$ the vector of the time-dependent expansion coefficients of $|\Psi(t)\rangle$ in the bare states. Invoking the RWA and neglecting all off-resonance terms, using Eq. (11.31), leads to a Schrödinger equation

$$i\hbar \frac{d}{dt} c(t) = \underline{\underline{\mathrm{H}}}(t) \cdot c(t) , \tag{11.33}$$

whose Hamiltonian is

$$\underline{\underline{\mathrm{H}}}(t) = -\frac{\hbar}{2} \begin{bmatrix} 0 & \Omega_{0,1}(t) & \ldots & \Omega_{0,n+m}(t) \\ \Omega_{1,0}(t) & 0 & \ldots & 0 \\ \ldots & \ldots & \ldots & \ldots \\ \Omega_{n+m,0}(t) & 0 & \ldots & 0 \end{bmatrix} . \tag{11.34}$$

This is an extension of the RWA three-level Hamiltonian to the case of many initial states. We have neglected the off-resonance terms which are assumed to be small compared to the level spacing $(\omega_i - \omega_j)$, $(i \neq j, i, j = 1, \ldots, n+m)$ and the Rabi frequencies [484]. The above contains no detuning parameters since the fields are assumed to be in resonance with the individual transition frequencies.

We now note that all but two of the eigenvalues of $\underline{\underline{\mathrm{H}}}(t)$, are zero. The two nonzero eigenvalues are

$$\lambda_{n+m}(t) = -\lambda_{n+m+1}(t) = \left(\sum_{j=1}^{n+m} |\Omega_{0,j}(t)|^2 \right)^{1/2} \cdot \frac{\hbar}{2} . \tag{11.35}$$

The zero eigenvalues correspond to *three* types of null adiabatic eigenvalue states, termed "initial null-eigenvector states" (INS), "mixed null-eigenvector states" (MNS) and "final null-eigenvector states" (FNS). In complete analogy to the three-state STIRAP, the eigenstates corresponding to these zero eigenvalues can be composed of "diads" of (initial, and final) energy eigenstates, written (in the interaction representation) as

$$\begin{aligned} |D^I_{kk'}(t)\rangle &= \Omega_{0,k'}(t)|k\rangle - \Omega_{0,k}(t)|k'\rangle , \quad k, k' = 1, \ldots, n , \\ |D^M_{kl}(t)\rangle &= \Omega_{0,l}(t)|k\rangle - \Omega_{0,k}(t)|l\rangle , \quad k = 1, \ldots, n ; \quad l = n+1, \ldots, n+m , \\ |D^F_{ll'}(t)\rangle &= \Omega_{0,l'}(t) l\rangle - \Omega_{0,l}(t)|l'\rangle , \quad l, l' = n+1, \ldots, n+m . \end{aligned} \tag{11.36}$$

Each such eigenstate executes a complete transfer of population from a single state $|j\rangle$ to a single state $|j'\rangle$. That these states correspond to the zero eigenvalue of the RWA Hamiltonian (Eq. (11.34)) can be explicitly verified by substitution. The construction in Eq. (11.36) generates $n\times(n-1)/2$ INS, $n\times m$ MNS, and $m\times(m-1)/2$ FNS, out of which only $n+m-1$ states are linearly-independent.

In order to affect the transfer, given the two arbitrary wave packets of Eq. (11.29) and the fields of Eq. (11.30), we need to take linear combinations of the above eigenstates. For example, the MNS states need to be combined such that they correlate at $t=t_{in}$ with the $|\Psi^i(t)\rangle$ wave packet and at $t=t_{end}$ with the $|\Psi^f(t)\rangle$ wave packet. The particular combination that satisfies this asymptotic condition is given (in the interaction representation) as

$$|D^M(t)\rangle \equiv \sum_{k,l} t_{kl}|D^M_{kl}(t)\rangle = \sum_{k=1}^{n}|k\rangle \sum_{l=n+1}^{n+m} t_{kl}\Omega_{0,l}(t) - \sum_{l=n+1}^{n+m}|l\rangle \sum_{k=1}^{n} t_{kl}\Omega_{0,k}(t)\,, \tag{11.37}$$

where the t_{kl} coefficients are chosen such that

$$\sum_{l=n+1}^{n+m} t_{kl}\Omega_{0,l}(t_{in}) \propto c^i_k\,, \quad \sum_{k=1}^{n} t_{kl}\Omega_{0,k}(t_{end}) \propto c^f_l\,. \tag{11.38}$$

Equations (11.38) can be satisfied by choosing t_{kl} (for $\Omega_{k,0}(t) \equiv \mathcal{O}_{k,0} f_\nu(t), \nu = S, P$) as

$$t_{kl} = \mathcal{O}_{k,0}\mathcal{O}_{l,0}\,, \quad \text{with} \quad \mathcal{O}_{k,0} = C c^i_k\,, \quad \mathcal{O}_{l,0}(t) = C' c^f_l\,, \tag{11.39}$$

where C, C' are chosen to guarantee that the magnitudes of the Rabi frequencies are small enough not to cause off-resonant transitions, yet large enough to satisfy the adiabaticity condition.

In order for the above description to be accurate, in addition to the neglect of the off-resonance terms we need to satisfy the *global* adiabaticity, or large pulse area, condition,

$$\frac{1}{|\max\{\mathcal{E}_i(t)\}|} \ll \tau_{pulse}\,. \tag{11.40}$$

This condition is satisfied if all the fields are strong enough. The second condition is that of *local* adiabaticity [431, 484]) discussed above in the context of our general introduction to adiabatic passage phenomena, namely,

$$\left(\underline{\underline{U}}^{-1}(t)\cdot\dot{\underline{\underline{U}}}(t)\right)_{k,l} \ll |\lambda_k(t)-\lambda_l(t)|\,, \tag{11.41}$$

where $\underline{\underline{U}}(t)$ is the transformation that diagonalizes $\underline{\underline{H}}(t)$. Equation (11.41) is satisfied only in some cases, namely when the field components $\mathcal{E}_j(t)$ change slowly enough with time.

We expect local adiabaticity to be violated when degenerate eigenstates are present in the spectrum of the RWA Hamiltonian. Although in the present case such degeneracy exists, due to the existence of the INS, MNS, and FNS types of zero eigenvalues discussed above, the nonadiabatic couplings between the MNS and INS/FNS states are nevertheless zero [448]. The vanishing of the nonadiabatic couplings stems from the fact that the INS and FNS of Eq. (11.36) are orthogonal to both the initial ($|\Psi^i(t_{in})\rangle = \sum_k c_k^i e^{-i\omega_k t_{in}}|k\rangle$) and the final ($|\Psi^f(t_{end})\rangle = \sum_l c_l^f e^{-i\omega_l t_{end}}|l\rangle$) wave packets. Hence the nonadiabatic coupling terms between the INS/FNS and the MNS are also zero and local adiabaticity is valid. Thus, the population transfer process from $|\Psi^i(t_{in})\rangle$ to $|\Psi^f(t_{end})\rangle$ via the MNS is completely oblivious to the presence of the INS or the FNS.

Figure 11.6 illustrates the validity of Eq. (11.37) for the following chain of population transfers:

$$|1\rangle \to c_4|4\rangle + c_5|5\rangle \to c_1|1\rangle + c_2|2\rangle + c_3|3\rangle \to c_4'|4\rangle + c_5'|5\rangle \,. \tag{11.42}$$

In the first link of the transfer chain, $|1\rangle \to c_4|4\rangle + c_5|5\rangle$, which is seen to exhibit 100% transfer probability, the multimode Rabi frequencies were chosen as, $\Omega_{0,1}(t) = \mathcal{O}_{0,1}\exp[-(t - t_0)^2/\tau^2]$ and $\Omega_{0,l}(t) = \mathcal{O}_{0,l}\exp[-t^2/\tau^2]$, $(l = 4, 5)$, with $t_0 = 2\tau$ being the delay between the pulses. The coefficients of Eq. (11.39) are chosen to be $C = C' = 100/\tau$. The next transfer link is realized by repeating the pulses with different central times and amplitudes $\mathcal{O}_{0,k}$, in accordance with Eq. (11.39).

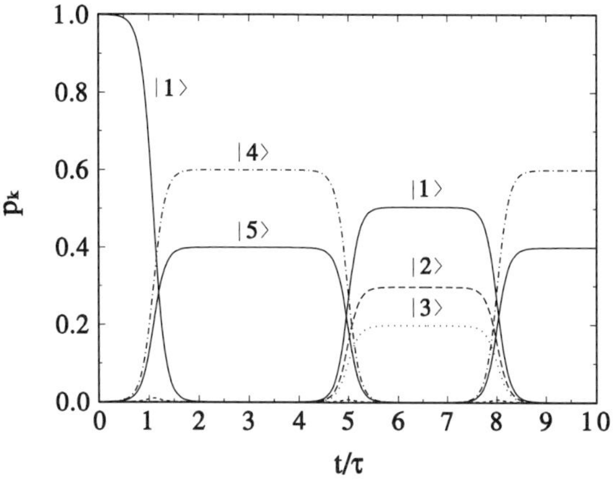

Figure 11.6 Sequence of population transfers illustrating the NQC scheme. Shown are complete transfers of the type: $|1\rangle \to c_4|4\rangle + c_5|5\rangle \to c_1|1\rangle + c_2|2\rangle + c_3|3\rangle \to c_4'|4\rangle + c_5'|5\rangle$. Taken from [483, Figure 2]. (Reprinted with permission. Copyright 2002 American Physical Society.)

11.3 The Degenerate Quantum Control Problem

In "degenerate quantum control" (DQC) we wish to control population transfer between wave packets

$$|\Psi^i(t_{in})\rangle = e^{-i\omega_i t_{in}} \sum_k c_k^i |k\rangle \longrightarrow |\Psi^f(t_{end})\rangle = e^{-i\omega_f t_{end}} \sum_{k'} c_{k'}^f |k'\rangle ,$$

whose constituents energy eigenstates are essentially-degenerate. By this term we mean that these states are degenerate in comparison to the widths of the pulses used in the transfer process. The problem is more demanding than the nondegenerate one, because it is no longer possible to use differences in energies to address different energy eigenstates components.

There have been a number of suggestions for solving this problem for various special cases [484, 486, 487]. Currently the most general approach is to perform the process in two steps [485]. In this approach, illustrated in Figure 11.7a, we first use a pair of counterintuitively ordered pulses to empty the population of the initial wave packet $|\Psi^i\rangle$, and transfer it to a single ("parking") state $|0\rangle$. We do so by using a set of nondegenerate auxiliary intermediate states $|j\rangle (j = 1, \ldots, n)$. We do not need to worry about individually addressing each initial energy eigenstates

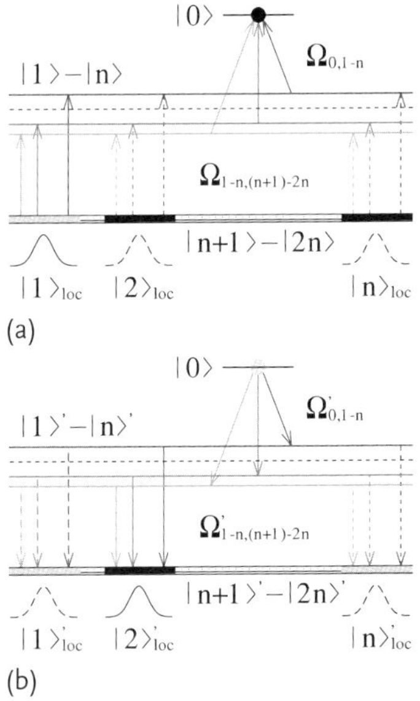

Figure 11.7 The DQC scheme. (a) The first step: Population in an initial wave packet $|\Psi^i\rangle$, also called $|1\rangle_{loc}$, composed of n nearly-degenerate eigenstates is transferred to a single state $|0\rangle$, by a two-photon adiabatic passage via n nondegenerate intermediate states. (b) The second step: population transfer by a time reversed process with different Rabi frequencies from $|0\rangle$ to the target wave packet $|\Psi^f\rangle$, also called $|2\rangle'_{loc}$, composed of another set of nearly-degenerate eigenstates. The process can be repeated to generate other $|n\rangle'_{loc}$ localized states. Taken from [485, Figure 1]. (Reprinted with permission. Copyright 2004 American Physical Society.)

so long as the transfer via the nondegenerate intermediate states is complete. This implies that both the Stokes pulse, characterized by the $\Omega_{0;1,\dots,n}$ Rabi frequencies, linking all the $|j\rangle$ states to the $|0\rangle$ state, and the pump pulse, characterized by the $\Omega_{n+1;1,\dots,n}$ Rabi frequencies, linking all the $|k\rangle$ states to all the $|j\rangle$ states, should be strong enough. In the second step (Figure 11.7b), the population in the parking state $|0\rangle$ is transferred to the target wave packet $|\Psi^f\rangle$, composed of the same (or different) set of nearly-degenerate eigenstates $|k'\rangle$. At this stage we do need to address the individual target eigenstates components and, due to the degeneracy of these states, we can only do so if the Rabi frequencies differ by some other attribute such as phase, polarization, or direction.

In order to see how this is done we summarize the above by writing the electric field used in the first step as

$$E(t) = E_S(t) + E_P(t) , \tag{11.43}$$

where

$$E_S(t) = \text{Re} \sum_{j=1}^{n} \hat{\epsilon}_{0,j} \mathcal{E}_{0,j}(t) e^{-i\omega_{0,j} t} , \tag{11.44}$$

$$E_P(t) = \text{Re} \sum_{j=1}^{n} e^{-i\omega_{j,D} t} \sum_{l=n+1}^{2n} \hat{\epsilon}_{j,l} \mathcal{E}_{j,l}(t) . \tag{11.45}$$

In the second step the form assumed by the Stokes and pump pulses is simply reversed. The Stokes and pump field components are characterized by the polarization directions $\hat{\epsilon}_{0,j}$ and $\hat{\epsilon}_{j,l}$ and the slowly varying amplitudes $\mathcal{E}_{0,j}(t)$ and $\mathcal{E}_{j,l}(t)$. The center frequencies of the field components are chosen to be in near resonance with the system's transition frequencies, $\omega_{0,j} = \omega_0 - \omega_j$ and $\omega_{j,D} = \omega_j - \omega_D$, where ω_D $(= \omega_i$ or $\omega_f)$ is the energy of the nearly-degenerate initial (final) states. As in the NQC case, we choose the field amplitudes associated with the discrete frequencies of each pulse to have a common time envelope, $\Omega_{j,i}(t) = \mathcal{O}_{j,i} f_\nu(t)$, $(\nu = S, P)$, with f_S preceding f_P in both steps.

The vector $c(t) = (c_0, c_1, \dots, c_n, c_{n+1}, \dots, c_{2n})^{\mathsf{T}}$ of the coefficients in the wave function expansion of the bare states in the interaction representation is obtained by solving Eq. (11.33), with the RWA Hamiltonian $\underline{\underline{\mathrm{H}}}(t)$ written in the compact form,

$$\underline{\underline{\mathrm{H}}}(t) = -\frac{\hbar}{2} \begin{pmatrix} 0 & \boldsymbol{\Omega}_0 & 0 \\ \boldsymbol{\Omega}_0^\dagger & 0 & \underline{\underline{\mathrm{H}}}_F \\ 0 & \underline{\underline{\mathrm{H}}}_F^\dagger & 0 \end{pmatrix} , \quad \underline{\underline{\mathrm{H}}}_F = \begin{pmatrix} \boldsymbol{\Omega}_1 \\ \dots \\ \boldsymbol{\Omega}_n \end{pmatrix} , \tag{11.46}$$

and

$$\begin{aligned} \boldsymbol{\Omega}_0 &= (\Omega_{0,1}(t), \dots, \Omega_{0,n}(t)) , \\ \boldsymbol{\Omega}_j &= (\Omega_{j,n+1}(t), \dots, \Omega_{j,2n}(t)) , \quad j = 1, \dots, n , \end{aligned} \tag{11.47}$$

with † denoting the Hermitian conjugation operation. The $\boldsymbol{\Omega}_{0,\dots,n}$ vectors of time-dependent Rabi frequencies are different in the two steps, because they control the population transfer between different wave packets.

We now show that in order to address arbitrary wave packets in any DQC step, the corresponding Rabi frequency vectors $\{\boldsymbol{\Omega}_1, \ldots, \boldsymbol{\Omega}_n\}$ must be *linearly-independent* of one another. When the linear independence is assured, $\mathcal{D} \equiv \det\left(\underline{\underline{\mathrm{H}}}_F\right) \neq 0$ and the Hamiltonian $\underline{\underline{\mathrm{H}}}(t)$ has just one zero eigenvalue, with the corresponding ("dark") eigenvector being given as $\boldsymbol{c} = (1, \boldsymbol{0}, \boldsymbol{x})^{\mathrm{T}}$, where $\boldsymbol{0}$ denotes an n-dimensional zero vector and $\boldsymbol{x}$ is an n-dimensional vector given as, $\boldsymbol{x} = -\underline{\underline{\mathrm{H}}}_F^{-1}\boldsymbol{\Omega}_0^{\dagger}$. Direct operation of $\underline{\underline{\mathrm{H}}}(t)$ on $\boldsymbol{c}$ immediately confirms that this is indeed a "dark" eigenvector.

By constructing the $\boldsymbol{\Omega}_0^{\dagger}$ vector of Eq. (11.47) to be proportional to the kth column of the $\underline{\underline{\mathrm{H}}}_F$ matrix, that is, $\Omega_{0,i}^* \propto \Omega_{i,k}$ $(k = 1, \ldots, n)$, we guarantee that the $\boldsymbol{c}$ vector correlates at the start of the first step (end of the second step) with $|k\rangle$ ($|k'\rangle$) states, and at the end of the first step (the start of the second step) with the $|0\rangle$ state. Hence, it follows from the linearity of the above equations that with the choice $\Omega_{0,i}^* \propto \sum_{k=1}^{n} c_k \Omega_{i,k}$, where $c_k = c_k^i$ (c_k^f) at the start of the first step (end of the second step), the "dark" eigenvector correlates at the start of the first step (end of the second step) with the $|\Psi^i\rangle$ ($|\Psi^f\rangle$) superposition state [485].

As we show below, there are a number of ways of guaranteeing the linear independence of the vectors of Rabi frequencies. Thus we can couple the essentially-degenerate states in such a way that each Rabi frequency has a different (complex) phase. This can be achieved for certain combinations of the dipole matrix elements and the phases and polarizations of the pulse's discrete frequency components. For example, in the CC of current directionality, discussed in Section 3.4.3, the linear independence is achieved by the different phases of the light beams used. In the control of the Al_3O isomerization discussed below, the linear independence is achieved if the dipole moment vectors $\boldsymbol{d}_{j,i}$ are pointing in different directions, so that independent polarization vector $\hat{\epsilon}_{j,i}$ directions can be used. Obviously, only a limited number of degrees of freedom can be gained in this way, so that other degrees of freedom, mostly based on the character of the excited states, must be used in systems with more degenerate levels.

To illustrate the versatility of the above DQC method we apply it to the control of isomerization processes of the Jahn–Teller distorted Al_3O molecule [488]. As shown in Figure 11.8, the Jahn–Teller effect results in the formation of three identical broken symmetry minima (isomers) corresponding to T-shaped configurations in which the distance between two of the aluminum atoms is smaller than their distance to the third aluminum atom [489]. This situation gives rise to three sets of highly stable wave packets localized about each T-shaped minimum. The in-plane tunneling motion of a localized wave packet from one such minimum to another is known as pseudorotation. Another important motion is the out-of-plane vibration of the oxygen atom relative to the Al_3 triangle.

Our control objective is to transfer population between the localized states of the three minima in as complete fashion as possible. The control objective falls under the DQC category because low-lying localized states appear in triplets, which are nearly three-fold degenerate, as discussed below. In order to simplify the dynamics, we consider partial or complete hindering of the overall rotation of the molecule, by for example, depositing it on a surface of some material or in a pocket of a

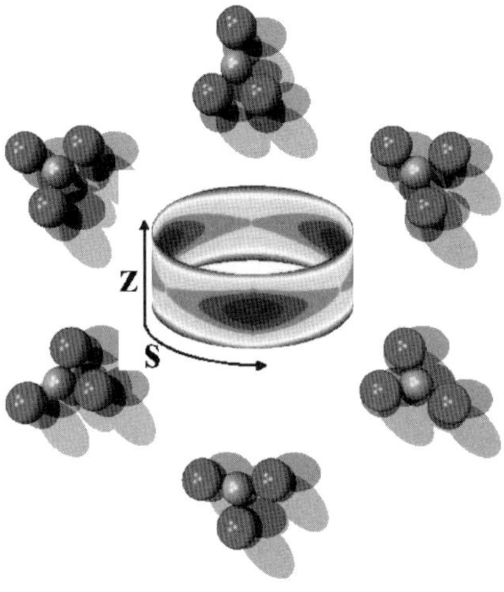

Figure 11.8 The three Jahn–Teller-distorted stable T-shaped configurations of the Al_3O molecule and the three equilateral "transition state" saddle points which separate them. In the center, we display the potential along s, the pseudorotation (isomerization), and z, the out-of-plane motion of O relative to the Al_3 plane. Taken from [485, Figure 2]. (Reprinted with permission. Copyright 2004 American Physical Society.)

larger molecule. As a result, the rotation levels become highly separated and can be neglected in the present analysis. Then the in-plane $(x-y)$ and out-of-plane (z) motions occur in well-defined laboratory directions that are, respectively, parallel and perpendicular to the surface. This allows us to choose the polarization directions of the field components that will couple with dipole moment components that are linearly-independent. As shown below, in order to do that we need to use states in which one of the out-of-plane vibrational modes is excited [485].

The simulation of the control dynamics proceeds by first obtaining, using *ab initio* methods [490], the potential surface and electric dipole moments in a two-dimensional (2D) subspace of the full (6D) configuration space of Al_3O, containing the pseudorotation (s) and the out-of-plane (z) modes (see Figure 11.8). The relevant eigenstates and eigenenergies are obtained using well established methodologies [491–493] for solving the 2D nuclear Schrödinger equation $[E_k - H]\phi_k(s,z)=0$.

The conical intersection with the first excited state occurs at sufficiently high energies [489] to allow neglect of Berry phase effects [494–497] for relatively low lying vibrational states [498], as used here. As a result, the two-dimensional eigenenergies are describable by two $\{v_s, v_z\}$ quantum numbers, essentially enumerating the number of nodes along the s and z directions in each T-shaped minimum (see also Figure 11.9). The low lying $\{v_s, v_z\}$ states appear as triplets, each triplet composed of two degenerate eigenstates and a third, nondegenerate, state.

The three eigenstates in the lowest ($\{0,0\}$) triplet are shown in Figure 11.9a. We take linear combinations of these states to form the initial (final) localized wave packets, $|\Psi^a\rangle = \sum_k c_k^a |k\rangle$, $(a = i, f; i, f = 1, 2, 3)$, displayed in Figure 11.9b. The three nondegenerate delocalized components of the $\{4,0\}$, $\{6,0\}$ and $\{4,1\}$

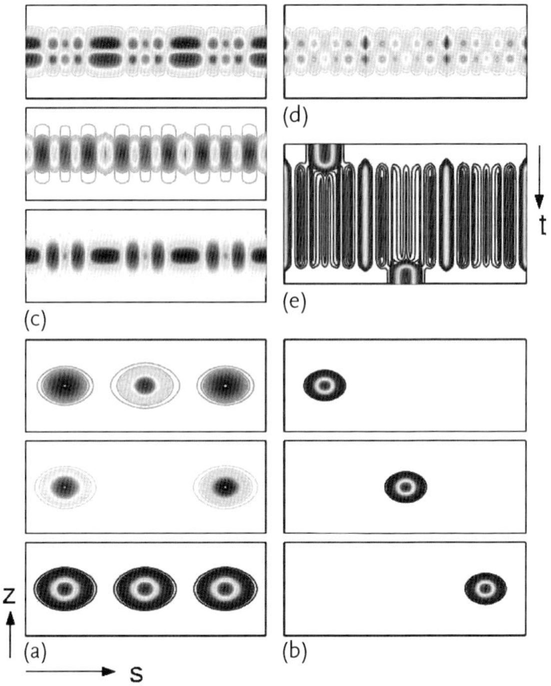

Figure 11.9 The DQC dynamics: (a) The three lowest $\{0,0\}$ energy eigenstates of Al_3O as a function of the s and z coordinates. (b) Superpositions of these eigenstates to form localized wave packets about each T-shaped minimum. (c) The $|j\rangle$ intermediate states. (d) The $|0\rangle$ parking state. (e) The time dependence of the DQC dynamics, showing the controlled transfer between two T-shaped minima. Taken from [485, Figure 3]. (Reprinted with permission. Copyright 2004 American Physical Society.)

triplets, shown in Figure 11.9c, are chosen as the $|j\rangle$ intermediate states. Finally, the nondegenerate component of the $\{6,1\}$ triplet, presented in Figure 11.9d, is used as the parking state $|0\rangle$.

The three nondegenerate intermediate states are coupled to the localized states $|\Psi^i\rangle$ by transitions along the x, y and z directions, respectively. The same states are coupled to the parking $|0\rangle$ state by transitions along the z, z and $y(x)$ directions, respectively. In this way one achieves the required linear independence of the $\boldsymbol{\Omega}_1$, $\boldsymbol{\Omega}_2$, and $\boldsymbol{\Omega}_3$ Rabi frequency vectors, due to the fact that the corresponding dipole matrix elements form linearly-independent vectors.

The Rabi frequencies for the two DQC steps for both the S and P pulses can be written as

$$\Omega_{j,k}(t) = \Omega_{j,k}^{\max}\left\{\exp\left[-\frac{(t+\tau)^2}{\tau^2}\right] + \exp\left[-\frac{(t-\tau)^2}{\tau^2}\right]\right\} ,$$
$$\Omega_{k,0}(t) = \Omega_{k,0}^{\max}\left\{\exp\left[-\frac{(t+3\tau)^2}{\tau^2}\right] + \exp\left[-\frac{(t-3\tau)^2}{\tau^2}\right]\right\} , \tag{11.48}$$

with $\Omega_{j,k}^{\max} = \Omega_{k,0}^{\max} = 0.4\,\text{cm}^{-1}$ and $\tau = 5\,\text{ns}$. The dipole moment matrix elements are of the order 10^{-3}–10^{-1} D. The duration of the laser pulses is roughly $8\tau \ll \tau_i$ [485], where τ_i is the tunneling time (in the μs time scale) from one minimum to the next, so that the $|\Psi^i\rangle$, $(i = 1, 2, 3)$ states remain localized in their corresponding T-shaped minima for sufficiently long times to be used as legitimate initial and final states.

In Figure 11.9e, we display the time evolution of the system, during the two steps of the DQC process. The z-integrated probability density $\int dz|\langle s, z|\Psi(t)\rangle|^2$ is presented as a function of s and t. We start with the localized $|\Psi^1\rangle$ state being initially populated. At the end of the first step, its population is parked on the excited delocalized $|0\rangle$ state, using as intermediate the $|j\rangle$, $(j = 1, 2, 3)$ states (which remain unpopulated throughout the process). In the parking state $|0\rangle$, the bulk of the probability density is symmetrically concentrated near the three saddle points. The parking time should be short enough so that no population is lost due to relaxation. Finally, at the end of the second step, the parked population is transferred to the localized $|\Psi^2\rangle$ state. Additional studies [447] have generalized the above DQC solution to achieving selective and complete population transfer to several energetically degenerate *continuum* channels.

11.4 Adiabatic Encoding and Decoding of Quantum Information

In addition to quantum control, the systems analyzed above are of interest in the context of quantum encoding/decoding of information [499], quantum computations [188], quantum communication [500], and the preparation of various quantum devices [501, 502]. We now discuss one of these applications, namely the use of adiabatic passage techniques for encoding and decoding of information. Contrary to quantum control, where we wish to transfer population from one known state to another, in the quantum decoding problem, we do not know the initial state. Rather we should devise a process that transfers population from each of the possible linear combinations of a set of energy eigenstates to a single final energy eigenstate whose nature can be easily determined.

We thus consider a situation [499] illustrated in Figure 11.10 in which the information is encoded as a superposition of N energy eigenstates, $|i_A\rangle (i = 1, \ldots, N)$, the M distinct phases of whose population amplitudes form a discrete set of "q-nits" [503]. (The term "q-nits" is the generalization of the term "q-bits" to nonbinary codes.) Decoding is performed by transferring populations using a pump pulse that couples each initial $|i_A\rangle$ state to a unique intermediate state $|i_B\rangle$, and a Stokes pulse that couples each $|i_B\rangle$ state to all the final $|k_C\rangle$ states $(k = 1, \ldots, L$ with $L \geq N)$. The Rabi frequencies coupling terms associated with the pump pulse are denoted as Ω_{ii}^{AB}, whereas those associated with the Stokes pulse are denoted as Ω_{ik}^{BC}.

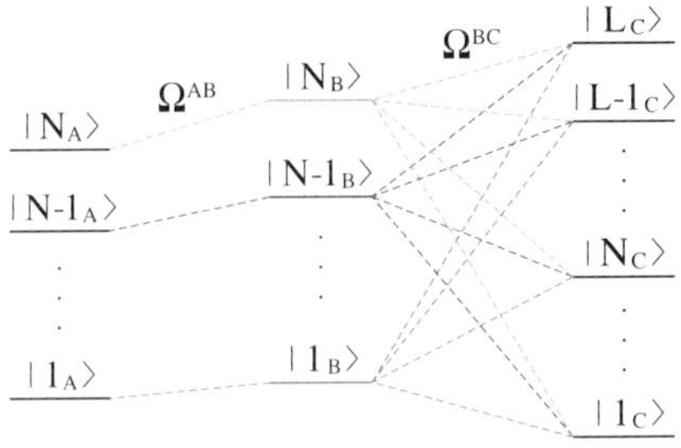

Figure 11.10 The adiabatic decoder. Quantum information is initially encoded as the phases of population amplitudes of a superposition of the $|1_A\rangle, \ldots, |N_A\rangle$ states. Depending on the code written on this superposition state, the information is decoded by transferring the population of this state to a specific state belonging to the $|1_C\rangle, \ldots, |L_C\rangle$ $(L > N)$ set. This is done adiabatically via the intermediate $|1_B\rangle, \ldots, |N_B\rangle$ states. Taken from [499, Figure 1]. (Reprinted with permission. Copyright 2004 Elsevier BV.)

These pulses execute population transfer according to a predetermined protocol of each linear combination of the $|i_A\rangle$ states, representing a number, to (predominantly) a single final $|k_C\rangle$ state. As in the above, we affect the transfer in a "counter-intuitive" fashion by having the Stokes pulse precede the pump pulse. Assuming a single, linear, polarization direction, the scalar components of the electric fields of the two pulses are given as

$$E_P(t) = \mathrm{Re} \sum_{k,l} \mathcal{E}_{k,l}^{AB}(t) e^{-i\omega_{k,l}^{AB} t} , \quad E_S(t) = \mathrm{Re} \sum_{k,l} \mathcal{E}_{k,l}^{BC}(t) e^{-i\omega_{k,l}^{BC} t} . \tag{11.49}$$

Given the field components, the Rabi frequencies are defined as

$$\Omega_{kl}^{\alpha}(t) \equiv \frac{d_{k,l}^{\alpha} \mathcal{E}_{k,l}^{\alpha}(t)}{\hbar} \equiv \mathcal{O}_{k,l}^{\alpha} f_\nu(t) , \quad \text{where} \quad d_{k,l}^{\alpha} = \langle k_{A(B)} | \mathbf{d} \cdot \hat{\epsilon} | l_{B(C)} \rangle ,$$
$$\nu = S, P , \quad \alpha = AB, CD . \tag{11.50}$$

As in the treatment of NQC and DQC, the time-dependent $0 < f_\nu(t) < 1$, $(\nu = S, P)$ functions determine the temporal properties of the Stokes and pump pulses.

The total Hamiltonian for this case is given as

$$H(t) = H_M - \frac{\hbar}{2} \sum_{i=1}^{N} \left[\Omega_{ii}^{AB}(t) e^{-i\omega_{i,i}^{AB} t} |i_B\rangle\langle i_A| + \text{h.c.} \right]$$
$$+ \frac{\hbar}{2} \sum_{i=1}^{N} \sum_{k=1}^{L} \left[\Omega_{ki}^{BC}(t) e^{-i\omega_{k,i}^{BC} t} |k_c\rangle\langle i_B| + \text{h.c.} \right] . \tag{11.51}$$

The first (H_M) term represents the diagonal material Hamiltonian; the second term represents the one-to-one resonant coupling of the initial states to the intermediate states by the pump pulse (whose frequency components are $\omega_{i,i}^{AB} \equiv \omega_{i_B} - \omega_{i_A}$);

the third term represents the resonant coupling of each intermediate state to every final state by the Stokes pulse, with frequency components $\omega_{k,i}^{BC} \equiv \omega_{k_C} - \omega_{i_B}$.

The vector of expansion coefficients $\boldsymbol{c} = (c_1, c_2, \ldots)^\mathrm{T}$ of the system wave function is obtained by solving the Schrödinger equation (Eq. (11.33)), whose time-dependent Hamiltonian is given in the RWA as

$$\underline{\underline{\mathrm{H}}}(t) = -\frac{1}{2}\begin{pmatrix} \mathbf{0} & \underline{\underline{\mathrm{H}}}^{AB\dagger} & \mathbf{0} \\ \underline{\underline{\mathrm{H}}}^{AB} & \mathbf{0} & \underline{\underline{\mathrm{H}}}^{BC\dagger} \\ \mathbf{0} & \underline{\underline{\mathrm{H}}}^{BC} & \mathbf{0} \end{pmatrix}, \tag{11.52}$$

where $H_{i,j}^{AB} = \hbar\Omega_{ii}^{AB}\delta_{ij}$ is a diagonal submatrix and $\mathrm{H}_{ik}^{BC} = \hbar\Omega_{ik}^{BC}$ is a full submatrix. We assume that the level energies are such that only the coupling of the initial states to the intermediate states by the pump pulse whose frequency components are $\omega_{i,i}^{AB}$ and the coupling of each intermediate state to every final state by the Stokes pulse, with frequency components $\omega_{k,i}^{BC}$ are sufficiently close to resonance to matter [504]. We can neglect the off-resonance terms under the assumption of sufficiently large level spacing, compared to the Rabi frequencies used [484].

$\underline{\underline{\mathrm{H}}}(t)$ has $2N + L$ eigenvalues, of which $2N$ at most are nonzero, $\lambda_{1,\ldots,2N}(t) \neq 0$. The remaining L eigenvalues are zero, $\lambda_{2N+1,\ldots,2N+L} = 0$, giving rise to dark states. By direct substitution into the null eigenvalue state $\underline{\underline{\mathrm{H}}}(t) \cdot \boldsymbol{d}_k(t) = 0$ $(k = 1, \ldots, L)$ relation we can verify that the expansion coefficients of the null eigenvalue states $\boldsymbol{d}_k(t)$ are given as

$$\begin{aligned} \boldsymbol{d}_1 &= \left(\frac{\Omega_{11}^{BC}}{\Omega_{11}^{AB}}, \ldots, \frac{\Omega_{N1}^{BC}}{\Omega_{NN}^{AB}}, 0, \ldots, 0, -1, 0, \ldots 0\right)^\mathrm{T}, \\ \boldsymbol{d}_2 &= \left(\frac{\Omega_{12}^{BC}}{\Omega_{11}^{AB}}, \ldots, \frac{\Omega_{N2}^{BC}}{\Omega_{NN}^{AB}}, 0, \ldots, 0, 0, -1, \ldots, 0\right)^\mathrm{T}, \\ &\vdots \\ \boldsymbol{d}_L &= \left(\frac{\Omega_{1L}^{BC}}{\Omega_{11}^{AB}}, \ldots, \frac{\Omega_{NL}^{BC}}{\Omega_{NN}^{AB}}, 0, \ldots, 0, 0, 0, \ldots, -1\right)^\mathrm{T}. \end{aligned} \tag{11.53}$$

In order to simplify the analysis, we assume, without loss of generality [448, 483], that $\Omega_{11}^{AB} = \Omega_{22}^{AB} = \ldots = \Omega_{NN}^{AB} = \Omega^{AB}$. With this simplification, the first N coefficients in the $\boldsymbol{d}_k(t)$ null eigenvectors are solely determined by the vector of the Stokes-field Rabi frequencies $\boldsymbol{\Omega}_{Sk} = (\Omega_{1k}^{BC}, \Omega_{2k}^{BC}, \ldots, \Omega_{Nk}^{BC})^\mathrm{T}$. Moreover, the $\boldsymbol{d}_k(t)$ states correlate on a one-to-one basis with the $|1_C\rangle, \ldots, |L_C\rangle$ final states. Thus, if we manage to initially populate only one of the $\boldsymbol{d}_k(t)$ states, AP would exclusively transfer the population to a single final $|k_C\rangle$ state at the end of the process.

If the transfer is exclusively to a single state, the decoding process is perfect. If the transfer is nonexclusive, but still predominantly to a single target state, we deem the decoding, though not perfect, adequate. This is the case in many trial cases: It is not possible to populate just a single dark state. In fact, the initial population of the $|1_A\rangle, \ldots, |N_A\rangle$ states is distributed among several $(k = 1, \ldots, L)$ null eigenvalue states. The distribution p_k is determined by the square of the projection of the vector of initial coefficients $\boldsymbol{c}_0 = (c_1, c_2, \ldots, c_N)^\mathsf{T}$ on the vector of Stokes-field Rabi frequencies $\boldsymbol{\Omega}_{Sk}$, that is, $p_k \propto |\boldsymbol{c}_0^\dagger \cdot \boldsymbol{\Omega}_{Sk}|^2$.

We can maximize the population p_i, transferred to the predetermined $|i_C\rangle$ state, by choosing $\boldsymbol{\Omega}_{Si} \propto \boldsymbol{c}_0$. At the same time, it is in principle possible to minimize transfer to undesired $|k_C\rangle (k \neq i)$ states, by requiring that the $\boldsymbol{\Omega}_{Sk}$ and $\boldsymbol{\Omega}_{Si}$ vectors be orthogonal to each other, $\boldsymbol{\Omega}_{Sk}^\dagger \cdot \boldsymbol{\Omega}_{Si} \approx 0$. Whereas, the first of the above conditions is easily satisfied, the second condition cannot be fully realized if $N < L$, because one cannot orthogonalize all the L vectors, $\boldsymbol{\Omega}_{Sk}$, in an $N(< L)$ dimensional space. Only if $N = L$ can their components be chosen such that all the L null eigenvalue states of Eq. (11.53) be mutually orthogonal. In that case the transfers to the final $|k_C\rangle$ states are both complete and exclusive. The results show that unless $N \ll L$, in most cases adequate, though not perfect, decoding can be realized.

As an example, we examine the transfer of numeric information stored in the form of $M = 2$ phases of the population coefficients of N equally populated initial levels $|i_A\rangle$. In this phase encoding, the binary number $0 = (0, 0, \ldots, 0)$, that is, $0 \cdot 2^0 + 0 \cdot 2^1 + \cdots + 0 \cdot 2^N$, is represented by the vector of amplitudes $c_1 = (1, 1, \ldots, 1)$ of the $|i_A\rangle$ states; the binary number $(1, 0, \ldots, 0)$, that is, $1 \cdot 2^0 + 0 \cdot 2^1 + \cdots + 0 \cdot 2^N = 1$, by $c_2 = (-1, 1, \ldots, 1)$, and so on. Because quantum states are known up to an overall phase, it is possible, using N levels, to encode in this manner only 2^{N-1} numbers.

Figure 11.11 illustrates the action of the adiabatic decoder. The magnitudes of all Rabi frequencies are $|\mathcal{O}_{ij}| = 30/\tau$, where $f_\nu(t) = \exp[-(t - t_\nu)^2/\tau^2]$, $(\nu = S, P)$, and $t_P - t_S = 2\tau$. The system decodes each of the binary phase stored number by transferring the population of the initial superposition state, representing this number, to predominantly *one* of the 16 final $|k_C\rangle$ eigenstates. The degree of exclusivity of the transfer is such that 5 other final states end up with 2.78 times less population, while the remaining 10 states being considerably less populated.

It is possible to obtain analytically the populations p_k of the final $|k_C\rangle$ states using the fact that $p_k \propto \left|\boldsymbol{c}_0^\dagger \cdot \boldsymbol{\Omega}_{Sk}\right|^2$, where $\boldsymbol{c}_0$ is the encoded initial vector $\boldsymbol{c}_0 = (c_1, c_2, \ldots, c_5)$, and $\boldsymbol{\Omega}_{Sk}$ is the Stokes Rabi frequency vector. For example, when encoding the number zero, $\boldsymbol{c}_0 = (1, 1, 1, 1, 1)^\mathsf{T}$, and choosing $\mathcal{O}_{S1} = (1, 1, 1, 1, 1)^\mathsf{T}$ and $\mathcal{O}_{S2} = (-1, 1, 1, 1, 1)^\mathsf{T}$, one obtains that $p_1 \propto |\boldsymbol{c}_0^\dagger \cdot \mathcal{O}_{S1}|^2 = 25$ and $p_2 \propto |\boldsymbol{c}_0^\dagger \cdot \mathcal{O}_{S2}|^2 = 9$, so that $p_1/p_2 = 25/9 \approx 2.78$, as in Figure 11.11. Thus the scalar products $\boldsymbol{c}_0^\dagger \cdot \mathcal{O}_{S1-16}$, yield all the probabilities, p_{1-16}; five of which are $\frac{9}{25} p_1$ and ten are $\frac{1}{25} p_1$. Given the normalization factor $n = 1 + 5(9/25) + 10(1/25) = 3.2$, one obtains that $p_1 = 1/n = 0.3125$, determining the populations p_k in all the states. We can increase the density of encoding [505], by using higher q-nit states, obtained by using $M > 2$ roots of the identity $\exp(i2\pi j/M)$. A system with N initial states and information stored in M possible phases gives rise to M^{N-1} numbers which can be decoded using the M^{N-1} final states.

The methodology used in decoding works equally well for encoding. The encoding process can be phrased as starting with a single state and transferring its population according to some preset protocol to a linear combination of other states. The method chosen for encoding, which is to time-reverse the decoding pulse se-

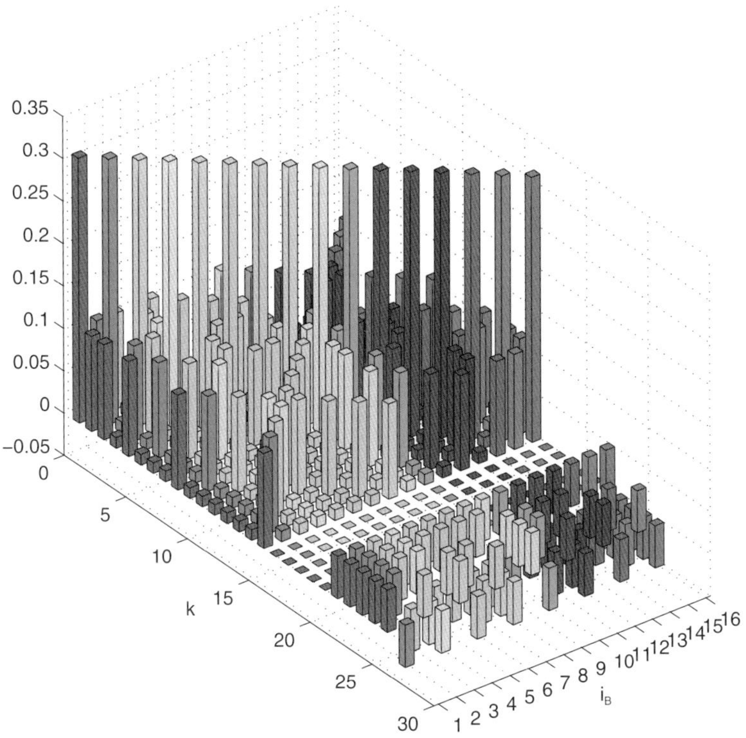

Figure 11.11 The results of the binary-to-hexadecimal decoder. In the back we see the populations of the $2^{N-1} = 16$ final levels transferred from superpositions of the $N = 5$ initial states, shown up front. The intermediate states, in the midregion, are unpopulated. The c_i ($i = 1$–5) amplitudes of the initial states are depicted by bars 1/10 in height (instead of the correct $1/\sqrt{5}$ values). The decoding of different binary numbers is manifested in that the bulk of the transferred population ends up in one final state, whose identity depends on the encoded number. Other final states are at least 2.78 times less populated. Taken from [499, Figure 2]. (Reprinted with permission. Copyright 2004 Elsevier BV.)

quence, is very effective [499] in that the population of the initial state is predominantly deposited in the encoded states. In fact, P_{tran}, the probabilities of transfer from each initial state to the correct linear combination of encoded states, as obtained by the time reversed method, is identical to the decoding process, that is, $P_{tran} = p_{max}$. Thus the scheme appears to behave like a complex but passive transmissive system.

11.5 Multistate Piecewise Adiabatic Passage

In this section we discuss a method of executing complete population transfers between quantum states in a piecewise manner using a train of short laser pulses. When done adiabatically, the method, is called "piecewise adiabatic passage" (PAP) [506–508]. The PAP process is related to the concepts of "coherent accumulation" [509, 510] and the Ramsey experiment [511], where a slow process is implemented in a piecewise manner using a train of short, mutually coherent, laser pulses. PAP is a special case of CCAP with a set of discrete frequencies discussed in Sections 12.1.2 and 12.2. The connection is due to the fact that a pulse composed of a set of discrete frequencies appears as a train of pulses in the time domain. PAP can be applied to a large class of problems because it benefits from the high peak powers and large spectral bandwidths afforded by femtosecond pulses. As we show below, the degree of population transfer is robust to a wide variation in the absolute and relative intensities, durations, and time ordering of the pulses.

Figure 11.12a,b shows a computation for an ordinary ("reference") STIRAP for $|1\rangle = 3s$, $|2\rangle = 3p$, and $|3\rangle = 3d$ electronic states of the Na atom, using the empirically derived [512] $d_{i,j}$ matrix elements. In Figure 11.12a,b we use two circularly polarized laser pulses with sine-squared $\varepsilon_{p(d)}$ pump (dump) field envelopes and temporal duration of 0.75 ps. The pulses are applied with a 0.6 ps delay and have central wavelengths $\lambda_p = 2\pi c/\omega_p = 589$ nm and $\lambda_d = 2\pi c/\omega_d = 819$ nm. As expected, we observe a complete transfer of population from state $|1\rangle$ to state $|3\rangle$, while never significantly populating state $|2\rangle$.

We now modify the above scenario by letting the material system evolve according to the above scheme until time t_1, at which point both laser fields are abruptly turned off. At some later time t_2, we abruptly turn both fields back on, making sure that the phase of each field at $t = t_2$ equals that at $t = t_1$. Since the coefficients $b_i(t)$ do not evolve during the zero-field period, $b_i(t > t_2)$ continue to follow the adiabatic passage with $b_i(t_1)$ serving as a new starting point. Repeating the turn-off/turn-on procedure many times allows us to execute the AP process in a controlled piecewise manner.

Figure 11.12c,d demonstrates the action of a pulse sequence obtained by turning off the original reference STIRAP fields at periodic intervals, while scaling up the $\varepsilon_{p(d)}(t)$ pump (dump) pulses during the "on" periods, as described in detail below. As shown, the resulting pulse sequence closely reproduces the effect of an ordinary AP. Moreover we find that this piecewise adiabatic passage (PAP) procedure is insensitive to the particular shape and strength of the pulses, as well as to the durations of the "off" periods.

A soft turn-on/turn-off version of PAP is shown in Figure 11.12e–h. In Figure 11.12e,f the pump and the dump laser fields are obtained by the spectral shaping of 75 fs pulses:

$$\varepsilon^{shaped}_{p(d)}(\omega) = \varepsilon^{unshaped}_{p(d)}(\omega) F_{p(d)}(\omega) , \tag{11.54}$$

where $F_{p(d)}(\omega) = \exp\{i A_{p(d)} \sin[\Delta T(\omega - \omega_{p(d)})]\}$ with $A_{p(d)} = \mp 1.2$, $\Delta T = 229$ fs. The resulting pulse sequence, shown in Figure 11.12e, is comprised of copies of the original transform limited femtosecond pulse of duration τ separated in time by ΔT, and with smoothly changing amplitudes defined by the value of A [513]. We see that the "soft" turn-on/turn-off PAP is as successful in bringing about the desired population transfer as the abrupt turn-on/turn-off version discussed above.

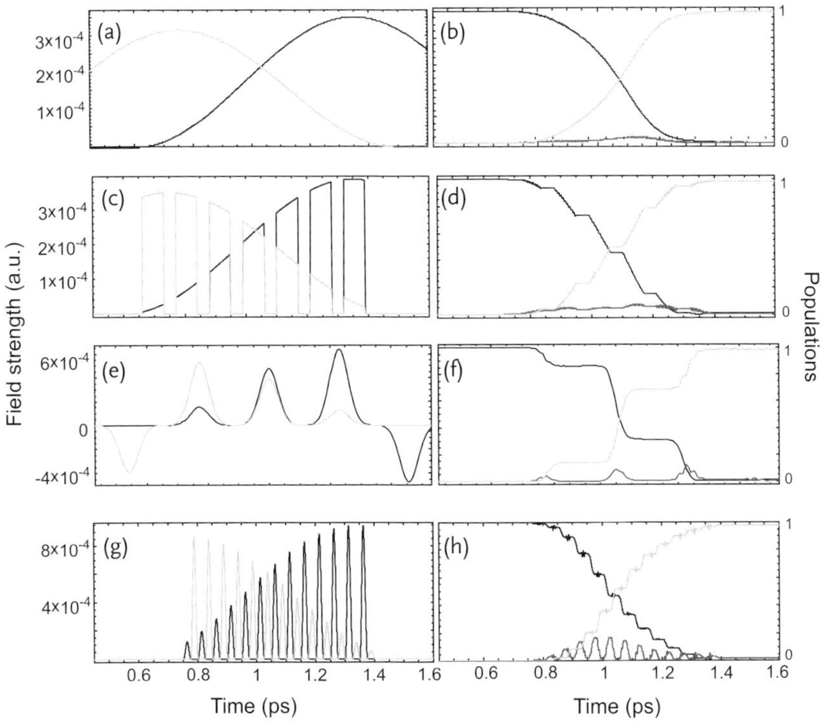

Figure 11.12 Four types of PAP processes: (a,c,e,g) are the pulse sequences where black is the pump, and light gray is the dump. (b,d,f,h) are the populations ($|b_i(t)|^2$) where black is for $i = 1$, dark gray is for $i = 2$, and light gray is for $i = 3$. (a,b): The reference AP. (c,d): PAP with sharp turn-ons and turn-offs. (e,f) and (g,h): PAP with smooth pulses. Taken from [506]. (Reprinted with permission. Copyright 2007 American Physical Society.)

It is instructive to treat each pump-dump segment of the combined pulse train as a deviation from the smoothly varying reference AP field (see Figure 11.12a). We do so by writing the interaction Hamiltonian as $H_{MR}(t) = H_0(t) + H_1(t)$, where $H_{0(1)}(t) \equiv -\boldsymbol{d} \cdot E_{0(1)}(t)$. Here $E_0(t) = \mathrm{Re}\{\varepsilon_0(t)\exp(-i\omega_0 t)\}$ is the reference field from the ordinary AP Hamiltonian and $E_1 = \mathrm{Re}\{\varepsilon_1(t)\exp(-i\omega_1 t)\}$ describes the deviation of the PAP field from it. We require $E_1(t)$ to have zero pulse area, that is, $\int_\tau dt\varepsilon_1(t) = 0$. This requirement makes the PAP evolution operator and the reference AP evolution operator be essentially the same during each segment [514, Figure 1]

$$S_n = \mathcal{T}\exp\left[-i\int_{t_n-\frac{\tau}{2}}^{t_n+\frac{\tau}{2}} H_0(t)dt - i\int_{t_n-\frac{\tau}{2}}^{t_n+\frac{\tau}{2}} H_1(t)dt\right]$$
$$\approx \mathcal{T}\exp\left[-i\int_{t_n-\frac{\tau}{2}}^{t_n+\frac{\tau}{2}} H_0(t)dt\right] \equiv S_{0n}\,, \qquad (11.55)$$

where $\mathcal{T}$ denotes the time ordering operator and S_{0n} is the evolution operator of the reference AP during the nth segment.

To simplify the method and maintain its robustness we replace the exact zero area requirement of H_1 with a coarse grained requirement, namely that $\int_{t_n-\tau/2}^{t_n+\tau/2} E(t)dt = E_0(t_n)\tau$. Constructed this way, the field of the pulse train is only defined by the coarse-grained $E_0(t)$ profile, and is insensitive to its short-time behavior. At the same time, since the pulse area of the reference AP satisfies the adiabaticity condition, so does the area of the PAP pulse train.

The behavior discussed here can also be viewed as a coherent accumulation discussed in [509, 510]. It is easy to see that when the target states are equidistant in energy the shaping of a pulse with multiple local frequencies results in a train of mutually coherent ultrashort pulses separated by the evolution period of the wave packet. Though each pulse in the train transfers only a small amount of population to the target superposition state, population of that state coherently accumulates [515–519] piece by piece, reaching 100% at the end of the interaction regardless of the total energy of the pulse train [450]. Thus, the process depicted in Figure 11.12g,h where the pump and the dump pulses do not overlap, can also be viewed as the driving of population by the pump into the intermediate level followed by its dump into the target state. The cumulative action of the train of coherent nonoverlapping pulses is insensitive to the exact intensity of the pump and dump trains, as long as their envelope mimics that of the reference AP.

A very informative, though, in general, incomplete, description of the system is given by following the trajectory of the amplitudes vector $\boldsymbol{a}_{x(y)(z)} \equiv |b_{1(2)(3)}|$ shown in Figure 11.13a. Each pump pulse, transferring populations between states $|1\rangle$ and $|2\rangle$, rotates $\boldsymbol{a}$ by a small angle $\alpha_P = \int_\tau \Omega_P^*(t)dt$ about the z-axis. Likewise, each dump pulse rotates $\boldsymbol{a}$ about the x-axis by an angle $\alpha_D = \int_\tau \Omega_D(t)dt$. These two rotations result in an overall rotation of $\boldsymbol{a}$ by an angle α_0 about an axis defined by the (θ_0, ϕ_0) polar and azimuthal angles, given in lowest-order expansion of α_P, α_D

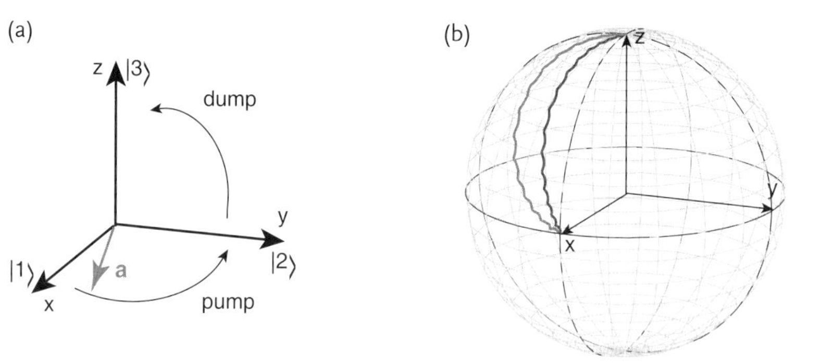

Figure 11.13 (a): Depiction of the $\boldsymbol{a} \equiv |b_1|, |b_2|, |b_3|$ amplitudes vector. (b): Two sample trajectories of the $\boldsymbol{a}$ vector during the PAP process. Both the pump and the dump sequences of small rotations rotate the vector by an overall angle of $10 \times 2\pi$. The left (right) trajectory corresponds to 50 (100) pulses in each sequence. Taken from [506, Figure 3]. (Reprinted with permission. Copyright 2007 American Physical Society.)

as,

$$\alpha_0 = \left[\frac{(\alpha_P^2 + \alpha_D^2)}{2}\right]^{\frac{1}{2}}, \quad \phi_0 = -\frac{\alpha_P}{2}, \quad \tan\theta_0 = \frac{\alpha_D}{\alpha_P}. \tag{11.56}$$

As $\Omega_{p(d)}(t)$ evolve, the otherwise stable (θ_0, ϕ_0) vector, which coincides with the null vector of our reference AP moves slowly from being aligned along the x-axis to being aligned along the z-axis. As shown in Figure 11.12, as long as the individual α_P and α_D are small, the $\boldsymbol{a}$ vector will faithfully follow the (θ_0, ϕ_0) null vector. Piecewise adiabaticity can be shown to be guaranteed as long as,

$$\Delta\theta_0 \ll \left[\frac{(\alpha_P^2 + \alpha_D^2)}{2}\right]^2. \tag{11.57}$$

This condition coincides, up to an insignificant numerical factor, with the adiabaticity condition, Eq. (11.15), of conventional AP. We thus expect the overall evolution to be adiabatic as long as Eq. (11.57) is satisfied and the α_P, α_D angles at each step are small. These expectations were verified numerically for various sequences of alternating rotations with slowly varying amplitudes. Figure 11.13b depicts two such trajectories differing by the number of pump-dump pulses and the pump-dump amplitudes. The trajectory starts at the state $|1\rangle$ and follows the AP route slightly outside the zeroth meridian (i.e., somewhat populating state $|2\rangle$) to coincide finally with state $|3\rangle$.

In piecewise adiabatic passage, the alternating operators for the pump and dump transitions may be of quite general nature, and may act in a Hilbert space that is much larger than that spanned by three states. The states themselves are not required to be nondegenerate, nor even energy eigenstates of the system. The only necessary requirements are that each pump operator transfers a fraction of population between an initial and some intermediate state, each dump operator transfers population between that intermediate and some desired target state, and that the coherence between the three states is preserved throughout the process.

The ability to alternate between the pump and the dump segments can be useful in minimizing deleterious effects of the Stark shifting of levels. This is particularly the case in the two-photon + one-photon (2 + 1) stimulated hyper-Raman adiabatic passage (STIHRAP [433, 434, 462, 520]) whose implementation with short laser pulses is considered very difficult due to the strong time-dependent AC Stark shifts of the levels. Although one can still transfer the population into the target level by detuning both the pump and the dump lasers from the exact resonance [520], the robustness of the method is significantly reduced since the conditions connecting the field-induced Stark shifts with the required frequency offsets are quite stringent.

The set of additional control knobs available in the PAP scheme includes the number of segments which comprise the pulse sequences, and the absolute phases of the pump and dump pulses. Similar to the Ramsey interference effect [511], if either one or both fields are detuned off the exact resonance, the probability

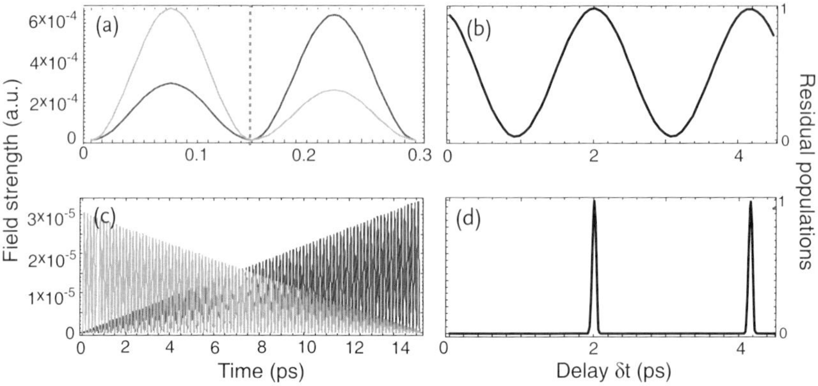

Figure 11.14 The turn-off periods dependence. (a,c): The field envelopes of the pump (dark gray) and dump (light gray) pulse sequences. The dotted line in panel (a) marks the onset of the turn-off period(s). (b,d): The final $|b_3|^2$ population as a function of the duration of the turn-off period(s). Taken from [506, Figure 2]. (Reprinted with permission. Copyright 2007 American Physical Society.)

transfer in PAP becomes sensitive to the duration of the turn-off periods. In Figure 11.14a we show PAP performed with just two pairs of (75 fs) pump and dump pulses inducing one-photon transitions between the levels 3s, 3p, and 3d of Na. Figure 11.14b depicts the final population of the state 3d as a function of the duration of the turn-off periods. Figure 11.14c,d shows the same results for the pulse trains composed of 100 segments with linearly increasing (decreasing) pump (dump) field strengths. The dump wavelength was set to $\sim$ 818 nm, 1 nm away from the exact 3p–3d resonance.

We first note that even as few as two pairs of pulses suffice to reproduce the effect of conventional AP. It can be also seen in Figure 11.14b and d, that there is a strong dependence of the final state population on the duration of the turn-off period(s). The sharp peaks seen in the latter case are analogous to the Ramsey fringes: the longer the pulse trains the narrower the fringes. This suggests that applying PAP to a metastable target state implemented with a frequency comb source [521] might present a useful alternative to the precision measurements of atomic transitions and frequency standards.

11.5.1
Multistate Piecewise Adiabatic Passage – Experiments

In this section we describe two experiments [507, 508] which demonstrate the use of multistate PAP in achieving a complete population transfer between quantum wave packets. One experiment deals with transfer between two energy eigenstates [507]; the other [508] is its extension to the multilevel target states case, in which one controls the amplitudes and relative phases of the fine structure doublet in atomic potassium. Both experiments display the theoretically expected features discussed above; namely that population transfer is complete; that the process is robust; and that it is insensitive to the exact value of the pulse intensities, delays and shapes, as long as one maintains adiabaticity. Of additional interest is that the multistate experiment uses the 1986 bichromatic perturbative control method discussed in Section 3.2 to verify the makeup (amplitudes and phases) of the target wave packets created in the process.

11.5.1.1 Chirped Adiabatic Passage

In addition to the case of AP of multilevel wave packets through intermediate states by two time delayed pulses considered in Section 11.1.1 above, it is also possible to execute AP without an intermediate state by sweeping, or "chirping", an instantaneous frequency of a single pulse or a train of pulses across a resonance [507]. The adiabatic crossing of the excited state by the ground state dressed with a photon of a given instantaneous frequency results in complete population transfer between those *two* states. By introducing local frequency chirp about each resonant transition frequency (Figure 11.15b), we can realize a *simultaneous* AP into multiple target states which make up the final wave packet. Figure 11.15b provides an intuitive illustration of the robustness of the method, not only with respect to the field amplitude and frequency, but also with respect to the dynamic level shifts associated with the presence of a strong polychromatic field. Crossings of the dressed ground state with multiple excited states will occur even if the latter are heavily perturbed

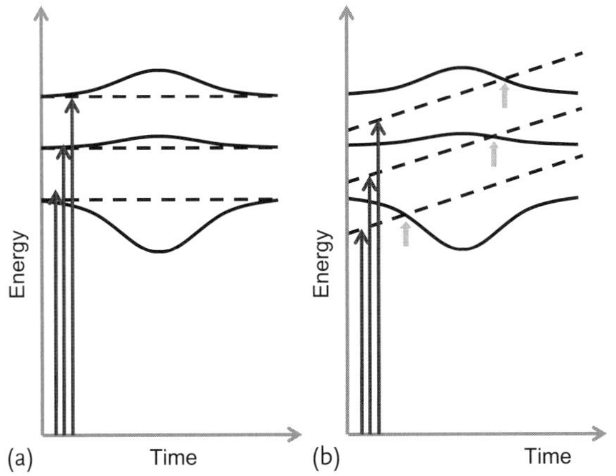

Figure 11.15 Illustration of the interaction of a multilevel system with polychromatic coherent radiation consisting of a number of discrete spectral components (black vertical arrows). Dashed lines represent the ground state "dressed" with a photon. Bare excited states of the system (solid lines) are dynamically shifted in energy due to the presence of nonresonant components of the laser field. This results in incomplete population transfer via nonadiabatic process (a). When the spectral components are simultaneously chirped in frequency (b), adiabatic transfer into a coherent superposition of excited states is executed. Different timing of the level crossings (gray arrows) complicates the control over the resulting wave packet, but does not affect the completeness of excitation. Taken from [507, Figure 1]. (Reprinted with permission. Copyright 2008 American Physical Society.)

by the off-resonance components of the driving field. Dynamic shifts of transition frequencies result in a slightly different timing of each adiabatic crossing. This discord between multiple APs may slightly complicate achieving the target wave packet, but it does not reduce the efficiency of the population transfer. This statement is verified in Figure 11.16 presenting numerical calculations of a PAP process with a train of ultrashort pulses in which one transfers populations from a single state to a superposition of two states.

In the multistate PAP transfer experiment [508] the probing of the resultant wave packet is done using the bichromatic coherent control scenario [21–25] described in Section 3.2. Applying that scenario to the present case we write the wave packet we wish to probe as

$$|\chi(t)\rangle = b_0(t)|4S_{1/2}\rangle + b_1(t)|4P_{1/2}\rangle + b_2(t)|4P_{3/2}\rangle \ . \tag{11.58}$$

Because of the low probe energy, the ionization probability is proportional to the population of the $5S_{1/2}$ state induced by the bichromatic transition from the $4P_{1/2}, 4P_{3/2}$ doublet. In the perturbative limit the population of the $5S_{1/2}$ state is given [21–25] as

$$P(t) \propto |b_1(t)|^2 |\epsilon(\omega_1)|^2 d(11) + |b_2(t)|^2 |\epsilon(\omega_2)|^2 d(22) + 2\mathrm{Re}\left[b_1(t) b_2^*(t) \epsilon(\omega_1) \epsilon^*(\omega_2) d(12)\right] \tag{11.59}$$

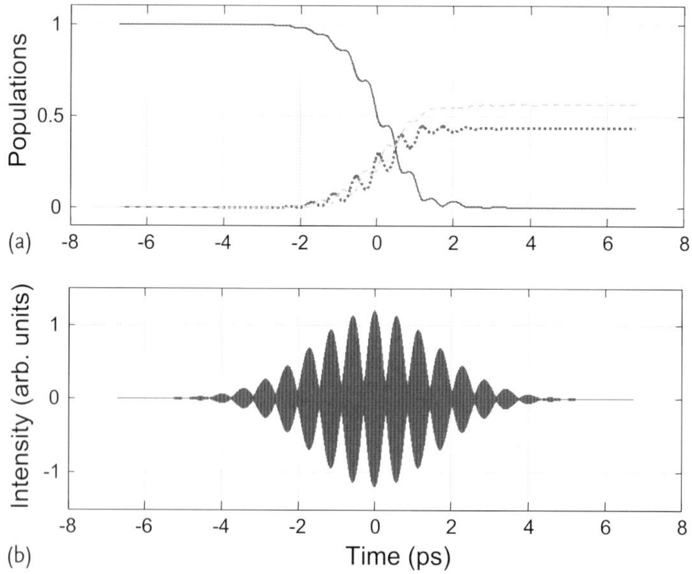

Figure 11.16 A numerical simulation of PAP for potassium atom. Population transfer is shown in (a) as a function of time. The solid line represents the ground state population while the dashed and dotted lines correspond to the populations of $4P_{1/2}$ and $4P_{3/2}$ excited states, respectively. The excitation field amplitude is plotted in (b) on the same time scale. Taken from [507, Figure 2]. (Reprinted with permission. Copyright 2008 American Physical Society.)

where $b_1(t)$ and $b_2(t)$ are the excited eigenstate amplitudes (Eq. (11.58)), $\epsilon(\omega_1)$ and $\epsilon(\omega_2)$ are *probe* field amplitudes at the resonant probe transition frequencies of $4P_{1/2} \rightarrow 5S_{1/2}$ and $4P_{3/2} \rightarrow 5S_{1/2}$, respectively, and $d(ij) = \langle\psi_j|\boldsymbol{d}\cdot\hat{\epsilon}|5S_{1/2}\rangle\langle 5S_{1/2}|\boldsymbol{d}\cdot\hat{\epsilon}|\psi_i\rangle$ with $\boldsymbol{d}$ being the transition-dipole operator.

The last term in Eq. (11.59) is the interference term which depends on the difference between the relative phase of the excited wave functions $b_{1,2}$ and the relative phase of the two resonant probe field components $\epsilon(\omega_{1,2})$. Since the latter is constant, the ionization probability, and therefore the ion signal recorded as a function of the pump-probe time delay, oscillates as $\cos[(\omega_1 - \omega_2)t + \alpha(12)]$, where $\alpha(12)$ is the material phase of Section 3.2 $\alpha(12) = \arg\{b_1^* b_2 d(12)\}$. Thus the relative amplitudes and phases of the $4P$ states populated in the process can be extracted from the oscillation contrast and phase of the ion signal [522].

11.5.1.2 Rabi Flopping

In order to calibrate the dipole matrix elements, the excited state populations and the excitation pulse area, the conditions were set such that the system exhibited separate Rabi flopping of each of the $4P$ states. In both cases, the spectral phase was kept flat across the whole spectrum of a pulse. With the available laser power, the experiment reached pulse areas of up to 3π for each transition. By fitting the data with a $\sin^2 A_{1,2}$ function, it was possible to calibrate A_1 and A_2, the areas of the $4P_{1/2}$ excitation pulse (D_1) and the $4P_{3/2}$ excitation pulse (D_2) as a function of the energy. Simultaneously, the magnitude of the ion signal corresponding to the pulse area of π was measured. These calibrations were then used for assessing the populations.

In order to execute an adiabatic passage into a single excited state, a quadratic spectral phase chirp $\varphi(\omega) = (\alpha/2)(\omega - \omega_{1,2})^2$ was introduced near one of the resonant frequencies $\omega_{1,2}$ (see Figure 11.17b), while blocking the pulse near the other. The frequency chirp was gradually increased until the population transfer displayed AP-like saturation with the intensity, as shown in Figures 11.18b,c for both the D_1 and D_2 transitions. This occurred at $\alpha = 270 \times 10^3\ \mathrm{fs}^2$. For both transitions, the ion signal peaks at a pulse area of $\approx \pi$ and stays relatively flat with increasing pulse energy.

Once the pulse area and chirp required to satisfy the adiabaticity conditions were determined, population transfer into a superposition of $4P_{1/2}$ and $4P_{3/2}$ states was executed. According to Eq. (11.59), if both states are populated coherently, quantum beats should be observed in the ionization signal as a function of the probe pulse delay. At the same time, interference between the two light pulses at the D_1 and D_2 frequencies results in the time domain in a train of pulses separated by the inverse of the $4P_{1/2}$ and $4P_{3/2}$ fine structure splitting (1.73 THz) that is, 578 fs. A numerical example of such pulse train is shown in Figure 11.16b.

Rabi flopping results when the spectral phase of pump pulses is kept flat across the whole double-peak spectrum. In the time domain, such flat spectral phase translates into a train of pulses with constant center frequency displaying no phase shift between consecutive pulses [450]. The resulting Rabi flopping between the

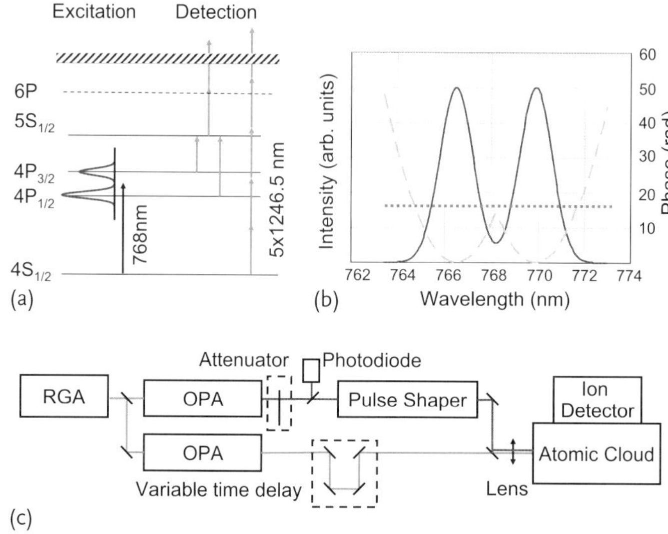

Figure 11.17 (a) The relevant quantum states of potassium. The double-peaked curve represents the shaped spectrum of the pump pulses. The gray arrows depict the possible ionized channels by the probe pulses. (b) The spectral shaping of the pump pulse whose local frequency chirping is centered about the two electronic resonances at 766.5 and 766.9 nm: Intensity (solid) and phase (dashed). The dash-dotted line represents the flat spectral phase used for a nonadiabatic excitation scheme. (c) The experimental setup. Two optical parametric amplifiers (OPA) are pumped by a Ti:sapphire regenerative amplifier (RGA) producing 130 fs 2 mJ pulses at 1 KHz repetition rate. One OPA is used to generate pump pulses of variable energy, controlled by an attenuator and measured by a fast photodiode. Each pulse in the train of pump pulses is shaped by a pulse shaper and is weakly focused on a cloud of Potassium atoms inside a vacuum chamber. Pulses belonging to the train of probe pulses are delayed by a variable time delay and tightly focused on the atomic cloud. Potassium atoms, ionized by the probe pulses, are accelerated towards and detected by a multichannel plate based ion detector. Taken from [508, Figure 3]. (Reprinted with permission. Copyright 2009 American Physical Society.)

ground state and the excited wave packet are shown in Figure 11.19a. The two-dimensional plot shows the ion signal as a function of the time delay (vertical axis) and effective pulse area (horizontal axis), which in the case of multiple excitation channels can be used as a convenient scale of the interaction strength, defined as

$$A_{eff} = \sqrt{A_1^2 + A_2^2}\,, \qquad (11.60)$$

where $A_i \equiv \int_{-\infty}^{\infty} \Omega_i(t)dt$, with Ω_i being the time-dependent Rabi frequencies for the ith transition.

The oscillations along the vertical axis indicate quantum beating between $4P_{1/2}$ and $4P_{3/2}$ states. For the flat spectral phase both experiments and computations (Figure 11.19a,c) demonstrate the strong dependence of the excitation on the effective pulse area. Rabi flopping between the ground state and the excited wave packet are manifested by the periodic reappearance of the beat signal as a function of the field strength. Thus, the high beating contrast at $A_{eff} = \pi$ disappears almost com-

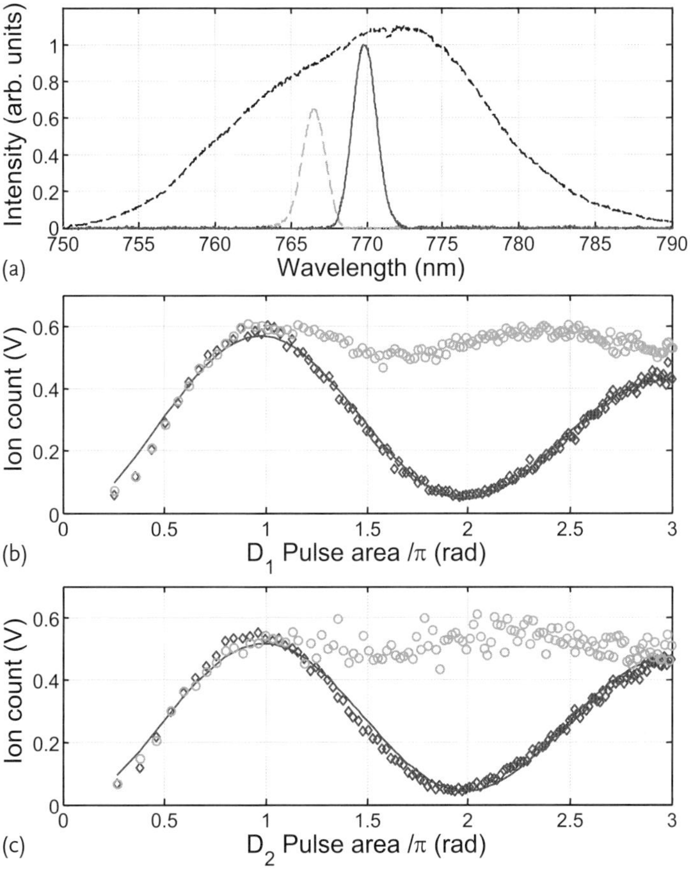

Figure 11.18 Adiabatic and nonadiabatic excitation of a single ($^4P_{1/2}$ or $^4P_{3/2}$) excited state. (a) Pump pulse spectra used to excite either D_1 or D_2 transitions separately (solid and dashed gray lines, respectively). Spectral profile of the initial unshaped femtosecond pulse is shown as dashed black curve. (b, c) Nonadiabatic (diamonds) and adiabatic (circles) population transfer into $^4P_{1/2}$ (b) and $^4P_{3/2}$ (c) states. Ion signal, proportional to the target state population, is plotted as a function of the corresponding pulse area, $A_{1,2}$. Solid lines show the anticipated $\sin^2(A_{1,2})$ dependence fitted to the experimental data. Taken from [508, Figure 4]. (Reprinted with permission. Copyright 2009 American Physical Society.)

pletely at $A_{eff} \approx 2\pi$, as shown by the vertical cross sections of the two-dimensional data of Figure 11.19b.

The effect of the off-resonance components is mainly to shift the resonance positions. For example, the $4P_{1/2}$ level is shifted due to the presence of an off-resonant D_2 pulse, whereas the $4P_{3/2}$ level is shifted by an off-resonant D_1 field. Since the shifts are in general unequal, the accumulated quantum phase of the two states, $|4S_{1/2}\rangle$ and $|4S_{3/2}\rangle$, depends on the energy of both pulses. The effect is reproduced by the numerical calculations shown in Figure 11.19c. Figure 11.19d demonstrates the degree to which the dependence of the $|b_1(t)|^2$ and $|b_2(t)|^2$ populations on

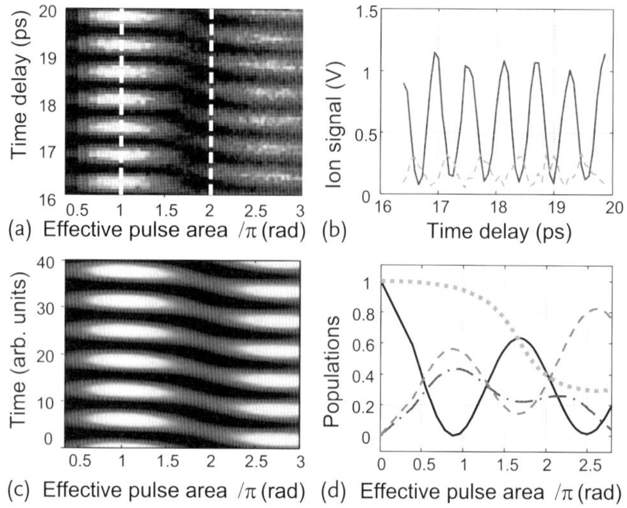

Figure 11.19 Nonadiabatic population transfer into a superposition of $4P_{1/2}$ and $4P_{3/2}$ states. The two-dimensional plots ((a) experiment, (c) calculation) show the ion signal as a function of the effective pulse area and pump-probe time delay. Oscillations along the vertical (time) axis reflect quantum beating due to the time evolution of the wave packet. The quantum beats are shown in (b) for pulse areas of π (solid) and 2π (dashed), corresponding to the dashed vertical white lines in (a). In plot (d), we present the calculated populations of the two excited states, $4P_{1/2}$ (dash-dotted) and $4P_{3/2}$ (dashed), and the relative phase (dots) between the corresponding wave functions, as a function of the effective pulse area of the excitation field. The ground state population is shown as a solid black line. Taken from [508, Figure 5]. (Reprinted with permission. Copyright 2009 American Physical Society.)

pulse energy deviates from the periodic Rabi oscillations behavior, once the effective pulse area exceeds π.

In contrast to Rabi flopping, when the pulses are chirped (Figure 11.17b), the observed quantum beats become insensitive to the pulse area (Figure 11.20a,c). Figure 11.20b shows that when the D_1 and D_2 frequency chirp equals 270×10^3 fs^2, the contrast of the experimentally observed quantum beats changes only little throughout the entire $A_{eff} \approx \pi$ to $A_{eff} \approx 3\pi$ range of pulse areas. This demonstrates the robustness of the global population transfer with respect to pulse energy changes, in agreement with the above theoretical discussion of the AP process.

The experiment demonstrated a threshold of pulse area for the onset of AP at $A_{eff} \approx \pi$. Beyond this value the *global* population transfer is complete for all pulse areas, as can be seen by comparing Figures 11.19d and 11.20d. In spite of the robustness of the overall transfer, as shown in Figure 11.20d, the ratio of populations of the two excited states varies somewhat with the excitation pulse energy. The dynamic Stark shift which causes this effect does not however affect the *relative phases* of the excited states, which, as shown in Figure 11.20b,d, remain constant with pulse area.

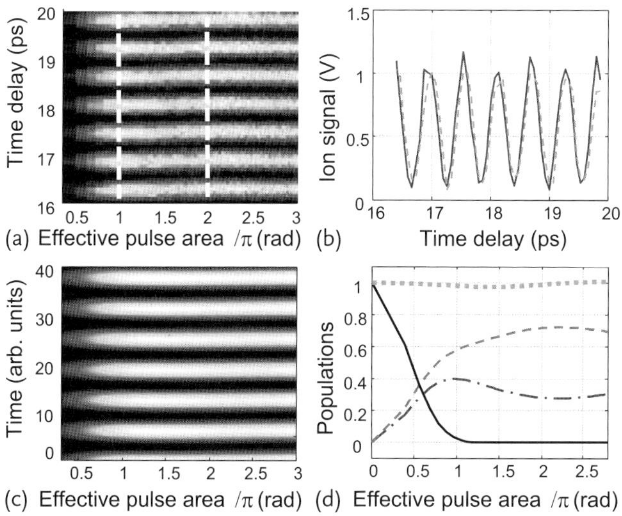

Figure 11.20 Adiabatic population transfer into a superposition of $4P_{1/2}$ and $4P_{3/2}$ states. The two-dimensional plots ((a) experiment, (c) calculation) show the ion signal as a function of the effective pulse area and pump-probe time delay. The oscillations along the vertical (time) axis reflect quantum beating due to the time evolution of the wave packet. The quantum beats are shown in (b) for pulse areas of π (solid) and 2π (dashed), corresponding to the dashed vertical white lines in (a). In plot (d), we present the calculated populations of the two excited states, $4P_{1/2}$ (dash-dotted) and $4P_{3/2}$ (dashed), and the relative phase (dots) between the corresponding wave functions, as a function of the effective pulse area of the excitation field. The ground state population is shown as a solid line. Taken from [508, Figure 6]. (Reprinted with permission. Copyright 2009 American Physical Society.)

The control over the relative phases of the two states was achieved by adding on top of the frequency chirp a constant phase to the D_1 or the D_2 pulses. The resulting phase profile, attained using the pulse shaper, is shown in Figure 11.21a for a relative phase of 0, $\pi/2$ and π radian. Two-dimensional energy-time scans for these three phase shifts are shown in Figure 11.21b. The vertical cross sections of each two-dimensional plot, displaying quantum beat patterns corresponding to $\approx$ π pulses, are plotted in Figure 11.21c. The phase of the oscillations, which directly reflects the relative phase of the $4P_{1/2}$ and $4P_{3/2}$ eigenstates, closely follows the extra phase shift introduced via the pulse shaper.

As discussed above and shown explicitly in [450], for the case of *negligible Stark shifts*, the state amplitudes at the end of the pulse are simply proportional to Rabi frequencies of the corresponding transitions. Thus, the distribution of populations among excited states (here, $4P_{1/2}$ and $4P_{3/2}$) can be controlled by changing the relative strength of the corresponding spectral components of the excitation field (here, the energies of D_1 and D_2 pulses). This simple strategy fails, however, once the dynamic Stark shifts become comparable to the energy bandwidth of the laser pulses. This situation is illustrated in Figure 11.15b, where one can see that un-

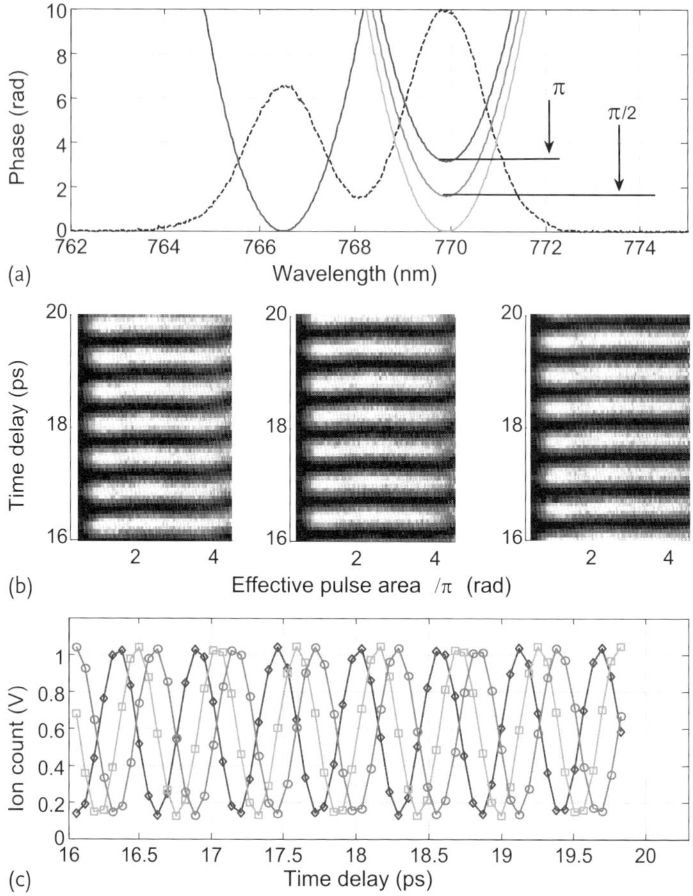

Figure 11.21 Experimental control over the quantum phase of an excited wave packet. (a) To control the relative phase of the eigenstates in the target superposition state, an extra constant phase shift of 0, $\pi/2$ and π radian is added to the spectral phase of D_1 pulses. The dashed line shows the double-peaked spectral amplitude of the pulse. The introduced phase shift results in a corresponding vertical shift of the detected quantum beats shown in (b). (c) Vertical cross sections of the two-dimensional scans in (b) at pulse area of π. Diamonds, squares, and circles corresponds to 0, $\pi/2$ and π phase shifts of D_1 pulse. Taken from [508, Figure 8]. (Reprinted with permission. Copyright 2009 American Physical Society.)

equal Stark shifts effectively change the time at which each adiabatic passage is executed. The farther these APs from being simultaneous, the bigger the deviation of the population distribution from the expected one.

It is possible to compensate for the detrimental effects of the Stark shifts as shown (computationally) in Figure 11.22 where the $\beta \equiv |b_1|^2/|b_2|^2$ ratio is plotted versus its target value for different control methods. In all calculations, the effective pulse area was set to π, and the local spectral chirps near D_1 and D_2

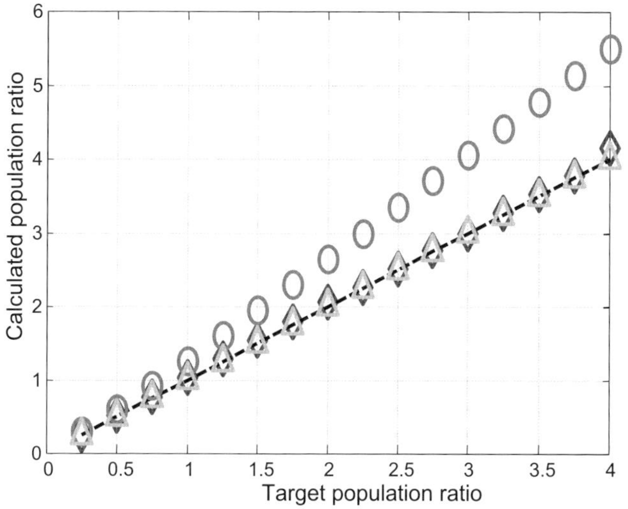

Figure 11.22 The quality of the population control with PAP. Numerical simulations demonstrate possible discrepancy between the achieved population ratio β of two $4P$ states (vertical axis) and its target value (horizontal axis), for different control methods (see text for symbol description). Taken from [508, Figure 9]. (Reprinted with permission. Copyright 2009 American Physical Society.)

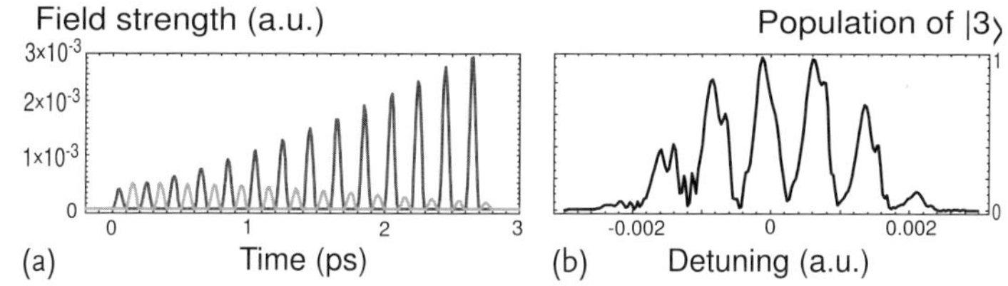

Figure 11.23 Field envelopes of the (2 + 1) PAP, and the final population of the target state as a function of the dump detuning.

lines was equal to $270 \times 10^3\ \text{fs}^2$. By accurately measuring populations of the excited states an iterative correction can be introduced into the relative energy of each spectral component of the excitation field, proportional to the deviation of the observed population distribution from its target shape. The numerical results of such an iterative procedure appear as triangles in Figure 11.22. We see that after only two iterations the calculated β is brought to within 0.2% of its target value! Experimental demonstration of the transfer and the completeness of transfer are given in Figures 11.23 and 11.24.

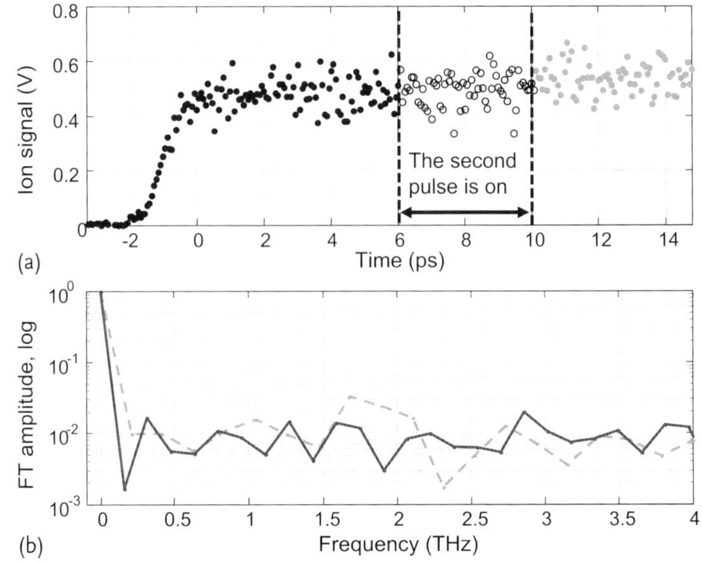

Figure 11.24 Demonstration of the population transfer completeness. (a) Ion signal as a function of the pump-probe time delay for a sequence of frequency chirped D_1 and D_2 pulses separated by 8 ps (the time when the second pulse is present is between two dashed vertical lines). Filled black (gray) dots correspond to the signal before (after) the arrival of the second pulse. (b) Fourier transform of the ion signal before (solid) and after (dashed) the arrival of the second pulse (log scale). The frequency of the quantum beating (1.73 THz) is where a strong peak is expected in the case when both states are populated. Taken from [508, Figure 7]. (Reprinted with permission. Copyright 2009 American Physical Society.)

11.6 Electromagnetically Induced Transparency

We saw that STIRAP uses interference effects generated by irradiation with two lasers to control characteristics of the time-dependent level populations. An alternative phenomenon, also associated with the existence of trapped states, utilizes interferences to control features of the absorption and emission of radiation. If, instead of applying the STIRAP pulse ordering, in which $\mathcal{E}_2(t)$ is applied before $\mathcal{E}_1(t)$ and allowed to decay as $\mathcal{E}_1(t)$ reaches its peak value, we maintain $\mathcal{E}_2(t) \gg \mathcal{E}_1(t)$ at all times, we produce a phenomenon termed "electromagnetically induced transparency" (EIT) [412, 413, 415, 416, 418, 431, 523–531]. In this phenomenon the absorption of radiation by the system at a specific frequency is suppressed. A simplistic picture of the suppression is that $\mathcal{E}_2(t) \gg \mathcal{E}_1(t)$, leads, following Eq. (11.21), to suppression of the absorption since in that case $\theta \approx 0$ at all times. Under these circumstances $b_1(t) \approx 1$, $b_0(t) \approx 0$, $b_2(t) \approx 0$. Hence the trapped state $|\lambda_1(t)\rangle \approx |E_1\rangle$ does not react to the light at all, becoming a so-called "dark" state. Qualitatively, this arises because the two fields $\mathcal{E}_1(t)$ (termed the probe laser) and $\mathcal{E}_2(t)$ (termed the coupling laser) have created competitive absorption pathways that destructively interfere with one another, canceling the absorption of light.

The formation of a "dark" state may be detected by allowing two beams of light of frequencies $\omega_1 = [E_0 - E_1]/\hbar$ and an $\omega_2 = [E_0 - E_2]/\hbar$ to transverse a dense medium of atoms or molecules. In the absence of the ω_2 beam, the ω_1 beam is readily absorbed due to the $\omega_1 = [E_0 - E_1]/\hbar$ resonance condition. By contrast, in the presence of the ω_2 beam, as long as $\mathcal{E}_2(t) \gg \mathcal{E}_1(t)$, neither beam will be absorbed [412].

11.6.1
EIT: a Resonance Perspective

EIT is, however, a much more dramatic effect than the general description given above would lead us believe. Not only is the absorption small, as the choice of the pulse ordering derived above implies, but the absorption actually goes to zero as a very sharp function of the frequency.

The EIT phenomenon is intimately related to the formation of *resonances*, that is, the broadening of spectral lines due to the interaction of bound states with a (radiative or nonradiative) continuum of states. When the broadening becomes comparable to the spacing between the lines, the resonances are said to *overlap*. This property can be controlled optically due to the splitting of *dressed states*, which depends on the intensity of the light field used to induce the Autler–Townes (AT) splitting [532]. Once the relative positions of the various levels are known, there is a well established theory [8, 90, 301, 302, 533] for dealing with interferences between overlapping resonances.

In order to understand how the interference between overlapping resonances is associated with the strong-field effects in EIT we simplify the problem by noting that since $\mathcal{E}_2(t) \gg \mathcal{E}_1(t)$, we can treat $\mathcal{E}_1(t)$ as a small perturbation and first obtain the adiabatic eigenstates resulting from the $\mathcal{E}_2(t)$-induced interaction between $|E_0\rangle$ and $|E_2\rangle$.

It was realized early on [8], following the analysis of Fano [302], that the interference between overlapping resonances can give rise to *dark states*. As shown in Figure 11.25 such dark states are characterized by the vanishing of photoabsorption cross sections at specific energies. Found experimentally a few years later [221, 534], dark states feature very highly in many applications in coherent optics and in particular in "lasing without inversion" [272, 418, 535, 536] and adiabatic passage phenomena [402, 430, 431, 444, 537].

The emergence of dark states as a result of overlapping resonances was at the heart of the original EIT idea of Harris *et al.* [412, 418]. However, only limited use of the general theory of overlapping resonances has been subsequently made. In particular, the Wigner–Weisskopf approximation [538], according to which the resonance width is independent of the energy ("unstructured continuum") was invariably assumed. The widths of the various levels were thus represented [530] as constant imaginary additions to the Hamiltonian matrix elements, or, in the density matrix description of the process, as decay and dephasing *rates*.

In complex molecules, continua have structures because the levels decay indirectly due to the existence of "tier-like" coupling schemes where levels belonging

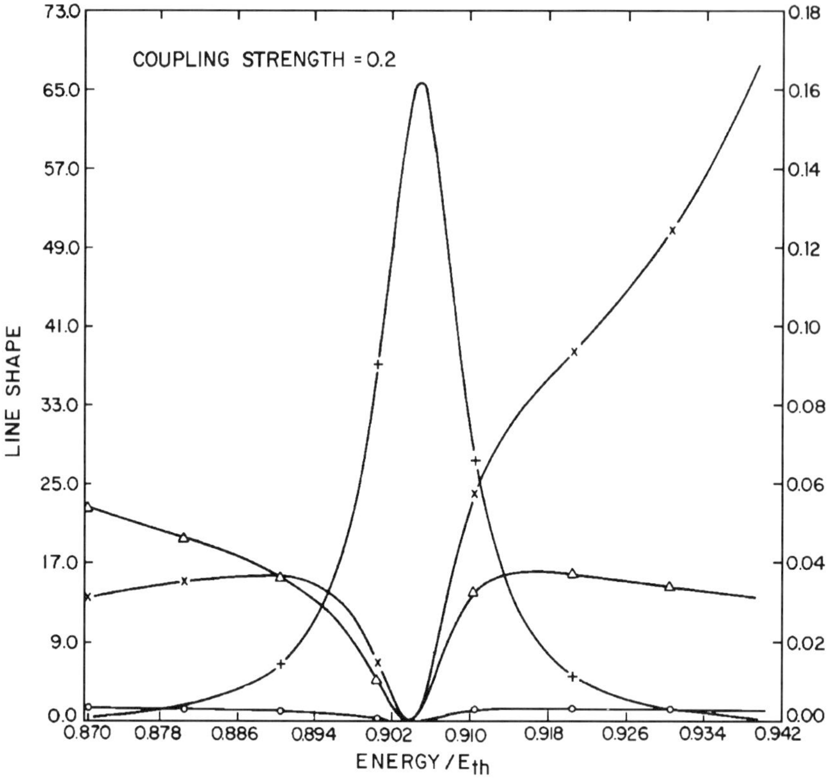

Figure 11.25 Resonance line-shapes in a model He-H_2 complex. The *Q* space composed of the He-H_2 complex in which the H_2 fragment is confined to the $j = 2$ rotational state, is coupled by the potential anisotropy to the *P* space composed of the He-H_2($j = 0$) manifold. Shown are four (v = 0, 1, 2, 3) vibrational resonances, with v = 0 (circles), v = 1 (triangles), v = 2 (plus signs), and v = 3 (crosses), in the vicinity of the center of the v = 2 resonance. The left-hand scale pertains to v = 2 and the right-hand scale to all other resonances. We see that although the resonances do not overlap appreciably (notice the difference between the v = 2 scale on the left-hand side and the right-hand scale pertaining to the "tails" of the other resonances), each of the v = 0, 1, 3 resonance exhibits a "hole" at the exact position of the v = 2 maximum. Taken from [8, Figure 4]. (Reprinted with permission. Copyright 1972 American Institue of Physics.)

to a given vibrational mode ("tier") are coupled to levels belonging to just one other tier, which are coupled to yet another limited set of levels, and so on, until one reaches the continuum tiers. The tier-like scheme thus leads to highly "structured" continua, which are in addition, often coupled to one another, leading to "multichannel" scattering [301] and dissociation [539].

Below we present a comprehensive theory of EIT, based on the overlapping resonances concepts [540, 541].

11.6.2
EIT as Emerging from the Interference between Resonances

We consider the situation illustrated in Figure 11.26 in which a ground state $|E_1\rangle$, a resonance state $|E_0\rangle$ (decaying radiatively or nonradiatively) coupled optically to a third state $|E_2\rangle$ (which can also decay radiatively or nonradiatively) by a strong guiding field

$$E_2(t) \equiv \hat{\boldsymbol{\varepsilon}}_2\varepsilon_2(t) = \hat{\boldsymbol{\varepsilon}}_2\mathrm{Re}\mathcal{E}_2(t)e^{-i\omega_2 t}\,, \tag{11.61}$$

is depicted, where $\hat{\boldsymbol{\varepsilon}}_2$ is the polarization vector. We probe the system by another laser pulse

$$E_1(t) \equiv \hat{\boldsymbol{\varepsilon}}_1\varepsilon_1(t) = \hat{\boldsymbol{\varepsilon}}_1\mathrm{Re}\mathcal{E}_1(t)e^{-i\omega_1 t}\,. \tag{11.62}$$

We first treat $\varepsilon_1(t)$ as a perturbation and obtain the adiabatic eigenstates resulting from the $\varepsilon_2(t)$-induced interaction between the $|E_0\rangle$ and $|E_2\rangle$ states. The extension to the strong $\varepsilon_1(t)$ pulse regime is presented in Section 11.6.4 below.

Therefore, temporarily neglecting $\varepsilon_1(t)$, we can expand the system wave function in just two states,

$$|\Psi(t)\rangle = b_0(t)|E_0\rangle e^{-iE_0t/\hbar} + b_2(t)|E_2\rangle e^{-iE_2t/\hbar}\,. \tag{11.63}$$

Using the expansion of Eq. (11.63) we obtain from the time-dependent Schrödinger equation

$$i\hbar\frac{\partial\Psi(t)}{\partial t} = H\Psi(t) = [H_M + H_{MR}]\Psi(t)\,, \tag{11.64}$$

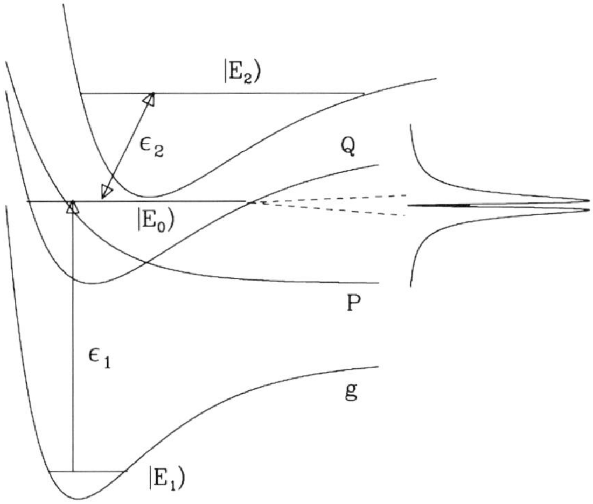

Figure 11.26 A ground state $|E_1\rangle$ is excited by a weak laser pulse ε_1 to a resonance state $|E_0\rangle \in Q$ decaying radiatively or nonradiatively to space P. The $|E_0\rangle$ state is coupled optically to a third state $|E_2\rangle$ by a strong guiding field ε_2 and undergoes as a result Autler–Townes splitting. As a result of the splitting and the decay, an EIT "hole" is formed at $E = E_0$. Taken from [540, Figure 1]. (Reprinted with permission. Copyright 2007 American Physical Society.)

where H_M is the material Hamiltonian and H_{MR} is the matter–radiation interaction, given in the dipole approximation as

$$H_{MR} = -\boldsymbol{d} \cdot \boldsymbol{E}(t) , \tag{11.65}$$

the usual set of ordinary coupled differential equations for the $\underline{\mathrm{b}} \equiv (b_0, b_2)^{\mathrm{T}}$ coefficients vector,

$$\frac{d}{dt}\underline{\mathrm{b}} = i\underline{\underline{\mathrm{H}}} \cdot \underline{\mathrm{b}}(t) , \tag{11.66}$$

where $\underline{\underline{\mathrm{H}}}$ is given in the "rotating-wave approximation" (RWA) [272] as

$$\underline{\underline{\mathrm{H}}} = \begin{pmatrix} 0 & \Omega_2^*(t)e^{i\Delta_2 t} \\ \Omega_2(t)e^{-i\Delta_2 t} & 0 \end{pmatrix} , \tag{11.67}$$

with the detuning and the Rabi frequency given respectively as

$$\Delta_2 \equiv \omega_2 - |E_0 - E_2|/\hbar , \quad \Omega_2(t) \equiv \boldsymbol{d}_2 \cdot \hat{\boldsymbol{\varepsilon}}_2 \mathcal{E}_2(t)/\hbar . \tag{11.68}$$

We now transform $\underline{\underline{\mathrm{H}}}$ to a form which does not contain the highly oscillatory $e^{-i\Delta_2 t}$ terms which might invalidate the adiabatic approximation. Multiplying Eq. (11.66) by a diagonal matrix, $\exp(i\underline{\underline{\Delta}}t/2)$, with $\underline{\underline{\Delta}}$ being the diagonal detuning matrix,

$$\underline{\underline{\Delta}} \equiv \begin{pmatrix} -\Delta_2 & 0 \\ 0 & \Delta_2 \end{pmatrix} , \tag{11.69}$$

results in the following,

$$\begin{aligned} \exp\left[i\left(\underline{\underline{\Delta}}t/2\right)\right] \cdot \frac{d}{dt}\underline{\mathrm{b}} &= \frac{d}{dt}\exp\left[i\left(\underline{\underline{\Delta}}t/2\right)\right] \cdot \underline{\mathrm{b}} - i\underline{\underline{\Delta}} \cdot \exp\left[i\left(\underline{\underline{\Delta}}t/2\right)\right] \cdot \frac{\underline{\mathrm{b}}}{2} \\ &= i\exp\left[i\left(\underline{\underline{\Delta}}t/2\right)\right] \cdot \underline{\underline{\mathrm{H}}} \cdot \exp\left[i\left(\underline{\underline{\Delta}}t/2\right)\right] \cdot \exp\left[i\left(\underline{\underline{\Delta}}t/2\right)\right] \cdot \underline{\mathrm{b}}(t) . \end{aligned} \tag{11.70}$$

We can eliminate the oscillatory terms by defining

$$\underline{\mathrm{c}} \equiv \exp\left[i\left(\underline{\underline{\Delta}}t/2\right)\right] \cdot \underline{\mathrm{b}} , \tag{11.71}$$

and obtain that

$$\frac{d}{dt}\underline{\mathrm{c}} = i\underline{\underline{\mathrm{H}'}} \cdot \underline{\mathrm{c}}(t) , \tag{11.72}$$

where

$$\underline{\underline{\mathrm{H}'}} = \begin{pmatrix} -\Delta_2/2 & \Omega_2^*(t) \\ \Omega_2(t) & \Delta_2/2 \end{pmatrix} . \tag{11.73}$$

We now build adiabatic solutions by diagonalizing Eq. (11.73) using a 2 × 2 unitary matrix

$$\underline{\underline{U}} = \begin{pmatrix} \cos\theta & e^{-i\phi_2}\sin\theta \\ -e^{i\phi_2}\sin\theta & \cos\theta \end{pmatrix} . \tag{11.74}$$

The θ and ϕ_2 angles of $\underline{\underline{U}}$ are given as

$$\tan\theta = \frac{|\Omega_2(t)|}{-\left(\Delta_2^2/4 + |\Omega_2(t)|^2\right)^{\frac{1}{2}} + \Delta_2/2} , \tag{11.75}$$

with ϕ_2 being the argument of Ω_2 (Eq. (11.68)). The corresponding diagonal eigenvalue matrix $\hat{\lambda}$ is composed of the two roots,

$$\lambda_{1,2}(t) = \pm\lambda(t) = \pm\left[\Delta_2^2/4 + |\Omega_2(t)|^2\right]^{\frac{1}{2}} . \tag{11.76}$$

Operating with $\underline{\underline{U}}^\dagger$ on Eq. (11.72) and defining $\underline{a} \equiv (a_1, a_2)^\mathsf{T} \equiv \underline{\underline{U}}^\dagger \cdot \underline{c}$ we obtain, by neglecting the nonadiabatic coupling matrix $\underline{\underline{A}} = \underline{\underline{U}} \cdot d\underline{\underline{U}}^\dagger/dt$, the adiabatic approximation for $\underline{a}$,

$$\frac{d}{dt}\underline{a} = i\hat{\lambda}(t) \cdot \underline{a}(t) , \tag{11.77}$$

whose solution is given by

$$\begin{aligned} a_{1,2}(t) &= \exp\left\{\pm i \int_0^t \lambda(t')dt'\right\} a_{1,2}(0) \\ &= \exp\left\{\pm i \int_0^t \left[\Delta_2^2/4 + |\Omega_2(t')|^2\right]^{\frac{1}{2}} dt'\right\} a_{1,2}(0) . \end{aligned} \tag{11.78}$$

We now introduce the (weak) $\varepsilon_1(t)$ pulse. Since $|E_1\rangle$ is the initially populated state, in the absence of $\varepsilon_1(t)$, neither the $|E_0\rangle$ or $|E_2\rangle$ states, nor the $|\lambda_1\rangle$ and $|\lambda_2\rangle$ adiabatic states can ever be populated. Thus, in the absence of the $\varepsilon_1(t)$ pulse, the only effect of the $\varepsilon_2(t)$ pulse is to change the spectrum of the Hamiltonian.

Assuming that the adiabatic condition indeed holds, the states seen by the $\varepsilon_1(t)$ pulse with $\varepsilon_2(t)$ on, are the adiabatic states $|\lambda_1\rangle$ and $|\lambda_2\rangle$, rather than the $|E_0\rangle$ and $|E_2\rangle$ material states. Using the definition of $\underline{a}$ (Eqs. (11.71) and (11.74)), we can write the adiabatic states, using the identity $\underline{b} = e^{-i\underline{\underline{\Delta}}t/2} \cdot \underline{\underline{U}} \cdot \underline{a}$, as

$$\begin{aligned} |\lambda_1(t)\rangle &= e^{i\int_0^t \lambda(t')dt' + i\Delta_2 t/2 - iE_0 t}\left\{\cos\theta\,|E_0\rangle + \sin\theta\, e^{-i\phi_2(t) - i\omega_{2,0}t}|E_2\rangle\right\} \\ |\lambda_2(t)\rangle &= e^{-i\int_0^t \lambda(t')dt' - i\Delta_2 t/2 - iE_0 t}\left\{-\sin\theta\, e^{i\phi_2(t)}|E_0\rangle + \cos\theta\, e^{-i\omega_{2,0}t}|E_2\rangle\right\} . \end{aligned} \tag{11.79}$$

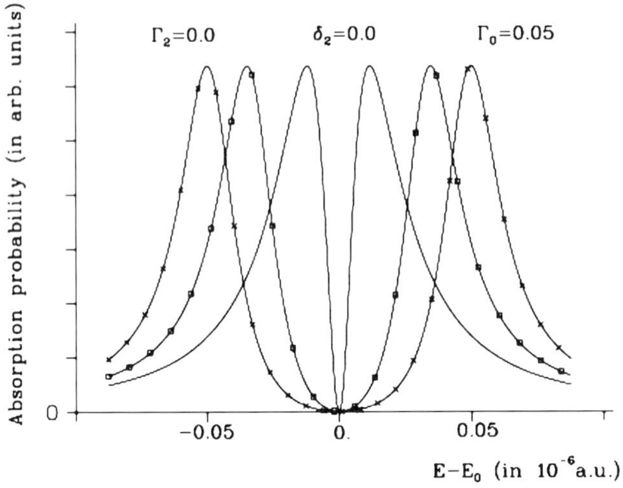

Figure 11.27 The formation of the EIT "hole" for unstructured continuum. $\overline{\Gamma}_0 = 0.05 \times 10^{-6}$ a.u., $\overline{\Gamma}_2 = 0$, and $\overline{\Delta}_2 = 0$. Shown is the line-shape at three different times: (cross) is at the peak of the pulse, $t = 0$, (square) is as the pulse begins to wane, $t = 0.75 \times 10^6$ a.u., and (full line) is at the tail of the pulse, $t = 1.5 \times 10^6$ a.u. A simple Gaussian pulse of the form $\mathcal{E}_2(t) = \mathcal{E}_0 e^{-t^2}$ was assumed. Taken from [540, Figure 3]. (Reprinted with permission. Copyright 2007 American Physical Society.)

Here $|\lambda_1(t)\rangle$ and $|\lambda_2(t)\rangle$ are obtained by setting either $\underline{a} = (a_1, 0)^{\mathsf{T}}$ or $\underline{a} = (0, a_2)^{\mathsf{T}}$ and $\omega_{2,0} \equiv (E_2 - E_0)/\hbar$.

When $\Delta_2 = 0$ (i.e., when ω_2 is exactly resonant with the $|E_2\rangle$ to $|E_0\rangle$ transition), it follows from Eq. (11.75) that $\theta = 3\pi/4$. If, in addition, we assume that the pulse has no chirp (i.e., that the phase of $\mathcal{E}_2(t)$, $\phi_2(t) = 0$), we have that

$$|\lambda_1(t)\rangle = e^{i\int_0^t |\Omega_2(t')|dt' - iE_0 t}\{|E_0\rangle - e^{-i\omega_{2,0}t}|E_2\rangle\}/\sqrt{2}\,,$$

$$|\lambda_2(t)\rangle = e^{-i\int_0^t |\Omega_2(t')|dt' - iE_0 t}\{|E_0\rangle| + e^{-i\omega_{2,0}t}|E_2\rangle\}/\sqrt{2}\,. \quad (11.80)$$

We see that the time evolution of the $|E_0\rangle$ component of $|\lambda_1(t)\rangle$ is governed by a "quasi-energy" of $E_0 - |\Omega_2(t)|$, whereas the time evolution of the $|E_0\rangle$ component of $|\lambda_2(t)\rangle$ is governed by a "quasi-energy" of $E_0 + |\Omega_2(t)|$. We say that the two levels are "Autler–Towns" split by an amount equal to $2|\Omega_2(t)|$.

We now consider the broadening of the adiabatic levels due to the decay channels considered in this section earlier. When this broadening is comparable to, or in excess of the $2|\Omega_2(t)|$ splitting (see Figure 11.27) the switch on of the $\varepsilon_1(t)$ pulse results in the simultaneous excitation of the two adiabatic eigenstates. As an example, we show in Figure 11.33 the original levels and adiabatic states for the case of Sr. The probability of a one-photon absorption to each scattering state $|E\rangle$ is given in first-order perturbation theory as ($\hbar = s$) hereafter in this section

$$P_{\mathrm{n}}(E) = |2\pi\epsilon_1(\omega_{E,1})d_{1,n}(E)|^2\,, \quad (11.81)$$

where $d_{1,\boldsymbol{n}}(E)$ are bound–free dipole matrix elements between states on the ground (g) and excited (e) electronic potential surfaces, given as

$$d_{1,\boldsymbol{n}}(E) = \langle E_1 | d_{e,g} | E, \boldsymbol{n}^- \rangle \,. \tag{11.82}$$

In Eq. (11.81), $\omega_{E,1} = (E - E_1)/\hbar$ is the transition frequency between $|E_1\rangle$ and $|E\rangle$, and $\epsilon_1(\omega)$ is the temporal Fourier transform of the pulse,

$$\epsilon_1(\omega) \equiv \frac{1}{2\pi} \int dt \varepsilon_1(t) \exp[-i\omega(z/c - t)] \,, \tag{11.83}$$

where z is the direction of propagation of the light and c is the velocity of light in vacuum. Assuming for simplicity that no direct transitions to the continuum occur, that is, that only the Q space of Eq. (7.46) is coupled radiatively to $|E_1\rangle$, we have that

$$d_{1,\boldsymbol{n}}(E) = \sum_s \langle E_1 | d_{e,g} | \phi_s \rangle \langle \phi_s | E, \boldsymbol{n}^- \rangle \,. \tag{11.84}$$

If we now identify the $|\phi_s\rangle$ bound states (which become resonances due to the interaction with the continuum) with the adiabatic states of Eq. (11.79), we can write that

$$d_{1,\boldsymbol{n}}(E) = \sum_{s=1,2} \langle E_1 | d_{e,g} | \lambda_s \rangle \langle \lambda_s | E, \boldsymbol{n}^- \rangle \equiv \sum_{s=1,2} \langle E_1 | d_{e,g} | \lambda_s \rangle \ell_s(E) \tag{11.85}$$

where we have restricted the treatment to the interaction of just two resonances. Using Eq. (11.79) and the fact that $\langle E_2 | \hat{\boldsymbol{\varepsilon}} \cdot \boldsymbol{d} | E_1 \rangle = 0$, we have that

$$d_{1,\boldsymbol{n}}(E) = d_{1,0} e^{-iE_0 t} \left\{ \cos\theta e^{i\Delta_2 t/2 + i\int_0^t \lambda(t')dt'} \ell_1(E) - \sin\theta e^{-i\Delta_2 t/2 - i\int_0^t \lambda(t')dt' + i\phi_2} \ell_2(E) \right\}, \tag{11.86}$$

where $d_{1,0} \equiv \langle E_1 | d_{e,g} | E_0 \rangle$.

When $\omega_2 = \omega_{2,0}$, that is, it is exactly on resonance, and $\phi_2 = 0$, this expression reduces to

$$d_{1,\boldsymbol{n}}(E) = \frac{1}{\sqrt{2}} e^{-iE_0 t} d_{1,0} \left\{ \ell_1(E) e^{i\int_0^t |\Omega_2(t')| dt'} - \ell_2(E) e^{-i\int_0^t |\Omega_2(t')| dt'} \right\} . \tag{11.87}$$

Using Eq. (7.53) we can write an exact expression for the amplitude of observing the $|\lambda_s\rangle$ states as

$$\ell_s(E) \equiv \langle \lambda_s | E, \boldsymbol{n}^- \rangle = \sum_{s'} \langle \lambda_s | [E - i\epsilon - Q\mathcal{H}Q]^{-1} | \lambda_{s'} \rangle \langle \lambda_{s'} | H | E, \boldsymbol{n}^-; 1 \rangle \,. \tag{11.88}$$

Following [8] we can write the $E - Q\mathcal{H}Q$ matrix of Eq. (7.57) in the two overlapping resonances case as

$$E - Q\mathcal{H}Q = \begin{pmatrix} E - E_0 - \lambda_1 - \Delta_{1,1} - i\Gamma_{1,1}/2 & -\Delta_{1,2} - i\Gamma_{1,2}/2 \\ -\Delta_{2,1} - i\Gamma_{2,1}/2 & E - E_0 - \lambda_2 - \Delta_{2,2} - i\Gamma_{2,2}/2 \end{pmatrix}, \tag{11.89}$$

where

$$\Gamma_{s,s'}(E) = 2\pi \sum_{\mathrm{n}} \langle \lambda_s | H | E, \boldsymbol{n}^-; 1\rangle \langle E, \boldsymbol{n}^-; 1 | H | \lambda_{s'} \rangle \tag{11.90}$$

and

$$\Delta_{s,s'}(E) = \mathrm{P}_v \int dE' \sum_{\mathrm{n}} \frac{\langle \lambda_s | H | E', \boldsymbol{n}^-; 1\rangle \langle E', \boldsymbol{n}^-; 1 | H | \lambda_{s'} \rangle}{E - E'} . \tag{11.91}$$

We obtain that the inverse matrix is given as

$$[E - Q\mathcal{H}Q]^{-1} = \frac{1}{\mathcal{D}} \begin{pmatrix} a & b \\ c & d \end{pmatrix}, \tag{11.92}$$

with

$$\begin{aligned} & a \equiv E - E_0 - \lambda_2 - \Delta_{2,2} - i\Gamma_{2,2}/2 , \quad b \equiv \Delta_{1,2} + i\Gamma_{1,2}/2 , \\ & c \equiv \Delta_{2,1} + i\Gamma_{2,1}/2 , \quad d \equiv E - E_0 - \lambda_1 - \Delta_{1,1} - i\Gamma_{1,1}/2 , \quad \mathcal{D} \equiv ad - cb . \end{aligned} \tag{11.93}$$

Using Eq. (11.88) we obtain that

$$\begin{aligned} \ell_1(E) &= \langle \lambda_1 | E, \boldsymbol{n}^- \rangle = \frac{1}{\mathcal{D}} [a \langle \lambda_1 | H | E, \boldsymbol{n}^-; 1\rangle + b \langle \lambda_2 | H | E, \boldsymbol{n}^-; 1\rangle] , \\ \ell_2(E) &= \langle \lambda_2 | E, \boldsymbol{n}^- \rangle = \frac{1}{\mathcal{D}} [c \langle \lambda_1 | H | E, \boldsymbol{n}^-; 1\rangle + d \langle \lambda_2 | H | E, \boldsymbol{n}^-; 1\rangle] . \end{aligned} \tag{11.94}$$

It follows from Eq. (11.79) that

$$\begin{aligned} \langle \lambda_1 | H | E, \boldsymbol{n}^-; 1\rangle &= e^{-i\int_0^t \lambda(t')dt' - i\Delta_2 t/2 + iE_0 t} \left\{ \cos\theta\, V_{0,\boldsymbol{n}} + \sin\theta\, e^{i\phi_2 + i\omega_{2,0}t} V_{2,\boldsymbol{n}} \right\} , \\ \langle \lambda_2 | H | E, \boldsymbol{n}^-; 1\rangle &= e^{i\int_0^t \lambda(t')dt' + i\Delta_2 t/2 + iE_0 t} \left\{ -\sin\theta\, e^{-i\phi_2} V_{0,\boldsymbol{n}} + \cos\theta\, e^{i\omega_{2,0}t} V_{2,\boldsymbol{n}} \right\} , \end{aligned} \tag{11.95}$$

where

$$V_{0,\boldsymbol{n}}(E) \equiv \langle E_0 | H | E, \boldsymbol{n}^-; 1\rangle , \quad V_{2,\boldsymbol{n}}(E) \equiv \langle E_2 | H | E, \boldsymbol{n}^-; 1\rangle . \tag{11.96}$$

Using Eq. (11.95) and neglecting the terms containing the highly oscillatory $e^{\pm i\omega_{2,0}t}$ factors, we obtain from Eq. (11.90) that

$$\begin{aligned} & \Gamma_{1,1} = \overline{\Gamma}_0 \cos^2\theta + \overline{\Gamma}_2 \sin^2\theta , \quad \Gamma_{2,2} = \overline{\Gamma}_0 \sin^2\theta + \overline{\Gamma}_2 \cos^2\theta , \\ & \Gamma_{1,2} = \left(-\overline{\Gamma}_0 + \overline{\Gamma}_2\right) e^{-2i\int_0^t \lambda(t')dt' - i\Delta_2 t + i\phi_2} \sin\theta\cos\theta , \end{aligned} \tag{11.97}$$

where

$$\overline{\Gamma}_i(E) \equiv 2\pi \sum_{n} |\langle E, \boldsymbol{n}^-; 1|H|E_i\rangle|^2 , \quad i = 0, 2 . \tag{11.98}$$

In the same manner from Eq. (11.91)

$$\Delta_{1,1} = \overline{\Delta}_0 \cos^2\theta + \overline{\Delta}_2 \sin^2\theta , \quad \Delta_{2,2} = \overline{\Delta}_0 \sin^2\theta + \overline{\Delta}_2 \cos^2\theta ,$$

$$\Delta_{1,2} = \left(-\overline{\Delta}_0 + \overline{\Delta}_2\right) e^{-2i\int_0^t \lambda(t')dt' - i\Delta_2 t + i\phi_2} \sin\theta\cos\theta , \tag{11.99}$$

where

$$\overline{\Delta}_i(E) \equiv \left(\frac{1}{2\pi}\right) \mathrm{P}_v \int dE' \frac{\overline{\Gamma}_i(E')}{E - E'} , \quad i = 0, 2 . \tag{11.100}$$

Hence from Eq. (11.94), and after defining $I \equiv e^{-i\int_0^t \lambda(t')dt' - i\Delta_2 t/2}$ and $F \equiv e^{iE_0 t}/\mathcal{D}$, we have

$$\ell_1(E) = F\Big[(aI\cos\theta - bI^*\sin\theta)V_{0,\boldsymbol{n}}$$
$$+ \left(aI\sin\theta\, e^{i\phi_2 + i\omega_{2,0}t} + bI^*\cos\theta\, e^{i\omega_{2,0}t}\right) V_{2,\boldsymbol{n}}\Big] ,$$

$$\ell_2(E) = F\Big[\left(cI^*\cos\theta - dI\sin\theta\right) V_{0,\boldsymbol{n}}$$
$$+ \left(cI^*\sin\theta\, e^{i\phi_2 + i\omega_{2,0}t} + dI\cos\theta\, e^{i\omega_{2,0}t}\right) V_{2,\boldsymbol{n}}\Big] . \tag{11.101}$$

After neglecting the highly oscillatory $e^{i\omega_{2,0}t}$ terms in Eq. (11.86), we further obtain

$$d_{1,\boldsymbol{n}}(E) = \frac{d_{1,0} V_{0,\boldsymbol{n}}}{\mathcal{D}} \left[E - E_0 + \lambda\cos 2\theta - \left(\overline{\Delta}_2 + i\overline{\Gamma}_2/2\right)\right] . \tag{11.102}$$

Thus, if state $|E_2\rangle$ is unstable, giving rise to the $i\overline{\Gamma}_2$ term, there is no real E for which the transition dipole matrix element vanishes. In case of $\Delta_2 = 0$ detuning and $\cos 2\theta = 0$, we have

$$d_{1,\boldsymbol{n}}(E) = \frac{1}{\mathcal{D}} d_{1,0} V_{0,\boldsymbol{n}} \left[E - E_0 - \left(\overline{\Delta}_2 + i\overline{\Gamma}_2/2\right)\right] . \tag{11.103}$$

When we use the explicit form of $\mathcal{D}$, as given by Eq. (11.93), and the values of $\Gamma_{i,j}$ as given by Eq. (11.97), we have

$$\mathcal{D} = \left[E - E_0 - \lambda - \overline{\Delta}_0\cos^2\theta - \overline{\Delta}_2\sin^2\theta - \frac{i}{2}(\overline{\Gamma}_0\cos^2\theta + \overline{\Gamma}_2\sin^2\theta)\right]$$
$$\times \left[E - E_0 + \lambda - \overline{\Delta}_0\sin^2\theta - \overline{\Delta}_2\cos^2\theta - \frac{i}{2}(\overline{\Gamma}_0\sin^2\theta + \overline{\Gamma}_2\cos^2\theta)\right]$$
$$- \left[\overline{\Delta}_0 - \overline{\Delta}_2 + \frac{i}{2}\left(\overline{\Gamma}_0 - \overline{\Gamma}_2\right)\right]^2 \sin^2\theta\cos^2\theta . \tag{11.104}$$

In case we irradiate exactly on resonance, $\Delta_2 = 0$, we also have that $\cos^2\theta = \sin^2\theta = 1/2$, which results in

$$\mathcal{D}(t) = \left(E - E_0 - \frac{1}{2}\overline{\Delta}_0 + \overline{\Delta}_2 - \frac{i}{4}\overline{\Gamma}_0 + \overline{\Gamma}_2\right)^2 - \left(\frac{1}{2}\overline{\Delta}_0 - \overline{\Delta}_2 + \frac{i}{4}\overline{\Gamma}_0 - \overline{\Gamma}_2\right)^2 - \lambda^2(t)\,. \tag{11.105}$$

11.6.2.1 Unstructured Continua

If we neglect the variation of $\Gamma_{0,2}(E)$ with energy, and assume that $E \gg 0$, we have that the integrand defining $\Delta_{0,2}$ is antisymmetric about E and is essentially zero at the integration limits, hence $\overline{\Delta}_i \approx 0$. Hence

$$|\mathcal{D}(t)|^2 = \frac{1}{4}\left\{\left[(E - E_0)^2 - \overline{\Gamma}_0\overline{\Gamma}_2/4 - |\Omega_2(t)|^2\right]^2 + \left[(E - E_0)\left(\overline{\Gamma}_0 + \overline{\Gamma}_2\right)\right]^2\right\}. \tag{11.106}$$

The channel specific probability of absorption of a photon of energy $E - E_1$ from state $|E_1\rangle$ is given, using Eqs. (11.103), (11.106), and (11.81) (assuming $\Delta_2 = 0$) as

$$P_n(E) = \frac{|2\pi d_{1,0} V_{0,n}\epsilon_1(\omega_{E,1})|^2\left[(E - E_0)^2 + \overline{\Gamma}_2^2/4\right]}{\left[(E - E_0)^2 - |\Omega_2(t)|^2 - \overline{\Gamma}_0\overline{\Gamma}_2/4\right]^2 + \left[(E - E_0)\left(\overline{\Gamma}_0 + \overline{\Gamma}_2\right)/2\right]^2}\,. \tag{11.107}$$

$P(E)$, the total probability for absorbing a photon of energy $E - E_1$ from state $|E_1\rangle$, given as $P_1(E) = \sum \boldsymbol{n} P_n(E)$, is

$$P_1(E) = \frac{2\pi\overline{\Gamma}_0\,|d_{1,0}\epsilon_1(\omega_{E,1})|^2\left[(E - E_0)^2 + \overline{\Gamma}_2^2/4\right]}{\left[(E - E_0)^2 - |\Omega_2(t)|^2 - \overline{\Gamma}_0\overline{\Gamma}_2/4\right]^2 + \left[(E - E_0)\left(\overline{\Gamma}_0 + \overline{\Gamma}_2\right)/2\right]^2}\,, \tag{11.108}$$

where we have used Eq. (11.98), according to which, $\overline{\Gamma}_0 = 2\pi\sum_n |V_{0,n}|^2$. We see that the basic form of the total photon-absorption probability remains essentially the same as in the single continuum case, with sums over channel specific widths and shifts replacing the single channel entities.

11.6.2.2 Structured Continua

When the variation of $\Gamma_{0,2}(E)$ with energy cannot be neglected we cannot assume that $\Delta_{0,2}(E)$ vanish. In that case we need to compute $\mathcal{D}$ according to Eq. (11.104). Assuming that $\Gamma_{0,2}(E)$ can be parametrized, for example, as a sum of Lorentzian functions

$$\overline{\Gamma}_i(E) = \sum_j A_{i,j}\frac{\gamma_{i,j}}{\left(E - e_{i,j}\right)^2 + \gamma_{i,j}^2/4}\,, \qquad i = 0, 2\,, \tag{11.109}$$

we have that

$$\overline{\Delta}_i(E) = \frac{1}{2\pi}\mathsf{P}_v \int_{-\infty}^{\infty} dE' \frac{\overline{\Gamma}_i(E')}{E - E'}$$
$$= \frac{1}{2\pi}\mathsf{P}_v \int_{-\infty}^{\infty} dE' \sum_j \frac{A_{i,j}\gamma_{i,j}}{\left(E' - e_{i,j} + \frac{i\gamma_{i,j}}{2}\right)\left(E' - e_{i,j} - \frac{i\gamma_{i,j}}{2}\right)(E - E')} . \tag{11.110}$$

Writing the P_v integral as the difference between a contour integral which includes the $E' = e_i - i\gamma_{i,j}/2$ poles but avoids the $E' = E$ and $E' = e_i + i\gamma_{i,j}/2$ poles, where the straight line segment is supplemented by a large semicircle on the lower half complex plane (whose contribution vanishes as $|E'| \to \infty$ because on it the integrand is proportional to $|E'|^{-3}$), and the small semicircle contribution about the $E' = E$ pole, one has that

$$\overline{\Delta}_i(E) = \sum_j \frac{-iA_{i,j}}{-i\left(E - e_{i,j} + i\gamma_{i,j}/2\right)} + \frac{iA_{i,j}\gamma_{i,j}}{2\left(E - e_{i,j} + i\gamma_{i,j}/2\right)\left(E - e_{i,j} - i\gamma_{i,j}/2\right)}$$
$$= \sum_j \frac{A_{i,j}(E - e_{i,j})}{(E - e_{i,j})^2 + \gamma_{i,j}^2/4} . \tag{11.111}$$

An illustration of a typical case of structured continua of the dressed AT split pair of states is given in Figure 11.28. We see that the shift function is antisymmetric about e_i, $i = 0, 2$. As a result, the probability of absorption, given as the square of $d_{1,n}(E)$ of Eq. (11.102), is no longer symmetric about the line center. Thus the appearance of asymmetry in the EIT absorption line-shape is a hallmark of a structured continuum.

11.6.3 Photoabsorption

We first present calculations of the photoabsorption of the probe pulse $\varepsilon_1(t)$ by the three level system depicted in Figure 11.26. In this system the ground state $|E_1\rangle$ is electric dipole coupled to an upper state ($|E_0\rangle$), that is in turn coupled by a strong optical pulse to a third state ($|E_2\rangle$). The (channel specific) photoabsorption probability from the ground state is given by $P_{\mathbf{n}}(E)$, and the total photoabsorption of the ε_1 pulse by $P_1(E)$.

In Figures 11.27–11.34 we plot the absorption line-shapes as a function of the detuning from the $E_1 - E_0$ resonance at different times in the history of the Gaussian pulse linking $|E_0\rangle$ to $|E_2\rangle$, with $t = 0$ being the pulse maximum. The $|E_0\rangle$ and $|E_2\rangle$ states are coupled nonradiatively to some continuum channels representing the P space.

We first analyze the situation for *unstructured continua*. Figure 11.27 shows the situation when only the $|E_0\rangle$ level (the one with the dipole allowed transition to the

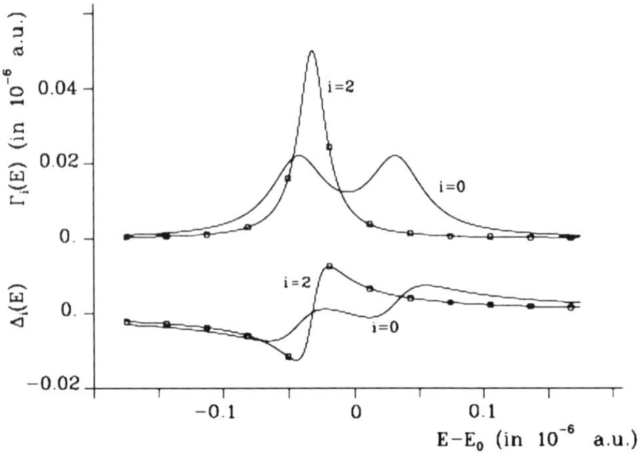

Figure 11.28 An example of the widths and shifts of a case of "highly structured" continua, characterized by $\gamma_{0,1} = \gamma_{0,2} = 0.02$, $\gamma_{2,1} = 0.01$, $e_{0,1} = 0$, $e_{0,2} = -0.03$, $A_{0,1} = A_{0,2} = 0.02$, and $A_{2,1} = 0.05$, coupled to the Autler–Townes split pair. Shown are $\overline{\Gamma}_0(E)$, $\overline{\Gamma}_2(E)$, $\overline{\Delta}_0(E)$, and $\overline{\Delta}_2(E)$. Taken from [540, Figure 2]. (Reprinted with permission. Copyright 2007 American Physical Society.)

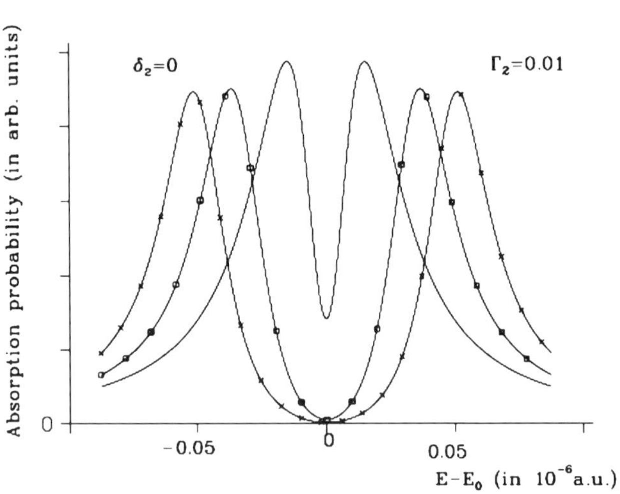

Figure 11.29 The same as in Figure 11.27 for a broadened $|E_2\rangle$ ($\overline{\Gamma}_2 = 0.01$) with zero detuning ($\Delta_2 = 0$). Taken from [540, Figure 4]. (Reprinted with permission. Copyright 2007 American Physical Society.)

ground state) is broadened ($\overline{\Gamma}_0 = 0.05 \times 10^{-6}$ a.u.). We assume no detuning ($\Delta_2 = 0$) of the center of the strong pulse connecting the $|E_0\rangle \leftrightarrow |E_2\rangle$ states. A *perfect* EIT dip is seen to arise. In contrast, Figure 11.29 shows the zero-detuning ($\Delta_2 = 0$) situation when both levels ($|E_0\rangle$ and $|E_2\rangle$) are broadened ($\overline{\Gamma}_0 = 0.05 \times 10^{-6}$ a.u., $\overline{\Gamma}_2 = 0.01 \times 10^{-6}$ a.u.). In this case, as clearly shown in Eqs. (11.107) and (11.108), the line-shape does not dip to zero. Figure 11.30 pertains to the case when both

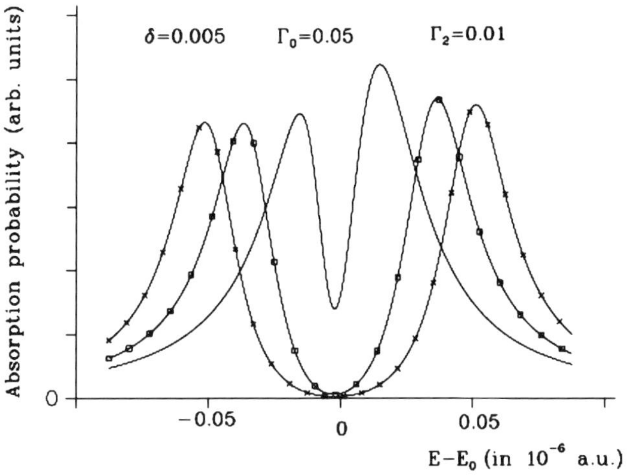

Figure 11.30 The same as in Figure 11.27 for a broadened $|E_2\rangle$ ($\overline{\Gamma}_2 = 0.01$) with finite detuning ($\Delta_2 = 0.005$). Taken from [540, Figure 5]. (Reprinted with permission. Copyright 2007 American Physical Society.)

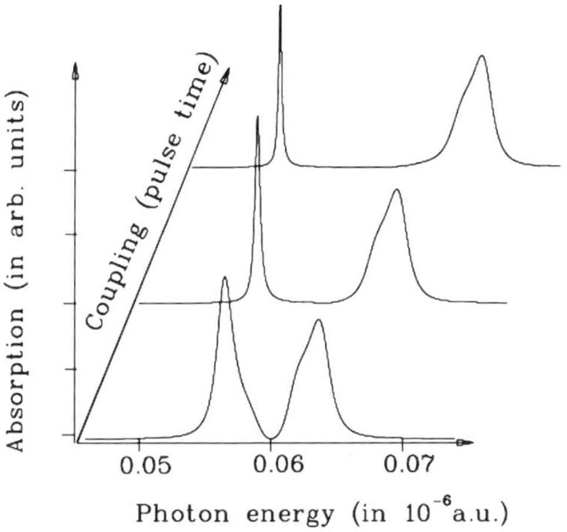

Figure 11.31 EIT for structured continuum ($A_{0,1} = 0.02$, $\gamma_{0,1} = 0.02$, $e_{0,1} = 0$, $A_{0,2} = 0.02$, $\gamma_{0,2} = 0.02$, $e_{0,2} = -0.02$, $A_{2,1} = 0.002$, $\gamma_{2,1} = 0.01$, $e_{2,1} = 0$) with no detuning ($\Delta_2 = 0$), as a function of the ε_2 coupling strength (the uppermost trace corresponds to the peak of the ε_2 pulse). In the two upper traces the first peak was divided by 2. All energies and widths are expressed in 10^{-6} a.u.

levels ($|E_0\rangle$ and $|E_2\rangle$) are broadened ($\overline{\Gamma}_0 = 0.05 \times 10^{-6}$ a.u., $\overline{\Gamma}_2 = 0.01 \times 10^{-6}$ a.u.) in the presence of detuning ($\Delta_2 = 0.005$). Here, the EIT does not dip to zero and the whole line-shape is asymmetrically biased to the blue by an amount which depends on the intensity of the $|E_0\rangle \leftrightarrow |E_2\rangle$ coupling field.

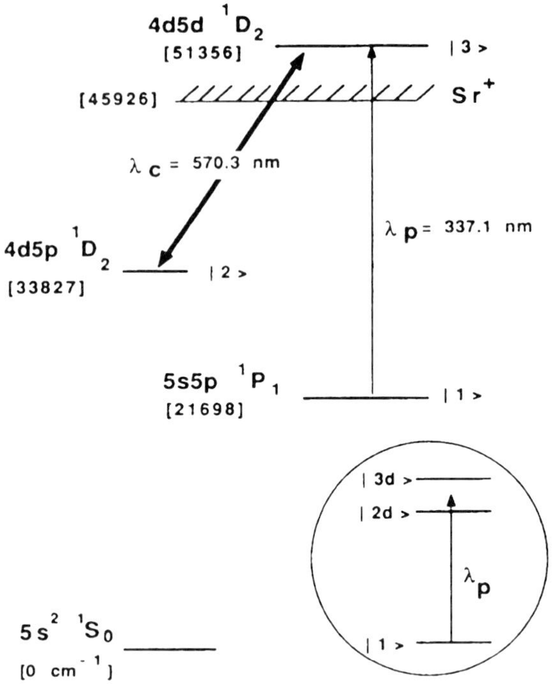

Figure 11.32 Scenario associated with Electromagnetically Induced Transparency (i.e., old levels, new levels $|\lambda_1\rangle$ and $|\lambda_2\rangle$, and associated fields). Note that the inset shows excitation to two adiabatic states, as discussed in the text. Here our levels $|E_0\rangle$, $|E_1\rangle$, and $|E_2\rangle$ are denoted $|3\rangle$, $|1\rangle$, and $|2\rangle$, respectively. The laser wavelengths are denoted λ_p and λ_c. Taken from [413, Figure 1]. (Reprinted with permission. Copyright 1991 American Physical Society.)

Experimental demonstrations of EIT abound. For example, Harris [413] showed that a gas of Sr atoms, which is normally opaque when irradiated with a laser operating at the ω_1 transition frequency, becomes transparent when accompanied by a strong laser operating at the ω_2 frequency. An example is shown in Figure 11.33. Here, applying a 570.3 nm laser beam resonant with the 4d5p$^1D_2$ to 4d5d$^1D_2$ transition increases the transmission of light at 337.1 nm (a wavelength which is near resonant with 5s5p$^1D_2$ to 4d5d$^1D_2$) by a factor of e^{19}, with nearly all of the Sr atoms remaining in their ground state. Experimental results (solid line) are compared to computations (dashed line) in Figure 11.33 where the significant increase in transmission with changes in the central frequency of the $\mathcal{E}_1(t)$ pulse is shown. Note the enormous increase in transmission at a detuning close to zero. Related results have been obtained by Stoicheff *et al.* [542] in atomic hydrogen. Also of interest is the use of EIT, by Kasapi [543], who was able to identify the presence of 0.3% ^{207}Pb in 99.7% ^{208}Pb by eliminating the absorption of the ^{208}Pb species, revealing the underlying traces of ^{207}Pb that still absorbed the incident radiation. The ability to distinguish the two isotopes of lead by this technique is based on the small energy differences in the spectra of the two systems.

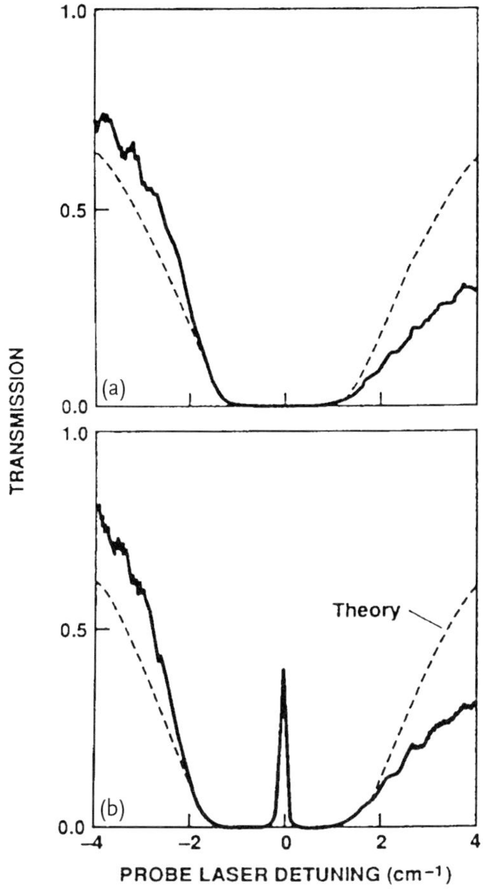

Figure 11.33 Transmission vs. probe laser detuning in Sr at a density of 5×10^{17} atoms/cm^3. (a) is without the ω_2 laser. (b) with the Rabi frequency $\Omega_{0,2} = 1.5\,\text{cm}^{-1}$. Taken from [413, Figure 3]. (Reprinted with permission. Copyright 1991 American Physical Society.)

Since these initial EIT experiments in dilute gases, EIT has also been observed in solids where the decoherence rates are considerably higher. Specifically, EIT has been seen in Y_2SiO_5 doped with Pr^{3+} [416, 544] and in solid Ruby [545]. The experimental results suggest that EIT persists as long as the inhomogeneous width associated with the transition between levels $|E_1\rangle$ and $|E_2\rangle$ is less than the Rabi frequencies coupling $|E_1\rangle$ with $|E_0\rangle$ and $|E_0\rangle$ with $|E_2\rangle$.

We now turn our attention to the *structured continua* case. Figure 11.31 displays the absorption of two AT split levels for structured continua of the type displayed in Figure 11.28 with no detuning ($\Delta_2 = 0$). Two asymmetric line-shapes separated by an almost complete transparency "hole" are seen to form. The two parts of the line-shape change, while being pushed apart, with one of them becoming much

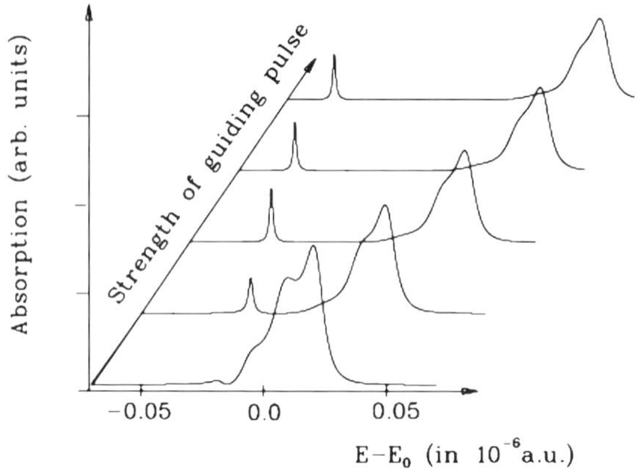

Figure 11.34 EIT for structured continuum with detuning ($\Delta_2 = 0.01$), as a function of the ε_2 coupling strength (the uppermost trace corresponds to the peak of the ε_2 pulse). The continuum parameters are as in Figure 11.31.

sharper, as the ε_2 coupling laser intensity increases. As shown above, the sharpening of one of the peaks does not occur in the unstructured continuum case.

In Figure 11.34 we show EIT for a structured continua *with* detuning ($\Delta_2 = 0.01$). The effect of the detuning in this case is quite dramatic, as it causes one of the EIT peaks to almost disappear while, as in Figure 11.31, making the line-shapes change substantially with increased ε_2 coupling strengths.

11.6.4
The Resonance Description of Slowing Down of Light by EIT

The slowing down of light associated with EIT was first observed by Hau *et al.* [296, 546, 547] The evidence for the effect in the Na_2 system is shown in Figure 11.35 where slowing down to speeds of a few meters per second is being demonstrated. Traditionally, the slowing down of light associated with EIT was attributed in a phenomenological way to the narrowing down of the dispersion curve due to the formation of the EIT transmission window. When a pulse propagates, the physically meaningful velocity is the "group velocity" $v_g = c/n_g$ where n_g is the group's index of refraction, given as

$$n_g = n + \omega \frac{dn}{d\omega} \tag{11.112}$$

where n is the "phase" index of refraction for each frequency component ω. As depicted in Figure 11.36, n changes sharply in the vicinity of an EIT window with a positive slope and n_g becomes large and positive, thereby greatly reducing $v_g = c/n_g$. As we saw above, the weaker the coupling laser, the sharper is the EIT transmission window and the larger is the positive slope of n as a function of ω. Thus, as we switch off the coupling laser, the group velocity gets smaller and smaller. Finally when the coupling laser strength approaches zero, the group velocity also

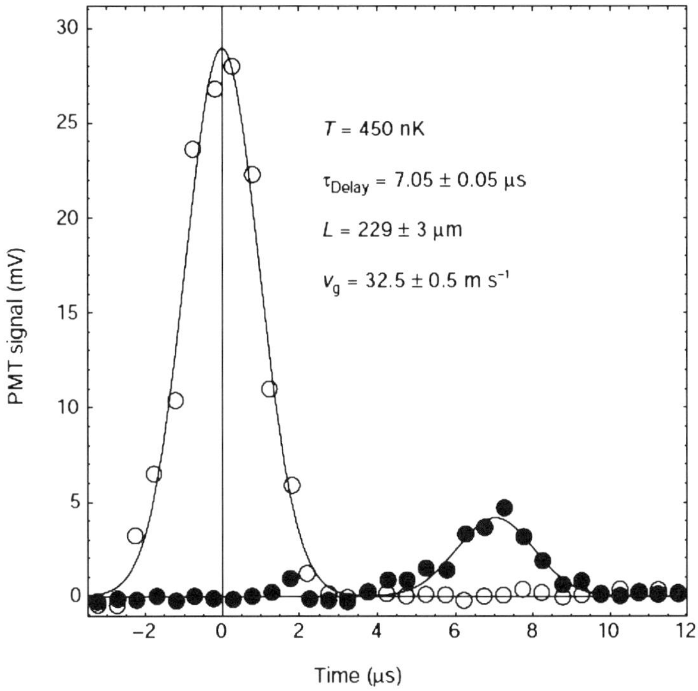

Figure 11.35 Experimental demonstration of the EIT slowing down of light relative to a reference freely propagating light pulse. Taken from [296, Figure 3]. (Reprinted with permission. Copyright 1999 Nature Publishing Group.)

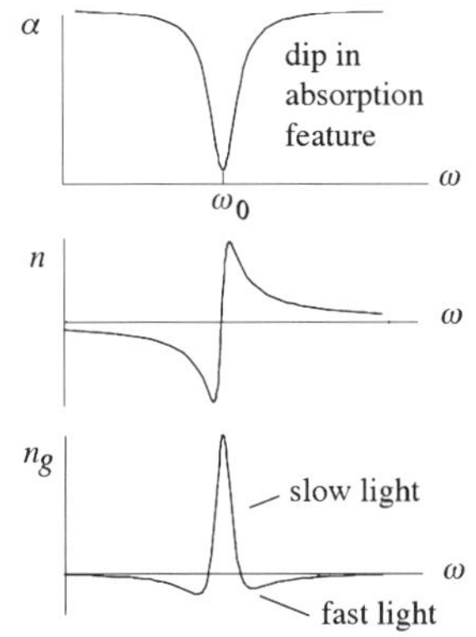

Figure 11.36 Dispersion curves in the vicinity of an EIT transmission window. From [548].

approaches zero and the light is said to have been "stopped". Of course light cannot be stationary, and what happens is that as the coupling laser is turned off, a STIRAP process takes place and a smooth population transfer between the ground state, $|E_1\rangle$, and the excited state, $|E_2\rangle$, occurs. The light energy thus gets stored in the matter medium. In the absence of decoherence, this light storage is coherent over the entire material ensemble. The light can therefore be "resurrected" by gradually switching back the coupling laser. This effect has been observed by several groups [549–551]. It has great potential as a means of storage of "quantum information".

We now turn to a more "mechanistic" description of the slowing down of light. As in all dispersion phenomena, the slowing down of light associated with EIT [296, 546, 547] originates in levels which are *off-resonance* with respect to the $\varepsilon_1(t)$ probe laser. The main difference is that the role of the off resonance levels is dramatically more important in EIT than in ordinary photoabsorption line-shapes because these levels occur very near the probe laser center frequency.

In order to see how this happens in detail, we go beyond the weak-field limit used so far to treat the effect of the $\varepsilon_1(t)$ probe pulse. We do so by incorporating $|E_1\rangle$ in the expansion of $\Psi(t)$ – the solution of the time-dependent Schrödinger equation (Eq. (11.64)). We thus have that

$$|\Psi(t)\rangle = b_1(t)|E_1\rangle e^{-iE_1t/\hbar} + \sum_{\mathrm{n}} \int dE\, b_{E,\boldsymbol{n}}(t)|E,\boldsymbol{n}^-\rangle e^{-iEt/\hbar} . \tag{11.113}$$

Substituting this expansion in Eq. (11.64), we obtain a set of ordinary differential equations (ODE) for the expansion coefficients $b_1(t)$ and $b_{E,\boldsymbol{n}}(t)$:

$$\frac{db_1(t)}{dt} = \frac{i}{\hbar}\sum_n \int dE\, b_{E,n}(t) d_{1,\boldsymbol{n}}(E)\mathcal{E}_1^*(t)e^{-i(\omega_{E,1}-\omega_1)t}$$

$$\frac{db_{E,\boldsymbol{n}}(t)}{dt} = \frac{i}{\hbar} b_1(t) d_{\boldsymbol{n},1}(E)\mathcal{E}_1(t)e^{i(\omega_{E,1}-\omega_1)t} , \tag{11.114}$$

where $\omega_{E,1} \equiv (E - E_1)/\hbar$. We now eliminate the continuum by solving the second part of Eq. (11.114), which yields

$$b_{E,\boldsymbol{n}}(t) = \frac{i}{\hbar}\int_{-\infty}^{t} dt' b_1(t') d_{\boldsymbol{n},1}(E)\mathcal{E}_1(t')e^{i\delta_E t'} , \tag{11.115}$$

where $\delta_E \equiv \omega_{E,1} - \omega_1$. Substituting Eq. (11.115) into the first part of Eq. (11.114) we obtain

$$\frac{db_1(t)}{dt} = \frac{\mathcal{E}_1^*(t)}{\hbar}\int_{-\infty}^{t} dt' b_1(t') F_1(t-t')\mathcal{E}_1(t') , \tag{11.116}$$

where $F_1(\tau)$, the *spectral autocorrelation* function

$$F_1(\tau) \equiv -\int d\delta_E \sum_{\mathrm{n}} |d_{1,\boldsymbol{n}}(E)|^2 e^{-i\delta_E\tau} , \tag{11.117}$$

is just the Fourier transform of $P_1(E)$, the total absorption probability from state $|E_1\rangle$,

$$P_1(E) \equiv \sum_{\mathrm{n}} |d_{1,\boldsymbol{n}}(E)|^2 . \tag{11.118}$$

For an unstructured continuum we can use the explicit form of $P_1(E)$ of Eq. (11.108) to obtain that

$$F_1(\tau) = \frac{|d_{1,0}|^2}{2\pi} \int d\delta_E \frac{-\overline{\Gamma}_0 \left[(E-E_0)^2 + \overline{\Gamma}_2^2/4\right] e^{-i\delta_E \tau}}{\left[(E-E_0)^2 - |\Omega_2(t)|^2 - \overline{\Gamma}_0 \overline{\Gamma}_2/4\right]^2 + \left[(E-E_0)\overline{\Gamma}_0 + \overline{\Gamma}_2/2\right]^2}. \tag{11.119}$$

In Figure 11.37 the $P_2(E)$ and $P_1(E)$ for unstructured continua are shown.

A sample $F_1(\tau)$ is shown in Figure 11.38, using which, one obtains with the aid of Eq. (11.116) the $b_1(t)$ coefficient (an example of which given in Figure 11.39). The

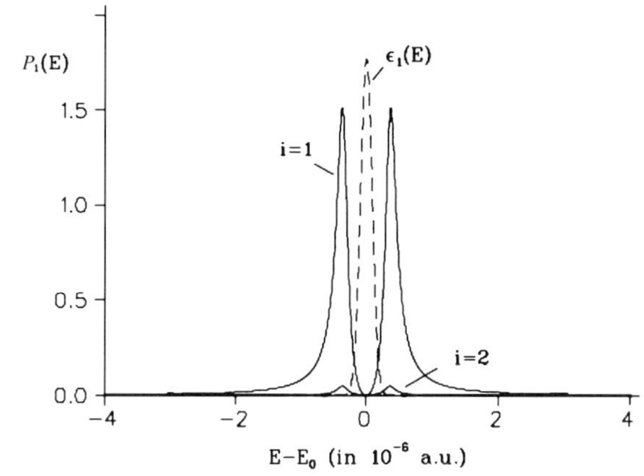

Figure 11.37 $P_2(E)$ and $P_1(E)$ for unstructured continua $\overline{\Gamma}_0 = 4 \times 10^{-6}$ a.u., $\overline{\Gamma}_2 = 10^{-8}$ a.u., and $\Delta_2 = 0$. Shown also is the probe laser line-shape $\epsilon_1(E)$.

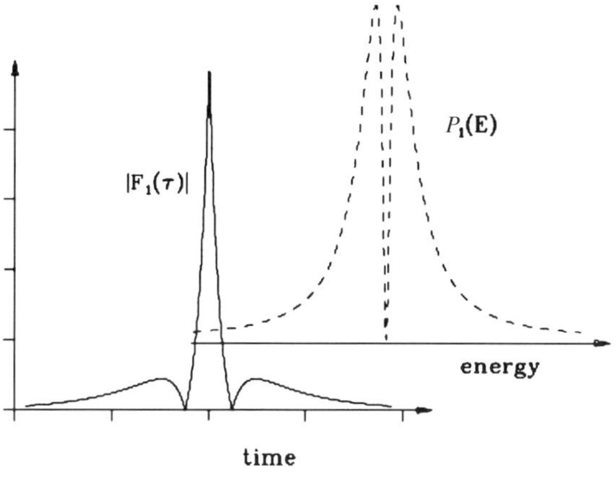

Figure 11.38 The spectral autocorrelation function $F_1(\tau)$ (full line) and the EIT line-shape (broken line) corresponding to it.

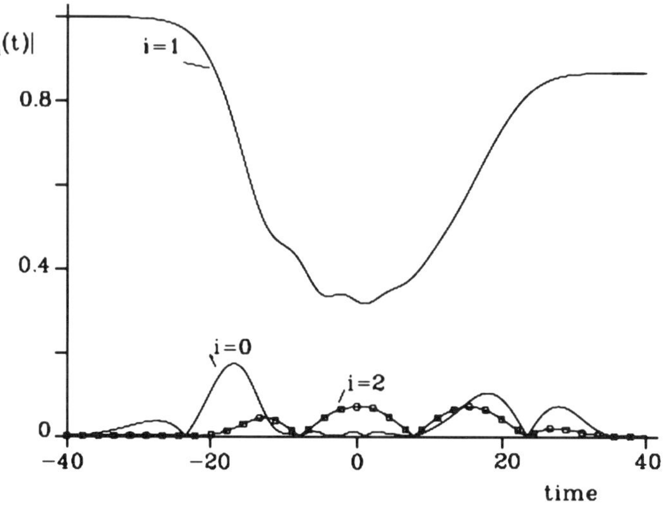

Figure 11.39 The $b_1(t)$, $b_0(t)$, and $b_2(t)$ expansion coefficients as functions of time and energy for the EIT line-shape and laser profile of Figure 11.37.

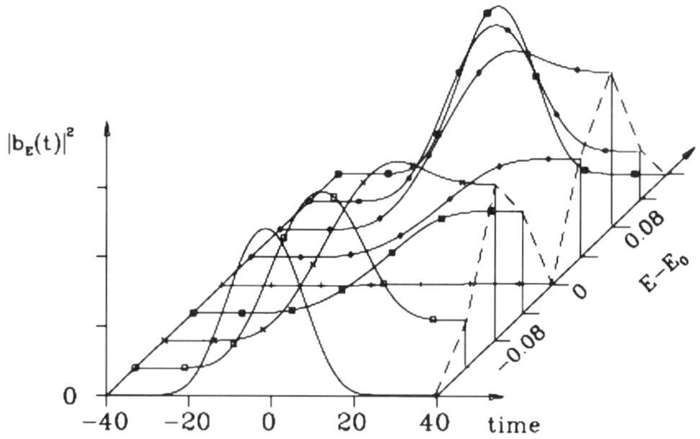

Figure 11.40 $|b_E(t)|^2$, the absolute value squared of the dressed states' expansion coefficients as a function of time, expressed in 10^6 a.u., and $E - E_0$, expressed in 10^{-6} a.u.

continuum preparation coefficients $b_{E,n}(t)$ (a sample is shown in Figure 11.40) are then readily obtained from Eq. (11.115).

We next turn our attention to calculating the (transient) populations of levels $|E_0\rangle$ and $|E_2\rangle$. These can be obtained using Eq. (11.113), as

$$\langle E_0|\Psi(t)\rangle = \sum_n \int dE\, b_{E,n}(t)\langle E_0|E, n^-\rangle e^{-iEt/\hbar} . \tag{11.120}$$

Noting that because of selection rules $|E_1\rangle$ is coupled exclusively by the dipole operator to $|E_0\rangle$, we have that

$$d_{1,n}(E) = d_{1,0}\langle E_0|E, \boldsymbol{n}^-\rangle\ . \tag{11.121}$$

Thus Eq. (11.120) can be written as

$$\begin{aligned}
\langle E_0|\Psi(t)\rangle &= \frac{1}{d_{1,0}}\sum_{n}\int dE\, b_{E,n}(t)d_{1,n}(E)e^{-iEt/\hbar} \\
&= \frac{ie^{-i(\omega_1+E_1/\hbar)t}}{\hbar d_{1,0}}\int_{-\infty}^{t} dt'\mathcal{E}_1(t')b_1(t')\sum_{n}\int dE\,|d_{n,1}(E)|^2\, e^{-i\delta_E(t-t')} \\
&= -\frac{ie^{-i(\omega_1+E_1/\hbar)t}}{d_{1,0}}\int_{-\infty}^{t} dt'\mathcal{E}_1(t')b_1(t')F_1(t-t')\ .
\end{aligned} \tag{11.122}$$

The calculation of $\langle E_2|\Psi(t)\rangle$ is more complicated and we need to repeat the procedure adopted in calculating $d_{1,n}(E)$. We thus write

$$\langle E_2|\Psi(t)\rangle = \sum_{n}\int dE\, b_{E,n}(t)\sum_{s=1,2}\langle E_2|\lambda_s\rangle\ell_s(E)e^{-iEt/\hbar}\ . \tag{11.123}$$

Using Eqs. (11.79), (11.97), (11.99), (11.101), and neglecting as before the highly oscillatory terms containing the $e^{\pm i\omega_{2,0}t}$ factors, we obtain that

$$\sum_{s=1,2}\langle E_2|\lambda_s\rangle\ell_s(E) = \frac{V_{2,n}}{\mathcal{D}}\left[E - E_0 - \lambda\cos 2\theta - \left(\overline{\Delta}_0 + i\overline{\Gamma}_0/2\right)\right] = \frac{d_{0,n}(E)}{d_{2,0}}\ , \tag{11.124}$$

where $d_{0,n}(E) \equiv \langle E_0|d_{e,g}|E, \boldsymbol{n}^-\rangle$ and $d_{2,0} \equiv \langle E_2|d_{e,g}|E_0\rangle$ and we have used the fact that the only state coupled to $|E_0\rangle$ by the $\varepsilon_2(t)$ pulse is $|E_2\rangle$. Thus,

$$\begin{aligned}
\langle E_2|\Psi(t)\rangle &= \frac{1}{d_{2,0}}\sum_{n}\int dE\, b_{E,n}(t)d_{0,n}(E)e^{-iEt/\hbar} \\
&= \frac{ie^{-i(\omega_1+E_1/\hbar)t}}{\hbar d_{2,0}}\int_{-\infty}^{t} dt'\mathcal{E}_1(t')b_1(t')\sum_{n}\int dE\, d_{0,n}(E)d_{n,1}(E)e^{i\delta_E t'} \\
&= -\frac{ie^{-i(\omega_1+E_1/\hbar)t}}{d_{2,0}}\int_{-\infty}^{t} dt'\mathcal{E}_1(t')b_1(t')F_{0,1}(t-t')\ ,
\end{aligned} \tag{11.125}$$

where $F_{0,1}(\tau)$ is the *spectral cross-correlation* function

$$F_{0,1}(\tau) \equiv -\int d\delta_E\, P_{0,1}(E)e^{-i\delta_E\tau}\ , \tag{11.126}$$

where

$$P_{0,1}(E) \equiv \sum_{n} d_{0,n}(E) d_{n,1}(E) . \tag{11.127}$$

We now turn our attention to the calculation of $\langle E', \boldsymbol{m}^-; 1|\Psi(t)\rangle$ – the amplitudes for populating the (radiative or nonradiative) $|E', \boldsymbol{m}^-; 1\rangle$ bare continuum states, that is, the eigenstates of PHP (see Eq. (7.44)), given as

$$\langle E', \boldsymbol{m}^-; 1|\Psi(t)\rangle = \sum_{n} \int dE\, b_{E,n}(t) \langle E', \boldsymbol{m}^-; 1|E, \boldsymbol{n}^-\rangle e^{-iEt/\hbar} . \tag{11.128}$$

With Eqs. (7.42), (11.88), and (11.95), neglecting the terms containing the highly oscillatory $e^{i\omega_{2,0}t}$ factor and using Eqs. (11.97) and (11.103), we have for the $\langle E', \boldsymbol{m}^-; 1|E, \boldsymbol{n}^-\rangle$ amplitudes

$$\langle E', \boldsymbol{m}^-; 1|E, \boldsymbol{n}^-\rangle = \frac{1}{E - i\epsilon - E'} \left\{ \frac{V_{m,0}(E') d_{1,n}(E)}{d_{1,0}} + \frac{V_{m,2}(E') d_{0,n}(E)}{d_{0,2}} \right\} . \tag{11.129}$$

Substituting in Eq. (11.128) and with Eq. (11.115) we finally have that

$$\langle E', \boldsymbol{m}^-; 1|\Psi(t)\rangle = \sum_{n} \int dE\, e^{-iEt/\hbar} b_{E,n}(t) \frac{1}{E - i\epsilon - E'} \times \left\{ \frac{V_{m,0}(E') d_{1,n}(E)}{d_{1,0}} + \frac{V_{m,2}(E') d_{0,n}(E)}{d_{0,2}} \right\} . \tag{11.130}$$

Using Eq. (11.115) we can show that

$$\langle E', \boldsymbol{m}^-; 1|\Psi(t)\rangle = i e^{-i\omega_1 t} \left\{ \frac{V_{m,0}(E')}{d_{1,0}} \int_{-\infty}^{t} dt' b_1(t') \mathcal{E}_1(t') G_1(E', t - t') + \frac{V_{m,2}(E')}{d_{0,2}} \int_{-\infty}^{t} dt' b_1(t') \mathcal{E}_1(t') G_{1,0}(E', t - t') \right\} . \tag{11.131}$$

where

$$\begin{aligned} G_1(E', \tau) &\equiv \int d\delta_E e^{-i\delta_E \tau} \frac{P_1(E)}{E - i\epsilon - E'} \\ &= \int d\delta_E e^{-i\delta_E \tau} P_1(E) \left[\frac{E - E' + i\epsilon}{(E - E')^2 + \epsilon^2} \right] \\ G_{0,1}(E', \tau) &\equiv \int d\delta_E e^{-i\delta_E \tau} \frac{P_{0,1}(E)}{E - i\epsilon - E'} \\ &= \int d\delta_E e^{-i\delta_E \tau} P_{0,1}(E) \left[\frac{E - E' + i\epsilon}{(E - E')^2 + \epsilon^2} \right] . \end{aligned} \tag{11.132}$$

The contribution of the free one-photon states to the absorption for unstructured continuum is shown in Figure 11.41.

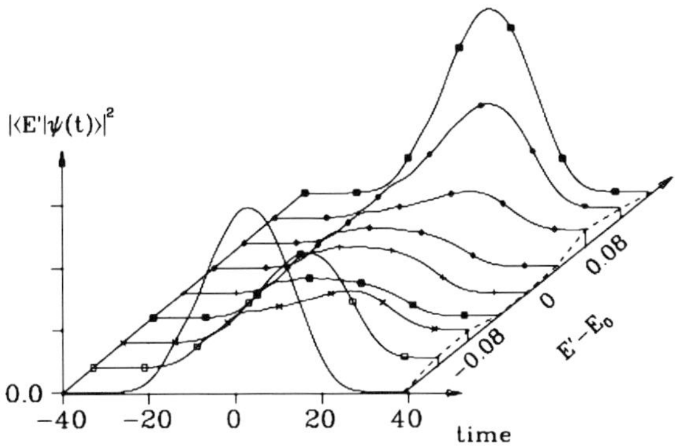

Figure 11.41 The contribution of the free one-photon states to the absorption for unstructured continuum. Plotted is $|\langle E'|\Psi\rangle|^2$ as a function of time and energy. $\overline{\Gamma}_0 = 0.4 \times 10^{-6}$ a.u., $\overline{\Gamma}_2 = 0.02 \times 10^{-6}$ a.u., and $\Delta_2 = 0$. The time axis (t) is expressed in 10^6 a.u. and the energy axis in 10^{-6} a.u.$|\langle E'|\Psi\rangle|$ is expressed in units of units of $\{1/E\}^{1/2}$ where E is expressed in a.u.

Equation (11.131) shows that it is possible to control the population of various continuum states by shaping the $\mathcal{E}_1(t)$ pulse. As demonstrated in Eq. (11.131), in the strong-field regime, the amplitude for observing a given $|E, \boldsymbol{m}^-; 1\rangle$ bare state is composed of two terms, one associated with the $G_1(E', \tau)$ autocorrelation function, and the other associated with the $G_{0,1}(E', \tau)$ cross-correlation function. These functions are respectively associated with the $P_1(E)$ absorption spectrum or with the (complex) $P_{0,1}(E)$ "Raman-type" spectrum. By changing the phase structure of $\mathcal{E}_1(t)$ it is thus possible to constructively interfere between the two terms which contribute to the population of a certain set of $|E, \boldsymbol{m}^-; 1\rangle$ states, while destructively interfering between the terms which contribute to the population of other $|E', \boldsymbol{m}'^-; 1\rangle$ bare states.

A demonstration of this type of coherent control was presented in a series of experiments conducted by Wollenhaupt and Baumert [552, 553] in which by varying the chirp or the phase of a train of pulses which comprise the $\mathcal{E}_1(t)$ field (in a method called SPODS – "selective population of dressed states"), they were able to control the relative kinetic energies E of continuum states of photoelectrons resulting from the ionization of K atoms.

We have thus presented a comprehensive theory of EIT in which both the structure and multiplicity of (coupled) continua are taken into account, which strongly emphasizes the fact that EIT is a manifestation of interferences in the continuum. As such, it is a property of the way the full continuum eigenfunctions are convoluted with the matter-radiation Hamiltonian and the initial bound states. The exact nature of this convolution depends on the type of spectroscopy used to probe the continuum; whether it is linear, or nonlinear, but the EIT line-shapes, and especially the EIT dips, are properties of the continuum.

12
Photodissociation Beyond the Weak-Field Regime

In this chapter we continue our discussion of control in moderately strong fields, focusing on the photodissociation of molecules by moderately strong pulses using approximate analytical approaches. The treatment of the dissociation of a molecule by a strong CW light is postponed to Section 15.1, where we show that the "dressed state" picture emerging from the quantum description of light is the most natural and useful way of thinking about the interaction of molecules with "truly" strong CW fields.

12.1
One-Photon Dissociation with Laser Pulses

We consider now the case of dissociation by the net absorption of just one photon, to be termed "one-photon dissociation". As in the weak-field domain (Chapter 3) the molecule is assumed to dissociate into two fragments as a result of the interaction with a laser pulse. It is convenient to parametrize the incident electric field (Eq. (1.7)) as

$$E(t) = 2\hat{\epsilon}\mathcal{E}(t)\cos(\omega_1 t) \; , \tag{12.1}$$

where we suppress the spatial z variable .

Assuming that the field is in near-resonance or on-resonance with transitions from the initial bound state $|E_1\rangle$ to the continuum, (see Figure 12.1a) we expand the full time-dependent wave function as:

$$|\Psi(t)\rangle = b_1(t)|E_1\rangle \exp(-iE_1 t/\hbar) + \sum_n \int dE\, b_{E,\boldsymbol{n}}(t)|E, \boldsymbol{n}^-\rangle \exp(-iEt/\hbar) \; . \tag{12.2}$$

Here we have suppressed the channel (q) index for convenience, assuming that it is contained in the $\boldsymbol{n}$ index. As usual, we insert Eq. (12.2) into the time-dependent Schrödinger equation, $i\hbar\partial\Psi/\partial t = H\Psi(t)$, and use the orthogonality of the eigenfunctions of H_M, to obtain an indenumerable set of first-order differential

Quantum Control of Molecular Processes, Second Edition. Moshe Shapiro and Paul Brumer

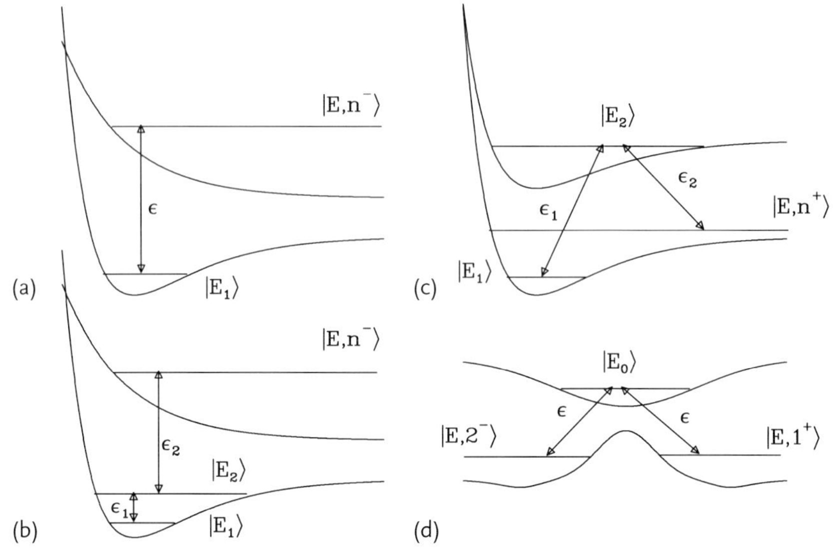

Figure 12.1 Energy levels and pulses pertaining to: (a) One-photon dissociation; (b) resonantly enhanced two-photon dissociation; (c) resonantly enhanced two-photon association; (d) laser catalysis.

equations that are analogous to Eq. (2.3):

$$\frac{db_1}{dt} = i\int dE \sum_n \Omega_{1,E,\boldsymbol{n}}(t) b_{E,\boldsymbol{n}}(t) \exp(-i\Delta_E t) , \tag{12.3a}$$

$$\frac{db_{E,\boldsymbol{n}}}{dt} = i\Omega^*_{1,E,\boldsymbol{n}}(t)\exp(i\Delta_E t) b_1(t) , \quad \text{for each } E \text{ and } \boldsymbol{n} . \tag{12.3b}$$

Here we have retained the "rotating wave" terms (see Eq. (2.13)) only. The detuning, Δ_E, is given by

$$\Delta_E \equiv \omega_{E,1} - \omega_1 , \quad \text{with} \quad \omega_{E,1} \equiv \frac{E - E_1}{\hbar} , \tag{12.4}$$

and $\Omega_{1,E,\boldsymbol{n}}(t)$ is the (time-varying) "Rabi frequency", defined as

$$\Omega_{1,E,\boldsymbol{n}}(t) \equiv \langle E_1 | \hat{\boldsymbol{\varepsilon}} \cdot \boldsymbol{d} | E, \boldsymbol{n}^- \rangle \mathcal{E}(t)/\hbar . \tag{12.5}$$

Unlike the treatment in the weak-field domain we do not assume that $b_1(t) \approx 1$ at all times. Rather, we integrate the $b_{E,\boldsymbol{n}}$ continuum coefficients of Eq. (12.3) over time, while imposing the boundary condition that the continuum states are empty at the start of the process, (i.e., $b_{E,\boldsymbol{n}}(t \to -\infty) = 0$), to get:

$$b_{E,\boldsymbol{n}}(t) = i \int_{-\infty}^{t} dt' \, \Omega^*_{1,E,\boldsymbol{n}}(t') b_1(t') \exp(i\Delta_E t') . \tag{12.6}$$

The state-specific photodissociation probability, $P_n(E)$, (Eq. 2.74), is the long-time probability, at fixed energy E, of observing a particular internal state $|\boldsymbol{n}\rangle$ of the dissociated fragments. Hence, using Eq. (12.5) and Eq. (12.6), we have that

$$P_n(E) = |b_{E,\boldsymbol{n}}(t \to \infty)|^2 = \left| \frac{1}{\hbar} \langle E_1 | \hat{\boldsymbol{\varepsilon}} \cdot \boldsymbol{d} | E, \boldsymbol{n}^- \rangle \int_{-\infty}^{\infty} dt' \mathcal{E}^*(t') b_1(t') \exp(i \Delta_E t') \right|^2 . \tag{12.7}$$

It follows that the final branching ratio between two product states is given by

$$\frac{P_n(E)}{P_m(E)} = \left| \frac{b_{E,\boldsymbol{n}}(\infty)}{b_{E,\boldsymbol{m}}(\infty)} \right|^2 = \left| \frac{\langle E_1 | \hat{\boldsymbol{\varepsilon}} \cdot \boldsymbol{d} | E, \boldsymbol{n}^- \rangle}{\langle E_1 | \hat{\boldsymbol{\varepsilon}} \cdot \boldsymbol{d} | E, \boldsymbol{m}^- \rangle} \right|^2 . \tag{12.8}$$

We see that in the strong-field domain, as in the weak-field regime, the relative probabilities of populating different asymptotic states at a fixed energy E are *independent of the attributes of the laser pulse*. This result, which coincides with that of perturbation theory (see Eq. (3.5)) holds true irrespective of the laser power, provided that only *one* initial state $|E_1\rangle$ is coupled via $\hat{\boldsymbol{\varepsilon}} \cdot \boldsymbol{d}$ to the continua (i.e., only one bound state appears in Eq. (12.2)). As in the weak-field domain, the situation changes completely if more than one bound state is effectively coupled to the continua.

We obtain a closed-form solution for the bound part of the problem by substituting Eq. (12.6) into Eq. (12.3a), giving a first-order integro-differential equation for b_1:

$$\frac{db_1}{dt} = \frac{-1}{\hbar^2} \int dE \sum_n |\langle E, \boldsymbol{n}^- | \hat{\boldsymbol{\varepsilon}} \cdot \boldsymbol{d} | E_1 \rangle|^2 \mathcal{E}(t) \int_{-\infty}^{t} dt' \mathcal{E}^*(t') \exp\left[-i\Delta_E(t-t')\right] b_1(t') . \tag{12.9}$$

Equation (12.9) can be solved numerically in a straightforward fashion. Nevertheless, it is instructive to analyze it in terms of $F_1(t-t')$, the "spectral autocorrelation function" [9, 11, 261, 554], defined as the Fourier transform of the absorption spectrum,

$$F_1(t-t') = \int dE\, P_1(E) \exp\left[-i\omega_{E,1}(t-t')\right] . \tag{12.10}$$

Here $P_i(E)$, the absorption spectrum from the ith state, where

$$P_i(E) \equiv \sum_n |\langle E, \boldsymbol{n}^- | \hat{\boldsymbol{\varepsilon}} \cdot \boldsymbol{d} | E_i \rangle|^2 . \tag{12.11}$$

Using this definition for $F_1(t-t')$, we can rewrite Eq. (12.9) as

$$\frac{db_1}{dt} = \frac{-\mathcal{E}(t)}{\hbar^2} \int_{-\infty}^{t} dt' \mathcal{E}^*(t') F_1(t-t') \exp\left[i\omega_1(t-t')\right] b_1(t') . \tag{12.12}$$

The value of the ground state coefficient at time t is seen to be determined by its past history at $t' < t$ through the "memory kernel" $\mathcal{E}(t)\mathcal{E}^*(t')F_1(t-t')$.

12.1.1
Slowly Varying Continuum

The simplest (though approximate) solution of Eq. (12.12) is obtained by assuming that all the continua are "flat", that is, that the bound-continuum matrix elements vary slowly with energy and can be replaced by their value at some average energy, say $E_L = E_1 + \hbar\omega_1$,

$$\sum_n |\langle E, \boldsymbol{n}^- | \hat{\boldsymbol{\varepsilon}} \cdot \boldsymbol{d} | E_1 \rangle|^2 \approx \sum_n |\langle E_L, \boldsymbol{n}^- | \hat{\boldsymbol{\varepsilon}} \cdot \boldsymbol{d} | E_1 \rangle|^2 . \tag{12.13}$$

This approximation, called the "flat continuum" or "slowly varying continuum approximation" (SVCA) [261, 555, 556] localizes the autocorrelation function in time, since by Eq. (12.10) and Eq. (12.13)

$$F_1(t - t') = 2\pi\hbar P_1(E_L)\delta(t - t') . \tag{12.14}$$

Suqbstituting Eq. (12.14) in Eq. (12.12) and integrating over E and t', we obtain that

$$\frac{db_1}{dt} = -\Omega_1^I(t) b_1(t) . \tag{12.15}$$

Hence,

$$b_1(t) = b_1(-\infty) \exp\left[- \int_{-\infty}^{t} \Omega_1^I(t')dt' \right] , \tag{12.16}$$

where $\Omega_1^I(t)$, the "imaginary Rabi frequency", is defined as

$$\Omega_1^I(t) \equiv \pi P_1(E_L)|\mathcal{E}(t)|^2/\hbar = \pi \sum_n |\langle E_L, \boldsymbol{n}^- | \hat{\boldsymbol{\varepsilon}} \cdot \boldsymbol{d} | E_1 \rangle \mathcal{E}(t)|^2/\hbar . \tag{12.17}$$

The factor of 1/2 relative to Eq. (12.14) arises because the integration over t' in Eq. (12.12) is carried out over the $[-\infty, t]$ range and not over the usual $[-\infty, +\infty]$ range.

It follows from Eq. (12.16) that a "slowly varying" continuum acts as an irreversible "perfect absorber", since in this approximation $b_1(t)$ decreases monotonically (though not necessarily as a simple exponential) with time.

In many cases the continuum may have structures that are narrower than the bandwidth of the pulse. Such structures may be due to either the natural spectrum of the molecular Hamiltonian [533, 557], or due to the interaction with the strong external field [264, 266–268, 558]. Under such circumstances we expect the SVCA approximation to break down, yielding nonmonotonic decay dynamics.

Given Eqs. (12.6) and (12.16) the amplitude $b_{E,\boldsymbol{n}}(t)$ is given by

$$b_{E,\boldsymbol{n}}(t) = i \int_{-\infty}^{t} dt' \Omega_{1,E,\boldsymbol{n}}^*(t') b_1(-\infty) \exp\left[- \int_{-\infty}^{t'} \Omega_1^I(t'')dt'' + i\Delta_E t' \right] . \tag{12.18}$$

12.1.2
Bichromatic Control

We now generalize this result by considering photoexcitation from a superposition state $b_1|E_1\rangle + b_2|E_2\rangle$ where $b_i \equiv b_i(-\infty)$. Assuming no transitions between levels 1 and 2, and using the SVCA, we can write an analytic formula for "bichromatic control" that goes beyond perturbation theory. Allowing the coefficients b_1 and b_2 to decay according to Eq. (12.16) we obtain that

$$b_{E,\boldsymbol{n}}(t \to \infty) =$$
$$\frac{i}{\hbar}\Bigg\{\langle E, \boldsymbol{n}^- |\hat{\boldsymbol{\varepsilon}} \cdot \boldsymbol{d}| E_1\rangle b_1 \int_{-\infty}^{\infty} dt' \mathcal{E}_1^*(t') \exp\left[i\Delta_{E,1}t' - \frac{\pi}{\hbar}P_1(E_L)\int_{-\infty}^{t'} |\mathcal{E}_1(t'')|^2 dt''\right]$$
$$+ \langle E, \boldsymbol{n}^- |\hat{\boldsymbol{\varepsilon}} \cdot \boldsymbol{d}| E_2\rangle b_2 \int_{-\infty}^{\infty} dt' \mathcal{E}_2^*(t') \exp\left[i\Delta_{E,2}t' - \frac{\pi}{\hbar}P_2(E_L)\int_{-\infty}^{t'} |\mathcal{E}_2(t'')|^2 dt''\right]\Bigg\}, \tag{12.19}$$

where $\Delta_{E,i} = \omega_{E,i} - \omega$, $i = 1, 2$. Therefore, the probability of observing a particular channel $\boldsymbol{n}$ is given as

$$P_{\mathrm{n}}(E) = \frac{4\pi^2}{\hbar^2}\Big|\langle E, \boldsymbol{n}^- |\hat{\boldsymbol{\varepsilon}} \cdot \boldsymbol{d}| E_1\rangle |\zeta_1(E)| e^{-i\theta_1(E)} b_1 + \langle E, \boldsymbol{n}^- |\hat{\boldsymbol{\varepsilon}} \cdot \boldsymbol{d}| E_2\rangle |\zeta_2(E)| e^{-i\theta_2(E)} b_2\Big|^2 . \tag{12.20}$$

where

$$\zeta_i(E) \equiv \frac{1}{2\pi}\int_{-\infty}^{\infty} dt \exp(-i\Delta_{E,i}t)\mathcal{E}_i(t) \exp\left[-\frac{\pi}{\hbar}P_i(E_L)\int_{-\infty}^{t} |\mathcal{E}_i(t')|^2 dt'\right]. \tag{12.21}$$

and $\theta_i(E)$ is the phase of $\zeta_i(E)$.

Thus, we obtain a form, which is correct (within the range of validity of the SVCA) for strong fields, which resembles the weak-field bichromatic control result of Eq. (3.12). The only difference is that instead of the Fourier transform of the electric field of the pulse, Eq. (12.20) depends on the Fourier transform of the product of the pulse electric field and the decaying factor $\exp[-(\pi/\hbar)P_i(E_L)\int_{-\infty}^{t}|\mathcal{E}_i(t')|^2 dt']$, which describes the depletion of the initial state(s) due to the action of the pulse.

12.1.3
Resonance

Returning to the case of excitation from a single $|E_1\rangle$ level, consider now the limit in which the continuum is composed of a *single* resonance at $E = E_s$. This means that the continuous spectrum is described by a single Lorentzian form, positioned at E_s and with full width at half maximum Γ_s:

$$P_1(E) = \frac{\overline{d_s^2}\,\Gamma_s^2/4}{(E - E_s)^2 + \Gamma_s^2/4}\,, \tag{12.22}$$

where $\overline{d_s^2}$ is an average dipole moment square that determines the height of the absorption curve and $2\pi\overline{d_s^2}\,\Gamma_s/4$ is the area under the absorption curve. Substituting this form into Eq. (12.10), and using the fact that $t \geq t'$, we have that

$$\exp\left[i\omega_1(t - t')\right] F_1(t - t') = \frac{\overline{d_s^2}\,\Gamma_s}{4} f_s^+(t) f_s^-(t')\,, \tag{12.23}$$

where

$$f_s^{\pm}(t) = \sqrt{2\pi}\exp[\mp i\chi_s t]\,, \tag{12.24}$$

with

$$\chi_s \equiv \Delta_s - i\frac{\Gamma_s}{2\hbar}\,, \quad \text{and} \quad \Delta_s \equiv \frac{E_s - E_1}{\hbar} - \omega_1\,. \tag{12.25}$$

Using Eq. (12.23) we can transform Eq. (12.12) into two coupled first-order differential equations,

$$\frac{db_1}{dt} = \frac{i}{4\hbar}\mathcal{E}(t)\overline{d_s^2}\,\Gamma_s f_s^+(t) B_s(t)\,, \tag{12.26a}$$

$$\frac{dB_s}{dt} = \frac{i}{\hbar}\mathcal{E}(t) f_s^-(t) b_1(t)\,. \tag{12.26b}$$

These two equations can be solved by reducing them to a form similar to the Schrödinger equation and using the WKB-like approximate solution [559, 560]. Specifically, we first obtain a second-order equation by differentiating, Eq. (12.26) to give

$$\frac{d^2 b_1}{dt^2} = \frac{d\ln\mathcal{E}(t)}{dt}\frac{db_1}{dt} - \frac{\pi|\mathcal{E}(t)|^2}{2\hbar^2}\overline{d_s^2}\,\Gamma_s b_1 + \frac{\mathcal{E}(t)}{4\hbar}\overline{d_s^2}\,\Gamma_s\chi_s f_s^+(t) B_s(t)\,, \tag{12.27}$$

where we have used the explicit form of f_s^+ (Eq. (12.24)). We thus obtain from Eqs. (12.26) and (12.27), that

$$\frac{d^2 b_1}{dt^2} = \left(\frac{d\ln\mathcal{E}(t)}{dt} - i\chi_s(t)\right)\frac{db_1}{dt} - \frac{\pi}{2\hbar^2}|\mathcal{E}(t)|^2\,\overline{d_s^2}\,\Gamma_s b_1\,, \tag{12.28}$$

where $\chi_s(t)$ is defined by equating Eqs. (12.27) and (12.28).

This second-order differential equation in time is homeomorphic to a one-dimensional time-independent Schrödinger equation in a spatial variable. This can be seen by defining

$$g_1(t) = -\frac{d\ln\mathcal{E}(t)}{dt} + i\chi_s(t)\,, \tag{12.29}$$

$$g_0(t) = \frac{\pi}{2\hbar^2} |\mathcal{E}(t)|^2 \overline{d_s^2 \Gamma_s} , \tag{12.30}$$

writing Eq. (12.28) as

$$\frac{d^2 b_1}{dt^2} + g_1(t)\frac{db_1}{dt} + g_0(t) b_1 = 0 , \tag{12.31}$$

and transforming $b_1(t)$, according to,

$$c(t) = \exp\left[\frac{1}{2}\int_{t^*}^{t} g_1(t')dt'\right] b_1(t) = \left(\frac{\mathcal{E}(t)}{\mathcal{E}(t^*)}\right)^{-\frac{1}{2}} \exp\left[\frac{i}{2}\int_{t^*}^{t} \chi_s(t')dt'\right] b_1(t) . \tag{12.32}$$

Doing so, we obtain a Schrödinger-like equation in $c(t)$,

$$\left[\frac{d^2}{dt^2} - W(t)\right] c(t) = 0 , \tag{12.33}$$

where

$$\begin{aligned} W(t) &= \frac{1}{2} g_1'(t) + \frac{1}{4} g_1^2(t) - g_0(t) \\ &= -\frac{d\mathcal{E}(t)}{2dt}\frac{d^2 \ln \mathcal{E}(t)}{dt^2} + \frac{1}{4}\left(\frac{d \ln \mathcal{E}(t)}{dt} - i\chi_s(t)\right)^2 - \frac{\pi}{2\hbar^2} |\mathcal{E}(t)|^2 \overline{d_s^2 \Gamma_s} . \end{aligned} \tag{12.34}$$

$W(t)$ is analogous to the term $2m\,(V(x) - E)\,/\hbar^2$ in the time-independent Schrödinger equation, with x being the coordinate, $V(x)$ is the potential, E is the energy and m is the mass of a particle. The quantity t^*, satisfying the $W(t^*) = 0$ equation, is defined, again in analogy with the time-independent Schrödinger equation, as the "turning point" (i.e., the point where $V(x^*) = E$).

If there is only one turning point, the solutions of Eq. (12.33) can be well approximated in terms of the uniform regular and irregular Airy functions Ai and Bi [559, 561],

$$c_{uni}(t) = \left[\frac{T(t)}{-W(t)}\right]^{\frac{1}{4}} \{C_a Ai[-T(t)] + C_b Bi[-T(t)]\} , \tag{12.35}$$

where the complex argument T is defined as

$$T(t) = \left[\frac{3}{2}\int_{t^*}^{t} (-W(t'))^{\frac{1}{2}}\, dt'\right]^{\frac{2}{3}} , \tag{12.36}$$

and C_a and C_b are constants determined by the initial conditions $b_s = 1,\ b_{E,n} = 0$.

Equation (12.35) is the exact solution of the equation,

$$\left[\frac{d^2}{dt^2} - W(t) - \eta(t)\right] c_{uni}(t) = 0 \tag{12.37}$$

where

$$\eta(t) = [T(t)]^{1/2} \frac{d^2}{dt^2} [T(t)]^{-1/2} . \tag{12.38}$$

Since $|\eta(t)| \ll |W(t)|$, c_{uni} is an excellent approximate solution to Eq. (12.33). If there is more than one turning point, the Airy functions can still be used (provided the turning points do not coalesce), by writing the solutions of Eq. (12.35) for each time interval containing a turning point and matching these solutions and their derivatives across the time intervals. Usually no more than two turning points exist.

The above equations can be simplified when $|T|$ is large because in that case we can use the asymptotic forms of the Ai and Bi functions [4] in Eq. (12.35),

$$\begin{aligned}
Ai(T) &\xrightarrow{|T|\to\infty} \frac{1}{2}\pi^{-\frac{1}{2}} T^{-\frac{1}{4}} \exp(-\zeta) , \quad |\arg T| < \pi \\
Bi(T) &\xrightarrow{|T|\to\infty} \pi^{-\frac{1}{2}} T^{-\frac{1}{4}} \exp(\zeta) , \quad |\arg T| < \frac{1}{3}\pi \\
Ai(-T) &\xrightarrow{|T|\to\infty} \pi^{-1/2} T^{-\frac{1}{4}} \sin\left(\zeta + \frac{\pi}{4}\right) , \quad |\arg T| < \frac{2}{3}\pi \\
Bi(-T) &\xrightarrow{|T|\to\infty} \pi^{-\frac{1}{2}} T^{-\frac{1}{4}} \cos\left(\zeta + \frac{\pi}{4}\right) , \quad |\arg T| < \pi ,
\end{aligned} \tag{12.39}$$

where $\zeta = (2/3)T^{3/2} = \int_{t^*}^{t} (-W(t'))^{1/2} dt'$.

For small enough Γ_s, $|\arg T| < (2/3)\pi$ and we can use the last two identities in Eq. (12.39) to obtain a (first-order) WKB approximation,

$$\begin{aligned}
c_{WKB}(t) = \left[-\pi^2 W(t)\right]^{-\frac{1}{4}} &\left\{ C_a \sin\left[\int_{t^*}^{t} (-W(t'))^{\frac{1}{2}} dt' + \frac{\pi}{4}\right]\right. \\
&\left. + C_b \cos\left[\int_{t^*}^{t} (-W(t'))^{\frac{1}{2}} dt' + \frac{\pi}{4}\right]\right\} .
\end{aligned} \tag{12.40}$$

Using the relation Eq. (12.32) we obtain the first-order WKB approximation that

$$\begin{aligned}
b_1(t) = &\left[\frac{\pi \mathcal{E}(t^*)}{\mathcal{E}(t)}\right]^{-\frac{1}{2}} [-W(t)]^{-\frac{1}{4}} \exp\left[-\frac{i}{2}\int_{t^*}^{t} \chi_s(t')dt'\right] \\
&\times \left\{ C_a \sin\left[\int_{t^*}^{t} (-W(t'))^{\frac{1}{2}} dt' + \frac{\pi}{4}\right]\right. \\
&\left. + C_b \cos\left[\int_{t^*}^{t} (-W(t'))^{\frac{1}{2}} dt' + \frac{\pi}{4}\right]\right\} .
\end{aligned} \tag{12.41}$$

For CW radiation $\mathcal{E}(t) = \mathcal{E}_0$ is a constant. Hence,

$$[-W]^{\frac{1}{2}} = \frac{1}{2}\left[\left(\Delta_s - i\frac{\Gamma_s}{2\hbar}\right)^2 + \frac{2\pi}{\hbar^2}|\mathcal{E}_0|^2\overline{d_s^2}\Gamma_s\right]^{\frac{1}{2}} . \tag{12.42}$$

For CW radiation (that is turned on at $t = 0$) we choose C_a and C_b such that $b_1(0) = 1$, and obtain that

$$b_1(t) = \exp\left[-\frac{i\Delta_s t}{2} - \frac{\Gamma_s t}{4\hbar}\right]\cos\left([-W]^{\frac{1}{2}}\, t\right) . \tag{12.43}$$

To obtain the CW transition amplitude to a level E in the continuum we use Eq. (12.6), according to which

$$b_{E,\boldsymbol{n}}(t) = i\Omega^*_{1,E,\boldsymbol{n}}\int_0^t dt' b_1(t')\exp(i\Delta_E t') = \frac{i}{\hbar}\langle E_s, \boldsymbol{n}^-|\hat{\boldsymbol{\varepsilon}}\cdot\boldsymbol{d}|E_1\rangle\mathcal{E}_0^*$$
$$\times\frac{e^{-\eta_{E,s}t}\left\{-\eta_{E,s}\cos\left[(-W)^{\frac{1}{2}}\, t\right] + (-W)^{1/2}\sin\left[(-W)^{\frac{1}{2}}\, t\right]\right\} + \eta_{E,s}}{\eta^2_{E,s} - W} , \tag{12.44}$$

where $\eta_{E,s} \equiv i(\Delta_s/2 - \Delta_E) + \Gamma_s/4\hbar$.

When we irradiate at the center of the resonance ($\Delta_s = 0$) and examine the amplitude for populating a continuum energy level at this center ($\Delta_E = 0$, see Eq. (12.4)), then $\eta_{E,S} = \Gamma_s/4\hbar$, the expression simplifies and the probability of observing the state $|E_s, \boldsymbol{n}^-\rangle$ is

$$P_{\mathbf{n}}(E_s, t) = |b_{E_s,\boldsymbol{n}}(t)|^2 = \left|\frac{\langle E_s, \boldsymbol{n}^-|\hat{\boldsymbol{\varepsilon}}\cdot\boldsymbol{d}|E_1\rangle\mathcal{E}_0^*}{2\pi|\mathcal{E}_0|^2\overline{d_s^2} - \Gamma_s/2}\right|^2$$
$$\times\left|e^{-\frac{\Gamma_s t}{4\hbar}}\left\{-\cos\left[(-W)^{\frac{1}{2}}\, t\right] + \left(\frac{\hbar(-W)^{\frac{1}{2}}}{\Gamma_s}\right)\sin\left[(-W)^{\frac{1}{2}}\, t\right]\right\} + 1\right|^2 . \tag{12.45}$$

Similar formulae were obtained in [3, 562].

We see that the transition displays damped Rabi-type oscillations between the initial state and the final continuum states where the damping is given by resonance *decay rate* $\Gamma_s/2\hbar$. Although the frequency of the Rabi oscillations is a function of the field strength $\mathcal{E}_0$, the branching ratio between channels is independent of the laser parameters, as pointed out, based on general grounds, in Eq. (12.8).

The above equations are only valid when $2\pi|\mathcal{E}_0|^2\overline{d_s^2} > \Gamma_s/2$. If $2\pi|\mathcal{E}_0|^2\overline{d_s^2} < \Gamma_s/2$, for example, when the field is weak or when the resonance width is very large, as in the flat unstructured continuum situation, another limit applies. Noting that in that case $(-W)^{\frac{1}{2}}$ of Eq. (12.42) becomes imaginary, we make use of the first of Eqs. (12.39), from which it follows that the initial state decays in a monotonic (exponential) fashion. It is easy to show that in that case one obtains the SVCA result of Eq. (12.16).

If there is more than one resonance, we follow the approach in Section 7.3.1, and express the bound-continuum dipole matrix elements $\langle E_1|\hat{\boldsymbol{\varepsilon}}\cdot\boldsymbol{d}|E,\boldsymbol{n}^-\rangle$ as a sum of resonances [556]:

$$\langle E_1|\hat{\boldsymbol{\varepsilon}}\cdot\boldsymbol{d}|E,\boldsymbol{n}^-\rangle = \sum_{s=1}^{N} \frac{i\,d_{s\boldsymbol{n}}\Gamma_s/2}{E-E_s+i\Gamma_s/2}\,. \tag{12.46}$$

The spectrum is now given by

$$P_1(E) = \sum_{\mathrm{n}} |\langle E_1|\hat{\boldsymbol{\varepsilon}}\cdot\boldsymbol{d}|E,\boldsymbol{n}^-\rangle|^2 = \sum_{s's} \frac{d_{s's}\Gamma_s\Gamma_{s'}/4}{(E-E_s+i\Gamma_s/2)(E-E_{s'}-i\Gamma_{s'}/2)}\,, \tag{12.47}$$

where $d_{s's} \equiv \sum_{\mathrm{n}} d_{s\boldsymbol{n}} d^*_{s'\boldsymbol{n}}$. If we only keep the diagonal ($s = s'$) terms, $P_1(E)$ becomes a sum of Lorentzians whereas the off-diagonal terms allow for interferences between overlapping resonances. Further, in the large Γ_s limit one obtains the SVCA [261]. An illustration of a typical spectrum obtained from Eq. (12.47) is given in Figure 7.18.

The Fourier transform of $P_1(E)$ (where, in Eq. (12.12), $t \geq t'$), now becomes

$$\exp\left[i\omega_1(t-t')\right] F_1(t-t') = 2\pi \sum_s \overline{d_s^2} \exp\left[-i\chi_s(t-t')\right] = \sum_s \overline{d_s^2} f_s^+(t) f_s^-(t')\,, \tag{12.48}$$

where

$$\overline{d_s^2} \equiv \sum_{s'} \frac{-i\,d_{s's}\Gamma_s\Gamma_{s'}/4}{E_s-E_{s'}-i(\Gamma_s+\Gamma_{s'})/2}\,, \tag{12.49}$$

and where $f_s^{\pm}$ are defined in Eq. (12.24).

Using Eq. (12.48), we can transform Eq. (12.9) into a discrete set of coupled differential equations:

$$\frac{db_1}{dt} = \frac{i}{\hbar}\mathcal{E}(t) \sum_s \overline{d_s^2} f_s^+(t) B_s(t)\,, \tag{12.50a}$$

$$\frac{dB_s}{dt} = \frac{i}{\hbar}\mathcal{E}^*(t) f_s^-(t) b_1(t)\,, \quad s = 1,\dots,N\,, \tag{12.50b}$$

which can be solved in a routine way using a variety of propagation methods. Once $b_1(t)$ is known, the continuum coefficient $b_{E,\boldsymbol{n}}$ can be computed by straightforward quadrature, using Eq. (12.6).

12.2 Computational Examples

Numerical studies allow us to explore aspects of these models for a number of molecular continua and pulse configurations. Consider first the effect of the pulse intensity on transition probabilities to a slowly varying continuum by considering a continuum composed of single broad Lorentzian of width $\Gamma_s = 2000\,\text{cm}^{-1}$, excited by a $120\,\text{cm}^{-1}$ wide pulse (i.e., a pulse of ~ 80 fs duration). The central frequency of the pulse is tuned to the center of the continuum ($\Delta_s = 0$) and the pulse peaks at $t = 0$.

In Figure 12.2a we show the $|b_{E,n}(t)|$ continuum coefficients as a function of time, at different intensities. The onset of off-resonance processes is typified by a nonmonotonic behavior: At off-pulse-center energies, the continuum coefficients rise and fall with the pulse, with the effect becoming more pronounced the further away from the line center the continuum energy levels are. In the far wings of the pulse the continuum coefficients are zero at the end of the pulse, giving rise to a pure transient, otherwise known as a "virtual" state. These results should be compared to the weak-field transients discussed in Section 2.1 and shown in Figure 2.2.

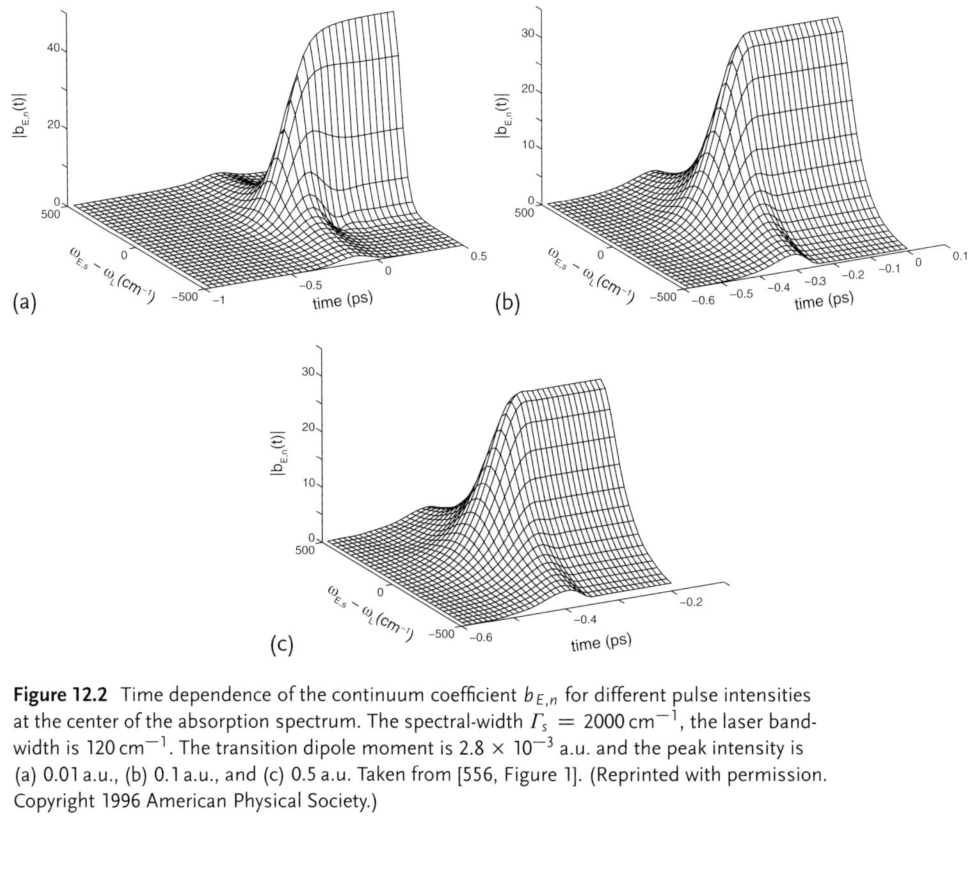

Figure 12.2 Time dependence of the continuum coefficient $b_{E,n}$ for different pulse intensities at the center of the absorption spectrum. The spectral-width $\Gamma_s = 2000\,\text{cm}^{-1}$, the laser bandwidth is $120\,\text{cm}^{-1}$. The transition dipole moment is 2.8×10^{-3} a.u. and the peak intensity is (a) 0.01 a.u., (b) 0.1 a.u., and (c) 0.5 a.u. Taken from [556, Figure 1]. (Reprinted with permission. Copyright 1996 American Physical Society.)

As we increase the field strength, the line-shapes of the photodissociation amplitudes (given as $|b_{E,n}(t = \infty)|$) broaden. This broadening is due to saturation of the continuum population which is greater for continuum states near the pulse center than at the pulse wings. Since a slowly varying continuum is an almost perfect absorber, as the pulse intensity increases the initial state $|E_1\rangle$ empties faster and the dissociation is over before any recurrence can occur. For example, in the 0.01 a.u. peak height case (Figure 12.2a), the continuum levels reach their final population by the time the pulse peaks (at $t = 0$). This time gets progressively shorter as the field strength is increased.

The situation is quite different for a *structured* continuum. Consider Figure 12.3, where the strong pulse induced transition to a narrow continuum ($\Gamma_s = 50\,\text{cm}^{-1}$) is displayed. The results show behavior that is intermediate between a "flat" continuum and a discrete set of levels. We see that "center line", $\omega_{E,1} - \omega_1 \approx 0$, continuum levels display recurrences, or Rabi oscillations, similar, though not identical, to those of a discrete two-level system. In contrast, continuum states at the pulse wings rise and fall smoothly with the pulse, as in the slowly varying continuum case.

Consider now narrowband continua composed of *two* distinct diffuse features (as shown in Figure 12.4a). In Figures 12.4b,c, as we switch on the pulse, the two initially separated lines begin to slowly merge. For moderate laser powers (Figure 12.4b), this merging is a signature of saturation: the continuum states at the

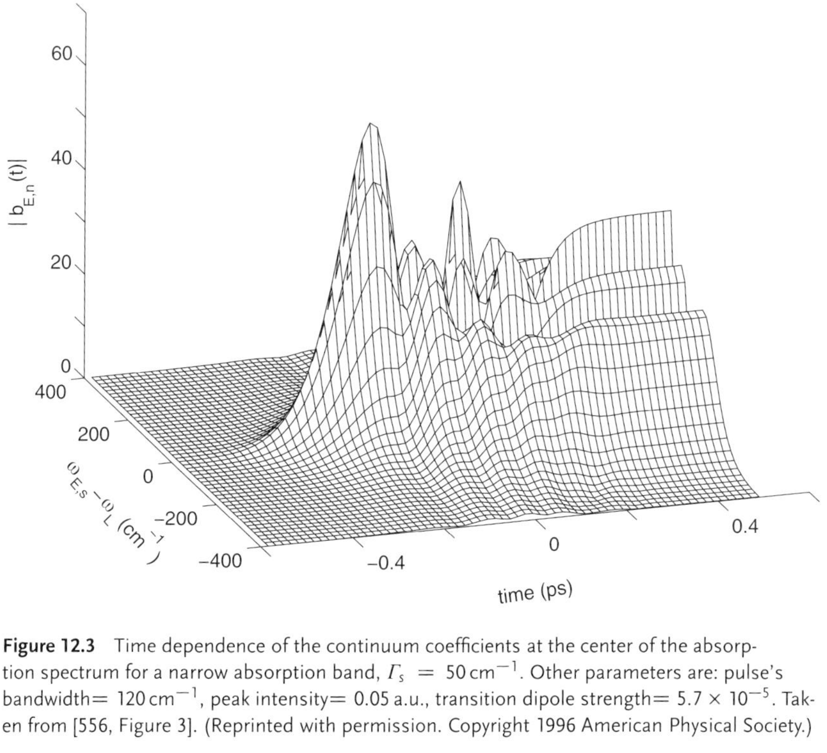

Figure 12.3 Time dependence of the continuum coefficients at the center of the absorption spectrum for a narrow absorption band, $\Gamma_s = 50\,\text{cm}^{-1}$. Other parameters are: pulse's bandwidth= $120\,\text{cm}^{-1}$, peak intensity= 0.05 a.u., transition dipole strength= 5.7×10^{-5}. Taken from [556, Figure 3]. (Reprinted with permission. Copyright 1996 American Physical Society.)

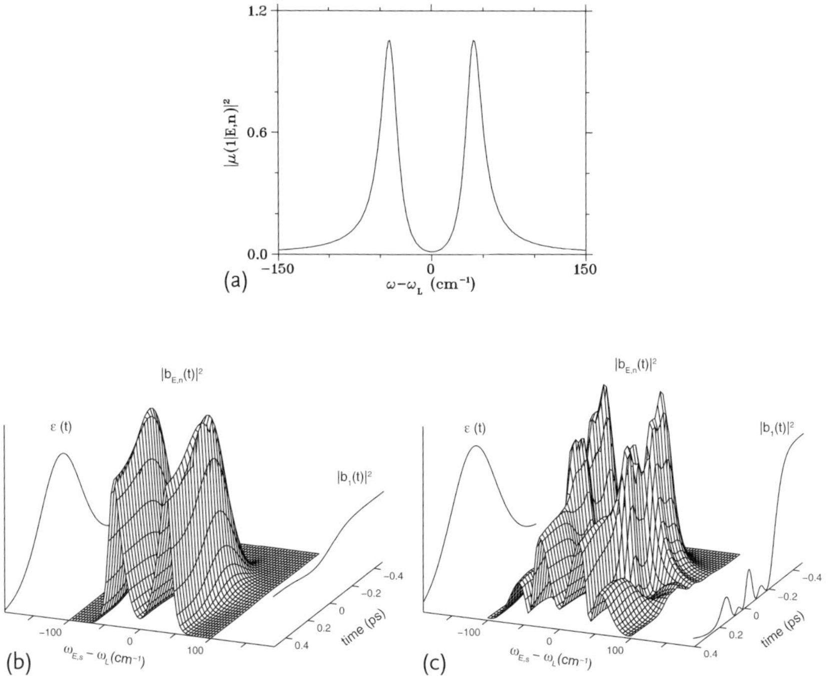

Figure 12.4 Time dependence of the continuum populations for a bound-continuum spectrum comprised of two ($s = 1, 2$) overlapping resonances. (a) The weak-field absorption spectrum. (b) $|b_{E,n}|^2$ as a function of t and E, for laser amplitude $\mathcal{E}_0 = 5 \times 10^{-3}$ a.u., (c) The same as in (b), but for $\mathcal{E}_0 = 5 \times 10^{-2}$ a.u. Also shown are the field envelope $\varepsilon(t)$ as a function of time, as well as $|b_1(t)|^2$. Note the direction of the time axis. Taken from [556, Figure 5]. (Reprinted with permission. Copyright 1996 American Physical Society.)

center of the absorption lines cease to rise while the population of continuum states between the line centers continues to increase. At higher laser intensities (Figure 12.4c), one sees the effect of the Rabi cycling: The populations of the continuum states at the line centers oscillate at a higher frequency than the populations at the wings of the lines. It may happen, as shown in Figure 12.4c, that continuum states at the line center execute a 2π-cycle and are empty at the end of the pulse, whereas continuum states away from the line centers execute only a π-cycling and are highly populated. Thus, under the action of the moderately strong laser pulse, the lines can be reversed: the absorption is effectively zero at the resonance (E_s) positions. In addition, the optically induced interference between the lines causes the formation of "dark states" [8, 221, 412, 563], that is, the cancellation of the absorption at the conclusion of the pulse, lying midway between the two line centers. As a result, we see three holes in the continuum populations in Figure 12.4c: two transparent lines at the center of the resonances due to 2π cycling of these continuum states, accompanied by a third transparent line residing midway between the resonances, due to destructive interference between the resonances. This effect is similar to EIT, discussed in Section 11.1.

13
Coherent Control Beyond the Weak-Field Regime: the Continuum

In this chapter we extend the discussion of control beyond the weak-field regime by considering transitions involving the continuum. We consider two-photon transitions from a bound state to a continuum (complete population transfer to a continuum); from a continuum to a bound state (photoassociation); and from one continuum to another (laser catalysis).

13.1
Control over Population Transfer to the Continuum by Two-Photon Processes

In this section we extend the treatment presented in Section 11.1 for three bound states to the case of resonantly enhanced two-photon dissociation. That is, we replace the final state $|E_2\rangle$ by a continuum of states $|E, \boldsymbol{n}^-\rangle$. The states are renumbered in order of increasing energy, $|E_1\rangle$, $|E_2\rangle$ and $|E, \boldsymbol{n}^-\rangle$ (see Figure 12.1b). If the behavior of such a system is similar to that of three-level systems, then complete population transfer to the continuum, hence complete ionization or dissociation, would become possible [261].

As in Section 11.1, we consider a molecule, initially ($t = 0$) in state $|E_1\rangle$, being excited to a continuum of states $|E, \boldsymbol{n}^-\rangle$, due to the combined action of *two* laser pulses of central frequencies ω_1 and ω_2. We assume, as depicted in Figure 12.1b, that ω_1 is in near-resonance with the transition from $|E_1\rangle$ to an intermediate bound state $|E_2\rangle$ and that ω_2 is in near-resonance with the transition from $|E_2\rangle$ to the continuum.

The total matter–radiation Hamiltonian is given in Eq. (11.2). We then solve the Schrödinger equation by expanding the total wave function as

$$|\Psi(t)\rangle = b_1(t)|E_1\rangle \exp(-iE_1t/\hbar) + b_2(t)|E_2\rangle \exp(-iE_2t/\hbar) + \sum_n \int dE\, b_{E,\boldsymbol{n}}(t)|E, \boldsymbol{n}^-\rangle \exp(-iEt/\hbar) , \quad (13.1)$$

and obtain a set of first-order differential equations, which is now of the form

$$\frac{db_1}{dt} = i\Omega_1^*(t)\exp(-i\Delta_1 t)b_2(t)\ , \tag{13.2a}$$

$$\frac{db_2}{dt} = i\Omega_1(t)\exp(i\Delta_1 t)b_1(t) + i\int dE \sum_{\text{n}} \Omega_{2,E,\boldsymbol{n}}(t)\exp(-i\Delta_E t)b_{E,\boldsymbol{n}}(t)\ , \tag{13.2b}$$

$$\frac{db_{E,\boldsymbol{n}}}{dt} = i\Omega_{2,E,\boldsymbol{n}}^*(t)\exp(i\Delta_E t)b_2(t)\ , \quad \text{for all } E \text{ and } \boldsymbol{n}\ , \tag{13.2c}$$

where

$$\Omega_1(t) \equiv \langle E_2|d_1\cdot\hat{\boldsymbol{\epsilon}}_1|E_1\rangle\mathcal{E}_1(t)/\hbar\ , \quad \Omega_{2,E,\boldsymbol{n}}(t) \equiv \langle E_2|d_2\cdot\hat{\boldsymbol{\epsilon}}_2|E,\boldsymbol{n}^-\rangle\mathcal{E}_2(t)/\hbar\ ,$$
$$\Delta_1 \equiv \frac{E_2 - E_1}{\hbar} - \omega_1\ , \quad \Delta_E \equiv \frac{E - E_2}{\hbar} - \omega_2\ . \tag{13.3}$$

We can eliminate the continuum equations by substituting the formal solution of Eq. (13.2c),

$$b_{E,\boldsymbol{n}}(t) = i\int_0^t dt'\,\Omega_{2,E,\boldsymbol{n}}^*(t')\exp(i\Delta_E t')b_2(t')\ , \tag{13.4}$$

into Eq. (13.2b) to obtain

$$\frac{db_2}{dt} = i\Omega_1(t)\exp(i\Delta_1 t)b_1(t) - \sum_n\int dE\,\Omega_{2,E,\boldsymbol{n}}(t)\exp(-i\Delta_E t)$$
$$\int_0^t dt'\,\Omega_{2,E,\boldsymbol{n}}^*(t')\exp(i\Delta_E t')b_2(t')\ . \tag{13.5}$$

By invoking the SVCA (Eq. (12.13)) we obtain the two-photon analog of Eq. (12.15):

$$\frac{db_2}{dt} = i\Omega_1(t)\exp(i\Delta_1 t)b_1(t) - \Omega_2^I(t)b_2(t)\ , \tag{13.6}$$

where

$$\Omega_2^I(t) = \pi\sum_n |\langle E_L,\boldsymbol{n}^-|\boldsymbol{d}_2\cdot\hat{\boldsymbol{\epsilon}}_2|E_2\rangle\mathcal{E}_2(t)|^2/\hbar\ . \tag{13.7}$$

coupled with Eq. (13.2c) gives a closed set of equations for $b_1(t)$ and $b_2(t)$.

13.1.1
The Adiabatic Approximation for a Final Continuum Manifold

Equations (13.2a) and (13.6), can be expressed as a 2×2 version of Eq. (11.8) where in this case

$$\underline{\text{b}} \equiv (\exp(i\Delta_1 t)b_1, b_2)^T\ , \tag{13.8}$$

and

$$\underline{\underline{H}} = \begin{pmatrix} \Delta_1 & \Omega_1^* \\ \Omega_1 & i\Omega_2^I \end{pmatrix} . \tag{13.9}$$

Assuming that Ω_1 is real, we obtain the adiabatic solutions to Eq. (11.8) by diagonalizing the $\underline{\underline{H}}$ matrix, as in Eq. (11.10).

The presence of the continuum, coupled with the SVCA, results in a complex symmetric $\underline{\underline{H}}$ matrix. Such matrices are diagonalizable using complex orthogonal matrices $\underline{\underline{U}}$, satisfying

$$\underline{\underline{U}}(t) \cdot \underline{\underline{U}}^T(t) = \underline{\underline{I}} . \tag{13.10}$$

Note that $\underline{\underline{U}}$ must be nonunitary on physical grounds in order to allow flux loss to the continuum.

In contrast to the three-level case discussed in Section 11.1, the 2×2 complex orthogonal $\underline{\underline{U}}$ matrix obtained here can be parameterized in terms of a single complex "mixing angle" α, where

$$\underline{\underline{U}} = \begin{pmatrix} \cos\alpha & \sin\alpha \\ -\sin\alpha & \cos\alpha \end{pmatrix} \tag{13.11}$$

and where

$$\alpha(t) = \frac{1}{2} \arctan\left(\frac{2\Omega_1}{i\Omega_2^I - \Delta_1} \right) \tag{13.12}$$

Operating with $\underline{\underline{U}}^T(t)$ on Eq. (11.8), and defining

$$\underline{a}(t) = \underline{\underline{U}}^T(t) \cdot \underline{b}(t) \tag{13.13}$$

we obtain that

$$\frac{d}{dt}\underline{a} = \left\{ i\underline{\underline{\lambda}}(t) + \underline{\underline{A}} \right\} \cdot \underline{a} , \tag{13.14}$$

with

$$\underline{\underline{A}} \equiv \frac{d\underline{\underline{U}}^T(t)}{dt} \cdot \underline{\underline{U}} = \begin{pmatrix} 0 & -\dot{\alpha} \\ \dot{\alpha} & 0 \end{pmatrix} . \tag{13.15}$$

Once again, the adiabatic solutions are given by

$$\underline{a}(t) = \exp\left\{ i \int_0^t \underline{\underline{\lambda}}(t')dt' \right\} \underline{a}(0) , \tag{13.16}$$

and with the elements of the diagonal eigenvalue matrix, $\underline{\underline{\lambda}}$, given this time as

$$\lambda_{1,2} = \frac{1}{2}\left\{ \Delta_1 + i\Omega_2^I \pm \left[(\Delta_1 - i\Omega_2^I)^2 + 4|\Omega_1|^2 \right] \right\} . \tag{13.17}$$

Using Eqs. (13.8) and (13.13), and imposing the initial condition, $\underline{b}(0) = (1, 0)$ we obtain for the $b_1(t)$ and $b_2(t)$ coefficients,

$$
\begin{aligned}
b_1(t) = & \left\{ U_{1,1}(t) \exp\left[i \int_0^t \lambda_1(t')dt' \right] U_{1,1}(0) \right. \\
& \left. + U_{2,1}(t) \exp\left[i \int_0^t \lambda_2(t')dt' \right] U_{2,1}(0) \right\} \exp(-i\Delta_1 t) , \\
b_2(t) = & \; U_{1,2}(t) \exp\left[i \int_0^t \lambda_1(t')dt' \right] U_{1,1}(0) \\
& + U_{2,2}(t) \exp\left[i \int_0^t \lambda_2(t')dt' \right] U_{2,1}(0) .
\end{aligned}
\tag{13.18}
$$

If both lasers are assumed to be off initially, that is, $\mathcal{E}_1(0) = \mathcal{E}_2(0) = 0$, we have that $\alpha(0) = 0$. Hence $U_{1,1}(0) = 1$, $U_{2,1}(0) = 0$, and

$$
\begin{pmatrix} b_1(t) \\ b_2(t) \end{pmatrix} = \begin{pmatrix} \exp(-i\Delta_1 t) \cos\alpha(t) \\ \sin\alpha(t) \end{pmatrix} \exp\left\{ i \int_0^t \lambda_1(t')dt' \right\} . \tag{13.19}
$$

Once $b_2(t)$ is known, the continuum coefficients $b_{E,n}(t)$ are obtained directly via Eq. (13.4).

As an example of this process, we consider the resonantly enhanced two-photon dissociation, by two laser pulses, of a molecule (e.g., Na_2) where the energy gap between the two bound states, $E_2 - E_1 = 20\,000\ \text{cm}^{-1}$, and the gap between E_2 and the continuum is $19\,000\ \text{cm}^{-1}$.

Assuming that $\underline{a}(t)$ is available, either via the adiabatic approximation or via an "exact" numerical computation, the probabilities to observe $|E_1\rangle$ and $|E_2\rangle$ as a function of time are given by

$$
\begin{aligned}
P_1 &\equiv |b_1|^2 = |a_1(t)\cos\alpha(t) + a_2(t)\sin\alpha(t)|^2 , \\
P_2 &\equiv |b_2|^2 = |-a_1(t)\sin\alpha(t) + a_2(t)\cos\alpha(t)|^2 ,
\end{aligned}
\tag{13.20}
$$

and the overall dissociation probability P_d is

$$
P_d \equiv 1 - (P_1 + P_2) . \tag{13.21}
$$

To examine the accuracy of the adiabatic approximation we compare, in Figure 13.1, the zeroth-order adiabatic solution for the probabilities with the exact numerical integration and with an iterative method [555] of solving Eq. (13.2). For this relatively slowly varying pulse and large detuning, the adiabatic solution is essentially exact. However, when going to shorter pulses the adiabatic approximation

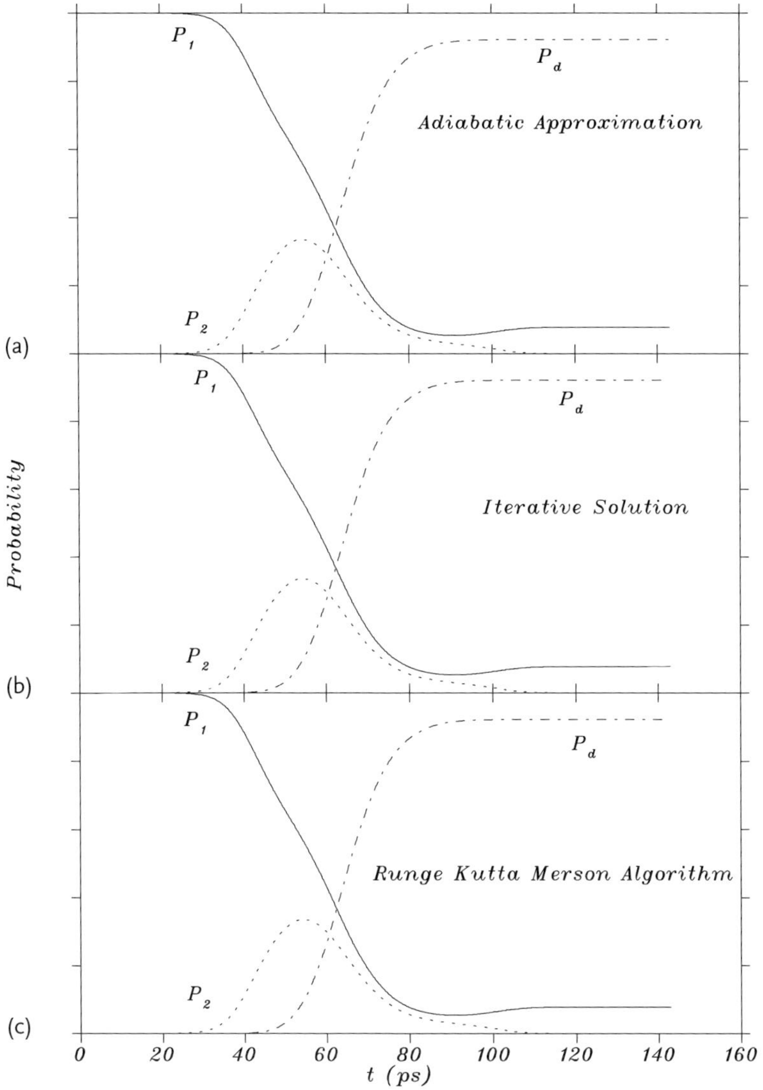

Figure 13.1 Populations of the initial (P_1), intermediate (P_2) and continuum (P_d) states vs. time, in the zero-order (adiabatic) approximation (a), and the exact solution, obtained by iterative corrections to the adiabatic approximation [555] (b), and by direct numerical integration (c). Pulses, lasting 100 ps, were coincident and detuned ($\Delta_1 = 10\,\text{cm}^{-1}$). Taken from [555, Figure 2]. (Reprinted with permission. Copyright 1996 American Institute of Physics.)

fails to properly partition the probability, at long times, between levels $|E_1\rangle$ and $|E_2\rangle$. Further, it fails to display the oscillations at intermediate times which are a result of the interference between the two adiabatic channels. This type of failure is expected because the shortening of the pulse induces rapid variation in the diagonalizing transformation, resulting in the breakdown of the adiabatic condition.

A different type of failure of the adiabatic approximation, unique to the case where the final state is in the continuum, is demonstrated in Figure 13.2. In this case the detuning Δ_1 is decreased with respect to the intermediate level to $1\,\mathrm{cm}^{-1}$ and the pulse duration is kept at 100 ps. Clearly, the adiabatic approximation fails to display the oscillations at intermediate times that result from the interference between two adiabatic channels. A reduction in detuning of the intermediate level

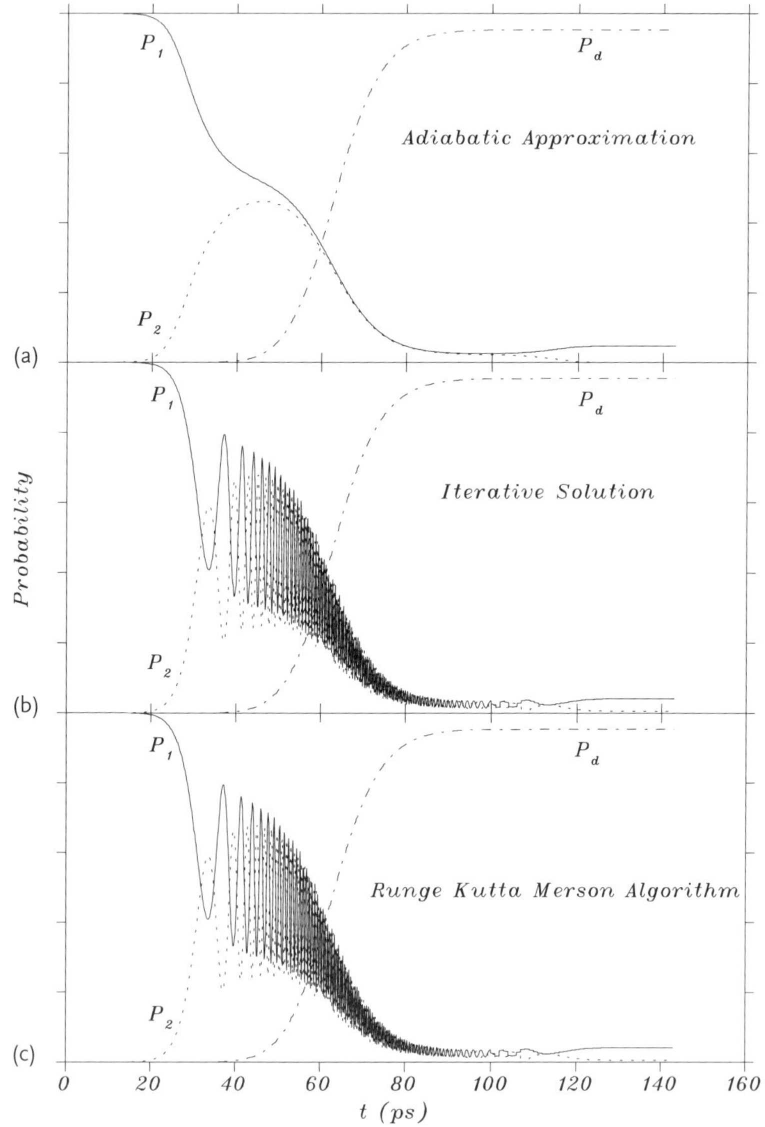

Figure 13.2 The same as in Figure 13.1 for $\Delta_1 = 1\,\mathrm{cm}^{-1}$. Taken from [555, Figure 5]. (Reprinted with permission. Copyright 1996 American Institute of Physics.)

does not affect the bound-state adiabatic condition, $\Omega^0 \Delta\tau \gg 1$, which appears to hold (as suggested by the large number of oscillations during pulse time). However, the more exact statement of the adiabatic condition (Eq. (11.15)) shows that the breakdown of the adiabatic approximation is due to the near divergence of the complex mixing angle α of Eq. (13.12) at small Δ_1, the complex nature of which is a result of the presence of the final continuum.

13.2 Pulsed Incoherent Interference Control

An extension of the three-level scheme discussed in the last section, through the addition of a third laser frequency, results in the pulsed version of incoherent interference control, discussed in Section 6.6.1. In that scenario two CW sources couple two bound states to the continuum, resulting in control over the branching between final fragment channels, that is, control over the selectivity of the process. The approach was particularly interesting because it was shown to be insensitive to the relative phase between the two light sources and hence did not require coherent lasers. Studies of CW-induced controlled photodissociation are, however, relatively limited in scope. The fact that the laser is always on prevents an analysis of the effects of laser parameters, such as the relative pulse orderings and the pulse shapes, which characterize pulsed sources, on the yield. In this section we extend the theory of incoherent interference control to the case of laser pulses and consider both selectivity of different final photodissociation channels, and conditions necessary to achieve high photodissociation yield.

Note that the approach introduced below relies exclusively on the computation of material matrix elements, as in the weak-field domain. As a result, one need only compute these matrix elements one time from which one then readily obtains dissociation rates and probabilities for a variety of pulse configurations and field strengths.

As shown in Figure 13.3, we consider a molecule (in this case Na_2), initially in state $|E_0\rangle$, which is excited by the combined action of two laser pulses, with central frequencies ω_0 and ω_1, to a continuum of states associated with two or more different product channels, at energy E. The frequency ω_0 is assumed to be in near-resonance with the transition to an intermediate bound state $|E_1\rangle$, and ω_1 to be in near-resonance with the transition frequency between $|E_1\rangle$ and the continuum. Thus ω_0 carries the system from $|E_0\rangle$ to $|E_1\rangle$, and ω_1 carries the system from $|E_1\rangle$ to the continuum. To avoid confusion, note that the energy levels are labeled somewhat differently than in the previous section.

In addition, as shown in Figure 13.3, the continuum is coupled to a third bound state $|E_2\rangle$ by a laser of central frequency ω_2. Basically, the three-level two-laser scheme described in Section 13.1 is being extended here to a four-level three-laser control scheme.

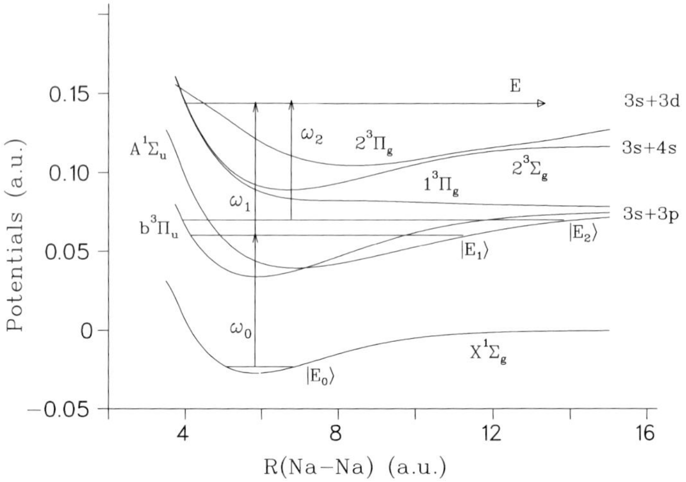

Figure 13.3 Incoherent interference control (IIC) scheme and potential energy curves for the $Na_2 \rightarrow Na(3s) + Na(3d)$, Na(4s), Na(3p) process. In this scheme an $\omega_0 + \omega_1$-photon excitation to the continuum interferes with an ω_2 photon from an initially unpopulated state. The two-photon absorption proceeds from an initial state, $|E_0\rangle$ (in Na_2 it is taken to be the $v = 5, J = 37$ state), via the $|E_1\rangle$ ($v = 35, J = 36, 38$) intermediate resonance belonging to the interacting $A^1\Sigma_u/^3\Pi_u$ electronic states. The ω_2 photon couples the continuum to the (initially unpopulated) $|E_2\rangle$ ($v = 93, J = 36$ or $v = 93, J = 38$) level of the $A^1\Sigma_u/^3\Pi_u$ electronic states. Taken from [270, Figure 1]. (Reprinted with permission. Copyright 1996 American Institute of Physics.)

The total Hamiltonian of the system is now given by

$$H_{tot} = H_M - 2\mathrm{Re}\{\boldsymbol{d}_0 \cdot \hat{\boldsymbol{\epsilon}}_0\mathcal{E}_0(t)\exp(i\omega_0 t) + \boldsymbol{d}_1 \cdot \hat{\boldsymbol{\epsilon}}_1\mathcal{E}_1(t)\exp(i\omega_1 t) + \boldsymbol{d}_2 \cdot \hat{\boldsymbol{\epsilon}}_2\mathcal{E}_2(t)\exp(i\omega_2 t)\}\ , \tag{13.22}$$

where H_M is the molecular Hamiltonian. The solution to the time dependent Schrödinger equation is obtained by expanding the total wave function $\Psi(t)$ in a basis set, this time composed of three bound states and a set of continuum states, $|E, \boldsymbol{n}^-\rangle$, as

$$|\Psi(t)\rangle = b_0|E_0\rangle\exp(-iE_0t/\hbar) + b_1|E_1\rangle\exp(-iE_1t/\hbar) + b_2|E_2\rangle\exp(-iE_2t/\hbar) + \sum_n \int dE\, b_{E,\boldsymbol{n}}(t)|E, \boldsymbol{n}^-\rangle\exp(-iEt/\hbar)\ . \tag{13.23}$$

Here $\boldsymbol{n} = \{\boldsymbol{m}, q\}$ where $q = 1, 2, \ldots$ denotes the product arrangement channel and $\boldsymbol{m}$ denotes the remaining quantum numbers other than the energy.

Substituting Eq. (13.23) into the time-dependent Schrödinger equation and using the orthogonality of the basis functions results in a set of first-order differential equations for the expansion coefficients which, in the rotating wave approximation, is given by:

$$i\hbar\frac{db_0}{dt} = -\mathcal{E}_0(t)d_{0,1}\exp(-i\Delta_1 t)b_1(t) \tag{13.24}$$

$$i\hbar\frac{db_1}{dt} = -\mathcal{E}_0^*(t)d_{1,0}\exp(i\Delta_1 t)b_0(t) - \mathcal{E}_1(t)\sum_n\int dE\,d(1|E,\boldsymbol{n})\exp(-i\Delta_{E,1}t)b_{E,\boldsymbol{n}}(t) \quad (13.25)$$

$$i\hbar\frac{db_2}{dt} = -\mathcal{E}_2(t)\sum_n\int dE\,d(2|E,\boldsymbol{n})\exp(-i\Delta_{E,2}t)b_{E,\boldsymbol{n}}(t) \quad (13.26)$$

$$i\hbar\frac{db_{E,\boldsymbol{n}}}{dt} = -\mathcal{E}_1^*(t)d(E,\boldsymbol{n}|1)\exp(i\Delta_{E,1}t)b_1(t) - \mathcal{E}_2^*(t)d(E,\boldsymbol{n}|2)\exp(i\Delta_{E,2}t)b_2(t)\,, \quad (13.27)$$

where

$$d_{1,0} \equiv \langle E_1|\boldsymbol{d}_0\cdot\hat{\boldsymbol{\epsilon}}_0|E_0\rangle\,,\quad d(E,\boldsymbol{n}|i) \equiv \langle E,\boldsymbol{n}^-|\boldsymbol{d}_i\cdot\hat{\boldsymbol{\epsilon}}_i|E_i\rangle\,,\quad i=1,2\,,$$
$$\Delta_1 \equiv \frac{E_1-E_0}{\hbar} - \omega_0\,,\ \Delta_{E,i} \equiv \frac{E-E_i}{\hbar} - \omega_i\,,\quad i=1,2\,. \quad (13.28)$$

Equation (13.27) shows that the contribution to the continuum amplitude $b_{E,\boldsymbol{n}}$ derives from excitations from levels $|E_1\rangle$ and $|E_2\rangle$. Below we assume $b_1(0) = b_2(0) = 0$. Hence state $|E_2\rangle$ is populated by the ω_2 pulse via the continuum state that was itself populated by the ω_1 pulse via $|E_1\rangle$. The two lowest-order routes to dissociation are $|E_0\rangle \to |E_1\rangle \to |E,\boldsymbol{n}^-\rangle$ and $|E_0\rangle \to |E_1\rangle \to |E,\boldsymbol{n}^-\rangle \to |E_2\rangle \to |E,\boldsymbol{n}'^-\rangle$. Contributions from these, and higher order terms, provide multiple pathways to the product in a given channel at energy E which interfere either constructively or destructively with one another, and afford the opportunity for quantum control.

To consider the nature of solutions to Eqs. (13.24)–(13.27), we substitute the formal solution of Eq. (13.27)

$$b_{E,\boldsymbol{n}}(t) = \frac{-1}{i\hbar}\left\{d(E,\boldsymbol{n}|1)\int_0^t dt'\mathcal{E}_1^*(t')\exp(i\Delta_{E,1}t')b_1(t') + d(E,\boldsymbol{n}|2)\int_0^t dt'\mathcal{E}_2^*(t')\exp(i\Delta_{E,2}t')b_2(t')\right\}\,, \quad (13.29)$$

into Eqs. (13.25) and (13.26) to obtain

$$i\hbar\frac{db_1}{dt} = -d_{1,0}\mathcal{E}_0^*(t)\exp(i\Delta_1 t)b_0(t) + \frac{\mathcal{E}_1(t)}{i\hbar}\sum_n\int dE|d(E,\boldsymbol{n}|1)|^2\exp(-i\Delta_{E,1}t)\int_0^t dt'\mathcal{E}_1^*(t')\exp(i\Delta_{E,1}t')b_1(t') + \frac{\mathcal{E}_1(t)}{i\hbar}\sum_n\int dE\,d(1|E,\boldsymbol{n})d(E,\boldsymbol{n}|2)\exp(-i\Delta_{E,1}t)\int_0^t dt'\mathcal{E}_2^*(t')\exp(i\Delta_{E,2}t')b_2(t')\,, \quad (13.30)$$

$$i\hbar\frac{db_2}{dt} = \frac{\mathcal{E}_2(t)}{i\hbar}\sum_n\int dE\, d(2|E,\boldsymbol{n})d(E,\boldsymbol{n}|1)\exp(-i\Delta_{E,2}t)$$
$$\int_0^t dt'\mathcal{E}_1^*(t')\exp(i\Delta_{E,1}t')b_1(t')$$
$$+\frac{\mathcal{E}_2(t)}{i\hbar}\sum_n\int dE|d(E,\boldsymbol{n}|2)|^2\exp(-i\Delta_{E,2}t)$$
$$\int_0^t dt'\mathcal{E}_2^*(t')\exp(i\Delta_{E,2}t')b_2(t')\,. \quad (13.31)$$

Invoking the slowly varying continuum approximation (SVCA) (Eq. (12.13)),

$$\sum_n d(i|E,\boldsymbol{n})d(E,\boldsymbol{n}|j) \approx \sum_n d(i|E_i+\hbar\omega_i,\boldsymbol{n})d(E_j+\hbar\omega_j,\boldsymbol{n}|j) \equiv \frac{\mathcal{I}_{i,j}}{\pi}\,,$$
$$i,j=1,2\,. \quad (13.32)$$

we have that

$$\int dE\sum_n d(i|E,\boldsymbol{n})d(E,\boldsymbol{n}|j)\exp[-i\Delta_{E,i}(t-t')] = 2\hbar\mathcal{I}_{i,j}\,\delta(t-t')\,. \quad (13.33)$$

Utilizing the SVCA greatly simplifies Eqs. (13.30) and (13.31), giving

$$i\hbar\frac{db_1}{dt} = -d_{1,0}\mathcal{E}_0^*(t)\exp(i\Delta_1 t)b_0(t) - i|\mathcal{E}_1(t)|^2\mathcal{I}_{1,1}b_1(t)$$
$$-i\mathcal{E}_1(t)\mathcal{E}_2^*(t)\mathcal{I}_{1,2}\exp(-i\Delta t)b_2(t)\,, \quad (13.34)$$

$$i\hbar\frac{db_2}{dt} = -i\mathcal{E}_2(t)\mathcal{E}_1^*(t)\mathcal{I}_{2,1}\exp(i\Delta t)b_1(t) - i|\mathcal{E}_2(t)|^2\mathcal{I}_{2,2}b_2(t)\,, \quad (13.35)$$

where $\Delta \equiv \Delta_{E,1} - \Delta_{E,2} = (E_2 - E_1)/\hbar + \omega_2 - \omega_1$. Note that the factor of two in Eq. (13.33) has been cancelled by a factor of one-half which arises from evaluating the integrals of $\delta(t-t')$ at the integration end point $t' = t$.

The resultant coupled equations are still not in a convenient form to solve numerically, or to examine from the point of the view of the adiabatic approximation. To this end we define a modified coefficient three-vector:

$$\underline{c} \equiv (c_0, c_1, c_2)^T \equiv (\exp(i\Delta_1 t)b_0, b_1, b_2\exp(-i\Delta t))^T\,. \quad (13.36)$$

Equations (13.34), (13.35) and (13.24) can then be rewritten as the matrix equation:

$$\frac{d}{dt}\underline{c}(t) = \frac{i}{\hbar}\underline{\underline{f}}(t)\cdot\underline{c}(t)\,, \quad (13.37)$$

where

$$\underline{\underline{f}}(t) = \begin{pmatrix} \hbar\Delta_1 & d_{0,1}\mathcal{E}_0 & 0 \\ d_{1,0}\mathcal{E}_0^* & i|\mathcal{E}_1|^2\mathcal{I}_{1,1} & i\mathcal{E}_1\mathcal{E}_2^*\mathcal{I}_{1,2} \\ 0 & i\mathcal{E}_1^*\mathcal{E}_2\mathcal{I}_{2,1} & i|\mathcal{E}_2|^2\mathcal{I}_{2,2} - \hbar\Delta \end{pmatrix}\,, \quad (13.38)$$

and the explicit time dependence of the field envelopes has been omitted for brevity. By writing

$$d_{0,1} = |d_{0,1}| \exp(i\phi_{0,1}) , \quad \mathcal{I}_{1,2} = |\mathcal{I}_{1,2}| \exp(i\phi_{1,2}) ,$$
$$\mathcal{E}_i(t) = |\mathcal{E}_i(t)| \exp[i\theta_i(t)] , \quad i = 0, 1, 2 , \tag{13.39}$$

we can factorize the $\underline{\underline{f}}(t)$ matrix as

$$\underline{\underline{f}}(t) = \underline{\underline{\hat{e}}}^{-1}(t) \cdot \underline{\underline{g}}(t) \cdot \underline{\underline{\hat{e}}}(t) \tag{13.40}$$

where $\underline{\underline{e}}(t)$ is a diagonal matrix containing the phase factors:

$$\underline{\underline{e}}(t) = \begin{pmatrix} 1 & 0 & 0 \\ 0 & \exp\left[i(\phi_{0,1} + \theta_0)\right] & 0 \\ 0 & 0 & \exp\left[i(\phi_{1,2} + \phi_{0,1} + \theta_0 + \theta_1 - \theta_2)\right] \end{pmatrix} , \tag{13.41}$$

and

$$\underline{\underline{g}}(t) = \begin{pmatrix} \hbar\Delta_1 & |d_{0,1}\mathcal{E}_0| & 0 \\ |d_{0,1}\mathcal{E}_0| & i|\mathcal{E}_1|^2\mathcal{I}_{1,1} & i|\mathcal{E}_1\mathcal{E}_2\mathcal{I}_{1,2}| \\ 0 & i|\mathcal{E}_1\mathcal{E}_2\mathcal{I}_{1,2}| & i|\mathcal{E}_2|^2\mathcal{I}_{2,2} - \hbar\Delta \end{pmatrix} . \tag{13.42}$$

The $\underline{\underline{e}}$ matrix contains all the information about the laser phases.

We now introduce a new three-component vector defined as

$$\underline{a}(t) \equiv (a_0, a_1, a_2)^T = \underline{\underline{e}}(t) \cdot \underline{c}(t) , \tag{13.43}$$

and combine Eq. (13.43) with Eq. (13.37) to obtain that

$$\frac{d}{dt}\underline{a}(t) = \frac{i}{\hbar}\left\{\underline{\underline{g}} + \underline{\underline{\dot{\theta}}}\right\} \cdot \underline{a}(t) , \tag{13.44}$$

where $\underline{\underline{\dot{\theta}}}$ is a diagonal time derivative phase matrix

$$\underline{\underline{\dot{\theta}}} = \hbar \begin{pmatrix} 0 & 0 & 0 \\ 0 & \dot{\theta}_0 & 0 \\ 0 & 0 & \dot{\theta}_0 + \dot{\theta}_1 - \dot{\theta}_2 \end{pmatrix} , \tag{13.45}$$

where $\dot{\theta} \equiv d\theta/dt$. Note that were it not for the variation of the laser phases with time, embodied in the $\underline{\underline{\dot{\theta}}}$ matrix, $\underline{a}$ would be independent of the laser phases.

Using Eqs. (13.36) and (13.43) we can express the full expansion coefficients b_1 and b_2 in terms of a_1 and a_2 as

$$b_1 = c_1 = a_1 \exp\left[-i(\phi_{0,1} + \theta_0)\right] , \tag{13.46}$$

$$b_2 = c_2 \exp(i\Delta t) = a_2 \exp\left[-i(\phi_{0,1} + \theta_0 + \phi_{1,2} + \theta_1 - \theta_2 - \Delta t)\right] . \tag{13.47}$$

After substituting these expressions for the $\underline{b}$ vector into Eq. (13.27), we obtain an explicit expression for the continuum coefficient:

$$i\hbar\frac{db_{E,n}}{dt} = -\left[d(E,\boldsymbol{n}|1)|\mathcal{E}_1(t)|\mathrm{a}_1(t) + d(E,\boldsymbol{n}|2)|\mathcal{E}_2(t)|e^{-i\phi_{1,2}}\mathrm{a}_2(t)\right] \cdot e^{i\Delta_{E,1}t - i(\phi_{0,1}+\theta_0+\theta_1)}, \tag{13.48}$$

in terms of which the total product probability P_q of producing product in channel q is given as

$$P_q = \sum_m \int dE\, P_{m,q}(E) = \sum_m \int dE\, |b_{E,m,q}(t\to\infty)|^2 . \tag{13.49}$$

The desired dissociation probability can therefore be found by first solving Eq. (13.44) and Eq. (13.48) and inserting the result into Eq. (13.49). If we assume that $\hat{\theta}$ is time independent, the $\underline{\underline{\dot{\theta}}}$ term drops out of Eq. (13.44) and it follows from Eq. (13.48) that the $b_{E,n}$ coefficients become independent of the ω_2 laser phase θ_2. Qualitatively this is so, because, as discussed in Section 6.6.1, in the $|E_0\rangle \to |E_1\rangle \to |E,\boldsymbol{n}^-\rangle \to |E_2\rangle \to |E,\boldsymbol{n}'^-\rangle$ excitation pathway, the stimulated emission of an ω_2 photon in the $|E,\boldsymbol{n}^-\rangle \to |E_2\rangle$ step contributes, in accord with Section 6.6.1, a factor of $\exp[-i\theta_2]$ to the molecular wave function that cancels exactly the $\exp[i\theta_2]$ phase factor accompanying the absorption of the same frequency photon in the $|E_2\rangle \to |E,\boldsymbol{n}'^-\rangle$ step.

In addition to the loss of the θ_2 phase dependence, when the laser phase matrix is time-independent, the dependence on the phases of the other lasers is also lost. This is because in this case the time-independent $\exp[-i(\phi_{0,1}+\theta_0+\theta_1)]$ term factors out of the expression for $b_{E,n}$ in Eq. (13.48), contributing unity when $|b_{E,\boldsymbol{n}}|^2$ is computed. As noted above, when the laser phase matrix is time-independent, $\underline{a}(t)$ of Eq. (13.44) becomes independent of all laser phases. Hence, using Eq. (13.49), it follows that in that case P_q is totally independent of the phases of all three lasers. As a consequence, for time-independent laser phases, we can control branching reactions in this scenario by using rather *incoherent* laser sources, a result similar to the situation with CW sources (see [426] and Section 6.6.2), hence the name "incoherent interference control" (IIC). However, as a related consequence, the laser phase is no longer a control variable. Rather, control knobs in IIC are mainly the $\omega_2 - \omega_1$ difference, and to a smaller extent the lasers intensities, pulse widths and pulse orderings.

When the laser phases are time-dependent we can gain insight into the phase dependence of the control by investigating the adiabatic states. We follow to some extent the analysis of Kobrak and Rice [427] who have considered a similar, though not identical, system in which population transfer between two bound states proceeds via an intermediate continuum, dominated by a single resonance. Rather than diagonalizing the entire $\underline{\underline{g}} + \underline{\underline{\dot{\theta}}}$ matrix of Eq. (13.44) (as Kobrak and Rice have done in their 3×3 case) which would yield highly complicated expressions, we only diagonalize the lower right 2×2 block of the $\underline{\underline{g}} + \underline{\underline{\dot{\theta}}}$ matrix.

We transform the $\underline{\underline{g}} + \underline{\underline{\dot{\theta}}}$ matrix of Eq. (13.42) as

$$\underline{\underline{h}}(t) = \underline{\underline{U}}^T(t) \cdot \left\{\underline{\underline{g}} + \underline{\underline{\dot{\theta}}}\right\} \cdot \underline{\underline{U}}(t) \tag{13.50}$$

where

$$\underline{\underline{U}} = \begin{pmatrix} 1 & 0 & 0 \\ 0 & & \\ 0 & \multicolumn{2}{c}{\underline{\underline{u}}} \end{pmatrix}, \tag{13.51}$$

with $\underline{\underline{u}}$ being a 2×2 complex orthogonal matrix, parametrized as in Eq. (13.11) by a complex mixing angle α. Here $\underline{\underline{u}}$ is the diagonalizing transformation of the 2×2 lower block of the $\underline{\underline{g}} + \underline{\underline{\dot{\theta}}}$ matrix which we denote $\underline{\underline{r}}(t)$:

$$\underline{\underline{r}}(t) = \begin{pmatrix} i|\mathcal{E}_1|^2\mathcal{I}_{1,1} + \hbar\dot{\theta}_0 & i|\mathcal{E}_1\mathcal{E}_2\mathcal{I}_{1,2}| \\ i|\mathcal{E}_1\mathcal{E}_2\mathcal{I}_{1,2}| & i|\mathcal{E}_2|^2\mathcal{I}_{2,2} - \hbar(\Delta - \dot{\theta}_0 - \dot{\theta}_1 + \dot{\theta}_2) \end{pmatrix}. \tag{13.52}$$

We have that

$$\underline{\underline{u}}^T \cdot \underline{\underline{r}} \cdot \underline{\underline{u}} = \underline{\underline{\lambda}} \tag{13.53}$$

with $\underline{\underline{\lambda}}$ being the 2×2 eigenvalue matrix

$$\lambda_{1,2} = \frac{1}{2}\Bigg\{ i|\mathcal{E}_1|^2\mathcal{I}_{1,1} + i|\mathcal{E}_2|^2\mathcal{I}_{2,2} + \hbar[2\dot{\theta}_0 + \dot{\theta}_1 - \dot{\theta}_2 - \Delta] \\ \pm \left[\left(i|\mathcal{E}_1|^2\mathcal{I}_{1,1} - i|\mathcal{E}_2|^2\mathcal{I}_{2,2} - \hbar[\dot{\theta}_1 - \dot{\theta}_2 - \Delta] \right)^2 - 4|\mathcal{E}_1\mathcal{E}_2\mathcal{I}_{1,2}|^2 \right]^{\frac{1}{2}} \Bigg\}. \tag{13.54}$$

With the above definitions of $\underline{\underline{U}}$ and $\underline{\underline{u}}$, the transformation of $\underline{\underline{g}} + \underline{\underline{\dot{\theta}}}$ (Eq. (13.50)) results in the following matrix:

$$\underline{\underline{h}}(t) = \begin{pmatrix} \hbar\Delta_1 & \cos\alpha|d_{0,1}\mathcal{E}_0| & \sin\alpha|d_{0,1}\mathcal{E}_0| \\ \cos\alpha|d_{0,1}\mathcal{E}_0| & \lambda_1 & 0 \\ \sin\alpha|d_{0,1}\mathcal{E}_0| & 0 & \lambda_2 \end{pmatrix}. \tag{13.55}$$

Allowing the $\underline{\underline{U}}$ matrix to operate on the expansion coefficients vector, $\underline{a}' \equiv \underline{\underline{U}} \cdot \underline{a}$, has the effect of transforming Eqs. (11.8) which describes the time evolution of the system, to:

$$\frac{d}{dt}\underline{a}'(t) = \frac{i}{\hbar}\left\{\underline{\underline{h}}(t) + \underline{\underline{A}}\right\} \cdot \underline{a}'(t), \tag{13.56}$$

where the nonadiabatic coupling matrix $\underline{\underline{A}}$ is defined in Eq. (11.13). Assuming that the $\mathcal{E}_0$ pulse (whose sole purpose is to excite $|E_0\rangle$ to $|E_1\rangle$, and in no way to control the process) is weak, and assuming the adiabatic approximation (in which we neglect $\underline{\underline{A}}$), we can easily solve Eq. (13.56) to obtain that

$$a'_0(t) = \exp(i\Delta_1 t)a'_0(0), \tag{13.57a}$$

$$a_1'(t) = \frac{i}{\hbar}\cos\alpha |d_{0,1}\mathcal{E}_0| a_0'(0) \int_0^t dt' \exp\left[i\Delta_1 t' + \frac{i}{\hbar}\int_0^{t'} dt'' \lambda_1(t'') \right] , \tag{13.57b}$$

$$a_2'(t) = \frac{i}{\hbar}\sin\alpha |d_{0,1}\mathcal{E}_0| a_0'(0) \int_0^t dt' \exp\left[i\Delta_1 t' + \frac{i}{\hbar}\int_0^{t'} dt'' \lambda_2(t'') \right] . \tag{13.57c}$$

When we use the back transformation from $\underline{a}'$ to $\underline{a}$, $\underline{a} = \underline{\underline{U}}^T \cdot \underline{a}'$, we obtain the desired solution

$$a_1(t) = \frac{i}{\hbar}|d_{0,1}\mathcal{E}_0| \int_0^t dt' \exp[i\Delta_1 t'] \left\{ \cos^2\alpha \exp\left[\frac{i}{\hbar}\int_0^{t'} dt'' \lambda_1(t'') \right] - \sin^2\alpha \exp\left[\frac{i}{\hbar}\int_0^{t'} dt'' \lambda_2(t'') \right] \right\} , \tag{13.58a}$$

$$a_2(t) = \frac{i}{\hbar}|d_{0,1}\mathcal{E}_0| \int_0^t dt' \exp[i\Delta_1 t'] \cos\alpha \sin\alpha \left\{ \exp\left[\frac{i}{\hbar}\int_0^{t'} dt'' \lambda_1(t'') \right] + \exp\left[\frac{i}{\hbar}\int_0^{t'} dt'' \lambda_2(t'') \right] \right\} , \tag{13.58b}$$

where we have used the fact that $a_0(t) = c_0(t)$, and that at $t = 0$ all the population resides in state $|E_0\rangle$, that is, that $c_0(0) = 1$.

In order to obtain the probabilities and branching ratios to various channels, given by Eq. (13.49), we can substitute the expressions for a_1 and a_2 from Eq. (13.58) into Eq. (13.48). Because of the dependence of $\lambda_{1,2}$ on $\dot{\theta}_{0,1,2}$ (see Eq. (13.54)), some dependence of P_q on the laser phases remains for time-dependent phases. A numerical solution [564] of Eq. (13.37) shows, however, that this dependence is tiny: the a_1 and a_2 coefficients obtained numerically were found to be unaffected (changes of less than 1% were computed), even when rapidly varying time-dependent phases were introduced.

As an example of pulsed IIC, we consider the situation depicted in Figure 13.3 of the two photon dissociation of Na_2 to yield the Na(3p) + Na(3s), Na(4s) + Na(3s) and Na(3d) + Na(3s) products. We denote the probabilities of their formation as P(3p), P(4s), and P(3d), and choose $|E_0\rangle$ as the $v = 0$, $J = 33$ level of the Na_2 ground $^1\Sigma_g$ state (the high J state is chosen to mimic thermal conditions in a high temperature heat-pipe [270].); $|E_1\rangle$ as the mixed state formed by spin–orbit coupling the $v = 33, J = 31, 33$ level of the $A^1\Sigma_u$ and the $v = 33$, $J = 32$ level of the $b^3\Pi_u$ electronic state [274]; and $|E_2\rangle$ as the $v = 93$, $J = 32$ or $v = 93$, $J = 34$ state of the $b^3\Pi_u$ state. The laser frequency $\omega_0 = 17\,844\,\text{cm}^{-1}$, that is, in close resonance with the $|E_0\rangle$ to $|E_1\rangle$ transition (a detuning of $-0.21\,\text{cm}^{-1}$); the $\omega_1 = 17\,712\,\text{cm}^{-1}$. The frequency ω_2 of the control laser is tuned over the range of 12 929–13 017 cm^{-1}, which corresponds to varying the detuning Δ from -43.8 to $+43.8\,\text{cm}^{-1}$. The laser pulse profiles in Eq. (13.22) are taken to be Gaussians of the following form,

$$\mathcal{E}_i(t) = \bar{\mathcal{E}}_i \exp\left[-\frac{(t-\tau_i)^2}{2\alpha_i^2}\right] , \quad (i = 0, 1, 2) , \tag{13.59}$$

where $\bar{\mathcal{E}}_i$ is the peak amplitude, and τ_i, α_i are the temporal center position and width of the ith pulse.

The numerical solutions to the original Eqs. (13.30)–(13.31) results are plotted in Figures 13.4 and 13.5 and contrasted with the experimental results of [270]. The experiments are made easier by the fact that the IIC control scenario does not require laser coherence, hence, generally available, nontransform limited nsec dye lasers can be used. In the experiment [270], the IIC scenario was studied for just two channels of the Na_2 two-photon dissociation process,

$$\text{Na(3s)} + \text{Na(3p)} \longleftarrow \text{Na}_2 \longrightarrow \text{Na(3s)} + \text{Na(3d)}$$

Two dye lasers pumped by a frequency-doubled Nd:YAG laser were used. One dye laser, of frequency ω_2 tuned between $13\,312\,\text{cm}^{-1}$ and $13\,328\,\text{cm}^{-1}$, was used to couple the continuum to the $|E_2\rangle$ vib-rotational state of the $A^1\Sigma_u/^3\Pi_u$

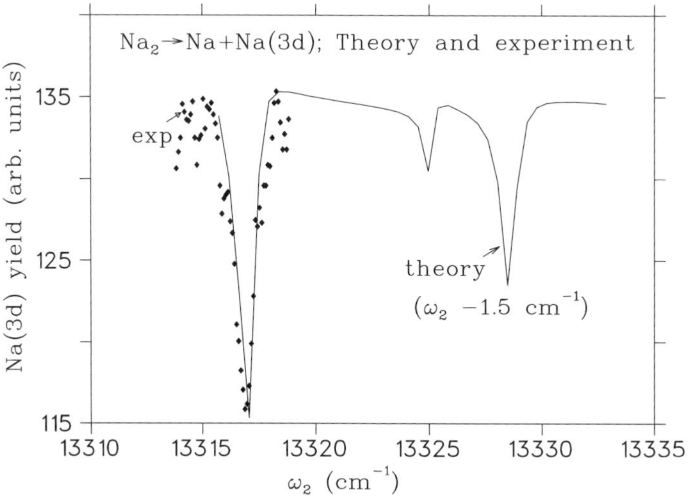

Figure 13.4 Comparison of the experimental and theoretical $Na_2 \rightarrow Na + Na(3d)$ yields as a function of ω_2. In the calculation, an intermediate $v = 33$, $J = 31, 33$ resonance is used and ω_1 is fixed at 17 720.7 cm^{-1}. The intensities of the two laser fields are $I(\omega_1) = 1.72 \times 10^8\,\text{W/cm}^2$ and $I(\omega_2) = 2.84 \times 10^8\,\text{W/cm}^2$. The ω_2 frequency axis of the calculated results was shifted by $-1.5\,\text{cm}^{-1}$ in order to better compare the predicted and measured line-shapes. Taken from [270, Figure 3]. (Reprinted with permission. Copyright 1996 American Physical Society.)

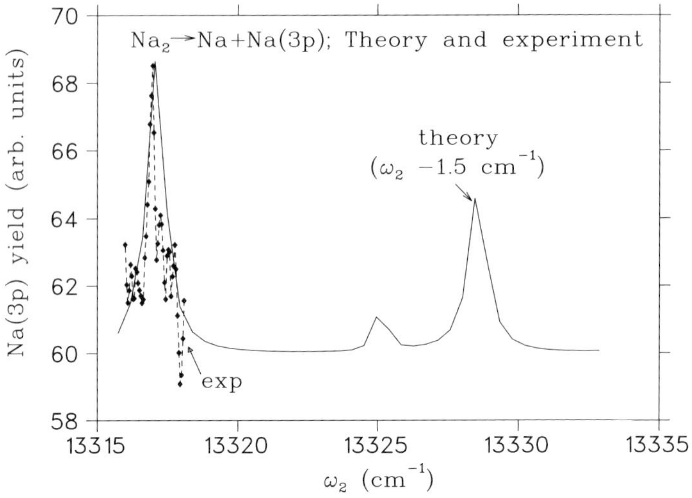

Figure 13.5 Comparison of the experimental and theoretical $Na_2 \rightarrow Na + Na(3p)$ yields as a function of ω_2, with parameters as in Figure 13.4. Taken from [270, Figure 4]. (Reprinted with permission. Copyright 1996 American Physical Society.)

mixed electronic state [271, 426] of Na_2. The other dye laser, of frequency $\omega_1 = 17\,474.12\,cm^{-1}$, was used to induce a two-photon dissociation of the $|E_0\rangle$ (v $= 5, J = 37$ ground state of Na_2), through intermediate resonances $|E_1\rangle$ (v $= 35, J = 38$ and v $= 35, J = 36$) of the $A^1\Sigma_u/^3\Pi_u$ mixed state. The ω_1 and ω_2 pulses, both of $\sim$ 5 ns duration with the stronger amongst them (ω_2) having an energy of $\sim$ 3.5 mJ, were made to overlap in a heat pipe containing Na vapor at 370 – 410°C. Spontaneous emission from the excited Na atoms (Na(3d) $\rightarrow$ Na(3p) and Na(3p) $\rightarrow$ Na(3s)) resulting from the Na_2 photodissociation, was detected and dispersed in a spectrometer and a detector with a narrow bandpass filter.

Figure 13.4 shows the experimental Na(3d) emission as a function of ω_2 at a fixed ω_1. Each point represents an average over a few hundred laser shots, each chosen to have an ω_2 pulse energy that deviates by less than 5% from 3.5 mJ (intensity of $\sim 10^7$ W/cm^2). In Figure 13.5 we plot the experimental Na(3p) yield obtained in the same experiment.

Also shown in Figures 13.4 and 13.5 are the theoretical calculations [270] described above. We see that the Na(3d) signal yield dips, and the Na(3p) signal peaks, as a function of the $\omega_2 - \omega_1$ frequency difference, yielding a $\sim$ 30% modulation in the Na(3p)/Na(3d) branching ratio. Considering the uncertainties in the theoretical potentials used [73, 270], the agreement between theory and experiment (especially in the Na(3d) signal) is remarkable. Additional computations [269, 558] suggest that the observed experimental substructures may be due to the excitation of numerous additional, as yet unassigned, thermally populated vib-rotational Na_2 energy levels.

13.3 Resonantly Enhanced Photoassociation

The set of Eqs. used in Section 13.1 for two-photon dissociation problems can be used to address another significant problem; that of resonantly enhanced *association*, depicted schematically in Figure 12.1c. The significance of photoassociation stems from the possibility of using it to form ultracold molecules, a topic of considerable interest. Those laser cooling schemes that work for atoms [353, 354, 565–567] tend to fail for molecules, mainly due to the presence of many near-resonance lines and the fact that other degrees of freedom, in addition to translation (rotations, vibrations, and so on), must be cooled. Though there are a number of viable suggestions to radiatively cool the internal degrees of freedom of molecules using light [568, 569], such methods do not address the issue of cooling the *translational* degrees of freedom.

Rather than cool warm molecules one can try to *synthesize* cold molecules by (photo) associating cold atoms. The molecules thus formed are expected to maintain the translational temperature of the recombining atoms, because the center of mass motion remains unchanged in the association process (save for the little momentum imparted by the photon). This idea was first proposed by Julienne *et al.* [570, 571] who envisioned a multistep association, first involving the continuum-to-bound excitation of translational continuum states of cold trapped atoms to an excited vibrational level in an excited electronic molecular state. This step was followed by a bound-bound spontaneous emission to the ground electronic state

An undesirable feature of this scheme is that the spontaneous nature of the second step allows the molecules to end up in a large range of vibrational levels. As a consequence, the use of *stimulated* emission [424, 469, 470, 472, 572–584], discussed below, is preferable because it allows population transfer to a particular final molecular state of interest. The process is most efficient if executed as an adiabatic passage from a continuum (APC) to the final molecular bound state of interest. As in bound-bound AP (Section 11.1) central to the APC is the formation of a photoassociation "dark state". This "light-dressed" state, once formed, is impervious to further actions of the light fields. When the light fields are applied as pulses, APC results due to the dark state changing its nature by "following" the makeup of the applied pulses from that of a scattering state of the associating atoms to that of a molecular bound state.

Following the theory discussed below, such dark states have been observed experimentally [388, 473, 474, 585]. In addition formation of ground state molecules by photoassociating a "Feshbach molecule" [362, 586] using STIRAP has been demonstrated [332, 388, 587–589], and shown to be a very promising way of creating high density of ultracold molecules in their ground internal state.

13.3.1
Theory of Photoassociation of a Coherent Wave Packet

In photoassociation the initial state is the scattering state and the goal is to transfer the population to the final *bound* state $|E_1\rangle$. We therefore consider a pair of colliding atoms described by scattering states $|E, \boldsymbol{n}^+\rangle$ as defined in Eq. (2.52), with $\boldsymbol{n}$ incorporating the quantum indices specifying the electronic states of the separated atoms and E being the total collision energy. As explained in conjunction with Eq. (2.52), the plus notation signifies, in contrast with the minus states that were previously used to describe dissociation processes, that the *initial* state of the fragments is known.

Following [469], we focus attention on a Λ-type system, shown in Figure 12.1c, subjected to the combined action of two laser pulses of central frequencies ω_1 and ω_2. Here ω_2 is in near-resonance with the transition from the $|E, \boldsymbol{n}^+\rangle$ continuum to an intermediate bound state $|E_2\rangle$ and ω_1 is in near-resonance with the transition from $|E_2\rangle$ to $|E_1\rangle$.

With the total Hamiltonian of the system given by Eq. (11.2) and the material wave function of the system expanded as in Eq. (13.1), we obtain a set of first-order differential equations for the expansion coefficients that is essentially identical to that of Eq. (13.2), except that the bound-continuum dipole matrix elements are now of the form $\langle E_2|\boldsymbol{d}_2\cdot\boldsymbol{\epsilon}_2|E, \boldsymbol{n}^+\rangle$, involving the $|E, \boldsymbol{n}^+\rangle$, rather than the $|E, \boldsymbol{n}^-\rangle$, states.

In the photoassociation case, contrary to the dissociation cases discussed above, the continuum is initially populated, that is, $b_{E,\boldsymbol{n}}(0) \neq 0$. Hence the formal solution of Eq. (13.2c) is now of the form:

$$b_{E,\boldsymbol{n}}(t) = b_{E,\boldsymbol{n}}(t=0) + i\int_0^t dt'\,\Omega^*_{2,E,\boldsymbol{n}}(t')\exp(i\Delta_E t')b_2(t'). \tag{13.60}$$

Substituting this solution into Eq. ((13.2b) gives

$$\begin{aligned}\frac{db_2}{dt} &= i\Omega_1(t)\exp(i\Delta_1 t)b_1(t) - \Gamma b_2 \\ &\quad + i\sum_n \int dE\,\Omega_{2,E,\boldsymbol{n}}(t)\exp(-i\Delta_E t)b_{E,\boldsymbol{n}}(t=0) \\ &\quad - \sum_n \int dE \int_0^t dt'\,\Omega_{2,E,\boldsymbol{n}}(t)\Omega^*_{2,E,\boldsymbol{n}}(t')\exp\left[-i\Delta_E(t-t')\right]b_2(t')\,,\end{aligned} \tag{13.61}$$

where Γ is the spontaneous emission rate.

If the molecular continuum is unstructured we can invoke the SVCA and replace the energy-dependent bound-continuum dipole matrix elements by their value at the pulse center, given (in the Λ configuration of Figure 12.1c) as $E_L = E_2 - \hbar\omega_2$. This is the case, for example, for Na_2 at threshold energies, where the bound-continuum dipole matrix elements vary with energy by less than 1% over a typical ns-pulse bandwidth. Within the SVCA, Eq. (13.61) becomes

$$\frac{db_2}{dt} = i\Omega_1(t)\exp(i\Delta_1 t)b_1(t) - \left[\Gamma + \Omega_2^I(t)\right] b_2(t) + iF(t)\,, \tag{13.62}$$

where

$$F(t) \equiv \mathcal{E}_2(t)\bar{d}_2(t)/\hbar\,, \tag{13.63}$$

with

$$\bar{d}_2(t) \equiv \sum_n \int dE\langle E_2|\boldsymbol{d}_2\cdot\boldsymbol{\epsilon}_2|E,\boldsymbol{n}^+\rangle\exp(-i\Delta_E t)b_{E,\boldsymbol{n}}(t=0)\,, \tag{13.64}$$

and where $\Omega_2^I(t)$ is defined as in Eq. (13.7). Equations (13.62) and (13.2) can be expressed in matrix notation as

$$\frac{d}{dt}\underline{\mathrm{b}} = i\left\{\underline{\underline{\mathrm{H}}}'\cdot\underline{\mathrm{b}}(t) + \underline{\mathrm{f}}\right\}\,, \tag{13.65}$$

where

$$\underline{\mathrm{f}}(t) = (0, F(t))^T\,, \tag{13.66}$$

with $\underline{\mathrm{b}}$ as defined in Eq. (13.8) and

$$\underline{\underline{\mathrm{H}}}' = \underline{\underline{\mathrm{H}}} + \begin{pmatrix} 0 & 0 \\ 0 & i\Gamma \end{pmatrix} \tag{13.67}$$

with $\underline{\underline{\mathrm{H}}}$ defined in Eq. (13.9).

The *net association rate* $R(t)$ is the rate of population-change in the bound manifold, given by $(d/dt)(|b_1|^2 + |b_2|^2)$. It can be written, using Eq. (13.65) and its complex conjugate, as

$$\begin{aligned} R(t) &= \frac{d}{dt}\left(|b_1|^2 + |b_2|^2\right) = \frac{d}{dt}|\underline{\mathrm{b}}|^2 = \underline{\mathrm{b}}^\dagger\cdot\left(\frac{d}{dt}\underline{\mathrm{b}}\right) + \left(\frac{d}{dt}\underline{\mathrm{b}}^\dagger\right)\cdot\underline{\mathrm{b}} \\ &= i\left\{\underline{\mathrm{b}}^\dagger\cdot(\underline{\underline{\mathrm{H}}}' - \underline{\underline{\mathrm{H}}}'^\dagger)\cdot\underline{\mathrm{b}} + \underline{\mathrm{b}}^\dagger\cdot\underline{\mathrm{f}} - \underline{\mathrm{f}}^\dagger\cdot\underline{\mathrm{b}}\right\} \\ &= 2\mathrm{Im}[F^*(t)b_2(t)] - 2\Omega_2'^I(t)|b_2(t)|^2\,. \end{aligned} \tag{13.68}$$

The first term in Eq. (13.68) represents the association rate,

$$R_{rec}(t) \equiv 2\mathrm{Im}[F^*(t)b_2(t)]\,, \tag{13.69}$$

and the second term is the back-dissociation rate,

$$R_{diss}(t) \equiv 2\Omega_2'^I(t)|b_2(t)|^2\,. \tag{13.70}$$

$$\Omega_2'^I(t) \equiv \Omega_2^I(t) + \Gamma\,, \tag{13.71}$$

see Eq. (13.9) and (13.67). As expected, the net association rate (Eq. (13.68)) is the difference between the association rate and the back-dissociation rates.

As in Eq. (11.10), we can solve Eq. (13.65) adiabatically by diagonalizing the $\underline{\underline{H}}'$ matrix. Operating with $\underline{\underline{U}}(t)$ on Eq. (13.65), with $\underline{a}(t)$ defined as in Eq. (13.13), we obtain that

$$\frac{d}{dt}\underline{a} = \left\{ i\underline{\underline{\lambda}}(t) + \underline{\underline{A}} \right\} \cdot \underline{a} + i\underline{g} \,, \tag{13.72}$$

where the source-vector $\underline{g}$ is given as

$$\underline{g}(t) = \begin{pmatrix} F(t)\,U_{1,2}(t) \\ F(t)\,U_{2,2}(t) \end{pmatrix} = \begin{pmatrix} F(t)\sin\theta(t) \\ F(t)\cos\theta(t) \end{pmatrix} . \tag{13.73}$$

Invoking the adiabatic approximation, we obtain from Eq. (13.72) that

$$\frac{d}{dt}\underline{a} = i\underline{\underline{\lambda}}(t) \cdot \underline{a}(t) + i\underline{g}(t) \,. \tag{13.74}$$

In the association process the initial conditions are such that

$$\underline{a}(t = 0) = 0 \,, \tag{13.75}$$

so that the adiabatic solutions are

$$\underline{a}(t) = \underline{\underline{v}}(t) \cdot \underline{q}(t) \,, \tag{13.76}$$

where

$$\underline{\underline{v}}(t) = \exp\left\{ i \int_0^t \underline{\underline{\lambda}}(t')dt' \right\} \,, \tag{13.77}$$

and

$$\underline{q}(t) = i \int_0^t \underline{\underline{v}}^{-1}(t') \cdot \underline{g}(t')dt' \tag{13.78}$$

with $\lambda_{1,2}$ given by Eq. (13.17).

Using Eq. (13.13) and Eq. (13.8), we obtain for the $b_1(t)$ and $b_2(t)$ coefficients in the adiabatic approximation:

$$\begin{aligned} b_1(t) = i \Bigg\{ &\cos\theta(t) \int_0^t \exp\left[i \int_{t'}^t \lambda_1(t'')dt'' \right] F(t')\sin\theta(t')dt' \\ &- \sin\theta(t) \int_0^t \exp\left[i \int_{t'}^t \lambda_2(t'')dt'' \right] F(t')\cos\theta(t')dt' \Bigg\} \exp(-i\Delta_1 t) \,, \\ b_2(t) = i \Bigg\{ &\sin\theta(t) \int_0^t \exp\left[i \int_{t'}^t \lambda_1(t'')dt'' \right] F(t')\sin\theta(t')dt' \\ &+ \cos\theta(t) \int_0^t \exp\left[i \int_{t'}^t \lambda_2(t'')dt'' \right] F(t')\cos\theta(t')dt' \Bigg\} \,. \end{aligned} \tag{13.79}$$

Given $b_2(t)$, the (channel-specific) continuum coefficients $b_{E,n}(t)$ are obtained directly via Eq. (13.60).

It is instructive to study the adiabatic solution when there is insignificant temporal overlap between the two laser pulses. Assuming in that case that the ω_2 pulse precedes the ω_1 pulse, we have during the ω_2 pulse that $\mathcal{E}_2 \gg \mathcal{E}_1$, hence by Eq. (13.17) $\lambda_1 \approx \Delta_1$, $\lambda_2 \approx i\left[\Gamma + \Omega_2^I\right]$ and $\theta(t) = 0$. Substituting these values into Eq. (13.79), gives, that during the ω_2 pulse (when the ω_1 pulse is off):

$$b_1(t) = 0\,; \quad b_2(t) = i\int_0^t \exp\left\{-\int_{t'}^t \left[\Gamma + \Omega_2^I(t'')\right] dt''\right\} F(t')dt'\,. \tag{13.80}$$

From Eq. (13.63) it is clear that the source term $F(t)$ is linearly proportional to the pulse amplitude. On the other hand, since $\Omega_2 > 0$ and $t' < t$, the $\exp\{-\int_{t'}^t [\Gamma + \Omega_2^I(t'')]dt''\}$ factor (describing dissociation back to the continuum) decays exponentially with increasing intensity. Thus, merely increasing the laser power does not necessarily increase the association yield. There exists some optimal intensity, beyond which the association probability decreases. Below we display some pulse configurations for a realistic case of photoassociation.

As an example of this formulation we consider pulsed photoassociation of a coherent wave packet of cold Na atoms [469]. The colliding atoms are described by an (energetically narrow) normalized Gaussian packet of $J = 0$ radial waves:

$$|\Psi(t=0)\rangle = \int dE\, b_E(t=0)|E, 3s + 3s\rangle\,, \tag{13.81}$$

where $|E, 3s + 3s\rangle$ are the translational Na–Na *s*-waves with the atoms in the 3s state, and b_E at time zero is taken as

$$b_E(t=0) = (\delta_E^2 \pi)^{-1/4} \exp\left\{-\frac{(E - E_{col})^2}{2\delta_E^2} + i\Delta_E t_0\right\}\,. \tag{13.82}$$

Here, t_0 denotes the instant of maximum overlap of the Na + Na wave packet with the $|E_2\rangle$ state. In the simulations, E_{col}, the mean collision energy, varies between $E_{col} = 0.006\,95–0.0695\ \mathrm{cm}^{-1} \approx 0.01–0.1$ K and the wave packet widths, δ_E, vary over the range $\delta_E = 10^{-4}–10^{-3}\ \mathrm{cm}^{-1}$. State $|E_1\rangle$ is chosen as the $(X^1\Sigma_g^+, \mathrm{v} = 0, J = 0)$ state and $|E_2\rangle$ as the $(A^1\Sigma_n^+, \mathrm{v}' = 34, J = 1)$ state, as shown in Figure 13.6. Thus, the combined effect of the two laser pulses is the transfer of population from the continuum to the ground vib-rotational state $(X^1\Sigma_g^+, \mathrm{v} = 0, J = 0)$, with the bound $(A^1\Sigma_n^+, \mathrm{v}' = 34, J = 1)$ state acting as an intermediate state. In order to minimize spontaneous emission losses we concentrate on the "counter-intuitive" [321] scheme where the "dump" pulse $\mathcal{E}_1(t)$ is applied *before* the $\mathcal{E}_2(t)$ "pump" pulse.

As discussed in Section 13.1, when either the final or initial state is in the continuum, the Rabi frequency is imaginary, which changes the range of validity of the adiabatic approximation. For example, it does not necessarily hold even for "large

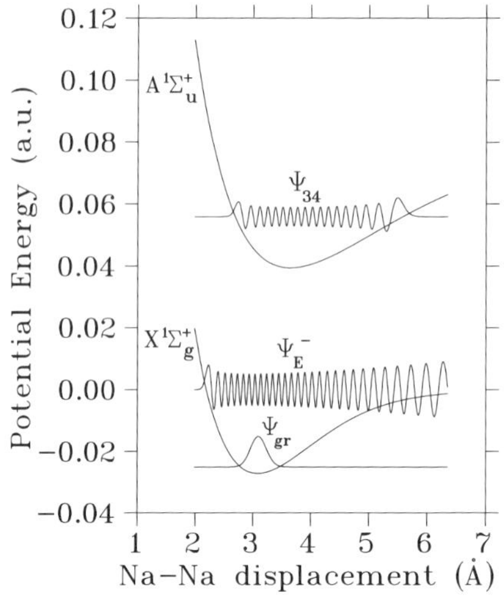

Figure 13.6 Potentials and vibrational wave functions used in the simulation of the Na + Na two-photon association. Taken from [469, Figure 2]. (Reprinted with permission. Copyright 1997 American Institute of Physics.)

area" $\int \Omega\, dt$ pulses [261, 555]. For example, as shown in Figure 13.2, in the presence of a continuum the adiabatic approximation tends to break down for small detunings. Despite this fact, we show below that with the proper choice of pulse parameters, it is possible to transfer the entire population contained in the continuum wave packet to the ground state, while keeping the intermediate state population low at all times. Moreover, for such pulse parameters the adiabatic solutions (Eq. (13.79)) are in perfect agreement with exact numerical solutions [469].

A typical population evolution is shown in Figure 13.7 which is obtained with pulse intensities of order 10^8 W/cm^2 and pulse durations of several ns. Clearly demonstrated is the completeness of the continuum-to-bound population transfer which proceeds with essentially no population in the intermediate state. These findings are very similar to the situation in the three bound states STIRAP process discussed in Section 11.1. In other words, at sufficiently high intensities, a *dark state* that is completely analogous to that of bound state STIRAP is formed in the photoassociation case as well. Experimental evidence for the formation of such dark states has been presented [388, 473] and will be discussed in greater detail below.

The pulse intensities used are sufficiently small to avoid unwanted photoionization, photodissociation and other strong-field parasitic processes. However, working with pulses requires that the atoms be sufficiently close to one another during the laser pulse that they can be recombined. In other words, the initial wave packet of continuum states considered here must be synchronized in time and in duration with the recombining pulses. It is therefore of interest to see whether it is possible to employ longer pulses (of lower intensity) in order to increase the absolute number of recombining atoms and the overall duty cycle of the process.

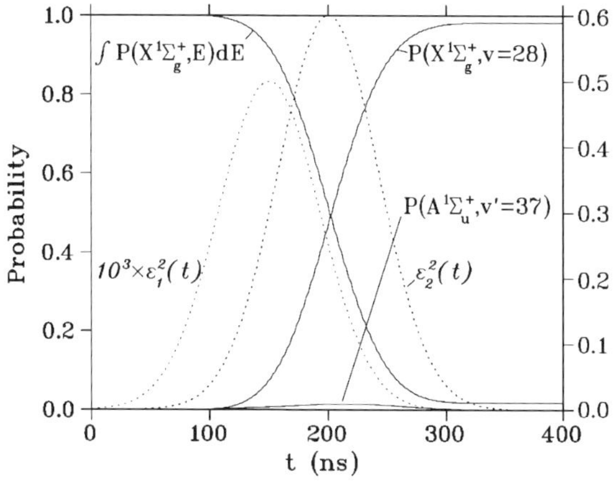

Figure 13.7 The appearance of a "dark state" in photoassociation via a "counterintuitive" pulse sequence. Shown, as a function of time, are integrated population of the wave packet of initial continuum states, the population of the v $= 34, J = 1$ intermediate state, and the population of the v $= 0, J = 0$ final ground state. Clearly seen is that at the intensities used there is essentially no population in the intermediate state, in complete similarity to the three bound states STIRAP process discussed in Section 11.1. Dashed lines are the intensity profiles of the two Gaussian pulses whose central frequencies are $\omega_1 = 18.143\,775\ \text{cm}^{-1}$ and $\omega_2 = 12.277\,042\ \text{cm}^{-1}$, (i.e., $\Delta_1 = \Delta_{E_{col}} = 0$). The maximum intensity of the dump pulse is $1.6 \times 10^8\ \text{W/cm}^2$ and that of the pump pulse is $3.1 \times 10^9\ \text{W/cm}^2$. Both pulses last 8.5 ns. The pump pulse peaks at $t_0 = 20$ ns, the peak time of the Na + Na wave packet. The dump pulse peaks 5 ns before that time. The initial kinetic energy of the Na atoms is $0.0695\ \text{cm}^{-1}$ (or 0.1 K). Taken from [469, Figure 4]. (Reprinted with permission. Copyright 1997 American Institute of Physics.)

Use of pulses of different intensity and different durations is illustrated for the "counterintuitive" scheme in Figure 13.8a,b, where the rates of association (R_{rec} of Eq. (13.69)), back-dissociation (R_{diss} of Eq. (13.70)), and the net association rate ($R(t)$ of Eq. (13.68)), are plotted as a function of time. A short-pulse case is shown in Figure 13.8a and a long-pulse case, with a more spread out wave packet, is shown in Figure 13.8b. Both Figures appear identical, though in Figure 13.8b the abscissa is scaled up by a factor of 10 and the ordinate is scaled down by a factor of 10.

The scaling behavior demonstrated in Figures 13.8 is due to the existence of exact scaling relations in Eq. (13.65). This scaling is obtained when the initial wave packet width and the pulse intensities are scaled down as

$$\delta_E \to \frac{\delta_E}{s}\,, \quad \mathcal{E}_1^0 \to \frac{\mathcal{E}_1^0}{s}\,, \quad \mathcal{E}_2^0 \to \frac{\mathcal{E}_2^0}{\sqrt{s}}\,, \tag{13.83}$$

and the pulse durations are scaled up as

$$\Delta t_{1,2} \to \Delta t_{1,2}\, s\,. \tag{13.84}$$

It follows from Eq. (13.7) that under these transformations

$$F(t) \to \bar{F}(t) = \frac{F\,(t/s)}{s}\,; \quad \Omega_{1,2}(t) \to \overline{\Omega_{1,2}}(t) = \frac{\Omega_{1,2}\,(t/s)}{s}\,,$$

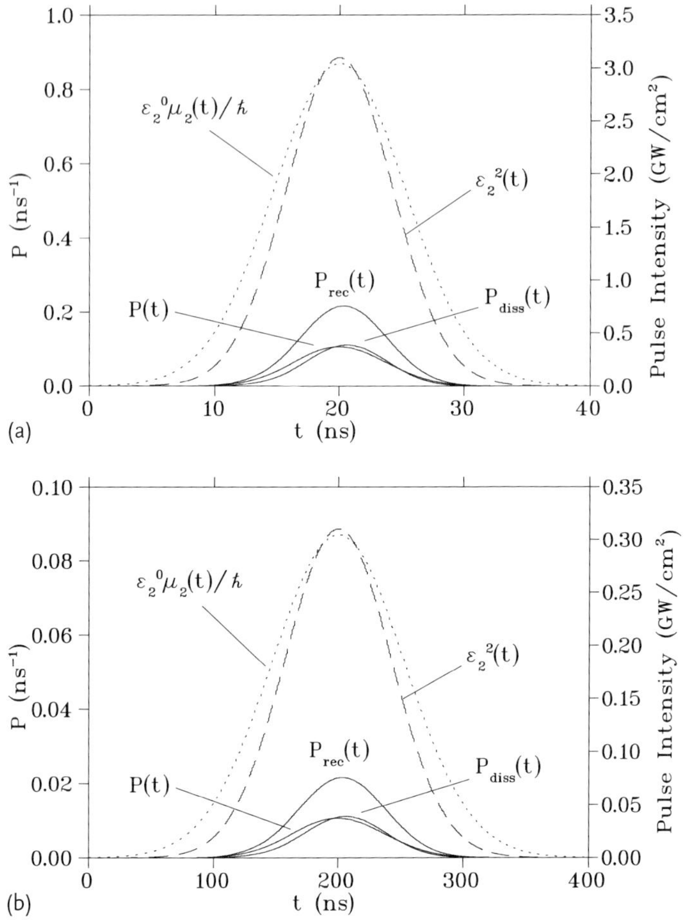

Figure 13.8 Rates of association, (P_{rec}) back-dissociation (P_{diss}) and total molecule formation (P) vs. t in the "counterintuitive" scheme. Dashed lines are pulse intensity profile, dotted lines denote the effective Rabi frequency $\mathcal{E}_2^0\bar{d}_2(t)/\hbar$, where $\mathcal{E}_2^0$ is the peak pulse intensity. (a) Initial wave packet width of $\delta_E = 10^{-3}\ \text{cm}^{-1}$ and other pulse parameters as in Figure 13.7. (b) The dynamics scaled by $s = 10$: Initial wave packet width of $\delta_E = 10^{-4}\ \text{cm}^{-1}$; both pulses lasting 85 ns; the pump pulse peaking at $t_0 = 200$ ns and the dump pulse peaking at 50 ns before that time. Peak intensity of the dump pulse is $1.6 \times 10^6\ \text{W/cm}^2$ and of the pump pulse is $3.1 \times 10^8\ \text{W/cm}^2$. Taken from [469, Figure 5]. (Reprinted with permission. Copyright 1997 American Institute of Physics.)

and Eq. (13.65) becomes

$$\frac{d}{dt/s}\underline{b}' = i\left\{\underline{\underline{H}}'(t/s)\cdot\underline{b}' + \underline{f}(t/s)\right\}\ , \tag{13.85}$$

where $\underline{b}'$ denotes the vector of solutions of the scaled equations. Thus, the scaled coefficients at time t are identical to the unscaled coefficients at times t/s.

One of the results of the above scaling relations is that the pulses' durations can be made longer and their intensities concomitantly scaled down, without changing the final population transfer yields. As noted above, lengthening of the pulses is beneficial because it causes more atoms to recombine within a given pulse.

There is a range of pulse parameters (such as the pulse area, $\Omega_{2,E_L}\Delta t_2$) that maximizes the association yield for a *fixed* initial wave packet. For both the "intuitive" and the "counterintuitive" schemes there is a clear maximum at a specific pulse area; merely increasing the pulse intensity does not lead to an improved association yield. We can attribute this behavior to the fact that the association rate (R_{rec} of Eq. (13.69)), increases linearly with increasing pulse intensity, whereas dissociation rate (R_{diss} of Eq. (13.70)) increases exponentially with the intensity. Hence, as long as the energetic width of the initial wave packet stays fixed, the association yield turns over with increasing pulse area. The turn-over point is different for the two pulse schemes: in the "counterintuitive" case it occurs at a much higher intensity (area).

The existence of a window of intensities for efficient association explains why it is not possible to increase the pulse durations *ad infinitum*, that is, to work with CW light. As $\Delta t_{1,2}$ increases, it follows from Eq. (13.83) that $|\mathcal{E}_2/\mathcal{E}_1|^2$ must also increase. Since $|\mathcal{E}_1|^2$ cannot vanish, $|\mathcal{E}_2|^2$ must diverge if one is to stay within the windows of intensities for efficient association in the CW limit. Hence radiative association as described in this section cannot take place in the CW regime.

There are various experimental demonstrations of photoassociation via two-photon transition as discussed above [590–596]. Evidence that counterintuitive pulse ordering results in large photoassociation cross sections has also been reported [332, 388, 473, 474, 585, 589] as discussed in greater detail below.

13.3.2
Photoassociation by the Consecutive Application of APC and STIRAP

One of the drawbacks of the APC scheme discussed above is that for many cases the Franck–Condon (FC) overlaps between the vibrational state excited from the continuum and the ground vibrational state are rather poor, necessitating (as in the Na_2 example above) the use of rather high powers. In order to overcome this problem we discuss, in this section, the execution of APC and bound-bound STIRAP *in tandem*.

As a specific example, consider the photoassociation of two ^{85}Rb atoms colliding on the Rb_2 $X^1\Sigma_g^+$-potential [597]. One of the possibilities illustrated in Figure 13.9a is to do so by first executing APC with a pair of laser pulses to transfer a fraction of the continuum population to $|E_1\rangle$, which is an (excited) vibrational state of the Rb_2 $X^1\Sigma_g^+$-state, via $|E_2\rangle$, which is an intermediate $A^1\Sigma_u/b^3\Pi_u$ vibrational state. The Rb_2 $X^1\Sigma_g^+$-state is chosen to have good FC factors with the $|E_2\rangle$ state. Later, a bound-bound STIRAP process is used to transfer the population from the $|E_1\rangle$ state to the ground vibrational state.

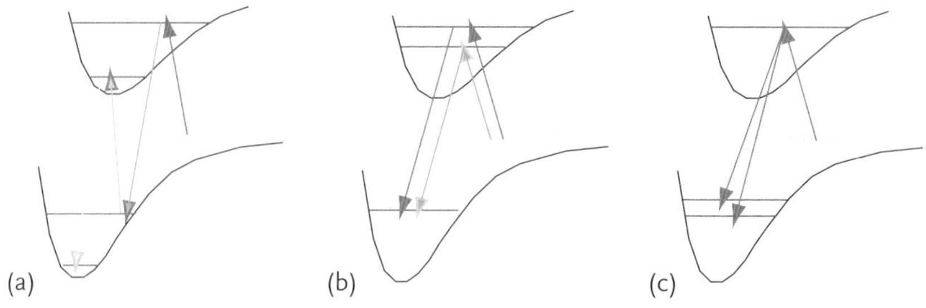

Figure 13.9 Three combinations of AP pairs discussed in the text. Taken from [597, Figure 1]. (Reprinted with permission. Copyright 2007 American Physical Society.)

Figure 13.10 shows the Born–Oppenheimer potentials for the Rb–Rb system. The $X^1\Sigma_g^+$ and $1^3\Sigma_u$ potentials, as well as the spin–orbit coupling terms, are taken from [599, 600].

Due to the presence of a resonance lying just a notch above the continuum threshold [601, 602], the computed low energy s-wave elastic cross section of two

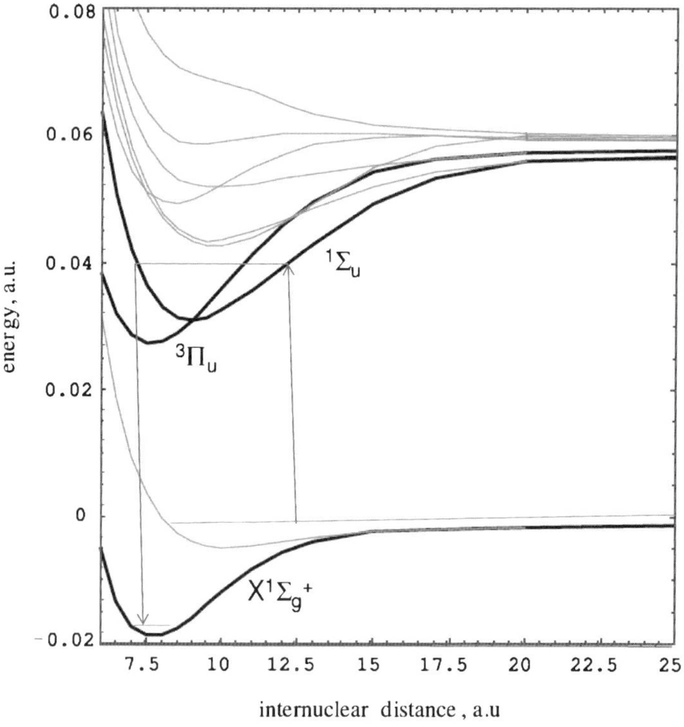

Figure 13.10 Black lines: Rb_2 Born–Oppenheimer potentials involved in the photoassociation calculations. Gray lines: Other Rb_2 potentials (not used in the calculation). The data shown in the figure are adopted from [598, Figure 2]. (Reprinted with permission. Copyright 1997 American Physical Society.)

^{85}Rb atoms on the $X^1\Sigma_g^+$ and $1^3\Sigma_u$ potentials is in access of 5.7×10^6 a.u.^{2} [603, 604]. As shown in Figure 13.11a, the scattering resonance enhances the photoassociation probability because it increases the amplitude of the continuum wave function in the inner region, thereby augmenting the $X^1\Sigma_g^+ - A^1\Sigma_u/b^3\Pi_u$ continuum-bound FC factors. The increase in the FC factors also means that we need to use much lower laser intensities in order to guarantee adiabaticity.

The dependence of the continuum-bound s-wave FC factors at collision energies of $E \approx 100$ μK on the bound $A^1\Sigma_u/b^3\Pi_u$ states energies is shown in Figure 13.11a. The APC scheme is expected to work best for transitions to bound states lying in the vicinity of the $A^1\Sigma_u/b^3\Pi_u(\mathrm{v} = 133, J = 1)$ level ($E = 9404\,\mathrm{cm}^{-1}$), for which

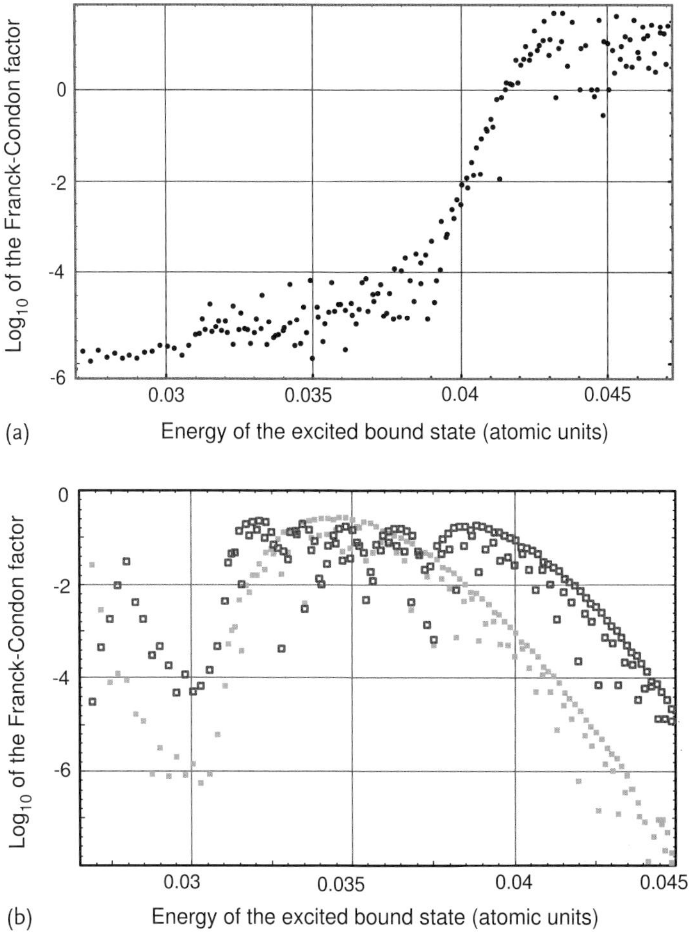

Figure 13.11 (a) FC factors for the $X^1\Sigma_g^+ - A^1\Sigma_u/b^3\Pi_u$ continuum-bound transitions. (b) FC factors for the $A^1\Sigma_u/b^3\Pi_u - X^1\Sigma_g^+$ bound–bound transitions, for the $\mathrm{v} = 0$ (filled gray) and $\mathrm{v} = 3$ (black) $X^1\Sigma_g^+$-vibrational states. Taken from [597, Figure 4]. (Reprinted with permission. Copyright 2007 American Physical Society.)

the continuum-bound FC factor is as high as 31.5 a.u. The choice of this level is also based on the availability of large area (and microsecond-long) Nd:YAG lasers at $\lambda = 1064$ nm – in resonance with the continuum-bound transitions to it.

Figure 13.12 shows an example of photoassociation of a Gaussian wave packet of Rb atoms described by Eq. (13.82) whose parameters are: $E_0 = 100\,\mu$K, $\delta_E = 70\,\mu$K, and $t_0 = 1150$ ps. The resulting continuum envelope $F_0(t)$ is shown in Figure 13.12a). In this calculation, the first pair of pulses transfers the entire population of the continuum wave packet to the X(v = 4, $J = 0$) state (of energy $E = -4001\,\text{cm}^{-1}$). The pulse durations are chosen so that their spectral widths will roughly coincide with that of the initial continuum wave packet. As shown in Figure 13.12, the final population of the X(v = 4, $J = 0$) state is 0.6. If there is no spontaneous emission of the intermediate state (i.e., $\Gamma = 0$), the population of the X(v = 4, $J = 0$) level reaches values as high as 0.9.

Having populated the X(v = 4, $J = 0$) state, we now perform a bound–bound STIRAP process (using the pair of pulses of Figure 13.9a) to execute a complete population transfer to the final, X(v = 0, $J = 0$), state. In the STIRAP process a pump pulse of center frequency of 11 261 cm^{-1} is followed after a delay of 600 ns by a dump pulse of center frequency of 11 507 cm^{-1}. These frequencies are in near-resonance with an intermediate $A^1\Sigma_u/b^3\Pi_u$ vibrational state of energy $E_1 = 7262\,\text{cm}^{-1}$. As a result of the process, the population of the X(v = 4, $J = 0$) state is completely transferred to the X(v = 0, $J = 0$) state.

In order to estimate the fraction of atoms photoassociated per pulse we need to multiply $P(E)$, the photoassociation probability of each colliding pair at energy E, by the number of collisions suffered by a given atom during the pulse. As illustrated in Figure 13.13, this number is calculated as follows: At a given energy E, the velocity of a given atom is $v = (2E/m)^{\frac{1}{2}}$ and the distance transversed by it during a pulse of Δt_2 duration is $v\Delta t_2$. The cross section for collision is πb^2 where b is the impact parameter, related to the J partial wave angular momentum as $b = (J + 1/2)/p = (J + 1/2)/(2mE)^{\frac{1}{2}}$. Hence, the number of collisions suffered

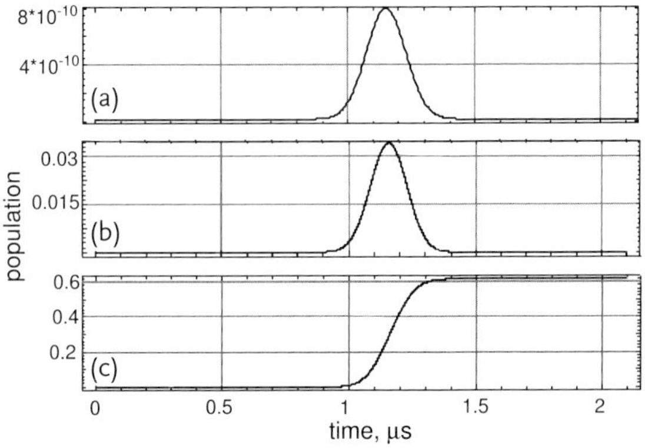

Figure 13.12 Photoassociation of a coherent wave packet. (a) $|F_0(t)|^2$; (b) the population of the intermediate $A^1\Sigma_u/b^3\Pi_u$ state; (c) the population of the X(v = 4, $J = 0$) target state. Taken from [597, Figure 5]. (Reprinted with permission. Copyright 2007 American Physical Society.)

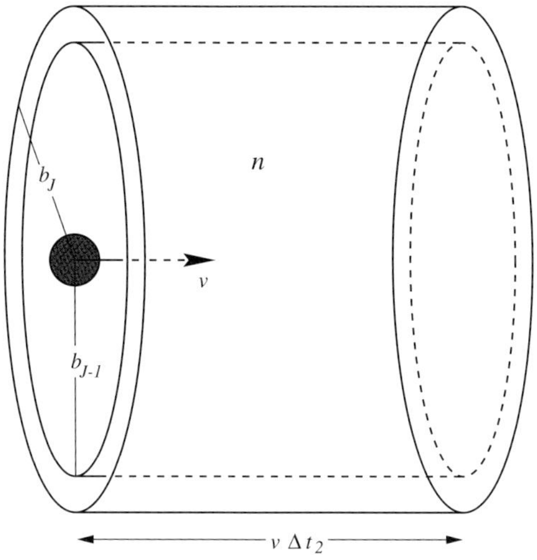

Figure 13.13 Calculation of the number of atoms in an ensemble of density n of impact parameter b_J that are in close enough vicinity to an atom moving at velocity v to be photoassociated during a pulse lasting Δt_2.

by the atom during the pulse is $N = n\pi b^2 v \Delta t_2$ where n is the density of atoms. Putting all this together we have for $J = 0$ that the fraction of atoms photoassociated per pulse is

$$f = \frac{P(E)\pi\left(\frac{2E}{m}\right)^{1/2}\Delta t_2}{8mE} = \frac{P(E)\pi\Delta t_2}{4m^{\frac{3}{2}}(2E)^{\frac{1}{2}}}\,. \tag{13.86}$$

For $\Delta t_2 = 750\,\text{ns}$, atomic density of $n = 10^{11}\,\text{cm}^{-3}$, collision energy of $E = 100\,\mu\text{K}$, and the ^{85}Rb–^{85}Rb reduced mass of $m = 1823 \times 85/2\,\text{a.u.}$, we have that $f \approx 2\times10^{-7}$ per pulse, in agreement with Figure 13.14f. Thus one needs to repeat the pulse sequence about 5×10^6 times in order to photoassociate the majority of the ensemble. Since many mode-locked laser sources operate at $\sim 10^6$ pulse/s and above, the above process can be completed in less than a minute.

13.3.3
Interference between Different Pathways

As in weak-field CC the introduction of a number of interfering pathways to the APC process is expected to enhance the population of a desired final state. Below, two simple interference schemes involving an initial continuum using the "double-Λ" [440] and the "tripod" schemes are examined.

The double-Λ scheme (Figure 13.9b) consists of performing a two-path photoassociation using two pairs of pulses. Two states, for example, $|2\rangle \equiv A^1\Sigma_u/b^3\Pi_u(\text{v} = 133,\ J = 1)$ and $|3\rangle \equiv A^1\Sigma_u/b^3\Pi_u(\text{v} = 136,\ J = 1)$, serve as intermediates leading to for example, the final $|1\rangle \equiv X(\text{v} = 4,\ J = 0)$ state. For simplicity we

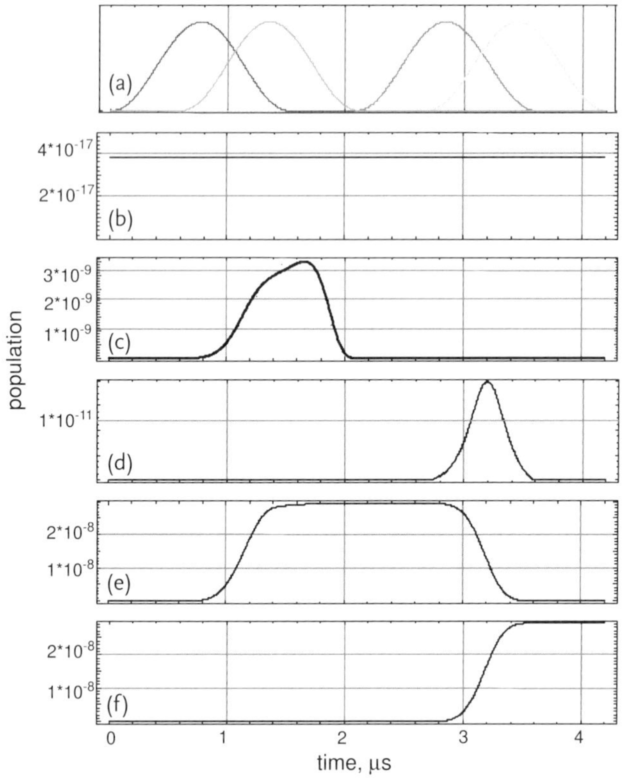

Figure 13.14 Photoassociation of the atomic ensemble in a trap. (a) Envelopes of the four laser pulses, unscaled; (b) $F_0(t)^2$; (c)–(f) Bound-state populations, weighted over the ensemble, for different bound states; (c) $A^1\Sigma_u/b^3\Pi_u$ (v = 133, J = 1), E = 9404 cm^{-1}; (d) $X^1\Sigma_g^+$ (v = 4, J = 0), E = −4001 cm^{-1}; (e) $A^1\Sigma_u/b^3\Pi_u$ (v = 35, J = 1), E = 7262 cm^{-1}; (f) $X^1\Sigma_g^+$ (v = 0, J = 0), E = −4236 cm^{-1}. Taken from [597, Figure 6]. (Reprinted with permission. Copyright 2007 American Physical Society.)

assume that $\Delta_{12} = \Delta_{13} = \Delta_b$, $\Delta_{E2} = \Delta_{E3} = \Delta_E$, that $\Omega_{12} = \Omega_{13} = W_b$, and that $\Omega_{E3} = \Omega_{E2}\exp[i\alpha]$. With these assumptions, we can write the time-dependent equations analogous to Eq. (13.2) as

$$\begin{aligned}
\dot{b}_1 &= i\,\Omega_b^*\,(b_2 + b_3)e^{-i\,\Delta_b\,t}\,,\\
\dot{b}_2 &= i\,\Omega_b\,b_1 e^{i\,\Delta_b\,t} - \Gamma\,(b_2 + e^{-i\alpha}b_3) + i\,\Omega_E\,F_0(t)\,,\\
\dot{b}_3 &= i\,\Omega_b\,b_1 e^{i\,\Delta_b\,t} - \Gamma\,(e^{i\alpha}b_2 + b_3) + i\,e^{i\alpha}\Omega_E\,F_0(t)\,.
\end{aligned} \tag{13.87}$$

Introducing the new amplitude $b_{exc} = b_2 + b_3$, we see that the process is analogous to the single pathway photoassociation governed by a combined continuum-bound α-independent Rabi frequency ranging between 0 and $2\Omega_E$. We find that the decay rate of b_{exc} does depend on α and is lower than the one governing the single pathway APC. If $\alpha = 0$, then the decay rate for b_{exc} is equal to $2\pi\Omega_E^2$ rather than $\pi(2\Omega_E)^2$.

The role of the relative phase in the double-λ AP was also checked [510]. Using intensities of $I_{b2} = 3.5\times10^3$ W/cm^2, and $I_{E2} = 5\times10^3$ W/cm^2, for the two transition involving the $A^1\Sigma_u/b^3\Pi_u(\mathrm{v} = 133,\ J = 1)$ intermediate state, it was found [510] that the maximal yield of AP in the double-λ passage is higher than that achieved in the single pathway passage with twice the field strength. The efficiency of double-λ AP depends on the relative phase between the four Rabi frequencies.

An alternative configuration which takes advantage of interfering pathways is the "tripod" configuration, shown in Figure 13.9c. Based on bound–bound studies [408, 411, 453, 504], one expects that the ratio between the Rabi frequencies of the two dump pulses would determine the branching ratio between the two bound X states. Calculations [510] performed on the tripod photoassociation of a wave packet in the $X^1\Sigma_g^+$-continuum to the $X^1\Sigma_g^+(\mathrm{v} = 4, J = 0)$ state via the $A^1\Sigma_u/b^3\Pi_u(\mathrm{v} = 133, J = 1)$ state using an additional dump pulse that couples the same intermediate $A^1\Sigma_u/b^3\Pi_u(\mathrm{v} = 133, J = 1)$ state to a different final state $X^1\Sigma_g^+(\mathrm{v} = 5, J = 0)$ $E = -3944\,\mathrm{cm}^{-1}$ confirm this expectation: the probability to populate either the $X^1\Sigma_g^+(\mathrm{v} = 4, J = 0)$ or $X^1\Sigma_g^+(\mathrm{v} = 5, J = 0)$ state is indeed proportional to the corresponding ratio between the Rabi frequencies squared.

13.3.4 Experimental Realizations

Dark states that arise in the adiabatic passage from the continuum, as discussed above, have been observed experimentally [473, 474, 605]. Similarly dark states associated with Feshbach molecules [362, 586] have been observed [388]. In Figure 13.15 we display the results of an experiment measuring the disappearance of the atomic Rb signal due to the photoassociation of Rb atoms to form Rb_2 molecule. In this experiment two CW lasers were used, one serving as a pump, and the other as a dump, to photoassociate Rb atoms in a Rb Bose–Einstein condensate to form a degenerate gas of Rb_2 ground state molecules in a specific vib-rotational state. As a signature for the decoupling of this coherent atom-molecule gas from the light field, a striking suppression ("dark state" formation) of photoassociation loss is observed. The experimentally observed dark state is similar to the theoretical one presented in Figure 13.7 in that the population in the intermediate state in both cases is negligible. In contrast to the theory displayed in Figure 13.7 the experiment involves CW light fields, hence no population transfer due to the temporal evolution of this dark state has been observed.

An important tool in performing PA to form ultracold molecules has been the use of "Feshbach molecules" as intermediates. A Feshbach molecule [362, 586, 606] is a scattering resonance (see Section 7.3.1) whose center energy is tuned magnetically to lie *below* the molecular dissociation limit. Feshbach molecules are bound states, but because of their very weak binding energy they usually dissociate upon collisions with other atoms or molecules. Photoassociation to form ultracold ground state molecules using "Feshbach molecules" as intermediates has been demonstrated by a number of groups [332, 388, 587, 589]. Winkler *et al.* [587] have used this technique to create ground state $^{87}Rb_2$ molecules, Ni *et al.* [388] have cre-

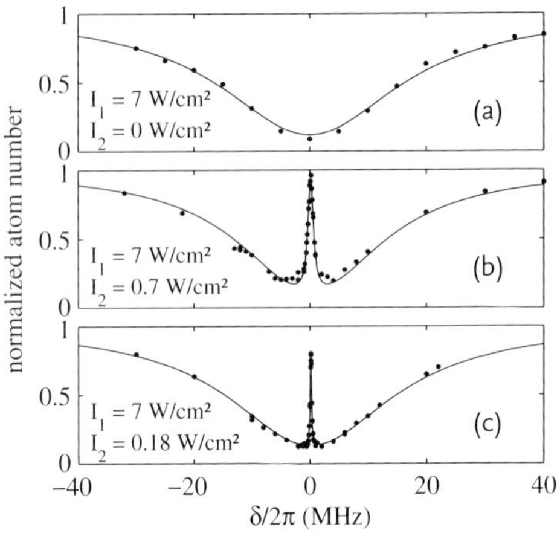

Figure 13.15 The atomic Rb signal as a function of δ – the detuning from resonance of the pump laser at three different dump laser intensities. (a) Dump intensity = 0. The concentration of the Rb atoms is depleted as a result of photoassociation of two Rb atoms to form a Rb_2 molecule as I_1 sweeps through the resonance. For (b) dump intensity = $0.7\,W/cm^2$ and (c) dump intensity = $0.18\,W/cm^2$, a narrow dark state, corresponding to the quenching of the photoassociation and the disappearance of the population in the intermediate molecular level, appears. This dark state becomes narrower and narrower as the intensity of the dump laser goes down. (Compare panel (b) to panel (c)). Taken from [473, Figure 1]. (Reprinted with permission. Copyright 2005 American Physical Society.)

ated $^{40}K^{87}Rb$ heteronuclear molecules, and Spiegelhalder *et al.* [332] have created an ensemble of $^{6}Li^{40}K$. In all these experiments one makes use of a STIRAP process (see Figure 13.16) to coherently transfer the extremely weakly bound $^{87}Rb_2$ or KRb Feshbach molecules to the vib-rotational ground state of the ground electronic state.

The KRb experiment [388] was remarkable in being able to produce an ensemble of molecules at relatively high density ($n = 10^{12}\,cm^{-3}$) at an (expansion-determined) translational temperature of 350 nK. The starting point of this experiment was an atomic degenerate gas mixture of fermionic ^{40}K atoms and bosonic ^{87}Rb atoms confined in an optical dipole trap. By using a 550 G magnetic field it was possible to lower the energy of a Feshbach resonance lying above the onset of the continuum to a truly bound state [607–610]. As many as 10^4 Feshbach molecules, whose binding energy of $0.767 \times 10^{-5}\,cm^{-1}$ can be detected directly using time-of-flight absorption imaging, were thus formed.

As shown in Figure 13.16a, the transfer scheme of the Feshbach molecules to the $a^3\Sigma, v = 0$ molecules involves three molecular levels, the initial state $|i\rangle$, the intermediate state $|e\rangle$, and the final state $|g\rangle$. These states are coupled by two laser fields. The pump laser drives the up transition to the $|e\rangle = |v' = 23\rangle$ vibrational level of the electronically excited $2^3\Sigma$ state. This state has good FC overlaps with both the weakly bound Feshbach molecule and the deeply bound ground vibrational state $|g\rangle$ which is coupled to it by the dump laser field. As shown in Figure 13.16a, the main contribution to the $\langle e|i\rangle$ FC integral occurs near the outer turning point of

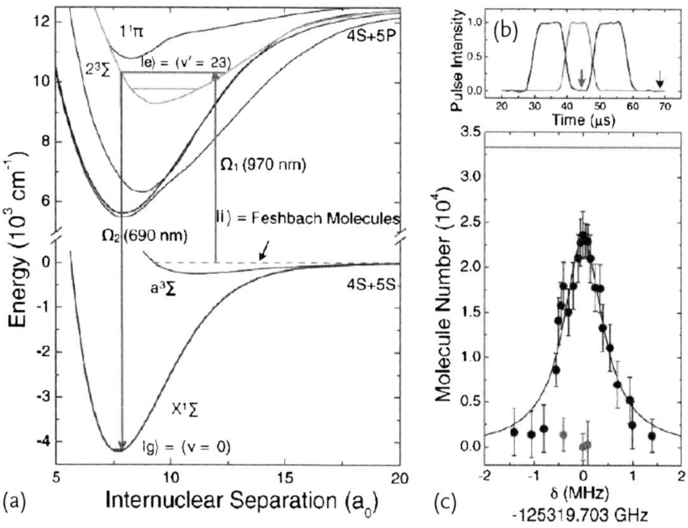

Figure 13.16 Two-photon coherent state transfer from weakly bound Feshbach molecules $|i\rangle$ to the absolute molecular ground state $|g\rangle$ (v $=$ 0, J $=$ 0 of $X^1\Sigma$). (a) Transfer scheme. Here, the intermediate state $|e\rangle$ is the v′ $=$ 23 level of the Ω $=$ 1 component of the electronically excited $2^3\Sigma$ potential. The chosen intermediate state lies just below the $1^1\Pi$ excited electronic potential, which provides the necessary triplet-singlet spin mixing to transfer predominantly triplet character Feshbach molecules to the vib-rotational ground state of the singlet electronic ground potential, $X^1\Sigma$. The vertical arrows are placed at the regions of greatest overlap of the up and down transitions. (b) Normalized Raman laser intensities versus time for the round trip STIRAP pulse sequence. Four transfers were performed each way using a maximum Rabi frequency of $2\pi \times 7$ MHz for the downward transition (black line) and a maximum Rabi frequency of $2\pi \times 4$ MHz for the upward transition (gray line). (c) The number of Feshbach molecules recovered after a round trip STIRAP transfer is plotted as a function of the detuning from a two-photon resonance. The round trip data were taken at the time indicated by the black arrow in (b). The gray data points show the remaining Feshbach molecule number when only one-way STIRAP is performed (at the time indicated by the gray arrow in (b)), where all Feshbach molecules are transferred to the ground state and are dark to the imaging light. The initial Feshbach molecule number is $3.3(4) \times 10^4$ (gray solid line), and the number after the round trip STIRAP is 2.3×10^4. The round trip efficiency is 69%, which suggests the one-way transfer efficiency is 83% and the number of the absolute ground state polar molecules is 2.7×10^4. Taken from [388, Figure 5]. (Reprinted with permission. Copyright 2008 American Association for the Advancement of Science.)

$|e\rangle$, whereas the $\langle g|e\rangle$ FC integral borrows its strength from the good overlap in the inner turning point region of $|e\rangle$.

In Figure 13.16b we see a typical time profiles of the pump and dump pulses for a full *round trip* composed of two STIRAP processes: The first process, corresponding to the $|i\rangle \rightarrow |e\rangle$ transfer, is followed after some delay by a process demonstrating good recovery of population in which the opposite $|e\rangle \rightarrow |i\rangle$ transfer is executed. True to the "counterintuitive" pulse ordering of STIRAP[321], the dump pulse precedes the pump pulse in the $|i\rangle \rightarrow |e\rangle$ transfer, with a reverse order in the $|e\rangle \rightarrow |i\rangle$ transfer. Figure 13.16c displays the number of Feshbach molecules recovered after one round trip STIRAP transfer as a function of the detuning from the two-photon resonance. The initial number of Feshbach molecules $3.3(4) \times 10^4$ (gray solid line), and the number after an on-resonance round trip STIRAP is 2.3×10^4. The round trip efficiency is thus 69%, suggesting that the one-way transfer efficiency is 83% and the number of the absolute ground state polar molecules is 2.7×10^4.

The above picture is supplemented in Figure 13.17a, taken also from [388], which displays the number of remaining Feshbach molecules as a function of the pump and dump frequency difference. Figure 13.17b demonstrates the decoupling of the system from the laser fields and the formation of a dark state involving a Feshbach molecule as the dump laser is scanned across one of the two-photon resonances shown in Figure 13.17a.

The above experiments clearly show that the field of photoassociation of ultracold atoms to form ultracold molecules analyzed theoretically above, has reached a maturity which would enable the production of ultracold molecules in large densities as a preparatory step for performing coherently controlled bimolecular reactions [333, 335, 336, 611, 612], a topic discussed in Chapter 8.

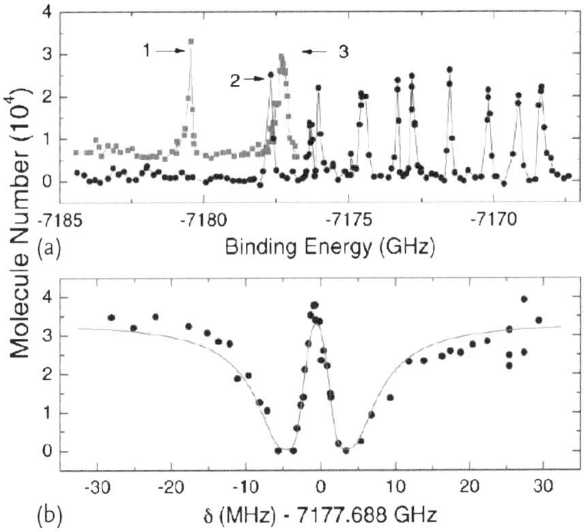

Figure 13.17 The $v = 0$ ground state level of the triplet electronic ground potential, $a^3\Sigma$. (a) Hyperfine and rotational states of the $a^3\Sigma$, $v = 0$ ground state molecule at a magnetic field of 546.94 G, observed using two-photon spectroscopy and scanning the frequency of the dump field. The measured number of remaining Feshbach molecules is plotted as a function of the frequency difference of the two laser fields. Shown are two sets of data, vertically offset for clarity, obtained using two different intermediate states, which are hyperfine and rotational states of the $v' = 10$ level of the electronically excited $2^3\Sigma$ potential. Peaks labeled 1 and 2 correspond to hyperfine states in the rotational ground state, and peak 3 corresponds to a rotationally excited state. (b) A precise determination of the energy and the transition dipole moments for individual states using the two-photon spectroscopy by scanning the up leg Ω_1 frequency. The measured number of remaining Feshbach molecules is plotted as a function of the two-photon detuning. The dark resonance shown here is for the triplet vib-rotational ground state corresponding to peak 2 in (a). Taken from [388, Figure 2]. (Reprinted with permission. Copyright 2008 American Association for the Advancement of Science.)

13.4 Laser Catalysis

Over the past two decades a number of scenarios for laser acceleration and suppression of dissociation processes and chemical reactions have been proposed [12, 613–637] that rely upon strong-field effects. The theme common to many of these schemes is quite different from that of coherent control in that the lasers affect chemical reactivity by modifying the molecular potential to produce so-called "dressed" potentials (see Chapter 15). Success is attained by designing dressed potentials that promote a given reaction. The main difference between laser enhancement of chemical reactions and ordinary photochemistry is that the former is designed so as to involve no net absorption of laser photons. The concept of "laser catalysis" [424, 637–640], that is, a process in which a laser field returns to its exact initial state after altering a reaction, is a refinement of such scenarios.

Consider first some of the factors affecting the design of such schemes. Ground electronic state based laser enhancement schemes [618–620, 622–624, 626, 627] rely on the induction of nuclear dipole moments to aid in promoting a desired reaction [12, 628–630]. For example, the use of IR radiation has been proposed to overcome reaction barriers on the ground electronic state [12, 628–630]. However, this proposal requires powers on the order of TW/cm^2. At these powers nonresonant multiphoton absorption, which invariably leads to ionization and/or dissociation, becomes dominant, drastically reducing the yield of the reaction of interest.

Continuum–continuum transitions involving *excited* electronic states [631] might be thought useful insofar as they ought to require less power than those occurring on the ground state, because in this case the laser can couple to strong electronic transition dipoles. However, in this case the continuum–continuum nuclear factors lead to smaller transition dipole matrix elements, and moreover, once the system is deposited on an unbounded excited electronic surface, it is impossible to prevent reaction on that surface and the resultant retention of the absorbed photon. Such a chain of events resembles that of conventional (weak-field) photochemistry where the laser is used to impart energy to the reaction and not to catalyze it.

Scenarios [424, 637–640] employing transitions between scattering states on the ground electronic surface and bound excited electronic states may reduce the above power requirements, primarily because of the involvement of the strong bound-continuum nuclear factors. For an excited surface possessing reaction well(s), such schemes give rise to "laser catalysis" [424, 637–640] described below because the reagents, once excited, remain in the transition state region and shuttle freely between the reactants' side and the products' side of the ground state barrier (see Figure 12.1d). If the energy available to the nuclei on the excited state is insufficient to break any bond, the system, not being able to escape the transition state region, eventually relaxes (radiatively or nonradiatively) back to the ground state. In doing so, it has *a priori* similar probabilities of landing on the products' side as on the reactants' side of the barrier. However, if the laser is strong enough, the stimulated radiative relaxation route, yielding back the same photon absorbed, overcomes the nonradiative channels, resulting in true laser catalysis.

We show below that the simple methodology developed in this book can account for many of the phenomena described above. The only required modification of the models discussed is to consider the coupling of a bound manifold to *two* continua (that of the reactants and that of the products) via laser pulses.

13.4.1 The Coupling of a Bound State to Two Continua by a Laser Pulse

As a model, we consider the $A + BC \rightarrow AB + C$ exchange reaction described by the smooth one-dimensional potential barrier along a reaction coordinate (Figure 13.18). The eigenstates of the system form a continuum of "outgoing" scattering states $|E, 1^+\rangle$ and $|E, 2^+\rangle$. In accordance with our general notation, the 1^+ and 2^+ indices are reminders that the reaction has originated in either arrangement channel 1, the $A + BC$ channel, or arrangement channel 2, the $AB + C$ channel. The situation is depicted in Figure 12.1d.

In accord with standard scattering theory [641], the asymptotic behavior of the $|E, 1^+\rangle$ and $|E, 2^+\rangle$ states is given by

$$\lim_{x \to -\infty} \langle x | E, 1^+ \rangle = \sqrt{\frac{m}{k_1 h}} \exp(i k_1 x) + R_1(E) \exp(-i k_1 x) , \tag{13.88}$$

$$\lim_{x \to +\infty} \langle x | E, 1^+ \rangle = T_1(E) \exp(i k_2 x) , \tag{13.89}$$

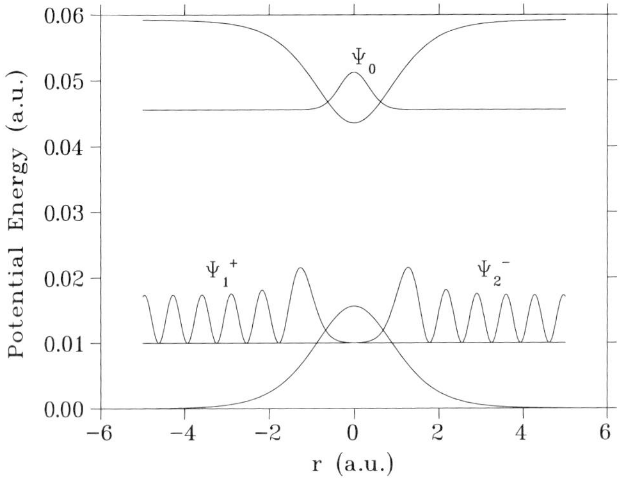

Figure 13.18 Eckart potentials and wave functions used in the simulation of the laser catalysis process. Potential parameters were $A = 0$ a.u., $B = 6.247$ a.u., $l = 4.0$ a.u. and $m = 1060.83$ a.u. Ψ_1^+ denotes the $|E, 1^+\rangle$ state of the text Ψ_2^- denotes the $|E, 2^-\rangle$ state of the text, and Ψ_0 denotes the $|E_0\rangle$ state of the text. Taken from [424, Figure 2]. (Reprinted with permission. Copyright 1998 American Physical Society.)

and

$$\lim_{x \to \infty} \langle x | E, 2^+ \rangle = \sqrt{\frac{m}{k_2 h}} \exp(-ik_2 x) + R_2(E) \exp(ik_2 x) , \tag{13.90}$$

$$\lim_{x \to -\infty} \langle x | E, 2^+ \rangle = T_2(E) \exp(-ik_1 x) , \tag{13.91}$$

where $k_{1,2} = \{\sqrt{2m[E - V(\mp\infty)]}\}/\hbar$. Here $R_i(E)$ denotes the amplitude for reflection, and $T_i(E)$ the amplitude for reaction when the system starts in channel i.

The laser catalysis scenario is shown in Figure 12.1d: Under the action of a laser pulse of central frequency ω, assumed to be in near-resonance with the transition from the continuum $|E, i^+\rangle$ to an intermediate bound state $|E_0\rangle$, population is transferred from states $|E, 1^+\rangle$ to a set of "incoming" scattering states $|E, 2^-\rangle$, with the asymptotic behavior

$$\lim_{x \to \infty} \langle x | E, 2^- \rangle = \sqrt{\frac{m}{k_2 h}} \exp(ik_2 x) + R_2^*(E) \exp(-ik_2 x) , \tag{13.92}$$

$$\lim_{x \to -\infty} \langle x | E, 2^- \rangle = T_2^*(E) \exp(ik_1 x) . \tag{13.93}$$

To address this problem, with the total Hamiltonian of the system given by Eq. (11.2), we expand the material wave function of the system as in Eq. (17.53):

$$|\Psi(t)\rangle = b_0 |E_0\rangle \exp(-iE_0 t/\hbar) + \int dE \left[b_{E,1}(t) |E, 1^+\rangle + b_{E,2}(t) |E, 2^+\rangle \right] \exp(-iEt/\hbar) , \tag{13.94}$$

where $|E_0\rangle$ and $|E, n^+\rangle$ satisfy the material Schrödinger equations

$$[E_0 - H_M] |E_0\rangle = [E - H_M] |E, n^+\rangle = 0 , \quad n = 1, 2 . \tag{13.95}$$

Substitution of the expansion of Eq. (13.94) into the time-dependent Schrödinger equation, and use of the orthogonality of the $|E_0\rangle$, $|E, 1^+\rangle$ and $|E, 2^+\rangle$ basis states, results in a set of first-order differential equations similar to Eq. (13.2):

$$\frac{db_0}{dt} = i \sum_{n=1,2} \int dE\, \Omega_{0,E,n}(t) \exp(i\Delta_E t) b_{E,n}(t) , \tag{13.96a}$$

$$\frac{db_{E,m}}{dt} = i\Omega^*_{0,E,m}(t) \exp(-i\Delta_E t) b_0(t) , \quad m = 1, 2 , \tag{13.96b}$$

where

$$\Omega_{0,E,n}(t) \equiv \langle E_0 | \hat{\boldsymbol{\varepsilon}} \cdot \boldsymbol{d} | E, n^+ \rangle \mathcal{E}(t)/\hbar , \quad n = 1, 2 , \quad \text{and} \quad \Delta_E \equiv \frac{E_0 - E}{\hbar} - \omega . \tag{13.97}$$

Substituting the formal solution of Eq. (13.96b), where t_0 denotes the initial time,

$$b_{E,n}(t) = b_{E,n}(t_0) + i \int_{t_0}^{t} dt' \Omega^*_{0,E,n}(t') \exp(-i\Delta_E t') b_0(t') , \tag{13.98}$$

into Eq. (13.96a), we obtain

$$\frac{d b_0}{d t} = i \sum_{n=1,2} \int d E\, \Omega_{0,E,n}(t) \exp(i \Delta_E t) b_{E,n}(t_0)$$
$$- \sum_{n=1,2} \int d E\, \Omega_{0,E,n}(t) \int_{t_0}^{t} d t' \,\Omega^*_{0,E,n}(t') \exp\left[i \Delta_E (t - t')\right] b_0(t') \,. \quad (13.99)$$

We now invoke the SVCA, which in the context of the two-photon Λ configuration of Figure 12.1d, reads

$$\sum_n |\langle E, n^- | \hat{\boldsymbol{\varepsilon}} \cdot \boldsymbol{d} | E_0 \rangle|^2 \approx \sum_n |\langle E_L, n^+ | \hat{\boldsymbol{\varepsilon}} \cdot \boldsymbol{d} | E_0 \rangle|^2 \,, \quad (13.100)$$

where $E_L = E_0 - \hbar\omega$. Upon substitution of Eqs. (13.3) and (13.100) into Eq. (13.99) we obtain that

$$\frac{d b_0}{d t} = i F(t) - \sum_{n=1,2} \Omega^I_n(t) b_0(t) \quad (13.101)$$

where

$$\Omega^I_n(t) \equiv \pi |\langle E_L, n^+ | \hat{\boldsymbol{\varepsilon}} \cdot \boldsymbol{d} | E_0 \rangle \mathcal{E}(t)|^2 / \hbar \,, \quad n = 1, 2 \,. \quad (13.102)$$

The source term $F(t)$ is given by

$$F(t) = \sum_{n=1,2} F_n(t) = \mathcal{E}(t) \sum_{n=1,2} \bar{d}_n(t)/\hbar \,, \quad (13.103)$$

where

$$\bar{d}_n(t) = \int d E \langle E_0 | \hat{\boldsymbol{\varepsilon}} \cdot \boldsymbol{d} | E, n^+ \rangle \exp(i \Delta_E t) b_{E,n}(t_0) \,, \quad n = 1, 2 \,. \quad (13.104)$$

As in the photoassociation case, we can obtain analytical solutions of Eq. (13.101):

$$b_0(t) = \nu(t) q(t) + b_0(t_0) \nu(t) \,, \quad (13.105)$$

where

$$\nu(t) = \exp\left\{ - \int_{t_0}^{t} \left[\Omega^I_1(t') + \Omega^I_2(t')\right] d t' \right\} \quad (13.106)$$

and

$$q(t) = i \int_{t_0}^{t} \frac{F(t')}{\nu(t')} d t' \,. \quad (13.107)$$

In the laser catalysis process, the initial conditions are such that $b_0(t_0) = 0$ and $b_{E,2}(t_0) = 0$ for all E. Therefore, we obtain for the $b_0(t)$ coefficient:

$$b_0(t) = i \int_{t_0}^{t} F_1(t') \exp\left\{-\int_{t'}^{t} [\Omega_1^I(t'') + \Omega_2^I(t'')]dt''\right\} dt' . \tag{13.108}$$

Given $b_0(t)$, the continuum population distributions $b_{E,1}(t)$ and $b_{E,2}(t)$ are obtained directly via Eq. (13.98). Typical potentials and eigenfunctions used to simulate one-photon laser catalysis are plotted in Figure 13.18.

As an illustration, consider laser catalysis with an Eckart potential [642, 643] for the ground state

$$V_{ground}(x) = V[\xi(x)] = -\frac{A\xi}{1-\xi} - \frac{B\xi}{(1-\xi)^2} ; \quad \xi = -\exp\left(\frac{2\pi x}{l}\right) , \tag{13.109}$$

with $A = 0,\ B = 6.247 \times 10^{-2}$ a.u., $l = 4.0$ a.u. and an inverted Eckart potential,

$$V_{excited}(x) = \mathcal{E} - V[\xi(x)] \tag{13.110}$$

having a well, for the excited state.

The particle's mass m was chosen as that of the $H + H_2 \rightarrow H_2 + H$ reaction, that is, $m = 1060.83$ a.u. The intermediate state was the $v = 0$ level of the inverse Eckart potential given in Eq. (13.110), with the same parameters as above [644, 645]. The resulting potential curves, similar to the $H + H_2$ reaction path, are plotted in Figure 13.18. Given these parameters, eigenfunctions and eigenenergies were obtained using the formulae of [424].

The initial state of the system is described by a normalized Gaussian wave packet of outgoing scattering states

$$|\Psi(t=0)\rangle = \int dE\, b_{E,1}(t=0)|E, 1^+\rangle \tag{13.111}$$

where $b_{E,1}(t = 0)$ is given by Eq. (13.82). Simulations were made for initial collision energies of $E_{col} = 0.005$–0.03 a.u. and wave packet widths $\delta_E = 10^{-4}$–10^{-3} cm^{-1}.

The time dependence of the expansion coefficients is shown in Figure 13.19. Initial collision energy for this calculation was 0.01 a.u. At this energy the nonradiative reaction probability is negligible. Here the effect of the laser pulse is to induce a near-complete ($> 99\%$) population transfer from the wave packet of $|E, 1^+\rangle$ states (localized to the left of the potential barrier) to a wave packet of $|E, 2^-\rangle$ states (localized to the right of the barrier), while keeping the population of the $|E_0\rangle$ states to a bare minimum. The latter serves to minimize spontaneous emission losses.

Due to the invariance of Eq. (13.101) upon the rescaling of Eq. (13.83), it is possible to freely vary the pulse durations and intensity, as long as the integrated pulse power $|\mathcal{E}^0|^2 \times \Delta t$ is kept fixed. Thus the dynamics remains the same if the time

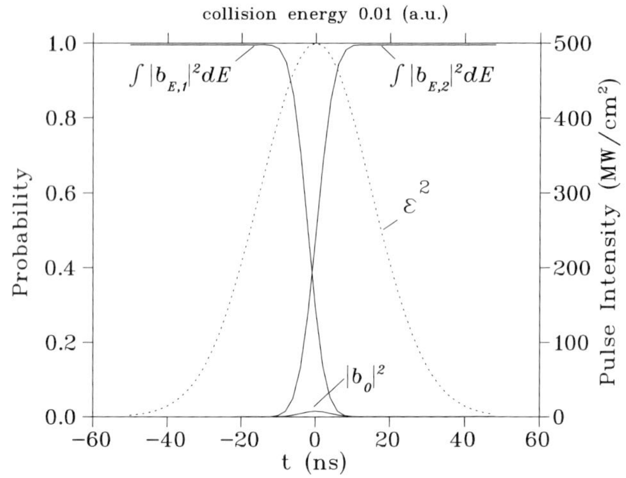

Figure 13.19 Integrated populations of incoming and outgoing continuum states and the population of the v $= 0$ intermediate state vs. time. The dashed line is the intensity profile of the Gaussian pulse whose maximum intensity is 5×10^8 W/cm^2. The FWHM of the pulse is 30 ns and its central frequency was chosen so that $\Delta_{E_{col}} = 0$. The initial reactant collision energy is 0.01 a.u. and initial wave packet width is $\delta_E = 10^{-3}$ cm^{-1}. Taken from [424, Figure 5]. (Reprinted with permission. Copyright 1998 American Physical Society.)

evolution of the system is scaled up by a factor of ten and the pulse intensity is scaled down by the same factor. The advantage of long pulses is that only reactants that collide during the laser pulse will react. One anticipates that longer pulses would increase the number of product molecules formed within a single pulse duration. The disadvantage of longer pulses is that the power requirements become increasingly more difficult to fulfill, because the peak power must go down exactly as $1/\Delta t$, whereas in most practical devices the power goes down much faster with increasing pulse durations.

As mentioned above, by keeping the population of the intermediate resonance low (as is the case in Figure 13.19) the spontaneous emission losses are effectively eliminated. In Figure 13.20 we plot the intermediate level population as a function of t at four different pulse intensities. Radiative reaction probability for all plotted intensities is near unity. However, it is evident that the intermediate state population throughout the process decreases with increasing pulse intensity. Thus, to avoid spontaneous emission losses, high pulse intensities should be used.

Calculated reactive line shapes (i.e., the reaction probability as a function of the pulse center frequency), at three pulse intensities are shown in Figure 13.21. The initial collision energy is 0.014 a.u., that is, slightly closer to the barrier maximum than before. At this energy the nonradiative reaction probability is about 9%. This causes the line-shapes to assume an asymmetric form due to the interference between the nonradiative tunneling pathway and the laser catalyzed pathway. We see that the reaction probability is enhanced for a positive (blue) detuning and suppressed for a negative (red) detuning.

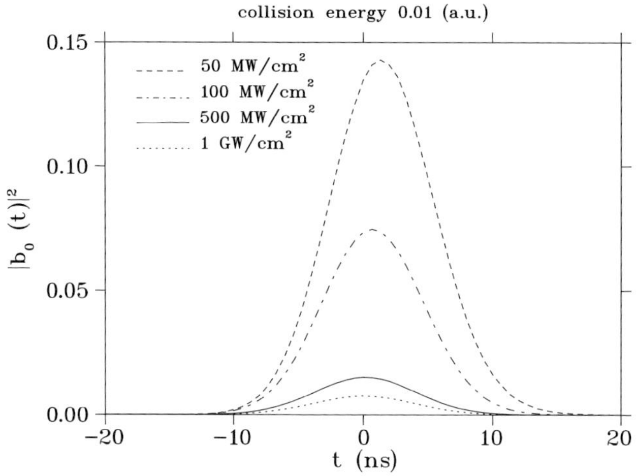

Figure 13.20 Intermediate state population vs. time at pulse intensities of 50, 100, 500 MW/cm^2 and 1 GW/cm^2. Taken from [424, Figure 7]. (Reprinted with permission. Copyright 1998 American Physical Society.)

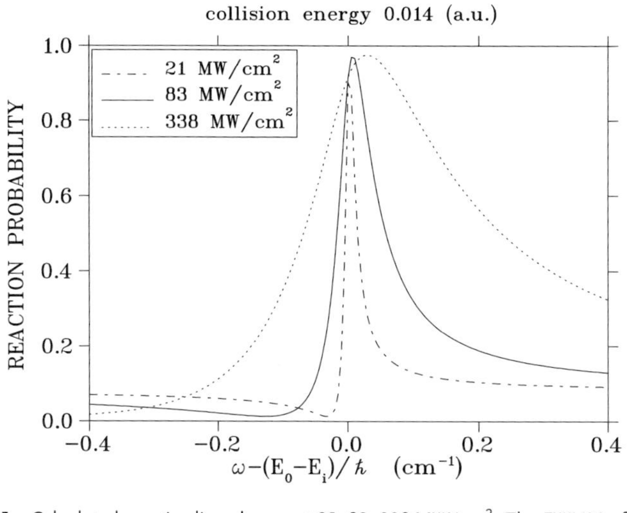

Figure 13.21 Calculated reactive line shapes at 21, 83, 338 MW/cm^2. The FWHM of the pulse is 20 ns. Reactants collision energy is 0.014 a.u. and the initial wave packet width is $\delta_E = 10^{-3}$ cm^{-1}. Taken from [424, Figure 8]. (Reprinted with permission. Copyright 1998 American Physical Society.)

The fact that there is a point where the reaction probability assumes the value of unity is best understood by adopting the "(photon) dressed states" picture, discussed extensively in Section 15.1 below. For this discussion it suffices to say that the only difference between the ordinary potential matrix and the dressed one is that in our (2×2) case, the (1, 1) diagonal matrix element of the dressed potential

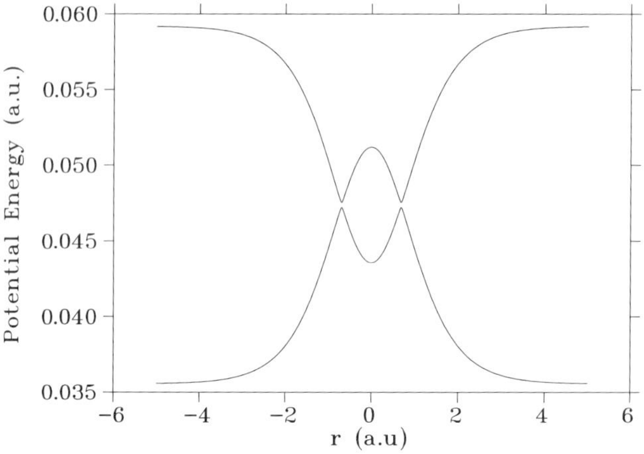

Figure 13.22 Dressed state potentials for the laser catalysis process at maximum pulse intensity. Initial kinetic energy is 0.01 a.u.

matrix is a sum of the material potential and the energy of the laser photon. The (2, 2) potential matrix element is just the excited state potential, and the off-diagonal matrix elements are the field-dipole coupling terms. When this matrix is diagonalized one obtains the two "field"–matter eigenvalues shown in Figure 13.22. As demonstrated in this figure, the ground field–matter eigenvalue assumes the shape of a double barrier potential and the excited eigenvalue – the shape of a double well potential. The separation between these eigenvalues increases as the coupling field strength is increased.

In the adiabatic approximation, particles starting out in the remote past in the ground state remain on the lowest eigenvalue at all times. These particles experience resonance scattering by a double barrier potential. It is known that in this situation there is one energy point with unity tunneling probability, irrespective of the details of the potential [646–651], This point occurs when the incident energy is near a bound state of the well contained within the barriers. Similar phenomena have been noted for semiconductor devices [646–648], in the context of the Ramsauer–Townsend effect [227] and for Fabry–Pérot interferometers [652]. Re-

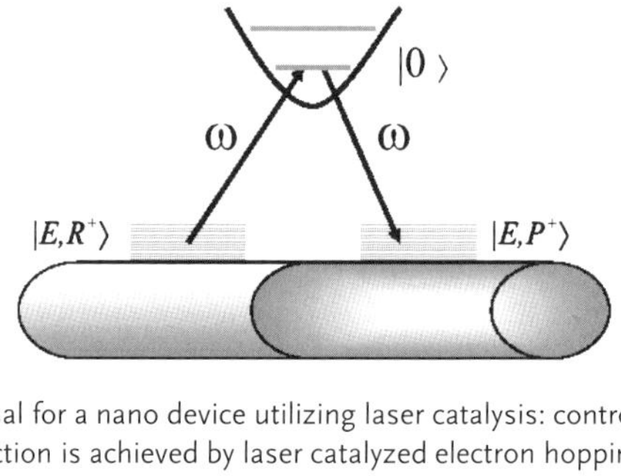

Figure 13.23 A proposal for a nano device utilizing laser catalysis: control over currents going through a nanojunction is achieved by laser catalyzed electron hopping above the junction barrier.

lated observations, linking the phenomenon of field-induced transparency with the emergence of a field-dressed double barrier potential, were made by Vorobeichik *et al.* [653].

The point where the tunneling is suppressed (i.e., when the tunneling probability of Figure 13.21 is zero) appears, in the dressed states picture, as a result of the breakdown of the adiabatic approximation: At the energy of tunneling suppression, the flux leaking to the excited double well eigenvalue interferes destructively with the flux remaining on the low double barrier potential.

An example for a proposal of a nano transistor in which control over currents going through a nanojunction between two different carbon nanotubes is achieved by laser catalyzed electron hopping above the junction barrier is shown in Figure 13.23.

14
Coherent Control of the Synthesis and Purification of Chiral Molecules

A molecule is said to be "chiral" if it does not coincide, or cannot be made to coincide using a simple rotation, with its mirror image. In such cases, the molecule and its mirror image are called "enantiomers", with one enantiomer being "right-handed" and the other enantiomer being "left-handed." A sufficient (though not necessary) condition for chirality is for the molecule to have at least one "asymmetric" carbon atom, that is, a carbon atom bonded to four different groups of atoms. Two enantiomers can be distinguished experimentally, for example, by their ability to rotate linearly polarized light in opposite directions. A mixture of the two enantiomers is called a "racemic" mixture or a "racemate".

The existence of enantiomers is one of the fundamental broken symmetries in nature [654–657]. It is also one of great practical importance because biological processes are often stereospecific, motivating a longstanding interest in asymmetric synthesis, that is, molecular processes that preferentially produce one of the enantiomeric pairs. In this chapter we explain how to use coherent control techniques to perform asymmetric synthesis [145, 658–660] using the strong electric dipole–electric field interaction. This is in sharp contrast with previous techniques [654, 661] where efforts were made to use the far weaker magnetic dipole interaction terms. As discussed below, the controlled transition between enantiomers has been treated using both classical as well as quantum light.

Section 14.1 introduces general rules under which this type of control is possible. Then, in Section 14.2 below, we consider the two-photon dissociation of a single quantum state of a $B{-}A{-}B'$ molecule to yield $BA + B'$ and $B + AB'$, where B and B' are enantiomers. Two results are demonstrated: (i) ordinary photodissociation of the nonchiral BAB' molecule with linearly polarized light yields identical cross sections for the production of the right-handed (B) and the left-handed (B') fragment; (ii) symmetry breaking can be induced by coherently controlling an interference term. The "control knobs" are the usual relative phases or the delay time between an excitation and a dissociation pulse.

In Section 14.3 below, we introduce a coherent control scheme for enhancing the fraction of one enantiomer, given a racemic mixture of molecules. This scheme, called "laser distillation", is of great practical interest because the effect obtained is substantial, even in the presence of decoherence. An alternative scheme, due to Manz, Fujimura and coworkers [662, 663] is briefly described in Section14.4. The use of adiabatic passage for chiral discrimination is discussed in Section 14.5

Quantum Control of Molecular Processes, Second Edition. Moshe Shapiro and Paul Brumer

14.1
Principles of Electric Dipole Allowed Enantiomeric Control

Below we discuss a number of scenarios for manipulating enantiomer populations. However, these scenarios are, presumably, a small subset of an entire class of scenarios capable of achieving this goal. The key issue then is to establish the general conditions under which the electric dipole electromagnetic field interaction may be used to attain selective control over the population of a desired enantiomer. These rules are established in this section.

Consider a molecule, described by the *total* Hamiltonian (including electrons and nuclei) H_{MT}. As above, H_{MT} has eigenstates describing the L and D enantiomers, denoted $|L_i\rangle$ and $|D_i\rangle$ $(i = 1, 2, 3 \ldots)$ that satisfy

$$\mathcal{I}|L_i\rangle = -|D_i\rangle\,; \quad \mathcal{I}|D_i\rangle = -|L_i\rangle\,, \tag{14.1}$$

where $\mathcal{I}$ is the operator that inverts all space fixed coordinates through the origin. Note that the choice of phase (here minus one) in Eq. (14.1) is arbitrary, and that neither $|L_i\rangle$ nor $|D_i\rangle$ have well-defined parity since they are not eigenstates of $\mathcal{I}$.

The dipole interaction of this molecule with an incident time-dependent electric field $\boldsymbol{E}(t)$ is described by the Hamiltonian:

$$H(\boldsymbol{E}) = H_{MT} - \boldsymbol{d} \cdot \boldsymbol{E}\,. \tag{14.2}$$

Here $\boldsymbol{d}$ is the total dipole operator, including both electron and nuclear contributions, and we have explicitly indicated the dependence of the Hamiltonian on the electric field. Consider now the effect of inversion on H. Noting that $\mathcal{I}$ operates on the coordinates of the molecule, that $\mathcal{I}^\dagger = \mathcal{I}$ and that $[H_{MT}, \mathcal{I}] = 0$, we have [664] that

$$\mathcal{I}H(\boldsymbol{E})\mathcal{I} = H(-\boldsymbol{E})\,, \tag{14.3}$$

where $H(-\boldsymbol{E}) = H_{MT} + \boldsymbol{d} \cdot \boldsymbol{E}$. Further, if we define $U(\boldsymbol{E})$ and $U(-\boldsymbol{E})$ as the propagators corresponding to dynamics under $H(\boldsymbol{E})$ and $H(-\boldsymbol{E})$, respectively, then

$$U(\boldsymbol{E})\mathcal{I} = \mathcal{I}U(-\boldsymbol{E})\,. \tag{14.4}$$

To expose the underlying principles allowing achiral light-induced asymmetric synthesis, consider irradiating a racemic mixture of D and L in its ground electronic state with an electric field $\boldsymbol{E}$ and examine the difference δ between the amount of D and L formed. We consider first the coherent process using transform limited light in the absence of collisions. Then, the difference δ is given by

$$\begin{aligned}\delta = \sum_i P_i \sum_j [|\langle D_j|U(\boldsymbol{E})|D_i\rangle|^2 + |\langle D_j|U(\boldsymbol{E})|L_i\rangle|^2] \\ - [|\langle L_j|U(\boldsymbol{E})|D_i\rangle|^2 + |\langle L_j|U(\boldsymbol{E})|L_i\rangle|^2]\,,\end{aligned} \tag{14.5}$$

where P_i is the probability of state $|L_i\rangle$ and $|D_i\rangle$ in the initial mixed state. (Since the initial state is a racemic mixture, the states $|L_i\rangle$ and $|D_i\rangle$ appear with equal probability.) If $\delta = 0$ then there is no control over the chirality in the scenario defined by $U(\mathbf{E})$.

To determine the conditions under which δ is nonzero, we rewrite Eq. (14.5) as

$$\delta = \sum_i P_i \sum_j [|\langle D_j | U(\mathbf{E}) | D_i \rangle|^2 - |\langle L_j | U(\mathbf{E}) | L_i \rangle|^2] \\ + [|\langle D_j | U(\mathbf{E}) | L_i \rangle|^2 - |\langle L_j | U(\mathbf{E}) | D_i \rangle|^2] \tag{14.6}$$

and recast the second and third terms using

$$|\langle L_j | U(\mathbf{E}) | L_i \rangle|^2 = |\langle D_j | \mathcal{I}^\dagger U(\mathbf{E}) \mathcal{I} | D_i \rangle|^2 = |\langle D_j | U(-\mathbf{E}) | D_i \rangle|^2 , \\ |\langle D_j | U(\mathbf{E}) | L_i \rangle|^2 = |\langle D_j | U(\mathbf{E}) \mathcal{I} | D_i \rangle|^2 = |\langle D_j | \mathcal{I} U(-\mathbf{E}) | D_i \rangle|^2 \\ = |\langle L_j | U(-\mathbf{E}) | D_i \rangle|^2 , \tag{14.7}$$

giving

$$\delta = \sum_i P_i \sum_j [|\langle D_j | U(\mathbf{E}) | D_i \rangle|^2 - |\langle D_j | U(-\mathbf{E}) | D_i \rangle|^2] \\ + [|\langle L_j | U(-\mathbf{E}) | D_i \rangle|^2 - |\langle L_j | U(\mathbf{E}) | D_i \rangle|^2] . \tag{14.8}$$

Equation (14.8), the essential result of this section, provides the general condition under which electric fields, assuming a dipole interaction, can break the right/left symmetry of the initial state, and result in enhanced production of a desired enantiomer. Specifically, the difference between the amount of D and L formed is seen to depend entirely on the difference between the molecular dynamics when irradiated by $\mathbf{E}$ and by $-\mathbf{E}$. Hence, we can state that *a necessary condition for nonzero enantiomeric excess, and the breaking of the left–right symmetry, is that the dynamics depend on the sign of the electric field.* Note that the fact that molecular dynamics can depend on the phase of the incident electric field is well substantiated [665, 666], but its utility for asymmetric synthesis is only evident from this result. Finally, note that the result is completely consistent with symmetry-based arguments that can usefully provide conditions under which δ must equal zero. For example, a racemic mixture of thermally equilibrated molecules is rotationally invariant. Hence any rotation that converts $\mathbf{E}$ to $-\mathbf{E}$ could not, in this case, result in enantiomeric control. In particular, in this case the sum over M_J (where M_J is the component of the reactant's total angular momentum along the direction of laser polarization) implicit in the sum over P_i in Eq. (14.8) would result in $\delta = 0$. By contrast, as discussed below, a racemic mixture of M_J polarized molecules irradiated with linearly polarized light [667] gives nonzero δ. New $\delta \neq 0$ examples emanating from Eq. (14.8) are also expected to display similar non traditional characteristics.

Both qualitative and quantitative applications of Eq. (14.8) are possible. Qualitatively, for example, a traditional scheme where the ground electronic state of L and D are incoherently excited to bound levels of an excited state, gives $\delta = 0$. This is because all processes connecting the initial and final $|L_i\rangle$ and $|D_i\rangle$ states, that is, contributions to the matrix elements in Eq. (14.8), are even in the power of the electric field. Hence, propagation under $\mathbf{E}$ and $-\mathbf{E}$ are identical. By contrast, consider

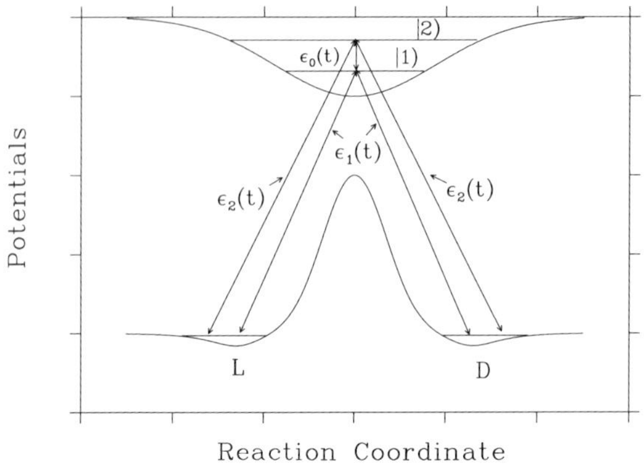

Figure 14.1 The "laser distillation" control scenario discussed in detail in Section 14.3. Two lasers, with pulse envelopes $\varepsilon_1(t)$ and $\varepsilon_2(t)$ couple, by virtue of the dipole operator, the states of the D and L enantiomers to two vib-rotational states $|1\rangle$ and $|2\rangle$ (denoted $|E_1\rangle$ and $|E_2\rangle$ in the text) in the excited electronic manifold. A third laser pulse with envelope $\varepsilon_0(t)$ couples the excited $|E_1\rangle$ and $|E_2\rangle$ states to one another. The system is allowed to absorb a photon and relax back to the ground state. After many such "excitation–relaxation" cycles, a significant enantiomeric excess is obtained, as explained in Section 14.3.

the four-level model scheme in Figure 14.1, and discussed in detail in Section 14.3. When $\varepsilon_0(t) \neq 0$ there exist processes connecting the initial and final $|L\rangle$ and $|D\rangle$ states that are of the form $|L\rangle \rightarrow |1\rangle \rightarrow |2\rangle \rightarrow |D\rangle$ and hence there are terms in Eq. (14.8) that are odd in the power of the electric field. One therefore anticipates the possibility of altering the enantiomeric excess using this combination of pulses, providing the basis for the control results reported later. Further, if $\varepsilon_0 = 0$ then the situation reverts to the case discussed above, where only processes even in the electric field contribute to transitions between the initial $|D\rangle, |L\rangle$ and final $|D\rangle, |L\rangle$ transitions, and hence control over the enantiomeric excess is lost. For this reason, the $\varepsilon_0(t)$ coupling laser is crucial to enantiomeric control. This qualitative picture is substantiated quantitatively later (Section 14.3).

What is required experimentally to achieve this kind of control is the ability to manipulate the phase of the electric field. One possible approach is to use ultrashort pulses [668, 669] that allow defining the overall electric field phase. Specific applications that are somewhat less experimentally demanding are discussed below.

14.2
Symmetry Breaking in the Two-Photon Dissociation of Pure States

Consider a molecule of the type BAB' where B and B' are enantiomers. This molecule possesses a plane of symmetry σ perpendicular to the B–A–B' axis. The operator corresponding to reflection across this plane is denoted σ_h. In order to coherently control the dissociation of this system, we take advantage of the existence of degenerate continuum states which do not possess this reflection symmetry. That is, these molecules possess degenerate continuum states $|E, \boldsymbol{n}, D^-\rangle$ and $|E, \boldsymbol{n}, L^-\rangle$ that correlate asymptotically with the dissociation of the right B' group and left B group, respectively. The collective quantum index $\boldsymbol{n}$ in the states $|E, \boldsymbol{n}, D^-\rangle$ and $|E, \boldsymbol{n}, L^-\rangle$ includes m, the magnetic quantum number of the B or B' fragment. These states are neither symmetric nor antisymmetric with respect to the reflection operator σ_h, although linear combinations of these states might possess this symmetry.

We now consider using the pump-dump scenario described in Section 3.6 above, for BAB' photodissociation. The application of this scenario to the chiral synthesis case is depicted schematically in Figure 14.2. Our aim is to control the relative yield of two product arrangement channels $B{-}A + B'$ and $B + A{-}B'$. That is, we consider $P_{q,\boldsymbol{n}}(E)$, with q labeling either the right ($q = D$) or left ($q = L$) handed product. As in Section 3.6, the product ratio $R_{DL;\boldsymbol{n}} = P_{D,\boldsymbol{n}}(E)/P_{L,\boldsymbol{n}}(E)$ is a function of the delay time $\Delta_d = (t_d - t_x)$ between pulses and the ratio $x = |c_2/c_3|$, the latter by varying the energy of the initial excitation pulse. Active control over the products $B + A - B'$ vs. $B' + A - B$, that is, a variation of $R_{DL;\boldsymbol{n}}$ with Δ_d and x, and hence control over left- vs. right-handed products, will result only if $P_{D,\boldsymbol{n}}(E)$ and $P_{L,\boldsymbol{n}}(E)$ have different functional dependences on the control parameters x and Δ_d.

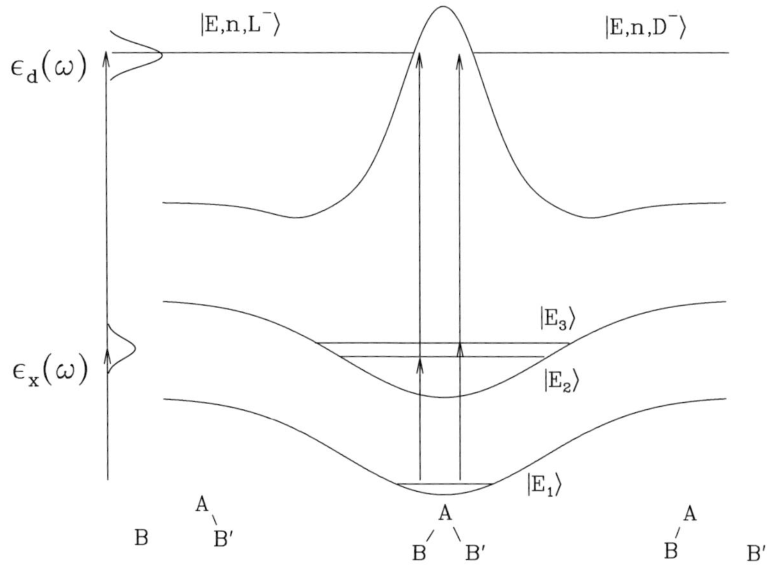

Figure 14.2 A schematic showing the controlled dissociation of the molecule $B{-}A{-}B'$ to yield the $B{-}A + B'$ or the $B + A{-}B'$ products, where B and B' are two enantiomers. The molecule is excited from an initial state $|E_1\rangle$ to a superposition of antisymmetric ($|E_2\rangle$) and symmetric ($|E_3\rangle$) vibrational states belonging to an excited electronic state, by an excitation pulse $\epsilon_x(\omega)$. After an appropriate delay time the molecule is dissociated by a second pulse $\epsilon_d(\omega)$, to the $|E, \boldsymbol{n}, D^-\rangle$ or $|E, \boldsymbol{n}, L^-\rangle$ continuum state.

To show that $P_{D,\boldsymbol{n}}(E)$ may differ from $P_{L,\boldsymbol{n}}(E)$ for the $B'AB$ case, note first that this molecule belongs to the C_s point group. This group possesses only one (hyper) plane of symmetry, denoted σ above, which is defined as the collection of points satisfying the requirement that the $B - A$ distance equals the $A - B'$ distance. Furthermore, we focus upon transitions between electronic states of the same representations, for example, A' to A' or A'' to A'' (where A' denotes the symmetric representation and A'' the antisymmetric representation of the C_s group). It is further assumed that the ground vibronic state belongs to the A' representation.

To obtain control, we choose the intermediate state $|E_3\rangle$ to be *symmetric*, and the intermediate state $|E_2\rangle$ to be *antisymmetric*, with respect to reflection in the σ hyperplane. Hence we must first demonstrate that it is possible to optically excite, simultaneously, both the symmetric $|E_3\rangle$ and antisymmetric $|E_2\rangle$ states from the ground state $|E_1\rangle$. Using Eq. (3.75) we see that this requires the existence of both a symmetric dipole component, denoted $\boldsymbol{d}_s$, and an antisymmetric component, denoted $\boldsymbol{d}_a$, with respect to reflection in the σ hyperplane, because, by the symmetry properties of $|E_3\rangle$ and $|E_2\rangle$,

$$\langle E_3|\boldsymbol{d}\cdot\hat{\boldsymbol{\epsilon}}|E_1\rangle = \langle E_3|\boldsymbol{d}_s\cdot\hat{\boldsymbol{\epsilon}}|E_1\rangle\,, \quad \langle E_2|\boldsymbol{d}\cdot\hat{\boldsymbol{\epsilon}}|E_1\rangle = \langle E_2|\boldsymbol{d}_a\cdot\hat{\boldsymbol{\epsilon}}|E_1\rangle\,. \tag{14.9}$$

We note that the coexistence of symmetric and antisymmetric components of the dipole moment is with respect to σ_h. Since the σ plane rotates with the molecule, the σ_h operation is said to be "body-fixed" (or "molecule-fixed"). Both the body-fixed symmetric $\boldsymbol{d}_s$ and the body-fixed antisymmetric $\boldsymbol{d}_a$ dipole moment components do occur in $A' \to A'$ electronic transitions whenever the geometry of a bent $B'-A-B$ molecule deviates considerably from the points on the σ hyperplane, characterized by the points of equidistance (C_{2v}) geometries (where $\boldsymbol{d}_a = 0$) (see Figure 14.3). The deviation of $\boldsymbol{d}_a$ from zero on the σ plane necessitates going beyond the Franck–Condon approximation, which assumes that the electronic dipole moment does not change as the molecule vibrates. (In the terminology of the theory of vibronic-transitions both symmetric and antisymmetric components can be nonzero due to a Herzberg–Teller intensity borrowing [670] mechanism).

Note also that the dipole moment operator, being a vector, must invert its sign under inversion $\mathcal{I}$. Hence, with respect to $\mathcal{I}$, the dipole moment is always *antisymmetric*. Thus, for the integrals in Eq. (14.9) to be nonzero also requires that $|E_3\rangle$ and $|E_1\rangle$ be of opposite symmetry with respect to inversion. Given the extant conditions

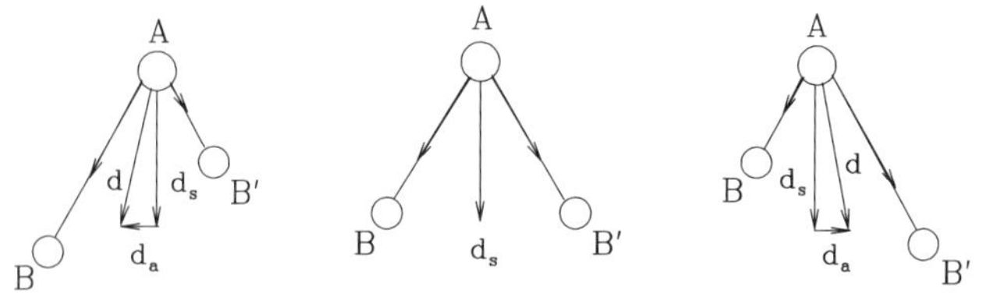

Figure 14.3 The emergence of an antisymmetric dipole component d_a in addition to the symmetric component d_s in a a bent BAB' triatomic molecule as a result of an asymmetric stretching vibration, assuming that the dipole is a vectorial sum of bond dipoles which are proportional to the bond lengths.

on the behavior of $|E_3\rangle$ and $|E_1\rangle$ with respect to the reflection σ_h, the symmetry requirements with respect to $\mathcal{I}$ are most easily accommodated through the rotational components of the $|E_3\rangle$ and $|E_1\rangle$ states.

Thus, the excitation pulse can create a superposition of $|E_2\rangle, |E_3\rangle$ consisting of two states of different reflection symmetry. The resultant superposition possesses no symmetry properties with respect to reflection [671].

We now show that the asymmetry created by this excitation of *nondegenerate* bound states translates into an asymmetry in the probability of populating the *degenerate* $|E, \boldsymbol{n}, D^-\rangle$, $|E, \boldsymbol{n}, L^-\rangle$ continuum states upon subsequent excitation. To do so we examine the properties of the bound–free transition matrix elements $\langle E, \boldsymbol{n}, q^- | \mathrm{d}_{e,g} | E_k\rangle$ that enter into the probability of dissociation (Eq. (3.77)). Note first that although the continuum states $|E, \boldsymbol{n}, q^-\rangle$ are asymmetric with respect to reflection, we can define symmetric and antisymmetric continuum eigenfunctions $|E, \boldsymbol{n}, s^-\rangle$ and $|E, \boldsymbol{n}, a^-\rangle$ via the relations

$$|E, \boldsymbol{n}, D^-\rangle \equiv |E, \boldsymbol{n}, s^-\rangle + |E, \boldsymbol{n}, a^-\rangle/\sqrt{2} \tag{14.10}$$

and

$$|E, \boldsymbol{n}, L^-\rangle \equiv |E, \boldsymbol{n}, s^-\rangle - |E, \boldsymbol{n}, a^-\rangle/\sqrt{2}\,, \tag{14.11}$$

using the fact that $\sigma_h|E, \boldsymbol{n}, D^-\rangle = |E, \boldsymbol{n}, L^-\rangle$.

Consider first the nature of the $\mathrm{d}_q(ij)$ that enter Eq. (3.79), prior to averaging over product scattering angles, and denoted $\mathrm{d}_q(ij; \hat{\boldsymbol{k}})$, where $\hat{\boldsymbol{k}}$ is the scattering direction. Since $|E_3\rangle$ is symmetric and $|E_2\rangle$ is antisymmetric, and adopting the notation $A_{s2} \equiv \langle E, \boldsymbol{n}, s^- | \mathrm{d}_a | E_2\rangle$, $S_{a3} \equiv \langle E, \boldsymbol{n}, a^- | \mathrm{d}_s | E_3\rangle$, and so on we have (see Eq. (3.79))

$$\begin{aligned}
\mathrm{d}_q(33; \hat{\boldsymbol{k}}) &= \sum{}'' \left[|S_{s3}|^2 + |A_{a3}|^2 \pm 2\mathrm{Re}(A_{a3} S_{s3}^*)\right] , \\
\mathrm{d}_q(22; \hat{\boldsymbol{k}}) &= \sum{}'' \left[|A_{s2}|^2 + |S_{a2}|^2 \pm 2\mathrm{Re}(A_{s2} S_{a2}^*)\right] , \\
\mathrm{d}_q(32; \hat{\boldsymbol{k}}) &= \sum{}'' \left[S_{s3} A_{s2}^* + A_{a3} S_{a2}^* \pm S_{s3} S_{a2}^* \pm A_{a3} A_{s2}^*\right] ,
\end{aligned} \tag{14.12}$$

where the plus sign applies for $q = D$, the minus sign applies for $q = L$, and $\mathrm{d}_q(23; \hat{\boldsymbol{k}}) = \mathrm{d}_q^*(32; \hat{\boldsymbol{k}})$. The double prime on the sum denotes a summation over all $q, \boldsymbol{n}$ other than the scattering angles and the product m, where m denotes the projection of the product angular momentum along the axis of laser polarization.

Equation (14.12) takes on a simpler form after angular averaging. The reason for this is that the overall parity of a state with respect to the inversion operation, $\mathcal{I}$, *must* change upon photon absorption since a photon has odd parity. As a result, if we have a single-photon absorption process in which the parity of a vibrational state is unchanged, then the parity of the rotational states must change, and vice versa. Close examination of Eq. (14.12) reveals that the S_{s3}^* term does not involve a change in the parity of the vibrational state, whereas the A_{a3} term does. As a result, the rotational wave functions associated with each term must have opposing parities and the angular integral of the product must vanish. (Note that Appendix 14.A discusses the form of these matrix elements and their products). The same goes for the $A_{s2}S^*_{a2}$ term. In a similar manner the $S_{s3}A^*_{s2} + A_{a3}S^*_{a2}$ term vanishes in the $d_q(32)$ interference term. By contrast, the $\pm S_{s3}S^*_{a2} \pm A_{a3}A^*_{s2}$ terms do not vanish upon angular integration since they correspond to final rotational states that have the same parity.

As a consequence, the net result is that, after angular averaging, Eq. (14.12) becomes

$$\begin{aligned} d_q(33) &= \sum{}' \left[|S_{s3}|^2 + |A_{a3}|^2\right] , \\ d_q(22) &= \sum{}' \left[|A_{s2}|^2 + |S_{a2}|^2\right] , \\ d_q(32) &= \sum{}' \pm \left[S_{s3}S^*_{a2} + A_{a3}A^*_{s2}\right] , \end{aligned} \tag{14.13}$$

where a single prime on the sum indicates that the sum over product m is not carried out.

These equations display two noteworthy features:

(i) $d_L(jj) = d_D(jj)$, $j = 2, 3$, that is, lacking interference, no discrimination between the left handed and right handed products is possible.

(ii) $d_L(23) \neq d_D(23)$, that is, laser controlled symmetry breaking, which depends upon $d_q(23)$ in accordance with Eq. (3.78), is possible. As noted below, this type of discrimination is possible only if we select of the direction of the angular momentum of the products (m-polarization).

To demonstrate the extent of expected control, as well as the effect of m summation, we considered a model of enantiomer selectivity, that is, HOH photodissociation in three dimensions, where the two hydrogens are assumed distinguishable,

$$H_aO + H_b \leftarrow H_aOH_b \rightarrow H_a + OH_b .$$

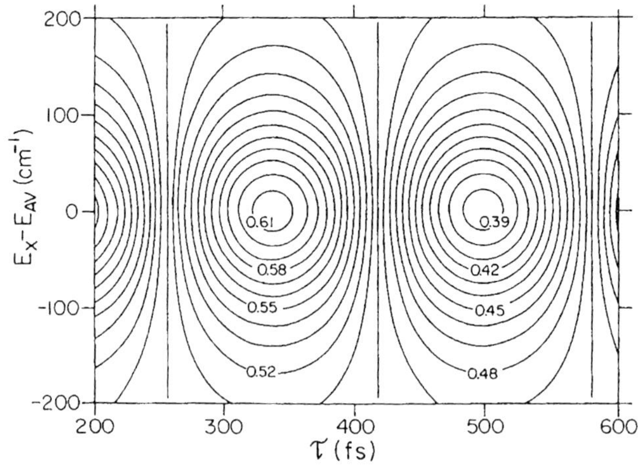

Figure 14.4 Contour plot of percent $H_aO + H_b$ (as distinct from $H_a + OH_b$) in H_aOH_b photodissociation. Ordinate is the detuning from $E_{av} = (E_3 - E_2)/2$ and the abscissa is the time delay between pulses. Taken from [25, Figure 9]. (Reprinted with permission. Copyright 2000 Elsevier.)

The computation is done using the formulation and computational methodology of [672]. In Appendix 14.A we briefly summarize the angular momentum algebra and some other details involved in performing this three dimensional quantum calculation on triatomic photodissociation. We also discuss the loss of control, in this scenario, with m averaging.

These equations, in conjunction with the artificial channel method [8] for computing the t matrix elements, were used to compute the ratio R_{DL} of the $H_aO + H_b$ (as distinct from the $H_a + OH_b$) product in a fixed m state. Specifically, Figure 14.4 demonstrates what happens when one first excites the superposition of symmetric plus asymmetric vibrational modes $[(1,0,0) + (0,0,1)]$ with $J_i = J_k = 0$ in the ground electronic state, and then photodissociates the molecule using vacuum UV light at 70 700 cm^{-1} to the B state using a pulse width of 200 cm^{-1}. The results show that varying the time delay between pulses allows for controlled variation of P_D from 61 to 39%. This variation is significant since it reveals the symmetry breaking arising within this scenario.

14.3
Purification of Racemic Mixtures by "Laser Distillation"

In Section 14.2 we showed that coherent control techniques can be used to direct the photodissociation of a BAB' molecule to yield an excess of a desired B or B' enantiomer. Throughout the treatment we assumed that prior to dissociation the BAB' molecule exists in single quantum state $|E_i, J_i, p_i\rangle$.

While the above process is of great scientific interest, practically speaking, we usually want to separate a *racemic mixture* of the B and B' enantiomers, that is, our initial state is a racemate. If we were to use the scenario of Section 14.2 to accomplish this one would have first to prepare the BAB' adduct in a pure state. Since the preparation of the BAB' adduct, and especially its separation from the BAB and $B'AB'$ adducts that would inevitably accompany it, is not a trivial task, it is preferable to find control methods that could separate the B and B' racemic mixture directly. In this section we outline a method that can achieve this much more ambitious task. The essential principles of this method remain the same as in Section 14.2, that is, excitation of a superposition of symmetric and antisymmetric states with respect to σ_h, the reflection operation.

Consider then a molecular system composed of a pair of stable nuclear configurations, denoted L and D, with L being the (distinguishable) mirror image of D. Note that the electronic Hamiltonian H_e commutes with σ_h, hence, the potential energy surfaces, which are the eigenvalues of the electronic Schrödinger equation at all nuclear configurations, must be symmetric with respect to σ_h.

Since L and D are assumed stable, it follows that the ground potential energy surface must possess a sufficiently high barrier at nuclear coordinates separating L and D such that the rate of interconversion between them by tunneling is negligible. By contrast, L and D need not be stable on an excited potential energy surface. To this end, we assume that there is at least one excited potential surface, denoted G, which possesses a potential well midway between the L and the D geometries (see also Figure 14.1). (A number of molecules expected to be of this type are tabulated in [657] and a number of examples are discussed below.) Hence, the interconversion between L and D on the excited surface G is expected to be very facile.

A direct consequence of the potential well midway between the L and the D geometries is the existence of stable vibrational eigenstates. Because of the symmetry of G, the vibrational eigenstates must be either symmetric or antisymmetric with respect to σ_h.

The procedure that we propose in order to enhance the concentration of a particular enantiomer when starting with a racemic mixture, that is, to "purify" the mixture, is as follows. The mixture of statistical (racemic) mixture of L and D is irradiated with a specific sequence of three coherent laser pulses, as described below. These pulses excite a coherent superposition of symmetric and antisymmetric vibrational states of G. After each pulse the excited system is allowed to relax back to the ground electronic state by spontaneous emission or any other nonradiative process. As shown below, by allowing the system to go through many irradiation and relaxation cycles the concentration of the selected enantiomer L or D is enhanced, depending on the laser characteristics. This scenario is called "laser distillation" of chiral enantiomers.

We note at the outset that detailed angular momentum considerations show that if the three incident lasers are of the same polarization then control results only if we do not average over M_J, the projection of the total angular momentum of the reactant along the z-axis (chosen as the direction of laser polarization). In particular, enantiomeric enhancement of one enantiomer from molecules in state M_J is exactly counterbalanced by enantiomeric enhancement of the other enantiomer by molecules in state $-|M_J|$. Hence, enantiomeric control in this scenario requires prior M_J selection of the molecules. This scenario is discussed below, but results are also provided for the case of three lasers of perpendicular polarization, where M_J averaging is nondestructive.

Consider then a molecule with Hamiltonian H_M, in the presence of a series of laser pulses. (In general we may deal with lasers which are not fully coherent but for simplicity we focus here on transform limited pulses of linearly polarized light). The treatment is in accord with Chapter 1, Eq. (1.5), where the interaction between the molecule and radiation is given by

$$H_{MR}(t) = -\boldsymbol{d} \cdot \boldsymbol{E}(t) = -2\boldsymbol{d} \cdot \sum_k \mathrm{Re}\left[\hat{\epsilon}_k \varepsilon_k(t) \exp(-i\omega_k t)\right] . \tag{14.14}$$

Here $\varepsilon_k(t)$ is the pulse envelope, ω_k is the central laser frequency and $\hat{\epsilon}_k$ is the polarization direction. Expanding $|\Psi(t)\rangle$ in eigenstates $|E_j\rangle$ of the molecular Hamiltonian [i.e., $H_M|E_j\rangle = E_j|E_j\rangle$]:

$$|\Psi(t)\rangle = \sum_j b_j \exp -i E_j t/\hbar |E_j\rangle , \tag{14.15}$$

and substituting Eq. (14.15) into the time-dependent Schrödinger equation gives the standard set of coupled equations:

$$\dot{b}_i = \frac{-i}{\hbar} \sum_j b_j \exp(-i\omega_{ji}t)\langle E_i | H_{MR}(t) | E_j \rangle \,, \tag{14.16}$$

where $\omega_{ji} = (E_j - E_i)/\hbar$.

As an example of an effective control scenario, consider the molecules D and L in their ground electronic states and in vib-rotational states $|E_D\rangle$ and $|E_L\rangle$, of energy $E_D = E_L$. We choose $\boldsymbol{E}(t)$ so as to excite the system to two eigenstates $|E_1\rangle$ and $|E_2\rangle$ of the electronically excited potential surface G. The states $|E_1\rangle$ and $|E_2\rangle$ are also coupled by an additional laser field (see Figure 14.1).

Specifically, we choose $\boldsymbol{E}(t)$ to be composed of three linearly polarized light pulses (all of the same polarization),

$$\boldsymbol{E}(t) = \sum_{k=0,1,2} 2\mathrm{Re}\left[\varepsilon_k(t)\exp(-i\omega_k t)\right]\hat{\epsilon}_k \,, \tag{14.17}$$

with ω_0 in near-resonance with $\omega_{2,1} \equiv (E_2 - E_1)/\hbar$, ω_1 is chosen to be near resonant with $\omega_{1,D} \equiv (E_1 - E_D)/\hbar$, and ω_2 near resonant with $\omega_{2,D} \equiv (E_2 - E_D)/\hbar$ (see Figure 14.1). In this case, only four molecular states are relevant and Eq. (14.15) becomes

$$\begin{aligned} |\Psi\rangle &= b_D(t)\exp -iE_D t/\hbar |E_D\rangle + b_L(t)\exp -iE_L t/\hbar |E_L\rangle \\ &\quad + b_1(t)\exp -iE_1 t/\hbar |E_1\rangle + b_2 \exp -iE_2 t/\hbar |E_2\rangle \,. \end{aligned} \tag{14.18}$$

Equation (14.16), in the rotating wave approximation, is then given by

$$\begin{aligned} \dot{b}_1 &= i\exp(i\Delta_1 t)\left[\Omega^*_{D,1}b_D + \Omega^*_{L,1}b_L\right] + i\exp(-i\Delta_0 t)\Omega_0^* b_2 \,, \\ \dot{b}_2 &= i\exp(i\Delta_2 t)\left[\Omega^*_{D,2}b_D + \Omega^*_{L,2}b_L\right] + i\exp(i\Delta_0 t)\Omega_0 b_1 \,, \\ \dot{b}_D &= i\exp(-i\Delta_1 t)\Omega_{D,1}b_1 + i\exp(-i\Delta_2 t)\Omega_{D,2}b_2 \,, \\ \dot{b}_L &= i\exp(-i\Delta_1 t)\Omega_{L,1}b_1 + i\exp(-i\Delta_2 t)\Omega_{L,2}b_2 \,, \end{aligned} \tag{14.19}$$

where $\Omega_{i,j}(t) \equiv d^{(j)}_{i,j}\varepsilon_j(t)/\hbar$, $\Omega_0 \equiv d^{(0)}_{2,1}\varepsilon_0(t)/\hbar$, $\Delta_j \equiv \omega_{j,D} - \omega_1$, $\Delta_0 \equiv \omega_{2,1} - \omega_0$, where $d^{(k)}_{i,j} \equiv \langle E_i | \boldsymbol{d}\cdot\hat{\boldsymbol{\varepsilon}}_k | E_j\rangle$, with $i = D, L$; $k = 0, 1, 2$ and $j = 1, 2$.

The essence of the laser distillation process lies in choosing the laser of central frequency ω_1 so that it excites the system to a state $|E_1\rangle$ which is *symmetric* with respect to the reflection operation σ_h, and to a state $|E_2\rangle$ which is *antisymmetric* with respect to σ_h. By contrast, $|E_D\rangle$ and $|E_L\rangle$ do not share these symmetries but are related to one another through reflection (i.e., $\sigma_h|E_D\rangle = |E_L\rangle$, $\sigma_h|E_L\rangle = |E_D\rangle$ whereas $\sigma_h|E_1\rangle = |E_1\rangle$, $\sigma_h|E_2\rangle = -|E_2\rangle$).

To consider the nature of the "Rabi frequencies" Ω in Eq. (14.19) we rewrite $|E_D\rangle$ and $|E_L\rangle$ in terms of a symmetric state $|S\rangle$ and an antisymmetric state $|A\rangle$:

$$\begin{aligned} |E_D\rangle &= |A\rangle + |S\rangle \,, \\ |E_L\rangle &= |A\rangle - |S\rangle \,. \end{aligned} \tag{14.20}$$

In addition to their symmetry properties with respect to σ_h, we choose the $|S\rangle$ and $|A\rangle$ states to be respectively symmetric and antisymmetric under the inversion operation $\mathcal{I}$. Coupled with the fact that the dipole operator must be antisymmetric with respect to $\mathcal{I}$, the relevant matrix elements satisfy the following relations:

$$\begin{aligned}
\langle 1|d^{(1)}|D\rangle &= \langle 1|d^{(1)}|A+S\rangle = \langle 1|d^{(1)}|A\rangle \,, \\
\langle 1|d^{(1)}|L\rangle &= \langle 1|d^{(1)}|A-S\rangle = \langle 1|d^{(1)}|A\rangle \,, \\
\langle 2|d^{(2)}|D\rangle &= \langle 2|d^{(2)}|A+S\rangle = \langle 2|d^{(2)}|S\rangle \,, \\
\langle 2|d^{(2)}|L\rangle &= \langle 2|d^{(2)}|A-S\rangle = -\langle 2|d^{(2)}|S\rangle \,.
\end{aligned} \tag{14.21}$$

That is,

$$\Omega_{D,1} = \Omega_{L,1} \,, \quad \Omega_{D,2} = -\Omega_{L,2} \,. \tag{14.22}$$

Given Eq. (14.22), Eq. (14.19) becomes

$$\begin{aligned}
\dot{b}_1 &= i\exp(i\Delta_1 t)\Omega^*_{D,1}[b_D + b_L] + i\exp(-i\Delta_0 t)\Omega_0^* b_2 \,, \\
\dot{b}_2 &= i\exp(i\Delta_2 t)\Omega^*_{D,2}[b_D - b_L] + i\exp(i\Delta_0 t)\Omega_0 b_1 \,, \\
\dot{b}_D &= i\exp(-i\Delta_1 t)\Omega_{D,1} b_1 + i\exp(-i\Delta_2 t)\Omega_{D,2} b_2 \,, \\
\dot{b}_L &= i\exp(-i\Delta_1 t)\Omega_{D,1} b_1 - i\exp(-i\Delta_2 t)\Omega_{D,2} b_2 \,.
\end{aligned} \tag{14.23}$$

The essence of optically controlled enantioselectivity in this scenario lies in Eq. (14.22) and the effect of these relationships on the dynamical equations for the level populations (Eq. (14.23)). Note specifically that the equation for $\dot{b}_D(t)$ is different than the equation for $\dot{b}_L(t)$, due to the sign difference in the last term in Eq. (14.23). Although not sufficient to ensure enantiomeric selectivity, the ultimate consequence of this difference is that populations of $|E_D\rangle$ and $|E_L\rangle$ after laser excitation are different when there is radiative coupling between levels $|E_1\rangle$ and $|E_2\rangle$.

Note, in accord with Section 14.1, the behavior of Eq. (14.23) under the transformation $\boldsymbol{E} \to -\boldsymbol{E}$. Specifically, changing $\boldsymbol{E}$ to $-\boldsymbol{E}$ means changing all $\varepsilon_j(t)$ to $-\varepsilon_j(t)$. Doing so, and defining $b_1' = -b_1$ and $b_2' = -b_2$ converts Eq. (14.23) into

$$\begin{aligned}
\dot{b}_1' &= i\exp(i\Delta_1 t)\Omega^*_{D,1}[b_D + b_L] - i\exp(i\Delta_0 t)\Omega_0^* b_2' \,, \\
\dot{b}_2' &= i\exp(i\Delta_2 t)\Omega^*_{D,2}[b_D - b_L] - i\exp(-i\Delta_0 t)\Omega_0 b_1' \,, \\
\dot{b}_D &= i\exp(-i\Delta_1 t)\Omega_{D,1} b_1' + i\exp(-i\Delta_2 t)\Omega_{D,2} b_2' \,, \\
\dot{b}_L &= i\exp(-i\Delta_1 t)\Omega_{D,1} b_1' - i\exp(-i\Delta_2 t)\Omega_{D,2} b_2' \,.
\end{aligned} \tag{14.24}$$

Clearly, Eq. (14.24) is the same as Eq. (14.23) barring the change of sign in the Ω_0 terms. Thus, the solution to Eq. (14.23) depends on the sign of $\boldsymbol{E}$ when $\varepsilon_0 \neq 0$. Hence, by the argument in Section 14.1, this scenario allows for chirality control when $\varepsilon_0(t) \neq 0$. For $\varepsilon_0(t) = 0$ the Eq. (14.24) is the same as Eq. (14.23) so that enantiomer control is not possible.

To obtain quantitative estimates for the extent of obtainable control we consider results for model cases assuming Gaussian pulses

$$\varepsilon_k(t) = \epsilon_k \exp\left[-\left(\frac{t-t_k}{\alpha_k}\right)^2\right], \quad (k = 0, 1, 2), \tag{14.25}$$

and system parameters $\langle 1|d^{(1)}|D\rangle = \langle 1|d^{(1)}|L\rangle = \langle 2|d^{(2)}|L\rangle = -\langle 2|d^{(2)}|D\rangle = 1\,\text{a.u.}$, $\langle 1|d^{(0)}|2\rangle = 1\,\text{a.u.}$, $\omega_{2,1} = 100\,\text{cm}^{-1}$ and $\Delta_0 = 0$. Figure 14.5 displays the final probabilities $P_D = |b_D(\infty)|^2$, $P_L = |b_L(\infty)|^2$ of population in $|E_D\rangle$ and $|E_L\rangle$, after a single pulse, for a variety of pulse parameters. Results are shown for various values of $\Delta_1 = \omega_{1,D} - \omega_1$ at various different pulse powers assuming that one starts solely with D, solely with L, or with a racemic mixture of both enantiomers. Clearly, for particular parameters, one can significantly enhance the population of one chiral enantiomer over the other. For example, for $\Delta_1 = -115\,\text{cm}^{-1}$, $\epsilon_0 = \epsilon_1 = 4.5 \times 10^{-4}$, a racemic mixture of D and L can be converted, after a single pulse, to a enantiomerically enriched mixture with predominantly D.

Control is strongly affected by the relative phase θ of the ε_1 and ε_0 fields, as shown in Figure 14.6. Here it is clear that changing θ by π interchanges the dynamical evolution of the L and D enantiomers.

Although not immediately obvious, this control scenario relies entirely upon quantum interference effects. To see this note that in the absence of an $\varepsilon_0(t)$ pulse, excitation from $|D\rangle$ or $|L\rangle$ to level $|E_i\rangle$, for example, occurs via one-photon excitation with $\varepsilon_i(t)$, $i = 1, 2$. In this case, as noted above, there is no chiral control. By

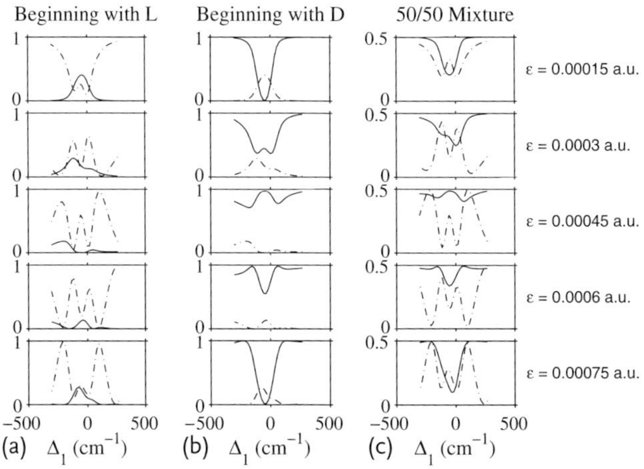

Figure 14.5 Probabilities of populating the $|E_D\rangle$ (solid lines) and $|E_L\rangle$ (dot-dash lines) after laser excitation, but prior to relaxation, as a function of the detuning Δ_1. Three different cases are shown, corresponding to three different initial conditions: (i) only state $|E_L\rangle$ is initially populated; (ii) only state $|E_D\rangle$ is initially populated; (iii) a statistical mixture made up of equal shares of the $|E_D\rangle$ and $|E_L\rangle$ states is initially populated. Results are shown for five different $\epsilon_1 = \epsilon_0 \equiv \epsilon$ laser peak electric fields, where Gaussian pulses are assumed with $\alpha_0 = \alpha_1 = 0.15$ ps, and $t_0 = t_1$.

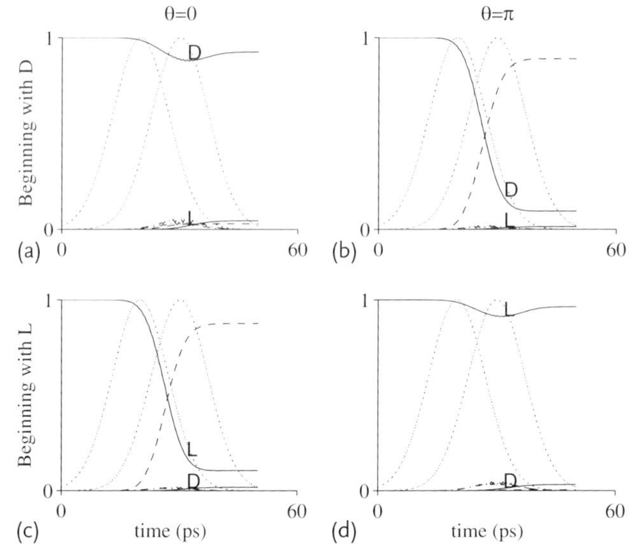

Figure 14.6 The time evolution of the enantiomeric populations for two different relative phases θ between the ϵ_1 and ϵ_0 beams. Solid line: population in the *D* or *L* enantiomer; dotted line: the ϵ_1 and ϵ_0 laser pulses; dashed line: excited state population in levels $|E_1\rangle + |E_2\rangle$.

contrast, with nonzero $\varepsilon_0(t)$, there is an additional (interfering) route to $|E_i\rangle$, that is, a two-photon route using $\varepsilon_j(t)$ excitation to level $|E_j\rangle$, $j \neq i$, followed by an $\varepsilon_0(t)$ induced transition from $|E_j\rangle$ to $|E_i\rangle$. The one- and two-photon routes interfere and, as implied in Section 3.4.3, allow for symmetry breaking transitions.

The computation, that results in Figure 14.5, which gives the result of a single pulse, provides input into a calculation of the overall result. In the overall process one starts with an incoherent mixture of N_D molecules of type *D* and N_L molecules of type *L*. In the first step the system is excited, as above, with a laser pulse sequence. In the second step, the system collisionally and radiatively relaxes so that all the population returns to the ground state to produce an incoherent mixture of $|E_L\rangle$ and $|E_D\rangle$. This pair of steps is then repeated until the populations of $|E_L\rangle$ and $|E_D\rangle$ reach convergence.

To obtain the result computationally note that the population after laser excitation, but before relaxation, consists of the weighted sum of the results of two computations: N_D times the results of laser excitation starting solely with molecules in $|E_D\rangle$, plus N_L times the results of laser excitation starting solely with molecules in $|E_L\rangle$. If $P_{D\leftarrow D}$ and $P_{L\leftarrow D}$ denote the probabilities of $|E_D\rangle$ and $|E_L\rangle$ resulting from laser excitation assuming the first of these initial conditions, and $P_{D\leftarrow L}$ and $P_{L\leftarrow L}$ for the results of excitation following from the second of these initial conditions, then the populations of $|E_D\rangle$ and $|E_L\rangle$ after laser excitation of the mixture are $N_D P_{D\leftarrow D} + N_L P_{D\leftarrow L}$ and $N_D P_{L\leftarrow D} + N_L P_{L\leftarrow L}$, respectively. The remainder of the population, $N_D[1 - P_{D\leftarrow D} - P_{L\leftarrow D}] + N_L[1 - P_{D\leftarrow L} - P_{L\leftarrow L}]$, is in the upper two levels $|E_1\rangle$ and $|E_2\rangle$. Relaxation from levels $|E_1\rangle$ and $|E_2\rangle$ then follows, with the excited population dividing itself equally between $|E_D\rangle$ and $|E_L\rangle$. The resultant populations $\mathcal{N}_D$ and $\mathcal{N}_L$ in ground state $|E_D\rangle$ and $|E_L\rangle$ is then:

$$\mathcal{N}_D = 0.5 N_D [1 + P_{D\leftarrow D} - P_{L\leftarrow D}] + 0.5 N_L [1 + P_{D\leftarrow L} - P_{L\leftarrow L}] ,$$
$$\mathcal{N}_L = 0.5 N_D [1 + P_{L\leftarrow D} - P_{D\leftarrow D}] + 0.5 N_L [1 + P_{L\leftarrow L} - P_{D\leftarrow L}] . \quad (14.26)$$

The sequence of laser excitation followed by collisional relaxation and radiative emission is then iterated to convergence. In the second step, for example, the populations in Eq. (14.26) are taken as the initial populations for two independent computations, one assuming a population of $\mathcal{N}_D$ in $|E_D\rangle$, with $|E_L\rangle$ unpopulated, and the second assuming a population of $\mathcal{N}_L$ in $|E_L\rangle$, with $|E_D\rangle$ unpopulated.

Clearly, convergence is obtained when the populations, post relaxation, are the same as those prior to laser excitation, that is, when $\mathcal{N}_D = N_D$, and $\mathcal{N}_L = N_L$. These conditions reduce to

$$N_D(1 - P_{D\leftarrow D} + P_{L\leftarrow D}) = N_L(1 + P_{D\leftarrow L} - P_{L\leftarrow L}) . \quad (14.27)$$

If the total population is chosen to be normalized ($N_D + N_L = 1$), then the final probabilities $\mathcal{P}_D, \mathcal{P}_L$ of populating states $|E_D\rangle$ and $|E_L\rangle$ are

$$\mathcal{P}_D = \frac{1 + P_{D\leftarrow L} - P_{L\leftarrow L}}{2 - P_{D\leftarrow D} + P_{L\leftarrow D} + P_{D\leftarrow L} - P_{L\leftarrow L}} ,$$
$$\mathcal{P}_L = \frac{1 - P_{D\leftarrow D} + P_{L\leftarrow D}}{2 - P_{D\leftarrow D} + P_{L\leftarrow D} + P_{D\leftarrow L} - P_{L\leftarrow L}} , \quad (14.28)$$

and the equilibrium enantiomeric branching ratio is simply,

$$R_{D,L} \equiv \frac{\mathcal{P}_D}{\mathcal{P}_L} = \frac{1 + P_{D\leftarrow L} - P_{L\leftarrow L}}{1 - P_{D\leftarrow D} + P_{L\leftarrow D}} . \quad (14.29)$$

Results for the converged probabilities for the cases depicted in Figure 14.5, are shown in Figure 14.7. The results clearly show substantially enhanced enantiomeric ratios at various choices of control parameters. For example, at $\epsilon_0 = \epsilon_1 = 1.5 \times 10^{-3}$, tuning Δ_1 to 50 cm^{-1} gives a preponderance of L whereas tuning to the $\Delta_1 = -125$ cm^{-1} gives more D.

Numerous other parameters in this system, such as the pulse shape, time delay between pulses, pulse frequencies and pulse powers, and so on can be varied to affect the final L to D ratio [658] resulting in a very versatile approach to asymmetric synthesis.

Finally, note that although only two ground state levels are included, the method applies equally well when a large number of ground state levels are included. In this case, relaxation will be amongst all of these ground state levels, but the proposed scenario, tuned to the above set of transitions, will "bleed" population from one M_J level of the desired enantiomer. As relaxation refills this level it will continue to be pumped over to the other enantiomer, with the overall effect that the major amount of the population will be transferred from one enantiomer to the other.

As a realization of the above scheme we now examine [673] the case of enantiomer control in dimethylallene, a molecule shown in Figure 14.8. Note that, at

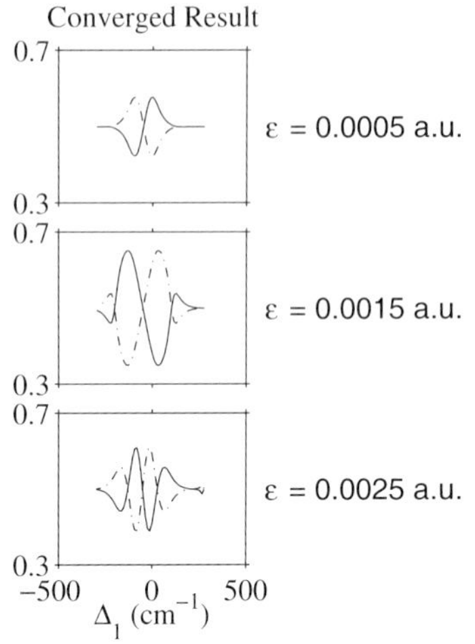

Figure 14.7 Results for laser distillation after a convergent series of steps comprised of radiative excitation, followed by collisional and radiative relaxation. Shown are the results at three different field strengths.

equilibrium in the ground state, the H–C–CH_3 groups at both ends of the molecule lie on planes that are perpendicular to one another, resulting a molecule that is chiral. By contrast, in the excited state, the C=C double bond breaks, allowing for rotation of one plane relative to the other. Cuts through the ground and first two excited state potential energy surfaces for this molecule along the α and θ coordinates (see Figure 14.8) are shown in Figure 14.9. The potentials show the features required for control in this scenario, that is, a minimum in the excited state potential surface at the geometry corresponding to the potential energy maximum on the ground state potential.

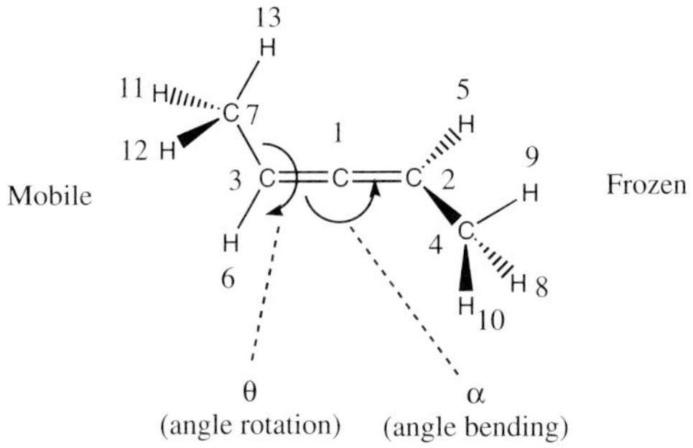

Figure 14.8 The geometry of the 1,3-dimethyl-allene and the two angles θ and α that were varied to scan the potential energy surface. Here θ is the dihedral angle between the H_3C–C=C and the C=C–CH_3 planes and α is the C–C–C bending angle, here shown by an arrow that brings the H_3C–C–H out of the plane of the paper. From [673, Figure 2]. (Reprinted with permission. Copyright 2001 American Chemical Society.)

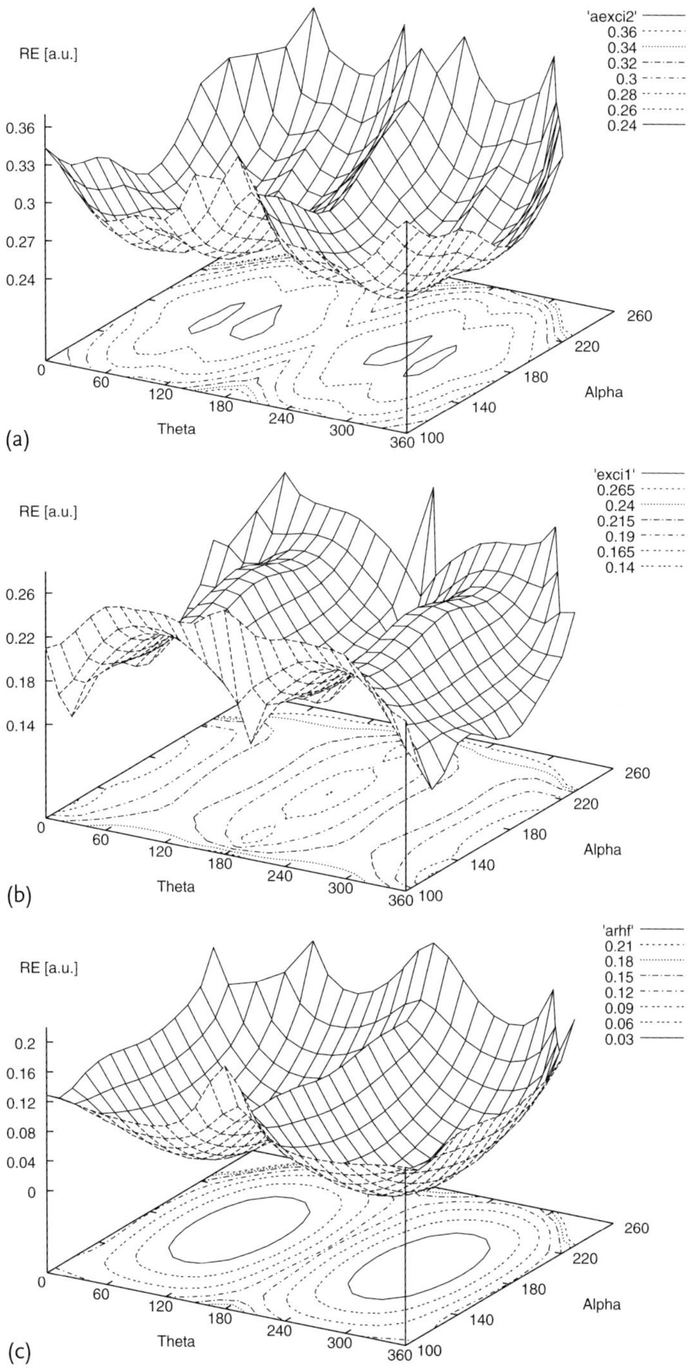

Figure 14.9 Potential energy surfaces for 1,3-dimethylallene. Only the in-plane surfaces for the ground and first two excited electronic states are shown. Taken from [673, Figure 4]. (Reprinted with permission. Copyright 2001 American Chemical Society.)

The results of a computation [659] on the control of L vs. D 1,3-dimethylallene are shown in Figure 14.10. Outstanding enantiomeric control over the dimethylallene enantiomers is evident for a wide variety of powers. For example, a most impressive result is achieved for $\Delta_1 = 0.0986\,\text{cm}^{-1}$ and $\epsilon_0 = 1.5 \times 10^{-4}$ a.u., $\epsilon_1 = \epsilon_2 = 4.31 \times 10^{-5}$ a.u., corresponding to laser powers of 7.90×10^8 and $6.52 \times 10^7\,\text{W/cm}^2$ respectively. Here a racemic mixture of dimethylallene in a specific J, M_J, λ state can be converted, after a series of pulses, to a mixture of dimethylallene, containing 92.7% of the D-dimethylallene in this state. (Here λ is the projection of the total angular momentum J along an axis fixed in the molecule.) Similarly, detuning to $\Delta_1 = -0.0986\,\text{cm}^{-1}$ results in a similar enhancement of L-dimethylallene. Slightly lower extremes of control are seen to be achievable for the two other laser powers shown. Further, control was achievable to field strengths down to $10^4\,\text{W/cm}^2$. Note, however, that this computation neglects the competitive process of internal conversion, discussed later below.

It is of some interest to note the character of the eigenstates $|E_1\rangle$ and $|E_2\rangle$ that contribute to these results; they are shown in Figure 14.11. Clearly they are states with considerable vibrational energy, so that they are broad enough in configuration space to overlap the ground electronic state, ground vibrational state wave functions. If this is not the case then the dipole matrix elements are too small to allow control at reasonable laser intensities.

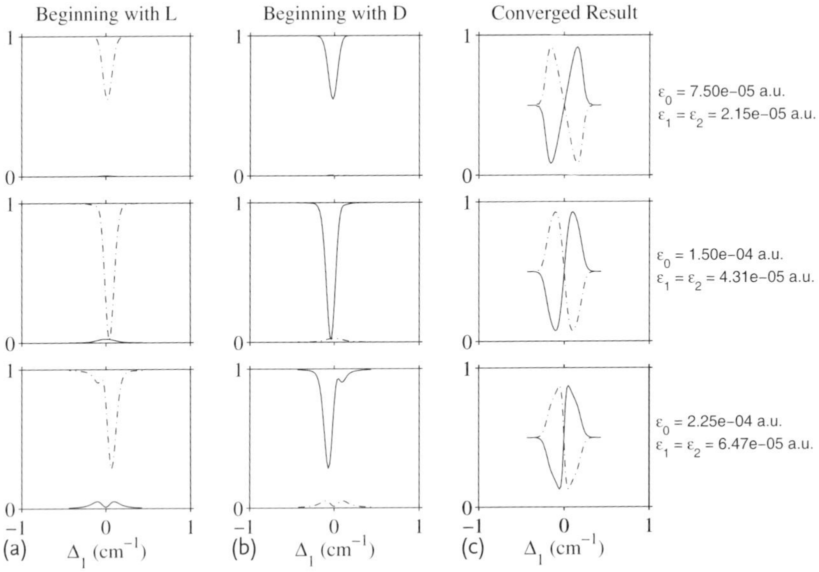

Figure 14.10 Control over dimethylallene enantiomer populations as a function of the detuning Δ_1 for various laser powers. The first column corresponds to probabilities of L (dot-dash curves) and D (solid curves) after a single laser pulse, assuming that the initial state is all L. The second column is similar, but for an initial state which is all D. (c) corresponds to the probabilities L and D after repeated excitation-relaxation cycles, as describe in the text. This is a corrected version of [659, Figure 2]. (Reprinted with permission. Copyright 2001 American Institute of Physics.)

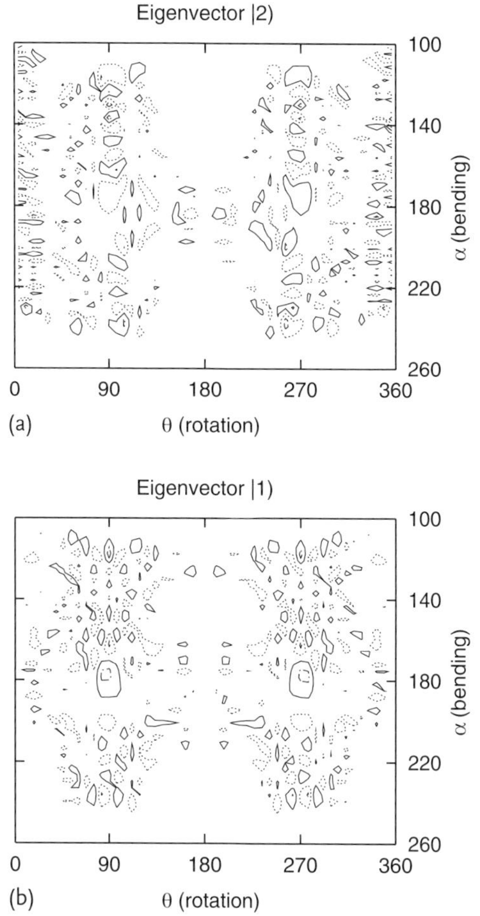

Figure 14.11 Contour plots of $|E_1\rangle$ and $|E_2\rangle$ where dash-dash lines $=$ 0.012 a.u., dot-dot lines $=$ 0.0004 a.u., solid lines $=$ -0.004 a.u. and dot-dash lines $=$ -0.012 a.u. Note that $|E_1\rangle$ is symmetric with respect to reflection and $|E_2\rangle$ is antisymmetric. Reflection here corresponds to changing ($\alpha \rightarrow 360° - \alpha$). Taken from [659, Figure 1]. (Reprinted with permission. Copyright 2001 American Institute of Physics.)

The primary experimental difficulty associated with this scenario is the requirement to isolate a particular subset of M_J levels, in order to avoid cancellation of M_J and $-|M_J|$ control. That is, from the viewpoint of the M_J structure, this scenario is associated with the level structure shown in Figure 14.12.

To remove this restriction we consider another scenario [667] where all of the three laser polarizations, $\hat{\epsilon}_0$, $\hat{\epsilon}_1$ and $\hat{\epsilon}_2$, are perpendicular to one another. This laser arrangement now allows for transitions between different M_J levels. The first few of these levels is shown in Figure 14.13. Under these circumstances, control survives averaging over M_J levels [667].

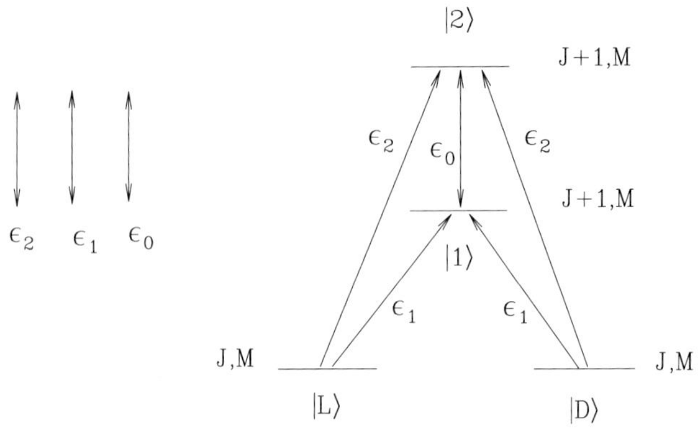

Figure 14.12 Schematic level diagram emphasizing the $M \equiv M_J$ features of the four level scheme in Figure 14.1.

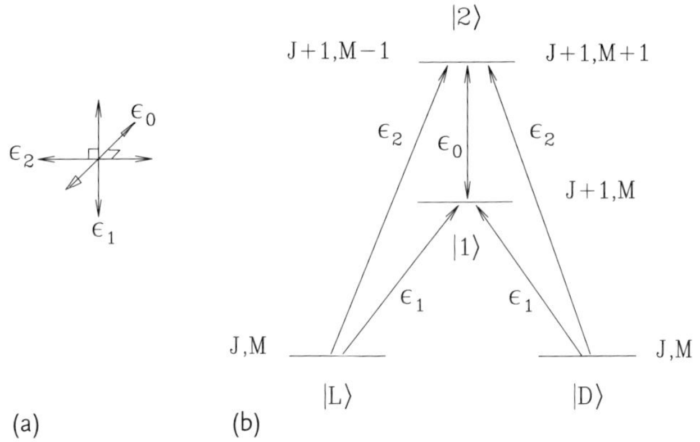

Figure 14.13 Schematic level diagram emphasizing the $M \equiv M_J$ coupling where three lasers of perpendicular polarization irradiate the D and L enantiomers. Only the first five levels that are coupled by these lasers are shown.

Sample results for the three laser case with perpendicular polarizations are shown in Figure 14.14, first row, where extensive control is evident. Here, even with M_J averaging, one can choose to convert the racemic mixture to over 90% of the *L* enantiomer, or of the *D* enantiomer, depending on the detuning. In this case the 1,3-dimethylallene was treated as an asymmetric top and averaging was carried out over all M_J levels.

A realistic model of dimethylallene control must also recognize the possibility of internal conversion to the ground state. In this process the electronically excited molecule undergoes a radiationless transition to the ground electronic state, leaving a highly vibrationally excited species. Only a few estimates or measurements of the internal conversion time scales for molecules are available [674, 675] and dimethylallene has not been explored. Further, after internal conversion one expects, in the dimethylallene case, that the excited molecule subsequently dissoci-

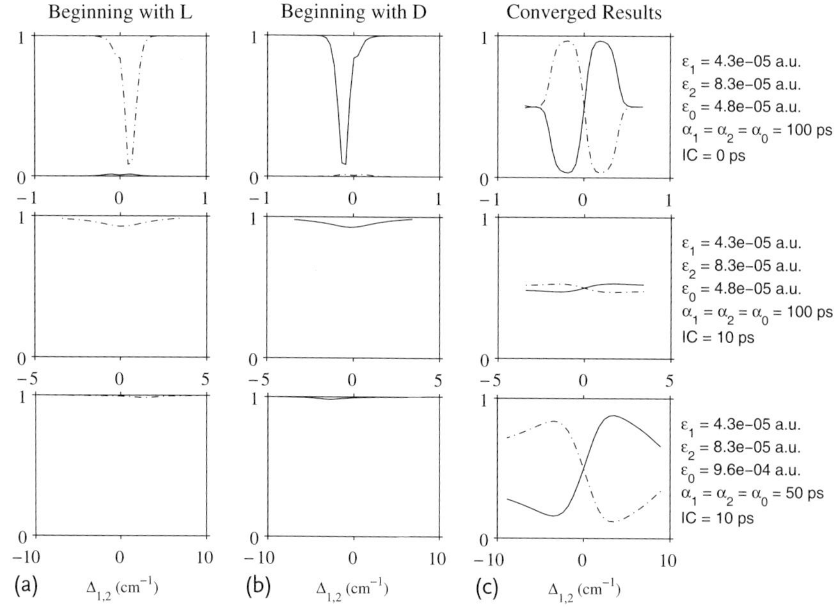

Figure 14.14 Control over dimethylallene enantiomer populations as a function of the detuning Δ_1 for various laser powers. The first column corresponds to probabilities of *L* (dot-dash curves) and *D* (solid curves) after a single laser pulse, assuming that the initial state is all *L*. The second column is similar, but for an initial state which is all *D*. The rightmost column corresponds to the probabilities *L* and *D* after repeated excitation-relaxation cycles, as described in the text. The first row corresponds to control using the laser parameters on the extreme right, in which there is no internal conversion; the second row uses the same laser parameters as does the first row, but with an internal conversion time of 10 ps; the bottom row shows results for an internal conversion time of 10 ps, but with the modified laser parameters shown.

ates, leaving molecular fragments that no longer participate in the control scenario. Hence, the process of internal conversion serves as a decoherence mechanism that can reduce control. Further decoherence effects, but on a slower time scale, would arise, for example, if the control was carried out in solution.

The second row in Figure 14.14 displays control with similar parameters as in the first row, but in the presence of a T_2 associated with decoherence chosen arbitrarily as 10 ps. Clearly, almost all of the control is lost. However, if the laser parameters are changed to those shown in Figure 14.14c, bottom-most column, then significant control is restored once again. In this case, however, the process occurs with the loss of considerable reactant population to dissociated dimethylallene. Additional studies designed to establish the relationship between the laser requirements for control, and the internal conversion rates published back in 2004 [676]. The possibility of alternate substituents to replace the Hydrogens is also of interest, as is the effect of changes to molecular structure to alter the radiative lifetime, the internal conversion rates, and so on.

Enantiomeric control is more difficult if the excited molecular potential energy surfaces do not posses an appropriate minimum at the σ_h hyperplane configurations (see Figure 14.1). In this case the method introduced in this section is not applicable. One may however be able to apply the laser distillation procedure by adding a molecule B to the initial L, D mixture to form weakly bound $L-B$ and $B-D$, which are themselves right and left handed enantiomeric pairs [677]. The molecule B is chosen so that electronic excitation of $B-D$ and $L-B$ forms an excited species G which has stationary ro-vibrational states which are either symmetric or antisymmetric with respect to reflection through σ_h. The species $L-B$ and $B-D$ now serve as the L and D enantiomers in the general scenario above and the laser distillation procedure described above then applies. Further, the molecule B serves as a catalyst that may be removed from the final product by traditional chemical means.

For example, L and D might be the left and right handed enantiomers of a chiral alcohol, and B is the ketone derived from this alcohol (see Figure 14.15). In this case, studies [677] of the electronic structure of the alcohol–ketone system indicate that there are weakly bound chiral alcohol–ketone minima in the ground electronic state, as desired. The particular advantage of using the ketone–alcohol complex is that the ketone, which is "recycled" after the conversion of one enantiomer to another, serves as a catalyst for the process.

The results in this chapter make clear that a chiral outcome, the enhancement of a particular enantiomer, can arise by coherently encoding quantum interference information in the excitation of a racemic mixture. The fact that the initial state displays a broken symmetry and that the excited state has states which are either symmetric or antisymmetric with respect to σ_h allows for the creation of a superposition state which does not have these transformation properties. Radiatively

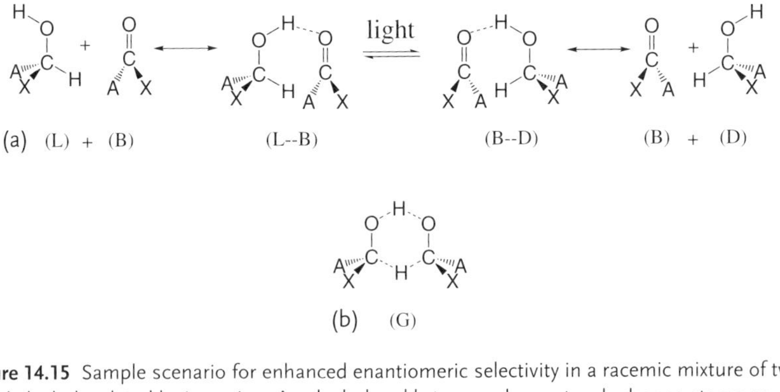

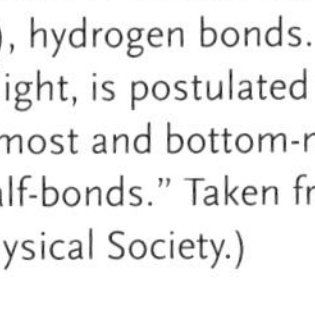

Figure 14.15 Sample scenario for enhanced enantiomeric selectivity in a racemic mixture of two chiral alcohols related by inversion. An alcohol and ketone exchange two hydrogen atoms so as to produce the ketone, but with an alcohol of reverse handedness. Here A and X are distinct organic groups and dashes denote, in (a), hydrogen bonds. The electronically excited species G, which is formed upon excitation with light, is postulated to be given by the structure at the bottom of the figure. In this case the topmost and bottom-most hydrogens are attached to the oxygens and carbons, respectively, by "half-bonds." Taken from [658, Figure 5]. (Reprinted with permission. Copyright 2000 American Physical Society.)

coupling the states in the superposition then allows for the transition probabilities from L and D to differ, allowing for depletion of the desired enantiomer.

14.4 Enantiomer Control: Oriented Molecules

An alternative way to introduce chirality into the interaction of matter with light using linearly polarized light has been introduced by Manz, Fujimura *et al.* [662, 663]. In this approach one first preorients the racemic mixture of D and L along some axis. Under these circumstances, there is a difference in the direction of the transition dipole moments of the left and right handed enantiomers. That is, matrix elements like $\langle E_i|\boldsymbol{d} \cdot \hat{\boldsymbol{\varepsilon}}_k|E_j\rangle$, are different for the two enantiomers $i = D$ or L. This distinction between L and D suffices to allow for the possibility of control over enantiomers.

As a simple example, consider the model [663] of Figure 14.16. The system is initially in a mixture of the ground vibrational state of D and L. The pump laser carries the system to a single excited vibrational state of the excited electronic state and the dump laser returns the system to an excited vibrational state of the ground elec-

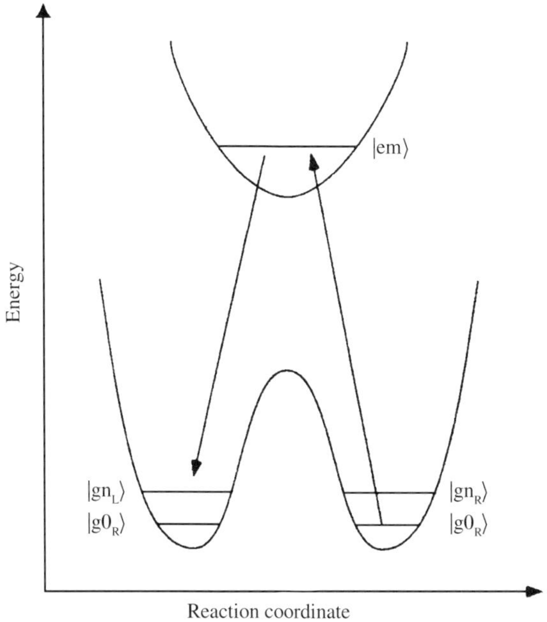

Figure 14.16 Model system for enantiomer control in accord with [663]. The notation is such that $|gn_R\rangle$ denotes the D enantiomer on the ground electronic state in vibrational state n, $|gn_L\rangle$ is the analogous state of the L enantiomer and $|em\rangle$ is the mth level of the excited electronic state. From [663, Figure 1]. (Reprinted with permission. Copyright 2002 American Institute of Physics.)

tronic state. This is then a pump-dump scenario, but transitions are solely between bound states.

To appreciate the essence of this control scenario, recall the results of applying a laser pulse $\varepsilon(t)$ to induce a transition between two bound states $|E_i\rangle$ and $|E_j\rangle$. We denote the dipole transition matrix element between these two states by $d_{i,j}$ and define $\kappa_{i,j} = 2d_{i,j}/\hbar$. Then it is well known [392] that complete population transfer between these levels can be accomplished by using a π-pulse, that is, a pulse of duration t satisfying

$$\int_{-\infty}^{t} \kappa_{i,j}\,\varepsilon(t')dt' = \pi \,. \tag{14.30}$$

Consider then the scenario in Figure 14.16, and suppose that one wishes to transfer population from L to D. Then, since $\langle E_i|\boldsymbol{d}\cdot\hat{\boldsymbol{\varepsilon}}_k|E_j\rangle$ differ for the L and D states, it is possible to choose a laser polarization such that this matrix element is zero for excitation of the ground state of D, but not L. Application of a π-pulse at this polarization will then transfer the ground state L population to the excited state. Application of a second π-pulse of different polarization can then transfer this population to the excited vibrational state of the ground electronic state of D, by now choosing a polarization that does not couple the excited state to L.

Sample results for the control over the oriented enantiomers of H_2POSH are shown in Figure 14.17. Clearly, as proposed, the method is very effective. In this case the primary experimental challenge is to orient the system prior to irradiation.

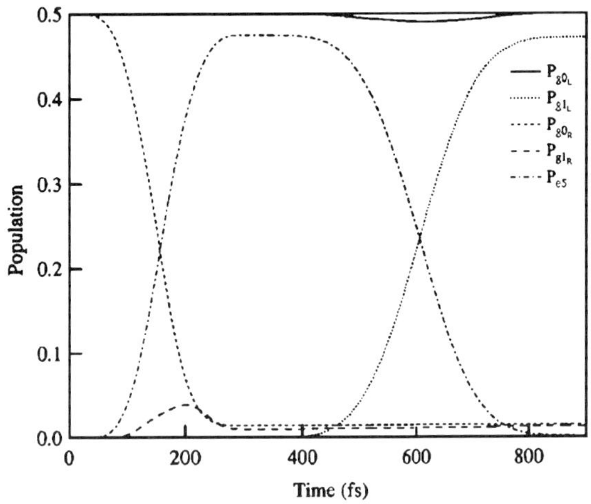

Figure 14.17 Sample computation of control in the pump-dump scenario for controlling chirality. The populations are defined in the upper right-hand corner of the figure. For example, P_{g0_R} denotes the population of $|g0_R\rangle$, and so on. Taken from [663, Figure 4]. (Reprinted with permission. Copyright 2002 American Institute of Physics.)

14.5 Adiabatic Purification of Mixtures of Right-Handed and Left-Handed Chiral Molecules

In this section we discuss the use of adiabatic passage techniques [678–684] to separate enantiomers [654–657, 664, 671, 685–690]. The techniques to be discussed here are distinct from the techniques discussed thus far [145, 658–663, 691–694]. Adiabatic passage chiral separation relies on the same basic property of the "laser distillation" method [658, 659, 694]: The difference between the *L* and *D* forms regarding the sign of some electric dipole matrix elements [658, 678].

The separation method discussed here constitutes the merging of one-photon vs. two-photon coherent control, discussed in Section 3.4.3, and AP, discussed in Section 11.1. As in the control of photo-current directionality discussed in Section 3.4.3, the interference between a one-photon and a two-photon pathway can, depending on the relative phases between the three laser fields used [658, 659, 694], result in the preferential excitation of one enantiomer (appearing as one of the localized minimum of Figure 14.18), relative to the other residing in the other local minimum.

If we ignore the parity-violating weak interactions in nuclei, whose effect is minuscule, the true eigenstates of chiral molecules do have well-defined parities, but these are not the states in which the molecules get trapped. Rather, chiral molecules

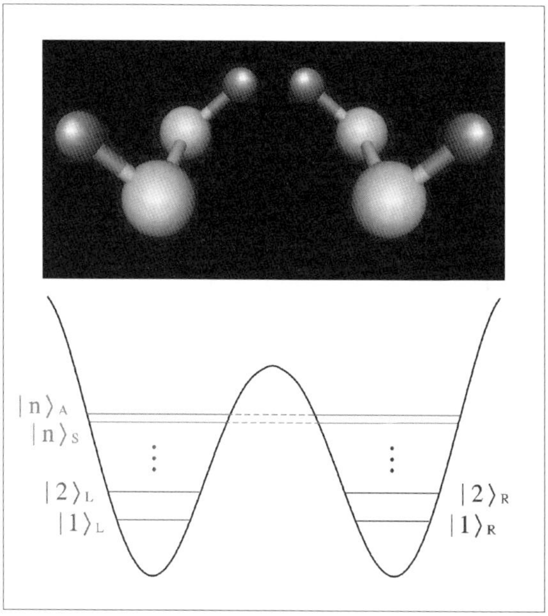

Figure 14.18 The double well potential energy curve for the torsional motion of the D_2S_2 molecule. The two enantiomers connected by this stereomutation torsional motion are displayed in the upper panel. Taken from [680, Figure 1]. (Reprinted with permission. Copyright 2003 American Institute of Physics.)

exist in broken-symmetry states that have ill defined parities, allowing for "cyclic population transfer" (CPT) [678–680] as explained below. An example of the symmetry broken minima is given in Figure 14.18. The states corresponding to these minima are linear combination of two eigenstates of opposite parities, which due to the height of the barrier for interconversion between enantiomers, are practically degenerate.

The "laser distillation method" [658], though potentially very effective for several molecules [659, 694], requires the continuous application of many laser pulses. It is thus of interest to see whether one can convert one enantiomer to another by applying a tailored pulse sequence only once. As discussed in Section 11.1, AP processes such as STIRAP are mainly sensitive to the energy levels and the magnitudes of transition dipole matrix elements, attributes which are identical for both enantiomers. In order to generate *phase sensitivity* in excitation by AP techniques of chiral molecules the cycle of the Λ system of STIRAP is closed by adding a third laser beam, thereby generating a Δ system displaying CPT [678]. More explicitly, the two-photon pathway $|1\rangle \leftrightarrow |2\rangle \leftrightarrow |3\rangle$ of the STIRAP Λ-system is supplemented by a one-photon process $|1\rangle \leftrightarrow |3\rangle$ that directly connects the initial and final states of the former Λ system (see Figure 14.19a).

Although closed loop systems involving three states have been studied previously [695, 696], only in chiral molecules is it possible to have a closed loop comprised of three electric dipole allowed one-photon transitions. Electric dipole allowed CPT is impossible for molecules with a well-defined parity (i.e., *achiral* molecules) because for such molecules, two states that are coupled radiatively by a one-photon process cannot be coupled by a two-photon process. Overall parity conservation dictates that the two states connected by a one-photon process must have opposite parities, whereas the same two states, if connected by a two-photon process, must have the same parities. These two conditions cannot be both satisfied; hence the two processes cannot couple the same two states.

It is of interest to note the similarity between this scenario and the generation and control of photocurrent directionality discussed in Section 3.4.3. In that scenario the one-photon process and the two-photon process excite two different final degenerate continuum states possessing opposite parities. These states then combine to form parity-broken states. The DC current formed is a result of interferences, depending on the phases of the light fields. The interference in that case only affects the directionality of the current, not the total population excited to the continuum. Indeed, when one averages over angles, the one-photon/two-photon interference term vanishes.

As depicted in Figure 14.19a, CPT is driven by three laser fields. The interference between the one-photon and two-photon pathways shown in Figure 14.19, renders the evolution of the system dependent on the total phase φ of the three (material + optical) coupling terms. Since the phases of the transition dipoles in the two enantiomers differ by π (see Eq. (14.33) below), the evolution of the two enantiomers under the action of the three fields is different.

14.5.1 Vibrational State Discrimination of Chiral Molecules

In order to demonstrate the method, consider the purification of a racemic mixture of the (transiently chiral) D_2S_2 molecule (shown schematically in Figure 14.18). In

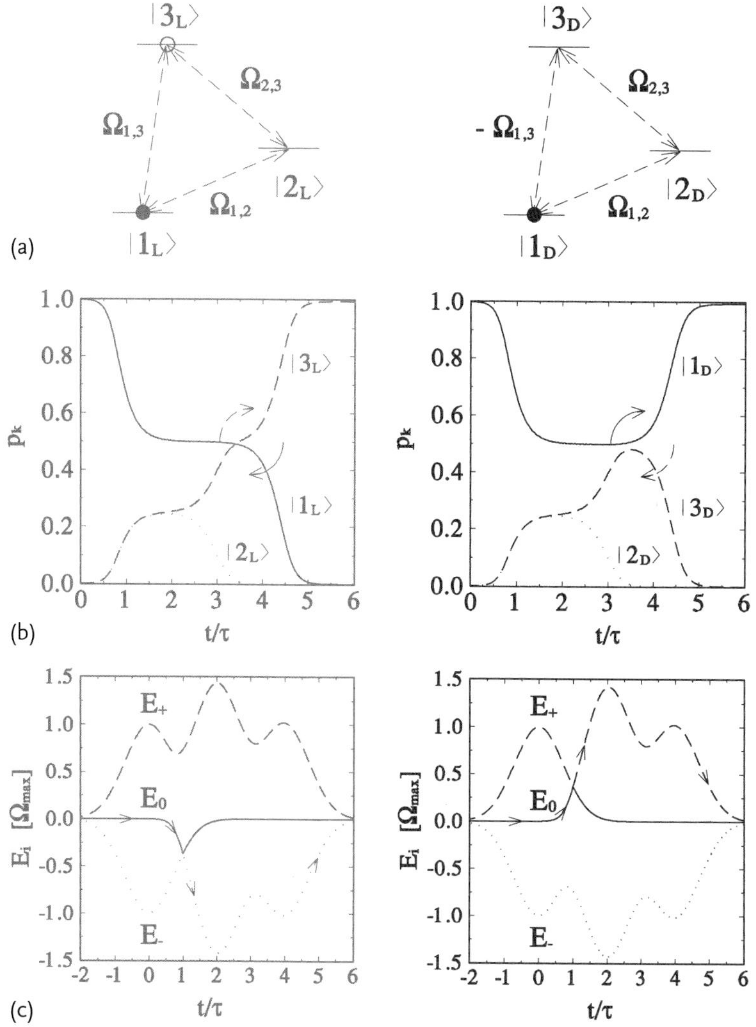

Figure 14.19 The enantio-discriminator: (a) A schematic plot of the CPT scheme, in which three levels of each enantiomer are resonantly coupled by three fields, as used in the discrimination step. (b) The time evolution of the population of the three levels: Both enantiomers start in the $|1\rangle$ state. At the end of the process the L enantiomer is transferred to the $|3_L\rangle$ state, while the D enantiomer remains in the initial $|1_D\rangle$ state. (c) The time dependence of the eigenvalues of the Hamiltonian of Eq. (14.35): The population initially follows the $|E_0\rangle \equiv |\lambda_0\rangle$ null eigenstate. At $t \approx \tau$ it crosses over diabatically to $|E_-\rangle \equiv |\lambda_-\rangle$ for one enantiomer and to $|E_+\rangle \equiv |\lambda_+\rangle$ for the other. ($|\lambda_{0,\pm}\rangle$ are defined in the text.) Taken from [680, Figure 2]. (Reprinted with permission. Copyright 2003 American Institute of Physics.)

this molecule, the pathway connecting the two enantiomers involves a relatively simple motion, that of the (hindered) rotation (torsion) of the D atoms about the S–S bond. The tunneling splitting of the lowest torsional states gives enantiomeric lifetimes of several msec [679, 680, 697]. Thus, although D_2S_2 is only transiently chiral, its chiral molecular structures remain stable for sufficiently long times to perform the excitation process.

According to this proposal, it is possible to achieve chiral purification of the racemic mixture of chiral D_2S_2 molecules in just two steps: In the first, "discrimination step", based on the CPT process, one enantiomer is excited, while leaving the other one in its initial state. In the second, "conversion step", the enantiomer excited in the first step is converted to its mirror-imaged form. Because of the completeness of the AP processes used in both steps, the racemic mixture of chiral molecules should be turned into a sample containing the enantiomer of our choice only, after the first-time completion of the two steps.

The two-step enantio-purification process makes use of five pairs of D_2S_2 rovibrational eigenstates. These vibrational states correspond to the combined torsional and S–D asymmetric stretching modes. The wave functions and energies of these states were calculated using potential and electric dipole surfaces obtained by *ab initio* methods [490]. The rotational states correspond to a rigid rotor. Within the pairs used,[2)] each eigenstate has an S/A (symmetric/antisymmetric) label denoting its parity.

The broken-symmetry states, denoted as $|k_{L,D}\rangle$, are linear combinations of the essentially-degenerate lower lying symmetric and antisymmetric eigenstates [657],

$$|k_{L,D}\rangle = \frac{1}{\sqrt{2}}(|k_S\rangle \pm |k_A\rangle)\,, \quad k = 1, \ldots, 4\,. \tag{14.31}$$

The higher lying $|5_S\rangle$ and $|5_A\rangle$ states are already split by $\Delta E^5_{S,A} = 0.38\,\text{cm}^{-1}$, and can be separately addressed using nsec laser pulses. This cannot be done for the essentially-degenerate lower lying $k = 1, \ldots, 4$ eigenstates, for which the splittings are much smaller than the bandwidths of most lasers in practical use.

As in Sections 11.2 and 11.3, the energy of state $|i\rangle$ is denoted by E_i, and the external electric field is chosen as a sum of components, each being in *resonance* with one of the $|i\rangle \leftrightarrow |j\rangle$ transition frequencies of interest,

$$\boldsymbol{E}(t) = \text{Re} \sum_{i \neq j} \hat{\epsilon} \mathcal{E}_{i,j}(t) e^{-i\omega_{i,j} t}\,, \tag{14.32}$$

where $\omega_{i,j} = \omega_i - \omega_j$, and $\hat{\epsilon}$ is the polarization direction. The total Hamiltonian has the same form as in Eq. (11.31).

The control scheme for the three-level discrimination step is given in Figure 14.19a. It is assumed that at sufficiently low temperature the D_2S_2 molecules in the racemic mixture reside initially in the ground $|1_L\rangle$ and $|1_D\rangle$ states. Therefore,

2) The $|1\rangle$, $|3\rangle$, $|4\rangle$ and $|5\rangle$ states correspond to torsional states with 0, 2, 3, 5 vibrational quanta, respectively. The $|2\rangle$ state is the first excited state of the asymmetric S–D stretching mode.

the objective at this step is to excite only one enantiomer, that is, transfer its population completely from the $|1\rangle$ to the $|3\rangle$ state, and keep the other in the $|1\rangle$ state [678]. Due to the essential degeneracy of the $|i_L\rangle$ and the $|i_D\rangle$ states for $i = 1, 2, 3$, the field $\boldsymbol{E}(t)$ excites simultaneously the $|i_\nu\rangle \leftrightarrow |j_\nu\rangle$, $(i \neq j = 1, 2, 3,\ \nu = L, D)$ transitions of both enantiomers [658, 694]. Therefore, one has to make use of the phase dependence of the CPT process. The fact that the electric dipole operator can only connect states of opposite parities results in

$$\Omega_{ij}^{L,D} = \left(\pm\left[\langle i_S|\mathrm{d}|j_A\rangle + \langle i_A|\mathrm{d}|j_S\rangle\right]\mathcal{E}_{ij}\right)/\hbar\ . \tag{14.33}$$

Hence, all the Rabi frequencies associated with the two enantiomers differ by a sign, that is, a phase factor of π, for transitions induced by the dipole component along the stereomutation coordinate. Since in the CPT processes the two enantiomers are influenced by the phase $\varphi^{L,D}$ of the products $\Omega_{12}^L\Omega_{23}^L\Omega_{31}^L$, $\Omega_{12}^D\Omega_{23}^D\Omega_{31}^D$, one obtains that $\varphi^L - \varphi^D = \pi$. This property is invariant to any arbitrary phase change in the individual wave functions of the states $|i_{L,D}\rangle$.

The vector component of the coefficients $c = (c_1, c_2, c_3)^{\mathrm{T}}$ of the expansion of the wave function for the discrimination step, is the solution of the Schrödinger equation

$$i\hbar\frac{d}{dt}c(t) = \underline{\underline{\mathrm{H}}}(t)\cdot c(t)\ , \tag{14.34}$$

with the RWA Hamiltonian after neglecting off-resonant terms given as

$$\underline{\underline{\mathrm{H}}}(t) = \frac{\hbar}{2}\begin{bmatrix} 0 & \Omega_{1,2}^*(t) & \Omega_{1,3}^*(t) \\ \Omega_{1,2}(t) & 0 & \Omega_{2,3}^*(t) \\ \Omega_{1,3}(t) & \Omega_{2,3}(t) & 0 \end{bmatrix}\ . \tag{14.35}$$

The phases of the complex Rabi frequencies $\Omega_{i,j}(t)$ are given as $\phi_{i,j} = \phi_{i,j}^{\mathrm{d}} + \phi_{i,j}^{E}$, where $\phi_{i,j}^{\mathrm{d}}$ are the phases of the dipole matrix elements $\mathrm{d}_{i,j}$, and $\phi_{i,j}^{E}$ are the phases of the electric field components $\mathcal{E}_{i,j}$. The evolution of the system is determined by the total phase $\varphi \equiv \phi_{1,2} + \phi_{2,3} + \phi_{3,1}$. This is most noticeable at the time $t = \tau$, for which the three Rabi frequencies are equal in magnitude, $|\Omega_{1,2}| = |\Omega_{1,3}| = |\Omega_{2,3}| = \Omega$.

Denoting the adiabatic eigenvalues of the Hamiltonian of Eq. (14.35) as λ_-, λ_0 and λ_+, it is easy to show that they undergo exact degeneracies (crossings) at $t = \tau$, with $\lambda_+ = 2\Omega$ and $\lambda_- = \lambda_0 = 2\Omega\cos(2\pi/3)$, for $\varphi = 0$, and $\lambda_- = -2\Omega$ and $\lambda_+ = \lambda_0 = -2\Omega\cos(2\pi/3)$, for $\varphi = \pi$. Depending on the polarizations of the fields chosen, one Rabi frequency $\Omega_{i,j}$ of the two enantiomers differ by a sign, making φ differ by π for the two. Therefore, as shown in the Figure 14.19c, the crossing occurs between different eigenvalues for different enantiomers, leading subsequently to their totally different dynamics (see Figure 14.19b).

The overall discrimination step works as follows: Start with a Stokes pulse $\mathcal{E}_{2,3}(t)$ that couples the $|2\rangle$ and $|3\rangle$ states and has the Rabi frequency $\Omega_{2,3}(t) = \Omega^{\max} f(t)$, where $\Omega^{\max} = 2\ \mathrm{ns}^{-1}$ and $f(t) = \exp[-t^2/\tau^2]$. At this stage of the process all the population resides in the $|\lambda_0\rangle$ adiabatic eigenstate. In the second stage one simultaneously adds two pump pulses of the Rabi frequencies $\Omega_{1,2}(t) = \Omega_{1,3}(t) = \Omega^{\max} f(t-2\tau)$ that couple the $|1\rangle \leftrightarrow |2\rangle$ and the $|1\rangle \leftrightarrow |3\rangle$ states. The phases of the optical fields are chosen so that $\varphi = 0$ for one enantiomer and, inevitably, $\varphi = \pi$ for the other. Therefore, the population, which has been following in both enantiomers the initial adiabatic level $|\lambda_0\rangle$, goes at $t = \tau$ smoothly through the crossing region and *diabatically* transfers to either the $|\lambda_-\rangle$ or the $|\lambda_+\rangle$ states, depending on whether $\varphi = 0$ or $\varphi = \pi$, that is, on the identity of the enantiomer.

After the crossing is complete, at $t > \tau$, the process becomes adiabatic again, with the enantiomer population residing fully in either $|\lambda_-\rangle$ or $|\lambda_+\rangle$. At this stage the $\mathcal{E}_{1,2}(t)$ pulse is slowly switched off while the $\mathcal{E}_{1,3}(t)$ field remains on. This is done by choosing $\Omega_{1,3}(t) = \Omega^{\max}\left(f(t-2\tau) + f(t-4\tau)\exp\{-it\Omega^{\max} f(t-6\tau)\}\right)$. As a result, the zero adiabatic eigenstate $|\lambda_0\rangle$ correlates adiabatically with state $|2\rangle$, which thus becomes empty after this process, while the occupied $|\lambda_+\rangle$ and $|\lambda_-\rangle$ adiabatic states correlate to, $|\lambda_\pm\rangle \to (|1\rangle \pm |3\rangle)/\sqrt{2}$.

The chirp, $\exp\{-it\Omega^{\max} f(t-6\tau)\}$, in the second term of $\Omega_{1,3}(t)$ causes a $\pi/2$ rotation in the $\{|1\rangle, |3\rangle\}$ subspace at $t \approx 5\tau$. As a result, state $|\lambda_+\rangle$ goes over to state $|3\rangle$ and state $|\lambda_-\rangle$ goes over to state $|1\rangle$, or vice versa, depending on φ. The net result of the adiabatic passage and the rotation is that one enantiomer returns to its initial $|1\rangle$ state and the other switches over to the $|3\rangle$ state. As shown in Figure 14.19b, the enantio-discriminator is very robust, with all the population transfer processes occurring in a smooth fashion.

After all the population in the L enantiomer has been successfully transferred to the $|3_L\rangle$ state, it is possible to use a combination of the solutions of NQC and DQC presented in Sections 11.2 and 11.3 to *convert* the $|3_L\rangle$ population to the D enantiomer. This is done by performing the $|3_L\rangle \to |4_D\rangle$ transfer with two in-parallel executed three-level STIRAP processes, using as intermediates the $|5_S\rangle$ and $|5_A\rangle$ states. Since no population is left in state $|1_L\rangle$, the center frequencies of the fields involved are tuned to guarantee that the D enantiomer remains intact in state $|1_D\rangle$.

Schematically, the dynamics follows the $|3_L\rangle \to \alpha|5_S\rangle + \beta|5_A\rangle \to |4_D\rangle$ pathway shown in Figure 14.20. The transfer is realized by simultaneously introducing two Stokes pulses $\mathcal{E}_{4,5S}(t)$ and $\mathcal{E}_{4,5A}(t)$ (of duration $\tau \gg (\omega_{5S} - \omega_{5A})^{-1}$), which resonantly couple each of the $|5_S\rangle$ and $|5_A\rangle$ states to the $|4_L\rangle$ and $|4_D\rangle$ state. After a delay of 2τ, two pump pulses, $\mathcal{E}_{3,5S}(t)$ and $\mathcal{E}_{3,5A}(t)$, are introduced. These pulses resonantly couple each of the $|5_S\rangle$ and $|5_A\rangle$ states to the $|3_L\rangle$ and $|3_D\rangle$ state. In this process only the $|4_D\rangle$ state is populated, while the $|3_D\rangle$ and $|4_L\rangle$ states, become empty at the end of the process. This is because the empty pair of states is coupled to the $|5_S\rangle$ and $|5_A\rangle$ states by vectors of Rabi frequencies Ω_i, that are orthogonal to analogous vectors of the populated pair of states, respectively. The chiral conversion of the excited enantiomer is achieved by choosing $\Omega_{4,5S}(t)$ to have the same sign as $\Omega_{3,5S}(t)$ and $\Omega_{4,5A}(t)$ to have an opposite sign to $\Omega_{3,5A}(t)$.

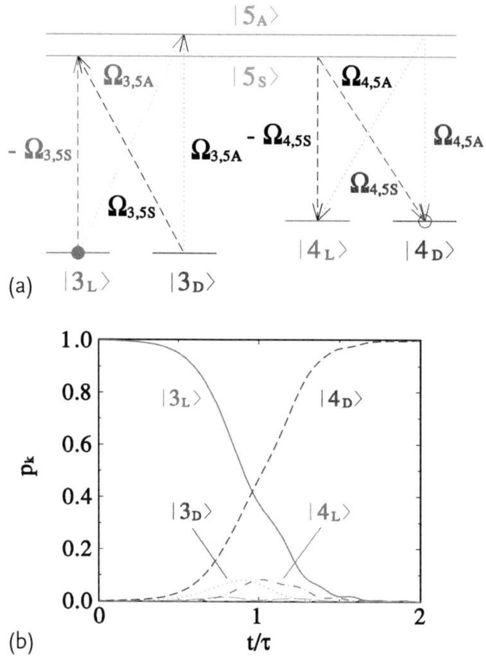

Figure 14.20 The enantio-converter scheme: (a) Population is transferred from the $|3_L\rangle$ state to the $|4_D\rangle$ state, while going through the superposition of $|5_{S,A}\rangle$ states. (b) The populations p_i of the broken-symmetry states, as a function of time, during the conversion step. Adapted from [680, Figures 3 and 4]. (Reprinted with permission. Copyright 2003 American Institute of Physics.)

The conversion step in the RWA and neglecting all off-resonant terms is described by the Hamiltonian,

$$\underline{\underline{H}}(t) = \frac{\hbar}{2}\begin{bmatrix} 0 & 0 & -\Omega^*_{3,5S} & \Omega^*_{3,5A} & 0 & 0 \\ 0 & 0 & \Omega^*_{3,5S} & \Omega^*_{3,5A} & 0 & 0 \\ -\Omega_{3,5S} & \Omega_{3,5S} & 0 & 0 & -\Omega^*_{4,5S} & \Omega^*_{4,5S} \\ \Omega_{3,5A} & \Omega_{3,5A} & 0 & 0 & \Omega^*_{4,5A} & \Omega^*_{4,5A} \\ 0 & 0 & -\Omega_{4,5S} & \Omega_{4,5A} & 0 & 0 \\ 0 & 0 & \Omega_{4,5S} & \Omega_{4,5A} & 0 & 0 \end{bmatrix}, \quad (14.36)$$

with the time-dependent wave function given by the

$$c(t) = (c_{3L}, c_{3D}, c_{5S}, c_{5A}, c_{4L}, c_{4D})^{\mathsf{T}}$$

vector of expansion coefficients after solving Eq. (14.34). This Hamiltonian matrix has four nonzero eigenvalues and two null eigenvalues, $\lambda_{1,2} = 0$, that correspond to two null states with coefficient vectors, $c_1(t) = (-d_+, -d_-, 0, 0, 2\Omega_{3,5S}, 0)^{\mathsf{T}}$, $c_2(t) = (-d_-, d_+, 0, 0, 0, 2\Omega_{3,5S})^{\mathsf{T}}$, where $d_\pm \equiv \Omega_{4,5S} \pm r\Omega_{4,5A}$, and $r \equiv \Omega_{3,5S}/\Omega_{3,5A}$.

These expressions imply that the system can follow two possible paths, of which only one results in chiral flipping of the initial state. The assumption that $r = 1$

and $r' = \Omega_{4,5S}/\Omega_{4,5A} = 1$ makes the null eigenstate $c_1(t_{ini})$ be the only eigenstate that correlates in the beginning of the process with the initial state $|3_L\rangle$, that is, the vector $c(t_{ini}) = (1,0,0,0,0,0)^\mathrm{T}$. At the end of the processes, this null state correlates with the vector $c_1(t_{end}) = (0,0,0,0,1,0)^\mathrm{T}$ for the $|4_L\rangle$ state, so the symmetry is preserved. On the other hand, if the phase of just one Stokes or one pump field component ($r = -1$ or $r' = -1$) is flipped, the system follows the null state c_2, which correlates at the end with the state $c_2(t_{end}) = (0,0,0,0,0,1)^\mathrm{T}$. The final population thus occupies the $|4_D\rangle$ state, with the opposite chirality, as required. We note that the transfer dynamics during the conversion step can be understood as the dynamics of two in-parallel executed three-level STIRAP processes, which is obvious when the conversion step is formulated in the $\{3S, 3A, 5S, 5A, 4S, 4A\}$ basis after inversion of Eq. (14.31).

Figure 14.20 shows the evolution of the calculated populations $p_i = |c_i|^2$. The process starts in the $|3_L\rangle$ state and ends in the $|4_D\rangle$ state. The Rabi frequencies are $\Omega_{3,5S}(t) = \Omega^{\max} f(t - 2\tau)$, $\Omega_{3,5A}(t) = 0.5\Omega^{\max} f(t - 2\tau)$, $\Omega_{4,5S}(t) = 0.4\Omega^{\max} f(t)$, $\Omega_{4,5A}(t) = -\Omega^{\max} f(t)$, with $\Omega^{\max} = 60\,\mathrm{ns}^{-1}$ and $f(t)$ as in the discrimination step. The transfer is complete and exclusive although, in the calculations presented in Figure 14.20, the $r = 2$ and $r' = -0.4$, are chosen thus demonstrating that the process is robust, as long as r and r' have the right relative signs. The schemes discussed have found applications for the detection and automatic repair of nucleotide base-pair mutations and targeted genome manipulation [698].

14.5.2 Spatial Separation of Enantiomers

Bruder *et al.* [682] have shown that the CCAP process discussed above can be used to *spatially* separate the two enantiomers. These authors drew an analogy, due to the existence of the vectorial nonadiabatic coupling terms, between this separation process and the Stern–Gerlach effect. Although the analogy with the magnetic Stern–Gerlach effect is highly seductive, as shown below, the effectiveness of the enantio-separation is due mainly to the scalar "light-induced potentials" (LIP) and thus has little to do with the pseudospin nature of the system. The nonadiabatic vector potential terms which bring about the pseudospin behavior, exist in fact for all, chiral and nonchiral, molecules. Moreover, the induced vector potential is identical for the right-handed and left-handed species, save for a cyclic permutation of the vibrational states. In contrast, the differences between the scalar LIP, created by the three inhomogeneous laser fields, are uniquely chiral.

In order to appreciate the feasibility of the optical separation of chiral molecules, consider wave packet and classical trajectories propagation for *strong-field* CCAP processes [683, 684]. Molecules of opposite handedness will be seen to follow different center of mass (CM) motions, with opposite enantiomers neatly separating in space. The fine details of the motion will be seen to depend on the internal states occupied by the molecules. Strong fields will be required to conduct the process in the practical mK domain. Weak fields, whose effects are describable by second-order perturbation theory [682], necessitate working in the, yet unrealistic, sub-μK

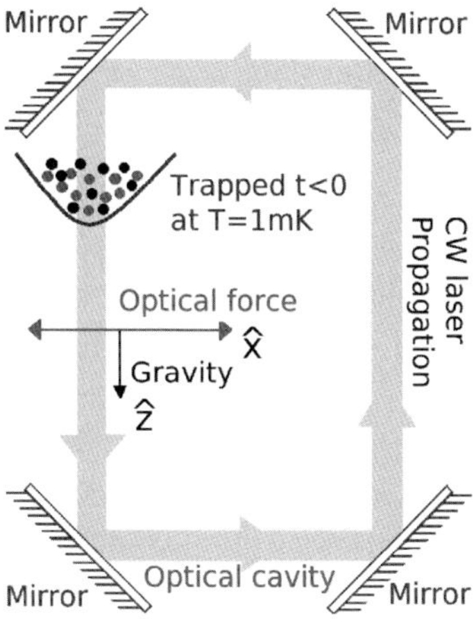

Figure 14.21 Separation via a cold molecular trap. Taken from [684, Figure 1]. (Reprinted with permission. Copyright 2010 American Institute of Physics.)

regime. Moreover, strong fields are necessary to maintain the adiabaticity which is at the heart of the CCAP process [678].

Consider a racemic mixture in an optical trap at temperatures of $\sim$ 1 mK, with the separating light's electric fields pointing in the horizontal direction. As shown in Figure 14.21, when the trapping potentials are lifted, the molecules free-fall in the vertical direction with the L/D enantiomers undergoing different parabolic trajectories as they separate in the horizontal direction. The second experimental configuration to consider is that of a molecular beam traveling at a very small angle relative to the laser propagation direction. In that case the LIP is applied in the vertical direction. The small angle between the propagation directions of the molecular and laser beams is adjusted so as to make the lasers' profiles follow the change in the CM position of the molecules as they separate.

14.5.3
Internal Hamiltonian and Dressed States

The total Hamiltonian of a (trapped) molecule in an inhomogeneous laser field, can be written [683, 684] as

$$H_{tot} = H_{CM}(\boldsymbol{r}) + H_{int}(\boldsymbol{r}) + U(\boldsymbol{r}) , \tag{14.37}$$

with $\boldsymbol{r}$ denoting the center of mass (CM) coordinate; H_{CM} – the CM kinetic energy operator; and $H_{int}(\boldsymbol{r})$ – the matter–radiation internal Hamiltonian. $U(\boldsymbol{r})$ includes all CM-dependent external potentials, that is, the trapping potential (if it exists), and the gravitational potential.

In order to derive the field-dressed states, first examine the internal Hamiltonian H_{int}. We consider a system composed of three "bare" states of each enantiomer, $|\alpha, n)$, $\alpha = L, D$; $n = 1, 2, 3$, where, $[E_n - H_{mat}]|\alpha, n) = 0$, with H_{mat} denoting the material (i.e., radiation-free) part of H_{int}. In the above, *round*, ( ), brackets and n, m indices are reserved for the notation of the bare states. The system is allowed to interact in a "*cyclic*" way with three, spatially displaced, inhomogeneous laser fields. These fields couple each $|\alpha, n)$ state to the two other $|\alpha, m \neq n)$ states of the same enantiomer. (Such a cyclic coupling scheme can only be attained for chiral molecules that lack a center of inversion [678].) We denote the relevant $m \leftrightarrow n$ Rabi frequency, whose sign (but not magnitude) depends on α, as $\Omega^{(\alpha)}_{m,n}$, where

$$\Omega^{(\alpha)}_{m,n} \equiv (\alpha, m|\boldsymbol{d}|\alpha, n) \cdot \mathcal{E}(\boldsymbol{r}, t)/\hbar \ . \tag{14.38}$$

Here $(\alpha, m|\boldsymbol{d}|\alpha, n)$ is the transition dipole matrix element between quantum states of the $\alpha = L, D$ enantiomer, and $\mathcal{E}(\boldsymbol{r}, t)$ is the radiation's electric field.

Assuming that the tunneling probabilities between the L and the D forms are negligible for all $|\alpha, n)$ states, one can write $\Psi^{(\alpha)}(\boldsymbol{r}, t)$, the solution of the time-dependent Schrödinger equation $i\hbar\partial\Psi^{(\alpha)}(\boldsymbol{r}, t)/\partial t = H_{tot}\Psi^{(\alpha)}(\boldsymbol{r}, t)$ for each enantiomer, as

$$\Psi^{(\alpha)}(\boldsymbol{r}, t) = \sum_{n=1}^{3} \psi^{(\alpha)}_n(\boldsymbol{r}, t)|\alpha, n) \ . \tag{14.39}$$

By invoking the rotating wave approximation, we can write the internal Hamiltonian of each enantiomer in the interaction representation as

$$H^{(\alpha)}_{int}(\boldsymbol{r}) = \sum_{n>m}^{3} \Omega^{(\alpha)}_{n,m}(\boldsymbol{r}) \exp(-i\Delta_{n,m}t)|\alpha, n)(\alpha, m| + \text{h.c.} \ , \tag{14.40}$$

where $\Delta_{n,m} \equiv (E_n - E_m)/\hbar - \omega_{n,m}$, with each $\omega_{n,m}$ being the carrier frequency of the laser field which couples the $|\alpha, n)$ state to the $|\alpha, m)$ state.

Using a unitary matrix $S^{(\alpha)}(\boldsymbol{r})$ we can diagonalize $H^{(\alpha)}_{int}$ to obtain $\lambda^{(\alpha)}_i(\boldsymbol{r})$ – the "field-dressed" eigenvalues, given as [678],

$$\lambda^{(\alpha)}_1 = \frac{1}{3}\left(\frac{2^{\frac{1}{3}}a}{c^{(\alpha)}} + \frac{c^{(\alpha)}}{2^{\frac{1}{3}}}\right) , \quad \lambda^{(\alpha)}_{2,3} = -\frac{1}{3}\left[\frac{(1 \pm i3^{\frac{1}{2}})a}{2^{\frac{2}{3}}c^{(\alpha)}} + \frac{(1 \mp i3^{\frac{1}{2}})c^{(\alpha)}}{2^{\frac{4}{3}}}\right] , \tag{14.41}$$

where

$$a \equiv 3\left(|\Omega_{1,2}|^2 + |\Omega_{2,3}|^2 + |\Omega_{3,1}|^2\right) , \quad b^{(\alpha)} \equiv 3^3 2\mathrm{Re}\,\mathcal{O}^{(\alpha)} ,$$
$$c^{(\alpha)} \equiv \left[b^{(\alpha)} + \sqrt{b^2 - 4a^3}\right]^{1/3} , \tag{14.42}$$

and $\mathcal{O}^{(\alpha)} \equiv \Omega^{(\alpha)}_{1,2}\Omega^{(\alpha)}_{2,3}\Omega^{(\alpha)}_{3,1}e^{-i\Sigma t}$, with $\Sigma \equiv \Delta_{12} + \Delta_{23} + \Delta_{31}$. There is a sign difference in $\mathcal{O}^{L/D}$ because it contains a product of an odd number of Rabi frequencies, each satisfying the L/D relation [678]

$$\Omega^{L}_{n,m} = -\Omega^{R}_{n,m} \ . \tag{14.43}$$

The "field-dressed" eigenstates, denoted as, $|\alpha, i\rangle$ $i = 1, 2, 3$, with *ordinary*, $\langle\ \rangle$, brackets and i, j indices reserved for the notation of the dressed states, can also be obtained analytically, as written explicitly below. Because of the assumed inhomogeneity of the fields, both $|\alpha, i\rangle$ and $\lambda_i^{(\alpha)}$ are functions of $\boldsymbol{r}$.

Now expand $\Psi^{(\alpha)}(\boldsymbol{r}, t)$ in terms of the dressed states, $\Psi^{(\alpha)}(\boldsymbol{r}, t) = \sum_i^3 \phi_i^{(\alpha)}(\boldsymbol{r}, t)$ $|\alpha, i\rangle$, with the vector of expansion coefficients $\underline{\underline{\phi}}^{(\alpha)} \equiv \left(\phi_1^{(\alpha)}, \phi_2^{(\alpha)}, \phi_3^{(\alpha)}\right)^T$ satisfying the matrix Schrödinger equation

$$i\hbar \frac{\partial}{\partial t} \underline{\phi}^{(\alpha)} = \underline{\underline{\mathrm{H}}}^{(\alpha)} \cdot \underline{\phi}^{(\alpha)} \tag{14.44}$$

where $\underline{\underline{\mathrm{H}}}^{(\alpha)}$, the matter-radiation Hamiltonian matrix, is given as [531, 682, 699]

$$\underline{\underline{\mathrm{H}}}^{(\alpha)}(\boldsymbol{r}) = \frac{1}{2m}\left(i\hbar\nabla_r + \underline{\underline{\mathrm{A}}}^{(\alpha)}(\boldsymbol{r})\right)^2 + \underline{\underline{\mathrm{V}}}^{(\alpha)}(\boldsymbol{r})\,. \tag{14.45}$$

Here,

$$\boldsymbol{A}_{i,j}^{(\alpha)}(\boldsymbol{r}) = i\hbar\langle\alpha, i|\nabla_r|\alpha, j\rangle \tag{14.46}$$

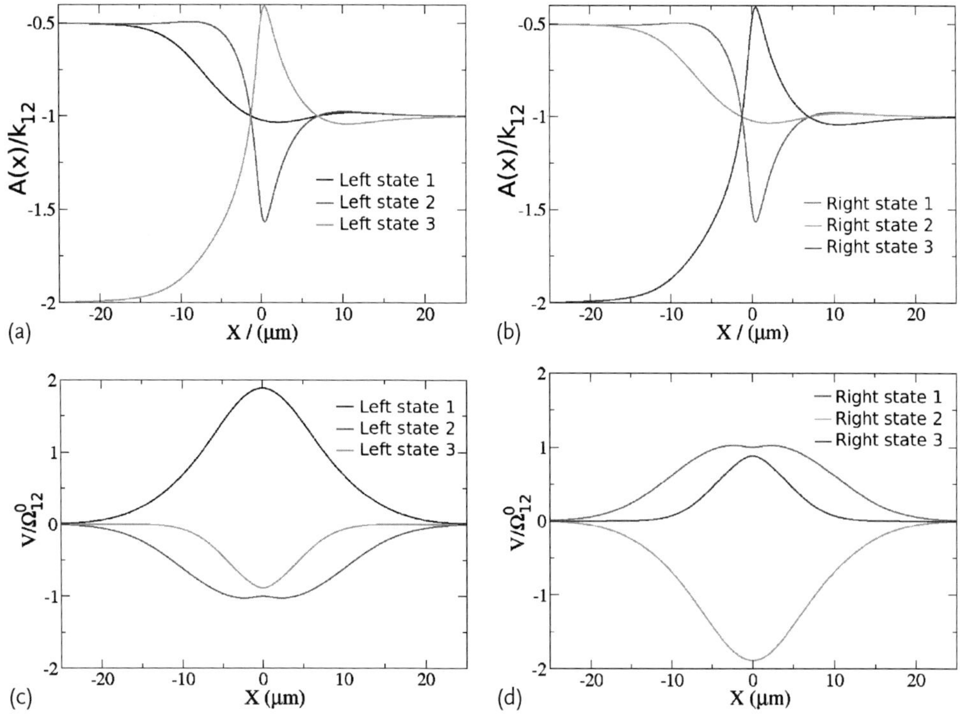

Figure 14.22 A sample of x-dependent LIPs for the $|\alpha, i\rangle$, $\alpha = L, D$; $i = 1, 2, 3$, dressed states. The vector potentials $\mathbf{A}_i^{(\alpha)}(x)$ are displayed in the top row; the scalar potentials $V_i^{(\alpha)}(x)$ are shown in the bottom row. Taken from [684, Figure 2]. (Reprinted with permission. Copyright 2010 American Institute of Physics.)

are the nonadiabatic matrix elements (displaying a "pseudo-magnetic" type behavior [682]), and

$$V_{i,j}^{(\alpha)}(\mathbf{r}) = \lambda_i^{(\alpha)}(\mathbf{r})\delta_{i,j} + \langle \alpha, i | U(\mathbf{r}) | \alpha, j \rangle \tag{14.47}$$

are the scalar (light-induced + trapping + gravitational) potential matrix elements.

The off-diagonal elements of the "coupled channels" set of equations (Eq. (14.44)) can be neglected, provided they are much smaller than the eigenvalue differences $|\lambda_i - \lambda_j|$ [699, 700]. One then obtains a set of three decoupled (*adiabatic*) Hamiltonians

$$H_i^{(\alpha)}(\mathbf{r}) = \frac{1}{2m}\left(i\hbar\nabla_r + A_i^{(\alpha)}(\mathbf{r})\right)^2 + V_i^{(\alpha)}(\mathbf{r}) , \quad i = 1, 2, 3 , \tag{14.48}$$

where $\mathbf{A}_i^{(\alpha)}(\mathbf{r}) = i\hbar\langle \alpha, i|\nabla_r|\alpha, i\rangle$ and $V_i^{(\alpha)}(\mathbf{r}) = \lambda_i^{(\alpha)}(\mathbf{r}) + \langle \alpha, i| U(\mathbf{r})|\alpha, i\rangle$. A sample of the scalar and vector potentials for three L and D states are given in Figure 14.22.

14.5.4 Laser Configuration

We now describe a set of adiabatic calculations demonstrating the enantiomeric spatial separation by the above method [683, 684], which assumes that all transitions are on resonance, that is, $\omega_{nm} = E_n - E_m$. In the context of the CCAP arrangement, this means that

$$\omega_{12} + \omega_{23} + \omega_{31} = 0 . \tag{14.49}$$

All laser fields are assumed to propagate in the $\hat{z}$ direction, and their transverse ($\hat{x}$ direction) profile to have a Gaussian form, hence the Rabi frequencies are also Gaussian functions of x

$$\Omega_{n,m}^{(\alpha)} = \Omega_{nm}^{\alpha,o} e^{-(x-x_{nm})^2/\sigma_{n,m}^2} e^{-ik_{nm}z/\hbar} . \tag{14.50}$$

Due to the cyclic arrangement, Eq. (14.49) also implies that $k_{nm} \equiv \omega_{nm}/c$ satisfy a *phase-matching* condition, $k_{12} + k_{23} + k_{31} = 0$.

With the above form for the laser fields, the dressed eigenstates are given for each enantiomer (the index α is suppressed) as,

$$|i\rangle = \frac{1}{\sqrt{N_i(x)}}\left\{e^{i(k_{12}+k_{23})z/\hbar} a_i(x)|1\rangle + e^{ik_{23}z/\hbar} b_i(x)|2\rangle + |3\rangle\right\} , \tag{14.51}$$

where (suppressing the x dependence)

$$a_i = \frac{-\bar{\Omega}_{23}^2 + \lambda_i^2}{\bar{\Omega}_{12}\bar{\Omega}_{23} + \bar{\Omega}_{13}\lambda_i} , \quad b_i = \frac{\bar{\Omega}_{13}\bar{\Omega}_{23} + \bar{\Omega}_{12}\lambda_i}{\bar{\Omega}_{12}\bar{\Omega}_{23} + \bar{\Omega}_{13}\lambda_i} . \tag{14.52}$$

N_i is a normalization factor, $N_i = |a_i|^2 + |b_i|^2 + 1$, and $\bar{\Omega}_{nm}(x) \equiv \Omega_{nm}(x, z)e^{ik_{nm}z/\hbar}$.

It can be shown that the $\hat{x}$ component of the vector potentials $\boldsymbol{A}_i$ is zero, hence, we obtain that,

$$\boldsymbol{A}_i = -\left[(k_{12} + k_{23})a_i^2 + k_{23}b_i^2\right]\hat{z} \,. \tag{14.53}$$

Thus, the diagonal vector potential satisfies the Coulomb gauge in which $\nabla_{\boldsymbol{r}} \cdot \boldsymbol{A}_i = 0$. The leading off-diagonal (nonadiabatic) coupling terms also point in the $\hat{z}$ direction and are given as

$$\boldsymbol{A}_{ij} = -[(k_{12} + k_{23})a_i a_j + k_{23}b_i b_j]\hat{z} \,. \tag{14.54}$$

14.5.5 Spatial Separation Using a Cold Molecular Trap

A proposal [683, 684] for the experimental implementation of the spatial separation of different enantiomers using a cold molecular trap is shown in Figure 14.21. In the absence of the separating laser field, the molecules constitute a racemic mixture composed of the $|\alpha, n)$; $\alpha = L, D$; $n = 1, 2, 3$, bare states. When the separating lasers are switched on the dressed states $|\alpha, i\rangle$; $\alpha = L, D$; $i = 1, 2, 3$ become the relevant ones. Each bare state $|\alpha, n)$ can be expressed as a linear combination of dressed states (the inverse of the unitary transformation of Eq. (14.51)),

$$|\Psi_n^{(\alpha)}(\boldsymbol{r}_{start})\rangle = |\alpha, n) = \sum_i S_{i,n}^{*(\alpha)}(\boldsymbol{r}_{start}) \cdot |\alpha, i\rangle \,. \tag{14.55}$$

At a time denoted as $t = 0$, one turns off the trapping potential, thereby releasing the molecules and allowing them to move under the joint influence of the gravitational and light-induced forces. For each dressed state, the classical equations of motion of the (x_i, y_i, z_i) CM coordinates and their conjugate momenta (p_{ix}, p_{iy}, p_{iz}) are given as

$$\dot{x}_i = \frac{p_{ix}}{m} \,, \quad \dot{y}_i = \frac{p_{iy}}{m} \,, \quad \dot{z}_i = \frac{p_{iz} - A_{iz}}{m} \tag{14.56}$$

and

$$\dot{p}_{ix} = \frac{\partial A_{iz}}{\partial x}\dot{z}_i - \frac{\partial V_i}{\partial x} \,, \quad \dot{p}_{iy} = \frac{\partial A_{iz}}{\partial y}\dot{z}_i - \frac{\partial V_i}{\partial y} \,, \quad \dot{p}_{iz} = mg \,. \tag{14.57}$$

Since in the adiabatic approximation the dressed states are decoupled from one another, these equations can be solved independently for each i states.

As pointed out above, the enantio-selectivity of the CCAP scenario is due to Eq. (14.43), arising from the sign difference between the L and the D electric dipole $(\alpha, m|\boldsymbol{d}|\alpha, n)$ matrix elements [678]. As shown in Figure 14.22, this sign difference mainly affects the scalar potential $V_i^{(\alpha)}(x, y)$. In order to minimize population losses in the higher ($i = 2, 3$) states, due for example to spontaneous emission and/or radiationless transitions, we only consider below ground electronic (vibrational) states. Hence, ω_{nm} are chosen to be in the mid-infrared range, corresponding to wavelengths of $\sim$ 5 μm. As a (transiently) chiral molecule, we consider D_2S_2 [680] and concentrate on the ν_5 vibrational mode whose fundamental absorption occurs at wavelengths of 5.137 μm. The rather weak transition dipoles for vibrational transitions ($\sim$ 0.025 D), are compensated for by placing the molecular sample *inside* the laser cavity. In this way the Rabi frequencies, Ω_{nm}, are enhanced by $\sqrt{Q}$. The Q-factor can reach values of $\sim$ 1000 for cavities with high quality reflective mirrors.

The parameters operating in this setup are explicitly,

$$|\Omega_{12}^{\circ}| = |\Omega_{13}^{\circ}| = |\Omega_{23}^{\circ}| = 1 \times 10^{-9}\,\text{a.u.} = 6.58\,\text{MHz} \tag{14.58}$$

and

$$\sigma_{13} = \sigma_{23} = 10\,\mu\text{m}\,, \quad \sigma_{12} = 7\,\mu\text{m}\,, \quad x_{13} = -x_{23} = 3\,\mu\text{m}\,, \quad x_{12} = 0\,. \tag{14.59}$$

In order to reduce the computational effort, especially for the quantum simulations, the spatial distribution of the CW lasers is chosen much wider in $\hat{y}$ than that in $\hat{x}$: $\sigma_y = 100$ μm. In this configuration, the potential energy surfaces $V_i(x, y)$ is much flatter in $\hat{y}$ than that in $\hat{x}$. The integration of Eqs. (14.56) and (14.57) is started at the following values: $x_i(t = 0) = x_o$, $y_i(t = 0) = y_o$, $z_i(t = 0) = z_o$, $\dot{x}_i(t = 0) = v_{xo}$, $\dot{y}_i(t = 0) = v_{yo}$ and $\dot{z}_i(t = 0) = v_{zo}$. At $t = 0$ the spread in space of the trapped molecules is assumed to be described by a Gaussian function of x and y,

$$\rho(x, y) = \left(\pi\sigma_r^2\right)^{-1/2} \exp\left(-(x^2 + y^2)/\sigma_r^2\right) \tag{14.60}$$

where $\sigma_r = 3$ μm. The spread in the initial translational velocity v_{xo}, v_{yo} and v_{zo} is assumed to follow a Boltzmann distribution,

$$\rho(v) \sim \exp\left(-\frac{0.5mv^2}{k_B T}\right)\,, \tag{14.61}$$

where $v = \sqrt{v_{xo}^2 + v_{yo}^2 + v_{zo}^2}$ and $T = 1$ mK. The spatial distribution in the $\hat{z}$ direction need not be specified because the calculations indicate that $\hat{z}$ direction of motion is almost that of a free-fall, that is, it is hardly affected by the LIPs. Therefore $z_o = 0$ is chosen.

It is possible to numerically propagate the CM motion for each of the six dressed states. In the $\hat{x}$ direction, it is sufficient to study the motion of the dressed states in the $x \in [-35$ μm, 35 μm] "observation window". Simulation performed on 2197 molecules show that after $t \approx 20$ μs it is possible to collect $|L, 1\rangle$ molecules at the edges of the observation window with essentially 100% purity! The separation of $|D, 1\rangle$ molecules is almost as good: molecules belonging to this state can be collected with 94% purity after a run time of $\sim$ 40 μs.

Sample trajectories demonstrating the separation of molecules of various states in the $\hat{x}$ direction are shown in Figure 14.23a. These trajectories originate at $v_{xo} =$

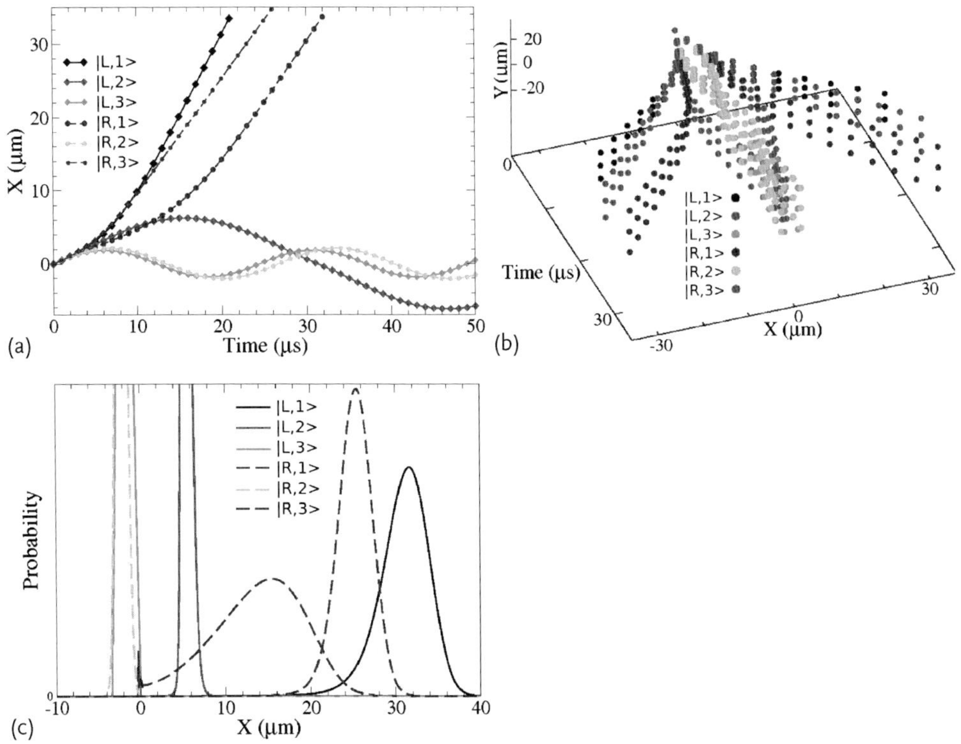

Figure 14.23 The CM motion for the trap configuration: (a) typical $\hat{x}$ trajectories for different states; (b) typical 3D trajectories for different states; (c) quantum wave packets at $t = 20\,\mu$s in $\hat{x}$. Taken from [684, Figure 3]. (Reprinted with permission. Copyright 2010 American Institute of Physics.)

$v_{yo} = 0.5\,\text{m/s}$ ($E_o = 1\,\text{mK}$) and $x_o = y_o = 0$. Three dimensional (3D) trajectories for sample states are exhibited in Figure 14.23b. These trajectories originate at $v_{xo}, v_{yo} = \pm 0.5\,\text{m/s}$ ($E_o = 1\,\text{mK}$) and $x_o = y_o = 0$. As evident from Figure 14.23b, the molecules do not fly beyond the $y \in [-40, 40]\,\mu$m region where the potentials are still flat. Hence, the $\hat{x}$ motion can be approximately decoupled from the $\hat{y}$ motion. Therefore, it is sufficient to perform quantum mechanical wave packet propagation in the $\hat{x}$ direction while using classical simulations for the $\hat{y}$ direction.

Results for the $\hat{x}$ motion wave packet propagation after 20 μs are given in Figure 14.23c. All six states are assumed to be given initially by the same Gaussian shaped wave packet whose average position is $x_o(t = 0) = 0$; average velocity – $v_{xo}(t = 0) = 0.5\,\text{m/s}$; and RMS velocity spread – $\Delta_{vx} = 0.25\,\text{m/s}$. In good agreement with the classical simulations of Figure 14.23b, the quantum wave packet calculations predict an excellent degree of spatial separation of the $|L, 1\rangle$ state from all other states.

It is possible to improve the above separation by starting out with a racemic mixture composed of a pair of *dressed* states. For example, the initial system can be prepared as a mixture of the $|L,3\rangle$ and $|D,3\rangle$ dressed states. After the trapping potential is turned off, the $|D,3\rangle$ molecules escape the observation window while the $|L,3\rangle$ molecules stay close to their initial positions. The parameters of the classical simulations are chosen to be identical to those of Eqs. (14.60) and (14.61). The parameters of the quantum simulations are $x_o(t=0)=0$, $v_{xo}(t=0)=0$ and $\Delta_{vx}=0.5\,\mathrm{m/s}$. The classical and quantum simulations agree with each other qualitatively, as shown in Figure 14.24, with the D enantiomer splitting into two parts which strongly separate from one another, while the L enantiomer essentially remaining in its initial place.

The spatial separation process is controlled by the competition between the light-induced forces and the initial (thermal) velocity spread. Clearly, the difference between the LIP of the L/D forms needs to be larger than the initial temper-

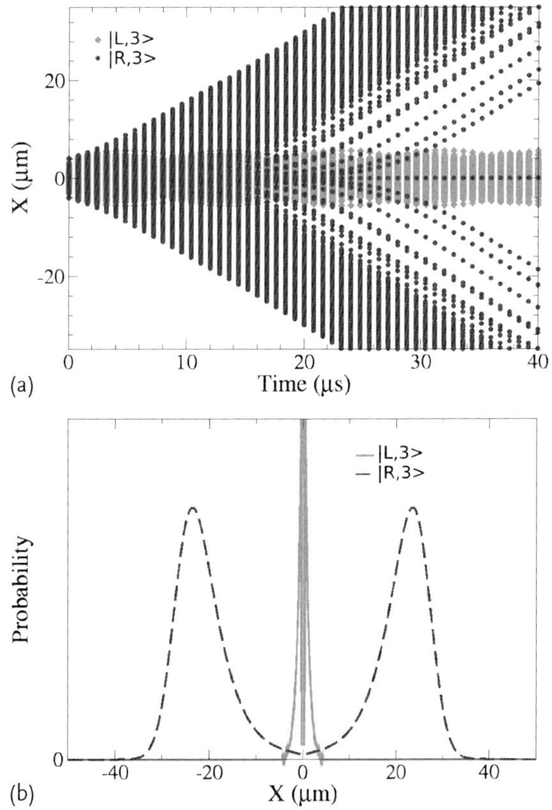

Figure 14.24 CM motion of a pair of $|L,3\rangle|D,3\rangle$ *dressed* states: (a) Classical calculations for 2197 molecular CM trajectories; (b) results of a quantum wave packet propagation at $t=22.5\,\mu\mathrm{s}$, $x_o(t=0)=0$, $v_{xo}(t=0)=0\,\mathrm{m/s}$ and $\Delta_{vx}=0.5\,\mathrm{m/s}$. Taken from [684, Figure 4]. (Reprinted with permission. Copyright 2010 American Institute of Physics.)

ature of the system. The choice of (an attainable) initial temperature of 1 mK ($\approx 3.2 \times 10^{-9}$ a.u.) means that the height difference of the scalar potentials between the two fastest moving states, $|L, 1\rangle$ and $|D, 1\rangle$, must be larger than this. In order to boost the effective height difference in the scalar LIPs an intracavity setup is used. Using a high Q factor of 1000, increases the Rabi frequency by a factor of 30. The effective height differences of the scalar LIPs increases to to 2.7×10^{-8} a.u. which is much larger than the 1 mK thermal energy.

14.A Appendix: Computation of $B-A-B'$ Enantiomer Selectivity

Section 14.2 requires the computation of the $B-A-B'$ photodissociation matrix elements in Eq. (14.12). To carry out this computation, and to understand the origin of the loss of control with m averaging, we sketch the details of the formalism for the model H_aOH_b system.

First, it is necessary to specify the relevant $\boldsymbol{n}$ and i quantum numbers that enter the bound–free matrix elements $\langle E, \boldsymbol{n}, q^-|\mathrm{d}_{e,g}|E_i\rangle$. For the continuum states, $\boldsymbol{n} = \{\hat{\boldsymbol{k}}, v, j, m\}$ where $\hat{\boldsymbol{k}}$ is the scattering direction, v and j are the vibrational and rotational product quantum numbers and m is the space-fixed z-projection of j. For the bound states, $|E_i\rangle$ actually denotes $|E_i, M_i, J_i, p_i\rangle$, where J_i, M_i and p_i are respectively, the bound state angular momentum, its space-fixed z-projection and its parity. The full (6-dimensional) bound–free matrix element can be written as a product of analytic functions involving $\hat{\boldsymbol{k}}$ and (three-dimensional) radial matrix elements:

$$
\begin{aligned}
&\langle E, \hat{\boldsymbol{k}}, v, j, m, q^-|\mathrm{d}_{e,g}|E_i, M_i, J_i, p_i\rangle = \\
&\left(\frac{\mu k_{vj}}{2\pi^2\hbar^2}\right)^{1/2} \sum_{J\lambda} (2J+1)^{1/2}(-1)^{M_i-j-m} D^J_{\lambda M_i}(\phi_k, \theta_k, 0) \\
&\times D^j_{-\lambda-m}(\phi_k, \theta_k, 0)\, t^{(q)}(E, J, \mathrm{v}, j, \lambda|E_i J_i p_i) \,. \qquad (14.A.1)
\end{aligned}
$$

Here the $D^J_{\lambda M_i}(\phi_k, \theta_k, 0)$ are the rotation matrices [326, 701], ϕ_k, θ_k are the scattering angles, μ is the reduced mass, k_{vj} is the momentum of the products and $t^{(q)}(E, J, \mathrm{v}, j, \lambda|E_i J_i p_i)$ is proportional to the radial partial wave matrix element [672] $\langle E, J, M, p, \mathrm{v}, j, \lambda, q^-|\mathrm{d}_{e,g}|E_i, M_i, J_i, p_i\rangle$. Here λ is the projection of J along the body fixed axis of the H–OH (c.m.) product separation.

The product of the bound–free matrix elements of Eq. (14.A.1), which enter Eq. (3.79), integrated over scattering angles and averaged over the initial [667]

$M_k(= M_i)$ quantum numbers, is

$$(2J_i+1)^{-1}\sum_{M_i}\int d\hat{\boldsymbol{k}}\langle E_k, M_i, J_k, p_k|\mathrm{d}_{e,g}|E,\hat{\boldsymbol{k}},\mathrm{v},j,m,q'^-\rangle$$
$$\times\langle E,\hat{\boldsymbol{k}},\mathrm{v},j,m,q^-|\mathrm{d}_{e,g}|E_i,M_i,J_i,p_i\rangle = (-1)^m\frac{8\pi\omega\mu}{\hbar^2(2J_i+1)}$$
$$\times\sum_{\mathrm{v}j}k_{\mathrm{v}j}\sum_{J\lambda J'\lambda'}[(2J+1)(2J'+1)]^{1/2}(-1)^{\{\lambda-\lambda'+J+J'+J_i\}}$$
$$\times\sum_{\ell=0,2}(2\ell+1)\begin{pmatrix}J & J' & \ell\\ \lambda & -\lambda' & \lambda'-\lambda\end{pmatrix}\begin{pmatrix}j & j & \ell\\ -\lambda & \lambda' & \lambda-\lambda'\end{pmatrix}$$
$$\times\begin{pmatrix}1 & 1 & \ell\\ 0 & 0 & 0\end{pmatrix}\begin{pmatrix}j & j & \ell\\ -m & m & 0\end{pmatrix}\begin{Bmatrix}1 & 1 & \ell\\ J & J' & J_i\end{Bmatrix}$$
$$\times\, t^{(q)}(EJ\mathrm{v}j\lambda p|E_iJ_ip_i)t^{(q')*}(EJ'\mathrm{v}j\lambda' p|E_kJ_kp_k)\,. \qquad (14.A.2)$$

Here $J_i = J_k$ has been assumed for simplicity. The $t^{(q)}$ matrix elements are computed with the artificial channel method [8].

Finally, we sketch the effect of a summation over product m states on symmetry breaking and chirality control. In this regard the three body model is particularly informative. Specifically, note that Eq. (14.12) provides $\mathrm{d}_q(32)$ in terms of products of matrix elements involving $|E,\boldsymbol{n},a^-\rangle$ and $|E,\boldsymbol{n},s^-\rangle$. Focus attention on those products that involve both wave functions, for example, terms like $S_{s3}S^*_{a2}$. These matrix element products can be written in the form of Eq. (14.A.2) where q and q' now refer to the antisymmetric or symmetric continuum states, rather than to channels D or L. Thus, for example, $S_{s3}S^*_{a2}$ results from using $|E,\boldsymbol{m},s^-\rangle$ in Eq. (14.A.1) to form S_{s3} and $|E,\boldsymbol{n},a^-\rangle$ to form S^*_{a2}. The resultant $S_{s3}S^*_{a2}$ has the form of Eq. (14.A.2) with $t^{(q)}$ and $t^{(q')}$ associated with the symmetric and antisymmetric continuum wave functions, respectively. Consider now the effect of summing over m. Standard formulae [672, 702] imply that this summation introduces a $\delta_{\ell,0}$ which, in turn forces $\lambda = \lambda'$ via the first and second $3j$ symbol in Eq. (14.A.2). However, it is possible to show that t matrix elements associated with symmetric continuum eigenfunctions and those associated with antisymmetric continuum eigenfunctions must have λ of different parities. Hence summing over m eliminates all contributions to Eq. (14.13) that involve both $|E,\boldsymbol{n},a^-\rangle$ and $|E,\boldsymbol{n},s^-\rangle$. Thus, we find after m summation:

$$\sum_m \mathrm{d}_L(23) = \sum_m \mathrm{d}_D(23) = 0\,. \qquad (14.A.3)$$

That is, control over the enantiomer ratio is lost upon m summation, both channels $q = D$ and $q = L$ having equal photodissociation probabilities.

15
Strong-Field Coherent Control

Photodissociation and control by strong electromagnetic fields, the topic of this chapter, is best treated by introducing states of the matter–radiation field, the so-called "dressed" states. This is most naturally done within the framework of a quantized electromagnetic fields, a topic introduced in Section 10.1.

15.1
Strong-Field Photodissociation with Continuous Wave Quantized Fields

In a photodissociation process induced by continuous wave (CW) radiation, we envision the molecule as existing initially ($t = -\infty$) in a state which does not interact with the field, that is, in an eigenstate $|E_i, N_i\rangle$ of H_f:

$$(E_i + N_i\hbar\omega_i - H_f)|E_i, N_i\rangle = 0 . \tag{15.1}$$

Here $|E_i, N_i\rangle = |E_i\rangle|N_i\rangle$ and where $|E_i\rangle$ is a bound state of H_M (defined in Eq. (2.1)) and $|N_i\rangle$ denotes a free radiation state with N_i photons in the $\boldsymbol{k}_i$ mode, whose frequency is ω_i. The molecule then interacts with the field, and we are interested in determining the photodissociation probability, defined as the probability of eventually (i.e., as $t \to \infty$) populating an eigenstate of H_f, where the molecule is dissociated. (Strictly speaking, as discussed in Chapter 2, the photodissociation probability is the probability of populating a (radiatively decoupled *and* materially decoupled) continuum eigenstate of $H_0 + H_R$, where H_0 is the free Hamiltonian of Eq. (2.39). However, as shown in Chapter 2, the nature of the incoming eigenstates of H_M, $|E, \boldsymbol{n}^-\rangle$, is such that the probability of populating a specific $|E, \boldsymbol{n}^-\rangle$ state at asymptotic times is identical to the probability of populating the $|E, \boldsymbol{n}; 0\rangle$ eigenstate of H_0.)

Following our strategy in the weak-field domain (see Chapter 3), we do not obtain the photodissociation probability by actually following the dynamics for long times. Rather, we calculate, at any given time, the transition probability to the particular fully interacting state that is guaranteed to evolve to the radiatively decoupled state of interest as $t \to \infty$.

The fully interacting incoming Hamiltonian eigenstates are denoted $|E, \boldsymbol{n}^-, \boldsymbol{N}^-\rangle$, where $\boldsymbol{N}$ is a vector of photon occupation numbers, $\boldsymbol{N} = (N_{\boldsymbol{k}_1}, N_{\boldsymbol{k}_2}, \ldots, N_{\boldsymbol{k}_m}, \ldots)$.

Quantum Control of Molecular Processes, Second Edition. Moshe Shapiro and Paul Brumer

These states are the strong-field analogs of the material incoming states, $|E, \boldsymbol{n}^-\rangle$, defined in Eq. (2.52). In order to find such fully interacting incoming states, we consider a particular radiatively decoupled state of interest in the distant future and evolve it backward in time to the present. This back-evolution is done as the field–matter interaction H_{MR} is switched on. To do so it is convenient to introduce the following notation:

$$|E, \boldsymbol{n}^-, \boldsymbol{N}^-, t\rangle \equiv \exp\left(-iHt/\hbar\right)|E, \boldsymbol{n}^-, \boldsymbol{N}^-\rangle = \exp\left(-iE_T t/\hbar\right)|E, \boldsymbol{n}^-, \boldsymbol{N}^-\rangle\,, \tag{15.2}$$

where $E_T = E + \sum_k N_k \hbar\omega_k$ is the total (matter + radiation) energy of the state $|E, \boldsymbol{n}^-, \boldsymbol{N}^-\rangle$, and

$$|E, \boldsymbol{n}^-, \boldsymbol{N}, t\rangle \equiv |E, \boldsymbol{n}^-, t\rangle|\boldsymbol{N}\rangle \equiv |E, \boldsymbol{n}^-\rangle|\boldsymbol{N}\rangle \exp\left(-iEt/\hbar\right)\,. \tag{15.3}$$

Note that the absence of the minus superscript in the state $|E, \boldsymbol{n}^-, \boldsymbol{N}, t\rangle$ implies that the molecule and radiation field are no longer coupled.

The desired fully interacting state is then

$$\begin{aligned}|E, \boldsymbol{n}^-, \boldsymbol{N}^-, t\rangle &= \lim_{t_1 \to \infty} \exp\left[-iH(t - t_1)/\hbar\right]|E, \boldsymbol{n}^-, \boldsymbol{N}^-, t_1\rangle \\ &= \lim_{t_1 \to \infty} \exp\left[-iH(t - t_1)/\hbar\right]|E, \boldsymbol{n}^-, t_1\rangle|\boldsymbol{N}\rangle\,.\end{aligned} \tag{15.4}$$

Equation (15.4) uses the fact that we can prove, in complete analogy to the proof given in the context of the material incoming states (see Eq. (2.57)), that as $t \to \infty$ the fully interacting states go over to the states wherein the matter and radiation are decoupled, but where the matter itself is interacting. That is,

$$|E, \boldsymbol{n}^-, \boldsymbol{N}^-, t\rangle \xrightarrow{t\to\infty} |E, \boldsymbol{n}^-, t\rangle|\boldsymbol{N}\rangle\,. \tag{15.5}$$

In turn, as the molecular fragments separate in the photodissociation process, these states go over to the noninteracting radiatively decoupled states,

$$|E, \boldsymbol{n}^-, t\rangle|\boldsymbol{N}\rangle \xrightarrow{t\to\infty} |E, \boldsymbol{n}; 0\rangle|\boldsymbol{N}\rangle \exp\left(-iEt/\hbar\right)\,. \tag{15.6}$$

As in the case of the incoming states for the pure material part (Eq. (2.52)), the $|E, \boldsymbol{n}^-, \boldsymbol{N}^-\rangle$ states (Eq. (15.2)) satisfy a modified Schrödinger equation:

$$\lim_{\epsilon \to +0}\left(E - i\epsilon + \sum_k N_k \hbar\omega_k - H\right)|E, \boldsymbol{n}^-, \boldsymbol{N}^-\rangle = 0\,. \tag{15.7}$$

Note that the total energy of the system is $E_T = E + \sum_k N_k \hbar\omega_k$, but neither E nor $\boldsymbol{N}$ are good quantum numbers at other than asymptotic times. That is, the molecule and radiation field exchange energy. As in the case of $\boldsymbol{n}^-$ for the pure material case (Eq. (2.52)), the $\boldsymbol{N}^-$ notation simply serves as a reminder that the incoming fully interacting states correlate at long times with the radiatively decoupled state, $|E, \boldsymbol{n}^-\rangle|\boldsymbol{N}\rangle$.

The proper limiting behavior is, as in the pure material case, achieved by adding $-i\epsilon$ to the energy (see Eq. (2.52)). This is equivalent to multiplying the radiation–matter interaction by a slowly decaying function to produce a time-dependent $H_{MR}(t)$, where

$$H_{MR}(t) \equiv H_{MR} \exp(-\epsilon t/\hbar) \xrightarrow{t\to\infty} 0 . \tag{15.8}$$

Since the state $|E, \boldsymbol{n}^-, \boldsymbol{N}^-, t\rangle$ contains the effect of the full Hamiltonian at time t, then the photodissociation amplitude $A(E, \boldsymbol{n}, \boldsymbol{N}, t|i, N_i)$ into the final state with energy E, internal quantum numbers $\boldsymbol{n}$ and radiation field described by $\boldsymbol{N}$, starting in the initial state $|E_i, N_i\rangle$ is simply the overlap between the radiatively decoupled initial state and the incoming fully interacting state. That is,

$$\begin{aligned} A(E, \boldsymbol{n}, \boldsymbol{N}, t|i, N_i) &= \langle E, \boldsymbol{n}^-, \boldsymbol{N}^-, t| E_i, N_i\rangle \\ &= \langle E, \boldsymbol{n}^-, \boldsymbol{N}^-| \exp(iHt/\hbar) |E_i, N_i\rangle , \end{aligned} \tag{15.9}$$

where the second equality in Eq. (15.9) arises from Eq. (15.2).

This expression is most readily evaluated for the slow-turn-on interaction of Eq. (15.8) through the formula:

$$e^{iHt/\hbar} = \lim_{\epsilon\to+0, t_1\to\infty} \left[1 - \frac{i}{\hbar}\int_t^{t_1} dt' e^{iHt'/\hbar} H_{MR}(t') e^{-iH_f t'/\hbar}\right] e^{iH_f t/\hbar} . \tag{15.10}$$

To obtain Eq. (15.10) we use the equality:

$$\begin{aligned} &-i\hbar\frac{d}{dt}\left[\exp(iHt/\hbar)\exp\left(-iH_f t/\hbar\right)\right] = \exp(iHt/\hbar)\, H_{MR}(t)\exp\left(-iH_f t/\hbar\right) \\ &+ \exp(iHt/\hbar)\, t\frac{dH_{MR}(t)}{dt}\exp\left(-iH_f t/\hbar\right) . \end{aligned}$$

Integrating both sides of this equation, and making the time dependence of the Hamiltonians explicit, gives:

$$\begin{aligned} &\exp(iH(t)t/\hbar)\exp\left(-iH_f t/\hbar\right) = \exp(iH(t_1)t_1/\hbar)\exp\left(-iH_f t_1/\hbar\right) \\ &- i/\hbar\left[\int_t^{t_1}\exp\left(iH(t')t'/\hbar\right) H_{MR}(t')\exp\left(-iH_f t'/\hbar\right) dt'\right. \\ &\left. + \int_t^{t_1}\exp\left(iH(t')t'/\hbar\right) t'\frac{dH_{MR}(t')}{dt'}\exp\left(-iH_f t'/\hbar\right) dt'\right] . \end{aligned} \tag{15.11}$$

Given the behavior of $H_{MR}(t)$ (Eq. (15.8)), the second term in the square brackets vanishes in the $\epsilon \to 0$ limit. Rearranging terms, we obtain Eq. (15.10) in the $\epsilon \to 0$ and $t_1 \to \infty$ limits.

Substituting Eq. (15.10) in Eq. (15.9) and using Eq. (15.7) gives:

$$A(E, \boldsymbol{n}, \boldsymbol{N}, t|i, N_i) = -\frac{i}{\hbar} \lim_{\epsilon \to +0} \lim_{t_1 \to \infty}$$

$$\times \langle E, \boldsymbol{n}^-, \boldsymbol{N}^- | \int_t^{t_1} dt' e^{i[E + i\epsilon/\hbar + \sum_k N_k \omega_k]t'} H_{MR} t' e^{-i(E_i/\hbar + N_i \omega_i)(t'-t)} | E_i, N_i \rangle$$

$$= \lim_{\epsilon \to +0} \frac{\langle E, \boldsymbol{n}^-, \boldsymbol{N}^- | H_{MR} t' | E_i, N_i \rangle \exp\left\{ i \left[(E + i\epsilon)/\hbar + \sum_k N_{\boldsymbol{k}} \omega_k \right] t \right\}}{E + i\epsilon - E_i + \sum_{\boldsymbol{k}} N_{\boldsymbol{k}} \hbar \omega_k - N_i \hbar \omega_i}, \tag{15.12}$$

where we have used the orthogonality between the bound and the continuum eigenstates of the material Hamiltonian, that is, $\langle E_i | E, \boldsymbol{n}^- \rangle = 0$.

The CW rate of transition $R(E, \boldsymbol{n}, \boldsymbol{N}|i, N_i)$, into the radiatively decoupled state $|E, \boldsymbol{n}^-, \boldsymbol{N}\rangle$, can be obtained as the rate of change of the photodissociation probability as the interaction is being slowly switched on,

$$R(E, \boldsymbol{n}, \boldsymbol{N}|i, N_i) = -\frac{d}{dt} |A(E, \boldsymbol{n}, \boldsymbol{N}, t|i, N_i)|^2 . \tag{15.13}$$

The minus sign is introduced in Eq. (15.13) because in our expressions for $|A(E, \boldsymbol{n}, \boldsymbol{N}, t|i, N_i)|^2$ the interaction is being switched off, rather than switched on. Using Eq. (15.13) we obtain that

$$R(E, \boldsymbol{n}, \boldsymbol{N}|i, N_i) = \lim_{\epsilon \to +0} \frac{2\epsilon}{\hbar} \frac{\exp\left(-\frac{2\epsilon t}{\hbar}\right) |\langle E, \boldsymbol{n}^-, \boldsymbol{N}^- | H_{MR} | E_i, N_i \rangle|^2}{\left(E - E_i + \sum_{\boldsymbol{k}} N_{\boldsymbol{k}} \hbar \omega_k - N_i \hbar \omega_i\right)^2 + \epsilon^2}$$

$$= \frac{2\pi}{\hbar} |\langle E, \boldsymbol{n}^-, \boldsymbol{N}^- | H_{MR} | E_i, N_i \rangle|^2 \, \delta \left(E - E_i + \sum_{\boldsymbol{k}} N_{\boldsymbol{k}} \hbar \omega_k - N_i \hbar \omega_i \right) . \tag{15.14}$$

The adiabaticity of the switch-off (i.e., the $\epsilon \to +0$ limit), has yielded a CW rate that is independent of time. Notice that the δ-function that appears in the transition rate expression guarantees that the transition rate is zero if the total energy is not conserved. By contrast, $A(E, \boldsymbol{n}, \boldsymbol{N}, t|i, N_i)$, the instantaneous transition amplitude, which is the overlap between a radiatively decoupled and a fully interacting state, permits a spread in the final energies observed.

Given the explicit form of H_{MR} (Eq. (10.18)), the expression for the transition rate assumes the form:

$$R(E, \boldsymbol{n}, \boldsymbol{N}|i, N_i) = \delta \left(E - E_i + \sum_{\boldsymbol{k}} N_{\boldsymbol{k}} \hbar \omega_k - N_i \hbar \omega_i \right) \sum_{\boldsymbol{k}} \frac{\pi \omega_k}{\varepsilon_0 V}$$

$$\times \left| \langle E, \boldsymbol{n}^-, \boldsymbol{N}^- | \hat{\boldsymbol{\varepsilon}}_k \cdot \boldsymbol{d} \left[\hat{a}_k \exp(i\boldsymbol{k} \cdot \boldsymbol{r} + i\phi_k) - \hat{a}_k^\dagger \exp(-i\boldsymbol{k} \cdot \boldsymbol{r} - i\phi_k) \right] | E_i, N_i \rangle \right|^2 . \tag{15.15}$$

Since we are dealing with strong fields, we can assume that N_i, the number of photons in the incident beam, is very large compared to unity. Using the properties of the creation and annihilation operators,

$$\hat{a}_k^\dagger|N_i\rangle = \begin{cases} (N_i+1)^{1/2}|N_i+1\rangle\,, & \boldsymbol{k}=\boldsymbol{k}_i\,,\\ 2^{1/2}|1_k\rangle|N_i\rangle\,, & \boldsymbol{k}\neq\boldsymbol{k}_i\,,\end{cases} \tag{15.16}$$

$$\hat{a}_k|N_i\rangle = \begin{cases} N_i^{1/2}|N_i-1\rangle\,, & \boldsymbol{k}=\boldsymbol{k}_i\,,\\ 0\,, & \boldsymbol{k}\neq\boldsymbol{k}_i\,,\end{cases} \tag{15.17}$$

we obtain when $N_i \gg 1$, that is, when $N_i+1\approx N_i$, that

$$\begin{aligned} R(E,\boldsymbol{n},\boldsymbol{N}|i,N_i) &= \frac{\pi\omega_i N_i}{\varepsilon_0 V}\delta\left(E-E_i+\sum_k N_k\hbar\omega_k - N_i\hbar\omega_i\right)\\ &\times\Big|\langle E,\boldsymbol{n}^-,\boldsymbol{N}^-|\hat{\boldsymbol{\varepsilon}}_i\cdot\boldsymbol{d}\,[|N_i-1\rangle\exp(i\omega_i z/c+i\phi_i)\\ &-|N_i+1\rangle\exp(-i\omega_i z/c-i\phi_i)]\,|E_i\rangle\Big|^2\,, \end{aligned} \tag{15.18}$$

where ϕ_i denotes $\phi(\omega_i)$ and z is the direction of propagation of the incident light beam. Here the contributions from the states $|1_k\rangle$ with $\boldsymbol{k}\neq\boldsymbol{k}_i$ are neglected since $N_i\gg 1$. Equation (15.18) applies to transitions to any combination of final photon number states (i.e., to any multiphoton process) and is correct to all orders of the radiation strength.

In the weak-field limit, and when only one photon is absorbed, we can approximate $|E,\boldsymbol{n}^-,\boldsymbol{N}^-\rangle$ by $|E,\boldsymbol{n}^-\rangle|N_i-1\rangle$. In addition, for visible light $\exp(i\omega_i z/c)$ is essentially constant over atomic and molecular dimensions (the dipole approximation). Under these conditions, with the initial state $|N_i\rangle$ we obtain from Eq. (15.15) the result, here called $R^{(1)}$:

$$R^{(1)}(E,\boldsymbol{n},N_i-1|i,N_i) = \frac{\pi\omega_i N_i}{\varepsilon_0 V}|\langle E,\boldsymbol{n}^-|\hat{\boldsymbol{\varepsilon}}_i\cdot\boldsymbol{d}|E_i\rangle|^2\,\delta(E-\hbar\omega_i-E_i)\,. \tag{15.19}$$

When we make the connection between $N_i^{1/2}$ and the incident radiation field envelope, we have

$$\mathcal{E}_i = i\left(\frac{\hbar\omega_i N_i}{2\epsilon_0 V}\right)^{1/2}\exp(i\omega_i z/c+i\phi_i)\,, \tag{15.20}$$

and we recover the weak-field expression, of Eq. (2.78),

$$R^{(1)}(E,\boldsymbol{n},N_i-1|i,N_i) = \frac{2\pi}{\hbar}|\mathcal{E}_i\langle E,\boldsymbol{n}^-|\hat{\boldsymbol{\varepsilon}}_i\cdot\boldsymbol{d}|E_i\rangle|^2\,\delta(E-\hbar\omega_i-E_i)\,. \tag{15.21}$$

15.1.1 The Coupled-Channels Expansion

In order to compute $R(E,\boldsymbol{n},\boldsymbol{N}|i,N_i)$ in the strong-field regime we must be able to evaluate the fully interacting wave functions $|E,\boldsymbol{n}^-,\boldsymbol{N}^-\rangle$. The "multichannel" aspect of the problem becomes quite involved for both dissociation of molecules [41, 42, 703–708] and the ionization of atoms [709–716] in the strong pulse regime.

General numerical methods for solving for the eigenfunctions of the radiatively coupled time-independent Schrödinger equation were developed by a number of research groups [703–708, 717, 718].

A powerful way of achieving this goal uses the "coupled-channels" expansion, a method widely used in calculations of scattering cross sections [6]. In the context of quantized matter–radiation problems, the coupled channels method amounts to expanding $|E, \boldsymbol{n}^-, \boldsymbol{N}^-\rangle$ in number states. Concentrating on the expansion in the ith mode, we write $|E, \boldsymbol{n}^-, \boldsymbol{N}^-\rangle$ as

$$|E, \boldsymbol{n}^-, \boldsymbol{N}^-\rangle = \sum_{N=N_i-m}^{N_i+m} |N\rangle\langle N|E, \boldsymbol{n}^-, \boldsymbol{N}^-\rangle \,. \tag{15.22}$$

Note that $\langle N|E, \boldsymbol{n}^-, \boldsymbol{N}^-\rangle$ are quantum states in the space of the material subsystem.

Using the orthogonality of the number states $\{|\boldsymbol{N}\rangle\}$ along with Eqs. (10.18), (15.16), (15.17), we transform the Schrödinger equation (Eq. (15.7)) into a set of coupled differential equations, the so-called "coupled-channels equations":

$$\begin{aligned}&\left[E_i + (N_i - N)\hbar\omega_i - H_M\right]\langle N|E, \boldsymbol{n}^-, \boldsymbol{N}^-\rangle = -i\left(\frac{\hbar\omega}{2\varepsilon_0 V}\right)^{1/2} \hat{\boldsymbol{\varepsilon}}_i \cdot \boldsymbol{d}\\ &\quad\times\left[(N+1)^{1/2}\exp\left(-i\omega_i z/c - i\phi_i\right)\langle N+1|E, \boldsymbol{n}^-, \boldsymbol{N}^-\rangle\right.\\ &\quad\left.-N^{1/2}\exp\left(i\omega_i z/c + i\phi_i\right)\langle N-1|E, \boldsymbol{n}^-, \boldsymbol{N}^-\rangle\right] ,\\ &N = N_i - m, \ldots, N_i + m \,.\end{aligned} \tag{15.23}$$

For a dissociation process in which only one net photon is absorbed, the final photon occupation number state is $|N_i - 1\rangle \equiv |0, 0, \ldots, N_i - 1, 0, \ldots\rangle$, that is, $E = E_i + \hbar\omega_i$. If $N \gg 1$, we can equate $(N+1)^{1/2} \approx N^{1/2} \approx N_i^{1/2}$ and using Eq. (15.20), we obtain that

$$\begin{aligned}&\left[E_i + (N_i - N)\hbar\omega_i - H_M\right]\langle N|E, \boldsymbol{n}^-, N_i - 1^-\rangle =\\ &\quad -\hat{\boldsymbol{\varepsilon}}_i \cdot \boldsymbol{d} \cdot \left[\mathcal{E}_i\langle N+1|E, \boldsymbol{n}^-, N_i - 1^-\rangle + \mathcal{E}_i^*\langle N-1|E, \boldsymbol{n}^-, N_i - 1^-\rangle\right] ,\\ &N = N_i - m, \ldots, N_i + m \,.\end{aligned} \tag{15.24}$$

This, or a similar, set of equations was applied to a large number of problems associated with molecular dissociation in intense laser fields [613, 719–721]. We now focus on the case of just two electronic states, defined as two eigenstates of the electronic Hamiltonian H_{el} (see Section 2.3.2):

$$\begin{aligned}&\left[H_{el}(\boldsymbol{y}|\boldsymbol{X}) - W_e(\boldsymbol{X})\right]\langle \boldsymbol{y}|e\rangle = 0 \,,\\ &\left[H_{el}(\boldsymbol{y}|\boldsymbol{X}) - W_{e'}(\boldsymbol{X})\right]\langle \boldsymbol{y}|e'\rangle = 0 \,.\end{aligned} \tag{15.25}$$

Here, $\boldsymbol{y}$ denotes the electronic coordinates, and $\boldsymbol{X}$ denotes all the nuclear coordinates $\{\boldsymbol{R}_\alpha\}$, $\alpha = 1, \ldots, n$, where n is the number of nuclei in the problem. The $W_j(\boldsymbol{X})$ eigenvalue is the potential experienced by the nuclei in the jth electronic state.

We expand each $\langle N|E, \boldsymbol{n}^-, N_i - 1^-\rangle$ component in the two electronic states,

$$\langle N|E, \boldsymbol{n}^-, N_i - 1^-\rangle = \big(|e\rangle\langle e| + |e'\rangle\langle e'|\big)\langle N|E, \boldsymbol{n}^-, N_i - 1^-\rangle = |e\rangle\langle N, e|E, \boldsymbol{n}^-, N_i - 1^-\rangle + |e'\rangle\langle N, e'|E, \boldsymbol{n}^-, N_i - 1^-\rangle\ ;$$

note that $\langle N, e|E, \boldsymbol{n}^-, N_i - 1^-\rangle$ and $\langle N, e'|E, \boldsymbol{n}^-, N_i - 1^-\rangle$ are quantum states in the space of the nuclear degrees of freedom.

Substituting in Eq. (15.24), we obtain, using Eq. (15.25), that

$$\begin{aligned}
&\left[E_i + (N_i - N)\hbar\omega_i - \sum_\alpha K_\alpha - W_e(\boldsymbol{X})\right]\langle N, e|E, \boldsymbol{n}^-, N_i - 1^-\rangle = -\hat{\boldsymbol{\varepsilon}}_i \cdot \boldsymbol{d}_{e,e'}(\boldsymbol{X}) \\
&\quad \cdot \left[\mathcal{E}_i\langle N+1, e'|E, \boldsymbol{n}^-, N_i - 1^-\rangle + \mathcal{E}_i^*\langle N-1, e'|E, \boldsymbol{n}^-, N_i - 1^-\rangle\right] , \\
&\left[E_i + (N_i - N)\hbar\omega_i - \sum_\alpha K_\alpha - W_{e'}(\boldsymbol{X})\right]\langle N, e'|E, \boldsymbol{n}^-, N_i - 1^-\rangle = -\hat{\boldsymbol{\varepsilon}}_i \cdot \boldsymbol{d}_{e',e}(\boldsymbol{X}) \\
&\quad \cdot \left[\mathcal{E}_i\langle N+1, e|E, \boldsymbol{n}^-, N_i - 1^-\rangle + \mathcal{E}_i^*\langle N-1, e|E, \boldsymbol{n}^-, N_i - 1^-\rangle\right] , \\
&N = N_i - m, \ldots, N_i + m\ ,
\end{aligned} \tag{15.26}$$

where K_α is the kinetic energy operator of the α nucleus, and $\hat{\boldsymbol{\varepsilon}}_i \cdot \boldsymbol{d}_{e,e'}(\boldsymbol{X}) \equiv \langle e|\hat{\boldsymbol{\varepsilon}}_i \cdot \boldsymbol{d}|e'\rangle$. We implicitly assume that the system has no permanent dipole moment, $\boldsymbol{d}_{e,e} = \boldsymbol{d}_{e',e'} = 0$.

The coupled channels expansion can be further simplified by introducing the (number state) rotating wave approximation (RWA), valid only when the field is of moderate intensity and the system is near resonance. As pointed out above, given an initial photon number state $|N_i\rangle$, the components of $|E, \boldsymbol{n}^-, N_i - 1^-\rangle$ of greatest interest for a one-photon transition are $\langle N_i|E, \boldsymbol{n}^-, N_i - 1^-\rangle$ and $\langle N_i \pm 1|E, \boldsymbol{n}^-, N_i - 1^-\rangle$. If $|E_i\rangle$ is the ground material state, then the $\langle N_i + m, e|E, \boldsymbol{n}^-, N_i - 1^-\rangle$, $m > 0$ components, and the $\langle N_i + m, e'|E, \boldsymbol{n}^-, N_i - 1^-\rangle$, $m \geq 0$ components are closed, since the total energy $E_i + N_i\hbar\omega_i$ is smaller than $E_j + (N_i + m)\hbar\omega_i$ for all E_j. This means that these components, though not necessarily zero at finite times, must vanish when the radiation and matter decouple at long times. For ω_i in the UV or visible range, the closed components are so far removed in energy from the initial state that for moderate field strengths they hardly affect $|E, \boldsymbol{n}^-, N_i - 1^-\rangle$. The same may be said about the $\langle e|\langle N_i - m|E, \boldsymbol{n}^-, N_i - 1^-\rangle$ open components. It is therefore reasonable in that case to only include the terms $\langle N_i, e|E, \boldsymbol{n}^-, N_i - 1^-\rangle$ and $\langle N_i - 1, e'|E, \boldsymbol{n}^-, N_i - 1^-\rangle$ in Eq. (15.26). Doing so, we obtain

$$\left[E_i - \sum_\alpha K_\alpha - W_e(\boldsymbol{X})\right]\langle N_i, e|E, \boldsymbol{n}^-, N_i - 1^-\rangle = -\,\mathcal{E}_i^*\hat{\boldsymbol{\varepsilon}}_i \cdot \boldsymbol{d}_{e,e'}(\boldsymbol{X})\langle N_i - 1, e'|E, \boldsymbol{n}^-, N_i - 1^-\rangle\ , \tag{15.27}$$

$$\left[E_i + \hbar\omega_i - \sum_\alpha K_\alpha - W_{e'}(\boldsymbol{X})\right]\langle N_i - 1, e'|E, \boldsymbol{n}^-, N_i - 1^-\rangle = -\,\mathcal{E}_i\hat{\boldsymbol{\varepsilon}}_i \cdot \boldsymbol{d}_{e',e}(\boldsymbol{X})\langle N_i, e|E, \boldsymbol{n}^-, N_i - 1^-\rangle\ . \tag{15.28}$$

Equations (15.27) and (15.28) constitute the number state RWA for moderate field photodissociation. They are nonperturbative within the basis set adopted, but are approximate in that they only incorporate a small number of number states and they neglect the contribution of all modes other than that of the incident beam.

An example for one of the first applications of the coupled channel equations with quantized fields for photodissociation problems is shown in Figure 15.1, where the dissociation of the IBr molecule by a two photon (visible + IR) process was studied [704]. The results of the calculations, shown in Figure 15.2, demonstrate how the strong IR photon broadens the transition ("power broadening") allowing the system to be dissociated even if the first photon is tuned substantially away from resonance. This illustrates how multiphoton transitions induced by strong fields [710] are less restricted by the need to be very close to an intermediate resonance, the situation described in Section 3.4.

The effect of strong infrared pulses on photodissociation dynamics involving the curve crossing situations shown schematically in Figure 15.3, which is akin to the laser catalysis scenario discussed in Chapter 13, was measured by Stolow *et al.* [722, 723]. In this experiment, the nonresonant dynamic Stark effect was used to control the electronic branching ratio and the avoided crossing induced predissociation in IBr. As shown in Figure 15.4 predissociation lifetimes could either be increased or decreased, and the branching ratios to the $I + Br^*(2P_{1/2})$ relative to the $I + Br(2P_{3/2})$

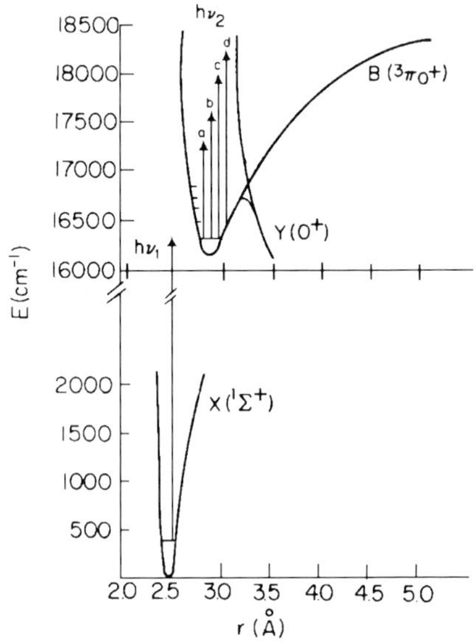

Figure 15.1 Energetics of the two color dissociation of IBr. The ground $X\,^1\Sigma^+$ potential curve and excited $Y(0^+)$ and $B^3\Pi_{0^+}$ states are shown. The intermediate level, at 16 333.03 cm^{-1} above the ground vibrational level of the molecule, is accessed by the weak $h\nu_1$ photon. Four $h\nu_2$ transitions were studied, and are marked as (a) 960 cm^{-1}, (b) 1282 cm^{-1}, (c) 1652 cm^{-1}, and (d) 1880 cm^{-1}. Taken from [704, Figure 1]. (Reprinted with permission. Copyright 1985 American Institute of Physics.)

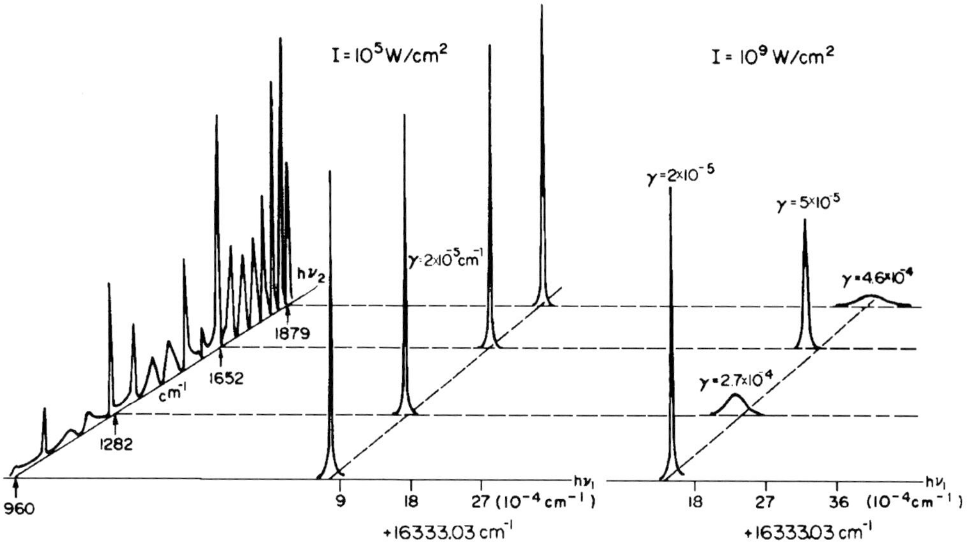

Figure 15.2 Effect of the strong laser field on the line shape for dissociation of an intermediate level at four $h\nu_2$ IR frequencies and at two intensities of the IR laser. The spectrum on the $h\nu_2$-axis (left hand side) is the computed IBr absorption spectrum in the weak-field limit, starting from 960 cm^{-1} above the intermediate level (which is 16 333.03 cm^{-1} above the ground vibrational level). We see that broadening of the 16 333.03 cm^{-1} line occurs for $I = 10^9$ W/cm^2 whenever the $h\nu_2$ photon is in near resonance with a strong predissociating line. Taken from [704, Figure 2]. (Reprinted with permission. Copyright 1985 American Institute of Physics.)

channels changed depending on the strength and timings of the short infrared pulses.

15.1.2
Number States vs. Classical Light

Note first that a pure number state is a state whose phase θ_k is evenly distributed between 0 and 2π. This is a consequence of the commutation relation [364] between $\hat{N}_k$ and $\hat{e}^{i\theta_k}$. Nevertheless, dipole matrix elements calculated between number states are (as all quantum mechanical amplitudes) well-defined complex numbers, and as such they have well-defined phases. Thus, the phases of the dipole matrix elements in conjunction with the mode phase ϕ_k (Eq. (10.18)) yield well-defined matter–radiation phases that determine the outcome of the photodissociation process. As in the weak-field domain, if only one incident radiation mode exists, the phase cancels out in the rate expression (Eq. (15.18)), provided that the RWA (Eqs. (15.27) and (15.28)) is adopted. However, in complete analogy with the treatment of weak-field control, if we operate on the material system with two or more radiation modes, the *relative* phase between them may have a pronounced effect on the fully interacting state, so that phase control is possible.

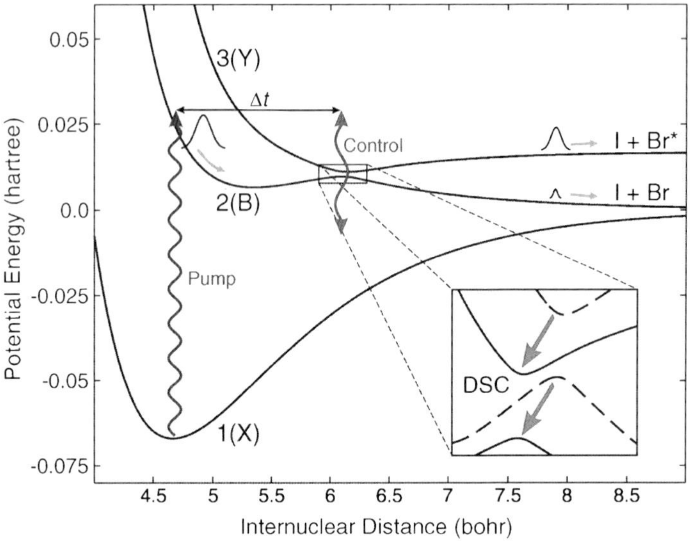

Figure 15.3 Schematics of the dynamic Stark control IBr dissociation experiment. As an excited state wave packet traverses a nonadiabatic crossing of states correlating to either the $I + Br(^2P_{3/2})$ or the $I + Br^* (^2P_{1/2})$ products, an ultrafast IR field is used to dynamically modify the adiabatic potential barrier (inset) via the Stark effect, mediating the reaction outcome. Because no transitions to other electronic states are involved, the system always remains on these two coupled potentials. Taken from [723, Figure 1]. (Reprinted with permission. Copyright 2006 Association for the Advancement of Science.)

In the weak-field limit it is possible to obtain the transition rates of Eq. (15.19) directly by solving Eq. (15.28) in first-order perturbation theory. This is done by assuming that $\langle N_i, e|E, \boldsymbol{n}^-, N_i - 1^-\rangle$ always remains larger than $\langle N_i - 1, e'|E, \boldsymbol{n}^-, N_i - 1^-\rangle$. It is therefore possible to neglect the right hand side of Eq. (15.27). Under these circumstances, Eq. (15.27) becomes identical to Eq. (2.1), whose solution is $|E_i\rangle$. This solution can be substituted in Eq. (15.28), which now becomes an inhomogeneous differential equation with a *known* source,

$$\left[E_i + \hbar\omega_i - \sum_\alpha K_\alpha - W_{e'}(\boldsymbol{X})\right]\langle e'|E, \boldsymbol{n}^-\rangle = -\mathcal{E}_i\hat{\boldsymbol{\varepsilon}}_i \cdot \boldsymbol{d}_{e',e}(\boldsymbol{X})\langle e|E_i\rangle \,. \qquad (15.29)$$

The continuum component $\langle e'|E, \boldsymbol{n}^-\rangle$ is therefore "driven" by the bound-state times the radiative coupling term. Driven equations of the type written here are implicitly used in all practical computational schemes developed for weak-field molecular photodissociation problems [8, 724].

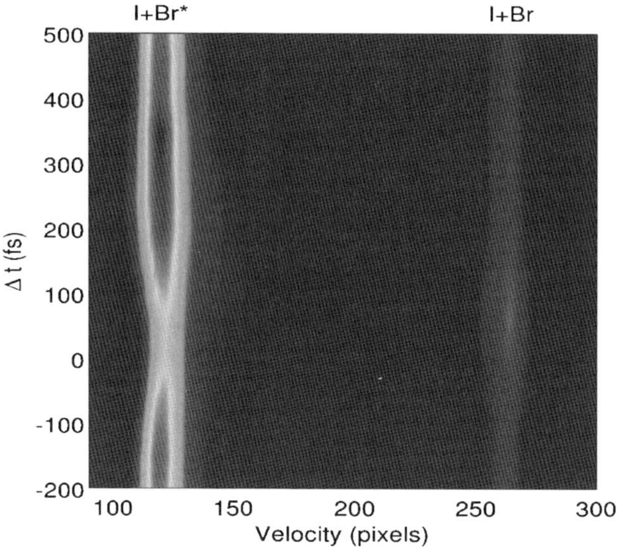

Figure 15.4 Experimental Iodine velocity distributions showing the two exit channels as a function of the IR pulse time delay Δt. By changing Δt, the branching fraction at the nonadiabatic crossing can be drastically altered. The smaller radius (velocity) corresponds to Br^* $(2P_{1/2})$; the higher, to $Br(2P_{3/2})$. At early and late delays, the field-free branching ratio is observed. Taken from [723, Figure 3]. (Reprinted with permission. Copyright 2006 Association for the Advancement of Science.)

15.2
Strong-Field Photodissociation with Pulsed Quantized Fields

Thus far we have treated CW number state light fields. The treatment of quantized fields in the *pulsed* domain is much more difficult computationally. This is due to the fact that representing a pulse requires *multimode* number states. Because of the presence of a large (or even a continuous) number of modes, coupled-channels expansions, such as the one introduced in the CW domain, cannot be used.

Over the years a number of computational methods have been developed to solve the problem of the interaction of a pulse of radiation with molecular and atomic systems [725–734, 736–740]. In order to keep the treatment as simple as possible, we only consider solving the problem in the adiabatic approximation, introduced in Chapter 11 for classical fields.

Replacing the electric field amplitude $\mathcal{E}_i$ by a time-dependent pulse envelope $\varepsilon_i(t)$, we can write the time-dependent Schrödinger equation, in the approximations that led to Eqs. (15.27) and (15.28), as two coupled equations of the form

$$\left[i\hbar\frac{d}{dt} - \sum_{\alpha} K_\alpha - W_e(\boldsymbol{X})\right]\langle N_i, e|E, \boldsymbol{n}^-, N_i - 1^-, t\rangle =$$
$$- \varepsilon_i^*(t)\hat{\boldsymbol{\varepsilon}}_i \cdot \boldsymbol{d}_{e,e'}(\boldsymbol{X})\langle N_i - 1, e'|E, \boldsymbol{n}^-, N_i - 1^-, t\rangle\ , \qquad (15.30)$$

$$\left[i\hbar\frac{d}{dt} + \hbar\omega_i - \sum_\alpha K_\alpha - W_{e'}(\boldsymbol{X})\right]\langle N_i - 1, e'|E, \boldsymbol{n}^-, N_i - 1^-, t\rangle =$$
$$- \varepsilon_i(t)\hat{\boldsymbol{\varepsilon}}_i \cdot \boldsymbol{d}_{e',e}(\boldsymbol{X})\langle N_i, e|E, \boldsymbol{n}^-, N_i - 1^-, t\rangle \, . \tag{15.31}$$

Defining a solution vector

$$\underline{\psi}(t) \equiv \begin{pmatrix} \langle N_i, e|E, \boldsymbol{n}^-, N_i - 1^-, t\rangle \\ \langle N_i - 1, e'|E, \boldsymbol{n}^-, N_i - 1^-, t\rangle \end{pmatrix} , \tag{15.32}$$

we can write Eqs. (15.30) and (15.31) in matrix notation as

$$\left[\left(i\hbar\frac{d}{dt} - \sum_\alpha K_\alpha\right)\underline{\underline{\hat{I}}} - \underline{\underline{W}}(t)\right] \cdot \underline{\psi}(t) = 0 \, , \tag{15.33}$$

where $\underline{\underline{\hat{I}}}$ is a 2×2 unity matrix and $\underline{\underline{W}}(t)$ is defined as

$$\underline{\underline{W}}(\boldsymbol{X}, t) \equiv \begin{pmatrix} W_e(\boldsymbol{X}) & -\varepsilon_i^*(t)\hat{\boldsymbol{\varepsilon}}_i \cdot \boldsymbol{d}_{e,e'}(\boldsymbol{X}) \\ -\varepsilon_i(t)\hat{\boldsymbol{\varepsilon}}_i \cdot \boldsymbol{d}_{e',e}(\boldsymbol{X}) & W_{e'}(\boldsymbol{X}) - \hbar\omega_i \end{pmatrix} . \tag{15.34}$$

By doing this, the problem has been transformed to a scattering problem on two potential energy surfaces, one being the ground surface $q = e$, and the other being a "field-dressed" surface, in which the excited state ($q = e'$) energy $W_{e'}$ has been lowered by the incident photon energy of $\hbar\omega_i$ to energy $W_{e'}(\boldsymbol{X}) - \hbar\omega_i$. The two surfaces are coupled by the transition dipole moment matrix element times the field.

15.2.1
Light-Induced Potentials

Following George *et al.* [618–620] we can gain insight into these equations by diagonalizing the $\underline{\underline{W}}$ matrix for every space-time $(\boldsymbol{X}, t)$ point, according to,

$$\underline{\underline{U}}^\dagger(\boldsymbol{X}, t) \cdot \underline{\underline{W}}(\boldsymbol{X}, t) \cdot \underline{\underline{U}}(\boldsymbol{X}, t) = \underline{\underline{\hat{\lambda}}}(\boldsymbol{X}, t) \, , \tag{15.35}$$

where $\underline{\underline{\hat{\lambda}}}(\boldsymbol{X}, t)$ is a 2×2 diagonal matrix of field-dressed adiabatic surfaces. These are the so-called "light-induced potentials" (LIP) [465, 618–620, 719–721, 741–743, 745–747], which are given as

$$\lambda_{1,2}(\boldsymbol{X}, t) = \frac{1}{2}\Big\{ W_e(\boldsymbol{X}) + W_{e'}(\boldsymbol{X}) - \hbar\omega_i$$
$$\pm \left[(W_e(\boldsymbol{X}) - W_{e'}(\boldsymbol{X}) + \hbar\omega_i)^2 + 4|\varepsilon_i(t)\hat{\boldsymbol{\varepsilon}}_i \cdot \boldsymbol{d}_{e',e}(\boldsymbol{X})|^2\right]^{1/2} \Big\} . \tag{15.36}$$

If, as in the classical field case (Eq. (11.16)), we neglect the nonadiabatic coupling terms, (in this case, the transformation to LIP entails nonadiabatic coupling terms arising from derivatives with respect to $\boldsymbol{X}$, embodied in the kinetic energy

K_α operators, *and* the time derivative), the problem decouples to two Schrödinger equations:

$$i\hbar \frac{d}{dt}\psi_i'(\boldsymbol{X},t) = \left[\sum_\alpha K_\alpha + \lambda_i(\boldsymbol{X},t)\right]\psi_i'(\boldsymbol{X},t)\,, \quad i = 1,2\,. \tag{15.37}$$

We see that the potentials have now been replaced by the $\lambda_i(\boldsymbol{X},t)$ LIP's. The solution to Eq. (15.33) is then obtained by back-transforming the solution of Eq. (15.37), that is, $\underline{\psi} = \underline{\underline{U}} \cdot \underline{\psi}'$.

As depicted in Figure 15.5, the situation (which is analogous to a radiation-free curve-crossing problem [748, 749]) is that of two ($\boldsymbol{X}$ and t dependent) eigenvalues that repel each other in the near crossing region. The stronger the electric field, the larger the repulsion between curves. Thus, in the pulsed case, as the field rises the "gap" between the two eigenvalues opens up, and then closes as the pulse falls. For weak fields, or for strong fields during the initial rise time of the pulse, if the system has been deposited (by an excitation pulse) on the excited potential surface it will stay on this surface and will dissociate to products linked to the original excited surface. If, however, the system reaches the crossing region at the maximum of the pulse, and the pulse is strong enough, the "gap" that opens up between the eigenvalues will force the system to remain on the higher eigenvalue and to dissociate to the products that are linked to the original ground potential surface.

This type of "strong-field control" was demonstrated theoretically by Suzor *et al.* [741, 741, 750] and Bandrauk *et al.* [719–721] in their study of the strong-field photodissociation of H_2^+, and experimentally by Zavriyev *et al.* [751–754] in their

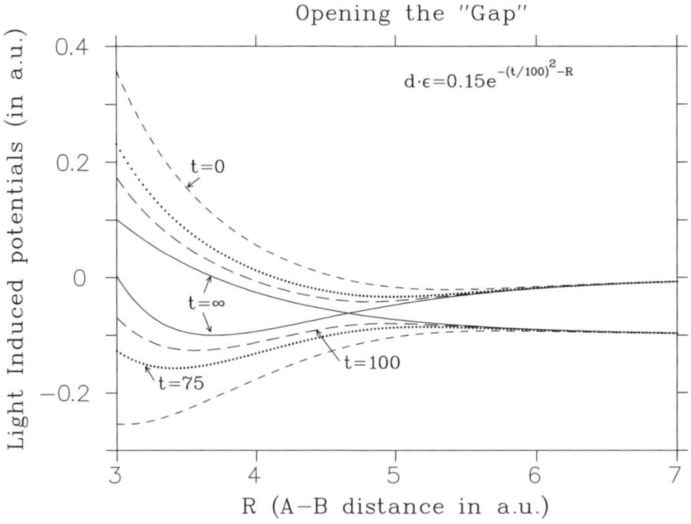

Figure 15.5 Opening of a gap between adiabatic states with the rise of a coupling pulse whose form is shown in the upper right corner.

study of the photodissociation of H_2 in intense laser fields. In these studies, another phenomenon, that of "bond softening", was demonstrated. Here [613, 750, 754, 755], due to the opening of the "gap", the lower eigenvalue is pushed down, thereby reducing the binding energy felt by the molecule. Moderate amounts of excitation can then be enough to give the molecule enough energy to cross the gap and be dissociated, provided it reaches the crossing point at the instant the "gap" opens. We see that an important element of strong-field control is, in addition to having a sufficiently strong field, the exact timings of the wave packet motion. This timing can be controlled by varying the time delay between the excitation pulse which "lifts" the wave packet to the excited state, and the "control" pulse, in charge of creating the "gap." This bond softening is observed at energies below the minimum of U_0^{LIP}. However, "bond hardening" or "vibrational trapping" occur in U_+^{LIP} where the molecule has an increased lifetime [635, 756, 757].

The extension of Eq. (15.33) to the three surface case [465, 745–747], shown in Figure 15.6, allows for adiabatic population transfer of entire *wave packets* (comprising many vibrational states) between electronic states, in complete analogy to the adiabatic population transfer in the three-state Λ system, treated in Section 11.1. As shown in Figure 15.6, the "bond softening" experienced by the $\lambda_i(X, t)$ LIP in the three-surface case results in a reduction in the barrier separating reactants and products of the nondiagonalized dressed potentials that is, $W_e(X)$ and $W_{e'}(X) - \hbar\omega_i$. In complete analogy with the three-level STIRAP problem, in the three surfaces case, a "counterintuitive" pulse sequence would be capable of inducing a complete population transfer between *wave packets*, each composed of many states.

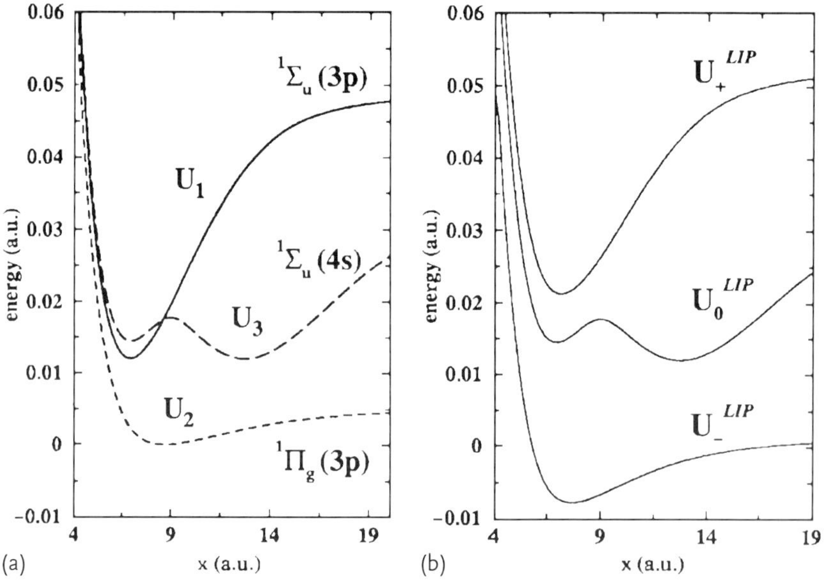

Figure 15.6 The "bond softening" experienced by the $\lambda_i(X, t)$ LIP in the three-surface case. Radiation free curves are shown in (a); and the LIP are shown in (b). Taken from [747, Figure 2]. (Reprinted with permission. Copyright 2001 Elsevier BV.)

15.3 Controlled Focusing, Deposition, and Alignment of Molecules

15.3.1 Focusing and Deposition

The light-induced potentials defined above were valid for both on-resonance, when $W_{e'}(\boldsymbol{X}) - W_e(\boldsymbol{X}) - \hbar\omega_i \approx 0$ and off-resonance processes, when this condition is not satisfied. Many multiphoton excitation experiments involve irradiating a molecule at a frequency that is far from any molecular transition or at energies far from that of the difference between potential energy surfaces. In this case, if the field is not too strong, we can assume that $(W_e(\boldsymbol{X}) + \hbar\omega_i - W_{e'}(\boldsymbol{X}))^2 \gg 4|\varepsilon_i(t)\hat{\boldsymbol{\varepsilon}}_i \cdot \boldsymbol{d}_{e',e}(\boldsymbol{X})|^2$ and expand the square root in Eq. (15.36) to obtain the following form for the two light-induced potentials,

$$\lambda_1(\boldsymbol{X}, t) = W_e(\boldsymbol{X}) + \frac{|\varepsilon_i(t)\hat{\boldsymbol{\varepsilon}}_i \cdot \boldsymbol{d}_{e',e}(\boldsymbol{X})|^2}{W_e(\boldsymbol{X}) + \hbar\omega_i - W_{e'}(\boldsymbol{X})},$$
$$\lambda_2(\boldsymbol{X}, t) = W_{e'}(\boldsymbol{X}) - \hbar\omega_i - \frac{|\varepsilon_i(t)\hat{\boldsymbol{\varepsilon}}_i \cdot \boldsymbol{d}_{e',e}(\boldsymbol{X})|^2}{W_e(\boldsymbol{X}) + \hbar\omega_i - W_{e'}(\boldsymbol{X})}. \tag{15.38}$$

We can rewrite the second term in Eqs. (15.38) as the product of a *dynamic polarizability tensor* of the molecule, given as

$$\underline{\underline{\alpha}}(\omega_i, \boldsymbol{X}) \equiv \frac{\boldsymbol{d}_{e,e'}(\boldsymbol{X}) \otimes \boldsymbol{d}_{e',e}(\boldsymbol{X})}{W_{e'}(\boldsymbol{X}) - W_e(\boldsymbol{X}) - \hbar\omega_i}, \tag{15.39}$$

times the polarization directions, $\hat{\varepsilon}_i$, times the field-squared. (The $\otimes$ symbol denotes the *outer product* of two vectors. For example, $\boldsymbol{a} \otimes \boldsymbol{b}$, where $\boldsymbol{a}$ and $\boldsymbol{b}$ each have x, y, and z components, is a 3×3 matrix whose elements are $a_x b_x$, $a_x b_y$, and so on). That is, the field-dependent component $\Delta W(\boldsymbol{X}, t)$ of the LIP felt by the system is given by

$$\Delta W(\boldsymbol{X}, t) = -\hat{\boldsymbol{\varepsilon}}_i \cdot \underline{\underline{\alpha}}(\omega_i, \boldsymbol{X}) \cdot \hat{\boldsymbol{\varepsilon}}_i |\varepsilon_i(t)|^2 . \tag{15.40}$$

This contribution may be viewed classically as being due to the interaction of an induced dipole, $\boldsymbol{d}^{ind}(t) = \underline{\underline{\alpha}}(\omega_i) \cdot \hat{\boldsymbol{\varepsilon}}_i \varepsilon_i(t)$ created by the field, with the field that created it, with the energy of interaction being given by $\Delta W(\boldsymbol{X}, t) = -\varepsilon_i(t)\hat{\boldsymbol{\varepsilon}}_i \cdot \boldsymbol{d}^{ind}$.

The only "nonclassical" aspect of the far off-resonance limit result, Eq. (15.40), is the dependence of the dynamic polarizability (Eq. (15.39)) on the nuclear coordinates $\boldsymbol{X}$. This dependence is in accord with qualitative classical expectations, because we expect different molecular configurations to be more easily polarizable than others. However, it is only quantum mechanics, via Eq. (15.39), that advises how to compute this shape-dependent polarizability.

This off-resonance LIP (Eq. (15.40)) gives rise to the so-called "dipole force". This force has been used in the focusing [758, 759] and trapping [760] of atoms passing through strong (usually standing wave) electromagnetic fields, and for nano-deposition (i.e. deposition on a nanometer size scale) of atoms on surfaces. In these experiments, the atoms are first cooled (e.g., by "laser cooling" [354]). They then pass through the optical standing waves, before impinging on a surface, where one observes the formation of periodic submicron atomic patterns. There are far fewer results for molecules (for which cooling is difficult), the most noteworthy being experiments [761, 762] and theory [763, 764] on the focusing of molecules using intense laser fields.

In order to better understand these effects, we consider an atom interacting with a standing wave field of the form

$$E_i(x, y, z, t) = 4\hat{\varepsilon}\mathcal{E}(z)\cos(ky)\cos(\omega_i t) , \tag{15.41}$$

where the polarization direction $\hat{\varepsilon}$ is assumed to lie along the z-axis. For nanodeposition cases, y is chosen to be parallel to the plane of the surface on which the atoms are to be deposited. It follows directly from Eq. (15.40) that an atom transversing this field will experience a periodic potential of the form,

$$\Delta W = -\alpha(\omega_i)\mathcal{E}^2(z)\cos^2(ky) , \tag{15.42}$$

where, as appropriate for atoms, we have suppressed the $\boldsymbol{X}$ (nuclear coordinate) dependence of ΔW and $\underline{\underline{\alpha}}(\omega_i)$. Further, the $\underline{\underline{\alpha}}$ tensor was replaced by an α scalar because, for atoms, the induced dipole is always parallel to the external field. The situation in molecules will be examined later.

The force acting on the atom in the direction parallel to the surface is given by

$$F_y = -\frac{\partial \Delta W}{\partial y} = -k\alpha(\omega_i)\mathcal{E}^2(z)\sin(2ky) . \tag{15.43}$$

For positive $\alpha(\omega_i)$ (i.e., when $W_{e'} - W_e - \hbar\omega_i > 0$), this force deflects the atoms towards the $y = n\pi/k,\ n = 0, 1, \ldots$ points, where the field intensity is maximal, and away from the $y = (n + 1/2)\pi/k,\ n = 0, 1, \ldots$ points, where the field intensity is zero. Close to one of the $y = n\pi/k$ stable points we can expand the force to yield,

$$F_y \approx \frac{\partial F_y}{\partial y}\left(y - \frac{n\pi}{k}\right) = -2k^2\alpha(\omega_i)\mathcal{E}^2(z)\left(y - \frac{n\pi}{k}\right) . \tag{15.44}$$

Thus, when the atom encounters the standing waves close to one of the $y \approx n\pi/k$ points, it experiences a harmonic restoring force proportional to $y - n\pi/k$. Assuming, for simplicity, a "flat-top" field profile in the z direction $\mathcal{E}(z) = \mathcal{E}$ for $0 < z < a$ and $\mathcal{E}(z) = 0$ for $z \leq 0$ or $z \geq a$, the frequency of the harmonic motion near one of the $y \approx n\pi/k$ points is given according to Eq. (15.44) as

$$\omega = k\mathcal{E}\left(\frac{2\alpha(\omega_i)}{M}\right)^{1/2} , \tag{15.45}$$

where M is the atomic mass. Since for harmonic potentials the period is independent of the degree of stretching of the oscillator relative to its equilibrium point, and provided that $|y - n\pi/k|$ is small enough for Eq. (15.44) to hold, it follows that all the atoms entering the field region will reach the $y = n\pi/k$ points after quarter of a period, that is, at a time equal to

$$\tau = \frac{\pi}{2\omega} = \frac{\pi}{2k\mathcal{E}}\left(\frac{M}{2\alpha(\omega_i)}\right)^{1/2} . \tag{15.46}$$

Therefore, if the incident atomic velocity v_z is normal to the surface, the best position to place the surface on which the atoms are to be deposited is at a distance of

$$\delta z = \mathrm{v}_z\tau = \frac{\mathrm{v}_z\pi}{2k\mathcal{E}}\left(\frac{M}{2\alpha(\omega_i)}\right)^{1/2} , \tag{15.47}$$

below the $z = 0$ line where the field starts. This will result in the best focusing of atoms at the $y = n\pi/k$, $n = 0, 1, \ldots$ points. This result, and a set of computed atomic trajectories leading to it, is demonstrated in Figure 15.7. Note, in particular, the focusing that occurs near 0.07 µs.

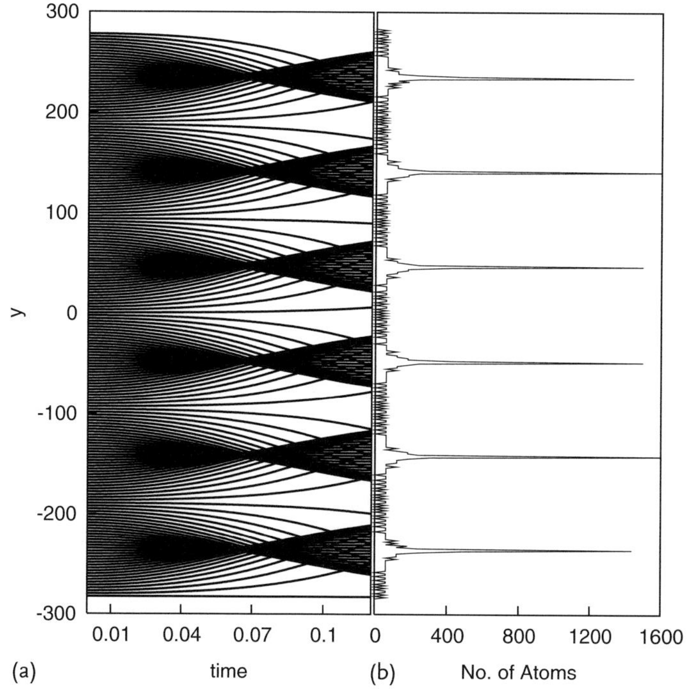

Figure 15.7 Trajectories and associated deposition for Rb atoms in the 16^2S Rydberg state passing through a field of intensity $I = 1.9 \times 10^7$ W/cm^2 and wavelength 188 495.6 nm. The ordinate shows the position along the y-axis, parallel to the surface. The abscissa of (a) shows the time in microseconds associated with the paths of trajectories, that are themselves shown as lines, incident on the surface. Note the visible "waist" associated with the focusing effect. (b) shows the intensity of the atoms incident on the surface at the focal waist. Adapted from [765]. (Reprinted with permission. Copyright 2004 American Chemical Society.)

The focusing noted above will deteriorate due to: (i) the failure of the harmonic approximation for atoms hitting y points for which Eq. (15.44) does not hold; (ii) a nonuniform v_z distribution; and (iii) nonzero v_y transverse velocity values, that is, deviations of the atomic directions from the normal. The last two effects can be minimized by cooling the atomic beam and selecting its directionality by using a sequence of slits.

The first effect cannot, however, be overcome in this manner. To estimate the fraction of atoms that will not focus well on the surface, we note that the range of harmonicity, which is the range of linearity of the $\sin 2ky$ function, is approximately $-1/4 < 2ky < 1/4$. This means that in each $[-\pi/2, \pi/2]$ interval, only a fraction of $\approx 1/(2\pi) \approx 16\%$ of the atoms will be tightly focused on the surface. This is so even if the surface is placed at the optimal distance of δz of Eq. (15.47) below the onset of the light field.

It is possible to overcome this lack of tight focusing by scattering away from the surface those atoms whose y impact parameter lies outside the $-1/(8k) < y < 1/(8k)$ tight-focusing range. One way of doing that is by changing the sign of $\alpha(\omega_i)$, thereby making the potential repulsive, for just these impact parameters. This can be accomplished [766], as explained below, by a straightforward application of the bichromatic control scenario of Section 3.2.1.

In essence, bichromatic control can be used to change the sign of $\alpha(\omega_i)$ as a function of y. That is, we prepare the molecular (or atomic) beam in a superposition of two energy eigenstates,

$$|\psi(t)\rangle = c_1|E_1\rangle \exp(-iE_1t/\hbar) + c_2|E_2\rangle \exp(-iE_2t/\hbar) \ . \qquad (15.48)$$

Instead of using the single field Eq. (15.41), we pass the molecules in this superposition state through a bichromatic standing wave field of the form:

$$\begin{aligned} \boldsymbol{E}(y,t) &= 4\hat{\boldsymbol{\varepsilon}}\left[\mathcal{E}_1^{(0)}\cos(k_1y)\cos(\omega_1t) + \mathcal{E}_2^{(0)}\cos(k_2y+\theta_F)\cos(\omega_2t)\right] \\ &\equiv \boldsymbol{E}(\omega_1) + \boldsymbol{E}(\omega_2) \ . \end{aligned} \qquad (15.49)$$

As in the usual bichromatic control scenario we use two laser frequencies which satisfy the $\omega_2 - \omega_1 = (E_1 - E_2)/\hbar$ relation, that is, we create two-pathway interference at the energy $E = E_1 + \hbar\omega_1 = E_2 + \hbar\omega_2$. Notice that as in the refractive index control discussed in Section 7.2, this energy is far off resonance from any molecular energy level.

The bichromatic off-resonance LIP obeys the same general relation to the field as does the monochromatic LIP, namely, $\Delta W(y) = -\boldsymbol{d}^{ind} \cdot \boldsymbol{E}(y,t)$. All that one need do is calculate the dipole induced in the material superposition state by the bichromatic field. Following our discussion of the control of refractive indices in Section 7.2, the induced dipole is given by

$$\begin{aligned} \boldsymbol{d}^{ind} &= \left[\underline{\underline{\alpha}}^{in}(\omega_1) + \underline{\underline{\alpha}}^{n}(\omega_1)\right]\cdot\boldsymbol{E}(\omega_1) + \left[\underline{\underline{\alpha}}^{in}(\omega_2) + \underline{\underline{\alpha}}^{n}(\omega_2)\right]\cdot\boldsymbol{E}(\omega_2) \\ &\quad + \underline{\underline{\alpha}}^{in}(\omega_{2,1}+\omega_1)\cdot\boldsymbol{E}(\omega_{2,1}+\omega_1) + \underline{\underline{\alpha}}^{in}(\omega_{2,1}-\omega_2)\cdot\boldsymbol{E}(\omega_{2,1}-\omega_2) \ , \end{aligned} \qquad (15.50)$$

where

$$E(\omega_{2,1} + \omega_1) = 4\hat{\varepsilon}\mathcal{E}_1^{(0)} \cos(k_1 y) \cos[(\omega_{2,1} + \omega_1)t]$$

and

$$E(\omega_{2,1} - \omega_2) = 4\hat{\varepsilon}\mathcal{E}_2^{(0)} \cos(k_2 y + \theta_F) \cos[(\omega_{2,1} - \omega_2)t] .$$

As explained in Section 7.2, $\underline{\underline{\alpha}}^{in}(\omega)$ is that part of the polarizability tensor that results from the interference between the two paths associated with the field-dressed energy E. It is a function of the $c_1^* c_2$ coherence between the two states that make up the superposition state in Eq. (15.48). By contrast, $\underline{\underline{\alpha}}^n(\omega)$ denotes the ordinary polarizability tensor, resulting from noninterfering terms. It is only a function of the populations, $|c_1|^2$ and $|c_2|^2$, of the $|E_1\rangle$ and $|E_2\rangle$ states.

The total polarizability at ω_i, $\underline{\underline{\alpha}}^{in}(\omega_i) + \underline{\underline{\alpha}}^n(\omega_i)$ is then a function of both θ_F and of θ_M, where $\theta_M = \arg(c_1/c_2)$. For example, Figure 15.8 shows the interference contribution to the $\alpha_{zz}(\omega)$ for rubidium at parameters indicated in the figure

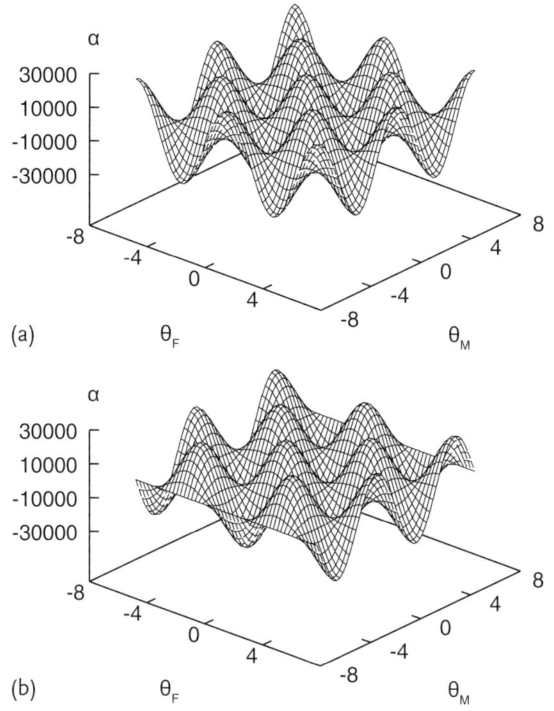

Figure 15.8 Interference contribution α^{in} to the polarizability (in a.u.) of Rb plotted against θ_M and θ_F for a superposition of the $|16s\rangle$ and $|16d\rangle$ atomic states. The two fields used are of intensity 19.1 W/cm^2 and 1.912×10^5 W/cm^2 at wavelengths 1782.53 and 1832.31 nm. (a) is the real component of α^{in} and (b) is the imaginary component. Adapted from [765]. (Reprinted with permission. Copyright 2004 American Chemical Society.)

caption. The range of $\alpha^{in}(\omega)$ is seen to be large and the $\alpha^n(\omega)$ contribution (not shown) is insignificant on the scale of Figure 15.8.

Assuming for simplicity that (as in atoms) the induced dipole moment is pointing in the direction of the field that is, only $\alpha_{zz}(\omega)$ is operative], we obtain the bichromatic LIP as

$$\Delta W(y) = -\boldsymbol{d}^{ind} \cdot \boldsymbol{E}(y,t) = V^n(y) + V^{in}(y)\,, \tag{15.51}$$

where the noninterfering contribution is given by

$$V^n(y) = -2\left[4\mathcal{E}_1^{(0)^2}\cos^2(k_1 y)\alpha^n(\omega_1) + 4\mathcal{E}_2^{(0)^2}\cos^2(k_2 y + \theta_F)\alpha^n(\omega_2)\right]\,, \tag{15.52}$$

and the interfering part is given by

$$\begin{aligned} V^{in}(y) = -2\Big[&4\mathcal{E}_1^{(0)^2}\cos(k_1 y)\cos(k_2 y + \theta_F)\alpha_r^{in}(\omega_1) \\ &+4\mathcal{E}_2^{(0)^2}\cos(k_1 y)\cos(k_2 y + \theta_F)\alpha_r^{in}(\omega_2)\Big]\,, \end{aligned} \tag{15.53}$$

where $\alpha_r^{in}(\omega_1)$ is the real part of the interference polarizability term (Eq. (7.38)). In deriving the above expression we have neglected the time-dependent parts of the LIP, emanating from the $\boldsymbol{E}(\omega_{2,1} + \omega_1)$ and $\boldsymbol{E}(\omega_{2,1} - \omega_2)$ terms, because these terms oscillate with high frequency and were found not to affect the trajectory of the atoms as they pass through the field.

Depending on the value of the $c_1^* c_2$ coherence, where c_1 and c_2 are the coefficients of the superposition state of Eq. (15.48), the $\alpha_r^{in}(\omega_1)$ term can either be positive or negative. Thus, we can control the magnitude of the negative polarizability, and hence the strength of the repulsive LIP that is added to the usual attractive LIP experienced in the monochromatic case when $\hbar\omega_i \ll W_{e'} - W_e$. As explained above, the controlled addition of a repulsive potential allows us to repel atoms that hit certain ranges of the y impact parameters. In particular, we can make the potential that is outside the $-1/(8k) < y < 1/(8k)$ tight-focusing range repulsive, thereby rejecting the atoms impinging at these impact parameters, while leaving the potential inside this range essentially harmonic. Hence, this affords the possibility of achieving much better focusing.

This is in fact what happens. Figure 15.9 shows the results in the presence and absence of interference contributions for a superposition comprised of two vibrational states of the N_2 molecule. Figure 15.9a,b shows the pattern of deposition, and the associated optical potential, for dynamics in the presence of $\Delta W(y) = V^{in}(y) + V^n(y)$. For comparison we show, in Figure 15.9c,d, the corresponding results assuming that there is no coherence between $|E_1\rangle$ and $|E_2\rangle$, that is, neglecting $V^{in}(y)$. In the absence of molecular coherence the optical potential is seen to be (Figure 15.9d) periodic, resulting in a series of short periodic deposition peaks (Figure 15.9c). By contrast, the inclusion of interference contributions (Figure 15.9a,b) result in significant enhancement and narrowing of peaks (FWHM of less than 4 nm) as well as the appearance of an aperiodic potential and associated aperiodic deposition pattern.

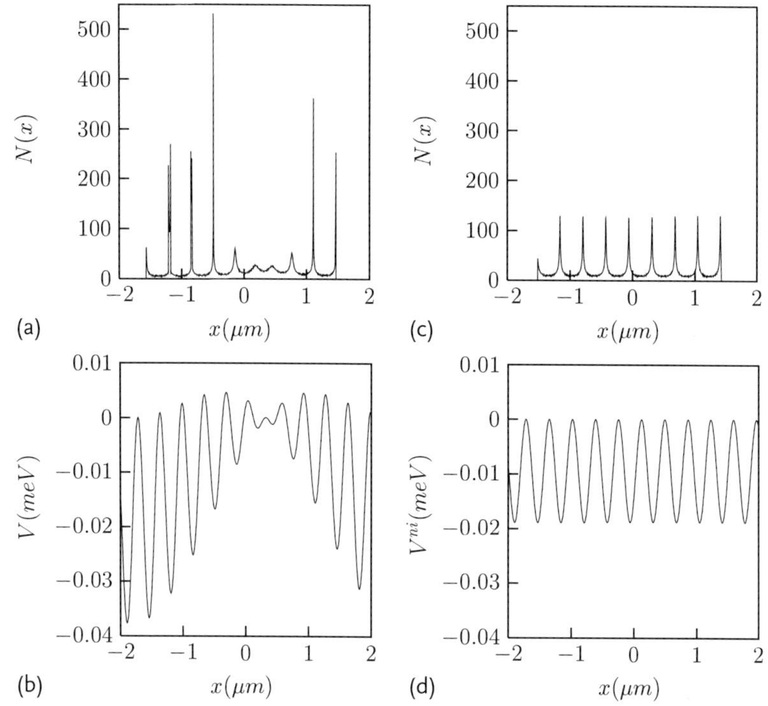

Figure 15.9 Molecular deposition of N_2 and associated optical potential for the initial superposition $0.2^{1/2}|0,0,0\rangle + 0.8^{1/2}|0,2,0\rangle$ due to $\Delta W(y) = V^{in}(y) + V^{n}(y)$ (panels (a) and (b)), and due to $\Delta W^{n}(y)$ only. (Here V^n is denoted V^{ni}) Here $|i, j, k\rangle$ denotes the state with vibrational quantum number i, rotational quantum number j and projection k of the total angular momentum along the z-axis. System parameters are $\mathcal{E}_2^0/\mathcal{E}_1^0 = 1.0 \times 10^4$, $\mathcal{E}_1^0 = 1.0 \times 10^2$ V/cm, $\lambda_1 = 0.628$ μm, $\lambda_2 = 0.736$ μm, $\theta_F = -2.65$ radian, and $t_{int} = 0.625$ μs, where t_{int} is the interaction time of the molecules with the field. Taken from [766, Figure 2]. (Reprinted with permission. Copyright 2000 American Physical Society.)

Control can also be achieved by changing the relative phase θ_F between the two light fields, defined by Eq. (15.49). Figure 15.10a,b shows significant differences in both the position and intensity of the peaks as a function of θ_F. By contrast, an analogous plot (not shown) where only V^n is included shows no variation in peak intensity as a function of θ_F. Similarly, Figure 15.10c,d shows the strong dependence of the deposition upon the magnitude of the coefficients of the created superposition.

15.3.2 Strong-Field Molecular Alignment

We now examine molecular *alignment* by strong laser fields [62, 767–790]. Molecular alignment was observed already in multiphoton ionization experiments of diatomic molecules by Normand *et al.* [767, 768]. It was explained in terms of the

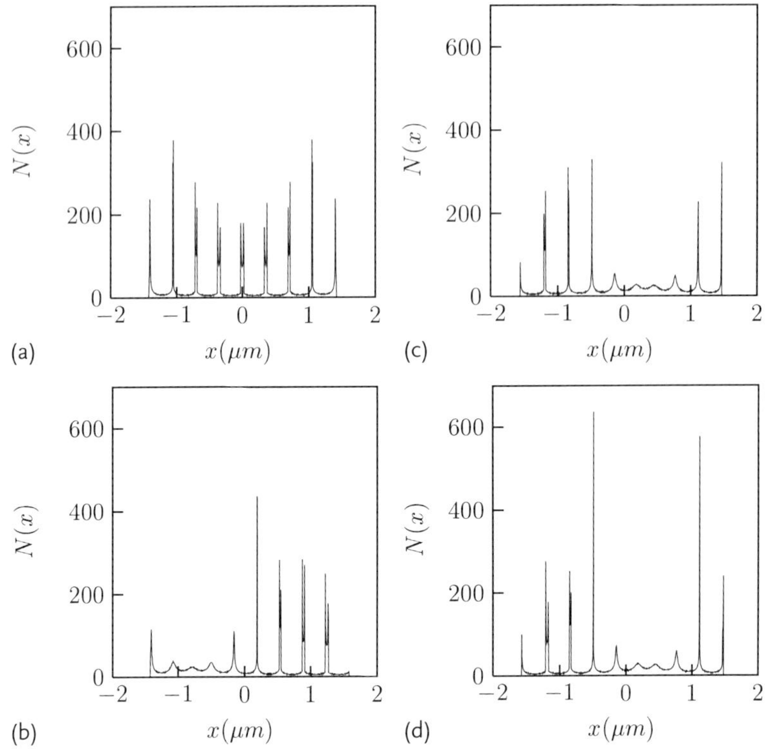

Figure 15.10 (a) and (b): Molecular deposition of N_2 associated with $0.8^{1/2}|0,0,0\rangle + 0.2^{1/2}|1,2,0\rangle$ for varying θ_F; that is, (a) $\theta_F = 0$, and (b) $\theta_F = 2.0$ rad. (c) and (d): Sample variation of deposition with changes in $|c_1|, |c_2|$: (c) $[0.99^{1/2}]|0,0,0\rangle + 0.1|0,2,0\rangle$ and (d) $0.4^{1/2}|0,0,0\rangle + 0.6^{1/2}|0,2,0\rangle$. Other parameters and notation are as in Figure 15.9. Taken from [766, Figure 3]. (Reprinted with permission. Copyright 2000 American Physical Society.)

off-resonance LIP by Friedrich and Herschbach [770], who showed, as explained below, that rather than rotate freely, molecules in the presence of the field execute a "pendular" like motion. Trapping and strong alignment of molecules [775, 777] was subsequently observed, as well as the acceleration of rotational motion of laser-aligned molecules by the "optical centrifuge" effect [791–793], which is also discussed below.

In recent years [776, 778–790] pulse shaping or a combination of two pulses have been used to achieve one-dimensional (1D) and three-dimensional (3D) alignment. One such method [776] consists of the use of an intense, elliptically polarized, nonresonant laser field to simultaneously force all three axes of a molecule (e.g., 3,4-dibromothiophene [776]) to align in space. Another method of generating (1D) alignment is to use the AC Stark effect [778]. The molecules are first field aligned by a slow (compared to the molecular rotational periods) "adiabatic" turn-on pulse which is then suddenly switched off to create partially revived [794, 795] field-free aligned wave packets.

Variants of this technique for achieving 3D alignment involve the use of two pulses [780, 783]. For example, one can 3D align asymmetric top molecules (such as sulphur dioxide [780]) by using two time-delayed, orthogonally polarized femtosecond laser pulses [780]. Alternatively, one can use a long "adiabatic" pulse, aligning molecules along their most polarizable axis, which is followed by an orthogonally polarized femtosecond pulse. In this way one creates wave packets which transiently become [794] 3D aligned [783]. The degree of alignment thus achieved (e.g., of iodobenzene [783] and 3,5-difluoroiodobenzene [788]) appears to be significantly better than that attainable by a single pulse.

The alignment effect is best explained using Eq. (15.40), noting that, in molecules, the induced dipole is not necessarily parallel to the field that induced it. In fact, the induced dipole can have three perpendicular components, d^X, d^Y and d^Z, in the X, Y, Z molecular-fixed coordinate system. Given these components, we can express $\hat{\boldsymbol{\varepsilon}}_i \cdot \boldsymbol{d}_{e',e}$, the projection of the transition dipole onto the laboratory-fixed z-axis, appearing in Eq. (15.38), as [701]

$$\begin{aligned}\hat{\boldsymbol{\varepsilon}}_i \cdot \boldsymbol{d} &= \mathrm{d}^Z \cos\Theta + \sin\Theta[-\mathrm{d}^X \cos\xi + \mathrm{d}^Y \sin\xi] \\ &= \mathrm{d}^Z \cos\Theta - \mathrm{d}^\perp \sin\Theta \cos(\xi + \beta)\,,\end{aligned} \tag{15.54}$$

where (see Figure 15.11), Θ is the polar angle of orientation of the molecular Z-axis relative to the laboratory z-axis (along which the field polarization $\hat{\varepsilon}_i$ lies); ξ is an azimuthal angle describing rotation of the molecule about the molecular Z-axis; and

$$\mathrm{d}^\perp \equiv \sqrt{|\mathrm{d}^X|^2 + |\mathrm{d}^Y|^2} \quad \text{and} \quad \tan\beta \equiv \frac{\mathrm{d}^Y}{\mathrm{d}^X}\,. \tag{15.55}$$

For brevity we have omitted the e', e subscripts.

Using Eq. (15.54) in Eq. (15.39) we have that

$$\hat{\boldsymbol{\varepsilon}}_i \cdot \underline{\underline{\alpha}}(\omega_i) \cdot \hat{\boldsymbol{\varepsilon}}_i = \frac{|\mathrm{d}^Z \cos\Theta - \mathrm{d}^\perp \sin\Theta \cos(\xi + \beta)|^2}{W_{e'} - W_e - \hbar\omega_i}\,. \tag{15.56}$$

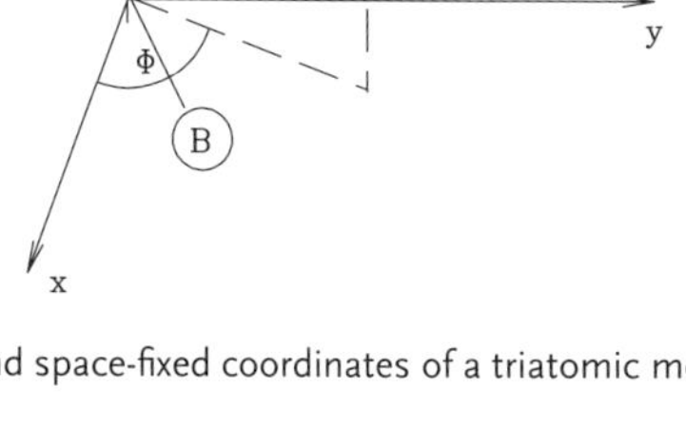

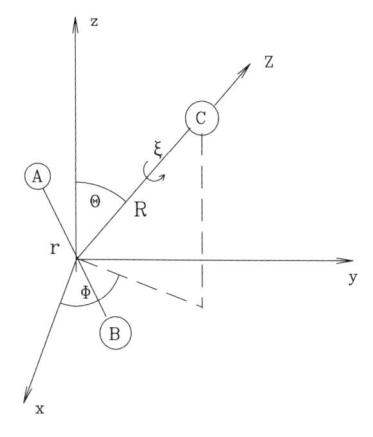

Figure 15.11 Body-fixed and space-fixed coordinates of a triatomic molecule.

Therefore, according to Eq. (15.40), the molecule feels a LIP ΔW that depends on the orientation of the molecular Z-axis relative to the field polarization and the azimuthal rotation angle of the molecular frame about the Z-axis.

Consider then the case of diatomic molecules in Σ electronic states. Here the e and e' electronic states have cylindrical symmetry about the Z-axis and the transition dipole matrix element $\boldsymbol{d}_{e,e'}$ is parallel to the molecular (Z-)axis. Hence only the d^Z component survives. In this case, the polarizability tensor reduces to $\underline{\underline{\alpha}}^{||}(\omega_i)$, which is

$$\underline{\underline{\alpha}}^{||}(\omega_i) = \frac{\boldsymbol{d}^Z \otimes \boldsymbol{d}^Z}{W_{e'} - W_e - \hbar\omega_i}\,, \tag{15.57}$$

and the LIP reduces to

$$\Delta W(t) = -\hat{\boldsymbol{\varepsilon}}_i \cdot \underline{\underline{\alpha}}^{||}(\omega_i) \cdot \hat{\boldsymbol{\varepsilon}}_i |\varepsilon_i(t)|^2 = -\frac{\left|\mathrm{d}^Z \varepsilon_i(t)\right|^2 \cos^2\Theta}{W_{e'} - W_e - \hbar\omega_i}\,. \tag{15.58}$$

The resultant force tends to align the molecule along the laboratory z-axis, that is, towards $\Theta = 0$.

The other diatomic extreme exists when the e electronic state is of Σ symmetry and the e' electronic state is of Π symmetry (i.e., it has one unit of angular momentum about the molecular axis). In that case the dipole matrix element $\boldsymbol{d}_{e,e'}$ is perpendicular to the molecular axis, that is, only the $\boldsymbol{d}^{\perp}_{e,e'}$ vector survives. The polarizability now reduces to $\underline{\underline{\alpha}}^{\perp}(\omega_i)$:

$$\underline{\underline{\alpha}}^{\perp}(\omega_i) = \frac{\boldsymbol{d}^{\perp} \otimes \boldsymbol{d}^{\perp}}{W_{e'} - W_e - \hbar\omega_i}\,, \tag{15.59}$$

and the LIP is

$$\Delta W(t) = -\frac{\left|\mathrm{d}^{\perp} \varepsilon_i(t)\right|^2 \sin^2\Theta \cos^2(\xi + \beta)}{W_{e'} - W_e - \hbar\omega_i}\,. \tag{15.60}$$

Averaging the $\cos^2(\xi + \beta)$ term of the rapidly revolving ξ angle yields 1/2, and we obtain for the LIP for this case that

$$\Delta W(t) = -\frac{\left|\mathrm{d}^{\perp} \varepsilon_i(t)\right|^2 \sin^2\Theta}{2(W_{e'} - W_e - \hbar\omega_i)}\,. \tag{15.61}$$

This potential tends to align the molecule *perpendicular* to the z-axis, that is, towards $\Theta = \pi/2$.

In general, more than two electronic states contribute to the polarizability and to the LIP. To include more than one excited state we write $\underline{\underline{\alpha}}(\omega_i)$ as

$$\underline{\underline{\alpha}}(\omega_i) = \sum_{e'} \frac{\boldsymbol{d}_{e,e'} \otimes \boldsymbol{d}_{e',e}}{W_{e'} - W_e - \hbar\omega_i}\,. \tag{15.62}$$

For diatomic molecules some of the transition dipole matrix elements will be parallel to the molecular axis and some will be perpendicular to it. Hence

$$\underline{\underline{\alpha}}(\omega_i) = \sum_{e'\|} \frac{\boldsymbol{d}_{e,e'} \otimes \boldsymbol{d}_{e',e}}{W_{e'} - W_e - \hbar\omega_i} + \sum_{e'\perp} \frac{\boldsymbol{d}_{e,e'} \otimes \boldsymbol{d}_{e',e}}{W_{e'} - W_e - \hbar\omega_i}, \tag{15.63}$$

and the LIP can be written as

$$\Delta W(t) = -|\varepsilon_i(t)|^2 \left[\alpha^{\|}(\omega_i)\cos^2\Theta + \frac{1}{2}\alpha^{\perp}(\omega_i)\sin^2\Theta\right], \tag{15.64}$$

where

$$\alpha^{\|}(\omega_i) \equiv \sum_{e'\|} \frac{|\mathrm{d}^Z_{e',e}|^2}{W_{e'} - W_e - \hbar\omega_i},$$

$$\alpha^{\perp}(\omega_i) = \sum_{e'\perp} \frac{|\mathrm{d}^{\perp}_{e',e}|^2}{W_{e'} - W_e - \hbar\omega_i}. \tag{15.65}$$

The LIP of Eq. (15.64) has two components, one attempting to align the molecule perpendicular to the field and one trying to align it parallel to the field. In diatomic molecules, the latter is usually much larger. Hence, the molecule attempts to align itself along the field direction. However, due to its initial rotational energy, the molecule cannot do so instantaneously. Since the dynamic polarizability is positive for $\hbar\omega_i \ll W_{e'} - W_e$, the LIP of Eq. (15.64) is usually negative, that is, it is purely attractive. Hence the initial effect of the LIP is to accelerate the rotational motion whenever Θ nears the potential minimum. However, in addition to the aligning force, the molecule feels the deflection force that draws it towards the high field region [796]. As it gets drawn more and more into the high field region it feels a greater and greater aligning force that further accelerates its rotational motion. The LIP is a function of both y and Θ so that these motions are no longer separable and energy can flow from the rotational motion to the translational motion and vice versa. It is therefore entirely possible that the molecule will lose enough kinetic energy in the Θ coordinate in favor of the y coordinate to have insufficient energy to execute a full rotation. At that point the motion (which resembles that of a pendulum) is called "pendular". Pendular motions are routinely observed in other situations, especially for molecules with a permanent dipole moment in a DC electric field [797].

The reverse can also happen, namely that the molecule will accelerate its rotational motion at the expense of the y motion and become trapped due to the y dependence of the LIP well. The probability of either phenomenon occurring depends on the degree of cooling of both the rotational motion and the translational motion of the molecule prior to its entrance into the high field region [770, 798]. If the temperature is sufficiently low, the field sufficiently high, and relaxation mechanisms to other degrees of freedom possible, trapping [760] and alignment [776] of molecules by highly off-resonance light fields becomes feasible. Once a molecule is trapped and aligned it can be manipulated to control a variety of processes, like enhancing reactivity [776, 799].

An interesting application of the optical alignment effect is the ability to accelerate the rotational motion and impart to the molecule hundreds of $\hbar$ units of angular momentum. In this case, one *rotates* the direction of polarization in space, causing a molecule to rotate with the field polarization onto which it is aligned. By accelerating the rotational motion of the field polarization one accelerates the rotational motion of the molecule which is aligned with it. This phenomenon, which was suggested by Ivanov *et al.* [791, 793] and demonstrated experimentally by Corkum *et al.* [792], is termed an "optical centrifuge."

To best understand how the optical centrifuge works, it is convenient to consider the field as polarized in the x–y plane, rather than along the z-axis as above. Here we reserve the z-axis for the direction of quantization of the molecular angular momentum, J.

Considering now $\hat{\boldsymbol{\varepsilon}}$ to lie in the x–y plane, and remembering that for diatomic molecules, ξ, the azimuthal angle describing rotation of the molecule about the molecular Z-axis (see Figure 15.11), can be chosen to be zero, the projections of $\boldsymbol{d}$ on x and y (instead of on z as was done in Eq. (15.54)), are given by [701],

$$\begin{aligned} \mathrm{d}^x &= \mathrm{d}^Z \sin\Theta \cos\Phi + \mathrm{d}^X \cos\Theta \cos\Phi - \mathrm{d}^Y \sin\Phi \,, \\ \mathrm{d}^y &= \mathrm{d}^Z \sin\Theta \sin\Phi + \mathrm{d}^X \cos\Theta \sin\Phi + \mathrm{d}^Y \cos\Phi \,, \end{aligned} \tag{15.66}$$

where (see Figure 15.11), Φ and Θ are the azimuthal and polar angle of orientation of the molecular z-axis relative to the laboratory z-axis (the axis of quantization).

Consider now a field whose linear polarization rotates in the x–y plane,

$$\boldsymbol{E}_i(t) = \varepsilon(t)\cos(\omega_i t)\,(\hat{\boldsymbol{x}}\cos\Phi_L(t) + \hat{\boldsymbol{y}}\sin\Phi_L(t)) \,, \tag{15.67}$$

where $\hat{\boldsymbol{x}}$ and $\hat{\boldsymbol{y}}$ are unit vectors in the x and y directions, respectively. We have that $\hat{\boldsymbol{\varepsilon}}_i \cdot \boldsymbol{d}$ is now given as

$$\begin{aligned} \hat{\boldsymbol{\varepsilon}}_i \cdot \boldsymbol{d} = \hat{\boldsymbol{\varepsilon}}_x \mathrm{d}^x + \hat{\boldsymbol{\varepsilon}}_y \mathrm{d}^y &= \mathrm{d}^Z \sin\Theta \cos(\Phi - \Phi_L(t)) \\ &+ \mathrm{d}^X \cos\Theta \cos(\Phi - \Phi_L(t)) - \mathrm{d}^Y \sin(\Phi - \Phi_L(t)) \,, \end{aligned} \tag{15.68}$$

Substituting Eq. (15.68) in Eqs. (15.39) and (15.40) the LIP is obtained as

$$\Delta W(t) = |\varepsilon_i(t)|^2 \frac{\left|(\mathrm{d}^Z \sin\Theta + \mathrm{d}^X \cos\Theta)\cos(\Phi - \Phi_L(t)) - \mathrm{d}^Y \sin(\Phi - \Phi_L(t))\right|^2}{W_{e'} - W_e - \hbar\omega_i} \,. \tag{15.69}$$

The d^z (parallel) term in the numerator creates a LIP which tends to align the molecule in the polarization (x–y) plane, whereas the second term tends to align the molecule along the z-axis – perpendicular to the polarization plane. In both cases the molecule will execute a nutational motion in which it rotates with the frequency of $\dot{\Phi}_L(t)$ about the z-axis and executes a pendular motion in Θ. For the parallel term of the LIP the equilibrium point for the pendular motion is $\Theta = \pi/2$, that is, in the x–y polarization plane. For the perpendicular part, the molecular pendulum moves about $\Theta = 0$, that is, perpendicular to the polarization plane.

In practice it is possible to rotate the direction of the polarization by building the linearly polarized field of Eq. (15.67) as a sum of two *circularly polarized* light fields whose phases vary (the phase variance with time is known as "chirp") in the opposite sense,

$$E_i(t) = \frac{\varepsilon(t)}{2}\left[\hat{x}\cos(\omega_i t + \Phi_L(t)) + \hat{y}\sin(\omega_i t + \Phi_L(t))\right] + \frac{\varepsilon(t)}{2}\left[\hat{x}\cos(\omega_i t - \Phi_L(t)) - \hat{y}\sin(\omega_i t - \Phi_L(t))\right] . \quad (15.70)$$

The $\Phi_L(t)$ chirp therefore determines the rate of rotation of the linearly polarized field. A molecule trapped in the minimum of the LIP would tend to follow faithfully the rotation of the linearly polarized field. As the rate of this rotation is accelerated so is the rate of the molecular rotation of the molecule.

The optical centrifuge is capable of imparting so much rotational energy to a diatomic molecule that the bond breaks. To see how this happens, note that the effective radial potential seen by a diatomic molecule is a sum of its "real" potential $V(R)$ and a centrifugal one associated with its angular motion. That is, it sees the effective potential

$$V_{eff}(R) = V(R) + \frac{\hbar^2 J(J+1)}{2\mu R^2} , \quad (15.71)$$

with a J dependent equilibrium distance $R_{eq}(J)$, given by the implicit equation,

$$R_{eq} = \left\{\frac{\hbar^2 J(J+1)}{\mu\, dV/dR(R_{eq})}\right\}^{\frac{1}{3}} .$$

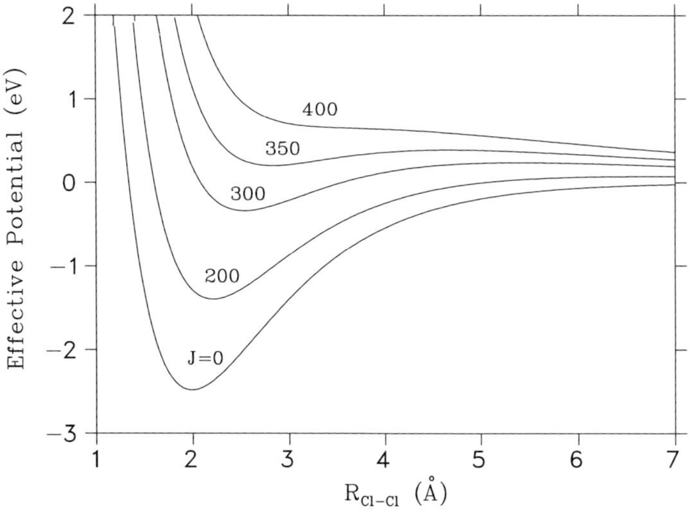

Figure 15.12 Effective potentials as a function of *J*.

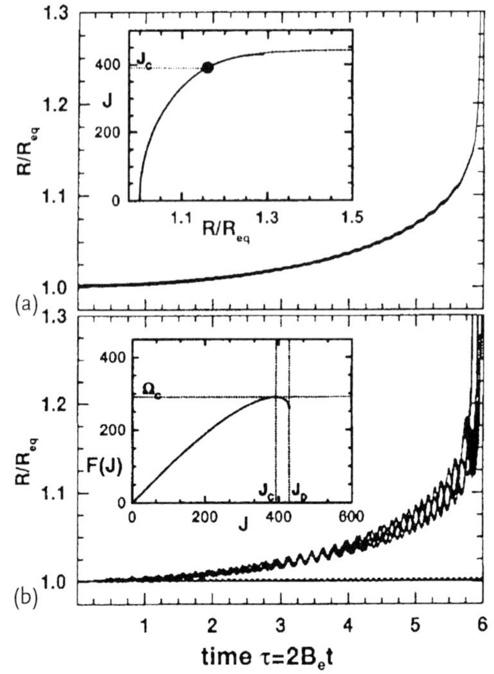

Figure 15.13 Sample trajectories for Cl_2. The pulse is 70 ps long (equal to one ground rotational period of Cl_2, corresponding to $\tau = 2\pi$ in dimensionless units), with 5.5 ps turn-on; the intensity is 1.7×10^{13} W/cm^2; $\omega_{final} = 0.02$ eV. J_C indicates the "critical" J at which the system breaks apart (which is implicit in Figure 15.12) (a) The "quiet" trajectory; the inset shows agreement between a numerical calculation (dashed line) and an analytic approximation (solid line). (b) Four sample trajectories $R(t)$ for Cl_2. Initially, $J = 30$ and a random orientation was chosen. Three trajectories show dissociation, and the fourth was not trapped. The inset shows $F(J) \equiv J/\mu R_{eq}^2(J)$. Taken from [791, Figure 1]. (Reprinted with permission. Copyright 1999 American Physical Society.)

Figure 15.12 displays the effective potentials for the Cl_2 molecule [793] and shows how the binding energy of the molecule is reduced with increasing J, until, at $J =$ 400 the effective potential is completely repulsive and the molecule dissociates.

Sample trajectories of the Cl_2 molecule on its way to dissociation are shown in Figure 15.13. We see a "quiet" trajectory which follows adiabatically the potential minimum as the system is rotated more and more by the LIP until it breaks apart, as well as "nonquiet" trajectories which deviate from the position of the minimum.

16
Coherent Control with Few-Cycle Pulses

Thus far we have dealt with relatively long pulses (duration of $\sim$ 30 fs and above) and used the light's *amplitude* and *phase* as two independent control variables. Control was very quantum mechanical in the sense that the light's phase was transferred to the material system, with control relying on interference between such phase-modified matter waves. Here we deal with ultrashort pulses where each pulse consists of just a few optical cycles ("few-cycle" pulses). In this case, the phase and the amplitude are no longer two independent variables, because any change in the phase (whose span is just a few radians) immediately changes the amplitude. Control now becomes much more classical because effects due to changes in amplitude can often be well explained by classical mechanics.

16.1
The Carrier Envelope Phase

In this section we define and discuss the so-called "carrier envelope phase". We do so by writing the field as

$$\varepsilon(z,t) = \mathrm{Re} \int_0^\infty d\omega |\epsilon(\omega)| \exp\left(i\omega n_\omega z/c - \omega t\right) , \tag{16.1}$$

where the main difference relative to our previous discussion (see Eq. 2.7) is the introduction of n_ω, the frequency-dependent refractive index of the medium. When the pulse bandwidth is very large the effects of "chromatic dispersion" (the dependence of n_ω) and hence of the phase velocity c/n_ω on the frequency becomes relevant: each frequency component travels at a slightly different speed. As a result, the group velocity of the pulse and the phase velocity of its center frequency ω_0 ("carrier wave") are not the same.

In order to quantify the "slippage" of the center frequency phase relative to the group velocity of the pulse envelope, it is customary [800–804] to rewrite Eq. (16.1) as

$$\varepsilon(z,t) = \mathrm{Re}\, \mathcal{E}(z,t) \exp\left(i\omega_0 n_0 z/c - \omega_0 t\right) , \tag{16.2}$$

Quantum Control of Molecular Processes, Second Edition. Moshe Shapiro and Paul Brumer

where n_0 is the refractive index at frequency ω_0, and $\mathcal{E}(z,t)$ is the pulse envelope, defined as

$$\mathcal{E}(z,t) = \int_0^\infty d\omega |\epsilon(\omega)| \exp\left\{i\left[(\omega n_\omega - \omega_0 n_0)z/c - (\omega - \omega_0)t\right]\right\} . \quad (16.3)$$

We now define ϕ_{CEP}, the carrier envelope phase (CEP) (depicted graphically in Figure 16.1), as the phase of the carrier wave relative to $z_{max}(t)$, the maximum of the $\mathcal{E}(z,t)$ pulse envelope. Subtracting $z_{max}(t)$ from z in Eq. (16.2), we have that

$$\varepsilon(z,t) = \mathrm{Re}\,\mathcal{E}(z,t)\exp[i\phi_{CEP}(z,t) - i\omega_0 t] , \quad (16.4)$$

where

$$\phi_{CEP}(z,t) = \omega_0 n_0[z - z_{max}(t)]/c . \quad (16.5)$$

As the pulse propagates through the medium, $z_{max}(t) = z_{max}(t=0) + ct/n_g$ moves at a slightly different velocity than the "fixed-carrier-phase" position $z(t) = z(t=0) + ct/n_0$. Specializing the expression to positions satisfying the "fixed-carrier-phase" condition eliminates the z dependence from ϕ_{CEP}, and we obtain that

$$\phi_{CEP}(t) = \omega_0 n_0[z(t) - z_{max}(t)]/c = \phi_{CEP}(t=0) + \omega_0 t\left(1 - \frac{n_0}{n_g}\right) . \quad (16.6)$$

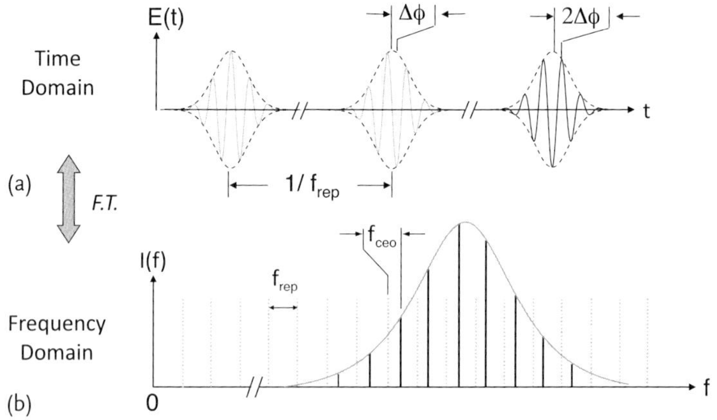

Figure 16.1 Time-frequency correspondence and relation between $\Delta\phi_{CEP}$ and f_{CEO}. (a) In the time domain, the relative phase between the carrier (solid) and the envelope (dotted) evolves from pulse to pulse by the amount $\Delta\phi_{CEP}$. (b) In the frequency domain, the elements of the frequency comb of a mode-locked pulse train are spaced by f_{rep}. The entire comb (solid) is offset from integer multiples (dotted) of f_{rep} by the CEO frequency $f_{CEO} = \Delta\phi_{CEP} f_{rep}/2\pi$. Taken from [802, Figure 1]. (Reprinted with permission. Copyright 2000 American Association for the Advancement of Science.)

In many cases the *shape* of the pulse envelope, $\mathcal{E}(z,t)$, as it moves through the medium, hence n_g, may not be completely constant in time, as it depends on various factors such as the intensity. Hence, $\phi_{CEP}(t)$ may deviate somewhat from the form of Eq. (16.6). This causes a problem because during the propagation of the pulse, the phase $\phi_{CEP}(t)$ changes by hundreds or even thousands of radians. Hence, any slight deviation from Eq. (16.6) may result, upon performance of the modulo operation, in a large error. The best solution is therefore to *measure* ϕ_{CEP}. This can for example be done for mode-locked laser pulses.

A mode-locked laser is generated from a single pulse circulating in a laser resonator. Every time this pulse hits the mirror that serves as the "output coupler", its attenuated replica is emitted. As a result, a sequence of pulses is generated with the time difference between each replica being the round trip time, $2Ln_g/c$, where L is the cavity length and n_g ($\approx n_0 + \omega_0(dn_0/d\omega)$) is the group refractive index in the laser medium. Viewed in frequency space this means that the train of pulses corresponds to an equidistant "comb" of frequencies, each differing from its closest neighbor by $f_{rep} = c/(2Ln_g)$, the pulse repetition rate.

To account for the large change in the CEP in each round trip we define f_{CEO}, the carrier envelope offset frequency (the CEO frequency) of a mode-locked laser as

$$f_{CEO} = \Delta\phi_{CEP} f_{rep}/2\pi \,, \tag{16.7}$$

where $\Delta\phi_{CEP}$ is the change (modulo 2π) in radians of the CEP per round trip in the resonator. The optical frequency of the nth member of a pulse train is therefore given as

$$\omega_n = 2\pi(f_{CEO} + n f_{rep}) = (\Delta\phi_{CEP} + 2n\pi)\, f_{rep} \,. \tag{16.8}$$

Due to the great accuracy to which f_{rep} is known, knowing ω_n is very useful for metrology as it serves as a standard against which all sorts of periodic sources can be calibrated. Basically one records the beat frequency of an unknown frequency against a range of comb frequencies. However in order to know ω_n accurately we first need to be able to measure f_{CEO} – the CEO frequency.

16.2 Coherent Control and the CEO Frequency Measurement

The CEO frequency can be measured using the one vs. two coherent control scenario, discussed in Section 3.4.3, termed in metrology the $f{-}2f$ interferometer. The ability to perform one vs. two interference using only a single pulse (or a train of pulses) stems from the large bandwidth of the few-cycle pulses. Such pulses often contain a "full octave", that is, a fundamental frequency and its second harmonic. If the pulse itself does not contain a full octave one can employ an auxiliary "white light continuum" produced by self-phase modulation [805, 806] of the pulse of interest, that *does* contain a full octave. In the self-phase modulation process, discussed below, there is an addition of frequency components to the pulse due to the nonlinear Kerr effect [807] which is the change in the refractive index with the pulse intensity. In the context of the determination of the CEO frequency, self-phase modulation can be used because, as shown below, in this process the pulse retains the same value of CEP as that of the input pulse.

The main difference between the one vs. two coherent control scenario considered in Section 3.4.3 and the present one is that in this case the energies of the final states that interfere are not the same. We can now use a nonlinear crystal to create $2\omega_n$, the second harmonic of ω_n, one of the comb frequencies, and interfere this second harmonic with the original pulse train. According to Eq. (16.8) the lowest frequency mismatch, giving rise to a time-dependent beat in the photodetector, would then be

$$\delta\omega = 2\omega_n - \omega_{2n} = 2(n + \Delta\phi_{CEP}) f_{rep} - (2n + \Delta\phi_{CEP}) f_{rep} = \Delta\phi_{CEP} f_{rep} \,. \tag{16.9}$$

Thus, a measurement of the lowest frequency of the one- vs. two-photon (time-dependent) beat signal gives f_{CEO} and the CEP phase change $\Delta\phi_{CEP}$ directly. In this way we can determine all the frequencies of the frequency comb, allowing for its use as a highly accurate frequency standard. The importance of this technique [800–804] was recognized by the awarding the 2005 Nobel prize in Physics to its inventors, John Hall [808] and Theodor Hänsch [809]. (These authors shared the prize with Roy Glauber for his contributions to quantum optics [366].)

16.3 The Recollision Model

As mentioned above, any change in the phase of a few-cycle pulse invariably involves change in the amplitude of the electric field that is felt by the material system that we wish to control. As a result "coherent control" becomes more "classical" in nature, since the effects of the *magnitude* of an electric field can often be well understood classically. Indeed, much of our understanding of single and multiple ionization, recombination and photoemission of high-harmonics under the action of strong laser pulses is derived from a semiclassical *recollision model* [810–813]. This model describes the process as a combination of tunnel ionization [814–818] and the motion of an electron under the influence of an oscillatory electric field, resulting in either the escape of the electron, in a phenomenon known as "above threshold ionization" (ATI) [819, 820], or in its recollision with other electrons, or in the emission of a photon, giving rise to "high-harmonics generation".

Thus we may think of four virtual steps:

16.3.1
Step 1: Tunnel Ionization

As shown in Figure 16.2, an electron bound to an ionic core subject to an alternating electric field sees an effective potential characterized by a strong attraction near the ion and a linear potential whose slope keeps changing sign due to the alternating electric field. Near the peak of a very strong pulse (intensities above 10^{13} W/cm^2), the effective potential barrier goes down, enabling the electron to tunnel through it. This is especially effective for a "quasistatic" IR or near IR light (e.g., wavelengths longer than $\sim$ 800 nm). Under such circumstances tunneling becomes quite probable, occurring during $\sim 300 \times 10^{-18}$ s ("attoseconds") around each field crest near the pulse peak.

Historically, semiclassical tunneling due to an alternating electric field was first analyzed [814–818] in the context of above threshold ionization [819] in which an electron absorbs many photons in the continuum, thereby raising its continuum kinetic energy by integer multiples of the photon's energy $\hbar\omega$. Specifically [814], the probability of tunneling into the continuum of a particle of mass m and charge q, residing initially in a bound state of energy E_i, by an electric field whose amplitude and frequency are E_0 and ω respectively, is

$$W = \exp\left[-\frac{2|E_i|f(\gamma)}{\hbar\omega}\right],$$

$$\text{where} \quad f(\gamma) = \left[1 + \frac{1}{2\gamma^2}\right]\sinh^{-1}\gamma - \frac{(1+\gamma^2)^{\frac{1}{2}}}{2\gamma};$$

$$\gamma = [2m|E_i|]^{1/2}\frac{\omega}{|q|E_0}. \tag{16.10}$$

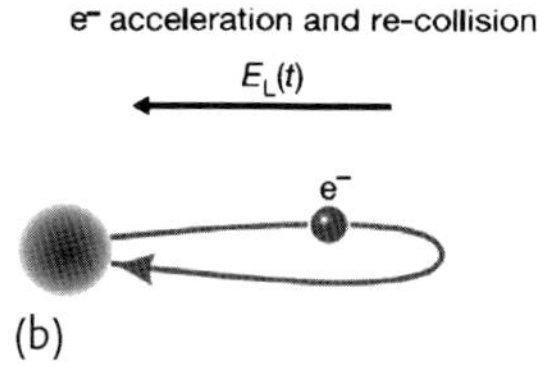

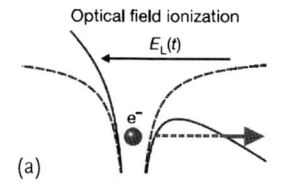

Figure 16.2 An illustration of the tunneling and the recollision process. (a) the barrier felt by the combined Coulomb potential and the dipolar interaction with the radiation's electric field. When the radiation's electric field is strong enough, tunneling may occur. (b) Acceleration of the liberated electron by the oscillating field may cause some of the liberated electrons to recollide with the remaining electrons of the parent ion. Adopted from [823, Figure 1]. (Reprinted with permission. Copyright 2003 Nature Publishing Group.)

γ is called the "Keldysh parameter". When $\gamma > 1$ the above tunneling ionization picture is justified, otherwise it is customary to think of ionization as following the multiphoton route.

16.3.2
Step 2: Classical "Swing" Motion

The liberated electron is rocked back and forth by the oscillating electric field, typically gaining a kinetic energy of $\sim$ 50–1000 eV during its first femtosecond of freedom. Because this motion can be estimated to involve the action of hundreds of photons, the use of classical mechanics is justified. Thus, following [821] we write the classical equation of motion of an electron (for which in a.u. – $m_e = 1$, $q = -1$), moving under the influence of an oscillating electric field pointing in the x-direction of amplitude E_0 and frequency ω, as

$$\frac{d^2x}{dt^2} = E_0 \cos(\omega(t - t_i) + \phi) , \tag{16.11}$$

where ϕ is the phase of the field at $t = t_i$ (the "moment of birth" into the continuum). The corresponding electron velocity is

$$\frac{dx}{dt} = -\frac{E_0}{\omega}[\sin(\omega(t - t_i) + \phi) - \sin\phi] , \tag{16.12}$$

and the electron displacement is

$$x(t) = -\frac{E_0}{\omega^2}\left\{\cos\phi - \cos\left[\omega(t - t_i) + \phi\right] - \omega(t - t_i)\sin\phi\right\} , \tag{16.13}$$

where we assumed that $dx(t_i)/dt = 0$ and that $x(t_i) = 0$. It follows from Eq. (16.12) that the electron's velocity is composed of two terms, the first is an oscillatory ("quiver") motion that follows the field's oscillations, and the second is a constant "drift" velocity, given by $E_0 \sin\phi/\omega$, whose value depends on the phase of the field at $t = t_i$.

16.3.3
Step 3: Recollision

Having gained a large amount of kinetic energy, some electrons escape, giving rise to ionization (and especially above threshold ionization [819]), while others turn back and recollide with the remaining unionized electrons that hover near the nucleus (or nuclei in the case of molecules), often liberating of additional electrons [822].

The kinetic energy acquired by the electron at the time of recollision can be calculated as follows: The occurrence of a recollision implies that $x(t_r) = 0$ for some $t_r > t_i$. Imposing this condition on Eq. (16.13), we obtain (as shown in Figure 16.3) that t_r is the point of intersection of the $\cos[\omega(t - t_i) + \phi]$ line with the

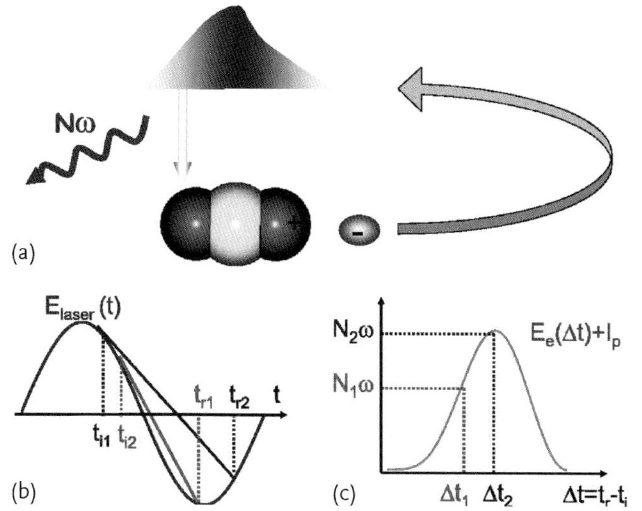

Figure 16.3 An illustration of how the recollision model explains high-harmonics generation. (a) An intense IR laser field liberates an electron. Within a fraction of the laser cycle, the electron is returned to the parent ion by the oscillating field. Recombination is accompanied by the emission of a high-harmonics photon. (b) The moment of ionization t_i determines the moment of recombination t_r. Solid curve shows one oscillation of the laser field E_{laser}. Straight lines connect the instants of ionization t_i and recombination t_r. (b) The time delay between ionization and recombination $\Delta t = t_r - t_i$ is mapped onto the harmonic number: $N\omega = E_e(\Delta t) + I_p$. Here, E_e is the energy of the returning electron and I_p is the ionization potential. Taken from [824, Figure 3]. (Reprinted with permission. Copyright 2009 National Academy of Sciences.)

$\cos\phi - \omega(t - t_i)\sin\phi$ straight line. Alternatively, if we know the time t_r at which recollision occurs, then we know the field phase ϕ at t_i, the "moment of birth" into the continuum, via the relation

$$\tan\phi = \frac{\cos(\omega\Delta t) - 1}{\sin(\omega\Delta t) - \omega\Delta t}, \tag{16.14}$$

where $\Delta t = t_r - t_i$.

This simple formula opens the way to a variety of control schemes with attosecond pulses discussed below. Basically what is required is control of the field's phase ϕ at which tunneling occurs. Such exquisite control is only possible for extremely short, few-cycle, pulses for which the electric field can be measured (by, e.g., attosecond streak camera techniques discussed below) and controlled.

As an illustration, we derive the conditions for the recolliding electron to attain maximum energy at the time it flies by the parent ion (which is when emission of a photon as discussed below can take place). Maximizing $|dx(t_r)/dt)|$, the electron velocity, at the moment of return subject to the Eq. (16.14) constraint (hence maximizing its kinetic energy at that instant) we obtain, that within the first cycle,

$$\left|\frac{dx(t_r)}{dt}\right|_{max} = 1.260\frac{E_0}{\omega}\,, \quad \text{at the points:} \quad \phi = 0.10\pi$$
$$\text{and} \quad \omega\Delta t = 1.30\pi\,. \tag{16.15}$$

The corresponding maximal electron kinetic energy is

$$\frac{1}{2}\left|\frac{dx(t_r)}{dt}\right|^2_{max} = 3.17\,U_p \tag{16.16}$$

where U_p is the cycle-average kinetic energy associated with the oscillatory component of the velocity (see Eq. (16.12)) of the free electron in an electromagnetic field (also called the pondermotive potential)

$$U_p = \frac{\overline{(dx/dt)^2}}{2} = \frac{(E_0/\omega)^2}{4}\,. \tag{16.17}$$

It follows from Eqs. (16.15) and (16.16) that the maximal energy (also known as the cutoff energy) is proportional to the peak electric field squared E_0^2 and the laser wavelength squared λ^2.

Figure 16.3 provides an illustration of Eq. (16.16). The maximum E_e kinetic energy acquired by the electron corresponds to the tunneling occurring at time t_{i1} of Figure 16.3 and the corresponding recombination occurring at time t_{r2}. For 800 nm radiation, the maximal kinetic energy of the recolliding electrons is estimated [810] to be acquired by electrons born at $t_{i1} \approx 1/20$ of the light's period after the peak of the electric field. The electron is estimated to hit the remaining electrons of the parent ion at $t_{r2} \approx 2/3$ of the light's period later.

16.3.4
Step 4: Emission of a Photon

Some of the recolliding electrons manage to lose all their energy by emitting a photon. Energy conservation dictates that if the system returns to its original material state, the energy of the emitted photon should equal the energy of the N photons absorbed, that is, that a photon is emitted at energy of $N\hbar\omega$, where ω is the frequency of the IR laser field. Thus, a series of high-harmonics is generated. Due to the large bandwidth generated in this process, if one is able to retain a well-defined phase between the high-harmonics, the emission described here, when viewed from the time-dependent point of view, corresponds to a sequence of attosecond pulses [825]. (Currently the term "attosecond" is used for subfemtosecond pulses whose duration is typically in the 10^{-16} s regime.)

The generation of a series of high-harmonics is viewed somewhat differently in the semiclassical recollision model: Although the electron gains energy in a continuous fashion from the field, the emission of a photon can take place only twice during each field cycle, when the electron flies by the parent ion. Thus emission occurs in bursts, each separated from the other by half the field's cycle. The frequencies corresponding to such a train of pulses form a sequence (or comb) of harmonics separated from the other by twice the field's frequency.

In isotropic media only the odd harmonics of the fundamental frequency are allowed; the even harmonics are not observed. (This is not strictly the case when the bandwidth of the IR pulse is so large that it contains a full octave, namely a fundamental frequency and its second harmonic, as discussed below). In quantum mechanics the fact that only odd harmonics appear is a result of parity conservation: Since each harmonic corresponds to the emission of a *single* photon, and a photon has odd parity, the continuum state from which emission occurs must have an opposite parity to that of the ground state. This can only be realized if that continuum state is a result of the absorption of an odd number of photons. In semiclassical mechanics (which is at the heart of the present recollision model), selection rules, such as the absence of the even harmonics, always arise as a result of interferences. In this case it is a result of a wave packet born at t_i flying by the $x = 0$ point during its *outward* motion, interfering with a wave packet born a portion of a cycle later flying by the $x = 0$ on its *inward* motion. For odd harmonics this interference creates a state of the opposite parity to that of the ground wave function, thereby allowing for the dipolar coupling between these states and the emission of a photon.

It follows from Eq. (16.16) that the highest photon energy (and therefore the highest harmonic) is given roughly as

$$N_{cutoff}\hbar\omega = 3.17U_p + I_p \tag{16.18}$$

where I_p is the ionization potential. Quantum mechanics does not follow these classical restrictions rigorously (see, e.g., [812, 813]), but observations do confirm the cutoff formula: at energies higher than this value, the probability of photon emission rapidly decreases. Clearly if we want to generate emission at a lesser high-harmonic order, we need to be able to change the field phase ϕ at which tunneling occurs. This, as discussed below, can be achieved by exciting the electron with an ultrashort few-cycle pulse, provided we can measure and control its electric field.

16.4 CEP Stabilization and Control

As mentioned above (Section 16.3), the phase of the field and hence the CEP, affects the conditions for recollision of an electron set free in the tunnel- or multiphoton–ionization processes [826]. The problem is that a train of laser pulses launches a number of electronic wave packets at different instants. Each of these follows a different trajectory because of the differences in the initial conditions of their motion, preventing precise control of strong-field-induced electronic dynamics. However, if the CEP can be controlled and especially set to zero [827], and the strength of the field just sufficient to reach the ionization threshold at the pulse center, a single isolated electronic wave packet can be formed.

The CEO frequency of a laser is influenced by a number of factors, such as the pump power or the tilt of the resonator mirrors. However, since as explained in Section 16.2, one is able to detect the CEP via $f-2f$ interferometry, one can use the interferograms to actively stabilize and control its value [823, 828–831]. The resultant laser is called CEO-stabilized or CEP-stabilized.

In addition to such active CEO frequency control schemes, it is possible to obtain frequency combs that naturally have a zero CEO frequency [827]. The method is based on the fact that a white light continuum pulse produced by a four-wave mixing self-phase modulation (SPM) [805, 806] scheme retains the same value of f_{CEO} as that of the input pulse. In order to see how this happens, consider ω_P, a new frequency added in the SPM process as a result of four-wave mixing,

$$\begin{aligned}\omega_P &= \omega_{n_1} + \omega_{n_2} - \omega_{n_3} ,\\ &= 2\pi\left[f_{CEO} + (n_1 + n_2 - n_3) f_{rep}\right] .\end{aligned} \tag{16.19}$$

We see that indeed any additional frequency generated by the above four-wave mixing process has the same CEO frequency as our original pulse train. We can now use any of the higher lying additional frequencies thus generated as a pump in a parametric down conversion process where the (seed) signal is ω_n, the original comb of frequencies. The idler comb of frequencies thus formed is given as

$$\omega_I = \omega_P - \omega_n = 2\pi(n_1 + n_2 - n_3 - n) f_{rep} , \tag{16.20}$$

that is, a pulse train with zero CEO frequency! The above process is currently being routinely used as a way of producing "self-phase-stabilized" train of ultrashort pulses [827].

16.4.1
The Attosecond Streak Camera

One can use an attosecond pulse to completely characterize the magnitude and phase of the field of a *single* few cycles IR pulse. Such a measurement may be achieved via the concept of the attosecond streak camera. The one-electron attosecond streak camera [832–839] is based on the idea that one can measure the phase of an IR field (or an attosecond field [834]) by performing an ionization process in combination with an ultrashort X-ray pulse.

As discussed in the recollision model of Section 16.3, the distribution of the electron's final momentum is a sensitive function of t_i, the electron's moment of birth into the continuum. In turn, t_i depends on ϕ, the electric field phase at that instant. Whereas in tunnel ionization there is a whole range of t_i's, these birth-times can be well defined if the ionization is performed with an auxiliary attosecond pulse. Examining the asymptotic momentum distribution of an ionized electron as a function of the delay between the two pulses thus allows "imaging" of the electric field of the IR few-cycle pulse. A similar idea can be used to characterize the phase of the attosecond pulse itself [834].

It is also possible to measure the correlation between two electrons [840] in this way. Using sequential double ionization of an inner shell electron knocking out an outer shell electron, one can measure $\theta_{1,2}$, the angle between the momenta of the escaping inner shell and outer shell electrons. $\theta_{1,2}$, which is a measure of electronic correlation, can be varied by controlling the birth time t_i of the first (inner shell) electron to be ionized. The net result is that a measurement of the kinetic energy distributions of two ionized electrons as a function of the delay between an attosecond and an IR pulse allows mapping out of the correlation between the inner shell and outer shell electrons.

16.5 Coherent Control of Sample Molecular Systems

16.5.1 One-Photon vs. Two-Photon Control with Few-Cycle Pulses

As mentioned above, a sufficiently broadband pulse might contain a full octave of frequencies; that is, some of the pulse's lower frequencies are accompanied by their second harmonics. Under such circumstances the one-photon vs. two-photon coherent control scenarios discussed extensively in Section 3.4.3 can be executed using frequencies contained within a *single* ultrashort pulse [841–843]. This indeed is the situation in many attosecond pulses produced in the process of electron recombination discussed in Section 16.3.

16.5.1.1 Backward-Forward Asymmetry in the Dissociative Photoionization of D_2.

We first consider the dissociative photoionization of $D_2 \to D + D^+ + e^-$ with few-cycle pulses experiment of Vrakking's group [842, 843], aimed at controlling the direction of recoil of the photofragments, namely whether it is the D^+ fragment that moves in the forward direction and the D fragment in the backward direction, or vice versa. Alternatively one can think of this experiment as the control of the motion of the remaining electron, whether it hops on the D^+ ion moving *forward*, or it hops on the D^+ ion moving *backward*. This experiment is very similar to that of DiMauro *et al.* [63, 755] which utilized two separate colors, at 1053 and 527 nm, one of them being the second harmonic of the other, to achieve coherent directionality control in the $D+H^+ \leftarrow DH^+ \to H+D^+$ process. It is also the process advocated in Section 10.3 to produce entangled atom-ion pairs in an optical lattice.

In interpreting their experiment, Vrakking's group [842] attributed the observed phase-controlled asymmetry, shown in Figure 16.4, to the detailed dynamics of electron–ion recollision induced dissociation. By computing wave packets on the various electronic states involved, they were able to show that the recolliding electron can localize about one D^+ or the other, in a phase-dependent manner. But since the two D^+ ions are completely identical to one another (i.e., the wave function should be properly symmetrized with respect to exchange between the two

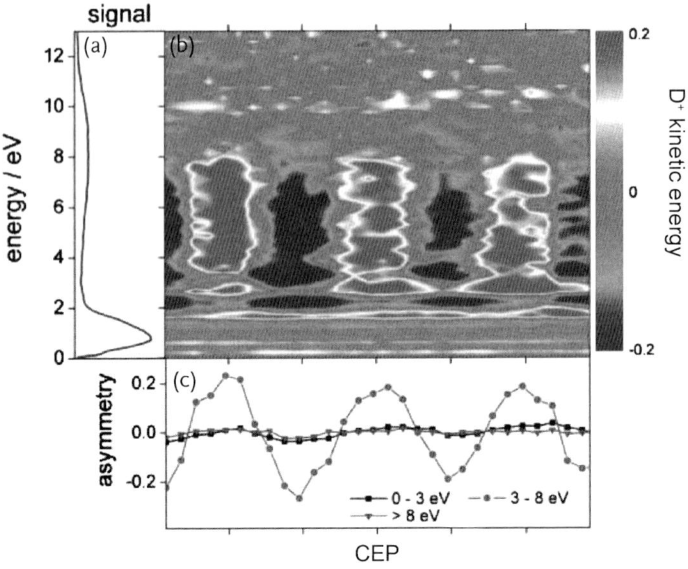

Figure 16.4 (a) D^+ kinetic energy spectrum for D_2 dissociation with 5 fs, 10^{14} W/cm^2 laser pulses without phase stabilization. (b) Map of forward-backward asymmetry parameter as a function of the D^+ kinetic energy and CEP (measured over a range of 6π with a step size of $\Delta\phi = 0.1\pi$. (c) Integrated asymmetry over several energy ranges versus CEP. Taken from [842, Figure 3]. (Reprinted with permission. Copyright 2006 American Association for the Advancement of Science.)

deuterons), this explanation sheds no light on why the electron chooses to localize about the forward moving deuteron, in preference to the backward moving other, or vice versa, depending on the CEP. That is, these computations do not give information on what happens in the *lab frame*: why with one CEP a D^+ will move in the forward direction, whereas with another CEP it is the D atom that moves in the forward direction.

This lab frame breaking of symmetry must arise from a strong-field variant of the one- vs. two-photon interference. As shown in Section 3.4.3, such an interference can break the symmetry in the lab frame, inducing an electron to move in the forward or backward lab-frame directions, irrespective of the exact nature of the electronic states involved, or the details of the recollision process. One- vs. two-photon interference is possible here because the bandwidth of the few cycles laser pulse contains both the frequency ω_n and its second harmonic $2\omega_n$. Support for this point of view is found in a computation [844] that demonstrated remarkable agreement between the perturbation theory based one- vs. two-photon interference mechanism and a full quantum simulation, for the setup used in this experiment [842, 843].

16.5.2
Control of the Generation of High-Harmonics

Use of many-cycle pulses leads, in an isotropic medium, to the generation of odd harmonics. This was shown above to result from parity conservation; that is, because each high-harmonic is composed of only one photon, the parity of the excited state from which this photon was emitted must be opposite to that of the final (ground) state. Since the excited state is populated via the multiphoton absorption of fundamental photons from the same ground state, the number of photons that can lead to the emission of just one photon must therefore be odd.

As mentioned above, a few-cycle pulse containing a full octave can however produce material states that break the backward-forward inversion symmetry, hence the above parity conservation symmetry considerations would not seem to apply. Indeed the observation of even harmonics generation when a fundamental frequency is mixed with its second harmonic [845] was explained in the context of the recollision model precisely in these terms. However, the production of symmetry broken states does not break the inherent inversion symmetry of the *operators* involved, such as that of the Hamiltonian (symmetric) or the dipole moment (antisymmetric). Thus the production of symmetry-broken states via the one-photon and two-photon interference, producing a linear combination of $\psi_{S,n}$, a symmetric state, and $\psi_{A,n}$ an antisymmetric n-harmonic generating state

$$\Psi_n = c_{S,n}\psi_{S,n} + c_{A,n}\psi_{A,n} \tag{16.21}$$

does not change the dipole selection rules with respect to emission to the (assumed symmetric) ground state, according to which

$$\langle \Psi_n|\hat{\epsilon}\cdot \boldsymbol{d}|\Psi_g\rangle = \left\{c^*_{S,n}\langle\psi_{S,n}| + c^*_{A,n}\langle\psi_{A,n}|\right\}\hat{\epsilon}\cdot\boldsymbol{d}|\Psi_g\rangle = c^*_{A,n}\langle\psi_{A,n}|\hat{\epsilon}\cdot\boldsymbol{d}|\Psi_g\rangle\ . \tag{16.22}$$

We see that although the broken symmetry state Ψ_n contains both symmetric and antisymmetric components, only one of these components (the antisymmetric one) matters as far as the dipole moment matrix elements go. The production of the other component by the one- vs. two-photon process does not seem to matter and does not explain by itself the formation of even harmonics.

As depicted in Figure 16.5, what the one- vs. two-photon interference actually does is to sever the link between the parity of the excited state and the parity of the high-harmonic produced by it. The presence of a fundamental frequency ω and its second harmonic 2ω, allows for the appearance of even harmonics produced by the absorption of an *odd* number of photons (giving rise to excited states of odd parity). For example, the 4th harmonic of ω (or the 2nd harmonics of 2ω) can result as shown in Figure 16.5 from the absorption of three photons: two ω-photons and one 2ω-photon. One might call this type of even harmonics emission with respect to both ω and 2ω, an "even-even" situation. This effect exists simultaneously with the more trivial "even-odd" situation in which the absorption of an odd number

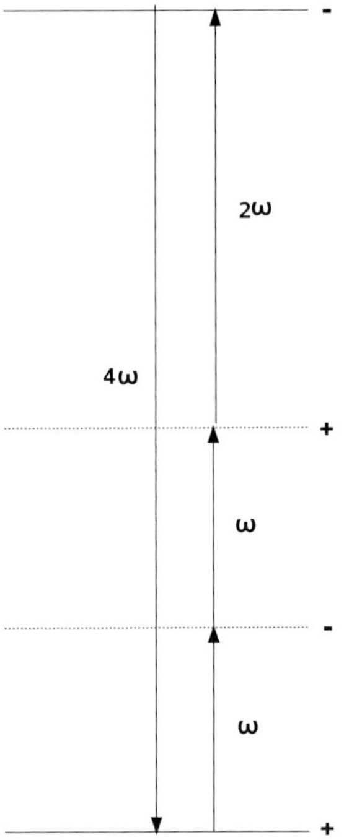

Figure 16.5 An illustration of the way even harmonics are generated due to the presence of a fundamental frequency ω and its second harmonic 2ω. Marked (as $\pm$) are the parities of the material states.

of the second harmonic 2ω-photons produces states of odd parity that emit odd harmonics of the 2ω-photons that coincide in frequency with *even* harmonics of the ω-photons. It is possible to control what type of even harmonics is produced by changing the phase between the one-photon field and the two-photon field. Control of the ratio between different high-harmonics orders of the same parity can also be achieved in this way as well as by using other coherent control [845, 846] and optimal control techniques [141, 847].

16.5.3
Control of Electron Transfer Processes

Electron transfer stands as one of the fundamental processes in molecular physics and chemistry [184]. For example, it is an integral component of the harpoon mechanism, a basic mechanism in elementary molecular reaction dynamics [848–850], where electron transfer occurs at large internuclear distances. Here we discuss a proposal [350] for using high-harmonic generation (HHG) to probe electron transfer dynamics within alkali halides, with NaI used as a specific example. The method extends high-harmonic probing to chemically relevant electronic dynamics.

Due to its eV-range energy, the electron liberated in HHG recollides with a subangstrom wave length, making the process extremely sensitive to the spatial structure of both the nuclei and electronic orbitals of the parent ion. With respect to nuclear degrees of freedom, recent experiments have shown HHG to be a sensitive probe of vibrational [851, 852] and rotational [853, 854] dynamics in molecules. A striking example of the high sensitivity to electronic degrees of freedom is the experimental realization of orbital tomography using high-harmonic emission, where the HOMO orbital of N_2 was reconstructed from a series of high-harmonic measurements [855], demonstrating that high-harmonic probing carries great potential for exploring electronic states of molecules. The ability to explore electron transfer dynamics within alkali halides, as discussed here, greatly extends this capability.

16.5.4 Electron Transfer in Alkali Halides

The two lowest Born–Oppenheimer potential energy curves of alkali halide diatomics display a generic characteristic. At large internuclear distances R, one potential curve corresponds to a covalent configuration M + X while the other corresponds to the ionic configuration $M^+ + X^-$; the two curves cross as R becomes smaller. Figure 16.6a shows the covalent and ionic curves for the case of NaI [856]. When an alkali halide in the ground electronic state at the equilibrium internuclear distance is photoexcited to the covalent state, the excited nuclear wave packet moves toward the crossing point, undergoing a nonadiabatic transition corresponding to electron transfer; that is, the valence electron moves from the M atom of the M + X species to the X atom to form $M^+ + X^-$. During the transfer, the electronic state is in a coherent superposition of electronic states. HHG emission from targets in coherent electronic superpositions was first studied with the focus on using the superposition state to tune and enhance the harmonic emission [857, 858], and more recently with the intention of imaging the electronic motion [859]. The study described here [350] considers the interesting case where the electronic coherence arises from the electronic-nuclear coupling present in alkali halides. The method has considerable advantages insofar as it provides an all-optical technique to probe the nonadiabatic electronic dynamics and provides for the simultaneous probing of a large range of momentum components of the electronic orbitals.

Consider NaI using the model and potential energy surfaces outlined in [856] and [860]. The pump laser pulse $E_p(t)$ is chosen to be

$$\varepsilon_p(t) = \mathcal{E}_p \exp\left[-4\ln 2t^2/\sigma_p^2\right]\cos(\omega_p t) \,. \tag{16.23}$$

With the pulse written in this way, $\sigma_p = 30$ fs is the full width at half maximum. The maximum electric field strength and central frequency are $\mathcal{E}_p = 1.7 \times 10^{-4}$ a.u. (intensity of 10^9 W/cm^2) and $\omega_p = 0.1373$ a.u. (328 nm). Figure 16.6 shows the calculated nuclear dynamics following excitation of the ground vibrational state on the ionic surface by the pump pulse. Figure 16.6b,c plots

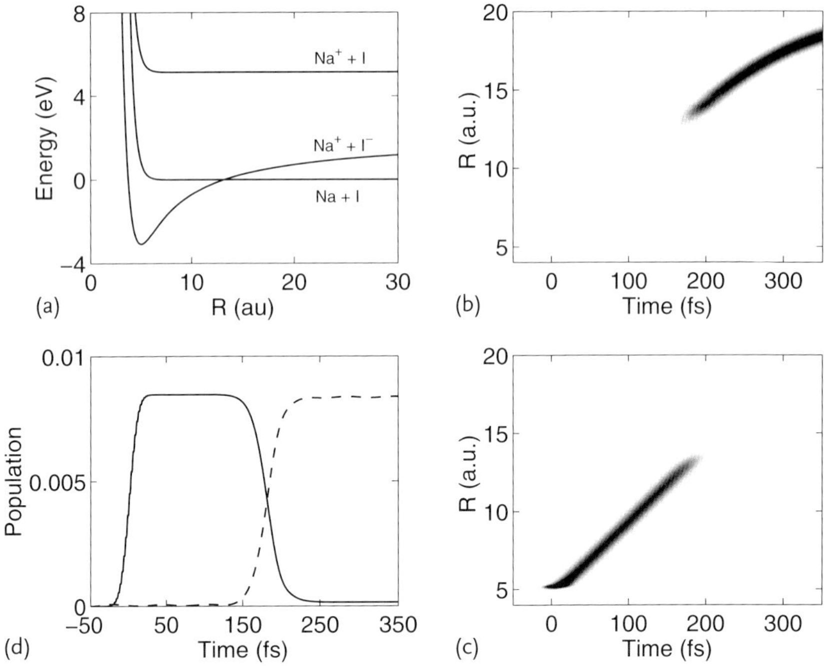

Figure 16.6 Nonadiabatic electron transfer dynamics of the excited wave packet. Relevant potential energy curves for electron transfer in the NaI system are shown in Panel (a). Panels (b) and (c) plot $|\chi_I(R,t)|^2$ and $|\chi_C(R,t)|^2$ respectively, that is, the time-dependent radial probability distributions on the ionic and covalent surfaces. (white corresponds to zero probability, black corresponds to maximum probability.) Panel (d) plots the electronic populations of the excited wave packet. The solid curve denotes the covalent state, the dashed curve denotes the ionic state. From [350, Figure 1]. (Reprinted with permission. Copyright 2008 American Physical Society.)

the time-dependent radial probability distributions $|\chi_I(R,t)|^2$ and $|\chi_C(R,t)|^2$ respectively, where R is the vibrational coordinate and the C and I subscripts refer to the covalent and ionic electronic states respectively. Figure 16.6d plots the time-dependent populations of both electronic states, $P_i(t) = \int dR|\chi_i(R,t)|^2$, $(i = I, C)$. Note that the initial ground state has been projected out of $\chi_I(R,t)$ before plotting these results in order to focus on the excitation portion of the total wave function. The nonadiabatic transition corresponding to electron transfer is clearly seen to occur over the range of 160–210 fs.

Consider now high-harmonic generation from the excited nuclear wave packet. The vector potential and electric field of the HHG pulse are set to

$$A(t) = -\frac{\mathcal{E}_H}{\omega_H}\exp\left[-4\ln 2(t-t_d)^2/\sigma_H^2\right]\cos(\omega_H t) \tag{16.24}$$

and $E_H(t) = -\partial A(t)/\partial t$, where t_d is the time delay between the excitation pulse and the HHG probe pulse, and the pulse is assumed to be polarized perpendicular to the molecular axis. Since the NaI and Na^+I^- species have relatively low ionization potentials ($\sim$ 5.14 eV) we choose a carrier wavelength of $\omega_H = 0.0143$ a.u. (3200 nm) for the HHG pulse in order to increase the recollision energy as much as possible prior to saturation. In order to avoid excessive time-averaging of the transition during the probing step, a single cycle pulse $\sigma_H = 2\pi/\omega_H$ (10.6 fs) with peak field strength of $\mathcal{E}_H = 0.02$ a.u. (1.4×10^{13} W/cm^2) is used.

The total wave function (vibrational and electronic) is written as

$$|\Psi_0(t)\rangle = |\chi_I(t)\rangle\, |\phi_I^{(n)}\rangle + |\chi_C(t)\rangle\, |\phi_C^{(n)}\rangle\,, \tag{16.25}$$

where $|\phi_j^{(n)}\rangle$ $(j = C, I)$ are n-electron states at the average bond length of the corresponding vibrational wave packet $|\chi_j(t)\rangle$. Note that the $|\phi_j^{(n)}\rangle$ are normalized eigenstates that depend parametrically on the average bond length, while the $|\chi_j(t)\rangle$ represent the time-dependent nuclear wave packets whose norms change as the vibrational wave packet moves from one surface to the other.

The strong-field approximation (SFA) is used to describe the ionization and continuum motion of the free electron [812, 816, 861, 862]. Although the SFA falls short of full quantitative prediction, it gives excellent qualitative insight and is sufficient to explore the main features of HHG probing of electron transfer in alkali halides.

In addition to the SFA, negligible depletion of the initial state during the HHG pulse is assumed, the nuclei are frozen during the HHG process, and include only the lowest electronic surface of the Na^+I ion following strong-field ionization from either the NaI or Na^+I^- state. With these assumptions, the dipole describing the HHG emission polarized perpendicular to the molecule is written, in atomic units ($\hbar = m_e = e = 1$), as

$$\begin{aligned}
d_{HHG}(t) =& \\
&- i \int_0^t dt' \int dk\, e^{-i\frac{1}{2}\int_{t'}^{t} d\tau [k + A(\tau)]^2}\, E_H(t') \\
&\times \Big[\langle \phi_I^{(n)} | \hat{r}_x | \phi^{(n-1)}; \boldsymbol{k}, t \rangle \langle \phi^{(n-1)}; \boldsymbol{k}, t' | \hat{r}_x | \phi_I^{(n)} \rangle \langle \chi_I(t_d) | \chi_I(t_d) \rangle e^{-iI_{p,I}(t-t')} \\
&+ \langle \phi_C^{(n)} | \hat{r}_x | \phi^{(n-1)}; \boldsymbol{k}, t \rangle \langle \phi^{(n-1)}; \boldsymbol{k}, t' | \hat{r}_x | \phi_C^{(n)} \rangle \langle \chi_C(t_d) | \chi_C(t_d) \rangle e^{-iI_{p,C}(t-t')} \\
&+ \langle \phi_C^{(n)} | \hat{r}_x | \phi^{(n-1)}; \boldsymbol{k}, t \rangle \langle \phi^{(n-1)}; \boldsymbol{k}, t' | \hat{r}_x | \phi_I^{(n)} \rangle \langle \chi_C(t_d) | \chi_I(t_d) \rangle e^{-iI_{p,C}t + iI_{p,I}t'} \\
&+ \langle \phi_I^{(n)} | \hat{r}_x | \phi^{(n-1)}; \boldsymbol{k}, t \rangle \langle \phi^{(n-1)}; \boldsymbol{k}, t' | \hat{r}_x | \phi_C^{(n)} \rangle \langle \chi_I(t_d) | \chi_C(t_d) \rangle e^{-iI_{p,I}t + iI_{p,C}t'} \Big]\,,
\end{aligned} \tag{16.26}$$

where $|\phi^{(n-1)}; \boldsymbol{k}, t\rangle \equiv |\phi^{(n-1)}\rangle\, |\boldsymbol{k}, t\rangle$, $|\phi^{(n-1)}\rangle$ is the $(n-1)$-electron state of the ion, $|\boldsymbol{k}, t\rangle$ are the single-electron plane wave continuum states defined as $\langle \boldsymbol{r} | \boldsymbol{k}, t\rangle = (2\pi)^{-(3/2)} e^{i[\boldsymbol{k}+A(t)]\cdot \boldsymbol{r}}$, and $\boldsymbol{r}$ is the spatial coordinate of the continuum electron.

The first two terms in Eq. (16.26) represent the harmonic emission from each electronic state individually – they would appear in identical form for an incoherent mixture of the $|\chi_I(t)\rangle\, |\phi_I^{(n)}\rangle$ and $|\chi_C(t)\rangle\, |\phi_C^{(n)}\rangle$ states. The second two terms represent the interference between these two states arising from the coherence between them. Note, significantly, that due to the $\langle \chi_C(t) | \chi_I(t) \rangle$ factor in the interference terms, the coherence is only manifested when the nuclear wave packets overlap.

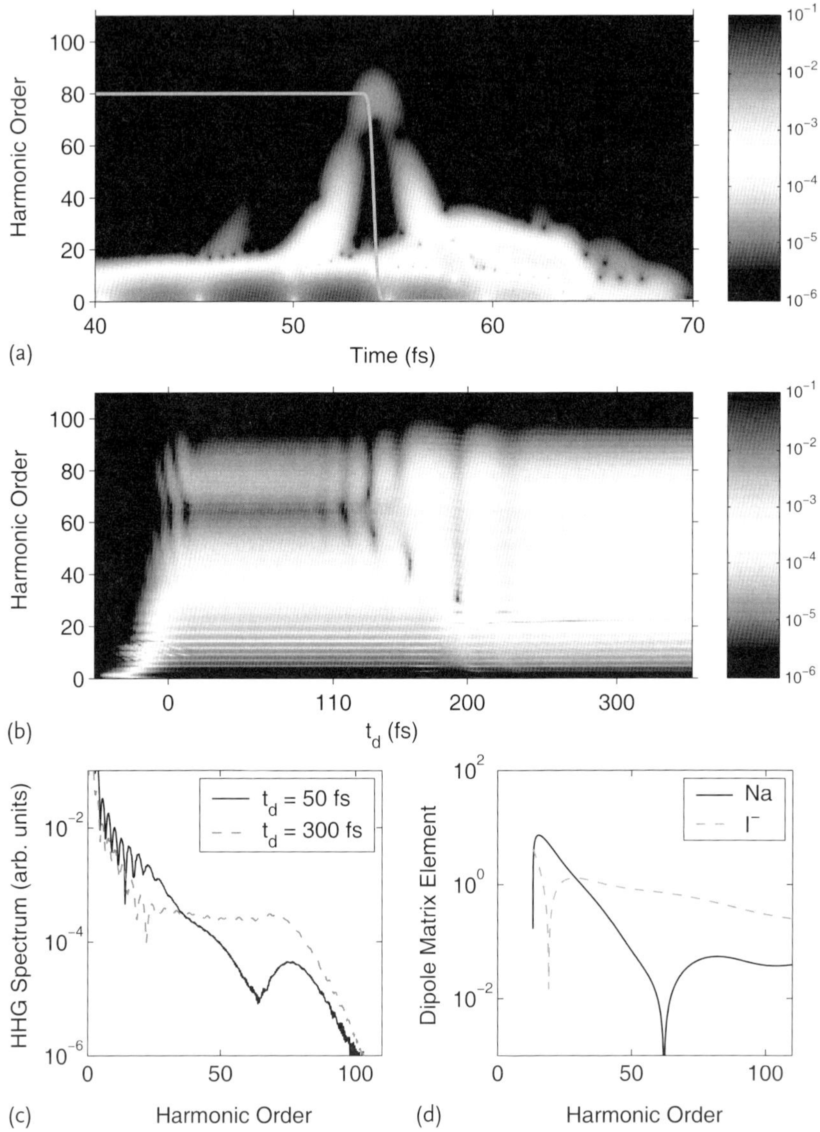

Figure 16.7 High-harmonic spectra probing the electron transfer process. (a) Time-frequency representation of $d_{HHG}(t)$ for $t_d = 50$ fs. The solid curve shows the filter $f(t)$ for this delay. (b) HHG spectrum as a function of t_d. (c) HHG spectra for $t_d = 50$ and 300 fs. (d) SFA recombination matrix elements plotted as a function of corresponding HHG emission energy E_{HHG}. From [350, Figure 2]. (Reprinted with permission. Copyright 2008 American Physical Society.)

Figure 16.7 plots emitted t_d-dependent HHG spectra as

$$\frac{\left|\mathcal{FT}\left[\frac{\partial^2 d_f(t)}{\partial t^2}\right]\right|^2}{\omega^4} = \frac{|\omega^2 \mathcal{FT}[d_f(t)]|^2}{\omega^4} = |\mathcal{FT}[d_f(t)]|^2 , \tag{16.27}$$

where $\mathcal{FT}$ denotes the Fourier Transform and $d_f(t)$ is the suitably time-filtered [350] value of $d_{HHG}(t)$. Figure 16.7b plots the spectra for all delays, while Figure 16.7c plots two representative spectra taken at $t_d = 50$ and 300 fs, before and after the electron transfer. Since a single-cycle pulse is used, the highest harmonics result from a single recollision wave packet and consequently display a continuous structure. Significantly, a large change in the HHG spectrum is seen to characterize the nuclear wave packet moving through the crossing region. The general character in the plateau region in Figure 16.7c is largely governed by the recombination matrix element $\langle \psi_j^H | \hat{r}_x | \boldsymbol{k}, t \rangle$ in the last step of the HHG process. For small t_d, when the valence electron is located on the Na atom, the recolliding electron recombines to the Na-3s orbital, while at later times the recombination involves the 5p orbital of I^-. Figure 16.7d plots $\langle \psi_j^H | \hat{r}_x | \boldsymbol{k}, t \rangle$ for both valence orbitals as a function of the harmonic order using the SFA relation between emitted energy and recollision momentum $E_{HHG} = k^2/2 + I_p$. The change in character of the harmonic emission is seen to directly reflect the change in the valence orbital structure.

In addition to this change in character of the plateau region, interference beats are seen in Figure 16.7b over the time interval where the nuclei pass through the crossing region, and readily provide a signature of the coherence present during the electron transfer process. The period of these interferences is seen to change as a function of t_d, a consequence of the fact that the phase of the interference terms in Eq. (16.26) is modulated by $I_{p,I}$, which varies as a function of t_d. Thus, the variation of the ionization potential of the ionic states becomes encoded in the interference beats as the nuclear wave packet undergoes the nonadiabatic transition.

This work demonstrates that HHG probing is capable of exposing the time evolving dynamics of electron transfer within alkali halides via the orbital structure and coherence during the electron transfer dynamics.

17
Case Studies in Optimal Control

Chapter 5 introduced the essential principles of optimal control. Here we describe a number of applications of the control of molecular processes, and emphasize (Section 17.4) the need for serious theoretical modeling of these experiments in order to properly extract the underlying control mechanisms.

An extensive review of experiments from an optimal control perspective, which is not the focus here, is provided in [97].

17.1
Creating Excited States

One of the earliest objectives of optimal control studies was to control the population of a specified vibrational state. Interest in this topic dates back to the early days of infrared multiphoton dissociation (IRMPD) [863], when researchers sought methods to preferentially populate a given vibrational state "before the deleterious effects of intramolecular vibrational relaxation (IVR) set in." In particular, the ideas discussed in Section 3.A were prevalent; that is, it was thought that one could "beat" out the rate of IVR by populating a given level while it remains "pure." However, as discussed in Section 3.A, the correct way to achieve this objective in the long-time limit is to focus on properties of energy eigenstates, and not on the time dependence of IVR.

The objective of populating specific energy levels using laser pulse sequences has since been realized in a number of ways. An example is the STIRAP approach discussed in Section 11.1. A number of alternative approaches are discussed below. We first consider the less general, though nontrivial, task of populating a target wave packet of states by a perturbative N-photon process. In particular we focus on the preparation of "bright" states, with the term "bright" to be defined precisely below.

Consider a molecule with Hamiltonian H_M, with energy eigenstates $|E_n\rangle$, of energy E_n, subjected to an optical pulse. Specifically, consider N-photon absorption by the molecule that is initially in state $|E_i\rangle$. We define two classes of states, *bright* states, $|\phi_s\rangle$, that are accessible by N photon absorption from $|E_i\rangle$, and *dark* states

Quantum Control of Molecular Processes, Second Edition. Moshe Shapiro and Paul Brumer

$|\chi_m\rangle$ that are not accessible by this process. That is, the dark states satisfy

$$\langle E_i|T^{(N)}(\omega)|\chi_m\rangle \approx 0 , \tag{17.1}$$

where $T^{(N)}(\omega)$ is the lowest-order N-photon transition operator (whose three-photon analog is given in Eq. (3.44)), given by

$$T^{(N)}(\omega) = \Pi_{k=1,2,\ldots,N-1}\left[\hat{\epsilon}\cdot \boldsymbol{d}(E_i - H_M + k\hbar\omega)^{-1}\right]\hat{\epsilon}\cdot \boldsymbol{d} . \tag{17.2}$$

For one-photon transitions $T^{(N)}(\omega) = \hat{\epsilon}\cdot \boldsymbol{d}$.

Whenever the lowest (Nth) order perturbation theory for the N-photon problem is valid, it is possible to generate a wave packet of *bright* states, assuming that such bright states exist. To see this, partition the excited state manifold into bright states $|\phi_s\rangle$, and dark states, $|\chi_m\rangle$. The eigenstates $|E_n\rangle$ of the molecular Hamiltonian H_M can therefore be written as

$$|E_n\rangle = \sum_s |\phi_s\rangle\langle\phi_s|E_n\rangle + \sum_m b_{m,n}|\chi_m\rangle . \tag{17.3}$$

Consider then excitation with an optical pulse of the form

$$\varepsilon(t) = \int d\omega\epsilon(\omega)\{\exp(-i\omega \boldsymbol{t}) + \text{c.c.}\} , \tag{17.4}$$

where $\boldsymbol{t} \equiv t - z/c$ and z is the direction of propagation, incident on the molecule in state $|E_i\rangle$. Analogous to Section 3.4.1, at the end of the pulse, and in the (Nth order) perturbative limit, the molecule will be in a superposition state of the form

$$\begin{aligned} |\psi(t)\rangle &= \frac{2\pi i}{\hbar}\sum_n \bar{\epsilon}^N(\omega_n)|E_n\rangle\langle E_n|T^{(N)}(\omega_n)|E_i\rangle \exp(-iE_n t/\hbar) \\ &= \frac{2\pi i}{\hbar}\sum_{n,s}\langle\phi_s|T^{(N)}(\omega_n)|E_i\rangle\bar{\epsilon}^N(\omega_n)|E_n\rangle\langle E_n|\phi_s\rangle \exp(-iE_n t/\hbar) , \end{aligned} \tag{17.5}$$

where $\omega_n \equiv (E_n - E_i)/(N\hbar)$, $\bar{\epsilon}(\omega_n) \equiv \epsilon(\omega_n)\exp(i\omega_n z/c)$ and where we have used Eq. (17.1) in the last equality.

If the frequency-width of the pulse is sufficiently large, (i.e., the pulse is sufficiently short in time) and ω_n is far from an intermediate resonance then we can invoke the SVCA (Eq. (12.13)). Alternatively, the power broadening may be large enough to smooth out the intermediate resonances. (Note that power broadening is the tendency of spectral absorption lines to broaden when measured using intense fields [285]). In this case, invoking the SVCA means assuming that $T^{(N)}(\omega_n)\bar{\epsilon}^N(\omega_n)$ varies more slowly with n than does $\langle E_n|\phi_s\rangle$. We can then take $T^{(N)}(\omega_n)\bar{\epsilon}^N(\omega_n)$ out of the sum in Eq. (17.5), approximate it by a constant $T^{(N)}(\bar{\omega})\bar{\epsilon}^N(\bar{\omega})$, and obtain,

$$\begin{aligned} |\psi(t)\rangle &= \frac{2\pi i}{\hbar}\sum_s\langle\phi_s|T^{(N)}(\bar{\omega})|E_i\rangle\bar{\epsilon}^N(\bar{\omega})\sum_n|E_n\rangle \exp(-iE_n t/\hbar)\langle E_n|\phi_s\rangle \\ &= \frac{2\pi i}{\hbar}\sum_s\langle\phi_s|T^{(N)}(\bar{\omega})|E_i\rangle\bar{\epsilon}^N(\bar{\omega})\exp(-iHt/\hbar)|\phi_s\rangle . \end{aligned} \tag{17.6}$$

It follows from Eq. (17.6) that immediately after the pulse (defined at $t = 0$) is off, the wave function is

$$|\psi(t)\rangle = \exp(-iHt/\hbar)\sum_s A_s|\phi_s\rangle\ , \tag{17.7}$$

where $A_s = (2\pi i/\hbar)\langle\phi_s|T^{(N)}(\bar{\omega})|E_i\rangle\bar{\epsilon}^N(\bar{\omega})$. Hence we see that in the Nth order perturbative regime, a sufficiently short pulse can indeed create a wave packet composed of pure bright states, at least at very short times after the end of the pulse.

This result can be generalized to the preparation of other types of "zero-order" states. For example, in accord with the objectives of "mode-selective chemistry" (Appendix 3.A) we may want to prepare a specific "local mode" vibrational state. To do so, consider [864] an M-level oscillator that has the "right" anharmonicity, that is, a system whose energy levels behave like

$$E_\mathrm{v} = E_0 + \hbar\mathrm{v}\left[\omega_{1,0} - \Delta^a(\mathrm{v}-1)\right]\ . \tag{17.8}$$

Here E_v is the vth vibrational energy level with wave function ϕ_v, $\omega_{1,0}$ is a harmonic frequency, and Δ^a is the anharmonicity constant. Under certain circumstances a system of this kind, initially in its ground state, and driven by a CW field

$$\varepsilon(t) = \epsilon_0\cos(\omega t)\ , \tag{17.9}$$

is equivalent to a two-level system. To see this, expand the wave function at time t, $\psi(t)$, as

$$\psi(t) = \sum_v a_\mathrm{v}\phi_\mathrm{v}\exp(-iE_\mathrm{v}t/\hbar)\ . \tag{17.10}$$

Substituting Eq. (17.10) into the time-dependent Schrödinger equation gives a set of coupled equations for the a_v coefficients of the form,

$$\begin{aligned} i\frac{da_\mathrm{v}}{d\tau} &= [(n-\mathrm{v}) - S_n]\mathrm{v}a_\mathrm{v} - F_0\left[(\mathrm{v}+1)^{1/2}a_{\mathrm{v}+1} + \mathrm{v}^{1/2}a_{\mathrm{v}-1}\right]\ ,\\ \mathrm{v} &= 0,1,2,\ldots,M\ , \end{aligned} \tag{17.11}$$

where $\tau \equiv \Delta^a t$ is a dimensionless time,

$$S_n \equiv n - 1 - \frac{\omega_{1,0}-\omega}{\Delta^a}\ , \tag{17.12}$$

and

$$F_0 \equiv \frac{d_{1,0}\epsilon_0}{2\hbar\Delta^a}\ , \tag{17.13}$$

is a dimensionless "Rabi frequency".

If $F_0 \ll 1$ then the above set of equations is identical to a two-level system,

$$\begin{aligned} i\frac{da_0}{d\tau} &= -F_0^{(n)}a_n\ ,\\ i\frac{da_n}{d\tau} &= S_n n a_n - F_0^{(n)}a_0\ , \end{aligned} \tag{17.14}$$

where

$$F_0^{(n)} = \frac{F_0^n n}{(n-1)!(n!)^{1/2}} . \tag{17.15}$$

These equations can be solved in the usual way to yield:

$$P_n(\tau) \equiv |a_n(\tau)|^2 = \left(\frac{F_0^{(n)}}{\Omega_n}\right)^2 \sin^2 \Omega_n \tau , \tag{17.16}$$

where

$$\Omega_n = \left\{\left(\frac{S_n n}{2}\right)^2 + \left(F_0^{(n)}\right)^2\right\}^{1/2} \tag{17.17}$$

is the n-photon Rabi frequency.

When ω is tuned to the n-photon resonance frequency,

$$\omega = \omega_{1,0} - \Delta^a(n-1) , \tag{17.18}$$

$S_n = 0$ and $P_n(\tau)$ reaches unity at the reduced time $\tau_p = \pi/(2\Omega_n) = \pi/[2F_0^{(n)}]$. Since $F_0 \ll 1$, then $\tau_p \gg 1$ and the time taken to attain complete population transfer to the nth level is very long. The same applies to the use of optimal pulses in populating rotational states [121] and to the use of adiabatic passage, discussed in Section 11.1. Further, it was found [864] that increasing F_0, which shortens τ_p, results in complete loss of selectivity. Basically, the power broadening increases so much that all n-photon resonances overlap.

Paramonov *et al.* [865–869] have solved this problem by using a short pulse of the form

$$\varepsilon(t) = \epsilon_0 \sin^2(\alpha t) \cos(\omega t) . \tag{17.19}$$

This pulse does not result in the coalescence of the n-photon resonances even when F_0 is increased because when $\alpha \ll \omega$ one obtains a set of equations that are similar to Eq. (17.11):

$$i\frac{da_v}{d\tau} = [(n-v) - S_n] v a_v - F_0 \sin^2\left(\frac{\alpha\tau}{\Delta^a}\right)\left[(v+1)^{1/2} a_{v+1} + v^{1/2} a_{v-1}\right] ,$$
$$v = 0, 1, 2, \ldots, M . \tag{17.20}$$

Hence [865] for a certain range of ω, F_0 and α, it is possible to preferentially excite any level n that one desires.

The pulse considered by Paramonov *et al.* [865] can be thought of as a sum of three "rectangular" pulses,

$$\varepsilon(t) = \frac{1}{4}\epsilon_0\{2\cos(\omega t) - \cos[(\omega+\alpha)t] - \cos[(\omega-\alpha)t]\} , \tag{17.21}$$

that have two frequencies in addition to the central frequency of Eq. (17.9). Thus, in addition to the shortness of the pulse, multiphoton excitation with $\sin^2(\alpha t)$ modulation is more efficient because it provides additional frequencies that help overcome frequency mismatches. In fact, a pulse of the type

$$\varepsilon(t) = \epsilon_0 \sin^{2m}(\alpha t)\cos(\omega t) \tag{17.22}$$

has also been considered by Paramonov and Savva [866]. It has, depending on the value of m and α, all the "right" resonance frequencies for a sequential multiphoton process, since $\varepsilon(t)$ can be rewritten as

$$\varepsilon(t) = \epsilon_0 \sum_{k=-m}^{m} C_k \cos[(\omega + 2k\alpha)t]\,. \tag{17.23}$$

Comparing $\omega + 2k\alpha$ to the level spacing $\omega_{1,0} - 2v\Delta^a$ we see that all the anharmonic frequencies, up to level m, are contained in the pulse if α is chosen equal to Δ^a.

The numerical simulations of Paramonov *et al.* [865, 866], were later confirmed by Manz *et al.* [125, 126, 129] using optimal control theory (OCT). These computations show that all of the frequencies in the pulse are important. However, the continuum of frequencies afforded by the leading and trailing edges of the pulse was found to be even more important for the overall process.

17.1.1
Using Prepared States

Having shown that it is possible to prepare bright states, or to prepare specific vibrational states in particular cases, we consider the utility of such states. If the task is merely to control the populations of stable molecules, then the discussion above demonstrates the possibility of doing so. Similarly, for example, Rabitz *et al.* [119, 475] have shown that it is possible to control the vibrational states of local bonds in a chain of harmonic oscillators.

Alternatively, such prepared excited states may prove useful photochemically under particular circumstances. This is especially true for local-mode type molecules [870–872]; that is, molecules for which vibrational eigenstates resemble localized excitation in individual bonds. As an example, in the case of HOD the large frequency difference between the OH and OD oscillators is such that intramolecular vibrational relaxation does not destroy the localized excitation. (Similar effects arise if one excites a resonance state that displays local behavior for example, see [873] for ABA type molecules.) As shown theoretically [874, 875], and confirmed experimentally [30–37], preparation of the OH stretch followed by an excitation laser leading to dissociation, gives a marked enhancement of the H atom photodissociation in many molecules.

Similarly, in the case of bimolecular reactions, Zare *et al.* [876], confirmed theoretical predictions and demonstrated experimentally [877–879] that by exciting *either* the OH *or* the OD bond in HOD, one can selectively enhance product formation in a subsequent H + HOD reaction. Specifically, when the OH bond is excited, the reaction yields H_2 + OD, whereas when the OD bond is excited, H reacts with HOD to form the HD + OH product. In these experiments, the OH was prepared either by overtone excitation [34, 35] to the fourth vibrational level $\nu = 4$, or by excitation to the $\nu = 1$ state by Raman pumping [81]. As yet to be verified experimentally is the computational prediction of Manz *et al.* [128, 129] that strong optimized pulses can also achieve selective excitation of higher lying vibrational states.

Finally, we note that vibrational excitation can also have an inhibitory effect, which also results in great selectivity, as shown for the B state photodissociation of HOD [880].

17.2
Optimal Control in the Perturbative Domain

In general, when perturbation theory applies, one can provide an elegant method for obtaining the optimal solution to problems such as those in photodissociation. Often, optimal fields derived in the perturbative domain [130–133] do not differ by much from the fields derived via the more general "brute-force" optimization methods.

To examine a particular case, consider the pump-dump control scenarios discussed in Section 3.6 and Chapter 5. Here a sequence of two pulses serves to first excite a molecule to a set of intermediate bound states with wave function $|\psi_i\rangle$ and energy E_i, and then to dissociate these states. We denote the two pulses ($k = 1, 2$) by

$$\begin{aligned}\varepsilon_k(t) &= \int d\omega \{\epsilon_k(\omega) \exp\left[-i\omega\left(t - \frac{z}{c}\right)\right] + \text{c.c.}\} \\ &= \int d\omega \{\bar{\epsilon}_k(\omega) \exp[-i\omega t] + \text{c.c.}\}\,, \quad k = 1, 2\,, \end{aligned} \tag{17.24}$$

where $\bar{\epsilon}_k(\omega) \equiv \epsilon_k(\omega) \exp(i\omega z/c)$ with z being the propagation direction. Here the $\varepsilon_k(t)$ peaks at time t_k (e.g., Eq. (3.72)) and τ is the delay time $\tau = t_2 - t_1$.

After the second pulse the probability $P_q(E)$ of observing a given product channel q at energy E is given by

$$P_q(E) = \frac{4\pi^2}{\hbar^4} \sum_n \left| \sum_i \bar{\epsilon}_1(\omega_{i,1}) d_{i,g} c_{i,1}(\tau) \bar{\epsilon}_2(\omega_{E,i}) d_{q,\boldsymbol{n};i}(E) \right|^2 , \tag{17.25}$$

where

$$d_{i,g} \equiv \langle E_i | \hat{\boldsymbol{\varepsilon}} \cdot \boldsymbol{d} | E_1 \rangle\,. \tag{17.26}$$

Here the $c_{i,1}$ coefficients are the eigenstate coefficients resulting from excitation with the first pulse, defined in Eq. (2.15), and the

$$d_{q,\boldsymbol{n};i}(E) \equiv \langle E, \boldsymbol{n}, q^- | \hat{\boldsymbol{\varepsilon}} \cdot \boldsymbol{d} | E_i \rangle \tag{17.27}$$

are the transition dipole matrix elements between the ith intermediate state and the scattering states.

The probability of observing a given q product channel irrespective of the energy is then

$$P_q = \int dE\ P_q(E)\ . \tag{17.28}$$

Equations (17.25) and (17.28) can be conveniently rewritten as

$$P_q = \sum_{i,j} d_{i,j}^{(q)} \varepsilon_{i,j} c_{i,1}(\tau) c_{j,1}^*(\tau)\ , \tag{17.29}$$

where $\omega_{i,j} = (E_i - E_j)/\hbar$,

$$d_{i,j}^{(q)} \equiv \frac{4\pi^2}{\hbar^4} \int dE \sum_n \bar{\epsilon}_2(\omega_{E,i}) \bar{\epsilon}_2^*(\omega_{E,j}) d_{q,\boldsymbol{n};i}(E) \mathrm{d}_{q,\boldsymbol{n};j}^*(E)\ , \tag{17.30}$$

and

$$\varepsilon_{i,j} \equiv \bar{\epsilon}_1(\omega_{i,1}) \bar{\epsilon}_1^*(\omega_{j,1}) d_{i,g} d_{j,g}^*\ . \tag{17.31}$$

We now consider maximizing either the probability P_q in a single channel q, or the selectivity of one channel in preference to another, that is, $P_{q_1} - P_{q_2}$. Optimization is carried out subject to the constraint of fixed average pulse power, that is,

$$J^{(k)} \equiv \int d\omega |\epsilon_k(\omega)|^2 \epsilon_k = I^{(k)}\ , \quad k = 1, 2\ . \tag{17.32}$$

Thus, we wish to maximize either

$$D^{(q)} \equiv P_q - \lambda_1 J^{(1)} - \lambda_2 J^{(2)}\ , \tag{17.33}$$

or

$$D^{(q_1,q_2)} \equiv P_{q_1} - P_{q_2} - \lambda_1 J^{(1)} - \lambda_2 J^{(2)}\ , \tag{17.34}$$

where λ_i, $i = 1, 2$ are Lagrange multipliers.

The resultant optimization equations can not be solved analytically. Numerically, it is convenient to expand each field in an orthonormal basis set $\{u_n(\omega, x)\}$, (e.g., harmonic oscillator eigenfunctions)

$$\bar{\epsilon}_1(\omega) = \sum_m a_m u_m(\omega, x_1)\ , \quad \bar{\epsilon}_2(\omega) = \sum_n b_n u_n(\omega, x_2)\ , \tag{17.35}$$

and solve for the set of a_m, b_n coefficients that optimize the desired target.

Defining

$$U_{mm',ij} \equiv u_m(\omega_{i,1}, x_1) u_{m'}(\omega_j, x_1) d_{i,g} d_{j,g}^*\ , \tag{17.36}$$

and

$$X^{(q)}_{nn',ij} \equiv \frac{4\pi^2}{\hbar^4} \sum_n \int dE\, u_n(\omega_{E,i}, x_2) u_{n'}(\omega_{E,j}, x_2) d_{q,n;i}(E) d^*_{q,n,j}(E) , \quad (17.37)$$

we have that

$$\varepsilon_{i,j} = \sum_{mm'} a_m a^*_{m'} U_{mm',ij} , \quad (17.38)$$

and

$$d^{(q)}_{i,j} = \sum_{nn'} b_n b^*_{n'} X^{(q)}_{nn',ij} . \quad (17.39)$$

Using Eqs. (17.29), (17.38) and (17.39), we can write P_q as a double bilinear form in the a_m and b_n coefficients,

$$P_q = \sum_{m,m'} a_m a^*_{m'} \sum_{n,n'} b_n b^*_{n'} Y^{(q)}_{mm',nn'}(\tau) , \quad (17.40)$$

where

$$Y^{(q)}_{mm',nn'}(\tau) \equiv \sum_{i,j} U_{mm',ij} X^{(q)}_{nn',ij} c_{i,1}(\tau) c^*_{j,1}(\tau) \exp(-i\omega_{i,j}\tau) . \quad (17.41)$$

With the availability of powerful time-independent computational techniques [8, 724] for both the bound–bound $d_{i,g}$ and bound–free $d_{q,n,j}(E)$ matrix elements, the $U_{mm',ij} X^{(q)}_{nn',ij}$ and hence the $Y^{(q)}_{mm',nn'}(\tau)$ matrices are calculable for many realistic systems.

The extrema of Eqs. (17.33) or (17.34), obtained via the relations

$$\frac{\partial D^{(q)}}{\partial a_m} = \sum_{m'} a^*_{m'} \left\{ \sum_{n,n'} b_n b^*_{n'} Y^{(q)}_{mm',nn'}(\tau) - \lambda_1 \delta_{m,m'} \right\} = 0 \quad (17.42)$$

and

$$\frac{\partial D^{(q)}}{\partial b_n} = \sum_{n'} b^*_{n'} \left\{ \sum_{m,m'} a_m a^*_{m'} Y^{(q)}_{mm',nn'}(\tau) - \lambda_2 \delta_{n,n'} \right\} = 0 , \quad (17.43)$$

result in a set of nonlinear equations in the field coefficients a_n, b_m. These equations can be solved iteratively as a set of linear equations by first defining two matrices

$$\left(\underline{\underline{B}}\right)_{m,m'} \equiv \sum_{n,n'} b_n b^*_{n'} Y^{(q)}_{mm',nn'}(\tau) , \quad (17.44)$$

$$\left(\underline{\underline{A}}\right)_{n,n'} \equiv \sum_{m,m'} a_m a^*_{m'} Y^{(q)}_{mm',nn'}(\tau) . \quad (17.45)$$

The iteration proceeds by assuming that

$$\underline{b} = \underline{b}^0 , \quad (17.46)$$

where $\underline{b}$ is a row vector composed of the b_n coefficients and $\underline{a}$ is a row vector composed of the a_m.

The assumption made in Eq. (17.46) reduces Eq. (17.42) to a set of linear algebraic eigenvalue equations,

$$\underline{a} \cdot \left(\underline{b}^0 \cdot \underline{\underline{Y}}^{(q)}_{mm'}(\tau) \cdot \underline{b}^{0\dagger} - \lambda_1 \underline{\underline{I}} \right) = 0 \,, \tag{17.47}$$

where $\underline{\underline{I}}$ is the identity matrix, and $\underline{\underline{Y}}^{(q)}_{mm'}(\tau)$ is the matrix of $Y^{(q)}_{mm',nn'}(\tau)$ coefficients for fixed m and m'. These equations are solved for the $\underline{\underline{a}}$ eigenvector matrix, out of which the $\underline{a}_1$ row of coefficients corresponding to the λ_1 eigenvalue that maximizes $D^{(q)}$ is chosen. These coefficients are then used to update the $\underline{\underline{A}}$ matrix (Eq. (17.45)), and to solve the eigenvalue equation for the $\underline{b}$ coefficients, which is

$$\underline{b} \cdot \left(\underline{\underline{A}} - \lambda_2 \underline{\underline{I}} \right) = 0 \,. \tag{17.48}$$

The $\underline{b}$ row corresponding to the λ_2 eigenvalue which maximizes the $D^{(q)}$ objective is then chosen to update $\underline{\underline{B}}$. The process is repeated until convergence.

The fields thus generated are still a function of the delay time τ, which is treated as a nonlinear parameter. One then solves Eq. (17.47) for every value of τ and obtains the optimal value of the time delay between the pulses as the time delay corresponding to the global maximum of $P_q(\tau)$.

Numerical results using this technique were obtained [881] for the pump-dump photodissociation of Na_2 to optimize the production of either Na(3s) + Na(3p) or Na(3s) + Na(4s). In this case optimal control often required pulses that were fairly heavily structured in laser phase and frequency. A more detailed study [881] indicated that this structure was necessary for the dissociation pulses, but not necessary for the excitation pulse. Further, as is often the case with OCT, pulses with very different structures were found to achieve similar control objectives in different ways.

Note that, if perturbation theory holds, then simple experimental techniques can be used to show control over a wide range of parameters. For example, consider the case of one- vs. three-photon control (Section 3.4.2) where the ratio of product in one channel to another $R_{qq'}(E)$ is given by Eq. (3.54). Measuring $R_{qq'}(E)$ at six values of the control parameters x and $\phi_3 - 3\phi_1$ suffices to provide the values of all the participating molecular matrix elements and their phases. Equation (3.54) can then be used to generate $R_{qq'}(E)$ at any other control parameter value, and hence to identify the optimum control point for the selected set of experimental conditions.

17.3 Adaptive Feedback Control

It may well be the case that we need to optimize laser pulses to get better results than those derived from the perturbative regime of coherent control. Judson and Rabitz [17] have suggested a method that circumvents the difficulties of applying

OCT to experiments by foregoing the theoretical step altogether and using experimental results directly. In essence, they propose using the experimental apparatus as an "analog computer".

In their approach, one irradiates a molecular sample of interest and measures the product distributions. These results are reported to a computer that runs a learning algorithm that is capable of recognizing patterns in the input-output measurement relationships. This algorithm then guides an iterative sequence of new experiments, each experiment being characterized by a different pulse structure. The iteration is facilitated by a cost functional, as in OCT, but the functional now only contains costs for the target state and for laboratory-related issues. The computer repeats the iteration and learning process until satisfactory convergence is reached. The advantage of this approach is that no solutions to the Schrödinger equation need be generated, nor do we need to know the molecular Hamiltonian.

This overall procedure is an example of an "adaptive feedback control" [882]. Experimentally, the key elements in the procedure are the pulse shaping device [164, 166–171], a rapid means of modifying the laser pulses, and a fast probe of the output (i.e., a rapid pump-probe duty cycle). In the Weiner–Heritage *discrete* pulse shaping scheme [164, 166, 171] the pulse is shaped by first dispersing it using a grating into a large number (typically 128) of frequency components that are made to transverse an array of the same number of cells filled with liquid crystals. By changing the voltage across each cell it is possible to change the dispersion properties of the liquid crystals in the cell, thereby imparting a controllable phase shift to each frequency component. All the components are then brought together by another grating, in what amounts to the inverse of the dispersion process, to form a pulse. The newly shaped pulse is a result of the beats between all the phase-shifted frequency components.

The *continuous* Warren pulse shaper [167–170], shown in Figure 17.1, works in essentially the same way, except that the liquid crystal array is replaced by an acousto-optical modulator in which a pattern of acoustic waves is used to (continuously) phase shift the dispersed components of the pulse. The great merit of both types of pulse shapers is that they allow the shape to be determined by a computer. This is done via the set of voltages applied to the (discrete or continuous) phase-shifting elements.

The learning algorithm recommended by Judson and Rabitz, and used thus far in adaptive feedback control, relies upon "genetic algorithms" [172]. These algorithms are global optimization methods based on several concepts from biological evolution. The first is the concept of a breeding population in which individuals who are more "fit" in some sense will have a higher chance of producing offsprings and of passing their genetic information on to successive generations. The second is the concept of crossover, in which a child's genetic material is a mixture of the genetic material of its parents. The third concept is that of mutation, where the genetic material is occasionally corrupted, leading to a certain degree of genetic diversity in the population.

The adaptive feedback control apparatus (Figure 17.1), consists of the molecules of interest, a laser whose pulse shape is defined by an acousto-optical modulator,

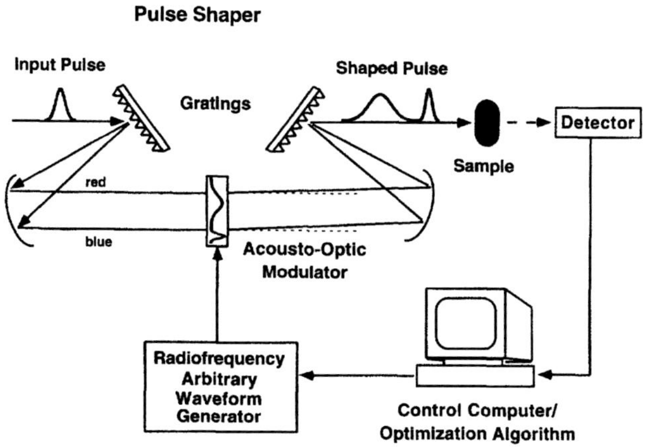

Figure 17.1 Schematic of the acousto-optic pulse-shaping feedback apparatus used in Adaptive Feedback experiments. Taken from [18, Figure 1]. (Reprinted with permission. Copyright 1997 Elsevier BV.)

controlled by a computer, and a measurement device that reads and feeds final population distributions (or other observables) back to the controlling computer. The genetic algorithm code runs on a controlling computer, supplying pulse sequences to the laser and receiving fitness values (the difference between the objective and an observed function of the molecular state) from the measurement device. Over several generations, the system as a whole will seek to optimize the fields.

In most applications, the genetic algorithm is implemented as follows. An individual (i.e., a single pulse sequence) is coded for by a "gene", which is a bit string of length N_{gene} that can be uniquely decoded to define the pulse sequence. A fitness function is defined that can discriminate between pulse sequences. For example, if we want to drive molecules into state j' we might choose the fitness function as $\sum_j (\delta_{j,j'} - \rho_j)^2$, where ρ_j is the observed population of state j. An initial population of individuals N_{pop} is formed by choosing N_{pop} bit strings, often initially at random. The fitness of these individuals is then evaluated.

Children of these first generation parents are then formed as follows. All the parents are ranked by fitness and the individuals with the highest fitness are placed directly into the second generation with no change. From the remaining parents, pairs of individuals are chosen and their genes are crossed over to form genes of the remaining second generation individuals. The crossover is effected by taking some subset of the bits from parent 1 and the complementary set of bits from parent 2, and combining them to form a new gene of child 1. The remaining bits from the two parents are combined to form the gene of child 2. Additionally, during replication, there is a small probability of a bit flip or mutation in a gene. This serves primarily to prevent premature convergence, in which a single very fit individual takes over the entire population.

Adaptive feedback control for molecules was first applied experimentally by Bardeen *et al.* [18] and by Yelin *et al.* [19] to two different problems. Using the setup of Figure 17.1, Bardeen *et al.* were able to optimize two objectives: the "efficiency" and "effectiveness" of the electronic excitation of the "IR125" dye molecule. The "efficiency" objective corresponded to maximizing the number of excited state molecules per integrated laser intensity. This goal was attained by constraining the objective through an integrated intensity penalty, as in Eq. (5.11). The "effectiveness" objective corresponded to maximizing the number of excited state molecules, irrespective of the integrated intensity of the pulse. This objective is obtained by removing the penalty constraint used in the efficiency objective.

The results displayed in Figures 17.2 –17.4 show that the most *efficient* result occurs when a relatively narrowband laser pulse of moderate intensity is applied at center frequency near the peak of the absorption spectrum of the dye molecule. The most *effective* result occurs when a *positively chirped* broadband laser pulse is applied. The positive chirp (i.e., an upward drift of the laser's central frequency with time) is helpful, because stimulated emission back to the ground state, which diminishes the number of excited state molecules, invariably occurs to the red of the absorbed photon. By rapidly shifting the laser center frequency more and more to the blue, one can successively eliminate the frequencies causing stimulated emission from the excited state shortly after it is formed by photon absorption.

Another early example for the use of computer-controlled Heritage–Weiner [164] phase shaping technique in conjunction with adaptive feedback control is to obtain a transform limited pulse from a less coherent pulse. By setting the target to maximize a two-photon absorption intensity, Silberberg *et al.* [19] were able, as shown in Figure 17.5, to modify the phases of the pulse in such a way as to narrow it down

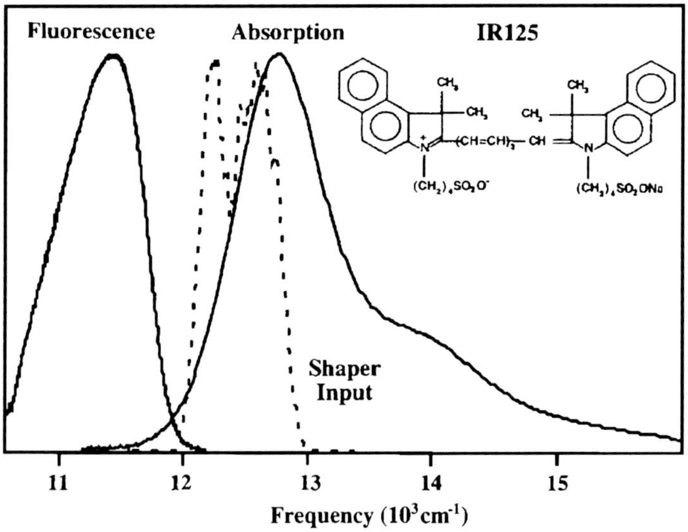

Figure 17.2 The molecular structure, and the absorption and fluorescence spectra of IR125 in methanol, along with the laser power spectrum before shaping (dashed line). Taken from [18, Figure 2]. (Reprinted with permission. Copyright 1997 Elsevier BV.)

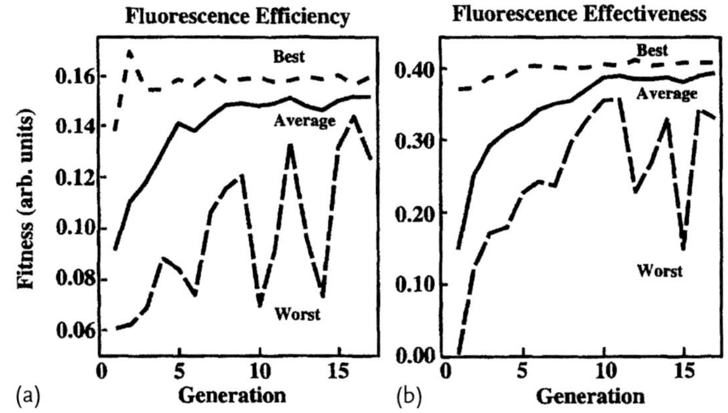

Figure 17.3 (a) The convergence of the genetic algorithm as measured by the behavior of populations at each generation. Results are shown for the best (small dashed), average (solid), and worst (large dashed) fitness in each generation, defined by the fluorescence efficiency (the ratio of fluorescence to laser power). One generation corresponds to approximately 30 experiments. (b) As in panel (a), but the fitness is now defined as the fluorescence effectiveness, proportional to the fluorescence power alone. Taken from [18, Figure 3]. (Reprinted with permission. Copyright 1997 Elsevier BV.)

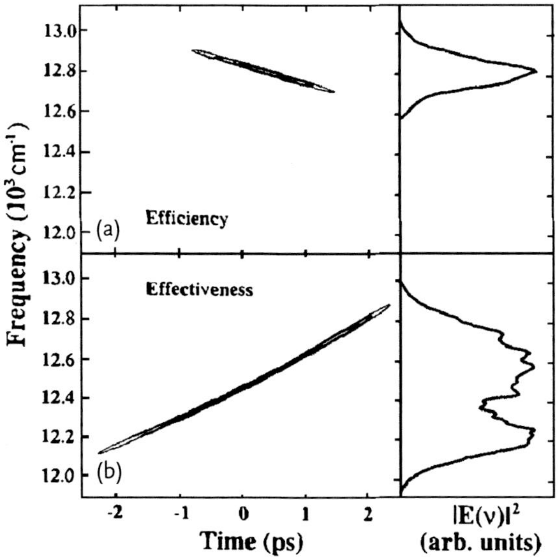

Figure 17.4 (a) A sample optimal pulse for the fluorescence efficiency as determined experimentally by the feedback algorithm. The left side is a plot of the experimentally determined Wigner transform which shows the intensity of the electric field as a function of time and frequency. The right side shows the spectrum $|E(\nu)|^2$. The efficiency does not appear to depend strongly on the laser chirp. (b) As in panel (a), but showing an optimal pulse for the fluorescence effectiveness. Taken from [18, Figure 5]. (Reprinted with permission. Copyright 1997 Elsevier BV.)

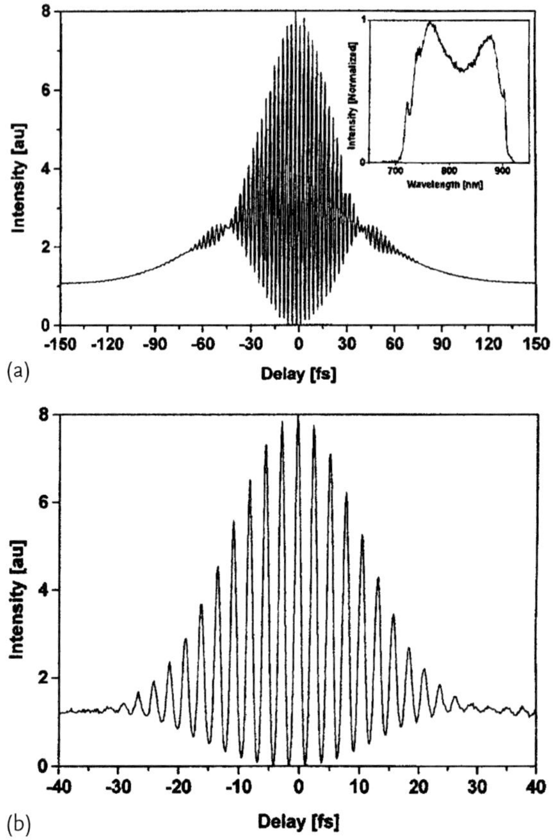

Figure 17.5 Interferometric autocorrelation traces of the uncompressed and the compressed pulses. (a) Uncompressed 80 fs pulses obtained directly from the Ti:sapphire laser (the power spectrum is shown in the inset). (b) Compressed pulses after 1000 iterations. The pulses were compressed to 14 fs. Taken from [19, Figure 2]. (Reprinted with permission. Copyright 1997 Optical Society of America.)

in time to essentially its transform limit. Here we show the autocorrelation trace of the pulse.

Gerber *et al.* [20] have used the same technique, coupled with their own version of an evolutionary algorithm, in a setup shown in Figure 17.6, to tailor femtosecond laser pulses to optimize the branching ratios of different organometallic photodissociation reaction channels. For example, they studied the photodissociation of $CpFe(CO)_2Cl$, (where Cp stands for cyclo-pentadiene, a pentagon made up of 4 CH groups and one CH_2 group). Gerber and coworkers were able to control two different bond cleaving reactions,

$$\text{channel A} \quad FeCl + \ldots \longleftarrow CpFe(CO)_2Cl \longrightarrow CO + CpFe(CO)Cl \quad \text{channel B}$$

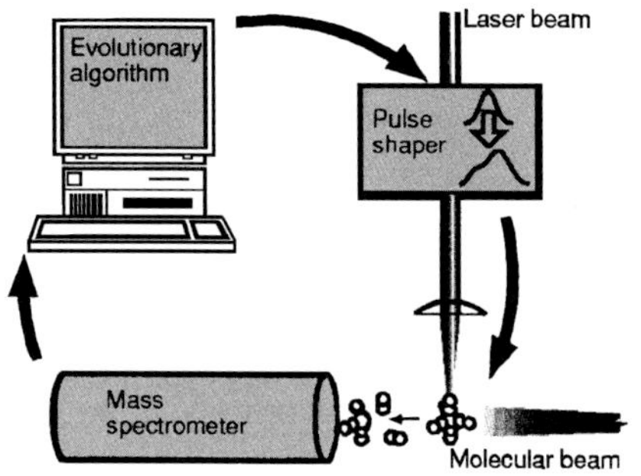

Figure 17.6 Schematic setup of the Gerber *et al.* experiment [20]. Femtosecond laser pulses are modified in a computer-controlled pulse shaper. Ionic fragments from molecular photodissociation are recorded with a reflection TOF mass spectrometer. This signal is used directly as feedback in the controlling evolutionary computer algorithm to optimize the branching ratios of photochemical reactions. Taken from [20, Figure 1]. (Reprinted with permission. Copyright 1998 American Association for the Advancement of Science.)

where the dots indicate a number of, as yet unknown, possible products. They have shown that the method works automatically and finds optimal solutions without prior knowledge of the molecular system and experimental environment. Sample results are shown in Figures 17.7 and 17.8 for the photodissociation of $CpFe(CO)_2Cl$. Here the ratio of $CpFeCOCl^+/FeCl^+$ was either maximized (solid blocks) or minimized (open blocks). The yields are shown at masses 91 and 184, with the parent ion shown at 212. The observed yield of the other ions is also shown, but was not included in the control protocol. Control over the desired ratio is quite evident.

Aspects of the laser fields that yield the optimized results are shown in Figure 17.9. Specifically, the autocorrelations $G_2(\tau) = \int [E(t-\tau) + E(t)]^4 dt$ are shown in Figure 17.9a for the experiment yielding the maximum in Figure 17.8 and in Figure 17.9b for the experiment yielding the minimum. Figure 17.9c shows a band-

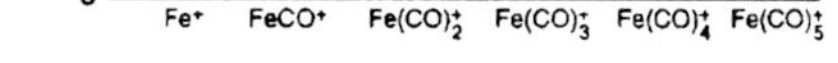

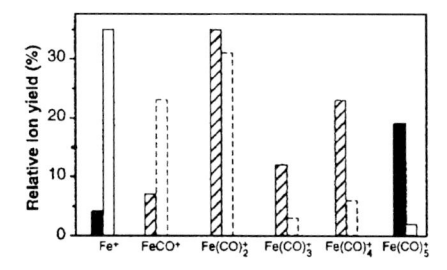

Figure 17.7 Relative $Fe(CO)_5$ photodissociation product yields. The yields are derived from the relative peak heights of the mass spectra. The ratio of $Fe(CO)_5^+/Fe^+$ is maximized (solid blocks) as well as minimized (open blocks) by the optimization algorithm, yielding a significantly different abundance of Fe^+ and $Fe(CO)_5^+$ in the two cases. The peak heights of all other masses [$Fe(CO)^+$ up to $Fe(CO)_4^+$] have not been included in the optimization procedure. Taken from [20, Figure 2]. (Reprinted with permission. Copyright 1998 American Association for the Advancement of Science.)

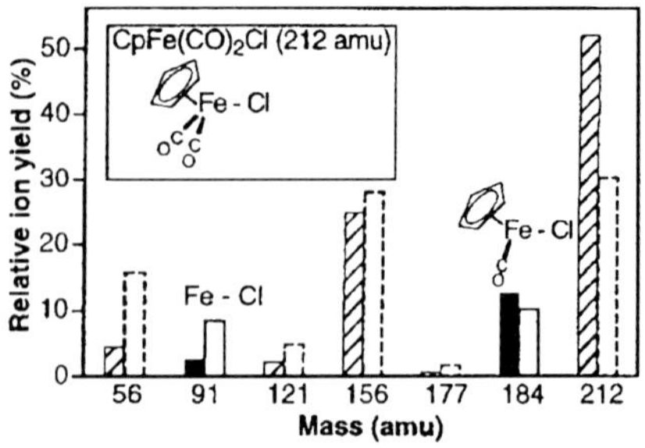

Figure 17.8 Adaptive Feedback Control over the products of the photodissociation of CO + CpFe(CO)Cl (mass 184) $\leftarrow$ $CpFe(CO)_2Cl$ $\rightarrow$ FeCl (mass 91) + ..., the ratio of $CpFeCOCl^+/FeCl^+$ is either maximized (solid blocks at masses 91 and 184) or minimized (open blocks at mass 91 and 184). Taken from [20, Figure 3]. (Reprinted with permission. Copyright 1998 American Association for the Advancement of Science.)

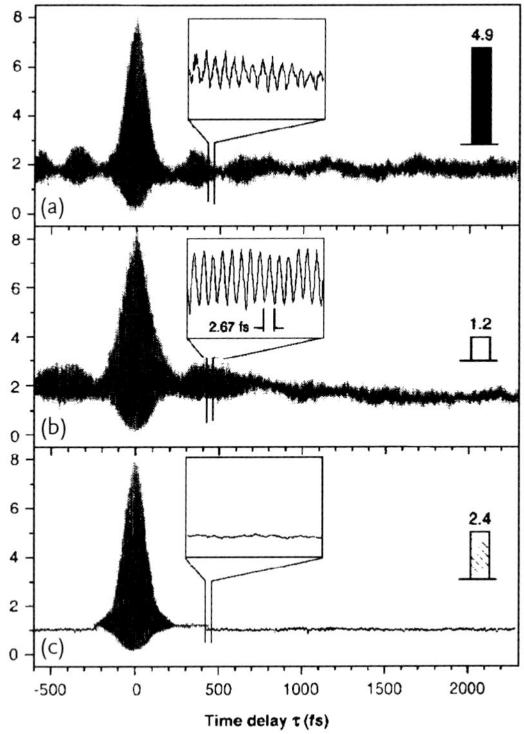

Figure 17.9 Aspects of the optimum laser pulses in the adaptive feedback control of the products of laser excitation of $CpFe(CO)_2Cl$. The autocorrelation $G_2(\tau)$ is shown for three different cases, as described in the text. The pulse shape differences in these three pulses is evident. Taken from [20, Figure 4]. (Reprinted with permission. Copyright 1998 American Association for the Advancement of Science.)

width limited laser autocorrelation for a laser that yields inferior results to that obtained with either of those shown in Figure 17.9a or b.

At present not much is known about the dynamics of the $CpFe(CO)_2Cl$ dissociation reaction as probed in Gerber's experiment, nor about those aspects of the pulse that enhance specific product production. Although the Weiner–Heritage pulse shaper only changes phases of different frequency components of the pulse, it is clear that this also has profound effect on the peak intensity of the pulse. As a result, it is conceivable that channel A involves a different multiphoton absorption (and ionization) route than does channel B. This means that the optimal pulse configurations may simply have excited the parent $CpFe(CO)_2Cl$ molecule to an energy region where dissociation to the channel of interest (A or B) is naturally preferred. Clearly, further work is needed, and is indeed ongoing, to clarify the details on the mechanism through which the yield is improved.

Additional studies have been carried out by Levis *et al.* [883] in which they demonstrated selective bond cleavage and rearrangement of chemical bonds having dissociation energies up to approximately 100 kcal/mol in molecules such as acetone, trifluoroacetone, and acetophenone. In particular, they showed control over the formation of CH_3CO from $(CH_3)_2CO$, CF_3 (or CH_3) from CH_3COCF_3, and $C_6H_5CH_3$

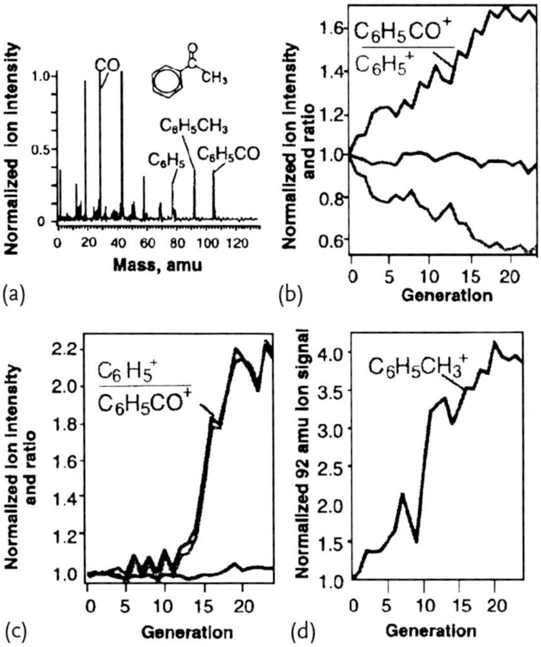

Figure 17.10 Adaptive feedback control over the products of laser excitation of acetophenone ($C_6H_5COCH_3$). (a) Shows the mass spectrum associated with products of the photoexcitation. (b)–(d) Show the intensity of the indicated ions and their ratio as a function of the "generation" of the adaptive feedback scheme. Taken from [883, Figure 5]. (Reprinted with permission. Copyright 2001 American Association for the Advancement of Science.)

(toluene) from $C_6H_5COCH_3$. In each case, ions associated with the products were measured. These experiments employed intense tailored laser pulses (on the order of 10^{13} W/cm^2) so that the system energy levels were dynamically Stark shifted into resonance with the laser.

A typical result is shown in Figure 17.10, where control over the products of photoexcitation of acetophenone ($C_6H_5COCH_3$) was sought. Figure 17.10b shows the results of maximizing the ratio of the $C_6H_5CO^+$ to $C_6H_5^+$ ion products and Figure 17.10c shows the result of minimizing this ratio. Clearly, control over the ion product is achieved within 20 or so generations. Furthermore, repeated experiments with different initial starting conditions yielded similar results. As with the experiments above, at this time there is no detailed understanding of the mechanism of the control fields, but work of this type is ongoing.

Finally, note that Motzkus and coworkers [884] have demonstrated that it is possible to control the internal conversion channel in a light-harvesting antenna complex of a photosynthetic purple bacterium. This implies the ability to control dynamics in a large complex system, as anticipated theoretically.

17.4
Analysis of Adaptive Feedback Experiments

As discussed above, optimal control takes advantage of engineering tools designed to optimize the system variables so as to maximize target cross sections within specific constraints. The procedure often results in complicated laser pulses, and the mechanisms that underlie the resultant control are difficult to discern. Efforts to experimentally identify the underlying dynamics, for example, by intelligently restricting the range of the optimal search have been relatively unenlightening. Similarly, the vast majority of experiments have not been modeled theoretically and hence have not yielded physical insight into the underlying mechanism. Two such studies have, however, been carried out for experiments done in the liquid phase, the medium of greatest chemical interest, and they are described below. In both cases, results show that despite the initial experimental analyses, neither of the experimental results imply control via coherent quantum interference effects, and neither of the experiments require other than simple qualitative classical explanations. The reasons for this can be traced back to the particular experimental apparatus, and to the particular optimization target chosen for the feedback algorithm.

17.4.1
trans–cis Isomerization in 3,3′-Diethyl-2,2′-thiacyanine Iodide

Isomerization between *cis* and *trans* configurations is ubiquitous in chemistry. Vogt *et al.* [885] examined control of *trans–cis* isomerization of 3,3′-diethyl-2,2′-thiacyanine iodide (NK88) in liquid methanol. The process takes place by excitation from the ground state to the excited state, followed by isomerization, as shown in

Figure 17.11. Using adaptive feedback, they demonstrated the ability to optimize the yield of the *cis* product by shaping a 60 nm wide fs laser source. Two cases were studied: "maximizing" pulses that increased the extent of *trans–cis* isomerization and "minimizing" pulses that minimized the amount of *trans–cis* isomerization: both are shown in Figure 17.12. The resultant optimal pulses consisting of a number of peaks as a function of time with an overall time width of 4 ps. Note that the minimizing pulse shows far more complex structure, which led to suggestions that complicated controlled coherent dynamics guides the dynamics away from the product *cis* state.

A subsequent detailed study of the dynamics, described below, used a simple one-dimensional model representing the *trans–cis* isomerization reaction coordinate, plus a decoherence term that modeled the remainder of the NK88 degrees of freedom as well as the liquid methanol environment. These computational results showed that the experimental results were consistent with control exercised through changes of the probability of excitation, rather than by control over the dynamics. That is, in the case of maximization of the isomerization, the pulse created significantly more of the excited state, whereas in the case of minimization, the pulse aimed to decrease the excited population. Coherent control was not evident.

The computation considered a model consisting of the system with Hamiltonian H_S, the bath H_B, system–bath H_{SB}, and system–electric field coupling described in the dipole approximation. The total Hamiltonian is given by

$$H = H_S + H_B + H_{SB} - \mathrm{d} \cdot \epsilon(t) \,, \tag{17.49}$$

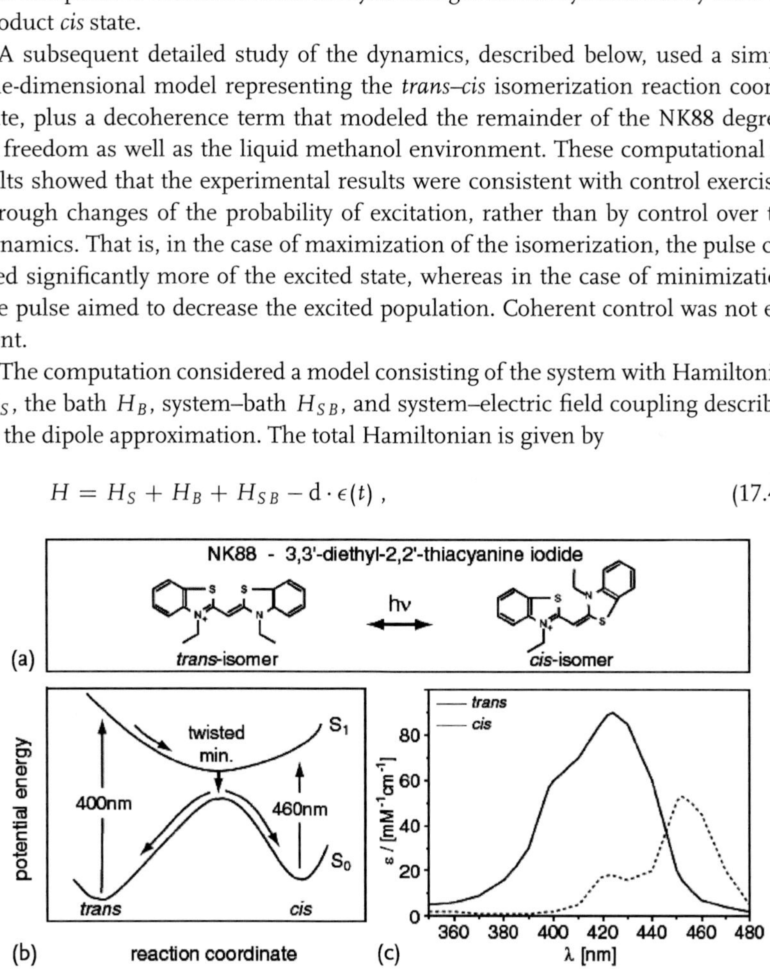

Figure 17.11 (a) Molecular structure of the two isomer configurations of the cyanine dye NK88. Irradiated by light of the proper wavelength, this molecule can undergo *trans–cis* isomerization. (b) Simplified potential energy surface. The reaction coordinate is the twist angle about the C–C double bond. (c) The ground state absorption spectra of the two isomers. While the *trans* isomer shows a broad ground state absorption centered at 420 nm, the absorption band of the *cis* isomer is redshifted with a maximum around 450 nm. Taken from [885, Figure 1]. (Reprinted with permission. Copyright 2005 American Physical Society.)

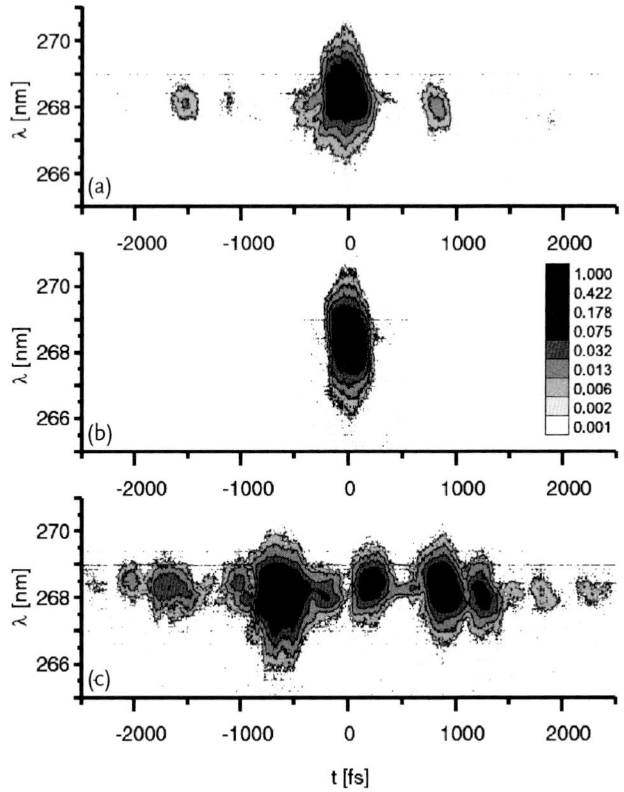

Figure 17.12 Optimal pulse shapes for (a) maximizing and (c) minimizing the ratio between the *cis* and *trans* isomers. For comparison the trace of the undiluted pulse is shown in (b). Taken from [885, Figure 4]. (Reprinted with permission. Copyright 2005 American Physical Society.)

where H_S describes the isomerization process via rotation in a one dimensional reaction coordinate ϕ, as shown in Figure 17.11, H_B represents all other degrees of freedom ("the bath"), d is the dipole moment, and $\epsilon(t)$ is the incident electric field at time t. The term H_{SB} couples the reaction coordinate to the remaining degrees of freedom.

In terms of the two participating electronic states, $V_g(\phi)$ and $V_e(\phi)$ are the ground and excited electronic state potential surfaces, and $V_{Ger}(\phi) = V_{eg}(\phi)$ is the coupling between them. In the adiabatic representation the ground state potential is a double well [885]. The simplest dynamics takes place by photoexcitation from the *trans* configuration to the excited electronic state followed by de-excitation to the *cis* and *trans* ground state via system–bath coupling. The bath is described as a set of harmonic oscillators of frequency ω_α and the system–bath coupling is bilinear, $H_{SB} = Q\sum_\alpha \hbar\kappa_\alpha(b_\alpha^\dagger + b_\alpha)$, where $b_\alpha^\dagger$ and b_α are the creation and annihilation operators pertaining to the αth harmonic oscillator. Parameters were chosen such that the electronic dephasing time around the Franck–Condon region of the *trans* configuration is $\sim$ 10 fs, and virtually complete relaxation from excited *trans* to stable *trans* and *cis* occurs within 5 ps. The former is a typical characteristic dephasing time whereas the latter is chosen to agree with experiment.

The dissipative dynamics of the system was evaluated using the Redfield equation with secular approximation, a particular type of master equation described in [178, 228]. In this approximation, the evolution of diagonal elements $\rho_{ii}(t)$ of the system density matrix is given by

$$\begin{aligned}\frac{\partial \rho_{ii}(t)}{\partial t} &= -i E(t)/\hbar \sum_m \left[\rho_{im}(t)\mu_{mi} - \mu_{im}\rho_{mi}(t)\right] \\ &\quad + \sum_{j\neq i} w_{ij}\rho_{jj}(t) - \rho_{ii}(t)\sum_{j\neq i} w_{ji}\,, \end{aligned} \tag{17.50}$$

where the transition probability is $w_{ji} = \Gamma^{+}_{ijji} + \Gamma^{-}_{ijji}$ and where each index denotes a state of the system, including the electronic and vibrational quantum numbers.

The evolution of the off-diagonal elements is described as

$$\begin{aligned}\frac{\partial \rho_{ij}(t)}{\partial t} &= -i\omega_{ij}\rho_{ij}(t) - \gamma_{ij}\rho_{ij}(t) \\ &\quad - i E(t)/\hbar \sum_m [\rho_{im}(t)\mu_{mj} - \mu_{im}\rho_{mj}(t)]\,, \end{aligned} \tag{17.51}$$

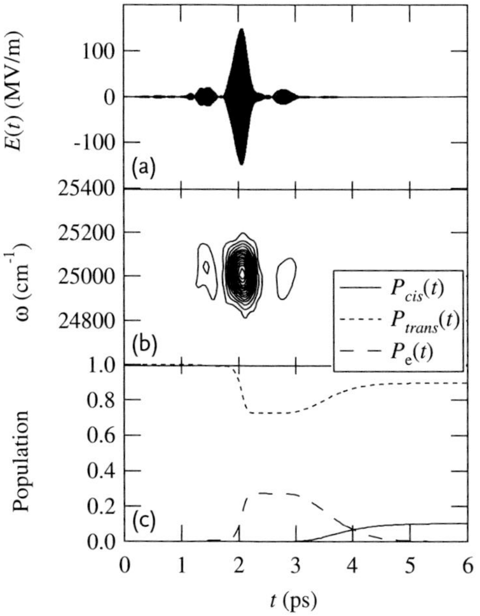

Figure 17.13 Time evolution under the optimized electric field with restriction on frequency and amplitude. (a) An electric field obtained by 64 iterations from random-initial phases Θ_i. (b) Time-frequency resolved spectrum. (c) Time evolution of populations. Taken from [887, Figure 1]. (Reprinted with permission. Copyright 2005 American Physical Society.)

with dephasing rate γ_{ij}. With the parameters chosen, the resultant vibrational dephasing time within the excited electronic state is $\approx$ 15 fs.

The adaptive feedback experiment was modeled by an electric field comprising 128 frequency values, where the phases of each frequency component are the optimization parameters; the frequency width is 200 cm^{-1}, and the time width is 2 ps. Specifically, the electric field function is taken to be

$$E(t) = \sum_{i=0}^{127} A \exp\left[-\left(\frac{t-t_0}{2\Delta t}\right)^2 - \left(\frac{\Omega_i - \Omega_0}{2\Delta\Omega}\right)^2\right] \cos(\Omega_i t + \Theta_i) , \qquad (17.52)$$

where $A = 5\,\text{MV/m}$, $t_0 = \Delta t = 2\,\text{ps}$, $\Omega_0 = 25\,000\,\text{cm}^{-1}$, $\Delta\Omega = 200\,\text{cm}^{-1}$, $\Omega_i = 24\,800 + 3.125\,i\,\text{cm}^{-1}$, and Θ_i are the optimization parameters. Note that in accord with experiment [885] the field is optimized by varying the phases Θ_i using an evolutionary algorithm, and the field amplitude, as well as the overall frequency width of the pulse, are constrained. Further, the algorithm is designed to simulate experimental conditions [886]. In the first study, consistent with experiment, both the frequency range and amplitude were constrained to the experimental range, ϵ.

A comparison of the fields obtained computationally shown in Figures 17.13 and 17.14, and those obtained experimentally (Figure 17.12), show excellent agreement. However, as noted above, all observed control was due to changes in the extent to which the molecule is excited. No coherent control scenario was evident.

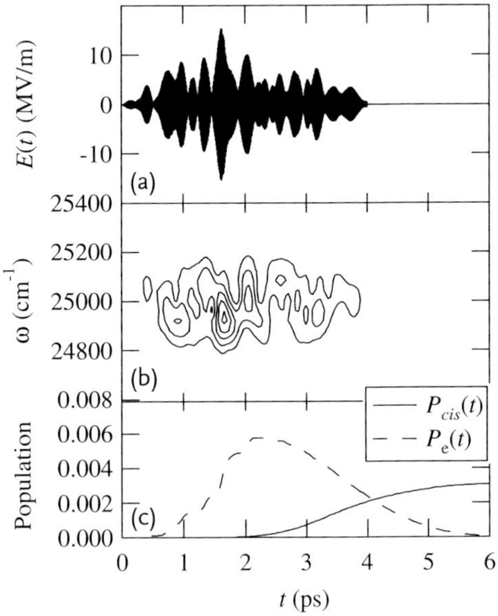

Figure 17.14 Time evolution of the system under the optimized electric field that is restricted in frequency and amplitude. In this case, the target of the control is the minimization of stable *cis*. (a) Optimized electric field. (b) Time-frequency resolved spectrum. (c) Time evolution of populations. Taken from [887, Figure 2]. (Reprinted with permission. Copyright 2005 American Physical Society.)

It should be emphasized that the beauty of this model of the isomerization experiment lies in its simplicity, focusing entirely on the essential features of the dynamics and of the decoherence. Alternate computational studies [888, 889] of the cyanine systems (which make no explicit reference to the experiments in [885]) suggest that complicated dynamics in such systems is *possible*. However, this one dimensional model + decoherence, with results in excellent agreement with experiment, shows that this adaptive feedback experiment does not reveal, or rely upon, complicated dynamical scenarios. This crucial attribute of the analysis of the experiment has been misconstrued in, for example, [97].

Where then is the "tool box" of control scenarios that would be expected to emerge from the optimization field, that is, scenarios such as those described in Chapters 3 and 7 above? Why are they not manifested in this adaptive feedback experiment? A second computational study shows that the answer relates to the practical restrictions involved in the experiment.

In particular, computational results [887] using a field with no frequency limitations, showed (see Figure 17.15) that pump-dump control requires a laser frequency width of at least 30 000 cm^{-1}, far greater than the width associated with the experimental apparatus. Results with this scenario were improved over those obtained with the experimentally restricted field; that is, the unconstrained pulse showed a final *cis* population of 0.36, as compared to the ratio of 0.16 obtained with the 60 nm wide pulse.

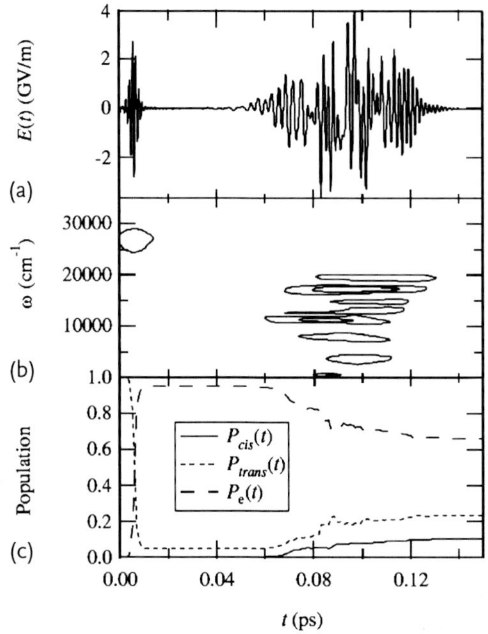

Figure 17.15 Short-time evolution of the system under the fully optimized pulse. After 0.15 ps, the electric field is essentially zero. (b) Time-frequency resolved spectrum of the pulse. (c) Time evolution of populations. Taken from [887, Figure 3]. (Reprinted with permission. Copyright 2005 American Physical Society.)

Of some interest is the fact that the decoherence here is sufficiently rapid so as to result in a modified pump-dump scenario wherein the second pulse operates well after the excited wave packet has decohered [890].

17.4.2
Controlled Stokes Emission vs. Vibrational Excitation in Methanol

A series of experiments [891–895] considered control of stimulated Raman scattering (SRS) in liquid methanol. The basic idea was to selectively excite one of two closely spaced Raman active modes of methanol associated with the symmetric and antisymmetric C–H stretch modes. The Raman modes were driven by a strong pump pulse with 150 fs bandwidth. This bandwidth is larger than the spacing of the excited Raman modes but smaller than the energies of the vibrational states themselves. This placed the experiment in the so-called nonimpulsive, or transient, regime where the molecules have enough time to vibrate during the pulse. By using an adaptive feedback algorithm to shape a strong pump pulse, significant control over the relative peak heights of the two Raman modes in the Stokes emission was demonstrated. Pulses were found that could either selectively enhance the emission from either mode, enhance both modes together, or completely suppress all emission. By assuming that the intensities of the peaks in the Stokes emission was directly proportional to the associated vibrational modes, it was concluded that selective vibrational excitation had been achieved.

In the absence of a firm theoretical footing, these experiments appeared to be an example of quantum or coherent control of mode-selective vibrational excitation in the liquid phase. This view had been temporarily upheld by a proposal for a coherent mechanism of Raman control in a closely related experiment using a double-Gaussian shaped pump pulse [895]. Unfortunately, this proposed mechanism was shown to be faulty [896], partly a consequence of the absence of theoretical support. Subsequently, a detailed theoretical analysis [897, 898] of these experiments, including both molecular excitation and nonlinear optical propagation of the pump and Stokes pulses, revealed a simple classical control mechanism.

The analysis of such experiments entails a complicated combination of the quantum mechanics of coherent control plus the use of Maxwell's equations to include the propagation of the classical electromagnetic field. This study revealed the presence of many control mechanisms capable of affecting the relative intensity of two Stokes peaks in the emitted Raman spectrum. All the identified mechanisms are third-order nonlinear optical effects [899], and include transient stimulated Raman scattering in a collisional environment, saturated Raman scattering, self- and cross-phase modulation, and focused beam effects. All mechanisms have the same clear physical interpretation: shaping the pump pulse controls the nonlinear optical response of the medium, which in turn controls the Stokes emission. Significantly, no coherent quantum interference effects were needed to explain the observed control. Indeed classical models that treat the vibrational modes as classical oscillators give the identical equations of motion for the vibrational excitations and the pump and Stokes fields. Interfering pathways of the type considered in coherent control would only contribute to fifth-order nonlinear optical emission, which contributes negligibly to the emission at the Stokes wavelengths.

Furthermore, although it was found that the vibrational populations are affected by the same control mechanisms that affect the Stokes spectra, the ratio of the Stokes spectra peak heights did not directly reflect the ratio of the level populations, as was assumed in the experiment. Second, the control was found to be completely incoherent at the molecular level. The reason for the incoherence was that no amplitude at the Stokes wavelengths was present in the initial pump pulse, hence forcing the Stokes mode to build up from noise (spontaneous emission or collisional excitation). These seeding processes are themselves incoherent, and prevented any chance for coherent dynamics. (Note that coherent dynamics could be induced if both the pump and Stokes frequencies have significant amplitude in the initial shaped pulse, suggesting again that the experimental bandwidth limitations seriously hampered the range of possible solutions available to the adaptive feedback algorithm.)

Three significant lessons emerge from these two studies. Two are rather obvious but ofttimes neglected – the first being that the optimization can be severely limited by constraints on the experimental apparatus, and the second is that optimization is geared towards optimizing a specific target (e.g., the ratio of Stokes vs. anti-Stokes lines as opposed to vibrational populations). The third, rather crucial result, is that optimization by a computer often leads to such complicated pulse shapes that detailed theoretical modeling is *necessary* in order to extract the underlying physical mechanisms leading to control.

17.5 Interference and Optimal Control

As this book has emphasized, there are two distinct paradigms for the quantum control of molecular processes: coherent control and optimal control. Coherent control is clearly based upon interfering pathways. Although not as manifestly evident, optimal control also relies upon the existence of multiple interfering pathways, as discussed below. Both of these paradigms bring their own insights to quantum control. In addition, they each motivate appropriate experiments in different technological domains. Thus, the energy-resolved viewpoint has primarily motivated nanosecond pulsed laser experiments, whereas the time-dependent perspective has primarily been used to devise and interpret ultrafast experiments . In this section we link the two approaches by using the energy resolved perspective to gain insight into pulse-shaped control [16]. We focus on photodissociation.

Consider a single bound molecular eigenstate $|E_1\rangle$ that is excited to the continuum. In accord with Eq. (2.2) the wave function $|\Psi(t)\rangle$ is of the form

$$|\Psi(t)\rangle = b_1(t)|E_1\rangle e^{-iE_1t/\hbar} + \sum_n \int dE\, b^{(1)}_{E,n,q}(t)|E, n, q^-\rangle e^{-iEt/\hbar} \,. \quad (17.53)$$

We can rewrite Eq. (17.53) in matrix notation as

$$|\Psi^{(1)}(t)\rangle = \int dE e^{-iEt/\hbar} \left(b^{(1)}_{E,\boldsymbol{n}_1,q}(t), b^{(1)}_{E,\boldsymbol{n}_2,q}(t), \dots,\right) \cdot \left(|E,\boldsymbol{n}_1,q^-\rangle, |E,\boldsymbol{n}_2,q^-\rangle, \dots\right)^T , \tag{17.54}$$

where the superscript $\boldsymbol{T}$ denotes the transpose operation that turns a row vector into a column vector, and $|\Psi^{(1)}(t)\rangle$ is the excited portion of the wave packet that originated from state $|E_1\rangle$. That is,

$$|\Psi^{(1)}(t)\rangle \equiv |\Psi(t)\rangle - b_1(t)|E_1\rangle e^{-iE_1t/\hbar} . \tag{17.55}$$

As shown in Chapters 3 and 12, one can not control the dynamics in such an excitation of a single initial state by shaping the laser pulse. That is, the ratio of products going to an individual product state is independent of the pulse shape.

Further, as discussed in Section 3.2, the inability to control the product ratio by shaping the pulse can be overcome by photodissociating not just one $|E_1\rangle$ bound state, but a superposition of several bound states $|E_i\rangle$ (as was done, for example, with bichromatic control). Such a superposition state can be created separately by an initial preparation pulse, as in the case of pump-dump control scenario (Sections 3.6 and 5.1). Alternatively, the superposition state can be created by the photolysis pulse itself (by e.g., a stimulated Raman process), provided that the bandwidth of the pulse is comparable to the energy spacings between the $|E_i\rangle$ levels.

Mathematically speaking, the goal of the control is the preparation of a single $|E,\boldsymbol{n},q^-\rangle$ state. If this is achieved, then complete control is guaranteed, by Eq. (2.66), insofar as only one fragment target state $|E,\boldsymbol{n};0\rangle$ will be populated as $t \to \infty$. In order to achieve the control target we consider preparing a whole array of wave packets, by, for example, starting with other initial states composed of the other system bound states $|E_i\rangle$. That is,

$$\underline{|\Psi_t\rangle} = \int dE e^{-iEt/\hbar} \underline{\underline{\mathrm{B}}} \cdot \underline{|E,\boldsymbol{n}\rangle} , \tag{17.56}$$

where

$$\underline{|\Psi_t\rangle}^T \equiv \left(|\Psi^{(1)}(t)\rangle, |\Psi^{(2)}(t)\rangle, |\Psi^{(3)}(t)\rangle, \dots\right) \tag{17.57}$$

$$\underline{\underline{\mathrm{B}}} \equiv \begin{pmatrix} b^{(1)}_{E,\boldsymbol{n}_1,q}, & b^{(1)}_{E,\boldsymbol{n}_2,q}, & b^{(1)}_{E,\boldsymbol{n}_3,q}, \dots \\ b^{(2)}_{E,\boldsymbol{n}_1,q}, & b^{(2)}_{E,\boldsymbol{n}_2,q}, & b^{(2)}_{E,\boldsymbol{n}_3,q}, \dots \\ & \dots & \\ & \dots & \\ & \dots & \end{pmatrix} , \tag{17.58}$$

and

$$\underline{|E,\boldsymbol{n}\rangle}^T \equiv \left(|E,\boldsymbol{n}_1,q^-\rangle, |E,\boldsymbol{n}_2,q^-\rangle, |E,\boldsymbol{n}_3,q^-\rangle, \dots\right) . \tag{17.59}$$

Using the results in Chapter 3, we see that the $\underline{\underline{\text{B}}}$ matrix factorizes as

$$\underline{\underline{\text{B}}} = \hat{\mathcal{E}}(E) \cdot \underline{\underline{\text{M}}}(E) \tag{17.60}$$

where

$$\underline{\underline{\text{M}}}(E) = \begin{pmatrix} \langle E_1|\hat{\boldsymbol{\varepsilon}} \cdot \boldsymbol{d}|E, \boldsymbol{n}_1, q^-\rangle^*, \langle E_1|\hat{\boldsymbol{\varepsilon}} \cdot \boldsymbol{d}|E, \boldsymbol{n}_2, q^-\rangle^*, \ldots \\ \langle E_2|\hat{\boldsymbol{\varepsilon}} \cdot \boldsymbol{d}|E, \boldsymbol{n}_1, q^-\rangle^*, \langle E_2|\hat{\boldsymbol{\varepsilon}} \cdot \boldsymbol{d}|E, \boldsymbol{n}_2, q^-\rangle^*, \ldots \\ \ldots \\ \ldots \\ \ldots \end{pmatrix} \tag{17.61}$$

and where $\hat{\mathcal{E}}(E)$ is a diagonal matrix, $[\hat{\mathcal{E}}(E)]_{ij} = \mathcal{E}_i(E)\delta_{ij}$, $\Delta_{E,i}$ is the detuning from the center laser frequency ω_L, i.e., $\Delta_{E,i} = (E - E_i)/\hbar - \omega_L$, with

$$\mathcal{E}_i(E) = \int_{-\infty}^{\infty} dt \varepsilon^*(t) e^{i\Delta_{E,i}t} b_i(t) \,. \tag{17.62}$$

Writing the array of possible wave functions produced as

$$\underline{|\Psi_t\rangle} = \int dE e^{-iEt/\hbar} \hat{\mathcal{E}}(E) \cdot \underline{\underline{\text{M}}}(E) \cdot \underline{|E, \boldsymbol{n}\rangle} \tag{17.63}$$

allows us to examine the possibility of taking different linear combinations of the components of the $\underline{|\Psi_t\rangle}$ vector so as to satisfy the control objectives of producing a single $|E, \boldsymbol{n}_i, q^-\rangle$ state. In this way different pathways starting with different precursor states leading to the same $|E, \boldsymbol{n}_i, q^-\rangle$ state will be seen to interfere to achieve the desired goal.

As an example, we consider a superposition state composed of the sum over the components of $\underline{|\Psi_t\rangle}$,

$$|\Psi'(t)\rangle = \sum_k \int dE e^{-iEt/\hbar} \mathcal{E}_k(E) \sum_j \text{M}_{k,j} |E, \boldsymbol{n}_j, q^-\rangle \,. \tag{17.64}$$

In the weak-field limit, the population and the phase of the initial levels can be assumed constant with time,

$$b_k(t) \approx b_k \equiv b_k(-\infty) \,, \tag{17.65}$$

in which case all the $\mathcal{E}_k(E)$ matrix elements factor as

$$\mathcal{E}_k(E) \approx b_k \int_{-\infty}^{\infty} dt \varepsilon^*(t) e^{i\Delta_{E,k}t} = 2\pi b_k \bar{\varepsilon}(\Delta_{E,k}) \,, \tag{17.66}$$

where

$$\bar{\varepsilon}(\omega) \equiv \frac{1}{2\pi} \int_{-\infty}^{\infty} dt \varepsilon^*(t) e^{i\omega t} \,. \tag{17.67}$$

Our objective to populate exclusively the ith fragment state $|E, \boldsymbol{n}_i, q^-\rangle$ can be realized in the weak-field domain by choosing the pulse shape that defines $|\Psi'(t)\rangle$ (Eq. (17.64)) to satisfy the condition

$$b_k \bar{\varepsilon}_i(\Delta_{E,k}) = \left(\underline{\underline{\mathrm{M}}}(E)^{-1}\right)_{i,k} . \tag{17.68}$$

This choice eliminates all but a *single* $|E, \boldsymbol{n}_i, q^-\rangle$ state in $\Psi'(t)$ given by Eq. (17.64).

Thus the control objective, the ith product state, is seen to be realized by starting out with an initial superposition of bound states,

$$|\Phi(t)\rangle = \sum_k b_k |E_k\rangle e^{-iE_k t/\hbar} , \tag{17.69}$$

and subjecting the system to the action of a pulse shaped according to Eq. (17.68). This allows for multiple-path interference between the various ways of generating the $|E, \boldsymbol{n}_i, q^-\rangle$ state. The weight of each pathway is chosen so as to cause destructive interference in the production of all the $|E, \boldsymbol{n}, q^-\rangle$ states but one, the $|E, \boldsymbol{n}_i, q^-\rangle$ state.

In general, control is incomplete because the pulse shaping conditions of Eq. (17.68) cannot be satisfied simultaneously for all energies. This can be seen by noting that the $(\underline{\underline{\mathrm{M}}}(E)^{-1})_{i,k}$ matrix element, which (for a single i) is a function of two variables, k and E, has to be equated to a product of a function of k, b_k, and a function of E, $\bar{\varepsilon}_i(\Delta_{E,k})$. In general, this equality cannot be satisfied. There are nevertheless important cases in which Eq. (17.68) can be satisfied. These are: either when $\underline{\underline{\mathrm{M}}}(E)$ does not vary too rapidly with E, or, conversely, when the $\langle E_j|\mu|E, \boldsymbol{n}, q^-\rangle$ matrix elements, which determine $\underline{\underline{\mathrm{M}}}(E)$ (and the absorption spectrum), span a very narrow range of energies (e.g., a narrow resonance).

The weak-field control discussed here must be achieved in two steps. First, it is necessary to create the $|\Phi(t)\rangle$ superposition state of Eq. (17.69). This state is then irradiated with the pulse satisfying Eq. (17.68). This is the essence of the weak-field pump-dump scenario. However, in the strong-field domain these two processes cannot be separated since the factorization of Eq. (17.66) does not hold. In that case the control conditions become

$$\mathcal{E}_{i,k}(E) = \left(\underline{\underline{\mathrm{M}}}(E)^{-1}\right)_{i,k} . \tag{17.70}$$

In this strong-field regime the $b_k(t)$ coefficients are embedded in $\mathcal{E}_k(E)$ (see Eq. (17.62)) and are themselves functions of $\varepsilon(t)$. Hence the problem is inherently nonlinear, necessitating an iterative solution. Nevertheless, the same interference mechanism outlined in the weak-field domain applies. The only difference is that the pulse shaping conditions are given implicitly via Eq. (17.70), rather than explicitly via Eq. (17.68), as in the weak-field domain.

We have elucidated the nature of pulse shaping control of photodissociation from the viewpoint of energy resolved coherent control theory. Clearly, when excitation is from a superposition of states, as in the vast majority of control scenarios, the role of the pulse shaping is to enhance a different set of interfering pathways for each control objective.

18
Coherent Control in the Classical Limit

The focus of this book has been on the effect of light on material systems. Nonlinear optics [272, 285] is a parallel field of endeavor in which the focus is on the reverse, that is, the effect of the material system on the incident and emergent light. Useful tools in this area have been developed within the framework of nonlinear response theory [899, 900], utilized below to reconsider coherent control and to examine control as the system approaches the classical limit.

We consider here the resonant symmetry breaking ω vs. 2ω scenario introduced in Section 3.4.3. Below, we review the essential physics of this control and extend it to the case of nonresonant excitation. The classical limit of these scenarios is then analyzed in a variety of ways. This approach is described in detail in [901].

18.1
The One-Photon vs. Two-Photon Scenario Revisited

18.1.1
Resonant Regime

Consider, once again, the one- vs. two-photon absorption coherent control scenario (Section 3.4.3). In this scenario a spatially symmetric system, initially prepared in one of its bound states, was photoexcited to an energy E in the continuum using a field $\varepsilon(t)$ given by

$$\varepsilon(t) = \epsilon_\omega \cos(\omega t + \phi_\omega) + \epsilon_{2\omega} \cos(2\omega t + \phi_{2\omega}) . \tag{18.1}$$

The 2ω component was chosen to be on resonance with the desired transition, transferring the population from the ground to the continuum state through one-photon absorption. The ω component couples the ground and continuum states through a two-photon resonant excitation. These two optical excitation pathways to the same final state interfere and create spatial anisotropy in the photodissociation. The origin of symmetry breaking was explained as follows: The potential of the system is invariant under spatial inversion or reflection and its Hamiltonian commutes with the parity operator. Hence, the system's bound eigenstates are of definite parity. In the dipole approximation, the one-photon transition connects states of opposite parity, while a two-photon process couples states of the same parity. Hence, the simultaneous resonant photoexcitation of the system through the two routes prepares a state in the continuum of the form

$$|\Psi\rangle = c_1 e^{-iEt/\hbar}|E, \text{odd}\rangle + c_2 e^{-iEt/\hbar}|E, \text{even}\rangle \,, \tag{18.2}$$

which is no longer a state of the parity operator. The average momentum $\langle p \rangle$ of the final state, for example, is:

$$\begin{aligned}\langle p \rangle &= |c_1|^2 \langle E, \text{odd}|p|E, \text{odd}\rangle + |c_2|^2 \langle E, \text{even}|p|E, \text{even}\rangle \\ &\quad + 2\text{Re}\left\{c_1^\star c_2 \langle E, \text{odd}|p|E, \text{even}\rangle\right\} .\end{aligned} \tag{18.3}$$

Since the momentum operator p is odd under inversion of the coordinates, only the third term contributes to symmetry breaking. Varying the relative phase $2\phi_\omega - \phi_{2\omega}$ between the ω and 2ω components $2\phi_\omega - \phi_{2\omega}$ controls the magnitude and sign of the interference, and hence gives control over the direction and magnitude of symmetry breaking.

The superposition in Eq. (18.2) describes the system after the interaction with the laser. The time dependence of the phases of the amplitudes in Eq. (18.2) arises from the normal, after the pulse, wave function evolution. Note that if the two final states excited are not degenerate then average values of properties induced in this scenario will oscillate in time with a period $|\hbar/\Delta E|$, determined by the energy difference ΔE between the two final states: the symmetry breaking effect produced will average out to zero in time. Hence, symmetry breaking that survives after the lasers are turned off can only be achieved in systems that have at least two degenerate excited states of opposite parity.

18.1.2
Off-Resonant Extension

The above remarks only hold true after the lasers have been switched off. When the lasers are on, it is possible to break the symmetry of any symmetric anharmonic system even if it lacks degenerate excited states of opposite parity.

To see this, note that while the system, initially in $|E_0\rangle$, is being driven by the laser field, the effect of the radiation is to create a superposition state of the form

$$|\Psi(t)\rangle = \sum_n |\Psi^{(n)}(t)\rangle \,, \tag{18.4}$$

where $|\Psi^{(0)}(t)\rangle = e^{-iE_0 t/\hbar}|E_0\rangle$ is the zeroth-order term and

$$|\Psi^{(n)}(t)\rangle = -\frac{i}{\hbar} e^{-\frac{i}{\hbar}H_0 t} \int_0^t dt' e^{\frac{i}{\hbar}H_0 t'} V(t')|\Psi^{(n-1)}(t')\rangle \,, \tag{18.5}$$

the nth order correction. Here H_0 is the system's Hamiltonian with eigenstates $\{|E_i\rangle\}$ and $V(t) = -d \cdot \varepsilon(t)$ is the radiation–matter interaction in dipole approximation. In this superposition, the time-dependent part of the phases within the created superposition state is determined not only by the energy of the states, as in Eq. (18.2), but also by the frequencies of the laser. The first-order term in the perturbation includes the main two frequencies of the laser (ω and 2ω). The second-order term includes the second harmonic (2ω and 4ω) as well as combinations of the two fundamentals; for example, $\omega \pm 2\omega$, etc. The essential point, and the origin of symmetry breaking, is that because of this mixing the wave function will contain terms of the form

$$|\Psi\rangle = c_1 e^{-i\Omega t/\hbar}|E_i, \text{odd}\rangle + c_2 e^{-i\Omega t/\hbar}|E_j, \text{even}\rangle\ , \tag{18.6}$$

whose phases oscillate in time at exactly the same rate. Hence the phase difference between them is zero even when the two states involved are not degenerate ($E_i \neq E_j$). As before, in order to create such a state, both an even and an odd ordered response to the radiation field are required and symmetry breaking results from the interference between the two processes.

For a system initially prepared in state $|E_0\rangle$ a detailed calculation of the net dipole induced off-resonance by the field in Eq. (18.1) gives third-order contributions in the field amplitude:

$$\begin{aligned}
\overline{\langle d\rangle} &= 2\text{Re}\left\{\overline{\langle\Psi^{(0)}(t)|d|\Psi^{(3)}(t)\rangle + \langle\Psi^{(1)}(t)|d|\Psi^{(2)}(t)\rangle}\right\}\\
&= \left(\frac{1}{\hbar}\right)^3 \frac{\epsilon_\omega^2\epsilon_{2\omega}}{4}\sum_{n,m,r}|d_{0r}d_{rm}d_{mn}d_{n0}|\\
&\quad\times\left[\left(\frac{1}{\omega_{r0}}+\frac{1}{\omega_{r0}-\omega}\right)\left(\frac{1}{\omega_{n0}-2\omega}+\frac{1}{\omega_{n0}+\omega}\right)\frac{1}{\omega_{m0}-\omega}\right.\\
&\quad+\left(\frac{1}{\omega_{r0}}+\frac{1}{\omega_{r0}+2\omega}\right)\frac{1}{\omega_{n0}+\omega}\frac{1}{\omega_{m0}+2\omega}\\
&\quad+\left(\frac{1}{\omega_{n0}}+\frac{1}{\omega_{n0}-2\omega}\right)\frac{1}{\omega_{r0}-\omega}\frac{1}{\omega_{m0}-2\omega}\\
&\quad\left.+\left(\frac{1}{\omega_{n0}}+\frac{1}{\omega_{n0}+\omega}\right)\left(\frac{1}{\omega_{r0}+2\omega}+\frac{1}{\omega_{r0}-\omega}\right)\frac{1}{\omega_{m0}+\omega}\right]\\
&\quad\times\cos(2\phi_\omega-\phi_{2\omega}+\alpha_{\text{rmn}})\ .
\end{aligned}\tag{18.7}$$

Here $d_{ij} = \langle E_i|d|E_j\rangle$, $\omega_{ij} = (E_i - E_j)/\hbar$, and α_{rmn} are the molecular phases defined by

$$\alpha_{\text{rmn}} = \alpha_{0r} + \alpha_{rm} + \alpha_{mn} + \alpha_{n0};\ d_{ij} = |d_{ij}|e^{i\alpha_{ij}} \tag{18.8}$$

with the convention $\alpha_{ij} = -\alpha_{ji}$. Equation (18.7) shows the broad applicability of the one- vs. two-photon scenario. While the system is being driven, an $\omega + 2\omega$ field induces symmetry breaking in any anharmonic symmetric system irrespective of the details of the system and of the driving frequency that is used. However, as emphasized below, this off-resonance effect dies out after the lasers are turned off.

Note that in the particular case of a harmonic oscillator $\overline{\langle d\rangle} = 0$. This can be verified by substituting the expressions for the transition frequencies ($\omega_{ij} = (i-j)\omega_0$)

and transition dipoles ($d_{ij} = qx_{ij} = q\sqrt{\hbar/2m\omega_0}(\sqrt{i}\delta_{i,j+1} + \sqrt{i+1}\delta_{i,j-1})$) for a harmonic oscillator into Eq. (18.7). Hence, the anharmonicities in the system's potential are an essential component of the phenomenon because they enable a nonlinear response to the laser field.

18.1.3
A Three-State Example

These ideas can be illustrated [901] by means of a simple numerical example. As a model, consider a system with three bound levels ($|E_0\rangle, |E_1\rangle, |E_2\rangle$) schematically shown in Figure 18.1 where the pairs $|E_0\rangle$ and $|E_1\rangle$, and $|E_1\rangle$ and $|E_2\rangle$, to have opposite parity such that $d_{ii} = 0$ and $d_{02} = 0$. Under the influence of the field given below (Eq. (18.10)) the wave function of the system evolves into a superposition

$$|\Psi(t)\rangle = \sum_{i=0}^{2} c_i(t)e^{-iE_it/\hbar}|E_i\rangle\,, \tag{18.9}$$

where the coefficients satisfy

$$\frac{\mathrm{d}}{\mathrm{d}\tau}\begin{pmatrix} c_0(\tau) \\ c_1(\tau) \\ c_2(\tau) \end{pmatrix} = i\alpha_0\varepsilon(\tau)\begin{pmatrix} 0 & e^{-i\tau} & 0 \\ e^{i\tau} & 0 & d_0e^{-i\delta\tau} \\ 0 & d_0e^{i\delta\tau} & 0 \end{pmatrix}\begin{pmatrix} c_0(\tau) \\ c_1(\tau) \\ c_2(\tau) \end{pmatrix},$$

where $\delta = (E_2 - E_1)/(\hbar\omega_0)$. This equation is integrated numerically with initial condition $|\Psi(t=0)\rangle = |E_0\rangle$ and the dynamics of the expectation value of the position operator, averaged over the period of the field, is followed. In the example below the transition dipole is assumed real, $\tau = t\omega_0$ is a dimensionless time, $\alpha_0 = d_{01}\epsilon_{2\omega}/(\hbar\omega_0)$, $d_0 = d_{12}/d_{01}$, and $\epsilon_0 = \epsilon_\omega/\epsilon_{2\omega}$. The laser used was a $\omega + 2\omega$ Gaussian pulse of width T_w and centered at T_c which, in dimensionless units, reads

$$\varepsilon(\tau) = \left[\cos\left(\frac{2\omega}{\omega_0}\tau + \phi_{2\omega}\right) + \epsilon_0\cos\left(\frac{\omega}{\omega_0}\tau + \phi_\omega\right)\right]e^{-\left(\frac{\tau - T_c}{T_w}\right)^2}\,. \tag{18.10}$$

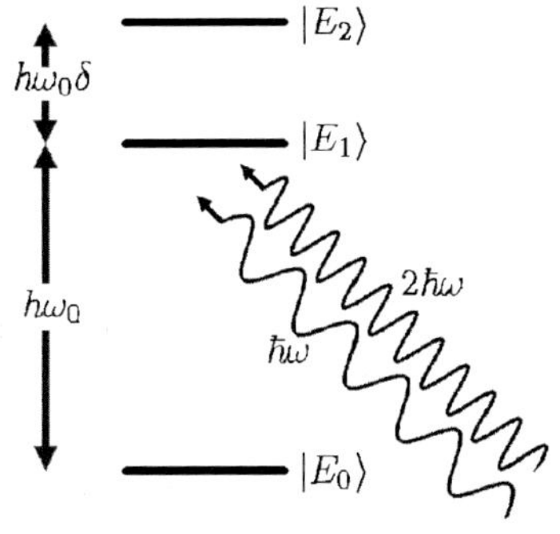

Figure 18.1 Schematic representation of the one vs. two scenario in a three-level system. Taken from [901, Figure 2.1].

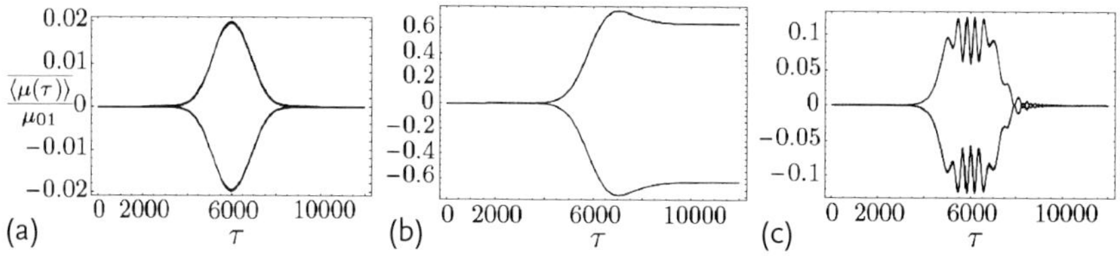

Figure 18.2 Net dipole induced by the 1 vs. 2 scenario in the three-level system schematically shown in Figure 18.1. The field used is defined in Eq. (18.10). Here, $T_w = 2000$, $T_c = 3T_w$ and $d_0 = 3$. The black and gray lines correspond to $2\phi_\omega - \phi_{2\omega} = 0$ and π, respectively. In panels (a) and (b) the two excited states are degenerate ($\delta = 0.0$), $\epsilon_0 = 15$ and $\alpha_0 = 10^{-3}$. In (a) the laser is slightly off-resonance ($\omega = 0.498\omega_0$) while (b) shows the resonant situation ($\omega = 0.5\omega_0$). In (c) the excited states are not degenerate ($\delta = 0.2$) and the lasers are off-resonance with $\omega = 0.59\omega_0$, $\alpha_0 = 0.12$ and $\epsilon_0 = 0.5$. Here, the y-axis labels are $\mu(\tau) \equiv d(\tau)$, $\mu_{01} \equiv d_{01}$. Taken from [901, Figure 2.2].

Figure 18.2 shows some representative results. When the two excited states in the system are degenerate the scenario can be applied under either off-resonant (Figure 18.2a) or resonant (Figure 18.2b) conditions. In the off-resonance case, phase-controllable net dipoles are induced while the system is being driven. However, as soon as the laser field is turned off, the symmetry breaking effect disappears. Under resonant conditions the scenario not only induces a permanent dipole in the system but the effect is an order of magnitude larger than its off-resonance counterpart. The case of excitation of nondegenerate states is shown in Figure 18.2c. In this case, it is still possible to break the symmetry of the system while the lasers are on.

18.1.4 Quantum Features

Two quantum mechanical features are manifested in this explanation of symmetry breaking: interference between two pathways, and parity effects. However, symmetry breaking has also been observed in classical mechanics [902]. Here quantum interference does not play a role and parity, which arises from the discontinuous operation of reflection, is not a conserved classical quantity [601, 903]. Further, unrelated classical and quantum arguments for symmetry breaking have been given in terms of the third-order nonlinear response of the system to the radiation field [56, 904, 905]. From this perspective, the system response mixes the laser frequencies and their harmonics in such a way as to generate a phase controllable nonoscillating (zero-harmonic (DC)) component in the photoinduced dipoles/currents.

Below we relate quantum and classical symmetry breaking, and discuss the classical limit of the ω vs. 2ω scenario. We address this issue by analytically considering the quantum-to-classical transition of the net dipole induced by an $\omega + 2\omega$ field in a quartic oscillator. This is the simplest model with a well-defined classical analog wherein induced symmetry breaking is manifested. To do so, we introduce a time-dependent perturbation theory approach in the Heisenberg representation that admits an analytic classical ($\hbar = 0$) limit [906] in the response of the oscillator to the field. As in many nonlinear response manipulations, the resultant expressions are cumbersome and lengthy. For this reason, we cite a source for these expressions, but focus on the more important qualitative content of the results.

18.2
The Quartic Oscillator

Consider [907] a charged particle confined in a bounding quartic potential that is being driven by an external radiation field $\varepsilon(t)$ in the dipole approximation. The anharmonic oscillator is defined by the Hamiltonian

$$H_0(x, p) = \frac{p^2}{2m} + \frac{1}{2}m\omega_0^2 x^2 + \lambda m \xi x^4 , \qquad (18.11)$$

where x, p and m denote the coordinate, momenta, and mass of the particle. Here $\lambda \in \{0, 1\}$, $\xi > 0$, with λ determining the strength of the anharmonicities in the potential. The total system Hamiltonian is

$$H(x, p, t) = H_0(x, p) - qx\varepsilon(t) , \qquad (18.12)$$

where q is the charge of the system and the field $\varepsilon(t)$ is given in Eq. (18.1). Since this system is bound, field-induced currents, as described in Section 3.4.3, are not possible. Rather, symmetry breaking is reflected in field-induced asymmetry in the average value of x.

In quantum mechanics, insight into spatial asymmetry induced by $\varepsilon(t)$ can be obtained via the long-time average of the position operator in the Heisenberg picture, that is, $\overline{\hat{x}_{\mathrm{H}}(t)}$. To obtain this quantity, consider a double perturbative expansion for the operator $\hat{x}_{\mathrm{H}}(t)$ in the anharmonicities and in the radiation–matter interaction. The anharmonicities are included to minimal order in a multiple-scale approximation [908, 909], while the interaction with the radiation field is taken into account up to third-order.

Given $\hat{x}_{\mathrm{H}}(t)$, the classical solution $x(t)$ can be neatly extracted [906] by first identifying the position and momentum operators, $\hat{x} = \hat{x}_{\mathrm{H}}(0)$ and $\hat{p} = \hat{p}_{\mathrm{H}}(0)$, in $\hat{x}_{\mathrm{H}}(t)$ with the classical position and momentum variables, $x = x(0)$ and $p = p(0)$, and then taking the limit $\hbar \to 0$, that is,

$$\lim_{\hbar \to 0, \hat{x} \to x, \hat{p} \to p} \hat{x}_{\mathrm{H}}(\hat{x}, \hat{p}, t) \to x(x(0), p(0), t) . \qquad (18.13)$$

With $\hat{x}_{\mathrm{H}}(t)$ and $x(t)$ one can then calculate quantum or classical averages for any initial state,

$$\langle \hat{x} \rangle(t) = \mathrm{Tr}(\hat{x}_{\mathrm{H}}(t)\hat{\rho}_0) , \quad \langle x \rangle_{\mathrm{c}}(t) = \mathrm{Tr}(x(t)\rho_0(x, p)) , \qquad (18.14)$$

respectively. Here, $\hat{\rho}_0$ is the density matrix of the quantum system at preparation time, while $\rho_0(x, p)$ is a classical phase space distribution of initial conditions.

Consider then the full evolution operator $\hat{U}(t)$, which provides $\hat{x}_{\rm H}(t) = \hat{U}^\dagger(t)\hat{x}\hat{U}(t)$. In the interaction picture $\hat{U}(t) = \hat{U}_0(t)\hat{U}_{\rm I}(t)$, where $\hat{U}_0(t) = \exp(-i\hat{H}_0 t/\hbar)$ is the evolution operator in the absence of the field while $\hat{U}_{\rm I}(t)$ captures the effects induced by $\hat{V}(t) = -q\hat{x}\varepsilon(t)$ on the dynamics. This allows a perturbative analysis to include the oscillator anharmonicity and a subsequent perturbation to include the external field. The former utilizes an exact operator solution for $\hat{x}_{\rm I}(t) = \hat{U}_0^\dagger(t)\hat{x}\hat{U}_0(t)$ to minimal order in a quantum multiple-scale perturbation theory [908, 909], a method that includes corrections to all orders in the anharmonicities.

The perturbative expansion for $\hat{U}_{\rm I}(t)$ is given by [910] $\hat{U}_{\rm I}(t) = \hat{U}_{\rm I}^{(0)} + \hat{U}_{\rm I}^{(1)}(t) + \hat{U}_{\rm I}^{(2)}(t) + \hat{U}_{\rm I}^{(3)}(t) + \cdots$, where $\hat{U}_{\rm I}^{(0)} = \hat{1}$ is the zeroth-order term and

$$\hat{U}_{\rm I}^{(n)}(t) = -\frac{i}{\hbar}\int_0^t dt'\,\hat{V}_{\rm I}(t')\,\hat{U}_{\rm I}^{(n-1)}(t') \quad (n \geq 1)\,, \tag{18.15}$$

the nth order correction. Here, $\hat{V}_{\rm I}(t) = -q\hat{x}_{\rm I}(t)\varepsilon(t)$ is the radiation–matter interaction in the interaction picture. Within this framework,

$$\hat{x}_{\rm H}(t) = \hat{U}_{\rm I}^\dagger(t)\hat{x}_{\rm I}(t)\hat{U}_{\rm I}(t) = \sum_{nm} U_I^{(m)\dagger}\hat{x}_I U_I^{(n)} \tag{18.16}$$

was calculated [907] up to third-order in the field. The result is composed of a huge number of oscillatory terms, only a subset of which has a zero-frequency (DC) term in the exponentials. These terms contribute to the net dipole, while the remaining oscillatory terms average out to zero in time.

Not all the resultant DC terms induce symmetry breaking. If the initial state is symmetric, only those terms for which $i + j$ is odd in $\hat{U}_{\rm I}^{(i)\dagger}\hat{x}_{\rm I}(t)\hat{U}_{\rm I}^{(j)}$ give a nonzero contribution to the expectation value. That is, symmetry breaking comes from the interference between an even and an odd order response to the field. This result is well known when the initial state is a parity eigenstate [55, 60, 61]. However, from this viewpoint this result relates to classical ideas like reflection symmetry, rather than that of changes of parity, a subtle but significant distinction since parity is nonclassical. (Being in a state of definite parity always implies reflection symmetry, but the converse is only true for pure states). From this viewpoint, the possibility of a classical nonzero limit to the $\omega + 2\omega$ control scenario need not be surprising.

This approach yields [907] a final operator expression for the net dipole:

$$\begin{aligned}\overline{\hat{x}_{\rm H}(t)} &= \overline{\hat{x}_{\rm I}(t)\hat{U}_{\rm I}^{(3)}(t)} + \overline{\hat{U}_{\rm I}^{(1)\dagger}(t)\hat{x}_{\rm I}(t)\hat{U}_{\rm I}^{(2)}(t)} + \text{h.c.}\\ &= \frac{q^3\epsilon_\omega^2\epsilon_{2\omega}}{16m^2\omega_0}\hat{\Gamma}(\hat{\mathcal{H}}, \omega, \lambda, \hbar)\cos(2\phi_\omega - \phi_{2\omega})\,,\end{aligned} \tag{18.17}$$

where the over-line denotes the time average. The lengthy expression for the operator $\hat{\Gamma}$ is given in [911]. Examination of this term shows that the only operator entering into $\hat{\Gamma}$ is $\hat{\mathcal{H}} = \hat{p}^2/(2m) + 1/2m\omega_0^2\hat{x}^2$, that is, $\hat{\Gamma}(\hat{\mathcal{H}}, \omega, \lambda, \hbar) =$

$\sum_{n=0}^{\infty} \Gamma_n(\omega, \lambda, \hbar)\hat{\mathcal{H}}^n$ where the Γ_n coefficients are c-numbers and that in the limit of zero anharmonicity, all symmetry breaking effects are lost, that is, $\lim_{\lambda\to 0} \hat{\Gamma}(\hat{\mathcal{H}}, \omega, \lambda, \hbar) = 0$. Hence, it is precisely because of the anharmonicities that the system can exhibit a nonlinear response to the laser, mix the frequencies of the field, and generate a zero-harmonic component in the response.

Equation (18.17) is a statement of the nonresonant control case, and the sign and magnitude of the induced dipole can be manipulated by varying the relative phase between the two frequency components of the laser. The form allows one to directly take the $\hbar \to 0$ limit in accord with Eq. (18.14), which is found to be analytic and nonzero, despite the fact that individual perturbative terms entering into Eq. (18.17) can exhibit singular behavior as $\hbar \to 0$. The result is of the form

$$\overline{x(t)} = \frac{q^3 \epsilon_\omega^2 \epsilon_{2\omega}}{16 m^2 \omega_0} \Gamma_c(\mathcal{H}, \omega, \lambda) \cos(2\phi_\omega - \phi_{2\omega}) , \tag{18.18}$$

where $\Gamma_c(\mathcal{H}, \omega, \lambda)$ is given in [911]. Hence, the field-induced interferences responsible for symmetry breaking do not disappear in the classical limit and are the source of classical control.

Significantly, in the quantum case the dipole can be divided into an $\hbar$-independent classical-like contribution $\overline{\hat{x}_c(t)} = \lim_{\hbar\to 0} \overline{\hat{x}_H(t)}$ and an entirely quantum mechanical part $\overline{\hat{x}_q(t)}$, so that:

$$\overline{\hat{x}_H(t)} = \overline{\hat{x}_c(t)} + \overline{\hat{x}_q(t)} . \tag{18.19}$$

When expectation values are calculated, as in Eq. (18.14), different initial states emphasize either $\overline{\hat{x}_c(t)}$ or $\overline{\hat{x}_q(t)}$ depending on the nature of the state. To demonstrate this and to expose the quantitative differences between quantum and classical symmetry breaking, the entire parameter space of Eqs. (18.17) and (18.18) can be explored. To do so requires a selection of initial state. Here the nth eigenstate of $\hat{\mathcal{H}}$ with energy $E_n = \hbar\omega_0(n + 1/2)$ is compared to a classical ensemble of particles with phase space density $\rho_0(x, p) = \delta(\mathcal{H} - E_n)/A$, where A is a normalization factor. For this density the classical ensemble average of $\overline{x(t)}$ coincides with the expectation value of the classical part of $\overline{\hat{x}_H}$, $\langle\overline{x(t)}\rangle_c = \langle\overline{\hat{x}_c(t)}\rangle$. This is true because $\langle\Gamma_c(\mathcal{H}, \omega, \lambda)\rangle_c = \Gamma_c(E_n, \omega, \lambda)$, in close analogy with the quantum case.

Figure 18.3 shows some representative results. The classical solution (Figure 18.3a), which in the parameter space shown is the same for all E_n, is composed of three broad resonances at $\omega = \left\{\frac{1}{2}(\omega_0 + \xi E_n),\ \omega_0 + \xi E_n,\ 2(\omega_0 + \xi E_n)\right\}$. In contrast, the quantum solutions exhibit a fine $\hbar$-dependent resonance structure, making quantum symmetry breaking look very different from the classical one. For example, the quantum $n = 0$ case (Figure 18.3b) differs dramatically from the classical solution except at small values for the anharmonicities. In all other cases, out of resonance and at small anharmonicities, the classical and quantum solutions resemble one another. The situation is very different near resonances where the quantum features can be dominant. Nevertheless, as the energy E_n of the initial state increases (the progression shown in Figure 18.3b–e) and becomes large with respect to $\hbar\omega_0$, the quantum resonances gradually merge together into three

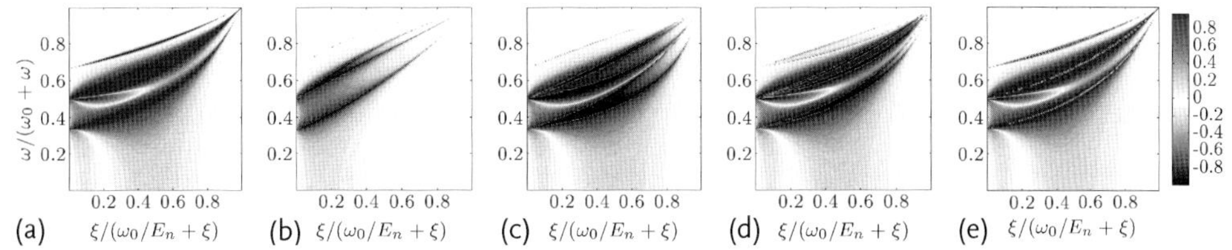

Figure 18.3 Net dipoles induced by an $\omega + 2\omega$ field in a quantum and classical anharmonic oscillator. The contour plots show the dependence of $(2/\pi)\arctan(16m^2\omega_0^4 E_n/q^3\epsilon_\omega^2\epsilon_{2\omega}\cos(2\phi_\omega - \phi_{2\omega})\langle\overline{\hat{x}_{\rm H}(t)}\rangle)$ on the anharmonicities of the potential (x-axis) and the frequency of the field (y-axis). The system is initially prepared in the nth eigenstate of $\hat{\mathcal{H}}$ with energy $E_n = \hbar\omega_0(n + 1/2)$. Panel (a) shows the classical part of the solution which in this parameter space is the same for all E_n. The remaining panels show the full quantum mechanical solution for: (b) $n = 0$; (c) $n = 1$; (d) $n = 3$; and (e) $n = 40$. The gray-scale code is given at the far right. Taken from [907, Figure 2]. (Reprinted with permission. Copyright 2009 American Physical Society.)

broad classical resonances. That is, the state progressively emphasizes the classical part of $\overline{\hat{x}_{\rm H}(t)}$, $\overline{\hat{x}_{\rm c}(t)}$, and quantum and classical symmetry breaking coincide.

Hence, laser control of symmetry breaking in this nonresonant case system is shown to correspond to the same physical phenomenon classically and quantum mechanically, with a common reliance on classically meaningful reflection symmetry arguments. From this perspective, the observed symmetry breaking is a consequence of field-driven interferences that do not vanish in the classical limit. In this sense the quantum interference contributions differ qualitatively from those in, for example, the double slit experiment.

18.3
Control in an Optical Lattice

The approach to the classical control limit in the resonant case has been computationally examined [350] in a proposed optical lattice scenario expected to be doable experimentally. The design allows one to explore both control as an effective $\hbar \to 0$ limiting process as well as examine effect of decoherence on quantum control. The computational results below also emphasize differences in quantum response in the domain of classically regular vs. chaotic dynamics.

The proposed system is a moving or shaken one-dimensional optical lattice [912–915] which (as shown below) can be viewed as a stationary spatially symmetric periodic potential interacting with a space homogeneous electric field. Both the $\hbar \to 0$ limit as well as the effect of decoherence are of interest, the latter by adding controlled amounts of decoherence through random momentum jumps induced by spontaneous emission. This allows an exploration of the effect of decoherence on a control scenario that can persist in the classical limit.

Consider an atom interacting with a longitudinally shaken one-dimensional optical lattice. The associated Hamiltonian is

$$H = \frac{P^2}{2m} + U f_1(t)\cos(2kx - \beta f_2(t)) , \tag{18.20}$$

where P is the atom momentum and m is its mass. The term U is the well depth created by the optical lattice with wave vector k. For an off-resonant interaction $U = \alpha I_0/4$ for where α is the atomic polarizability and I_0 is the peak lattice intensity. The pulse shapes $f_1(t)$ and $f_2(t)$ describe the temporal envelope and spatial shaking motion of the lattice, respectively, with β controlling the shaking strength. Such spatial motion of the lattice is experimentally achieved by applying a phase modulation to one of the counterpropagating beams that generate the optical potential [912]. Here we consider two-frequency driving fields with frequencies ω and 2ω:

$$f_2(t) = \cos(\omega t + \phi'_{rel}) + s\cos(2\omega t) , \tag{18.21}$$

where s controls the ratio of field amplitudes ($s = 1$ in all calculations below) and ϕ_{rel} is the relative phase between the driving fields. The atomic wave function $\Psi(x, t)$ satisfies the time-dependent Schrödinger equation.

Investigating the classical limit requires taking the limit as $\hbar \to 0$. To this end an effective Planck's constant, which can be varied by changing the lattice parameters, can be defined by casting the problem in reduced units. By rescaling the coordinates as

$$\theta = 2kx , \tag{18.22}$$

$$P_\theta = P\frac{2k}{\omega m} , \tag{18.23}$$

$$\tau = \omega t , \tag{18.24}$$

and defining

$$\mathcal{U} = \left(\frac{2k}{\omega}\right)^2 \frac{U}{m} , \tag{18.25}$$

the Hamiltonian becomes

$$\mathcal{H} = \frac{{P_\theta}^2}{2} + \mathcal{U} f_1(\tau)\cos(\theta - \beta f_2(\tau)) , \tag{18.26}$$

where $\mathcal{H} = (2k/\omega)^2 H/m$. The classical equations of motion are then

$$\dot{\theta} = \frac{\partial \mathcal{H}}{\partial P_\theta} = P_\theta , \tag{18.27}$$

$$\dot{P}_\theta = -\frac{\partial \mathcal{H}}{\partial \theta} = \mathcal{U} f_1(\tau) \sin(\theta - \beta f_2(\tau)) . \tag{18.28}$$

The associated quantum mechanical equations allow for the introduction of an effective Planck's constant, $\hbar_e$. Specifically, in the reduced units the Schrödinger equation becomes

$$i\hbar_e \frac{\partial \Psi(\theta,\tau)}{\partial \tau} = \left[-\frac{\hbar_e^2}{2} \frac{\partial^2}{\partial \theta^2} + \mathcal{U} f_1(\tau) \cos(\theta - \beta f_2(\tau)) \right] \Psi(\theta,\tau) , \tag{18.29}$$

where

$$\hbar_e = \frac{\hbar (2k)^2}{\omega m} \tag{18.30}$$

is the effective $\hbar$ in the reduced units. The term $\hbar_e$ can be tuned independently of $\mathcal{U}$ and β by varying the lattice parameters k and ω. Hence, the $\hbar_e \to 0$ limit can, in principle, be accessed experimentally. Analogous experimental control over the effective Planck's constant has been reported [916, 917] in the context of cold Cesium atoms in an amplitude-modulated standing wave of light.

In addition to the $\hbar_e$ dependence of the isolated system dynamics, one can also consider the effect of decoherence on the control, the most predominant of which is spontaneous emission.

18.3.1 Equivalence with Dipole Driving

This setup is related to a traditional one vs. two coherent control scenario through a gauge transformation. Specifically, consider a moving reference frame defined by the coordinate transformation

$$z = \theta - \beta f_2(\tau) \tag{18.31}$$

and employ the gauge transformation

$$\Psi'(z,\tau) = e^{\frac{izA(\tau)}{\hbar_e}} \Psi(z,\tau) , \tag{18.32}$$

where

$$A(\tau) = \beta \dot{f}_2(\tau) . \tag{18.33}$$

The Schrödinger equation takes the form $i\hbar_e \partial \Psi'(z,\tau)/\partial \tau = \mathcal{H}'(\tau)\Psi'(z,\tau)$, where

$$\mathcal{H}'(\tau) = \frac{P_z^2}{2} + \mathcal{U} f_1(\tau) \cos(z) + z\varepsilon(\tau) , \tag{18.34}$$

$P_z = \dot{z} + \beta \dot{f}_2(\tau)$ is the momentum conjugate to z, and

$$\varepsilon(\tau) = \beta \ddot{f}_2(\tau) . \tag{18.35}$$

18.3.2
Computational Results

As an example, consider the envelope

$$f_1(\tau) = \sin^2\left(\frac{\pi\tau}{2\sigma_\tau}\right) , \tag{18.36}$$

where $\sigma_\tau = 100$, with an added "absolute" phase, ϕ_{abs}, of the shaking motion of the lattice, defined by

$$\mathcal{H} = \frac{P_\theta{}^2}{2} + \mathcal{U} f_1(\tau)\cos(\theta - \beta f_2(\tau + \phi_{abs})) . \tag{18.37}$$

Here ϕ_{abs} determines the temporal shift between the envelope $f_1(\tau)$ and the underlying oscillations $f_2(\tau)$. Computations below assume $\mathcal{U} = 0.1$ and $\beta = 1$.

Classical and quantum results are compared below. In doing so all classical and quantum initial conditions are related by

$$\Psi_0(\theta) = \sqrt{\mathcal{D}_c(\theta)} , \tag{18.38}$$

where $\Psi_0(\theta)$ is the initial quantum wave function, $\mathcal{D}_c(\theta)$ is the distribution of initial classical trajectory positions, which all have zero initial momentum $P_\theta = 0$. With this definition $|\Psi_0(\theta)|^2 = \mathcal{D}_c(\theta)$, that is, the initial quantum and classical spatial probability distributions are the same. The initial quantum states are chosen to be real and have zero initial momentum.

The classical dynamics of this system exhibits chaos [96, 918]. This suggests an examination of the quantum/classical transition for regular and chaotic regions of the phase space. To this end, three initial states were examined: a spatially uniform state with

$$\mathcal{D}_c^{(u)}(\theta) = 1/(2\pi) \tag{18.39}$$

and two additional initial states localized in the regular $\mathcal{D}_c^{(r)}(\theta)$ and chaotic $\mathcal{D}_c^{(c)}(\theta)$ regions, respectively.

Consider first control in the classical system. Results are shown in the first column of Figure 18.4, which shows the final average momentum $P_{\theta,\text{avg}}$ (averaged over all initial trajectories) as a function of ϕ_{rel} and ϕ_{abs}. The top row presents results for the uniform distribution, while the middle and bottom plots correspond to the regular and chaotic distributions, respectively. The scale used for all control maps is shown in the bottom right of Figure 18.4.

In all cases, regions of control where a nonzero average momentum has been imparted to the atoms are seen. This nonzero momentum can be either positive (right-moving atoms) or negative (left-moving atoms) depending on the particular choice of phases of the driving lattice. The classical control displays a strong ϕ_{rel} dependence, showing that this traditional coherent control strategy (i.e., the dependence of the outcome of a driven process on the relative phase of two driving

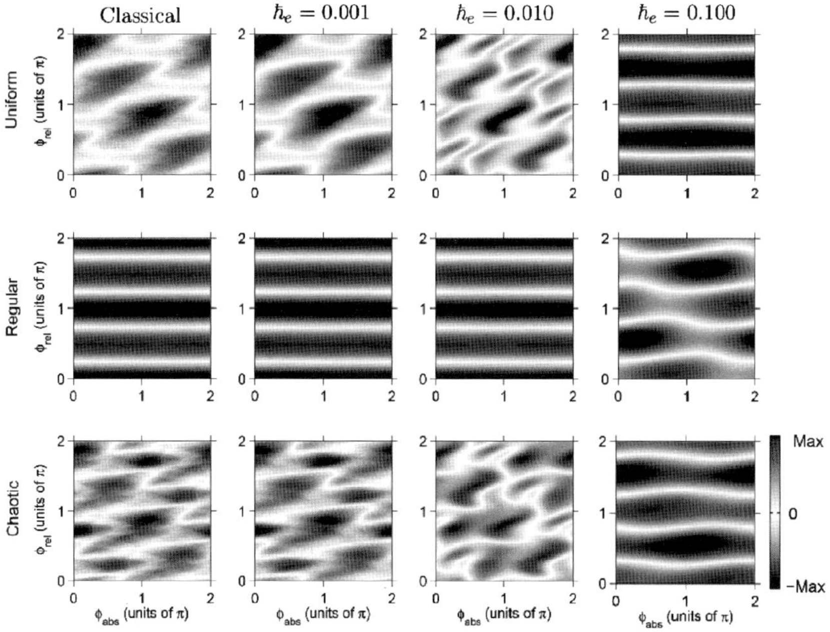

Figure 18.4 Control dynamics. Final average momentum for the classical and quantum systems. The top row is for the uniform initial state, while the middle and bottom row correspond to the regular and chaotic initial state. The first column plots the classical results, while the remaining columns plot the quantum results for three values of $\hbar_e = 0.001$, 0.01, and 0.1. Here $\phi_{rel} = (\phi'_{rel} + \pi)$. Taken from [389, Figure 1]. (Reprinted with permission. Copyright 2009 American Physical Society.)

frequencies) survives in classical mechanics. Note that for regular dynamics the results are independent of ϕ_{abs}, whereas the chaotic regions show a strong ϕ_{abs} dependence. Such ϕ_{abs} dependence arises in the chaotic dynamics since changing ϕ_{abs} is changing the relative delay between the pulse envelope and the underlying oscillations, slightly perturbing the peak force attained in each cycle of the driving field. Since chaotic dynamics is extremely sensitive to small perturbations, a strong ϕ_{abs} dependence results.

Of particular interest is the quantum control dynamics as $\hbar_e \to 0$. This is shown in Figure 18.4 for three values of $\hbar_e$ (0.1, 0.01, and 0.001) for the three initial states considered. Plotted are the expectation value $\langle P_\theta \rangle$ of the final momentum distribution as the control measure, that is, the quantum analog of $P_{\theta,\text{avg}}$.

Several points are notable: (i) both quantum and classical results show significant control and differ noticeably at the larger $\hbar_e$ values, (ii) As $\hbar_e \to 0$ the quantum result approaches the classical result. In particular, the ϕ_{rel} phase dependence of the control dynamics survives in the classical limit, as noted above, where it is identical in quantum and classical mechanics, (iii) as $\hbar_e \to 0$ the quantum result displays the underlying ϕ_{abs} dependence associated with the regular (second row in Figure 18.4) or chaotic (third row) classical behavior.

Not all properties of the quantum scenario match the classical scenario perfectly in the $\hbar_e \to 0$ limit. For example, in the small $\hbar_e$ limit (e.g., at $\hbar_e = 0.0001$) the final quantum momentum distribution $\mathcal{D}_q(P_\theta) = |\Psi(P_\theta)|^2$ (Figure 18.5b) shows highly oscillatory behavior that is not seen classically (Figure 18.5a). These oscillations, arising from quantum interference effects, are superimposed onto the underlying classical distribution and have little or no influence on the average control dynamics. Adding weak decoherence via spontaneous emission (1 momentum jump per cycle of the driving field and of magnitude $\delta P_\theta = \hbar_e$) completely suppresses the oscillations (Figure 18.5d) and gives the classical distribution, which itself remains unchanged by addition of the decoherence (Figure 18.5c). Similar results were obtained with alternate types of decoherence such as spatial jitter, random amplitude fluctuations or introducing a small initial temperature. Analogous observations of rapid quantum interferences destroyed by small decoherence were made by Ballentine *et al.* in a study of the quantum-to-classical transition of Hyperion, a moon of Saturn [919].

Since this small amount of decoherence destroys the quantum interference features and leads to the classical distributions, and since phase control is nonetheless present in the classical dynamics, small amounts of decoherence in this scenario removes a fundamentally quantum feature (oscillations in the momentum distribution) but is not detrimental, in the small $\hbar_e$ limit, to the utility of the coherent

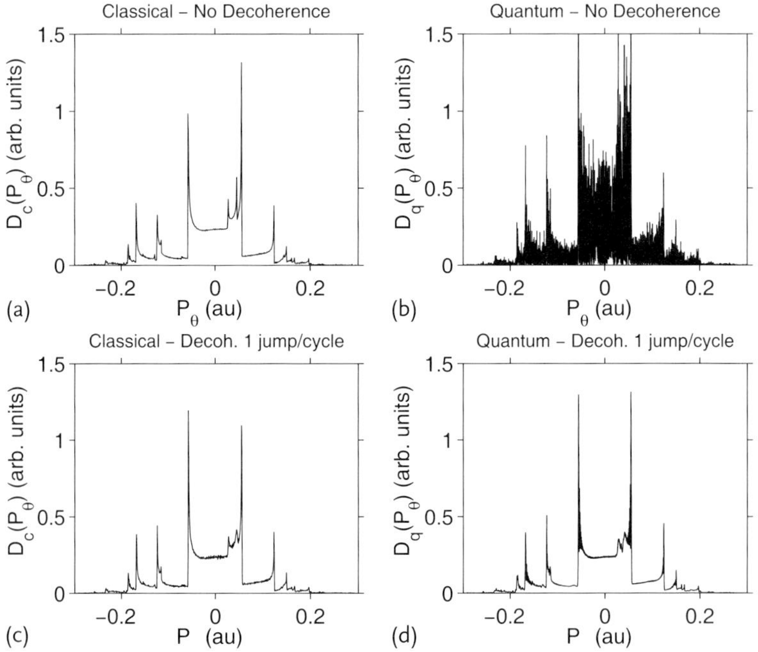

Figure 18.5 Classical and quantum momentum distribution (where $\hbar_e = 10^{-4}$) with (panels (c) and (d)) and without (panels (a) and (b)) decoherence. Taken from [389, Figure 3]. (Reprinted with permission. Copyright 2009 American Physical Society.)

control scheme. Thus, spontaneous emission is not a form of decoherence that destroys the 1 vs. 2 control in the optical lattice. Further, the control is seen to survive in the classical limit.

The results outlined in this chapter are of fundamental significance to the general area of coherent control. First, they emphasize the fact that quantum based coherent control scenarios can persist in the classical limit, albeit that the numerical values of the control can be vastly different in the quantum and classical regimes. This correspondence arises because the quantum interference terms are driven by the external laser fields. The expectation is that any quantum control scenario that can be cast as a nonlinear response problem falls into this class. What remains to be explored are (i) the circumstances under which the quantum results are well approximated by the classical dynamics [920], and (ii) the range of scenarios that are fully quantum; that is, scenarios that rely upon fully quantum effects – such as entanglement – in control of bimolecular processes at fixed energy (Section 8.1), and hence do not survive classically.

Appendix
Common Notation Used in the Book (in Order of Appearance)

H_M	Hamiltonian of the material system
$H_{MR}(t)$	Hamiltonian describing the interaction of material system with an incident external radiation field
$\boldsymbol{E}(\boldsymbol{r}, t)$	Incident classical electric field
$\boldsymbol{d} \equiv \sum_j q_j \boldsymbol{r}_j$	Molecular dipole moment
$\boldsymbol{E}(z, t)$	Incident classical electric field with assumption that all of its modes propagate in the same z direction
$\tau = t - z/c$	Retarded time
$\epsilon(\omega) = \lvert\epsilon(\omega)\rvert \exp(i\phi(\omega))$	Frequency spectrum of electric field $\boldsymbol{E}(z, t)$; $\boldsymbol{E}(z, t) = \hat{\epsilon} \int_{-\infty}^{\infty} d\omega \bar{\epsilon}(\omega) \exp(-i\omega\tau)$
$\bar{\epsilon}(\omega)$	$\epsilon(\omega) \equiv \epsilon(\omega) \exp(\omega z/c)$; $\boldsymbol{E}(z, t) = \hat{\epsilon} \int_{-\infty}^{\infty} d\omega \hat{\epsilon}(\omega) \exp(-i\omega\tau)$
$\lvert\Psi(t)\rangle, \lvert\psi(t)\rangle$	System wavefunction (depends on the context in the text)
$\omega_{m,n} \equiv (E_m - E_n)/\hbar$	Transition frequency between levels E_n and E_m
$d_{m,n} \equiv \langle E_m \lvert \hat{\epsilon} \cdot \boldsymbol{d} \rvert E_n \rangle$	Matrix element of the projection of the transition dipole operator $\boldsymbol{d}$ along the electric field direction $\hat{\epsilon}$
RW term	Rotating wave term
CRW term	Counter rotating wave term
RWA	Rotating wave approximation
$\mu = m_A m_B/(m_A + m_B)$	Reduced mass
$K_R = -[\hbar^2/(2\mu)]\nabla_R^2$	Kinetic energy operator in $\boldsymbol{R}$ in the coordinate representation
$\boldsymbol{k}$	Wave vector; $\lvert\boldsymbol{k}\rvert = 2\pi/\lambda$
$S_{\boldsymbol{n},\boldsymbol{m}}(E)$	Scattering matrix at energy E
$\lvert E, \boldsymbol{n}; 0\rangle$	Eigenfunctions of the "free" (noninteracting) material system Hamiltonian; $\langle E', \boldsymbol{m}; 0 \lvert E, \boldsymbol{n}; 0\rangle = \delta(E - E')\delta_{\boldsymbol{n},\boldsymbol{m}}$

Quantum Control of Molecular Processes, Second Edition. Moshe Shapiro and Paul Brumer

$\|E, \boldsymbol{n}\rangle$	Eigenfunctions of the fully interacting material system Hamiltonian; $\langle E', \boldsymbol{m}\|E, \boldsymbol{n}\rangle = \delta(E - E')\delta_{\boldsymbol{n},\boldsymbol{m}}$
$\|E, \boldsymbol{n}^+\rangle, \|E, \boldsymbol{n}^-\rangle$	*Outgoing* (labeled "+") and *Incoming* (labeled "−") full scattering states; $\langle E', \boldsymbol{n}^-\|E, \boldsymbol{m}^+\rangle = \delta(E - E')S_{\boldsymbol{n},\boldsymbol{m}}(E)$
$\|e\rangle$	Born–Oppenheimer electronic state
$d_{e,g} = \langle e\|\hat{\epsilon} \cdot \boldsymbol{d}\|g\rangle$	Matrix element of the projection of the transition dipole operator $\boldsymbol{d}$ along the electric field direction $\hat{\epsilon}$ for different electronic states g and e
$A_{\boldsymbol{n}}(E\|i)$	Photodissociation probability amplitude from the state i into the state characterized by $\boldsymbol{n}$ at energy E
$P_{\boldsymbol{n}}(E\|i) = \|A_{\boldsymbol{n}}(E\|i)\|^2$	Photodissociation probability from the state i into the state characterized by $\boldsymbol{n}$ at energy E
$\sigma_{\boldsymbol{n}}(E\|i)$	Photodissociation cross section from the state i into the state characterized by $\boldsymbol{n}$ at energy E
$U(t) = \exp(-iHt/\hbar)$	Evolution operator for the system Hamiltonian H
$F(t, t_0) = \langle \Psi(t_0)\|\Psi(t)\rangle$	Autocorrelation function for the system with wavefunction $\|\Psi(t)\rangle$
$I_\epsilon(E)$	Pulse-modulated absorption spectrum
$P_q(E)$	Probability of forming the product with chemical identity q
$R_{q,q'}(E) = P_q(E)/P_{q'}(E)$	Branching ratio between two channels q and q' at energy E
$P_q = \int dE\, P_q(E)$	Energy averaged $P_q(E)$
$R_{q,q'} = P_q/P_{q'}$	Branching ratio between P_q and $P_{q'}$
$V^I(t)$	Interaction Hamiltonian in the interaction representation
$\|\psi^I(t)\rangle$	Wavefunction in the interaction representation
$\mathcal{V}^{(n)}(t)$	nth order perturbation term in the interaction representation
D	Two-photon absorption operator
T	Three-photon absorption operator
$P_q(E, \hat{\boldsymbol{k}})$	Probability of photodissociation into channel q with energy E at angles $\hat{\boldsymbol{k}} = (\theta_k, \phi_k)$
$\hat{\eta}$	Polarization vector
$\|\gamma\rangle, E_\gamma$	Eigenstates and eigenvalues of the full system Hamiltonian $H = H_A + H_B + H_{AB}$
$\|\kappa\rangle, E^{(\kappa)}$	"Zeroth-order" eigenstates and eigenvalues of the sum of the decoupled Hamiltonians $H_A + H_B$

P, Q	Feshbach partitioning operators (projectors) onto the subspaces spanned by $\|\beta\rangle$ and $\|\kappa\rangle$ states, respectively
$V(\kappa\|\beta) = \langle\kappa\|QHP\|\beta\rangle$	Coupling term
Δ	Energy shift due to the coupling
$\boldsymbol{\Gamma}$	Decay rate due to the coupling
$\mathcal{H}$	Effective Hamiltonian
$\rho_s(t)$	System density matrix
$S(t) = \mathrm{Tr}(\rho_s^2(t))$	Purity of the system state
$\boldsymbol{E}^{\alpha}(t)$	Kraus operator
$\rho^{W}(q, p)$	Wigner representation of density matrix ρ
$A^{W}(q, p)$	Wigner representation of any operator A
$\sigma_j^{+} = \|j\rangle\langle 0\|, \sigma_j^{-} = \|0\rangle\langle j\|$	Transfer operators
$\langle O(t)\rangle$	Expectation value of the system property O
T_i	Phenomenological relaxation times in the energy representation
$\Omega(t)$	Rabi frequency
$\hat{a}^{\dagger}, \hat{a}$	Creation and annihilation operators, respectively
$\|+\rangle, \|-\rangle$	Spin half states
$I(t) = \|\langle\Psi_0\|\Psi(t)\rangle\|^2$	Survival probability of finding the state $\|\Psi_0\rangle$ at time t
$V_{eff}(R)$	Effective potential
$\langle a(t_1)a^*(t_2)\rangle$	Correlation function of the value a for times t_1 and t_2
$D_{\hat{k},q,i}(E, E_i J_i M_i, \omega_2, \omega_1)$	Probability amplitude for resonantly enhanced two-photon dissociation
$b_{E_f}^{(N)}$	N-photon transition amplitude between two bound states
$\boldsymbol{d}^{ind}(\boldsymbol{r}, t)$	Induced dipole moment
$\boldsymbol{P}(\boldsymbol{r}, t)$	Polarization vector
$\underline{\underline{\chi}}(t)$	Susceptibility tensor
$n(\omega)$	Complex refractive index
$\underline{\underline{\alpha}}(t)$	Polarizability tensor
$\|\phi_s\rangle$	"Zeroth-order" bound states (resonances)
$\|E, \boldsymbol{n}^{-}; 1\rangle$	"Zeroth-order" continuum states
$V(s\|E, \boldsymbol{n})$	Coupling between bound and continuum states
Δs	Energy shift of $\|\phi_s\rangle$ resonance
Γ_s	Decay rate of $\|\phi_s\rangle$ resonance
$\hat{J}$	Angular momentum operator
M_J	Projection of the angular momentum on the quantization (z) axis

P_j	Occupational probability for rotational quantum number j
$\|E, q, \boldsymbol{m}; 0\rangle$	Asymptotic states of $A + B$ (labeled q)
$\|E, q', \boldsymbol{n}; 0\rangle$	Asymptotic states of $C + D$ (labeled q')
$P_E(\boldsymbol{n}, q'; \boldsymbol{m}, q)$	Probability of transition from $\|E, q, \boldsymbol{m}; 0\rangle$ to $\|E, q', \boldsymbol{n}; 0\rangle$
$\sigma_E(\boldsymbol{n}, q'; \boldsymbol{m}, q)$	Cross section of forming $\|E, q', \boldsymbol{n}; 0\rangle$, having initiated the scattering in $\|E, q, \boldsymbol{m}; 0\rangle$
$\sigma_E(q'; \boldsymbol{m}, q)$	Cross section for scattering into arrangement q' independent of the product internal state $\boldsymbol{n}$
$\sigma_E(q', \theta, \phi; \boldsymbol{m}, q)$	Differential cross section into angles (θ, ϕ) for scattering into arrangement q' independent of the product internal state $\boldsymbol{n}$
σ^R	Cross section for Reactive Scattering
σ^{NR}	Cross section for Nonreactive Scattering
PI	Penning ionization
AI	Associative ionization
STIRAP	Stimulated Raman adiabatic passage
CPT	Coherent population trapping
σ^{PI}	Cross section for penning ionization
σ^{AI}	Cross section for associative ionization
$P(f, q'; i, q)$	Probability of producing product in final state $\|E, q', f; 0\rangle$ having started in the initial state $\|E, q, i; 0\rangle$
$\sigma = \boldsymbol{S}_{q'}^{\dagger} \boldsymbol{S}_q$	Optimization matrix for scattering problem
$\rho(\boldsymbol{r}, t)$	Charge density at the point $\boldsymbol{r}$ at time t
$\boldsymbol{B}(\boldsymbol{r}, t)$	Magnetic field
$\boldsymbol{j}(\boldsymbol{r}, t)$	Charge current
$\boldsymbol{A}(\boldsymbol{r}, t)$	Vector electromagnetic potential
$\Phi(\boldsymbol{r}, t)$	Scalar electromagnetic potential
$\boldsymbol{E}_L$	Longitudinal component of electric field $\boldsymbol{E}$
$\boldsymbol{E}_R$	Transverse component of electric field $\boldsymbol{E}$
H_R	Hamiltonian of the electromagnetic field
V_C	Coulomb potential
$\boldsymbol{A}_{\boldsymbol{k}}$	k^{th} expansion coefficient of $\boldsymbol{A}$
$Q_{\boldsymbol{k}}, P_{\boldsymbol{k}}$	Canonical coordinate and momentum, respectively, of the kth mode of the electromagnetic field
$\hat{a}_{\boldsymbol{k}}^{\dagger}, \hat{a}_{\boldsymbol{k}}$	Creation and annihilation operators, respectively, of the kth mode of the electromagnetic field
$\hat{N}_{\boldsymbol{k}} = \hat{a}_{\boldsymbol{k}}^{\dagger} \hat{a}_{\boldsymbol{k}}$	Number operator of the kth mode of the electromagnetic field, having eigenstates $\|N_{\boldsymbol{k}}\rangle$ and eigenvalues $N_{\boldsymbol{k}}$

$\|\alpha\rangle$	Coherent states of the electromagnetic field; solutions of $\hat{a}\|\alpha\rangle = \alpha\|\alpha\rangle$
$H_f = H_M + H_R$	"Radiatively decoupled" Hamiltonian; $H = H_M + H_R + H_{MR} = H_f + H_{MR}$
$\|M_i\rangle$	Matter (material system) state
$\|R_i\rangle$	Radiation field state
$\|\Psi^f\rangle$	Full matter + radiation state
$\langle R_i\|R_j\rangle$	Radiation overlap matrix element
$\|\mathcal{C}\rangle$	Schrödinger "cat-state"
$\|E, f^-\rangle$	Forward direction state
$\|E, b^-\rangle$	Backward direction state
$\underline{\underline{H}}$	Hamiltonian matrix
$\underline{\underline{A}}(t) = (d\underline{\underline{U}}^\dagger(t)/dt)\underline{\underline{U}}(t)$	Nonadiabatic coupling matrix
$\boldsymbol{E}_S(t)$	Stokes laser pulse
$\boldsymbol{E}_P(t)$	Pump laser pulse
NQC	Nondegenerate quantum control
DQC	Degenerate quantum control
q-nit	Generalization of the term q-bit for nonbinary codes
AP	Adiabatic passage
PAP	Piecewise adiabatic passage
STIHRAP	Stimulated hyper-Raman adiabatic passage
EIT	Electromagnetically induced transparency
AT splitting	Autler–Townes splitting
$\|\chi_i(t)\rangle$	Adiabatic states
$n_g = n + \omega(dn/d\omega)$	Group index of refraction
$v_g = c/n_g$	Group velocity
$F_i(\tau)$	Spectral autocorrelation function
SPODS	Selective population of dressed states
$\underline{b}(t), \underline{c}(t)$	Vectors of expansion coefficients (depending on the context)
SVCA	Slowly varying continuum approximation ("flat continuum")
$\overline{d^2}, \overline{\Delta}, \overline{\Gamma}$	Averaged d^2, Δ, Γ quantities, respectively
IIC	Incoherent interference control
APC	Adiabatic passage from a continuum
$R(t)$	Net association rate
FWHM	Full width at half maximum
$\mathcal{I}$	Inversion operator
$\|L\rangle, \|E, \boldsymbol{n}, L^-\rangle$ and $\|D\rangle, \|E, \boldsymbol{n}, D^-\rangle$	L and D enantiomer states, respectively
$\|E, \boldsymbol{n}, s^-\rangle, \|E, \boldsymbol{n}, a^-\rangle$	Symmetric and antisymmetric continuum eigenfunctions, respectively
P_L, P_D	Populations of L and D enantiomers, respectively

CPT	Cyclic population transfer
CCAP	Coherently controlled adiabatic passage
LIP	Light induced potential
$\boldsymbol{N} = (N_{\boldsymbol{k}1}, N_{\boldsymbol{k}2}, \ldots, N_{\boldsymbol{k}_m}, \ldots)$	Vector of photon occupation numbers
$\|E_i, N_i\rangle = \|E_i\rangle\|N_i\rangle$	Material + field eigenstate of H_f
$\|E, \boldsymbol{n}^-, \boldsymbol{N}^-\rangle$	Fully interacting *incoming* material + field Hamiltonian eigenstates
$\|E, \boldsymbol{n}^-, \boldsymbol{N}^-, t\rangle$	Time dependent fully interacting *incoming* material + field Hamiltonian eigenstates
$\|E, \boldsymbol{n}^-, t\rangle$	Time dependent *incoming* material Hamiltonian eigenstates
$A(E, \boldsymbol{n}, \boldsymbol{N}, t\|i, N_i)$	Photodissociation amplitude into the final state with energy E, internal quantum numbers $\boldsymbol{n}$ and radiation field described by $\boldsymbol{N}$, starting in the initial state $\|E_i, N_i\rangle$
$R(E, \boldsymbol{n}, \boldsymbol{N}, t\|i, N_i)$	Rate of transition into the radiatively decoupled state
H_{el}	Electronic Hamiltonian
CEP	Carrier-envelope phase
CEO	Carrier-envelope offset
ATI	Above threshold ionization
$U_p = \overline{(dx/dt)^2/2}$	Ponderomotive potential
I_p	Ionization potential
HHG	High harmonic generation
SFA	Strong field approximation
IRMPD	Infra-red multiphoton dissociation
IVR	Intramolecular vibrational relaxation
$T^{(N)}(\omega)$	Lowest order N-photon transition operator
OCT	Optimal control theory
GA	Genetic algorithm
H_S	System Hamiltonian
H_B	Bath Hamiltonian
H_{SB}	System–Bath coupling
SRS	Stimulated Raman scattering
$\hbar_e = \hbar(2k)^2/(\omega m)$	Effective Planck's constant

References

1 D'Alessandro, D. (2008) *Introduction to Quantum Control and Dynamics*, Chapman and Hall, New York.

2 Rice, S.A. and Zhou, M. (2000) *Optical Control of Molecular Dynamics*, John Wiley & Sons, Inc., New York.

3 Cohen-Tannoudji, C., Diu, B., and Laloë, F. (1977) *Quantum Mechanics*, vol. 1, John Wiley & Sons, Inc., New York.

4 Abramowitz, M. and Stegun, I.A. (1965) *Handbook of Mathematical Functions*, Dover, New York.

5 Pauling, L. and Wilson, E.B. (1935) *Introduction to Quantum Mechanics*, McGraw Hill, New York.

6 Bernstein, R.B. (ed.) (1979) *Atom-Molecule Collision Theory, A Guide for the Experimentalist*, Plenum Press, New York.

7 Manolopoulos, D.E., D'Mello, M., and Wyatt, R.E. (1990) *J. Chem. Phys.*, **93**, 403.

8 Shapiro, M. (1972) *J. Chem. Phys.*, **56**, 2582.

9 Shapiro, M. (1993) *J. Phys. Chem.*, **97**, 7396.

10 Heller, E.J. (1981) in *Potential Energy Surfaces, Dynamics Calculations*, (ed. D.G. Truhlar), Plenum Press, New York.

11 Heller, E.J. (1981) *Acc. Chem. Res.*, **14**, 368.

12 Kulander, K.C. and Orel, A.E. (1981) *J. Chem. Phys.*, **74**, 6529.

13 A. Bartana, U.B., Ruhman, S., and Kosloff, R. (1994) *Chem. Phys. Lett.*, **219**, 211.

14 Shapiro, M. and Brumer, P. (1986) *J. Chem. Phys.*, **84**, 540.

15 Brumer, P. and Shapiro, M. (1989) *Chem. Phys.*, **139**, 221.

16 Shapiro, M. and Brumer, P. (2001) *J. Phys. Chem.*, **105**, 2897.

17 Judson, R.S. and Rabitz, H. (1992) *Phys. Rev. Lett.*, **68**, 1500.

18 Bardeen, C.J., Yakovlev, V.V., Wilson, K.R., Carpenter, S.D., Weber, P.M., and Warren, W.S. (1997) *Chem. Phys. Lett.*, **280**, 151.

19 Yelin, D., Meshulach, D., and Silberberg, Y. (1997) *Opt. Lett.*, **22**, 1793.

20 Assion, T., Baumert, T., Bergt, M., Brixner, T., Kiefer, B., Seyfried, V., Strehle, M., and Gerber, G. (1998) *Science*, **282**, 919.

21 Brumer, P. and Shapiro, M. (1986) *Chem. Phys. Lett.*, **126**, 541.

22 Brumer, P. and Shapiro, M. (1992) *Ann. Rev. Phys. Chem.*, **43**, 257.

23 Shapiro, M. and Brumer, P. (1994) *Int. Rev. Phys. Chem.*, **13**, 187.

24 Shapiro, M. and Brumer, P. (1997) *Trans. Faraday Soc.*, **93**, 1263.

25 Shapiro, M. and Brumer, P. (2000) in *Advances in Atomic, Molecular and Optical Physics*, (eds B. Bederson and H. Walther), vol. 42, Academic Press, San Diego.

26 Shapiro, M. and Bersohn, R. (1980) *J. Chem. Phys.*, **73**, 3810.

27 Shapiro, M. (1986) *J. Phys. Chem.*, **90**, 3644.

28 Brumer, P. and Shapiro, M. (1999) in *Coherent Control in Atoms, Molecules and Semiconductors*, (eds W. Pötz and W.A. Schröder), Kluwer, Dordrecht.

29 Shapiro, M. and Brumer, P. (1987) *Faraday Discuss. Chem. Soc.*, **82**, 177.

Quantum Control of Molecular Processes, Second Edition. Moshe Shapiro and Paul Brumer

30 Crim, F.F. (1984) *Ann. Rev. Phys. Chem.*, **35**, 647.
31 Ticich, T.M., Likar, M.D., Dubal, H., Butler, L.J., and Crim, F.F. (1987) *J. Chem. Phys.*, **87**, 5820.
32 Likar, M.D., Baggott, J.E., Sinha, A., Ticich, T.M., Wal, R.L.V., and Crim, F.F. (1988) *J. Chem. Soc. Faraday Trans.*, **84**, 1483.
33 Sinha, A., Wal, R.L.V., and Crim, F.F. (1989) *J. Chem. Phys.*, **91**, 2929.
34 Crim, F.F. (1990) *Science*, **249**, 1387.
35 Wal, R.L.V., Scott, J.L., and Crim, F.F. (1991) *J. Chem. Phys.*, **94**, 1859.
36 Arutsi-Parpar, T., Schmid, R.P., Li, R.J., Bar, I., and Rosenwaks, S. (1997) *Chem. Phys. Lett.*, **268**, 163.
37 Bar, I. and Rosenwaks, S. (2001) *Int. Rev. Phys. Chem.*, **20**, 711.
38 Roman, P. (1965) *Advanced Quantum Theory*, Addison-Wesley, Boston.
39 Shapiro, M., Hepburn, J.W., and Brumer, P. (1988) *Chem. Phys. Lett*, **149**, 451.
40 Chan, C.K., Brumer, P., and Shapiro, M. (1991) *J. Chem. Phys.*, **94**, 2688.
41 Chelkowski, S. and Bandrauk, A.D. (1991) *Chem. Phys. Lett.*, **186**, 284.
42 Bandrauk, A.D., Gauthier, J.M., and McCann, J.F. (1992) *Chem. Phys. Lett.*, **200**, 399.
43 Szöke, A., Kulander, K.C., and Bardsley, J.N. (1991) *J. Phys. B.*, **24**, 3165.
44 Zuo, T. and Bandrauk, A.D. (1996) *Phys. Rev. A*, **54**, 3254.
45 Chen, C., Yin, Y.Y., and Elliott, D.S. (1990) *Phys. Rev. Lett.*, **64**, 507.
46 Chen, C., Yin, Y.Y., and Elliott, D.S. (1990) *Phys. Rev. Lett.*, **65**, 1737.
47 Park, S.M., Lu, S.P., and Gordon, R.J. (1991) *J. Chem. Phys.*, **94**, 8622.
48 Lu, S.P., Park, S.M., Xie, Y., and Gordon, R.J. (1992) *J. Chem. Phys.*, **96**, 6613.
49 Kleiman, V.D., Zhu, L., Li, X., and Gordon, R.J. (1995) *J. Chem. Phys.*, **102**, 5863.
50 Zhu, L., Kleiman, V.D., Li, X., Lu, S., Trentelman, K., and Gordon, R.J. (1995) *Science*, **270**, 77.
51 Zhu, L., Suto, K., Fiss, J.A., Wada, R., Seideman, T., and Gordon, R.J. (1997) *Phys. Rev. Lett.*, **79**, 4108.
52 Wang, X., Bersohn, R., Takahashi, K., Kawasaki, M., and Kim, H.L. (1996) *J. Chem. Phys.*, **105**, 2992.
53 Nagai, H., Ohmura, H., Ito, F., Nakanaga, T., and Tachiya, M. (2006) *J. Chem. Phys.*, **124**, 034304.
54 Gordon, R.J., Zhu, L.C., and Seideman, T. (1999) *Acc. Chem. Res.*, **32**, 1007.
55 Kurizki, G., Shapiro, M., and Brumer, P. (1989) *Phys. Rev. B*, **39**, 3435.
56 Baranova, B.A., Chudinov, A.N., and Zel'dovitch, B.Y. (1990) *Opt. Commun.*, **79**, 116.
57 Haché, A., Kostoulas, Y., Atanasov, R., Hughes, J.L.P., Sipe, J.E., and van Driel, H.M. (1997) *Phys. Rev. Lett.*, **78**, 306.
58 Schafer, K.J. and Kulander, K.C. (1992) *Phys. Rev. A*, **45**, 8026.
59 Potvliege, R.M. and Smith, P.H.G. (1992) *J. Phys. B*, **25**, 2501.
60 Atanasov, R., Haché, A., Hughes, L.P., van Driel, H.M., and Sipe, J.E. (1996) *Phys. Rev. Lett.*, **76**, 1703.
61 Dupont, E., Corkum, P.B., Liu, H.C., Buchanan, M., and Wasilewski, Z.R. (1995) *Phys. Rev. Lett.*, **74**, 3596.
62 Charron, E., Giusti-Suzor, A., and Mies, F.H. (1995) *J. Chem. Phys.*, **103**, 7359.
63 Sheehy, B., Walker, B., and DiMauro, L.F. (1995) *Phys. Rev. Lett.*, **74**, 4799.
64 Aubanel, E.E. and Bandrauk, A.D. (1994) *Chem. Phys. Lett.*, **229**, 169.
65 Ehlotzky, F. (2001) *Phys. Rep.*, **345**, 175.
66 Asaro, C., Brumer, P., and Shapiro, M. (1988) *Phys. Rev. Lett.*, **60**, 1634.
67 Tannor, D.J. and Rice, S.A. (1985) *J. Chem. Phys.*, **83**, 5013.
68 Tannor, D.J. and Rice, S.A. (1988) *Adv. Chem. Phys.*, **70**, 441.
69 Tannor, D.J. (2007) *Introduction to Quantum Mechanics: A Time Dependent Perspective*, University Science Press, Sausalito.
70 Levy, I., Shapiro, M., and Brumer, P. (1990) *J. Chem. Phys.*, **93**, 2493.
71 Abrashkevich, D.G., Shapiro, M., and Brumer, P. (1998) *J. Chem. Phys.*, **108**, 3585.
72 Seideman, T., Shapiro, M., and Brumer, P. (1989) *J. Chem. Phys.*, **90**, 7132.
73 Schmidt, I. (1987) Ph.D. Thesis, Kaiserslautern University.

The Na-Na potential curves and the relevant electronic dipole moments were taken from this thesis.
74 Papanikolas, J.M., Williams, R.M., Kleiber, P.D., Hart, J.L., Brink, C., Price, S.D., and Leone, S.R. (1995) *J. Chem. Phys.*, **103**, 7269.
75 Uberna, R., Amitay, Z., Qian, C.X.W., and Leone, S.R. (2001) *J. Chem. Phys.*, **114**, 10311.
76 Shapiro, M. and Brumer, P. (1993) *J. Chem. Phys.*, **98**, 201.
77 Tannor, D.J., Kosloff, R., and Rice, S.A. (1986) *J. Chem. Phys.*, **85**, 5805.
78 Rosenwaks, S. (2009) *Vibrationally Mediated Photodissociation*, Royal Society of Chemistry, Cambridge.
79 Pimentel, G.C. and Coonrod, J.A. (1987) *Opportunities in Chemistry Today and Tomorrow*, National Academy Press, Washington.
80 Bar, I., Cohen, Y., David, D., Rosenwaks, S., and Valentini, J.J. (1990) *J. Chem. Phys.*, **93**, 2146.
81 Bar, I., Cohen, Y., David, D., Rosenwaks, S., and Valentini, J.J. (1991) *J. Chem. Phys.*, **95**, 3341.
82 Wardlaw, D., Brumer, P., and Osborn, T.A. (1982) *J. Chem. Phys.*, **76**, 4916.
83 Parson, W.W. (2007) *Modern Optical Spectroscopy*, Springer, New York.
84 Christopher, P.S., Shapiro, M., and Brumer, P. (2005) *J. Chem. Phys.*, **123**, 064813.
85 Christopher, P.S., Shapiro, M., and Brumer, P. (2006) *J. Chem. Phys.*, **125**, 124310.
86 Christopher, P.S., Shapiro, M., and Brumer, P. (2006) *J. Chem. Phys.*, **124**, 184107.
87 Gerbasi, D., Sanz, A.S., Christopher, P.S., Shapiro, M., and Brumer, P. (2007) *J. Chem. Phys.*, **126**, 124307.
88 Savolainen, J., Fanciulli, R., Dijkhuizen, N., Moore, A.L., Hauer, J., Backup, T., Motzkus, M., and Herek, J.L. (2008) *Proc. Natl. Acad. Sci. USA*, **105**, 7641.
89 deGroot, M., Field, R.W., and Buma, J. (2009) *Proc. Natl. Acad. Sci. USA*, **106**, 2510.
90 Feshbach, H. (1962) *Ann. Phys. (NY)*, **19**, 287.
91 Feshbach, H. (1967) *Ann. Phys. (NY)*, **43**, 410.
92 Gerbasi, D. (2004) Ph.D. Dissertation, University of Toronto.
93 Thanopulos, I., Brumer, P., and Shapiro, M. (2010) *J. Chem. Phys.*, **133**, 154111.
94 Raab, A., Worth, G.A., Meyer, H.D., and Cederbaum, L.S. (1999) *J. Chem. Phys.*, **110**, 936.
95 Nuernberger, P., Vogt, G., Brixner, T., and Gerber, G. (2007) *Phys. Chem. Chem. Phys.*, **9**, 2470.
96 Brumer, P. (1990) *Encyclopedia of Modern Physics*, (ed. R.A. Meyers), Academic Press, New York.
97 Brif, C., Chakrabarti, R., and Rabitz, H. (2010) *New J. Phys.*, **12**, 075008.
98 Kosloff, R., Rice, S.A., Gaspard, P., Tersigni, S., and Tannor, D.J. (1989) *Chem. Phys.*, **139**, 201.
99 Rice, S.A. (1997) *Adv. Chem. Phys.*, **101**, 213.
100 Rice, S.A. and Zhao, M. (2000) *Optical Control of Molecular Dynamics*, John Wiley & Sons, Inc., New York.
101 Zewail, A.H. (1988) *Science*, **242**, 1645.
102 Zewail, A.H. and Bernstein, R.B. (1988) *Chem. Eng. News*, **66**(45), 24.
103 Rosker, M.J., Dantus, M., and Zewail, A.H. (1988) *J. Chem. Phys.*, **89**, 6113.
104 Rosker, M.J., Dantus, M., and Zewail, A.H. (1988) *J. Chem. Phys.*, **89**, 6128.
105 Bowman, M.J., Dantus, M., and Zewail, A.H. (1989) *Chem. Phys. Lett.*, **161**, 297.
106 Herek, J.L., Materny, A., and Zewail, A.H. (1994) *Chem. Phys. Lett.*, **228**, 15.
107 Baumert, T., Grosser, M., Thalweiser, R., and Gerber, G. (1990) *Phys. Rev. Lett.*, **64**, 733.
108 Baumert, T., Grosser, M., Thalweiser, R., and Gerber, G. (1991) *Phys. Rev. Lett.*, **67**, 3753.
109 Baumert, T., Helbing, J., and Gerber, G. (1997) *Adv. Chem. Phys.*, **101**, 47.
110 Tersigni, S., Gaspard, P., and Rice, S.A. (1990) *J. Chem. Phys.*, **93**, 1670.
111 Amstrup, B., Carlson, R.J., Matro, A., and Rice, S.A. (1991) *J. Phys. Chem.*, **95**, 8019.
112 Zhao, M. and Rice, S.A. (1991) *J. Chem. Phys.*, **95**, 2465.

113 Rice, S.A. (1991) in *Mode Selective Chemistry*, (eds J. Jortner, R.D. Levine and B. Pullman), Kluwer, Dodrecht.

114 Tannor, D.J. and Jin, Y. (1991) *Mode Selective Chemistry*, (eds J. Jortner, R.D. Levine and B. Pullman), Kluwer, Dordrecht.

115 Rice, S.A. (1992) *Science*, **258**, 412.

116 Tannor, D.J. (1994) *Molecules in Laser Fields*, (ed. A.D. Bandrauk), Marcel Dekker, New York.

117 Shi, S., Woody, A., and Rabitz, H. (1988) *J. Chem. Phys.*, **88**, 6870.

118 Peirce, A.P., Dahleh, M., and Rabitz, H. (1988) *Phys. Rev. A*, **37**, 4950.

119 Shi, S. and Rabitz, H. (1989) *Chem. Phys.*, **139**, 185.

120 Shi, S. and Rabitz, H. (1990) *J. Chem. Phys.*, **92**, 364.

121 Judson, R.S., Lehmann, K.K., Rabitz, H., and Warren, W.S. (1990) *J. Mol. Struct.*, **223**, 425.

122 Dahleh, M., Peirce, A.P., and Rabitz, H. (1990) *Phys. Rev. A*, **42**, 1065.

123 Yao, K., Shi, S., and Rabitz, H. (1990) *Chem. Phys.*, **150**, 373.

124 Warren, W.S., Rabitz, H., and Dahleh, M. (1993) *Science*, **259**, 1581.

125 Jakubetz, W., Manz, J., and Mohan, V. (1989) *J. Chem. Phys.*, **90**, 3686.

126 Jakubetz, W., Just, B., Manz, J., and Schreier, H.J. (1990) *J. Phys. Chem.*, **94**, 2294.

127 Hartke, B., Kolba, E., Manz, J., and Schor, H.H.R. (1990) *Ber. Bunsenges. Phys. Chem.*, **94**, 1312.

128 Combariza, J.E., Just, B., Manz, J., and Paramonov, G.K. (1991) *J. Phys. Chem.*, **95**, 10351.

129 Combariza, J.E., Daniel, C., Just, B., Kades, E., Kolba, E., Manz, J., Malisch, W., Paramonov, G.K., and Warmuth, B. (1992) in *Isotope Effects in Gas Phase Chemistry*, (ed. J.A. Kaye), ACS Symposium Series, vol. 502, American Chemical Society, Washington, DC.

130 Shapiro, M. and Brumer, P. (1992) *J. Chem. Phys.*, **97**, 6259.

131 Averbukh, I. and Shapiro, M. (1993) *Phys. Rev. A*, **47**, 5086.

132 Shapiro, M. and Brumer, P. (1993) *Chem. Phys. Lett.*, **208**, 193.

133 Kohler, B., Krause, J.L., Raski, F., Wilson, K.R., Yakovlev, V.V., Whitnell, R.M., and Yan, Y. (1995) *Acc. Chem. Res.*, **28**, 133.

134 Kosloff, D. and Kosloff, R. (1983) *J. Comput. Phys.*, **52**, 35.

135 Kosloff, R. and Kosloff, D. (1983) *J. Chem. Phys.*, **79**, 1823.

136 Kosloff, R. (1988) *J. Phys. Chem.*, **92**, 2087.

137 Kosloff, R. and Tal-Ezer, H. (1986) *Chem. Phys. Lett.*, **127**, 223.

138 Leforestier, C., Bisseling, R.H., and Cerjan, C. *et al.* (1991) *J. Comput. Phys.*, **94**, 59.

139 Meshulach, D., Yelin, D., and Silberberg, Y. (1997) *Opt. Commun.*, **138**, 345.

140 Meshulach, D. and Silberberg, Y. (1998) *Nature*, **396**, 239.

141 Winterfeldt, C., Spielmann, C., and Gerber, G. (2008) *Rev. Mod. Phys.*, **80**, 117, and references therein.

142 Dantus, M. and Lozovoy, V.V. (2004) *Chem. Rev.*, **104**, 1813.

143 Lozovoy, V.V., Zhu, X., Gunaratne, T.C., Harris, D.A., Shane, J.C., and Dantus, M. (2008) *J. Phys. Chem. A*, **112**, 3789.

144 Zhu, X., Gunaratne, T.C., Lozovoy, V.V., and Dantus, M. (2009) *J. Phys. Chem. A*, **113**, 5264.

145 Shapiro, M. and Brumer, P. (1991) *J. Chem. Phys.*, **95**, 8658.

146 Scherer, N.F., Ruggiero, A.J., Du, M., and Fleming, G.R. (1990) *J. Chem. Phys.*, **93**, 856.

147 Scherer, N.F., Carlson, R.J., Matro, A., Du, M., Ruggiero, A.J., Romero-Rochin, V., Cina, J.A., Fleming, G.R., and Rice, S.A. (1991) *J. Chem. Phys.*, **95**, 1487.

148 Blanchet, V., Bouchene, M.A., Cabrol, O., and Girard, B. (1995) *Chem. Phys. Lett.*, **233**, 491.

149 Blanchet, V., Nicole, C., Bouchene, M.A., and Girard, B. (1997) *Phys. Rev. Lett.*, **78**, 2716.

150 Blanchet, V., Bouchene, M.A., and Girard, B. (1998) *J. Chem. Phys.*, **108**, 4862.

151 Nicole, C., Bouchene, M.A., Zamith, S., Melikechi, N., and Girard, B. (1999) *Phys. Rev. A*, **60**, R1755.

152 Kinrot, O., Averbukh, I.S., and Prior, Y. (1995) *Phys. Rev. Lett.*, **75**, 3822.

153 Warmuth, C., Tortschanoff, A., Milota, F., Knopp, G., Shapiro, M., Prior, Y., Averbukh, I., Schleich, W., Jakubetz, W., and Kauffmann, H.F. (2000) *J. Chem. Phys.*, **112**, 5060.

154 Katsuki, H., Chiba, H., Girard, B., Meier, C., and Ohmori, K. (2006) *Science*, **311**, 1589.

155 Ohmori, K., Katsuki, H., Chiba, H., Honda, M., Hagihara, Y., Fujiwara, K., Sato, Y., and Ueda, K. (2006) *Phys. Rev. Lett.*, **96**, 093002.

156 Katsuki, H., Chiba, H., Girard, B., Meier, C., and Ohmori, K. (2009) *Ann. Rev. Phys. Chem.*, **60**, 487.

157 Hosaka, K., Shimada, H., Chiba, H., Katsuki, H., Teranishi, Y., Ohtsuki, Y., and Ohmori, K. (2010) *Phys. Rev. Lett.*, **104**, 180501.

158 Leichtle, C., Schleich, W.P., Averbukh, I.S., and Shapiro, M. (1998) *J. Chem. Phys.*, **108**, 6057.

159 Parker, J. and Stroud, C.R. (1986) *Phys. Rev. Lett.*, **56**, 716.

160 Yeazell, J.A. and Stroud, C.R. (1988) *Phys. Rev. Lett.*, **60**, 1494.

161 Alber, G., Ritch, H., and Zoller, P. (1986) *Phys. Rev. A*, **34**, 1058.

162 Henle, W.A., Ritch, H., and Zoller, P. (1987) *Phys. Rev. A*, **36**, 683.

163 Fedorov, M.V. and Movsesian, A.M. (1988) *J. Opt. Soc. Am. B*, **5**, 850.

164 Weiner, A.M., Heritage, J.P., and Thurston, R.N. (1986) *Opt. Lett.*, **11**, 153.

165 Weiner, A.M. and Heritage, J.P. (1987) *Phys. Rep. Appl.*, **22**, 1619.

166 Weiner, A.M., Leaird, D.E., Wiederrecht, G.P., and Nelson, K.A. (1990) *Science*, **247**, 1317.

167 Haner, M. and Warren, W.S. (1988) *Appl. Phys. Lett.*, **52**, 1459.

168 Melinger, J.S., Gandhi, S.R., Hariharan, A., Tull, J.X., and Warren, W.S. (1992) *Phys. Rev. Lett.*, **68**, 2000.

169 Melinger, J.S., Gandhi, S.R., Hariharan, A., Goswami, D., and Warren, W.S. (1994) *J. Chem. Phys.*, **101**, 6439.

170 Hillegas, C.W., Tull, J.X., Goswami, D., Strickland, D., and Warren, W.S. (1994) *Opt. Lett.*, **19**, 737.

171 Weiner, A.M. (2000) *Rev. Sci. Instrum.*, **71**, 1929.

172 Goldberg, D.E. (1993) *Genetic Algorithms in Search Optimization and Machine Learning*, Addison-Wesley, Reading.

173 Schwefel, H.P. (1995) *Evolution and Optimum Seeking*, John Wiley & Sons, Inc., New York.

174 Harel, G. and Akulin, V.M. (1999) *Phys. Rev. Lett.*, **82**, 1.

175 Omnes, R. (1994) *The Interpretation of Quantum Mechanics*, Princeton University Press, Princeton.

176 Schlosshauer, M. (2008) *Decoherence and the Quantum to Classical Transition*, Springer, New York.

177 Wilkie, J. and Brumer, P. (1997) *Phys. Rev. A*, **55**, 27.

178 Blum, K. (1981) *Density Matrix Theory and Applications*, Plenum Press, New York.

179 Jiang, X.P. and Brumer, P. (1993) *Chem. Phys. Lett.*, **208**, 179.

180 Pattanayak, A. and Brumer, P. (1997) *Phys. Rev. Lett.*, **79**, 4131.

181 Zurek, W.H. (1993) *Prog. Theor. Phys.*, **89**, 281.

182 Joos, E. (1996) in *Decoherence and the Appearance of the Classical World*, (eds D. Giulini, E. Joos, C. Kiefer, J. Kupsch, I-O. Stamatescu, and H.D. Zeh), Springer-Verlag, New York.

183 Weiss, E.A., Katz, G., Goldsmith, R.H., Wasielewski, M.R., Ratner, M.A., Kosloff, R., and Nitzan, A. (2006) *J. Chem. Phys.*, **124**, 074501.

184 May, V. and Kuhn, O. (2000) *Charge and Energy Transfer Dynamics in Molecular Systems*, Wiley-VCH Verlag GmbH, Berlin.

185 Breuer, H.P. and Petruccione, F. (2002) *The Theory of Open Quantum Systems*, Oxford University Press, Oxford.

186 Cohen-Tannoudji, C., Dupont-Roc, J., and Grynberg, G. (1992) *Atom-Photon Interactions*, John Wiley & Sons, Inc., New York.

187 Bellac, M.L. (2006) *Quantum Physics*, Cambridge Univ. Press, Cambridge. This book provides a basic introduction to the Lindblad form.

188 Nielsen, M.A. and Chuang, I.L. (2000) *Quantum Computation and Quantum Information*, Cambridge University Press, Cambridge.

189 Lidar, D.A., Bihary, Z., and Whaley, K.B. (2001) *Chem. Phys.*, **268**, 35.

190 Lindblad, G. (1976) *Commun. Math. Phys.*, **48**, 119.

191 Kraus, K. (1983) *States, Effects and Operations*, Springer, New York.

192 Alicki, R. and Fannes, M. (2001) *Quantum Dynamical Systems*, Oxford University Press, Oxford.

193 Lendi, K. (2003) *Lecture Notes in Physics*, **622**, 31.

194 Biswas, A., Shapiro, M., and Brumer, P. (2010) *J. Chem. Phys.*, **133**, 014103.

195 Huang, H. and Rossky, P.J. (2004) *J. Chem. Phys.*, **120**, 11380.

196 Prezhdo, O.V. and Rossky, P.J. (1998) *Phys. Rev. Lett.*, **81**, 5294.

197 Franco, I., Shapiro, M., and Brumer, P. (2008) *J. Chem. Phys.*, **128**, 244905.

198 Franco, I., Shapiro, M., and Brumer, P. (2008) *J. Chem. Phys.*, **128**, 244906.

199 Iwaki, L.K. and Dlott, D.L. (2000) *J. Phys. Chem. A*, **104**, 9109.

200 Engel, G.S., Calhoun, T.R., Reed, E.L., Ahn, T.K., Mancal, T., Cheng, Y.C., Blankenship, R.E., and Fleming, G.R. (2007) *Nature*, **446**, 782.

201 Collini, E., Wong, C.Y., Wilk, K.E., Curmi, P.M.G., Brumer, P., and Scholes, G.D. (201) *Nature*, **463**, 644.

202 Franco, I., Shapiro, M., and Brumer, P. (2007) *Phys. Rev. Lett.*, **99**, 126802.

203 Bardeen, C.J., Wang, Q., and Shank, C.V. (1995) *Phys. Rev. Lett.*, **75**, 3410.

204 Brixner, T., Damrauer, N.H., Niklaus, P., and Gerber, G. (2001) *Nature*, **414**, 57.

205 Elran, Y. and Brumer, P. (2004) *J. Chem. Phys.*, **121**, 2673.

206 Caldeira, A.O. and Leggett, A.J. (1983) *Physica (Amsterdam) A*, **121**, 587.

207 Unruh, W.G., and Zurek, W.H. (1992) *Phys. Rev. D*, **40**, 1071.

208 Hu, B.L., Paz, J.P., and Zhang, Y. (1989) *Phys. Rev. D*, **45**, 2843.

209 Hu, B.L., Paz, J.P., and Zhang, Y. (1993) *Phys. Rev. D*, **47**, 1576.

210 Jaffe, C. and Brumer, P. (1985) *J. Chem. Phys.*, **82**, 2330.

211 Gong, J. and Brumer, P. (1999) *Phys. Rev. E*, **60**, 1643.

212 Eckhardt, B., Hose, G., and Pollak, E. (1989) *Phys. Rev. A*, **39**, 3776.

213 Christoffel, K.M. and Brumer, P. (1985) *Phys. Rev. A*, **33**, 1309.

214 Han, H. and Brumer, P. (2005) *J. Chem. Phys.*, **122**, 144316.

215 Wu, L.A., Bharioke, A., and Brumer, P. (2008) *J. Chem. Phys.*, **129**, 041105.

216 Shapiro, M. and Brumer, P. (1995) *J. Chem. Phys.*, **103**, 487.

217 Wang, X., Levy, D.H., Rubin, M.B., and Speiser, S. (2000) *J. Phys. Chem. A*, **104**, 6558.

218 Caruso, F., Chin, A.W., Datta, A., Huelga, S.F., and Plenio, M.B. (2009) *J. Chem. Phys.*, **131**, 105106.

219 Rebentrost, P., Mohseni, M., Kassal, I., Lloyd, S., and Aspuru-Guzik, A. (2009) *New. J. Phys.*, **11**, 033003.

220 Chin, A., Datta, A., Caruso, F., Huelga, S., and Plenio, M.B. (2010) *New J. Phys.*, **12**, 065002.

221 Arimondo, E. and Orriols, G. (1976) *Lett. Nuovo Cim. Soc. Ital. Fis.*, **17**, 333.

222 Brandes, T. and Renzoni, F. (2000) *Phys. Rev. Lett.*,**85**,4148.

223 Emary, C. (2007) *Phys. Rev. B*, **76**, 245319.

224 Shuang, F. and Rabitz, H. (2004) *J. Chem. Phys.*, **121**, 9270.

225 Shuang, F., Rabitz, H., and Dykman, H. (2007) *Phys. Rev E*, **75**, 02110.

226 Kumar, P. and Malinovskaya, S. (2010) *J. Mod. Phys.*, **57**, 1243.

227 Taylor, J.R. (1972) *Scattering Theory*, John Wiley & Sons, Inc., New York.

228 Redfield, R.G. (1957) *IBM J. Res. Dev.*, **1**, 19.

229 Flores, S.C. and Batista, V.S. (2004) *J. Phys. Chem. B*, **108**, 6745.

230 Prokhorenko, V., Nagy, A.M., and Miller, R.J.D. (2005) *J. Chem. Phys.*, **122**, 184502.

231 Prokhorenko, V.J., Nagy, A.M., Waschuk, S.A., Brown, C.S., Birge, R.R., and Miller, R.J.D. (2006) *Science*, **313**, 1257.

232 van der Walle, P., Milder, M.T.W., Kuipers, L., and Herek, J.L. (2009) *Proc. Natl. Acad. Sci. USA*, **106**, 7714.

233 Katz, G., Ratner, M.A., and Kosloff, R. (2010) *New J. Phys.*, **12**, 015003.

234 Kosloff, R., Ratner, M., Katz, G., and Khasin, M. (2011) *Proc. Chem. of Solvay Conference*, in press.

235 Palma, G.M., Suominen, K.A., and Ekert, A.K. (1996) *Proc. R. Soc. A*, **452**, 567.

236 Gordon, G., Erez, N., and Kurizki, G. (2007) *J. Phys. B*, **40**, S75.

237 Gordon, G. (2009) *J. Phys. B*, **42**, 223001.

238 Han, H. and Brumer, P. (2005) *Chem. Phys. Lett.*, **406**, 237.

239 Shapiro, M. and Brumer, P. (1989) *J. Chem. Phys.*, **90**, 6179.

240 Allen, L. and Stroud, C.R. (1982) *Phys. Rep.*, **91**, 1. Where the case of $\phi = 0$ and $T_1 = T_2 = \infty$, is discussed.

241 Macomber, J.D. (1976) *The Dynamics of Spectroscopic Transitions*, John Wiley & Sons, Inc., New York.

242 Cao, J., Messina, M., and Wilson, K.R. (1997) *J. Chem. Phys.*, **106**, 5239.

243 Cao, J., Bardeen, C.J., and Wilson, K.R. (2000) *J. Chem. Phys.*, **113**, 1898.

244 Demirplak, M. and Rice, S.A. (2002) *J. Chem. Phys.*, **116**, 8028.

245 Demirplak, M. and Rice, S.A. (2006) *J. Chem. Phys.*, **125**, 194517.

246 Schirrmeister, D.H. and May, V. (1997) *Chem. Phys.*, **220**, 1.

247 Schirrmeister, D.H. and May, V. (1998) *Chem. Phys. Lett.*, **297**, 383.

248 Gong, J. and Rice, S.A. (2004) *J. Chem. Phys.*, **120**, 5117.

249 Batista, V.S. and Brumer, P. (2002) *Phys. Rev. Lett.*, **89**, 143201.

250 Guillar, V., Batista, V.S., and Miller, W.H. (2000) *J. Chem. Phys.*, **113**, 9510.

251 Campolieti, G. and Brumer, P. (1992) *J. Chem. Phys.*, **96**, 5969.

252 Sterling, M., Zadoyan, R., and Apkarian, V.A. (1996) *J. Chem. Phys.*, **104**, 6497.

253 Gershgoren, E., Vala, J., Kosloff, R., and Ruhman, S. (2001) *J. Phys. Chem. A*, **105**, 5081.

254 Ibrahim, H., Hejjas, M., Fushitani, M., and Schwentner, N. (2009) *J. Phys. Chem. A*, **113**, 7439.

255 Branderhorst, M.P.A., Londero, P., Wasylczyk, P., Brif, C., Kosut, R.L., Rabitz, H., and Walmsley, I.A. (2008) *Science*, pp. 320, 638.

256 Herzberg, G. (1950) *Molecular Spectra and Molecular Structure: I. Spectra of Diatomic Molecules*, 2nd edn, Van Nostrand, Princeton.

257 Bartram, D. and Ivanov, M. (2010) *Phys. Rev. A*, **81**, 043405.

258 Jiang, X.P., Shapiro, M., and Brumer, P. (1996) *J. Chem. Phys.*, **104**, 607.

259 Jiang, X.P. and Brumer, P. (1991) *Chem. Phys. Lett.*, **180**, 222.

260 Jiang, X.P. and Brumer, P. (1991) *J. Chem. Phys.*, **94**, 5833.

261 Shapiro, M. (1994) *J. Chem. Phys.*, **101**, 3849.

262 Camparo, J.C. and Lambropoulos, P. (1997) *Phys. Rev. A*, **55**, 552.

263 Karapanagioti, N.E., Xenakis, D., Charalambidis, D., and Fotakis, C. (1996) *J. Phys. B*, **29**, 3599.

264 Knight, P.L., Lauder, M.A., and Dalton, B.J. (1990) *Phys. Rep.*, **190**, 1.

265 Nakajima, T., Elk, M., Jian, Z., and Lambropoulos, P. (1994) *Phys. Rev. A*, **50**, R913.

266 Halfmann, T., Yatsenko, L.P., Shapiro, M., Shore, B.W., and Bergmann, K. (1998) *Phys. Rev. A*, **58**, R46.

267 Cavalieri, S., Eramo, R., Fini, L., Materazzi, M., Faucher, O., and Charalambidis, D. (1998) *Phys. Rev. A*, **57**, 2915.

268 Baldwin, K.G.H., Bott, M.D., Bachor, H.A., and Chapple, P.B. (2000) *J. Opt. B*, **2**, 470.

269 Chen, Z., Shapiro, M., and Brumer, P. (1995) *J. Chem. Phys.*, **102**, 5683.

270 Shnitman, A., Sofer, I., Golub, I., Yogev, A., Shapiro, M., Chen, Z., and Brumer, P. (1996) *Phys. Rev. Lett.*, **76**, 2886.

271 Chen, Z., Shapiro, M., and Brumer, P. (1993) *J. Chem. Phys.*, **98**, 8647.

272 Scully, M.O. and Zubairy, M.S. (1997) *Quantum Optics*, Cambridge University Press, Cambridge.

273 Schubert, M. and Wilhelmi, B. (1986) *Nonlinear Optics and Quantum Electronics*, John Wiley & Sons, Inc., New York.

274 Chen, Z., Brumer, P., and Shapiro, M. (1995) *Phys. Rev. A*, **52**, 2225.

275 Brumer, P., Chen, Z., and Shapiro, M. (1994) *Isr. J. Chem.*, **34**, 137.

276 Chen, Z., Brumer, P., and Shapiro, M. (1992) *Chem. Phys. Lett.*, **198**, 498.

277 Zhang, Q., Keil, M., and Shapiro, M. (2003) *J. Opt. Soc. Am. B*, **20**, 2255.

278 Pratt, S.T. (1996) *J. Chem. Phys.*, **104**, 5776.

279 Wang, F., Chen, C., and Elliott, D.S. (1996) *Phys. Rev. Lett.*, **77**, 2416.

280 Luc-Koenig, E., Aymar, M., Millet, M., Lecomte, J.M., and Lyras, A. (2000) *Eur. Phys. J. D*, **10**, 205.
281 Georgiades, N.P., Polzik, E.S., and Kimble, H.J. (1996) *Opt. Lett.*, **21**, 1688.
282 Dudovich, N., Dayan, B., Faeder, S.M.G., and Silberberg, Y. (2001) *Phys. Rev. Lett.*, **86**, 47.
283 Meshulach, D. and Silberberg, Y. (1999) *Phys. Rev. A*, **60**, 1287.
284 Dayan, B., Pe'er, A., Friesem, A., and Silberberg, Y. (2004) *Phys. Rev. Lett.*, **93**, 023005.
285 Boyd, R.W. (1992) *Nonlinear Optics*, Academic Press, Boston.
286 Dudovich, N., Oron, D., and Silberberg, Y. (2002) *Nature*, **418**, 512.
287 Xu, X.G., Konorov, S.O., Hepburn, J.W., and Milner, V. (2008) *Opt. Lett.*, **33**, 1177.
288 Konorov, S.O., Xu, X.G., Hepburn, J.W., and Milner, V. (2009) *J. Chem. Phys.*, **130**, 234505.
289 Li, H., Harris, D.A., Xu, B., Wrzesinski, P.J., Lozovoy, V.V., and Dantus, M. (2008) *Opt. Express*, **16**, 5499.
290 Pestov, D., X, W., Murawski, R.K., Ariunbold, G.O., Sautenkov, V.A., and Sokolov, A.V. (2008) *J. Opt. Soc. Am. B*, **25**, 768.
291 Pegoraro, A.F., Stolow, A., Ridsdale, A., Moffatt, D.J., and Pezacki, J.P. (2009) *Opt. Express*, **17**, 2984.
292 Pastirk, I., Zhu, X., Lozovoy, V.V., and Dantus, M. (2007) *Appl. Opt.*, **46**, 4041.
293 Bonen, K.D. and Kresin, V.V. (1997) *Electric Dipole Polarizabilities of Atoms, Molecules and Clusters*, World Scientific, Singapore.
294 McCullough, E., Shapiro, M., and Brumer, P. (2000) *Phys. Rev. A*, **61**, 041801.
295 McCullough, E. (1997) M.Sc. Dissertation, University of Toronto.
296 Hau, L.V., Harris, S.E., Dutton, Z., and Behroozi, C.H. (1999) *Nature*, **397**, 594.
297 Scully, M.O. (1991) *Phys. Rev. Lett.*, **67**, 1855.
298 Fiss, J.A., Khachatrian, A., Zhu, L., Gordon, R.J., and Seideman, T. (1999) *Faraday Discuss. Chem. Soc.*, **113**, 61.
299 Seideman, T. (1999) *J. Chem. Phys.*, **111**, 9168.
300 Gordon, R.J., Zhu, L., and Seideman, T. (2001) *J. Phys. Chem. A*, **105**, 4387.
301 Levine, R.D. (1969) *Quantum Mechanics of Molecular Rate Processes*, Clarendon Press, Oxford.
302 Fano, U. (1961) *Phys. Rev.*, **124**, 1866.
303 Goldstein, H. (1980) *Classical Mechanics*, 2nd edn, Addison-Wesley, Reading.
304 McCauley, J.C. (1997) *Classical Mechanics*, Cambridge University Press, Cambridge.
305 Brumer, P. (1981) *Adv. Chem. Phys.*, **47**, 201.
306 Pattanayak, A. and Brumer, P. (1996) *Phys. Rev. Lett.*, **77**, 59.
307 Blümel, R., Fishman, S., and Smilansky, U. (1986) *J. Chem. Phys.*, **84**, 2604.
308 Casati, G. and Chirikov, B. (1995) *Quantum Chaos: Between Order and Disorder*, Cambridge University Press, Cambridge.
309 Gong, J. and Brumer, P. (2001) *Phys. Rev. Lett.*, **86**, 1741.
310 Blümel, R. and Reinhardt, W.P. (1997) *Chaos in Atomic Physics*, Cambridge University Press, Cambridge.
311 Gong, J. and Brumer, P. (2001) *J. Chem. Phys.*, **115**, 3590.
312 Haake, F. (1992) *Quantum Signatures of Chaos*, Springer-Verlag, Berlin.
313 Krause, J.L., Shapiro, M., and Brumer, P. (1990) *J. Chem. Phys.*, **92**, 1126.
314 Shapiro, M. and Brumer, P. (1996) *Phys. Rev. Lett.*, **77**, 2574.
315 Holmes, D., Shapiro, M., and Brumer, P. (1996) *J. Chem. Phys.*, **105**, 9162.
316 Abrashkevich, A.G., Shapiro, M., and Brumer, P. (1998) *Phys. Rev. Lett.*, **81**, 3789.
317 Abrashkevich, A.G., Shapiro, M., and Brumer, P. (1999) *Phys. Rev. Lett.*, **82**, 3002.
318 Brumer, P., Abrashkevich, A.G., and Shapiro, M. (1999) *Faraday Discuss. Chem. Soc.*, **113**, 291.
319 Brumer, P., Bergmann, K., and Shapiro, M. (2000) *J. Chem. Phys.*, **113**, 2053.
320 Abrashkevich, A.G., Shapiro, M., and Brumer, P. (2001) Coherent control of atom-diatom reactive scattering: Isotopic variants of $h + h_2$ in three dimensions. *Chem. Phys.*, **267**, 81.
321 Bergmann, K., Theuer, H., and Shore, B.W. (1998) *Rev. Mod. Phys.*, **70**, 1003.

322 Unanyan, R.G., Fleischhauer, M., Bergmann, K., and Shore, B.W. (1998) *Opt. Commun.*, **155**, 144.

323 Gong, J., Shapiro, M., and Brumer, P. (2003) *J. Chem. Phys.*, **118**, 2626.

324 Gong, J.B. (2001) Ph.D. Dissertation, University of Toronto.

325 Schatz, G.C. and Kuppermann, A. (1976) *J. Chem. Phys.*, **65**, 4642.

326 Edmonds, A.R. (1960) *Angular Momentum in Quantum Mechanics*, Princeton University Press, Princeton.

327 Pack, R.T. and Parker, G.A. (1987) *J. Chem. Phys.*, **87**, 3888.

328 Tamir, M. and Shapiro, M. (1975) *Chem. Phys. Lett.*, **31**, 166.

329 Miller, W.H. (1969) *J. Chem. Phys.*, **50**, 407.

330 Krems, R.V. (2008) *Phys. Chem. Chem. Phys.*, **10**, 4079.

331 Carr, L.D., DeMille, D., Krems, R.V., and Ye, J. (2009) *New J. Phys.*, **11**, 055409.

332 Spiegelhalder, F., Trenkwalder, A., Naik, D., Kerner, G., Wille, E., Hendl, G., Schreck, F., and Grimm, R. (2010-01-28) E-print cond-mat 1001.5253.

333 Knoop, S., Ferlaino, F., Mark, M., Berninger, M., Schöbel, H., Nägerl, H., and Grimm, R. (2009) *Nat. Phys.*, **5**, 227.

334 Knoop, S., Ferlaino, F., Berninger, M., Mark, M., Nägerl, H., Grimm, R., D'Incao, J.P., and Esry, B. (2009) *Phys. Rev. Lett.*, **104**, 053201.

335 Ospelkaus, S., Ni, K.K., Wang, D., deMiranda, M.H.G., Neyenhuis, B., Quemener, G., Julienne, P.S., Bohn, J.L., Jin, D.S., and Ye, J. (2010) *Science*, **327**, 853.

336 Ni, K.K., Ospelkaus, S., Wang, D., Quemener, G., Neyenhuis, B., de Miranda, M.H.V., Bohn, J.L., Ye, J., and Jin, D.S. (2010) *Nature*, **464**, 1324.

337 Siska, P.E. (1993) *Rev. Mod. Phys.*, **65**, 337.

338 Arango, C.A., Shapiro, M., and Brumer, P. (2006) *Phys. Rev. Lett.*, **97**, 193202.

339 Arango, C.A., Shapiro, M., and Brumer, P. (2006) *J. Chem. Phys.*, **125**, 094315.

340 Mori, M., Watanabe, T., and Katsuura, K. (1964) *J. Phys. Soc. Jpn.*, **19**, 380.

341 Heinz, M., Vewinger, F., Schneider, U., Yatsenko, L., and Bergmann, K. (2006) *Opt. Commun.*, **264**, 248.

342 Knappe, S., Wynands, R., Kitching, J., Robinson, H.G., and Hollberg, L. (2001) *J. Opt. Soc. Am. B*, **18**, 1545.

343 Westphal, P., Horn, A., Koch, S., Schmand, J., and Andra, J.J. (1996) *Phys. Rev. A*, **54**, 4577.

344 Westphal, P., Horn, A., Koch, S., Schmand, J., and Andra, J.J. (1976) *Lett. Nuovo Cim.*, **17**, 333.

345 Herrera, F. (2008) *Phys. Rev. A*, **78**, 054702.

346 Spanner, M. and Brumer, P. (2007) *Phys. Rev. A*, **76**, 013408.

347 Zhang, J.Z.H. and Miller, W.H. (1989) *J. Chem. Phys.*, **91**, 1528.

348 Halavee, U. and Shapiro, M. (1976) *J. Chem. Phys.*, **64**, 2826.

349 Wong, J.K.C. and Brumer, P. (1979) *Chem. Phys. Lett.*, **68**, 517.

350 Spanner, M. and Brumer, P. (2008) *Phys. Rev. A*, **78**, 033840.

351 Gong, J.B. and Brumer, P. (2010) *J. Chem. Phys.*, **132**, 054306.

352 Smith, E.T., Dhirani, A.A., Koborowski, D.A., Rubenstein, R.A., Roberts, T.D., Yao, H., and Pritchard, D.E. (1998) *Phys. Rev. Lett.*, **81**, 1996.

353 Phillips, W.D. (1998) *Rev. Mod. Phys.*, **70**, 721.

354 Chu, S. and Wieman, C. (1989) *J. Opt. Soc. Am. B*, **6(11)**, 2020.
Special issue on laser cooling and trapping.

355 Weitz, M., Young, B.C., and Chu, S. (1994) *Phys. Rev. Lett.*, **94**, 2563.

356 Bethlem, H.L., van Roij, A.J., Jongma, R.T., and Meijer, G. (2002) *Phys. Rev. Lett.*, **88**, 133003.

357 Raizen, M.G. (2009) *Science*, **324**, 1403.

358 Frishman, E., Shapiro, M., and Brumer, P. (1999) *J. Chem. Phys.*, **110**, 9.

359 Shapiro, M. and Brumer, P. (2002) *Phys. Rev. A*, **66**, 052308.

360 Frishman, E., Shapiro, M., and Brumer, P. (1999) *J. Phys. Chem. A*, **103**, 10333.

361 Zeman, V., Shapiro, M., and Brumer, P. (2004) *Phys. Rev. Lett.*, **92**, 133204.

362 Regal, C.A., Ticknor, C., Bohn, J.L., and Jin, D.S. (2003) *Nature*, **424**, 47.

363 Jackson, J.D. (1962) *Classical Electrodynamics*, John Wiley & Sons, Inc., New York.

364 Loudon, R. (1983) *The Quantum Theory of Light*, 2nd edn, Clarendon Press, Oxford.

365 Schrödinger, E. (1926) *Naturwissenschaften*, **14**, 664.

366 Glauber, R.J. (1963) *Phys. Rev.*, **131**, 2766.

367 Greenstein, G. and Zajonc, A.G. (2005) *The Quantum Challenge: Modern Research on the Foundations of Quantum Mechanics*, 2nd edn, Jones and Bartlett, Sudbury.

368 Mandel, L. and Wolf, E. (1965) *Rev. Mod. Phys.*, **37**, 231.

369 Gong, J., Zeman, V., and Brumer, P. (2002) *Phys. Rev. Lett.*, **89**, 109301.

370 Bruss, D. and Leuchs, G. (2007) *Lectures on Quantum Information*, Wiley-VCH Verlag GmbH, Weinheim.

371 Buchleitner, A., Viviescas, C., and M. Tiersch, E. (2010) *Entanglement and Decoherence*, Springer, Berlin.

372 Horodecki, R., Horodecki, P., Horodecki, M., and Horodecki, K. (2009) *Rev. Mod. Phys.*, **81**, 865.

373 Zhao, Z., Chen, Y.A., Zhang, A.N., Yang, T., Briegel, H.J., and Pan, J.W. (2004) *Nature*, **430**, 54.

374 Häffner, H., Hansel, W., Roos, C.F., Benhelm, J., and Chekalkar, D. *et al.* (2005) *Nature*, **438**, 643.

375 Blinov, B.B., Moehring, D.L., Duan, L.M., and Monroe, C. (2004) *Nature*, **428**, 153.

376 Luo, L., Hayes, D., Manning, T.A., Matsukevich, D.N., Maunz, P., Olmschenk, S., Sterk, J.D., and Monroe, C. (2009) *Fortschr. Phys.*, **57**, 1133.

377 Perina, J. (1984) *Quantum Statistics of Linear and Nonlinear Phenomena*, D. Riedl, Dodrecht, Boston.

378 Raimond, J.M., Brune, M., and Haroche, S. (2001) *Rev. Mod. Phys.*, **73**, 565.

379 Opatrný, T. and Kurizki, G. (2001) *Phys. Rev. Lett.*, **86**, 3180.

380 Solano, E., Agarwal, G.S., and Walther, H. (2003) *Phys. Rev. Lett.*, **90**, 027903.

381 Bayer, M., Hawrylak, P., Hinzer, K., Fafard, S., Wasilewski, M.K.Z.R., Stern, O., and Forchel, A. (2003) *Science*, **291**, 451.

382 Král, P., Thanopulos, I., and Shapiro, M. (2005) *Phys. Rev. A*, **72**, 020303(R).

383 Král, P. and Tománek, D. (1999) *Phys. Rev. Lett.*, **82**, 5373.

384 Král, P. and Sipe, J. (2000) *Phys. Rev. B*, **61**, 5381.

385 Yurke, B. and Stoler, D. (1986) *Phys. Rev. Lett.*, **57**, 13.

386 Schleich, W., Peringo, M., and Kien, F.L. (1991) *Phys. Rev. A*, **44**, 2172.

387 Schrödinger, E. (1935) *Naturwissenschaften*, **23**, 807,823,844. English translation: (1980) *Proc. Am. Philos. Soc.*, **124**, 323.

388 Ni, K.K., Ospelkaus, S., de Miranda, M.H.G., Pe'er, A., Neyenhuis, B., Zirbel, J.J., Kotochigova, S., Julienne, P.S., Jin, D.S., and Ye, J. (2008) *Science*, **322**, 231.

389 Spanner, M., Franco, I., and Brumer, P. (2009) *Phys. Rev. A*, **80**, 053402.

390 Král, P. (1990) *Phys. Rev. A*, **42**, 4177.

391 Abragam, A. (1961) *The Principles of Nuclear Magnetism*, Clarendon Press, Oxford.

392 Allen, L. and Eberly, J.H. (1975) *Optical Resonance and Two-Level Atoms*, John Wiley & Sons, Inc., New York.

393 Grischkowski, D.G. (1970) *Phys. Rev. Lett.*, **24**, 866.

394 Grischkowski, D.G. and Armstrong, J.A. (1972) *Phys. Rev. A*, **6**, 1566.

395 D. G. Grischkowski, E.C. and Armstrong, J.A. (1973) *Phys. Rev. Lett.*, **31**, 422.

396 Grischkowski, D.G. (1973) *Phys. Rev. A*, **7**, 2096.

397 Takatsuji, M. (1971) *Phys. Rev. A*, **4**, 808.

398 Brewer, R.G. and Hahn, E.L. (1975) *Phys. Rev. A*, **11**, 1641.

399 Cook, R.J. and Shore, B.W. (1979) *Phys. Rev. A*, **20**, 539.

400 Kuzmin, M.V. and Sazonov, V.N. (1980) *Zh. Fiz.*, **79**, 1759.

401 Hioe, F.T. and Eberly, J.H. (1981) *Phys. Rev. Lett.*, **12**, 838.

402 Oreg, J., Hioe, F.T., and Eberly, J.H. (1984) *Phys. Rev. A*, **29**, 690.

403 Liedenbaum, C., Stolte, S., and Reuss, J. (1989) *Phys. Rep.*, **178**, 1.

404 Broers, B., van Linden van den Heuvell, H.B., and Noordam, L.D. (1992) *Phys. Rev. Lett.*, **69**, 2062.

405 Carroll, C.E. and Hioe, F.T. (1992) *Phys. Rev. Lett.*, **68**, 3523.

406 Carroll, C.E. and Hioe, F.T. (1995) *Phys. Lett. A*, **199**, 145.

407 Gaubatz, U., Rudecki, P., Schiemann, S., and Bergmann, K. (1990) *J. Chem. Phys.*, **92**, 5363.

408 Coulston, G. and Bergmann, K. (1992) *J. Chem. Phys.*, **96**, 3467.

409 Shore, B.W., Martin, J., Fewell, M.P., and Bergmann, K. (1995) *Phys. Rev. A*, **52**, 566.

410 Martin, J., Shore, B.W., and Bergmann, K. (1995) *Phys. Rev. A*, **52**, 583.

411 Theuer, H., Unanyan, R.G., Habschied, C., Klein, K., and Bergmann, K. (1999) *Opt. Express*, **4**, 77.

412 Harris, S.E., Field, J.E., , and Imamoglu, A. (1990) *Phys. Rev. Lett.*, **64**, 1107.

413 Boller, K.J., Imamoglu, A., and Harris, S.E. (1991) *Phys. Rev. Lett.*, **66**, 2593.

414 Field, J.E., Hahn, K.H., and Harris, S.E. (1991) *Phys. Rev. Lett.*, **67**, 3062.

415 Harris, S.E. (1993) *Phys. Rev. Lett.*, **70**, 552.

416 Ichimura, K., Yamamoto, K., and Gemma, N. (1998) *Phys. Rev. A*, **58**, 4116.

417 Takeoka, M., Fujishima, D., and Kannari, F. (2001) *Jpn. J. Appl. Phys. 1*, **40**, 137.

418 Harris, S.E. (1989) *Phys. Rev. Lett.*, **62**, 1033.

419 Scully, M.O., Zhu, S.Y., and Gavrielides, A. (1989) *Phys. Rev. Lett.*, **62**, 2813.

420 Kocharovskaya, O. (1992) *Phys. Rep.*, **219**, 175.

421 Cohen, J.L. and Berman, P.R. (1997) *Phys. Rev. A*, **55**, 3900.

422 deJong, F.B., Spreeuw, R.J.C., and van Linden van den Heuvell, H.B. (1997) *Phys. Rev. A*, **55**, 3918.

423 Feynman, R.P., Vernon, F.L.J., and Hellwarth, R.W. (1957) *J. App. Phys.*, **28**, 49.

424 Vardi, A. and Shapiro, M. (1998) *Phys. Rev. A*, **58**, 1352.

425 Vardi, A. and Shapiro, M. (2001) *Comments At. Mol. Phys.*, **2**, 233.

426 Chen, Z., Brumer, P., and Shapiro, M. (1993) *J. Chem. Phys.*, **98**, 6843.

427 Kobrak, M.N. and Rice, S.A. (1998) *J. Chem. Phys.*, **109**, 1.

428 Kobrak, M.N. and Rice, S.A. (1998) *Phys. Rev. A*, **57**, 1158.

429 Kuklinski, J.R., Gaubatz, U., Hioe, F.T., and Bergmann, K. (1989) *Phys. Rev. A*, **40**, 6741.

430 Shore, B.W. (1995) *Contemp. Phys.*, **36**, 15.

431 Vitanov, N.V., Fleischhauer, M., Shore, B.W., and Bergmann, K. (2001) *Adv. Atom. Mol. Opt. Phys.*, **46**, 55.

432 Vitanov, N.V. and Stenholm, S. (1997) *Opt. Commun.*, **135**, 394.

433 Guérin, S. and Jauslin, H.R. (1998) *Eur. Phys. J. D*, **2**, 99.

434 Yatsenko, L.P., Guerin, S., Halfmann, T., Shore, B.W., and Bergmann, K. (1998) *Phys. Rev. A*, **58**, 4683.

435 Band, Y. and Julienne, P.S. (1991) *J. Chem. Phys.*, **94**, 5291.

436 Band, Y. and Julienne, P.S. (1992) *J. Chem. Phys.*, **96**, 3339.

437 Glushko, B. and Kryzhanovsky, B. (1992) *Phys. Rev. A*, **46**, 2823.

438 Fleischhauer, M. and Manka, A.S. (1996) *Phys. Rev. A*, **54**, 794.

439 Vitanov, N.V. and Stenholm, S. (1997) *Phys. Rev. A*, **56**, 1463.

440 Vitanov, N.V. and Stenholm, S. (1999) *Phys. Rev. A*, **60**, 3820.

441 Shore, B.W., Bergmann, K., Oreg, J., and Rosenwaks, S. (1991) *Phys. Rev. A*, **44**, 7442.

442 Marte, P., Zoller, P., and Hall, J.L. (1991) *Phys. Rev. A*, **44**, 4118.

443 Malinovsky, V.S. and Tannor, D.J. (1997) *Phys. Rev. A*, **56**, 4929.

444 Theuer, H. and Bergmann, K. (1998) *Eur. Phys. J. D*, **2**, 279.

445 Law, C. and Eberly, J. (1998) *Opt. Express*, **2**, 368.

446 Nakajima, T. (1999) *Phys. Rev. A*, **59**, 559.

447 Thanopulos, I. and Shapiro, M. (2006) *Phys. Rev. A*, **74**, 031401.

448 Král, P., Thanopulos, I., and Shapiro, M. (2007) *Rev. Mod. Phys.*, **79**, 53.

449 Kyoseva, E.S. and Vitanov, N.V. (2006) *Phys. Rev. A*, **73**, 023420.

450 Shapiro, E.A., Milner, V., and Shapiro, M. (2009) *Phys. Rev. A*, **79**, 023422.

451 Esslinger, T., Sander, F., Weidemüller, M., Hammerich, A., and Hänsch, T.W. (1996) *Phys. Rev. Lett.*, **76**, 2432.

452 Kulin, S., Saubamea, B., Peik, E., Lawall, J., Leduc, M., and Cohen-Tannoudji, C. (1997) *Phys. Rev. Lett.*, **78**, 4185.

453 Kobrak, M.N. and Rice, S.A. (1998) *Phys. Rev. A*, **57**, 2885.

454 Unanyan, R., Vitanov, N., and Stenholm, S. (1998) *Phys. Rev. A*, **57**, 462.

455 Paspalakis, E., Protopapas, M., and Knight, P.L. (1997) *Opt. Commun.*, **142**, 34.

456 Vitanov, N.V. and Stenholm, S. (1997) *Phys. Rev. A*, **56**, 741.

457 Yatsenko, L.P., Unanyan, R.G., Halfmann, T., and Shore, B.W. (1997) *Opt. Commun.*, **135**, 406.

458 Carroll, C.E. and Hioe, F.T. (1996) *Phys. Rev. A*, **54**, 5147.

459 Unanyan, R., Vitanov, N.V., Shore, B.W., and Bergmann, K. (2000) *Phys. Rev. A*, **61**, 043408.

460 Yatsenko, L.P., Halfmann, T., Shore, B., and Bergmann, K. (1999) *Phys. Rev. A*, **59**, 2926.

461 Kylstra, N.J., Paspalakis, E., and Knight, P.L. (1998) *J. Phys. B*, **31**, L719.

462 Guérin, S., Yatsenko, L.P., Halfmann, T., Shore, B.W., and Bergmann, K. (1998) *Phys. Rev. A*, **58**, 4691.

463 Yatsenko, L.P., Shore, B.W., Halfmann, T., Bergmann, K., and Vardi, A. (1999) *Phys. Rev. A*, **60**, R4237.

464 Rickes, T., Yatsenko, L.P., Steurwald, S., Halfmann, T., Shore, B.W., Vitanov, N.V., and Bergmann, K. (2001) *J. Chem. Phys.*, **113**, 534.

465 Garraway, B.M. and Suominen, K.A. (1998) *Phys. Rev. Lett.*, **80**, 932.

466 Kallush, S. and Band, Y.B. (2000) *Phys. Rev. A*, **61**, 041401.

467 Solá, I.R., Santamaríblubbia, J., and Malinkovsky, V.S. (2000) *Phys. Rev. A*, **61**, 043413.

468 Rodriguez, M., Garraway, B.M., and Suominen, K.A. (2000) *Phys. Rev. A*, **62**, 53413.

469 Vardi, A., Abrashkevich, D., Frishman, E., and Shapiro, M. (1997) *J. Chem. Phys.*, **107**, 6166.

470 Vardi, A., Shapiro, M., and Bergmann, K. (1999) *Opt. Express*, **4**, 91.

471 Javanainen, J. and Mackie, M. (1998) *Phys. Rev. A*, **58**, R789.

472 Mackie, M., Kowalski, R., and Javanainen, J. (2000) *Phys. Rev. Lett.*, **84**, 3803.

473 Winkler, K., Thalhammer, G., Theis, M., Ritsch, H., Grimm, R., and Denschlag, J.H. (2005) *Phys. Rev. Lett.*, **95**, 063202.

474 Dumke, R., Weinstein, J.D., Johanning, M., Jones, K., and Lett, P. (2005) *Phys. Rev. A*, **72**, 041801.

475 Beumee, J.G.B. and Rabitz, H. (1990) *J. Math. Phys.*, **31**, 1253.

476 Aubanel, E.E. and Bandrauk, A.D. (1997) *Can. J. Chem.*, **107**, 1441.

477 de Araujo, L.E.E. and Walmsley, I.A. (1999) *J. Phys. Chem. A*, **103**, 10409.

478 Huang, G.M., Tarn, T.J., and Clark, J.W. (1983) *J. Math. Phys.*, **24**, 2608.

479 Grischkowski, D.G., Loy, M.M.T., and Liao, P.F. (1975) *Phys. Rev. A*, **12**, 2514.

480 Kuhn, A., Coulston, G.W., He, G.Z., Schiemann, S., Bergmann, K., and Warren, W.S. (1992) *J. Chem. Phys.*, **96**, 4215.

481 Unanyan, R., Fleischhauer, M., Shore, B., and Bergmann, K. (1998) *Opt. Commun.*, **155**, 144.

482 Král, P., Fiurášek, J., and Shapiro, M. (2001) *Phys. Rev. A*, **64**, 023414.

483 Král, P., Amitay, Z., and Shapiro, M. (2002) *Phys. Rev. Lett.*, **89**, 063002.

484 Shore, B.W. (1990) *The Theory of Coherent Atomic Excitation*, vol. 1, John Wiley & Sons, Inc., New York.

485 Thanopulos, I., Král, P., and Shapiro, M. (2004) *Phys. Rev. Lett.*, **92**, 113003.

486 Kis, Z., Karpati, A., Shore, B.W., and Vitanov, N.V. (2004) *Phys. Rev. A*, **70**, 053405.

487 Morris, J.R. and Shore, B.W. (1983) *Phys. Rev. A*, **27**, 906.

488 Bunker, P. and Jensen, P. (1998) *Molecular Symmetry and Spectroscopy*, NRC Research Press, Ottawa.

489 Boldyrev, A.T. and von R. Schleyer, P. (1991) *J. Am. Chem. Soc.*, **113**, 9045.

490 Frisch, M.J. (2001) *et al. Gaussian 98. Revision A.11*, Gaussian, Inc., Pittsburgh, PA.

491 Mayer, R. and Günthard, H.H. (1968) *J. Chem. Phys.*, **49**, 1510.

492 Mayer, R. and Günthard, H.H. (1970) *J. Chem. Phys.*, **50**, 353.

493 Hartchcock, M.A. and Laane, J. (1985) *J. Phys. Chem*, **89**, 4231.

494 Berry, M.V. (1984) *Proc. R. Soc. A*, **392**, 45.

495 Mead, C.A. (1992) *Rev. Mod. Phys.*, **64**, 51.

496 Yarkony, D.R. (1996) *Rev. Mod. Phys.*, **68**, 985.

497 Yarkony, D.R. (2001) *J. Phys. Chem. A*, **105**, 6277.
498 Kendrick, B. (1997) *Phys. Rev. Lett.*, **79**, 2431.
499 Král, P. and Shapiro, M. (2004) *Chem. Phys. Lett.*, **393**, 488.
500 Duan, L.M., Lukin, M.D., Cirac, J.I., and Zoller, P. (2001) *Nature*, **414**, 413.
501 Weinstein, Y.S., Pravia, M.A., Fortunato, E.M., Lloyd, S., and Cory, D.G. (2001) *Phys. Rev. Lett.*, **86**, 1889.
502 Vandersypen, L.M.K., Steffen, M., Breyta, G., Yannoni, C., Sherwood, M., and Chuang, I. (2001) *Nature*, **414**, 883.
503 Kaszlikowski, D., Gnacinski, P., Zukowski, M., Miklaszewski, W., and Zeilinger, A. (2001) *Phys. Rev. Lett.*, **85**, 4418.
504 Král, P. and Shapiro, M. (2002) *Phys. Rev. A*, **65**, 043413.
505 Braunstein, S.L. and Kimble, H.J. (2000) *Phys. Rev. A*, **61**, 042302.
506 Shapiro, E.A., Milner, V., Menzel-Jones, C., and Shapiro, M. (2007) *Phys. Rev. Lett.*, **99**, 033002.
507 Zhdanovich, S., Shapiro, E.A., Shapiro, M., Hepburn, J.W., and Milner, V. (2008) *Phys. Rev. Lett.*, **100**, 103004.
508 Zhdanovich, S., Shapiro, E.A., Hepburn, J.H., Shapiro, M., and Milner, V. (2009) *Phys. Rev. A*, **80**, 063405.
509 Stowe, M.C., Cruz, F.C., Marian, A., and Ye, J. (2006) *Phys. Rev. Lett.*, **96**, 153001.
510 Pe'er, A., Shapiro, E.A., Stowe, M.C., Shapiro, M., and Ye, J. (2007) *Phys. Rev. Lett.*, **98**, 113004.
511 Ramsey, N.F. (1950) *Phys. Rev.*, **78**, 695.
512 Ralchenko, Y., Kramida, A., and Reader, J. (2010) *NIST Atomic Spectra Database*, http://www.nist.gov/pml/data/asd.cfm.
513 Wollenhaupt, M., Prakelt, A., Sarpe-Tudoran, C., Liese, D., Bayer, T., and Baumert, T. (2006) *Phys. Rev. A*, **73**, 063409.
514 Dykhne, A.M. and Yudin, G.L. (1978) *Sov. Phys. Usp.*, **21**, 549.
515 Dhar, L., Rogers, J.A., and Nelson, K.A. (1994) *Chem. Rev.*, **94**, 157.
516 Kawashimaa, H., Wefers, M.M., and Nelson, K.A. (1996) *Physica B: Condens. Matter*, **219**, 734.
517 Marian, A., Stowe, M.C., Lawall, J.R., Felinto, D., and Ye, J. (2004) *Science*, **36**, 2063.
518 Felinto, D., Bosco, C.A.C., Acioli, L.H., and Vian, S.S. (2003) *Opt. Commun.*, **215**, 69.
519 Soares, A.A. and de Araujo, L.E.E. (2007) *Phys. Rev. A*, **76**, 043818.
520 Böhmer, K., Halfmann, T., Yatsenko, L.P., Shore, B.W., and Bergmann, K. (2001) *Phys. Rev. A*, **64**, 023404.
521 Ye, J. and Cundiff, S.T. (eds) (2005) *Femtosecond Optical Frequency Comb: Principle, Operation and Applications*, Springer, New York.
522 Zamith, S., Bouchene, M.A., Sokellb, E., Nicolec, C., Blanchet, V., and Girard, B. (2000) *Eur. Phys. J. D*, **12**, 255.
523 Harris, S.E., Field, J.E., and Immamoglu, A. (1992) *Phys. Rev. Lett.*, **64**, 1107.
524 Scully, M. (1992) *Phys. Rep.*, **219**, 191.
525 Harris, S.E. (1997) *Phys. Today*, **50**(7), 36.
526 Lukin, M.D., Hemmer, P., and Scully, M.O. (2000) *Adv. Atm. Mol. Opt. Phys.*, **42**, 347.
527 Matsko, A.B., Kocharovskaya, O., Rostovtsev, Y., Welch, G.R., and Scully, M.O. (2001) *Adv. Atm. Mol. Opt. Phys.*, **46**, 191.
528 Takeoka, M., Fujishima, D., and Kannari, F. (2001) *Japan. J. Appl. Phys. 1*, **40**, 137.
529 Lukin, M.D. (2003) *Rev. Mod. Phys.*, **75**, 457.
530 Fleischhauer, M., Imamoglu, A., and Marangos, J.P. (2005) *Rev. Mod. Phys.*, **77**, 633.
531 Ruseckas, J., Juzeliūnas, G., Öhberg, P., and Fleischhauer, M. (2005) *Phys. Rev. Lett.*, **95**, 010404.
532 Autler, S.H. and Townes, C.H. (1955) *Phys. Rev.*, **100**, 703.
533 Shapiro, M. (1998) *J. Phys. Chem.*, **102**, 9570.
534 Alzetta, G., Gozzini, A., Moi, L., and Orriols, G. (1976) *Nuovo Cim. Soc. Ital. Fis. B*, **36**, 5.
535 Kocharovskaya, O.A. and Khanin, Y.I. (1988) *JETP Lett.*, **48**, 630.
536 Gornyi, M.B., Matisov, B.G., and Rozhdestvenskii, Y.V. (1989) *Sov. Phys. JETP*, **68**, 728.

537 Gaubatz, U., Rudecki, P., Schiemann, S., and Bergmann, K. (1990) *J. Chem. Phys.*, **92**, 5363.

538 Weisskopf, V. and Wigner, E.P. (1930) *Zeitschr. Phys.*, **63**, 54.

539 Shapiro, M. and Bersohn, R. (1982) *Ann. Rev. Phys. Chem.*, **33**, 409.

540 Shapiro, M. (2007) *Phys. Rev. A*, **75**, 013424.

541 Shapiro, M. (2006) *Isr. J. Chem.*, **47**, 233.

542 Hakuta, K., Marmet, L., and Stoicheff, B.P. (1991) *Phys. Rev. Lett.*, **66**, 596.

543 Kasapi, A. (1996) *Phys. Rev. Lett.*, **77**, 1035.

544 Ham, B.S., Shahriar, M.S., and Hemmer, P.R. (1997) *Opt. Lett.*, **22**, 1138.

545 Zhao, Y., Wu, C., Ham, B.S., Kim, M.K., and Awad, E. (1997) *Phys. Rev. Lett.*, **79**, 641.

546 Kash, M.M., Sautenkov, V.A., Zibrov, A.S., Hollberg, L., Welch, G.R., Lukin, M.D., Rostovtsev, Y., Fry, E.S., and Scully, M.O. (1999) *Phys. Rev. Lett.*, **82**, 5229.

547 Bigelow, M.S., Lepeshkin, N.N., and Boyd, R.W. (2003) *Science*, **301**, 200.

548 Gauthier, D. and Boyd, R.W. http://www.phy.duke.edu/research/photon/

549 Liu, C., Dutton, Z., Behroozi, C., and Hau, L. (2001) *Nature*, **409**, 490.

550 Phillips, D.F., Fleischhauer, A., Mair, A., Walsworth, R.L., and Lukin, M.D. (2001) *Phys. Rev. Lett.*, **86**, 783.

551 Ginsberg, N.S., Garner, S.R., and Hau, L.V. (2007) *Nature*, **445**, 623.

552 Wollenhaupt, M. and Baumert, T. (2006) *J. Photochem. Photobiol. A*, **180**, 248.

553 Bayer, T., Wollenhaupt, M., Sarpe-Tudoran, C., and Baumert, T. (2009) *Phys. Rev. Lett.*, **102**, 023004.

554 Abrashkevich, A.G. and Shapiro, M. (1996) *J. Phys. B*, **29**, 627.

555 Vardi, A. and Shapiro, M. (1996) *J. Chem. Phys.*, **104**, 5490.

556 Frishman, E. and Shapiro, M. (1996) *Phys. Rev. A*, **54**, 3310.

557 Shapiro, M., Vrakking, M.J.J., and Stolow, A. (1999) *J. Chem. Phys.*, **110**, 2465.

558 Chen, Z., Shapiro, M., and Brumer, P. (1994) *Chem. Phys. Lett.*, **228**, 289.

559 Bohm, D. (1955) *Quantum Theory*, Prentice Hall, Englewood Cliffs.

560 Child, M.S. (1991) *Semiclassical Mechanics with Molecular Applications*, Oxford University Press, Oxford.

561 Amos, D.E. *Sandia National Laboratories, Computer routine library – slatec/ Complex Airy and subsidiary routines.*

562 Billy, N., Girard, B., Gonédard, G., and Vigue, J. (1987) *Mol. Phys.*, **61**, 65.

563 Arimondo, E. (1996) *Prog. Opt.*, **35**, 257.

564 Chen, Z., Brumer, P., and Shapiro, M. (1997) *Chem. Phys.*, **217**, 325.

565 Chu, S., Bjorkholm, J.E., Ashkin, A., and Cable, A. (1986) *Phys. Rev. Lett.*, **57**, 314.

566 Aspect, A., Cohen-Tannoudji, C., Dalibard, J., Heidemann, A., and Solomon, C. (1986) *Phys. Rev. Lett.*, **57**, 1688.

567 Lawall, J., Bardou, F., Shimizu, K., Leduc, M., Aspect, A., and Cohen-Tannoudji, C. (1994) *Phys. Rev. Lett.*, **73**, 1915.

568 Bahns, J.T., Stwalley, W.C., and Gould, P.L. (1996) *J. Chem. Phys.*, **104**, 9689.

569 Bartana, A., Kosloff, R., and Tannor, D.J. (1993) *J. Chem. Phys.*, **99**, 196.

570 Thorsheim, H.R., Weiner, J., and Julienne, P.S. (1987) *Phys. Rev. Lett.*, **58**, 2420.

571 Band, Y.B. and Julienne, P.S. (1995) *Phys. Rev. A*, **51**, R4317.

572 Coté, R. and Dalgarno, A. (1997) *Chem. Phys. Lett.*, **279**, 50.

573 Julienne, P.S., Burnett, K., Band, Y.B., and Stwalley, W.C. (1998) *Phys. Rev. A*, **58**, R797.

574 Mackie, M. and Javanainen, J. (1999) *Phys. Rev. A*, **60**, 3174.

575 Vardi, A. and Shapiro, M. (2000) *Phys. Rev. A*, **62**, 025401.

576 Vardi, A., Shapiro, M., and Anglin, J.R. (2002) *Phys. Rev. A*, **65**, 027401.

577 Drummond, P.D., Kheruntsyan, K.V., Heinzen, D.J., and Wynar, R.H. (2002) *Phys. Rev. A*, **65**, 063619.

578 Ling, H., Pu, H., and Seaman, B. (2004) *Phys. Rev. Lett.*, **93**, 250403.

579 Gabbanini, C., Fioretti, A., Lucchesini, A., Gozzini, S., and Mazzoni, M. (2000) *Phys. Rev. Lett.*, **84**, 2814.

580 Kemmann, M., Mistrik, I., Nussmann, S., Helm, H., Williams, C.J., and Julienne, P.S. (2004) *Phys. Rev. A*, **69**, 022715.

581 Luc-Koenig, E., Vatasescu, M., and Masnou-Seeuws, F. (2004) *Eur. Phys. J. D*, **31**, 239.

582 Koch, C.P., Palao, J.P., Kosloff, R., and Masnou-Seeuws, F. (2004) *Phys. Rev. A*, **70**, 013402.

583 Salzmann, W., Poschinger, U., Wester, R., Weidemüller, M., Merli, A., Weber, S.M., Sauer, F., Plewicki, M., Weise, F., Esparza, A.M., Wöste, L., and Lindinger, A. (2001) *Phys. Rev. A*, **73**, 023414.

584 Brown, B.L., Dicks, A.J., and Walmsley, I.A. (2006) *Phys. Rev. Lett.*, **96**, 173002.

585 Hulet, R.G. (2002) Coherence in photoassociation of a quantum degenerate gas, in *Cold Molecules and Bose–Einstein Condensates*, Les Houches, France.

586 Strecker, K.E., Partridge, G.B., and Hulet, R.G. (2003) *Phys. Rev. Lett.*, **91**, 080406.

587 Winkler, K., Lang, F., Thalhammer, G., v. d. Straten, P., Grimm, R., and Denschlag, J.H. (2007) *Phys. Rev. Lett.*, **98**, 043201.

588 Lang, F., Winkler, K., Strauss, C., Grimm, R., and Denschlag, J.H. (2008) *Phys. Rev. Lett.*, **101**, 133005.

589 Danzl, J.G., Haller, E., Gustavsson, M., Mark, M.J., Hart, R., Bouloufa, N., Dulieu, O., Ritsch, H., and Nägerl, H.C. (2008) *Science*, **321**, 1062.

590 Marvet, U. and Dantus, M. (1995) *Chem. Phys. Lett.*, **245**, 393.

591 Gross, P. and Dantus, M. (1997) *J. Chem. Phys.*, **106**, 8013.

592 Fioretti, A., Comparat, D., Crubellier, A., Dulieu, O., Masnou-Seeuws, F., and Pillet, P. (1998) *Phys. Rev. Lett.*, **80**, 4402.

593 Fioretti, A., Comparat, D., Drag, C., Amiot, C., O. Dulieu, F.M.S., and Pillet, P. (1998) *Eur. Phys. D*, **5**, 389.

594 Nikolov, A.N., Eyler, E.E., Wang, X.T., Li, J., Wang, H., Stwalley, W.C., and Gould, P.L. (1999) *Phys. Rev. Lett.*, **82**, 703.

595 Wynar, R., Freeland, R.S., Han, D.J., Ryu, C., and Heinzen, D.J. (2000) *Science*, **287**, 1016.

596 McKenzie, C., Denschlag, J.H., Häffner, H., Browaeys, A., de Araujo, L.E.E., Fatemi, F.K., Jones, K.M., Simsarian, J.E., Cho, D., Simoni, A., Tiesinga, E., Julienne, P.S., Helmerson, K., Lett, P.D., Rolston, S.L., and Phillips, W.D. (2001) *Phys. Rev. Lett.*, **88**, 120403.

597 Shapiro, E.A., Shapiro, M., Pe'er, A., and Ye, J. (2007) *Phys. Rev. A*, **75**, 013405.

598 Spiegelmann, F., Pavolini, D., and Daudey, J.P. (1989) *J. Phys. B: At. Mol. Opt. Phys.*, **22**, 2465.

599 Edvardsson, D., Lunell, S., and Marian, C.M. (2003) *Mol. Phys.*, **101**, 2381.

600 Marinescu, M., Sadeghpour, H.R., and Dalgarno, A. (1994) *Phys. Rev. A*, **49**, 982.

601 Landau, L.D. and Lifshitz, E.M. (1981) *Quantum Mechanics: Non-Relativistic Theory*, Butterworth-Heinemann, Oxford.

602 Weiner, J., Bagnato, V.S., Zilio, S., and Julienne, P.S. (1999) *Rev. Mod. Phys.*, **71**, 1.

603 Roberts, J.L., Claussen, N.R., J. P. Burke, J., Greene, C.H., Cornell, E.A., and Wieman, C.E. (1998) *Phys. Rev. Lett.*, **81**, 5109.

604 Vogels, J.M., Tsai, C.C., Freeland, R.S., Kokkelmans, S.J.J.M.F., Verhaar, B.J., and Heinzen, D.J. (1997) *Phys. Rev. A*, **56**, R1067.

605 Lang, F., Strauss, C., Winkler, K., Takekoshi, T., Grimm, R., and Denschlag, J.H. (2009) *Faraday Discuss.*, **142**, 271–282.

606 Pilch, K., Lange, A.D., Prantner, A., Kerner, G., Ferlaino, F., Nägerl, H., and Grimm, R. (2009) *Phys. Rev. A*, **79**, 042718.

607 Köhler, T., Goral, K., and Julienne, P.S. (2006) *Rev. Mod. Phys.*, **78**, 1311.

608 Ospelkaus, C., Ospelkaus, S., Humbert, L., Ernst, P., Sengstock, K., and Bongs, K. (2006) *Phys. Rev. Lett.*, **97**, 120402.

609 Zirbel, J.J., Ni, K.K., Ospelkaus, S., Incao, J.P.D., Wieman, C.E., Ye, J., and Jin, D.S. (2008) *Phys. Rev. Lett.*, **100**, 143201.

610 Zirbel, J.J., Ni, K.K., Ospelkaus, S., Nicholson, T.L., Olsen, M.L., Julienne, P.S., Wieman, C.E., Ye, J., and Jin, D.S. (2008) *Phys. Rev. A*, **78**, 013416.

611 Markiewisz, M., Kraemer, T., Waldburger, P., Herbig, J., Chin, C., Nägerl, H., and Grimm, R. (2007) *Phys. Rev. Lett.*, **99**, 113201.

612 Knoop, S., Ferlaino, F., Berninger, M., Mark, M., Nägerl, H., Grimm, R.,

D'Incao, J.P., and Esry, B. (2010) *Phys. Rev. Lett.*, **104**, 053201.
613 Bandrauk, A.D. (ed.) (1994) *Molecules in Laser Fields*, Marcel Dekker, New York.
614 Fedorov, M.V., Kudrevatova, O.V., Makarov, V.P., and Samokhin, A.A. (1975) *Opt. Commun.*, **13**, 299.
615 Kroll, N.M. and Watson, K.M. (1973) *Phys. Rev. A*, **8**, 804.
616 Kroll, N.M. and Watson, K.M. (1976) *Phys. Rev. A*, **13**, 1018.
617 Gerstein, J.I. and Mittleman, M.H. (1976) *J. Phys. B.*, **9**, 383.
618 Yuan, J.M., George, T.F., and McLafferty, F.J. (1976) *Chem. Phys. Lett.*, **40**, 163.
619 Yuan, J.M., Laing, J.R., and George T.F. (1977) *J. Chem. Phys.*, **66**, 1107.
620 George, T.F., Yuan, J.M., and Zimmermann, I.H. (1977) *Discuss. Faraday Soc.*, **62**, 246.
621 DeVries, P.L. and George, T.F. (1979) *Discuss. Faraday Soc.*, **67**, 129.
622 Lau, A.M.F. and Rhodes, C.K. (1977) *Phys. Rev. A*, **16**, 2392.
623 Lau, A.M.F. (1976) *Phys. Rev. A*, **13**, 139.
624 Lau, A.M.F. (1981) *Phys. Rev. A*, **25**, 363.
625 Dubov, V.S., Gudzenko, L.I., Gurvich, L.V., and Iakovlenko, S.I. (1977) *Chem. Phys. Lett.*, **45**, 351.
626 Bandrauk, A.D. and Sink, M.L. (1978) *Chem. Phys. Lett.*, **57**, 569.
627 Bandrauk, A.D. and Sink, M.L. (1981) *J. Chem. Phys.*, **74**, 1110.
628 Orel, A.E. and Miller, W.H. (1978) *Chem. Phys. Lett.*, **57**, 362.
629 Orel, A.E. and Miller, W.H. (1979) *Chem. Phys. Lett.*, **70**, 4393.
630 Orel, A.E. and Miller, W.H. (1980) *Chem. Phys. Lett.*, **73**, 241
631 Light, J.C. and Altenberger-Siczek, A. (1979) *J. Chem. Phys.*, **70**, 4108.
632 Foth, H.J., Polanyi, J.C., and Telle, H.H. (1982) *J. Phys. Chem.*, **86**, 5027.
633 Ho, T., Laughlin, C., and Chu, S.I. (1985) *Phys. Rev. A*, **34**, 122.
634 Chu, S.I. and Yin, R. (1987) *J. Opt. Soc. Am. B*, **4**, 720.
635 Yao, G. and Chu, S.I. (1992) *Chem. Phys. Lett.*, **197**, 413.
636 Matusek, D.R., Ivanov, M.Y., and Wright, J.S. (1996) *Chem. Phys. Lett.*, **258**, 255.
637 Shapiro, M. and Zeiri, Y. (1986) *J. Chem. Phys.*, **85**, 6449.
638 Seideman, T., Krause, J.L., and Shapiro, M. (1990) *Chem. Phys. Lett.*, **173**, 169.
639 Seideman, T., Krause, J.L., and Shapiro, M. (1991) *Faraday Discuss. Chem. Soc.*, **91**, 271.
640 Seideman, T. and Shapiro, M. (1991) *J. Chem. Phys.*, **94**, 7910.
641 Rodberg, L.S. and Thaler, R.M. (1967) *Introduction to the Quantum Theory of Scattering*, Academic Press, New York.
642 Eckart, C. (1930) *Phys. Rev.*, **35**, 1303.
643 Eyring, H., Walter, J., and Kimball, G.E. (1944) *Quantum Chemistry*, John Wiley & Sons, Inc., New York.
644 Liu, B. (1973) *J. Chem. Phys.*, **58**, 1925.
645 Truhlar, D.G. and Horowitz, C.J. (1978) *J. Chem. Phys.*, **68**, 2466.
646 Chang, L.L., Esaki, L., and Tsu, R. (1974) *Appl. Phys. Lett.*, **24**, 593.
647 Kan, S.C. and Yariv, A. (1990) *J. Appl. Phys.*, **67**, 1957.
648 Sa'ar, A., Kan, S.C., and Yariv, A. (1990) *J. Appl. Phys.*, **67**, 3892.
649 Yamamoto, H., Kanie, Y., Arakawa, M., and Taniguchi, K. (1990) *Appl. Phys. A*, **50**, 577.
650 Cai, W., Zheng, T.F., Hu, P., Lax, M., Shun, K., and Alfano, R. (1990) *Phys. Rev. Lett.*, **65**, 104.
651 Chao, K.A., Willander, M., and Galperin, Y.M. (1994) *Phys. Scr. T*, **54**, 119.
652 Yariv, A. (1991) *Optical Electronics*, 4th edn, Saunders College, Philadelphia.
653 Vorobeichik, I., Lefebvre, R., and Moiseyev, N. (1998) *Eur. Phys. Lett.*, **41**, 111.
654 Barron, L.D. (1982) *Molecular Light Scattering and Optical Activity*, Cambridge University Press, Cambridge.
655 Woolley, R.G. (1975) *Adv. Phys.*, **25**, 27.
656 Barron, L.D. (1986) *Chem. Soc. Rev.*, **15**, 189.
657 Quack, M. (1989) *Angew. Chem. Int. Ed.*, **28**, 571.
658 Shapiro, M., Frishman, E., and Brumer, P. (2000) *Phys. Rev. Lett.*, **84**, 1669.
659 Gerbasi, D., Shapiro, M., and Brumer, P. (2001) *J. Chem. Phys.*, **115**, 5349.
660 Brumer, P., Frishman, E., and Shapiro, M. (2001) *Phys. Rev. A*, **65**, 015401.
661 Salam, A. and Meath, W.J. (1998) *Chem. Phys.*, **228**, 115.

662 Fujimura, Y., Gonalez, L., Hoki, K., Manz, J., and Ohtsuki, Y. (1999) *Chem. Phys. Lett.*, **306**, 1.

663 Hoki, K., Gonzalez, L., and Fujimura, Y. (2002) *J. Chem. Phys.*, **116**, 2433.

664 Maierle, C.S. and Harris, R.A. (1998) *J. Chem. Phys.*, **109**, 3713.

665 Shirley, J.H. (1965) *Phys. Rev. B*, **138**, 979.

666 Brown, A. and Meath, W.J. (1998) *J. Chem. Phys.*, **109**, 9351.

667 Shapiro, M., Frishman, E., and Brumer, P. (2003) *J. Chem. Phys.*, **119**, 7237.

668 Apolonski, A., Poppe, A., Tempea, G., Spielmann, C., Udem, T., Holzwarth, R., Hansch, T.W., and Krausz, E. (2000) *Phys. Rev. Lett.*, **85**, 740.

669 Raman, C., Weinacht, T.C., and Bucksbaum, P.H. (1997) *Phys. Rev. A*, **55**, R3995.

670 Hollas, J.M. (1982) *High Resolution Spectroscopy*, Butterworths, London.

671 Harris, R.A., Shi, Y., and Cina, J.A. (1994) *J. Chem. Phys.*, **101**, 3459.

672 Segev, E. and Shapiro, M. (1980) *J. Chem. Phys.*, **73**, 2001.

673 Deretey, E., Shapiro, M., and Brumer, P. (2001) *J. Phys. Chem. A*, **105**, 9509.

674 Hayashi, M., Mebel, A.M., Liang, K.K., and Lin, S.H. (1998) *J. Chem. Phys.*, **108**, 2044.

675 Lochbrunner, S., Schults, T., Schmitt, M., Shaffer, J.P., Zgierski, M.Z., and Stolow, A. (2001) *J. Chem. Phys.*, **114**, 2519.

676 Gerbasi, D., Brumer, P., Thanopulos, I., Kràl, P., and Shapiro, M. (2004) *J. Chem. Phys.*, **120**, 11557.

677 Brumer, Y., Shapiro, M., Brumer, P., and Baldridge, K.K. (2002) *J. Phys. Chem. A*, **106**, 9512.

678 Král, P. and Shapiro, M. (2001) *Phys. Rev. Lett.*, **87**, 18300.

679 Král, P., Thanopulos, I., Shapiro, M., and Cohen, D. (2003) *Phys. Rev. Lett.*, **90**, 033001.

680 Thanopulos, I., Král, P., and Shapiro, M. (2003) *J. Chem. Phys.*, **119**, 5105.

681 Brumer, P. and Shapiro, M. (2004) *Chiral Photochemistry*, (eds Y. Inoue and V. Ramamurthy), Marcel Dekker, New York.

682 Li, Y., Bruder, C., and Sun, C.P. (2007) *Phys. Rev. Lett.*, **99**, 130403.

683 Li, X. and Shapiro, M. (2010) *J. Chem. Phys.*, **132**, 041101.

684 Li, X. and Shapiro, M. (2010) *J. Chem. Phys.*, **132**, 194315.

685 Morrison, R.T. and Boyd, R.N. (1992) *Organic Chemistry*, 6th edn, Prentice Hall, Englewood Cliffs.

686 Walker, D.C. (ed.) (1979) *Origins of Optical Activity in Nature*, Elsevier, Amsterdam.

687 Bodenhöfer, K., Hiermann, A., Seemann, J., Koppenhoefer, G.G.B., and Göpel, W. (1997) *Nature*, **387**, 577.

688 McKendry, R., Theoclitou, M.E., Rayment, T., and Abell, C. (1998) *Nature*, **391**, 566.

689 Knowles, W.S. (2002) *Angew. Chem. Int. Ed.*, **41**, 1997.

690 Zepik, H., Shavit, E., Tang, M., Jensen, T.R., Kjaer, K., Boldach, G., Leiserowitz, L., Weissbuch, I., and Lahav, M. (2002) *Science*, **295**, 1266.

691 Hoki, K., Ohtsuki, Y., and Fujimura, Y. (2001) *J. Chem. Phys.*, **114**, 1575.

692 Ohta, Y., Hoki, K., and Fujimura, Y. (2002) *J. Chem. Phys.*, **116**, 7509.

693 Hoki, K., Gonzalez, L., and Fujimura, Y. (2002) *J. Chem. Phys.*, **116**, 8799.

694 Frishman, E., Shapiro, M., Gerbasi, D., and Brumer, P. (2003) *J. Chem. Phys.*, **119**, 7237.

695 Unanyan, R., Yatsenko, L.P., Bergmann, K., and Shore, B.W. (1997) *Opt. Commun.*, **139**, 48.

696 Fleischhauer, M., Unanyan, R., Shore, B.W., and Bergmann, K. (1999) *Phys. Rev. A*, **59**, 3751.

697 Gottselig, M., Luckhaus, D., Quack, M., Stohner, J., and Willeke, M. (2000) *Helv. Chim. Acta*, **84**, 1846.

698 Thanopulos, I. and Shapiro, M. (2005) *J. Am. Chem. Soc.*, **127**, 14434.

699 Zhu, S.L. *et al.* (2006) *Phys. Rev. Lett.*, **97**, 240401.

700 Sun, C.P. and Ge, M.L. (1990) *Phys. Rev. D*, **41**, 1349.

701 Zare, R.N. (1988) *Angular Momentum*, John Wiley & Sons, Inc., New York.

702 Balint-Kurti, G.G. and Shapiro, M. (1981) *Chem. Phys.*, **61**, 137.

703 Bandrauk, A.D. and Atabek, O. (1989) *Adv. Chem. Phys.*, **73**, 823.

704 Shapiro, M. and Bony, H. (1985) *J. Chem. Phys.*, **83**, 1588.

705 Seideman, T. and Shapiro, M. (1988) *J. Chem. Phys.*, **88**, 5525.

706 Seideman, T. and Shapiro, M. (1990) *J. Chem. Phys.*, **92**, 2328.

707 Chu, S.I. (1979) *Chem. Phys. Lett.*, **64**, 178.

708 Chu, S.I. (1980) *Chem. Phys. Lett.*, **70**, 205.

709 Crance, M. and Aymar, M. (1980) *J. Phys. B*, **13**, L421.

710 Deng, Z. and Eberly, J.H. (1984) *Phys. Rev. Lett.*, **53**, 1810.

711 Grobe, R. and Eberly, J.H. (1992) *Phys. Rev. Lett.*, **68**, 2905.

712 Grobe, R. and Eberly, J.H. (1993) *Phys. Rev. A*, **48**, 623.

713 Grobe, R. and Eberly, J.H. (1993) *Phys. Rev. A*, **48**, 4664.

714 Grobe, R. and Eberly, J.H. (1993) *Laser Phys.*, **3**, 323.

715 Rzazewski, K. and Grobe, R. (1986) *Phys. Rev. A*, **33**, 1855.

716 Piraux, B., Bhatt, R., and Knight, P.L. (1990) *Phys. Rev. A*, **41**, 6296.

717 Atabek, O., Lefebvre, R., and Jacon, M. (1980) *J. Chem. Phys.*, **72**, 2670.

718 Blank, R. and Shapiro, M. (1994) *Phys. Rev. A*, **50**, 3234.

719 Bandrauk, A.D., Aubanel, E.E., and Gauthier, J.M. (1994) in *Molecules in Laser Fields*, (ed. A.D. Bandrauk), Marcel Dekker, New York.

720 Aubanel, E.E. and Bandrauk, A.D. (1992) *Chem. Phys. Lett.*, **197**, 419.

721 Bandrauk, A.D., Aubanel, E.E., and Gauthier, J.M. (1993) *Laser Phys.*, **3**, 381.

722 Sussman, B.J., Ivanov, M.Y., and Stolow, A. (2005) *Phys. Rev. A*, **71**, 051401R.

723 Sussman, B.J., Townsend, D., Ivanov, M.Y., and Stolow, A. (2006) *Science*, **314**, 278–281.

724 Schinke, R. (1992) *Photodissociation Dynamics*, Cambridge University Press, Cambridge.

725 Heather, R. and Metiu, H. (1988) *J. Chem. Phys.*, **88**, 5496.

726 Seel, M. and Domcke, W. (1991) *J. Chem. Phys.*, **95**, 7806.

727 Abou-Rachid, H., Nguyen-Dang, T.T., Chaudhury, R.K., and He, X. (1992) *J. Chem. Phys.*, **97**, 5497.

728 Kulander, K.C., Shafer, K.J., and Krause, J.L. (1992) in *Atoms in Intense Laser Fields*, Academic, New York.

729 Smyth, E.S., Parker, J.S., and Taylor, K.T. (1998) *Comput. Phys. Commun.*, **114**, 1.

730 Peskin, U. and Moiseyev, N. (1993) *J. Chem. Phys.*, **99**, 4590.

731 Peskin, U. and Moiseyev, N. (1994) *Phys. Rev. A*, **49**, 3712.

732 Peskin, U., Alon, O.E., and Moiseyev, N. (1994) *J. Chem. Phys.*, **100**, 7310.

733 Peskin, U., Kosloff, R., and Moiseyev, N. (1994) *J. Chem. Phys.*, **100**, 8849.

734 Moiseyev, N. (1995) *Comments At. Mol. Phys.*, **31**, 87.

735 Alon, O. and Moiseyev, N. (1995) *Chem. Phys.*, **196**, 499.

736 Pont, M. and Shakeshaft, R. (1997) in *Photon and Electron Collisions with Atoms and Molecules*, (eds P.G. Burke and C.J. Joachain), Plenum, New York.

737 Chelkowski, S., Zuo, T., Atabek, O., and Bandrauk, A.D. (1995) *Phys. Rev. A*, **52**, 2977.

738 Meier, C., Engel, V., and Manthe, U. (1998) *J. Chem. Phys.*, **109**, 36.

739 Kornberg, M.A. and Lambropoulos, P. (1999) *J. Phys. B*, **32**, L603.

740 Tanner, G., Richter, K., and Rost, J.M. (2000) *Rev. Mod. Phys.*, **72**, 498.

741 Giusti-Suzor, A., He, X., Atabek, O., and Mies, F.H. (1990) *Phys. Rev. Lett.*, **64**, 515.

742 Giusti-Suzor, A. and Mies, F.H. (1992) *Phys. Rev. Lett.*, **68**, 3869.

743 Wunderlich, C., Kobler, E., Figger, H., and Hansch, T.W. (1997) *Phys. Rev. Lett.*, **78**, 2333.

744 Wunderlich, C., Kobler, E., Figger, H., and Hansch, T.W. (2000) *Phys. Rev. A*, **62**, 023401.

745 Solá, I.R., Santamaria, J., and Malinovsky, V.S. (2000) *Phys. Rev. A*, **61**, 043413.

746 Solá, I.R., Chang, B.Y., Santamaria, J., Malinovsky, V.S., and Krause, J.L. (2000) *Phys. Rev. Lett.*, **85**, 4241.

747 Malinovsky, V.S. and Krause, J.L. (2001) *Chem. Phys.*, **267**, 47.

748 Zener, C. (1932) *Proc. R. Soc. A*, **137**, 696.

749 Landau, L.D. (1932) *Phys. Z. Sowjetunion*, **2**, 46.

750 Charron, E., Giusti-Suzor, A., and Mies, F.H. (1993) *Phys. Rev. Lett.*, **71**, 692.

751 Verschuur, J.W.J., Noordam, L.D., and van Linden van den Heuvell, H.B. (1989) *Phys. Rev. A*, **40**, 4383.

752 Zavriyev, A., Bucksbaum, P.H., Muller, H.G., and Schumacher, D.W. (1990) *Phys. Rev. A*, **42**, 5500.

753 Allendorf, S.W. and Szöke, A. (1991) *Phys. Rev. A*, **44**, 518.

754 Zavriyev, A. and Bucksbaum, P.H. (1994) in *Molecules in Laser Fields*, (ed. A.D. Bandrauk), Marcel Dekker, New York.

755 Giusti-Suzor, A., Mies, F.H., Dimauro, L.F., Charron, E., and Yang, B. (1995) *J. Phys. B*, **28**, 309.

756 Zavriyev, A., Bucksbaum, P.H., Squier, J., and Saline, F. (1993) *Phys. Rev. Lett.*, **70**, 1077.

757 Pegarkov, A.I. and Rapoport, L.P. (1988) *Opt. Spectrosc.*, **65**, 55.

758 Timp, G., Behringer, R.E., Tennat, D.M., Cunningham, J.E., Prentiss, M., and Berggren, K.K. (1992) *Phys. Rev. Lett.*, **69**, 1636.

759 Berggren, K., Prentiss, M., Timp, G.L., and Behringer, R.E. (1994) *J. Opt. Soc. Am. B*, **11**, 1166.

760 Miller, J.D., Cline, R.A., and Heinzen, D.J. (1993) *Phys. Rev. A*, **47**, R4567.

761 Stapelfeldt, H., Sakai, H., Constant, E., and Corkum, P.B. (1997) *Phys. Rev. Lett.*, **79**, 2787.

762 Sakai, H., Tarasevitch, A., Danilov, J., Stapelfeldt, H., Yip, R.W., Ellert, C., Constant, E., and Corkum, P.B. (1998) *Phys. Rev. A*, **57**, 2794.

763 Seideman, T. (1997) *J. Chem. Phys.*, **106**, 2881.

764 Seideman, T. (1997) *Phys. Rev. A*, **56**, R17.

765 Nguyen, N.A., Dey, B.K., Shapiro, M., and Brumer, P. (2004) *J. Phys. Chem. A*, **108**, 7878.

766 Dey, B.K., Shapiro, M., and Brumer, P. (2000) *Phys. Rev. Lett.*, **85**, 3125.

767 Normand, D., Lompre, L.A., and Cornaggia, C. (1992) *J. Phys. B*, **25**, L497.

768 Schmidt, M., Normand, D. and Cornaggia, (1994) C. *Phys. Rev. A*, **50**, 5037.

769 Dietrich, P., Strickland, D.T., Laberge, M., and Corkum, P.B. (1993) *Phys. Rev. A*, **47**, 2305.

770 Friedrich, B. and Herschbach, D.R. (1995) *Phys. Rev. Lett.*, **74**, 4623.

771 Bandrauk, A. and Aubanel, E.E. (1995) *Chem. Phys.*, **198**, 159.

772 Seideman, T. (1995) *J. Chem. Phys.*, **103**, 7887.

773 Kim, W. and Felker, P.M. (1996) *J. Chem. Phys.*, **104**, 1147.

774 Haberland, H. and Issendorff, B.V. (1996) *Phys. Rev. Lett.*, **76**, 1445.

775 Takekoshi, T., Patterson, B.M., and Knize, R.J. (1998) *Phys. Rev. Lett.*, **81**, 5105.

776 Larsen, J.J., Hald, K., Bjerre, N., Stapelfeldt, H., and Seideman, T. (2000) *Phys. Rev. Lett.*, **85**, 2470.

777 Rosca-Pruna, F. and Vrakking, M.J.J. (2001) *Phys. Rev. Lett.*, **87**, 153902.

778 Underwood, J.G., Spanner, M., Ivanov, M.Y., Mottershead, J., Sussman, B.J., and Stolow, A. (2003) *Phys. Rev. Lett.*, **90**, 223001.

779 Stapelfeldt, H. and Seideman, T. (2003) *Rev. Mod. Phys.*, **75**, 543.

780 Lee, K.F., Villeneuve, D.M., Corkum, P.B., Stolow, A., and Underwood, J.G. (2006) *Phys. Rev. Lett.*, **97**, 173001.

781 Poulsen, M., Ejdrup, T., Stapelfeldt, H., Hamilton, E., and Seideman, T. (2006) *Phys. Rev. A*, **73**, 033405.

782 Horn, C., Wollenhaupt, M., Krug, M., Baumert, T., de Nalda, R., and Banares, L. (2006) *Phys. Rev. A*, **73**, 031401.

783 Kumarappan, V., Viftrup, S., Holmegaard, L., Bisgaard, C., and Stapelfeldt, H. (2007) *Phys. Scr.*, **76**, C63–C68.

784 Fleischer, S., Averbukh, I.S., and Prior, Y. (2006) *Phys. Rev. A*, **74**, 041403.

785 Fleischer, S., Averbukh, I.S., and Prior, Y. (2007) *Phys. Rev. Lett.*, **99**, 093002.

786 Viftrup, S., Kumarappan, V., Trippel, S., Stapelfeldt, H., Hamilton, E., and Seideman, T. (2007) *Phys. Rev. Lett.*, **99**, 143602.

787 Fleischer, S., Averbukh, I.S., and Prior, Y. (2008) *J. Phys. B*, **41**, 074018.

788 Viftrup, S., Kumarappan, V., Holmegaard, L., Bisgaard, C., Stapelfeldt, H., Artamonov, M., Hamilton, E., and Seideman, T. (2009) *Phys. Rev. A*, **79**, 023404.

789 Dimitrovski, D., Abu-Samha, M., Madsen, L.B., Filsinger, F., Meijer, G., Küpper, J., Holmegaard, L., Kalhoj, L., Nielsen, J.H., and Stapelfeldt, H. (2011) *Phys. Rev. A*, **83**, 023405.

790 Hansen, J.L., Holmegaard, L., Kalhoj, L., Kragh, S.L., Stapelfeldt, H., Filsinger, F., Meijer, G., Küpper, J., Dimitrovski, D., Abu-Samha, M., Martiny, C.P.J., and Madsen, L.B. (2011) *Phys. Rev. A*, **83**, 023406.

791 Karczmarek, J., Wright, J., Corkum, P.B., and Ivanov, M.Y. (1999) *Phys. Rev. Lett.*, **82**, 3420.

792 Villeneuve, D.M., Aseyev, S.A., Dietrich, P., Spanner, M., Ivanov, M.Y., and Corkum, P.B. (2000) *Phys. Rev. Lett.*, **85**, 542.

793 Spanner, M. and Ivanov, M.Y. (2001) *J. Chem. Phys.*, **114**, 3456.

794 Averbukh, I.S. and Perel'man, N.F. (1991) *Sov. Phys. Usp.*, **34**, 572.

795 Vrakking, M.J.J., Villeneuve, D.M., and Stolow, A. (1996) *Phys. Rev. A*, **54**, R37.

796 Gershnabel, E. and Averbukh, I.S. (2010) *Phys. Rev. Lett.*, **104**, 153001.

797 Loesch, H.J. and Müller, J. (1993) *J. Phys. Chem.*, **97**, 2158.

798 Doyle, J.M. (1994) *Bull. Am. Phys. Soc.*, **39**, 1166.

799 Orr-Ewing, A.J. and Zare, R.N. (1994) *Ann. Rev. Phys. Chem.*, **45**, 315.

800 Xu, L., Spielmann, C., Poppe, A., Brabec, T., Krausz, F., and Hänsch, T.W. (1996) *Opt. Lett.*, **21**, 2008.

801 Telle, H.R., Steinmeyer, G., Dunlop, A.E., Stenger, J., Sutter, D.H., and Keller, U. (1999) *Appl. Phys. B*, **69**, 327.

802 Jones, D.J., Diddams, S.A., Ranka, J.K., Stentz, A., Windeler, R.S., Hall, J.L., and Cundiff, S.T. (2000) *Science*, **288**, 635.

803 Apolonski, A., Poppe, A., Tempea, G., Spielmann, C., Udem, T., Holzwarth, R., Hänsch, T.W., and Krausz, F. (2000) *Phys. Rev. Lett.*, **85**, 740.

804 Ye, J. and Cundiff, S.T. (2005) *Femtosecond Optical Frequency Comb: Principle, Operation, and Applications*, Springer, New York.

805 Shimizu, F. (1967) *Phys. Rev. Lett.*, **19**, 1097.

806 Alfano, R.R. and Shapiro, S.L. (1970) *Phys. Rev. Lett.*, **24**, 592,

807 Buckland, E.L. and Boyd, R.W. (1996) *Opt. Lett.*, **21**, 1117.

808 Hall, J.L. (2006) *Rev. Mod. Phys.*, **78**, 1279.

809 Hänsch, T.W. (2006) *Rev. Mod. Phys.*, **78**, 1297.

810 Corkum, P.B. (1993) *Phys. Rev. Lett.*, **71**, 1994.

811 Kulander, K.C., Schafer, K.J., and Krause, K.L. (1993) *Super-Intense Laser-Atom Physics*, (eds B. Piraux and A.L. Huillier, and K. Rzazewski), vol. 316, Series B: Physics, NATO Advanced Studies Institute, Plenum, New York.

812 Lewenstein, M., Balcou, P., Ivanov, M.Y., Huillier, A.L., and Corkum, P.B. (1994) *Phys. Rev. A*, **49**, 2117.

813 Schafer, K.J., Yang, B., DiMauro, L.F., and Kulander, K.C. (1993) *Phys. Rev. Lett.*, **70**, 1599.

814 Keldysh, L.V. (1965) *Sov. Phys. JETP*, **20**, 1307.

815 Ammosov, M., Delone, N., and Krainov, V. (1986) *Zh. Eksp. Teor. Fiz.*, **91**, 2008, English translation: (1986) Sov. Phys. JETP **64**, 1191.

816 Faisal, F.H.M. (1973) *J. Phys. B.*, **6**, L89.

817 Reiss, H.R. (1980) *Phys. Rev. A*, **22**, 1786.

818 Szöke, A. (1988) in *Atomic and Molecular Processes with Short Intense Laser Pulses*, (ed. A.D. Bandrauk), Plenum, New York.

819 Agostini, P., Fabre, F., Mainfray, G., Petite, G., and Rahman, N. (1979) *Phys. Rev. Lett.*, **42**, 1127.

820 Becker, W., Grasbon, F., Kopold, R., Milosevic, D.B., Paulus, G.G., and Walther, H. (2002) *Adv. Atom. Mol. Opt. Phys.*, **48**, 35.

821 LaGattuta, K.J. and Cohen, J.S. (1998) *J. Phys. B: At. Mol. Opt. Phys.*, **31**, 5281.

822 Yudin, G.L. and Ivanov, M.Y. (2001) *Phys. Rev. A*, **63**, 033404.

823 Baltuska, A., Udem, T., Uiberacker, M., Hentschel, M., Goulielmakis, E., Gohle, C., Holzwarth, R., Yakovlev, V.S., Scrinzi, A., Hänsch, T.W., and Krausz, F. (2003) *Nature*, **421**, 611.

824 Smirnova, O., Patchkovskii, S., Mairesse, Y., Dudovich, N., and Ivanov, M.Y. (2009) *Proc. Natl. Acad. Sci. USA*, **106**, 16556.

825 Paul, P.M., Toma, E.S., Breger, P., Mullot, G., Auge, F., Balcou, P., Muller, H.G., and Agostini, P. (2001) *Science*, **292**, 1689.

826 Nakajima, T. and Watanabe, S. (2006) *Phys. Rev. Lett.*, **96**, 213001.

827 Baltuska, A., Fuji, T., and Kobayashi, T. (2002) *Phys. Rev. Lett.*, **88**, 133901.

828 Telle, H.R., Steinmeyer, G., Dunlop, A.E., Stenger, J., Sutter, D.H., and Keller, U. (1999) *Appl. Phys. B*, **69**, 327.

829 Brabec, T. and Krausz, F. (2000) *Rev. Mod. Phys.*, **72**, 545.

830 Helbing, F.W., Steinmeyer, G., and Keller, U. (2003) *IEEE J. Sel. Top. Quantum Electron.*, **9**, 1030.

831 Vozzi, C., Cirmi, G., Manzoni, C., Benedetti, E., Calegari, F., Sansone, G., Stagira, S., Svelto, O., Silvestri, S.D., Nisoli, M., and Cerull, G. (2006) *Opt. Express*, **14**, 10109.

832 Dietrich, P., Krausz, F., and Corkum, P.B. (2000) *Opt. Lett.*, **25**, 16.

833 Drescher, M., Hentschel, M., Kienberger, R., Tempea, G., Spielmann, C., Reider, G.A., Corkum, P.B., and Krausz, F. (2001) *Science*, **291**, 1923.

834 Hentschel, M., Kienberger, R., Spielmann, C., Reider, G.A., Milosevic, N., Brabec, T., Corkum, P., Heinzmann, U., Drescher, M., and Krausz, F. (2001) *Nature*, **414**, 509.

835 Itatani, J., Quere, F., Yudin, G.L., Ivanov, M.Y., Krausz, F., and Corkum, P.B. (2002) *Phys. Rev. Lett.*, **88**, 173903.

836 Drescher, M., Hentschel, M., Kienberger, R., Uiberacker, M., Yakovlev, V., Scrinzi, A., Westerwalbesloh, T., Kleineberg, U., Heinzmann, U., and Krausz, F. (2002) *Nature*, **419**, 803.

837 Goulielmakis, E., Yakovlev, V.S., Cavalieri, A.L., Uiberacker, M., Pervak, V., Apolonski, A., Kienberger, R., Kleineberg, U., and Krausz, F. (2007) *Science*, **317**, 769.

838 Goulielmakis, E., Schultze, M., Hofstetter, M., Yakovlev, V.S., Gagnon, J., Uiberacker, M., Aquila, A.L., Gullikson, E.M., Attwood, D.T., Kienberger, R., Krausz, F., and Kleineberg, U. (2008) *Science*, **320**, 1614.

839 Eckle, P., Pfeiffer, A.N., Cirelli, C., Staudte, A., Dörner, R., Muller, H.G., Büttiker, M., and Keller, U. (2008) *Science*, **322**, 1525.

840 Emmanouilidou, A., Staudte, A., and Corkum, P.B. (2010) *New J. Phys.*, **12**, 103024.

841 Fortier, T.M., Roos, P.A., Jones, D.J., Cundiff, S.T., Bhat, R.D., and Sipe, J.E. (2004) *Phys. Rev. Lett.*, **92**, 147403.

842 Kling, M.F., Siedschlag, C., Verhoef, A.J., Khan, J.I., Schultze, M., Uphues, T., Ni, Y., Uiberacker, M., Drescher, M., Krausz, F., and Vrakking, M.J.J. (2006) *Science*, **312**, 246.

843 Sansone, G., Kelkensberg, F., Kling, M.F., Siu, W.K., Ghafur, O., Johnsson, P., Zherebtsov, S., Znakovskaya, I., Uphues, T., Benedetti, E., Ferrari, F., Lèpine, F., Swoboda, M., Remetter, T., L'Huillier, A., Nisoli, M., and Vrakking, M.J.J. (2009) in *Utrafast Phenomena XVI*, (eds P. Corkum, S. Silvestri, K.A. Nelson, E. Riedle, and R.W. Schoenlein), Springer Series in Chemical Physics, Springer, Berlin.

844 Starace, A.F., Pronin, E.A., Frolov, M.V., and Manakov, N.L. (2009) *Phys. Rev. A*, **80**, 063403.

845 Dudovich, N., Smirnova, O., Levesque, J., Mairesse, Y., Ivanov, M.Y., Corkum, P.B., and Villeneuve, D.M. (2006) *Acta Phys. Hung. B*, **26**, 359.

846 Christov, I.P., Bartels, R., Kapteyn, H.C., and Murnane, M.M. (2001) *Phys. Rev. Lett.*, **86**, 5458.

847 Christov, I.P., Murnane, M.M., and Kapteyn, H.C. (1997) *Phys. Rev. Lett.*, **78**, 1251.

848 Polanyi, M. (1932) *Atomic Reactions*, Williams and Norgate, London.

849 Herschbach, D.R. (1993) in *Nobel Lectures in Chemistry 1981–1990*, (ed. B.G. Malmström), World Scientific, Singapore.

850 Levine, R.D. (2005) *Molecular Reaction Dynamics*, Cambridge University Press, Cambridge.

851 Baker, S., Robinson, J., Haworth, C.A., Teng, H., Smith, R.A., Chirila, C.C.,

Lein, M., Tisch, J.W.G., and Marangos, J.P. (2006) *Science*, **312**.
852 Wagner, N.L., Wuest, A., Christov, I.P., Popmintchev, T., Zhou, X., Murnane, M.M., and Kapteyn, H.C. (2006) *Proc. Natl. Acad. Sci. USA*, **103**, 13279.
853 de Nalda, R., Heesel, E., Lein, M., Hay, N., R. Velotta, E.S., Castillejo, M., and Marangos, J.P. (2004) *Phys. Rev. A*, **69**, 031804R.
854 Itatani, J., Zeidler, D., Levesque, J., Spanner, M., Villeneuve, D.M., and Corkum, P.B. (2005) *Phys. Rev. Lett.*, **94**, 123902.
855 Itatani, J., Levesque, J., Zeidler, D., Niikura, H., Pepin, H., Kieffer, J.C., Corkum, P.B., and Villeneuve, D.M. (2004) *Nature*, **432**, 867.
856 Engel, V. and Metiu, H. (1989) *J. Chem. Phys.*, **90**, 6116.
857 Watson, J.B., Sanpera, A., Chen, X., and Burnett, K. (1996) *Phys. Rev. A*, **53**, R1962.
858 Sanpera, A., Watson, J.B., Lewenstein, M., and Burnett, K. (1996) *Phys. Rev. A*, **54**, 4320.
859 Niikura, H., Villeneuve, D.M., and Corkum, P.B. (2005) *Phys. Rev. Lett.*, **94**, 083003.
860 Rose, T.S., Rosker, M.J., and Zewail, A.H. (1988) *J. Chem. Phys.*, **88**, 6672.
861 Keldysh, L.V. (1964) *Zh. Eksp. Teor. Fiz.*, **47**, 1945.
862 Reiss, H.R. (1992) *Prog. Quant. Electr.*, **16**, 1.
863 Bagratashvili, V.N., Letokhov, V.S., Makarov, A.A., and Ryabov, E.A. (1984) *Laser Chem.*, **4**, 311.
864 Larsen, D.M. and Bloembergen, N. (1976) *Opt. Commun.*, **17**, 254.
865 Paramonov, G.K. and Savva, V.A. (1983) *Phys. Lett. A*, **97**, 340.
866 Paramonov, G.K. and Savva, V.A. (1984) *Chem. Phys. Lett.*, **107**, 394.
867 Paramonov, G.K., Savva, V.A., and Samson, A.M. (1985) *Infrared Phys.*, **25**, 201.
868 Paramonov, G.K. (1990) *Chem. Phys. Lett.*, **169**, 573.
869 Paramonov, G.K. (1991) *Phys. Lett. A*, **152**, 191.
870 Lawton, R.T. and Child, M.S. (1979) *Mol. Phys.*, **37**, 1799.
871 Child, M.S. and Halonen, H.S. (1984) *Adv. Chem. Phys.*, **57**, 1.
872 Jaffe, C. and Brumer, P. (1980) *J. Chem. Phys.*, **73**, 5645.
873 Joseph, T. and Manz, J. (1986) *Mol. Phys.*, **57**, 1149.
874 Zhang, J. and Imre, D. (1988) *Chem. Phys. Lett.*, **149**, 233.
875 Schinke, R., Wal, R.L.V., Scott, J.L., and Crim, F.F. (1991) *J. Chem. Phys.*, **94**, 283.
876 Bronikowski, M.J., Simpson, W.R., Girard, B., and Zare, R.N. (1991) *J. Chem. Phys.*, **95**, 8647.
877 Schatz, G.C., Colton, M.C., and Grant, J.L. (1984) *J. Phys. Chem.*, **88**, 2971.
878 Clary, D.C. (1991) *J. Chem. Phys.*, **95**, 7298.
879 Troya, D., González, M., and Schatz, G.C. (2001) *J. Chem. Phys.*, **114**, 8397.
880 Shapiro, M. (1992) in *Isotope Effects in Gas Phase Chemistry*, ACS Symposium Series, (ed. J.A. Kaye), vol. 502, American Chemical Society, Washington, DC.
881 Paci, J., Shapiro, M., and Brumer, P. (1998) *J. Chem. Phys.*, **109**, 8993.
882 Astrom, K.J. (1987) *Proc. IEEE*, **75**, 185.
883 Levis, R.J., Menkir, G.M., and Rabitz, H. (2001) *Science*, **292**, 709.
884 Herek, J.L., Wohlleben, W., Cogdell, R.J., Zeidler, D., and Motzkus, M. (2002) *Nature*, **417**, 533.
885 Vogt, G., Krampert, G., Niklaus, P., Nuernberger, P., and Gerber, G. (2005) *Phys. Rev. Lett.*, **94**, 068305.
886 Baumert, T., Brixner, T., Seyfried, V., Strehle, M., and Gerber, G. (1997) *Appl. Phys. B*, **65**, 779.
887 Hoki, K. and Brumer, P. (2005) *Phys. Rev. Lett.*, **95**, 168305.
888 Hunt, P.A. and Robb, M.A. (2005) *J. A. Chem. Soc.*, **127**, 5720.
889 Improta, R. and Santoro, P. (2005) *J. Chem. Theory Comput.*, **1**, 215.
890 Hoki, K. and Brumer, P. (2009) *Chem. Phys. Lett.*, **468**, 23.
891 Weinacht, T.C., White, J.L., and Bucksbaum, P.H. (1999) *J. Phys. Chem. A*, **103**, 10166.
892 Pearson, B.J., White, J.L., Weinacht, T.C., and Bucksbaum, P.H. (2001) *Phys. Rev. A*, **63**, 063412.
893 Weinacht, T.C. and Bucksbaum, P.H. (2002) *J. Opt. B Quantum Semiclass. Opt.*, **4**, R35.

894 White, J.L., Pearson, B.J., and Bucksbaum, P.H. (2004) *J. Phys. B: At. Mol. Opt. Phys.*, **37**, L399.

895 Pearson, B.J. and Bucksbaum, P.H. (2004) *Phys. Rev. Lett.*, **92**, 243003.

896 Pearson, B.J. and Bucksbaum, P.H. (2005) *Phys. Rev. Lett.*, **94**, 209901.

897 Spanner, M. and Brumer, P. (2006) *Phys. Rev A*, **73**, 023809.

898 Spanner, M. and Brumer, P. (2006) *Phys. Rev A*, **73**, 023810.

899 Mukamel, S. (1995) *Principles of Nonlinear Optical Spectroscopy*, Oxford University Press, Oxford.

900 Butcher, P.N. and Cotter, D. (1990) *The Elements of Nonlinear Optics*, Cambridge Univ. Press, Cambridge.

901 Franco, I. (2007) Ph.D. Dissertation, University of Toronto.

902 Flach, S., Yevtushenko, O., and Zolotaryuk, Y. (2000) *Phys. Rev. Lett.*, **84**, 2358.

903 Kibble, T.W.B. and Berkshire, F.H. (2004) *Classical Mechanics*, 5th edn, Imperial College Press, London.

904 Goychuk, I. and Hänggi, P. (1998) *Europhys. Lett.*, **43**, 503.

905 Denisov, S., Flach, S., Ovchinnikov, A., Yevtushenko, O., and Zolotaryuk, Y. (2002) *Phys. Rev. E*, **66**, 041104.

906 Osborn, T.A. and Molzahn, F.H. (1995) *Ann. Phys. (NY)*, **241**, 79.

907 Franco, I. and Brumer, P. (2006) *Phys. Rev. Lett.*, **97**, 040402.

908 Bender, C.M. and Bettencourt, L.M.A. (1996) *Phys. Rev. Lett.*, **77**, 4114.

909 Bender, C.M. and Bettencourt, L.M.A. (1996) *Phys. Rev. D*, **54**, 7710.

910 Fetter, A.L. and Walecka, J.D. (2003) *Quantum Theory of Many Particle Systems*, Dover Publications, Mineola, NY.

911 EPAPS Document No. E-PRLTAO-97-033631 for the expression for $\hat{\Gamma}$ and Γ_c. For more information on EPAPS, see http://www.aip.org/pubservs/epaps.html.

912 Schiavoni, M., Sanchez-Palencia, L., Renzoni, F., and Grynbert, G. (2003) *Phys. Rev. Lett.*, **90**, 094101.

913 Jones, P.H., Goonasekera, M., and Renzoni, F. (2004) *Phys. Rev. Lett.*, **93**, 073904.

914 Gommers, R., Bergamini, S., and Renzoni, F. (2005) *Phys. Rev. Lett.*, **95**, 073003.

915 Gommers, R., Denisov, S., and Renzoni, F. (2006) *Phys. Rev. Lett.*, **96**, 240604, and references therein.

916 Steck, D.A., Oskay, W.H., and Raizen, M.G. (2001) *Science*, **293**, 274.

917 Steck, D.A., Oskay, W.H., and Raizen, M.G. (2002) *Phys. Rev. Lett.*, **88**, 120406.

918 Gong, J. and Brumer, P. (2005) *Ann. Rev. Phys. Chem.*, **56**, 1.

919 Wiebe, N. and Ballentine, L.E. (2005) *Phys. Rev. A*, **72**, 022109.

920 Kryvohuz, M. and Cao, J.S. (2006) *Phys. Rev. Lett.*, **96**, 030403. This reference clearly shows that the classical and quantum nonlinear response can be vastly different.

Subject Index

Quantum Control of Molecular Processes, Second Edition. Moshe Shapiro and Paul Brumer